Antioxidant Activity of Essential Oils

Antioxidant Activity of Essential Oils

Editors

Delia Mirela Tit
Simona Gabriela Bungau

MDPI • Basel • Beijing • Wuhan • Barcelona • Belgrade • Manchester • Tokyo • Cluj • Tianjin

Editors

Delia Mirela Tit
Pharmacy
University of Oradea
Oradea
Romania

Simona Gabriela Bungau
Pharmacy
University of Oradea
Oradea
Romania

Editorial Office
MDPI
St. Alban-Anlage 66
4052 Basel, Switzerland

This is a reprint of articles from the Special Issue published online in the open access journal *Antioxidants* (ISSN 2076-3921) (available at: www.mdpi.com/journal/antioxidants/special_issues/antioxidant_essential_oil).

For citation purposes, cite each article independently as indicated on the article page online and as indicated below:

LastName, A.A.; LastName, B.B.; LastName, C.C. Article Title. *Journal Name* **Year**, *Volume Number*, Page Range.

ISBN 978-3-0365-6949-9 (Hbk)
ISBN 978-3-0365-6948-2 (PDF)

Contents

About the Editors

Delia Mirela Tit

Delia Mirela Tit is a former student of the University of Oradea, becoming a full professor in 2020, as a result of her activity as a member of the teaching staff, which started in 2005. She earned a diploma of Doctor of Medicine in 2014 and presented her Habilitation Thesis in 2019, both also at the University of Oradea. She is a supervisor at PhD level in the field of Pharmacy, having over 100 works indexed in the Web of Science and 4 books in publishing houses in Romania. Her main areas of interest are pharmaceutics, public health and sustainability.

Simona Gabriela Bungau

Simona Gabriela Bungau became a full professor in 2005, having conducted continuous teaching and research activities. She started her work in 1992 and passed through all the stages of a university teaching career. After obtaining two Bachelor's degrees (in Chemistry-physics and Pharmacy), she obtained her Doctorate (PhD) in Chemistry at "Babes-Bolyai" University in Cluj, Romania, in 2003. Her extensive research activity is highlighted by her number Web of Science indexed publications - over 400 - and over 10 books published in Romanian publishing houses and 2 in international publishing houses. Her initial fields of interest were Chemistry and Analytical Chemistry, which then expanded to Pharmacology, Botany, Public health, Ecology and Sustainability/Environmental protection. Obtaining the qualification of PhD supervisor in the field of Pharmacy came naturally, in 2015, by presenting her Habilitation Thesis at "Carol Davila" University of Medicine and Pharmacy, Bucharest, Romania. Participating in over 20 international and national research projects, numerous Erasmus teaching programs, and societies in her native country and abroad, Mrs. Bungau continues her activity of over 30 years at the same university, namely, the University of Oradea, Romania.

Preface to "Antioxidant Activity of Essential Oils"

In recent years, the high efficacy of herbal antioxidant products in various diseases has been reported. Numerous essential oils (EOs) with antioxidant properties must be mentioned here, as their use as natural antioxidants is now a field of wide interest—especially in food science and medicine. EOs are complex products, usually consisting of over 50 components at varying concentrations. The EO substances described as having antioxidant activity represent an important fraction of the total compounds, often being indicated as a potential source in the discovery and development of new bioactive compounds, with medical, pharmaceutical, cosmetic, and other uses. Additionally, EOs and their constituents have been studied as alternative additives in the food industry, emphasizing their advantages over synthetic antioxidants that often have negative effects on health.

This Special Issue collects original research articles and reviews dealing with all aspects of the antioxidant activity of EOs:

·The chemistry and mechanisms that are the basis of EOs' antioxidant activity;

·Methods used to measure antioxidant activity;

·In vitro and in vivo antioxidant activity;

·The pharmaceutical, cosmetic, and food applications of bioactive compounds from EOs and their mechanisms, focusing on their antioxidant activity;

·The testing of combinations of oils and combinations with other antioxidant compounds in order to increase their antioxidant potential;

·The antioxidant activity of innovative formulations, such as active packaging and nano-/microparticles containing EOs.

With 18 papers published, we consider this book as a valuable reward for all of us, authors and editors. In this regard, the guest editors also wish to express their gratitude to all those involved in the success of the aforementioned volume, with the hope of continued fruitful collaboration with all authors who have succeeded in publishing in this Special Issue, as well as potential authors who works in this rich subject.

Delia Mirela Tit and Simona Gabriela Bungau
Editors

antioxidants

MDPI

Editorial

Antioxidant Activity of Essential Oils

Delia Mirela Tit [1,2] and Simona Gabriela Bungau [1,2,*]

1 Department of Pharmacy, Faculty of Medicine and Pharmacy, University of Oradea, 410028 Oradea, Romania
2 Doctoral School of Biomedical Sciences, University of Oradea, 410087 Oradea, Romania
* Correspondence: sbungau@uoradea.ro

In recent few years, the high efficacy of herbal antioxidant products in various diseases has been reported. Numerous essential oils (EOs) with antioxidant properties must be mentioned here, their use as natural antioxidants being a field of real interest—especially in food science and medicine. EOs are complex products, usually consisting of over 50 components in varying concentrations. The EO substances described as having antioxidant activity represent an important fraction of the total compounds, often being indicated as a potential source in the discovery and development of new bioactive compounds, with medical, pharmaceutical, cosmetic, and other uses. Additionally, EOs and their constituents have been studied as alternative additives in the food industry, emphasizing their advantages over synthetic antioxidants that often have negative effects on health.

This Special Issue collected original research articles and reviews dealing with all aspects of the antioxidant activity of EOs:

- The chemistry and mechanisms at the basis of EOs' antioxidant activity;
- Methods used to measure antioxidant activity;
- In vitro and in vivo antioxidant activity;
- The pharmaceutical, cosmetic, and food applications of bioactive compounds from EOs and their mechanisms, focusing on their antioxidant activity;
- The testing of combinations of oils and combinations with other antioxidant compounds in order to increase their antioxidant potential;
- The antioxidant activity of innovative formulations, such as active packaging and nano/microparticles containing EOs.

A summary of the present information on this topic is provided by the 15 original research articles and two reviews gathered in this Special Issue.

In one review, Sharma et al. show that *Embelia ribes* may prove to be an efficient medicinal tool in the treatment of many illnesses, due to the abundance and variety of bioactive compounds that demonstrate antioxidant, wound-healing, anti-diabetic, antiviral, antibacterial, antifungal, anti-obesity, cardioprotective, and antifertility action, among other interesting pharmacological properties, such as that related to diseases of the central nervous system [1].

Another review, published by Agarwal et al., provides a thorough overview of the characteristics, content, and mechanism of action of citrus EOs in relation to a variety of health-related conditions. The soothing, calming, mood-lifting, and cheer-enhancing benefits of citrus EOs are coupled with their pleasant smell. Their dispersion generates a feeling of freshness and cleanliness, helps relieve tension and anxiety, and boosts mood and physical and emotional energy [2].

The effects of aromatherapy with EOs were also tested by Mot et al., in a study determining the chemotype, antioxidant activity, and prospective aromatherapy uses for *Salvia officinalis* (common sage) EO, in a medical setting. According to this investigation, EO has a limited antioxidant capability, but hospitalized patients may find that inhaling the oil's high borneol content increases their comfort/pleasure [3].

Citation: Tit, D.M.; Bungau, S.G. Antioxidant Activity of Essential Oils. *Antioxidants* **2023**, *12*, 383. https://doi.org/10.3390/antiox12020383

Received: 30 January 2023
Accepted: 2 February 2023
Published: 5 February 2023

In silico methods have been used by several authors, alongside classic in vitro methods, to test the antioxidant action and safety of the EOs/bioactive compounds studied. The study conducted by Karakoti et al. revealed some intriguing biological properties of three Vitex species (*V. agnus-castus*, *V. negundo*, and *V. trifolia*), particularly as natural phytotoxic agents and antioxidants, which supports the utilization of this plant species in traditional medicine beyond the domain of crop protection [4]. In the EO of *Allium sativum* cultivated in Peru, Herrera-Calderon et al. determined the presence of two oxygenated terpenes (α-bisabolol and an unknown constituent with the formula $C_{22}H_{42}O_4$), not reported in other studies, associated by the authors with the high antioxidant power of the studied oil. Additionally, in silico studies (on nicotinamide adenine dinucleotide phosphate oxidase) showed that α-bisabolol had the highest docking rating and displayed excellent stability. According to the ADMET prediction, all garlic components can be administered orally and topically without causing any toxicity [5]. According to the results of the research carried out by Minchán-Herrera et al., volatile *Valeriana pilosa* components may contribute to the observed antioxidant effect by functioning as potential CYP2C9 gene and xanthine oxidase inhibitors [6]. Shah et al. reported, for the first time, the pharmacological activities of *Scutellaria edelbergii* EOs. The results indicate the presence of the bioactive constituent methyl 7-abieten-18-oate, which has the potential to function as an antioxidant, efficient painkiller, and anti-inflammatory agent, in addition to being a prospective candidate molecule for use against microorganisms. The anti-inflammatory ability of methyl 7-abieten-18-oate to inhibit cyclooxygenase-2 enzyme activity has been further validated by computational research, and its absorption, distribution, metabolism, excretion, and toxicity (ADMET) properties support the use of this molecule for additional investigation in a clinical trial [7].

Cytotoxic assays against *Artemia salina* were performed in three studies. New information is presented by Cascaes et al. regarding the antioxidant activity, chemical composition, and early toxicity of EOs from the Brazilian Amazonian species Duguetia and Xylopia (Annonaceae). When compared with EO derived from D. riparia, which had low toxicity or was non-toxic, EOs obtained from *D. echinophora*, *X. frutescens*, and *X. emarginata* showed significant toxicity. The EOs of *X. frutescens* and *X. emarginata* had the strongest ability to neutralize 2,2-diphenyl-1-picrylhydrazyl (DPPH) and 2,2′-azino-bis(3-ethylbenzothiazoline-6-sulfonic acid (ABTS) radicals. The primary components of these EOs may be predominantly responsible for the observed antioxidant capacity, but the influence of minor constituents should also be mentioned [8]. *Myrcia sylvatica* and *Myrciaria floribunda* (Myrtaceae) EOs were studied by de Moraes et al. According to preliminary toxicity studies, *M. floribunda* EO was moderately toxic compared with *A. salina*, whereas *M. sylvatica* EO was very toxic. Furthermore, *M. floribunda* EO demonstrated a stronger ability to block the DPPH radical [9]. The first study on the chemical content, antioxidant properties, and initial toxicity of the Amazonian plant extract known as *Croton campinarensis* is presented by da Costa et al. The EO aromatic profile was represented by terpenes, with a prevalence of sesquiterpene hydrocarbons (87.95%). A significant suppression of DPPH radicals was seen in the Trolox Equivalent Antioxidant Capacity assessment. According to the initial cytotoxicity assay against *A. salina*, the EO of *C. campinarensis* has a lethal concentration of 20.84 ± 4.84 $\mu g \cdot mL^{-1}$, making it toxic [10].

In vitro and in vivo studies were performed to test the biological effects of *Artemisia judaica L.* and *A. visnaga* L. EOs and *Heracleum persicum* oil nanoemulsion. Mohamed et al. carried out phytochemical investigation of *Artemisia judaica* L. for its antioxidant and anti-inflammatory properties, high concentration of oxygenated monoterpenes and cinnamate derivatives, and its therapeutic effect in the treatment of skin wounds [11]. *A. visnaga* L's chemical content and antioxidant capacities were investigated by Kamal et al. Although in vitro studies indicated a low antioxidant potential, *A. visnaga* L. EO supplementation considerably enhanced antioxidant capacity, as evidenced by an increase in the antioxidant enzyme activities of catalase, superoxide dismutase, and plasma glutathione peroxidase, and a decrease in the levels of 3,4-methylenedioxyamphetamine (MDA), according to an in vivo investigation on Swiss albino mice [12]. Bashlouei et al. prepared *Heracleum persicum* oil nanoemulsion (HAE-NE) and investigated its biological properties against healthy human fibroblasts from the foreskin and human breast cancer cells. In the liver, kidney, and jejunum of mice, the HAE-NE at 1.5, 2.5, and 3.5 g/concentration increased caspase 3 and accelerated the sub-G1 peak of the cell cycle with no harmful effects. The results indicate that HAE-NE presented the ability to be turned into therapeutic medications and might represent an environmentally friendly nanotherapeutic alternative for use in food, cosmetics, and pharmaceutical applications [13].

Potential applications in the aquaculture sector were tested by Magara et al., by evaluating the regulation of antioxidant defense against oxidative stress in the rainbow trout, *Oncorhynchus mykiss*, fed fish meal supplemented with a supercritical extract of basil (F1-BEO). Increased F1-BEO supplementation in fish meal (1–3% *w/w*) led to decreased glutathione levels and the failure of numerous important antioxidant enzymes. Furthermore, the levels of MDA indicate that fish fed diets containing 0.5–2% (*w/w*) F1-BEO supplements had sufficient oxidative stress defense supported by antioxidant pathways over the experimental period. When trout are administered highly substituted meals with F1-BEO for prolonged periods of time, the decline in crucial stress-shielding molecules raises the alarm for possible oxidative damage [14].

To improve the quality attributes of Marchigiana Burgers, the oxidative stability, color characteristics, microbiological profile, and fatty acid content of burgers treated with and without a combination of EOs (*Origanum vulgare var. hirtum* and *Rosmarinus officinalis*) were all evaluated by Fusaro et al. in their study on young Marchigiana bulls. The use of EO had an impact on the amount of thiobarbituric acid reactive material ($p < 0.05$). EOS, however, had no impact on the microbial profile or color characteristics while storing meat [15]. Manzur et al. evaluated the antioxidant and antipathogenic properties of the commercial EOs from orange—*Citrus sinensis* (L.) Osbeck—and concluded that these oils may serve as natural and secure substitutes to increase the period of validity of foods by preventing contamination and oxidation with pathogens that ruin food. Additionally, sweet orange EOs may offer an inventive dual strategy for food conservation [16]. The rosemary extracts derived from raw materials, post-distillation materials, and post-supercritical CO_2 extraction ($scCO_2$) materials were all initially compared by Luca et al., using a phytochemical and multi-biological method. According to the authors, terpene-rich extracts (EO, $scCO_2$) that are typically utilized as aroma-active substances or food preservatives (antioxidants) can be replaced with byproducts [17].

The articles collected here offer fresh perspectives on the expanding knowledge and research possibilities in the creation of new nutraceutical/adjunctive approaches, taking advantage of the antioxidant potential of the bioactive chemicals found in EOs.

Conflicts of Interest: The authors declare no conflict of interest.

References

1. Sharma, V.; Gautam, D.N.S.; Radu, A.-F.; Behl, T.; Bungau, S.G.; Vesa, C.M. Reviewing the traditional/modern uses, phytochemistry, essential oils/extracts and pharmacology of *Embelia ribes* Burm. *Antioxidants* **2022**, *11*, 1359. [CrossRef] [PubMed]
2. Agarwal, P.; Sebghatollahi, Z.; Kamal, M.; Dhyani, A.; Shrivastava, A.; Singh, K.K.; Sinha, M.; Mahato, N.; Mishra, A.K.; Baek, K.-H. Citrus essential oils in aromatherapy: Therapeutic effects and mechanisms. *Antioxidants* **2022**, *11*, 2374. [CrossRef] [PubMed]
3. Mot, M.-D.; Gavrilaș, S.; Lupitu, A.I.; Moisa, C.; Chambre, D.; Tit, D.M.; Bogdan, M.A.; Bodescu, A.-M.; Copolovici, L.; Copolovici, D.M.; et al. *Salvia officinalis* L. essential oil: Characterization, antioxidant properties, and the effects of aromatherapy in adult patients. *Antioxidants* **2022**, *11*, 808. [CrossRef] [PubMed]
4. Karakoti, H.; Mahawer, S.K.; Tewari, M.; Kumar, R.; Prakash, O.; de Oliveira, M.S.; Rawat, D.S. Phytochemical profile, in vitro bioactivity evaluation, in silico molecular docking and ADMET study of essential oils of three *Vitex* species grown in Tarai Region of Uttarakhand. *Antioxidants* **2022**, *11*, 1911. [CrossRef] [PubMed]
5. Herrera-Calderon, O.; Chacaltana-Ramos, L.J.; Huayanca-Gutiérrez, I.C.; Algarni, M.A.; Alqarni, M.; Batiha, G.E.-S. Chemical constituents, in vitro antioxidant activity and in silico study on NADPH oxidase of *Allium sativum* L. (garlic) essential oil. *Antioxidants* **2021**, *10*, 1844. [CrossRef] [PubMed]

6. Minchán-Herrera, P.; Ybañez-Julca, R.O.; Quispe-Díaz, I.M.; Venegas-Casanova, E.A.; Jara-Aguilar, R.; Salas, F.; Zevallos-Escobar, L.; Yáñez, O.; Pino-Rios, R.; Calderon, P.B.; et al. *Valeriana pilosa* roots essential oil: Chemical composition, antioxidant activities, and molecular docking studies on enzymes involved in redox biological processes. *Antioxidants* **2022**, *11*, 1337. [CrossRef]
7. Shah, M.; Bibi, S.; Kamal, Z.; Al-Sabahi, J.N.; Alam, T.; Ullah, O.; Murad, W.; Rehman, N.U.; Al-Harrasi, A. Bridging the chemical profile and biomedical effects of *Scutellaria edelbergii* essential oils. *Antioxidants* **2022**, *11*, 1723. [CrossRef]
8. Cascaes, M.M.; De Moraes, Â.A.B.; Cruz, J.N.; Franco, C.d.J.P.; Silva, R.C.E.; Nascimento, L.D.d.; Ferreira, O.O.; Anjos, T.O.d.; de Oliveira, M.S.; Guilhon, G.M.S.P.; et al. Phytochemical profile, antioxidant potential and toxicity evaluation of the essential oils from *Duguetia* and *Xylopia* species (*Annonaceae*) from the Brazilian Amazon. *Antioxidants* **2022**, *11*, 1709. [CrossRef]
9. De Moraes, Â.A.B.; Ferreira, O.O.; da Costa, L.S.; Almeida, L.Q.; Varela, E.L.P.; Cascaes, M.M.; de Jesus Pereira Franco, C.; Percário, S.; Nascimento, L.D.d.; de Oliveira, M.S.; et al. Phytochemical Profile, Preliminary Toxicity, and Antioxidant Capacity of the Essential Oils of *Myrciaria floribunda* (H. West ex Willd.) O. Berg. and *Myrcia sylvatica* (G. Mey) DC. (Myrtaceae). *Antioxidants* **2022**, *11*, 2076. [CrossRef]
10. Da Costa, L.S.; de Moraes, Â.A.B.; Cruz, J.N.; Mali, S.N.; Almeida, L.Q.; do Nascimento, L.D.; Ferreira, O.O.; Varela, E.L.P.; Percário, S.; de Oliveira, M.S.; et al. First report on the chemical composition, antioxidant capacity, and preliminary toxicity to *Artemia salina* L. of *Croton campinarensis* secco, A. Rosário & PE Berry (Euphorbiaceae) essential oil, and in silico study. *Antioxidants* **2022**, *11*, 2410. [CrossRef]
11. Mohammed, H.A.; Qureshi, K.A.; Ali, H.M.; Al-Omar, M.S.; Khan, O.; Mohammed, S.A.A. Bio-evaluation of the wound healing activity of *Artemisia judaica* L. as part of the plant's use in traditional medicine; phytochemical, antioxidant, anti-inflammatory, and antibiofilm properties of the plant's essential oils. *Antioxidants* **2022**, *11*, 332. [CrossRef]
12. Kamal, F.Z.; Stanciu, G.D.; Lefter, R.; Cotea, V.V.; Niculaua, M.; Ababei, D.C.; Ciobica, A.; Ech-Chahad, A. Chemical composition and antioxidant activity of *Ammi visnaga* L. essential oil. *Antioxidants* **2022**, *11*, 347. [CrossRef] [PubMed]
13. Bashlouei, S.G.; Karimi, E.; Zareian, M.; Oskoueian, E.; Shakeri, M. *Heracleum persicum* essential oil nanoemulsion: A nanocarrier system for the delivery of promising anticancer and antioxidant bioactive agents. *Antioxidants* **2022**, *11*, 831. [CrossRef] [PubMed]
14. Magara, G.; Prearo, M.; Vercelli, C.; Barbero, R.; Micera, M.; Botto, A.; Caimi, C.; Caldaroni, B.; Bertea, C.M.; Mannino, G.; et al. Modulation of antioxidant defense in farmed rainbow trout (*Oncorhynchus mykiss*) fed with a diet supplemented by the waste derived from the supercritical fluid extraction of basil (*Ocimum basilicum*). *Antioxidants* **2022**, *11*, 415. [CrossRef] [PubMed]
15. Fusaro, I.; Cavallini, D.; Giammarco, M.; Serio, A.; Mammi, L.M.E.; De Matos Vettori, J.; Lanzoni, L.; Formigoni, A.; Vignola, G. Effect of diet and essential oils on the fatty acid composition, oxidative stability and microbiological profile of Marchigiana burgers. *Antioxidants* **2022**, *11*, 827. [CrossRef] [PubMed]
16. Manzur, M.; Luciardi, M.C.; Blázquez, M.A.; Alberto, M.R.; Cartagena, E.; Arena, M.E. *Citrus sinensis* essential oils as innovative antioxidants and an antipathogenic dual strategy in food preservation against spoilage bacteria. *Antioxidants* **2023**, *12*, 246. [CrossRef]
17. Luca, S.V.; Zengin, G.; Sinan, K.I.; Korona-Glowniak, I.; Minceva, M.; Skalicka-Woźniak, K.; Trifan, A. Value-added compounds with antimicrobial, antioxidant, and enzyme-inhibitory effects from post-distillation and post-supercritical CO_2 extraction by-products of rosemary. *Antioxidants* **2023**, *12*, 244. [CrossRef]

antioxidants

Article

Citrus sinensis Essential Oils an Innovative Antioxidant and Antipathogenic Dual Strategy in Food Preservation against Spoilage Bacteria

Marilina Manzur [1,†], María C. Luciardi [1,2,†], M. Amparo Blázquez [3,*], María R. Alberto [1,2], Elena Cartagena [1,2] and Mario E. Arena [1,2,*]

1 INBIOFAL (Instituto de Biotecnología Farmacéutica y Alimentaria) CONICET, Avenida Kirchner 1900, Tucumán 4000, Argentina
2 Facultad de Bioquímica, Química y Farmacia, Universidad Nacional de Tucumán (UNT), Ayacucho 471, Tucumán 4000, Argentina
3 Departament de Farmacologia, Facultat de Farmàcia, Universitat de València, Avd. Vicent Andrés Estellés s/n, 46100 Burjasot, Spain
* Correspondence: amparo.blazquez@uv.es (M.A.B.); mario.arena@fbqf.unt.edu.ar (M.E.A); Tel.: +54-381-4856367 (M.E.A.)
† These authors contributed equally to this work.

Abstract: The present study evaluates the chemical compositions and antioxidant and antipathogenic properties of commercial orange (*Citrus sinensis* (L.) Osbeck) essential oils obtained using the cold-press method (EOP) and the cold-press method followed by steam distillation (EOPD). The chemical compositions of the volatilizable fractions, determined by gas chromatography-mass spectrometry, were similar in both samples. A relatively large amount of γ-terpinene was found in the EOPD (1.75%) as compared to the EOP (0.84%). Monoterpene hydrocarbons with limonene (90.4–89.8%) followed by myrcene (3.2–3.1%) as the main compounds comprised the principal phytochemical group. The non-volatile phenolics were eight times higher in the EOP than in the EOPD. Several assays with different specificity levels were used to study the antioxidant activity. Although both essential oils presented similar reducing capacities, the radical elimination ability was higher for the EOP. Regarding the antipathogenic properties, the EOs inhibited the biomass and cell viability of *Staphylococcus aureus* and *Pseudomonas aeruginosa* biofilms. Furthermore, both EOs similarly attenuated the production of elastase, pyocyanin, and quorum-sensing autoinducers as assessed using Gram-negative bacteria. The EOP and EOPD showed important antioxidant and antipathogenic properties, so they could represent natural alternatives to extend the shelf life of food products by preventing oxidation and contamination caused by microbial spoilage.

Keywords: sweet orange; cold pressing; hydrodistillation; scavenging activity; reducing capacity; virulence factors; quorum sensing

Citation: Manzur, M.; Luciardi, M.C.; Blázquez, M.A.; Alberto, M.R.; Cartagena, E.; Arena, M.E. *Citrus sinensis* Essential Oils as an Innovative Antioxidants and Antipathogenic Dual Strategy in Food Preservation against Spoilage Bacteria. *Antioxidants* **2023**, *12*, 246. https://doi.org/10.3390/antiox12020246

Academic Editors: Delia Mirela Tit and Simona Bungau

Received: 4 January 2023
Revised: 16 January 2023
Accepted: 18 January 2023
Published: 21 January 2023

1. Introduction

Different industries, such as the pharmaceutical, sanitary, cosmetic, and food industries, have paid attention to essential oils (EOs) to improve the shelf life and quality of products due to their potent antimicrobial and antioxidant activities [1,2]. As a result, the application of naturally produced antimicrobial compounds, such as EOs extracted from plants, has received significant attention [3–6]. EOs are mixtures of 20–100 different plant secondary metabolites with significant chemical variability [7]. This chemical variability is due to the variable ecological and geographical conditions, the age of the plant, the harvesting time, and the different extraction methodologies. The variations in the chemical profiles of EOs may influence their biological activity [8]. In numerous cases, the EOs' bioactivities are attributed to one or two principal components. However, the major constituents sometimes do not represent the overall activity [8,9].

Several EOs, such as *Citrus* EOs, have obtained the "GRAS" (Generally Regarded as Safe) category from the US Food and Drug Administration, given their favorable safety profiles [8], which is why studies on the biological activities of *Citrus* essential oils are increasing. *Citrus* spp. have been extensively investigated for their EOs, although their biological activities are still under study [3]. *Citrus* species belonging to the Rutaceae family are among the most commercially significant crops cultivated in tropical and subtropical climate regions [10]. The orange is one of the top-rated citrus fruits, and orange production accounts for more than 50% of global citrus production [11]. *Citrus* essential oils are particularly fascinating since they can be used as antioxidants because of their ability to protect organisms and tissues from the damage inflicted by reactive oxygen species and as flavoring agents [12]. They are rich sources of bioactive compounds; about 85–99% of the components are volatile and include a mixture of monoterpenes, sesquiterpenes, and oxygenated derivatives (aldehydes, ketones, acids, alcohols, and esters) [13]. The EOs of citrus fruits of various species have shown various biological activities, such as antibacterial, antiviral, fungicidal, and antioxidant effects. Therefore, these EOs can be used as a safer alternative to synthetic preservatives [14,15].

Another critical problem in the food industry is the tolerance of foodborne pathogens to various environmental stressors used as preservation methods (heat, cold, salt, and acid conditions), as well as the pathogens' ability to form biofilms on biotic or abiotic surfaces. The biofilm allows bacteria to contaminate surfaces in contact with food and transfer onto them [16]. Several foodborne disease outbreaks have been associated with biofilms [17], which has become a significant challenge to food production [18,19]. In biofilm formation, quorum sensing (QS) enables a phenotypic change in bacteria, whereby sessile biofilm bacteria show increased resistance to many biocides, disinfectants, and antibiotics [20]. Therefore, inhibiting QS and the virulence factors controlled by it is a primary health objective.

On the other hand, microorganisms have an innate ability to produce reactive oxygen species to promote and maintain their redox cycle and enhance their microbial attachment by forming biofilms. Consequently, oxidative stress is a fundamental driving force for bacteria to transfer from the planktonic (free-living) state to the biofilm layer [21]. Thus, some authors have related the antioxidant property of a sample to its ability to reduce biofilm production by pathogenic bacteria, suggesting the use of antioxidant compounds as an alternative method to treat, prevent, and eradicate biofilms [21–23].

Considering these challenges of the food industry, this work aims to determine the chemical compositions of two orange essential oils obtained industrially using different methodologies: the cold-pressed method (EOP) and cold-pressed method followed by steam distillation (EOPD). Moreover, we attempt to determine their potential as antioxidant and antimicrobial agents against planktonic cells and as antipathogenic agents active against biofilm and other virulence factors controlled by QS of *Pseudomonas aeruginosa* and *Staphylococcus aureus*, two significant pathogenic food spoilage bacteria.

2. Materials and Methods

2.1. Sample

The oranges (*Citrus sinensis* (L.) Osbeck) were cultivated in Entre Rios, Argentina, in 2019, and their Eos (commercial samples) were provided by the Litoral Citrus Company.

Cold-pressed EOs (EOPs) represent 99.9% of the industrial and commercial EOs produced, which are obtained by applying cold pressure to orange peels. The cold-pressing of EOs followed by steam distillation (EOPD) was applied to the liquid discharged from the cold-pressed oil that did not separate in the initial centrifugation process.

2.2. Gas Chromatography-Mass Spectrometry

The gas chromatography–mass spectrometry (GC-MS) analysis was carried out with an Agilent 5973N apparatus equipped with a capillary column (95% dimethylpolysiloxane–5% diphenyl), HP-5MS UI (30 m in length and 0.25 mm i.d., with a 0.25 mm film thickness).

Here, 2 µL of a mixture containing 20 µL of EO samples in 0.5 mL of dichloromethane (99%, Fisher Scientific, Hampton, NH, USA) was injected. The column temperature program was 60 °C for 5 min, with 3 °C/min increases to 180 °C, then 20 °C/min increases to 280 °C, which was maintained for 10 min. The carrier gas was helium at a flow rate of 1 mL/min. Split mode injection (ratio 1:30) was employed. Mass spectra were taken over the m/z 30–500 range with an ionizing voltage of 70 eV [24]. The identification of EO components was based on matching their mass spectra peaks with those from the NIST 2005 Mass Spectral Library. The experimental values for Kovats retention indices (RIs), relative to C8–C30 n-alkanes, were determined compared to those from the available literature [25]. They were used as an additional tool to support the MS findings. The percentile presence of components in EO samples was calculated from the peak areas obtained in the area percentage reports (standard processing of chromatograms without replicates), without correction factors, using the normalization method.

2.3. *Total Phenolic Content*

The total phenolic content of the samples was measured spectrophotometrically based on the Folin–Ciocalteu method [22].

2.4. *Antioxidant Capacity*

2.4.1. Phosphomolybdenum Total Antioxidant Activity Assay

The total antioxidant activity of the samples was evaluated using the phosphomolybdenum method, according to Zengin et al. [26]. The sample solution (dil 1/100 DMSO) was combined with 1 mL of reagent solution (0.6 M sulfuric acid, 28 mM sodium phosphate, and 4 mM ammonium molybdate). A control experiment without samples was conducted identically (control). After 90 min of incubation at 95 °C, the absorbance was read at 695 nm. The total antioxidant capacity was calculated from a standard curve (6–50 µg/mL) of ascorbic acid (Biopack, Buenos Aires, Argentina). The results are expressed as ascorbic acid equivalents (AEs).

2.4.2. Nitric Oxide (NO) Scavenging Activity Method

Different concentrations of samples (15–150 µL/mL), sodium nitroprusside (99%, Sigma-Aldrich, St. Louis, MO, USA) (100 mM final concentration), and phosphate buffer (0.2 M, pH 7.4) at a final volume of 300 µL were incubated at 37 °C for 15 min under a light. A control experiment without samples was conducted in an identical manner (control). Then, the reaction mixtures were mixed with Griess reagent, and the absorbance of the chromophore formed was measured at 550 nm after 5 min [22]. The % nitric oxide scavenging activity was calculated using the following equation:

$$(\%)\ \text{Scavenging} = ((\text{DOcontrol} - \text{DOsample})/\text{DOcontrol}) \times 100$$

where DOcontrol is the absorbance of the mixture reaction containing all reagents except the test compounds and DOsample is the absorbance of the mixture reaction containing the test compounds. The SC_{50} (concentration necessary to scavenge 50% of radical) was calculated using a regression curve (scavenging concentration vs. sample concentration). From an ascorbic acid standard curve (25–200 µg/mL), the ascorbic acid equivalent antioxidant capacity (AEAC) of the samples was calculated.

2.4.3. ABTS Radical Scavenging Method

This assay was performed as described by Re et al. [27] with slight modifications. Samples at different volumes (diluted 1/20 with methanol) were mixed with 200 µL of ABTS radical solution (98%, Sigma-Aldrich, St. Louis, MO, USA) and made up with 96% ethanol to a final volume of 300 µL. A control experiment without samples was conducted identically (control). After a 1 h incubation period at room temperature, the absorbance was recorded at 734 nm using a microplate reader. The percentage of radical scavenging activity was calculated in the following way:

$$(\%)\ \text{Scavenging} = (\text{DOcontrol} - \text{DOsample})/\text{DOcontrol}) \times 100$$

where DOcontrol is the absorbance of the mixture reaction containing all reagents except the test compounds and DOsample is the absorbance of the mixture reaction containing the test compounds. The SC_{50} was calculated using a regression curve. From a Trolox standard curve (2–8 μg/mL), the Trolox equivalent antioxidant capacity (TEAC) of the samples was calculated.

2.4.4. Cupric-Reducing Antioxidant Capacity (CUPRAC) Method

This assay was determined using the method of Sadeer et al. [28]. The reaction mixture consisted of copper (II) chloride solution (10 mM), neocuproine (98%, Sigma-Aldrich, St. Louis, MO, USA) (7.5 mM), ammonium acetate buffer (1 M, pH = 7), and sample dilutions to reach a final volume of 820 μL. A control experiment without samples was conducted in an identical manner (control). The test tubes were incubated at room temperature (20 to 25 °C) for 30 min. The absorbance at 450 nm was monitored against a blank without neocuproine. The results are expressed as TEAC values using the calibration curve (5–20 μg/mL).

2.5. *Bacterial Growth Conditions*

The strains were obtained from the American Type Culture Collection (ATCC) and Laboratory of Research of Added Value of Regional Products and Foods (LVP) of INBIOFAL (Instituto de Biotecnología Farmacéutica y Alimentaria).

Two *P. aeruginosa* strains were used (ATCC 27853 as a reference and HT5, a multi-antibiotic-resistant strain isolated from a patient with food poisoning). These strains were cultured at 37 °C in Luria–Bertani (LB) medium. Additionally, two *S. aureus* strains were used (ATCC 6538 and HT1 methicillin-resistant). These strains were cultured at 37 °C in Müller–Hinton (MH) medium.

In a microtiter plate, 20 μL of each sample solution (1, 5, 10, 20, and 40 mg/mL) as mixed to arrive at final concentrations of 0.1, 0.5, 1, 2, and 4 mg/mL in wells ($n = 8$) with 180 μL of each strain suspension (OD 0.12 ± 0.01 at 560 nm) from the exponential-phase culture. A vehicle that dissolves EO (DMSO/water, 1:1) was used as the positive control for growth, and the antibiotic ciprofloxacin at low concentration (5 μg/mL) was used as the negative control. The growth was determined at 560 nm (Power Wave XS2, Biotek, Winooski, VT, USA) after 24 h of incubation at 37 °C.

2.6. *Biofilm Formation Assay*

After 24 h of incubation of bacterial cultures prepared as indicated above, the biofilms were stained with 200 μL of an aqueous crystal violet (pa-grade; Cicarelli, Santa Fe, Argentina) solution (0.1%, *w/v*) for 20 min [24,29]. After washing with water, the liquid in the wells was discarded and the material that remained fixed to the polystyrene (biofilm) was washed with distilled water. The crystal violet that adhered to the biofilm in each well was stained using 200 μL of absolute ethanol, and the absorbance was measured at 595 nm using a microtiter plate reader (Multiskan Go, Thermo, Waltham, MA, USA). The biofilm biomass inhibition was calculated relative to the biofilm production in the untreated control culture.

2.7. *Biofilm Metabolic Activity Assay*

Cell viability measured as the bacterial metabolic activity in the biofilm was assessed using the 3-[4,5-dimethylthiazol-2-yl]-2,5-diphenyltetrazolium bromide (MTT) (97.5%, Sigma-Aldrich, St. Louis, MO, USA) reduction assay with some modifications [24,29]. The bacterial biofilm is formed by incubating in all wells 200 μL of each bacterial suspension (OD 560 nm, 0.09 ± 0.02) for 24 h at 37 °C. Subsequently, the bacterial culture is discarded, and the already-formed biofilm remains adhered to the walls. Then, 180 μL of PBS (pH 6.5) and 20 μL of each sample are added to the solution (n = 8) per well (final concentrations of 0.1, 0.5, 1, 2, and 4 mg/mL per well), which are then incubated again at 37 °C for 24 h and washed with PBS. To determine bacterial survival, 100 μL of MTT solution (0.5 mg/mL) is added to each well and incubated for 3 or 6 h at 37 °C. If the compound is degraded, the formed purple formazan dissolves in the DMSO and the absorbance is measured at 570 nm. The controls used were the EO vehicle and ciprofloxacin (5 μg/mL).

2.8. Elastase Activity and Pyocyanin Quantification

In the cell-free culture, supernatants of each *P. aeruginosa* strain cultivated in the presence and absence of EOs and limonene, pyocyanin, and elastase activities were quantified as described by Díaz et al. [30]. The elastolytic activity in the supernatants was evaluated using the elastin–Congo red conjugate (Sigma–Aldrich, St. Louis, MO, USA) at 495 nm. At the same time, the pyocyanin activity was determined using the chloroform–HCl extraction method and was quantified via absorbance measurements at 520 nm. DMSO-treated cultures were used as controls, and each test was assessed for statistical significance (n = 3).

2.9. Quantification of N-Acyl Homoserine Lactones (AHL)

Autoinducers are measured for their QS inhibition using the β-galactosidase activity assay, using the reporter strain *P. aeruginosa* qsc 119, a mutant donated by P. Greenberg [31], which is incapable of producing its own AHL. This strain responds to exogenous active signal molecules generated by wild-type *P. aeruginosa* strains by producing β-galactosidase. Consequently, the activity of β-galactosidase is directly related to the concentration of AHL [31]. The AHLs were determined according to a previously reported method [29], using a cell-free culture supernatant obtained from *P. aeruginosa* (ATCC 27853 or HT5) grown individually (n = 8) in LB medium in the presence of final concentrations of 4.0, 2.0, 1.0, 0.5, and 0.1 mg/mL of orange EOs and limonene for 24 h. The antibiotic azithromycin was used at a low concentration (5 μg/mL) as a positive control for QS. The β-galactosidase activity was measured using the Miller test [32].

2.10. Statistical Analysis

Differences in the mean values were evaluated using an analysis of variance (ANOVA). Tukey's test was used for all pair-wise multiple comparisons of groups. In all analyses, p values < 0.05 were considered statistically different (Statistix 7.1, 2002).

3. Results

3.1. Chemical Composition and Antioxidant Activity

In both commercial orange oils (EOP and EOPD), high amounts of monoterpene hydrocarbons were found (96.11 and 97.08%, respectively). The main monoterpene was limonene (90.41 and 89.78%, respectively), followed by myrcene (3.19 and 3.05%, respectively). However, in the EOPD (Figure 1 and Table 1), γ-terpinene and α-pinene appeared in amounts greater than 1% (1.75 and 1.12%, respectively). Among the oxygenated monoterpenes, the major compound was linalool in both commercial essential oils (0.55 and 0.83%, respectively). In addition, linalyl acetate and sabinene hydrate were only present in respectively EOP, while in respectively EOPD, another compound, mentha-2,8-dien-1-ol, was identified. Small amounts of the sesquiterpenes, both hydrocarbons, and oxygenated sesquiterpenes were found. Minimum qualitative and quantitative differences in this phytochemical group were observed in respectively EOP and EOPD (0.55 vs. 0.49%, 0.12 vs. 0.34%, and 0.12 vs. 0.03%, respectively). Concerning the sesquiterpene hydrocarbons, δ-elemene and γ-muurolene were only found in the EOPD, whereas germacrene D, caryophyllene oxide, and nootkatone were detected among the sesquiterpenes and oxygenated sesquiterpenes in the EOP (Figure 2 and Table 1). In addition, the sesquiterpene hydrocarbon valencene showed a three-fold higher concentration in EOP than EOPD (Table 1). Other constituents such as hexadecanal, hexadecanoic acid, and tetracosane were only found in EOP.

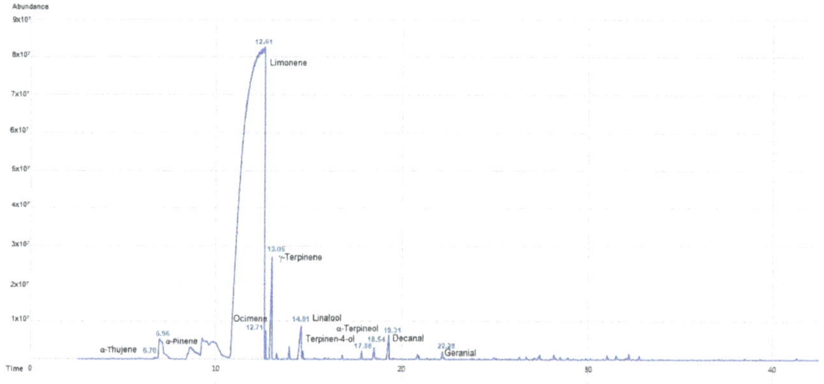

Figure 1. GC chromatogram of orange essential oil obtained using the cold-pressed method followed by steam distillation (EOPD).

Table 1. Chemical constituents of essential oils obtained from fruit peels of *Citrus sinensis* (L.) Osbeck (Rutaceae).

Compounds	RI [a]	KI [b]	Identification Methods	Relative Content (%)	
				EOP	EOPD
Monoterpene hydrocarbons				**96.111**	**97.080**
α-thujene	926	930	RI, MS	0.017	0.060
α-pinene	933	939	RI, MS	0.952	1.119
Sabinene	972	975	RI, MS	0.140	0.076
β-pinene	976	979	RI, MS	0.500	0.935
Myrcene	989	990	RI, MS	3.188	3.049
Limonene	1041	1039	RI, MS	90.408	89.781
Ocimene	1063	1050	RI, MS	0.061	0.153
γ-terpinene	1069	1059	RI, MS	0.838	1.752
Terpinolene	1088	1088	RI, MS	0.007	0.155
Oxygenated monoterpenes				**1.238**	**1.457**
Sabinene hydrate	1074	1070	RI, MS	0.006	-
Linalool	1101	1095	RI, MS	0.551	0.831
Mentha-2,8-dien-1-ol	1121	1122	RI, MS	-	0.020
cis-limonene oxide	1134	1136	RI, MS	0.011	0.017
trans-limonene oxide	1138	1142	RI, MS	0.029	0.012
Citronellal	1153	1153	RI, MS	0.078	0.050
Terpinen-4-ol	1176	1177	RI, MS	0.034	0.109
α-terpinol	1188	1188	RI, MS	0.077	0.200
Neral	1239	1238	RI, MS	0.125	0.064
Carvone	1241	1243	RI, MS	0.017	0.052
Linalyl acetate	1254	1257	RI, MS	0.001	-
Geranial	1269	1267	RI, MS	0.221	0.011
α-terpinyl acetate	1345	1349	RI, MS	0.011	0.007
Citronellyl acetate	1350	1352	RI, MS	0.005	0.012
Neryl acetate	1361	1361	RI, MS	0.027	0.032
Geranyl acetate	1380	1381	RI, MS	0.019	0.023
Limonen-10-yl-acetate	1405	1395	RI, MS	0.026	0.017

Table 1. *Cont.*

Compounds	RI [a]	KI [b]	Identification Methods	Relative Content (%)	
				EOP	EOPD
Sesquiterpene hydrocarbons				**0.552**	**0.487**
δ-elemene	1333	1338	RI, MS	-	0.016
α-copaene	1370	1376	RI, MS	0.037	0.042
β-cubebene	1384	1388	RI, MS	0.035	0.025
β-elemene	1385	1390	RI, MS	0.025	0.052
β-caryophyllene	1411	1419	RI, MS	0.043	0.034
β-copaene	1422	1432	RI, MS	0.054	0.024
α-bergamotene	1430	1434	RI, MS	0.011	0.008
α-humulene	1446	1454	RI, MS	0.009	0.016
β-farnesene	1452	1456	RI, MS	0.023	0.016
Germacrene D	1474	1480	RI, MS	0.034	-
γ-muurolene	1474	1479	RI, MS	-	0.057
Valencene	1486	1496	RI, MS	0.169	0.057
Bicyclogermacrene	1489	1500	RI, MS	0.012	0.011
α-muurolene	1492	1500	RI, MS	0.009	0.007
α-farnesene	1502	1505	RI, MS	0.034	0.069
δ-cadinene	1516	1523	RI, MS	0.055	0.049
Germacrene B	1549	1561	RI, MS	0.002	0.004
Oxygenated sesquiterpenes				**0.122**	**0.034**
Caryophyllene oxide	1574	1583	RI, MS	0.008	-
β-sinensal	1690	1699	RI, MS	0.053	0.011
α-sinensal	1746	1756	RI, MS	0.038	0.023
Nootkatone	1794	1806	RI, MS	0.023	-
Other compounds				**0.818**	**0.716**
Octanol	1076	1068	RI, MS	0.019	0.062
Nonanal	1104	1100	RI, MS	0.100	0.088
Decanal	1203	1201	RI, MS	0.517	0.431
Octanol acetate	1210	1213	RI, MS	0.016	0.014
2*E*-decenal	1259	1263	RI, MS	0.009	0.012
Undecanal	1302	1306	RI, MS	0.024	0.018
2*E*,4*E*-decadienal	1311	1315	RI, MS	0.005	0.009
Methyl-*N*-methyl anthranilate	1399	1406	RI, MS	-	0.009
Dodecanal	1403	1408	RI, MS	0.093	0.059
2*E*-dodecenal	1461	1466	RI, MS	-	0.006
Hexadecanal	1808	1817	RI, MS	0.009	-
Hexadecanoid acid	1961	1960	RI, MS	0.019	-
Tricosane	2283	2300	RI, MS	0.005	0.008
Tetracosane	2385	2400	RI, MS	0.002	-
Total VOCs				**98.841**	**99.774**

[a] RI: Retention index relative to C8–C30 n-alkane on HP-5MS column; [b] KI: Kovats retention index; VOCs: volatilizable organic compounds; EOP: essential oil obtained by cold-pressed method; EOPD: essential oil obtained by cold-pressed followed by steam distillation.

Different assays were carried out to evaluate the antioxidant properties of the EOP and EOPD. The ABTS assay based on the scavenging of a stable free radical (ABTS•+), CUPRAC assay based on the capacity of antioxidants to reduce ions copper, nitrite assay based on the scavenging of free radicals focused on nitrogen (•NO), and phosphomolybdenum assay based on the reduction of Mo(VI) to Mo(V). *C. sinensis* oils showed significant antioxidant potential in various experimental models by scavenging free radical and nitrogen species and reducing metals. The antioxidant results and phenolic content (Table 2 and Figure 3) showed that although both oils presented similar reducing capacity levels, the radical scavenging ability was higher for the EOP (SC_{50} 8 and 65.5 µL/mL for ABTS and NO, respectively) than EOPD. This could be related to their chemical compositions, since although both have similar terpene contents, the total polyphenol concentration of EOP is eight times higher than for EOPD. In addition, EOP contains greater amounts of oxygenated compounds, mainly the antioxidants nootkatone (not found in EOPD) and valencene (0.169 vs. 0.057).

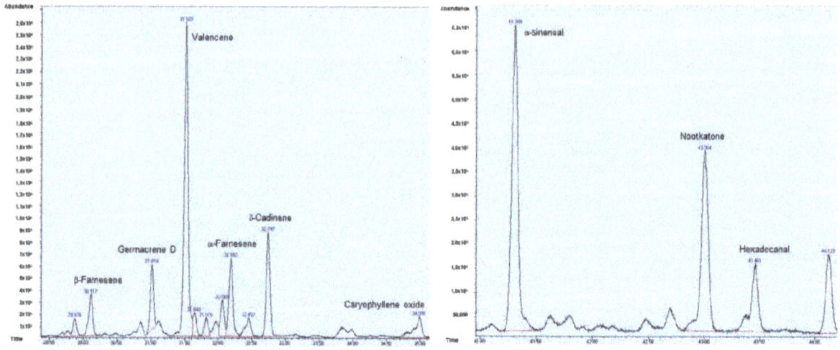

Figure 2. GC chromatogram of the sesquiterpene fractions of orange essential oil obtained using the cold-pressed method (EOP).

Table 2. Antioxidant activity of *Citrus sinensis* oils by total phenolic composition.

Oils	Phenolic Compounds	Reducing Capacity		Scavenging Capacity	
		Phosphomolybdenum Assay	CUPRAC Assay	Nitric Oxide Radical	ABTS Radical
	µg GAE/mL EO	mg AE/mL EO	mg TEAC/mL EO	mg AEAC/mL EO	mg TEAC/mL EO
EOP	84.80 ± 7.20	245.11 ± 18.60	0.55 ± 0.02	2.10 ± 0.45	0.55 ± 0.10
EOPD	10.53 ± 1.20 *	257.12 ± 10.60	0.52 ± 0.006	0.35 ± 0.08 *	0.07 ± 0.004 *

GAE: Gallic acid equivalent; AE ascorbic acid equivalent; AEAC: ascorbic acid equivalent antioxidant capacity; TEAC: Trolox equivalent antioxidant capacity. Results are expressed as means ± standard deviations (n = 3). EOP: Essential oil obtained by cold-pressed method; EOPD: essential oil obtained by cold-pressed method followed by steam distillation. Data are presented as means ± SEMs (n = 3); * indicates significant differences between samples, according to Tukey's test ($p < 0.05$).

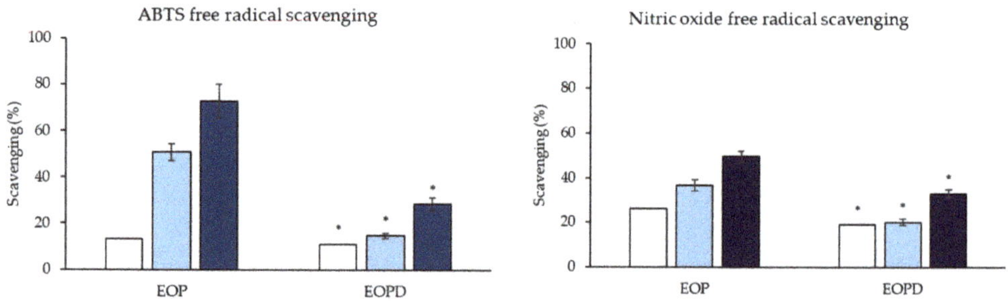

Figure 3. Free radical scavenging activity levels of different essential oil concentrations. EOP: Essential oil obtained by cold-pressed method; EOPD: essential oil obtained by cold-pressed method followed by steam distillation. Concentrations assayed for ABTS in EOP at 1.5 (□), 8 (■), and 12 (■) µL/mL and EOPD at 12 (□), 15 (■), and 30 (■) µL/mL. Concentrations assayed for nitric oxide in EOP at 15 (□), 30 (■), and 60 (■) µL/mL and EOPD at 30 (□), 60 (■), and 150 (■) µL/mL. Data are presented as means ± SEMs (n = 3); * represents significant differences between the oils, according to Tukey's test ($p < 0.05$).

3.2. Planktonic Growth and Total Biofilm Formation by S. aureus and P. aeruginosa in the Presence of Orange Oils

The EOP, EOPD, and their main component, limonene, moderately inhibited the planktonic growth of both *S. aureus* strains (Figure 4). However, the natural orange products inhibited bacterial biofilm formation by more than 50% for both bacteria at all concentrations assayed. The EOP's inhibition values ranged from 89 to 57% for the concentration range of 4 to 0.1 mg/mL, while the EOPD's inhibition values ranged from 90 to 53%. A dose-dependent effect until 2 mg/mL was observed for both strains. Limonene, for its part, had a lower inhibitory effect range of 46 to 25% at the same range of concentrations (Figure 4).

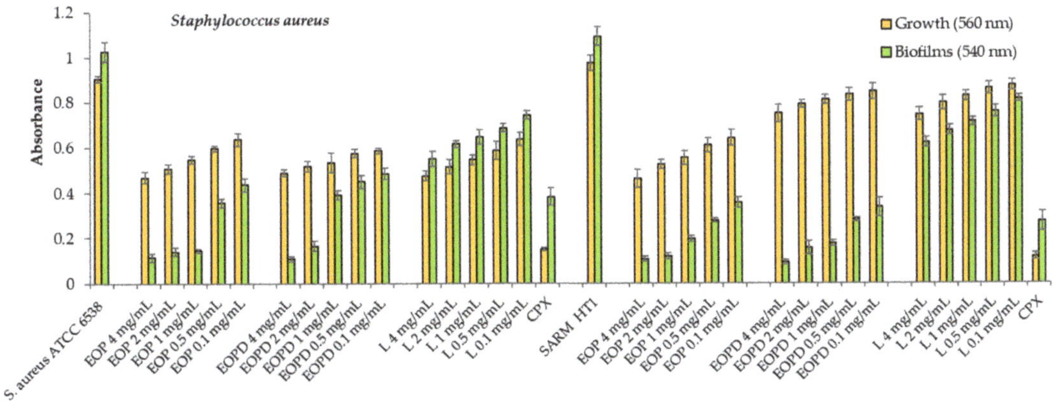

Figure 4. Growth and biofilms of *Staphylococcus aureus* (ATCC 6538 and methicillin-resistant HT1 strains). EOP: Essential oil obtained by cold-pressed method; EOPD: essential oil obtained by cold-pressed method followed by steam distillation; L: Limonene; CPX: ciprofloxacin. Data are presented as means ± SEMs (n = 8) of three independent experiments. All experiments showed significant differences compared to respective controls ($p < 0.05$).

The orange EOs reduced the growth of the *P. aeruginosa* strains from 31 to 6% in a concentration range of 4 to 0.1 mg/mL. In comparison, limonene caused decreases of 26 to 4% (Figure 5). However, all samples significantly inhibited the development of the *P. aeruginosa* ATCC 27853 biofilms. The EOP produced inhibition effects of 69 to 37%, the EOPD produced inhibition effects of 87 to 36%, and limonene showed the lowest inhibitory effects (46–33%) at the tested concentrations. Concerning the HT5 strain, the EOP and EOPD produced 77 to 50% inhibition rates for the concentration range of 4 to 0.1 mg/mL. On the other hand, the limonene caused inhibition rates of 51 to 25% in the same range of concentrations (Figure 5). The biofilm formation decreases were dose-dependent for both strains.

No significant ($p > 0.05$) differences were observed between the EOs (EOP and EOPD) based on the planktonic growth and biofilm biomass formation of the different bacterial species studied, except for the biofilm formation of the *P. aeruginosa* ATCC 27853 strain in the presence of 4 mg/mL of EO.

3.3. S. aureus and P. aeruginosa Biofilm Metabolic Activity in the Presence of Orange Oils

The tested products moderately inhibited the metabolic activity of *S. aureus* in the phenotype biofilm (Table 3). The EOP, EOPD, and limonene diminished the viability rates of the ATCC 6538 strain in the concentration range of 4 to 0.1 mg/mL by 42 to 27%, 48 to 30%, and 38 to 17%, respectively. The effects on the methicillin-resistant strain were 37–21%, 41–23%, and 42–26% for the EOP, EOPD, and limonene, respectively, at the tested concentrations. Likewise, the EOP, EOPD, and the main constituent of both, limonene, significantly inhibited the metabolic activity of *P. aeruginosa* in the phenotype biofilm (Table 3). At 4 mg/mL, the EOP, EOPD, and limonene inhibited the cell viability rates by 68, 65, and 36% for the ATCC 27853 strain; and by 57, 56, and 41% for the strain HT5, respectively. These results show a dose-dependent effect for all strains.

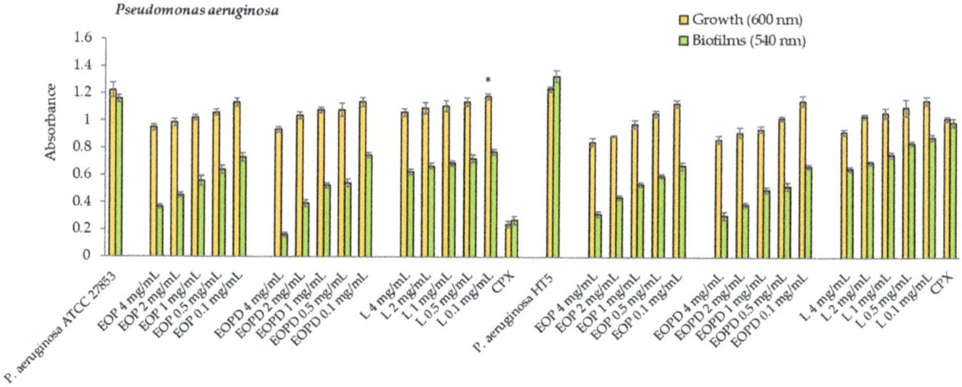

Figure 5. Growth and biofilms of *Pseudomonas aeruginosa* (ATCC 27853 and HT5 strains). EOP: Essential oil obtained by cold-pressed method; EOPD: essential oil obtained by cold-pressed method followed by steam distillation; L: limonene; CPX: ciprofloxacin. Data are presented as means ± SEMs (n = 8) of three independent experiments. All experiments show significant differences compared to the respective controls ($p < 0.05$), except the sample with asterisk (*).

Table 3. Biofilm metabolic activity (%) rates of *Staphylococcus aureus* and *Pseudomonas aeruginosa* strains.

Samples	*Staphylococcus aureus*		*Pseudomonas aeruginosa*	
	ATCC 6538	**HT1**	**ATCC 27853**	**HT5**
Control	100 ± 2.07	100 ± 1.97	100 ± 1.57	100 ± 2.00
EOP				
4 mg/mL	58.93 ± 1.19	63.09 ± 5.19	31.56 ± 7.00	43.21 ± 4.96
2 mg/mL	59.57 ± 1.97	70.14 ± 1.96	43.52 ± 3.88	49.23 ± 6.65
1 mg/mL	64.06 ± 3.30	73.22 ± 5.57	57.40 ± 3.62	58.87 ± 3.27
0.5 mg/mL	69.71 ± 3.00	76.59 ± 1.00	67.97 ± 2.78	61.62 ± 6.05
0.1 mg/mL	72.78 ± 1.88	78.97 ± 1.69	79.45 ± 2.35	66.41 ± 4.79
EOPD				
4 mg/mL	49.84 ± 2.25	59.06 ± 1.95	34.49 ± 7.04	43.96 ± 7.55
2 mg/mL	59.28 ± 1.35	63.32 ± 6.28	49.25 ± 2.64	50.55 ± 3.03
1 mg/mL	63.83 ± 2.07	69.57 ± 2.88	56.64 ± 3.80	57.92 ± 2.67
0.5 mg/mL	65.76 ± 5.25	73.22 ± 1.21	72.93 ± 2.19	61.59 ± 4.90
0.1 mg/mL	70.67 ± 1.57	76.57 ± 2.16	80.67 ± 1.58	71.80 ± 1.91
Limonene				
4 mg/mL	61.78 ± 3.57	58.29 ± 1.82	64.20 ± 3.73	58.56 ± 3.14
2 mg/mL	66.13 ± 3.36	60.70 ± 2.62	68.45 ± 2.66	66.50 ± 2.61
1 mg/mL	72.41 ± 1.94	63.14 ± 1.68	72.13 ± 1.70	71.30 ± 2.00
0.5 mg/mL	74.73 ± 2.16	70.61 ± 4.31	77.54 ± 1.52	78.99 ± 3.53
0.1 mg/mL	83.55 ± 2.14	74.47 ± 4.37	88.23 ± 1.59	84.74 ± 1.70
Ciprofloxacin 5 µg/mL	55.08 ± 2.81	55.16 ± 2.73	19.82 ± 3.12	64.22 ± 0.80

EOP: Essential oil obtained by cold-pressed method; EOPD: essential oil obtained by cold-pressed method followed by steam distillation. Data are presented as means ± SEMs (n = 8) of three independent experiments. All experiments show significant differences compared to respective controls ($p < 0.05$).

The EOP and EOPD exerted a similar ($p > 0.05$) action on the cell metabolic activity in a preformed biofilm of the selected strains.

3.4. Elastase, Pyocyanin, and Autoinducer Production by *P. aeruginosa* in the Presence of Orange Oils

The orange essential oils and limonene significantly inhibited the enzyme elastase, a key virulence factor of *P. aeruginosa* in the spread of infection (Figure 6). For both strains, the effect was significant and greater than 45%, even at the lowest concentration tested (0.1 mg/mL), reaching an elastolytic activity level close to 70% at 4 mg/mL for all natural orange products. No statistical differences were observed for elastase activity between the essential oils ($p > 0.05$). Likewise, the dose-dependent effect on the elastase production was only notable for the HT5 strain.

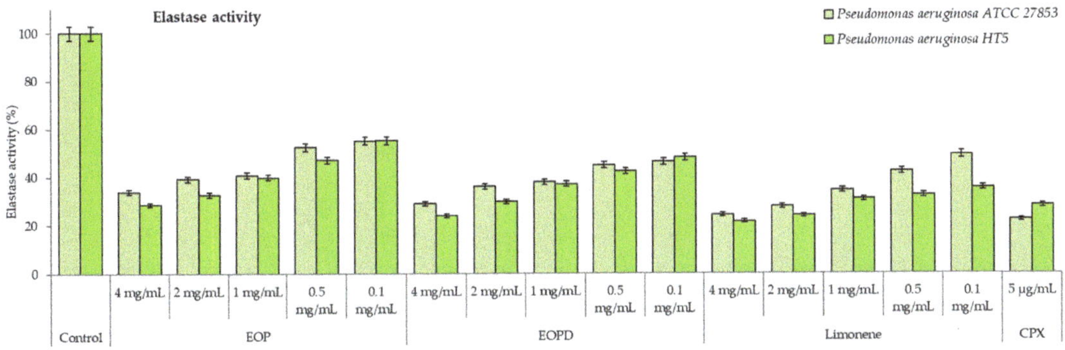

Figure 6. Elastase activity of *Pseudomonas aeruginosa* ATCC 27853 and HT5 strains. EOP: Essential oil obtained by cold-pressed method; EOPD: essential oil obtained by cold-pressed method followed by steam distillation; L: limonene; CPX: ciprofloxacin. Data are presented as means ± SEMs ($n = 8$) of three independent experiments. All experiments show significant differences compared to respective controls ($p < 0.05$).

Concerning the QS-dependent pyocyanin production of *P aeruginosa*, the EOP, EOPD, and limonene significantly inhibited the production of this toxic pigment. For both strains at the maximum concentration, the effects of EOP and EOPD equaled 65%, while for limonene the rate was 57%. At the lowest concentration, the effects were still significant, at close to 40% for both oils and less than 30% for limonene (Table 4). The results show a clear dose-dependent decrease in this virulence factor. No significant variations were observed in the production of pyocyanin between the essential oils ($p > 0.05$).

Table 4. Inhibition (%) of the production of the *Pseudomonas aeruginosa* virulence factor pyocyanin.

Sample	*P. aeruginosa* ATCC 27853	*P. aeruginosa* HT5
EOP 4 mg/mL	62.30 ± 2.05	68.02 ± 1.03
EOP 0.1 mg/mL	37.12 ± 3.00	40.07 ± 2.14
EOPD 4 mg/mL	65.21 ± 1.03	69.19 ± 0.00
EOPD 0.1 mg mL	41.42 ± 2.11	43.11 ± 4.08
Limonene 4 mg/mL	57.45 ± 0.05	58.23 ± 3.01
Limonene 0.1 mg/mL	30.00 ± 1.10	23.44 ± 3.23

EOP: Essential oil obtained by cold-pressed method; EOPD: essential oil obtained by cold-pressed method followed by steam distillation. Data are presented as means ± SEMs ($n = 3$) of three independent experiments. All experiments show significant differences compared to respective controls ($p < 0.05$).

As seen in Figure 7, both sweet orange EOs and limonene can reduce the β-galactosidase activity of *P. aeruginosa*, which depends on the QS mechanism. This fact indicates that EOP,

EOPD, and limonene can inhibit the production of AHLs in both strains and can interrupt the bacterial communication. In *P. aeruginosa* ATCC 27853, inhibition rates of 56 to 29% and 63 to 39% by EOP and EOPD, respectively, were observed in the concentration range of 4 to 0.1 mg/mL. Meanwhile, the principal constituent of the orange essential oils, limonene, had a lower inhibitory effect range of 35 to 13% for the same range of concentrations. For the multiresistant *P. aeruginosa* HT5, the EOP and EOPD caused 47 to 32% and 52 to 43% inhibition effects, respectively, for the 4 to 0.1 mg/mL concentration range. The effects of both EOs on the AHL production were dose-dependent; however, at the low concentrations, the EOPD showed greater inhibition of both strains ($p < 0.05$). In comparison, limonene reduced the production of autoinducers from 30 to 17% in the same range of concentrations.

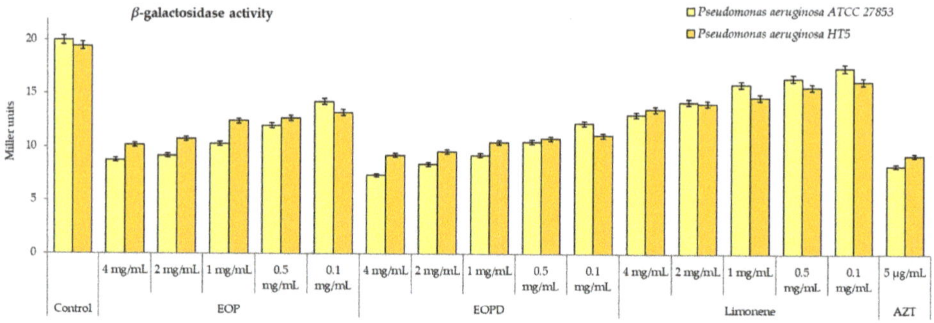

Figure 7. The β-galactosidase activity of *Pseudomonas aeruginosa* ATCC 27853 and HT5 strains. EOP: Essential oil obtained by cold-pressed method; EOPD: essential oil obtained by cold-pressed method followed by steam distillation; L: limonene; AZT: azithromycin. Data are presented as means ± SEMs (n = 8) of three independent experiments. All experiments show significant differences compared to respective controls ($p < 0.05$).

4. Discussion

The chemical compositions of *C. sinensis* EOs vary depending on the extraction method used. Nevertheless, they also depend on the region where the sample was taken and the cultivar was analysed. The composition results of the volatilizable fraction of orange EOs agreed with previous studies in which limonene was the main component, with an average concentration range of 75 to 97.3% [33–35]. In concordance with our results, myrcene (3.2–3.1%) was the second most abundant monoterpene hydrocarbon in *C. sinensis* EOs from Argentina, the United States, and Nepal [34,35]. Additionally, the monoterpenes α-pinene and γ-terpinene were observed in appreciable amounts. However, in *C. sinensis* EOs from Tunisia, β-pinene (1.45–1.82%) was the second most abundant monoterpene hydrocarbon, followed by α-pinene and sabinene [33]. The second main group contained the oxygenated monoterpenes with linalool as the main compound [34,35].

The cold-pressing technique is a unique process applied to extract edible oils from various oily seeds, kernels, peels, and fruits. This technique is important in producing specialty oils with typical characteristic aroma compounds and functional and nutritional compositions. Cold-pressed oils are preferred for their desirable flavour (α- and β-sinensal) characteristics, with their antioxidant components (phenolic compounds, nootkatone, valencene) having therapeutic effects [11,36]. Indeed, the cold-pressing method is common and one of the best techniques for isolating EOs. Using this method, pressure is applied to the plant sample without enhancing the temperature. This extraction technique has significant advantages compared to other extraction techniques, such as the lower energy costs and the fact it does not require a solvent or advanced equipment. The extraction methods influence the EO composition; as reported by González-Mas et al. [37], steam distillation could result in the loss of certain compounds with high molecular weights and low volatility, such as certain phenolic compounds. This observation is consistent with our results, where the concentration of phenolic compounds was eight times higher in the EOP than in the EOPD.

In addition, several studies have reported a relationship between *Citrus* EOs' compositions and their antioxidant properties [38]. Some authors assigned the antioxidant activity of *Citrus* EOs to the volatilizable fractions, particularly monoterpenes, while other researchers attributed it to the presence of phenolic compounds [13,39]. Based on the results obtained in this work with *C. sinensis* oils, we suggest that the free radical scavenging activity is conferred to antioxidant compounds such as polyphenols and valencene (present in higher concentrations in EOP) and nootkatone (only present in EOP). Meanwhile, their reducing capacity is attributed to the terpenes present in similar proportions in both oils. Frasinetti et al. [40] attributed the antioxidant capacity of the bitter orange, sweet orange, lemon, and mandarin EOs to the presence of monoterpenes, the main compounds found in these oils. Moreover, some reports have shown that monoterpenes, such as limonene, α- and γ-terpinene, terpinolene, geraniol, β-pinene, myrcene, α-terpineol, and linalool, were mainly responsible for the antioxidant potential of many plant oils, including *Citrus* oils [41–44]. In addition, oxygenated monoterpenes with different functional groups, such as phenols, alcohols, aldehydes, ethers, esters, and ketones, contribute significantly to the antioxidant properties of the *Citrus* EOs [41]. Raspo et al. [34] explained that these minor components or the synergy between them might be the cause of the antioxidant potential of *Citrus* EOs.

On the other hand, Noshad et al. [45] stated that phenolic compounds act as electron donors in free radical reactions and are often correlated with the antioxidant effects of EOs. In addition, Bonilla and Sobral [46] reported that the phenolic components of cinnamon EOs are capable of quenching reactive oxygen species to delay lipid oxidation. Moreover, the activities of monoterpene phenols have been associated with their phenolic structures and redox properties, which perform a fundamental function in the decomposition of peroxides, in addition to free radical neutralization [47].

In agreement with the present results, Raspo et al. [34] found that Argentinian orange EOs extracted by hydrodistillation showed antioxidant capacity. The authors used ABTS (16 mg Trolox equivalent/mL), DPPH (8 mg Trolox equivalent/mL), CUPRAC (3.5 mg Trolox equivalent/mL), and FRAP assays (0.15 mmol AA/mL). To our knowledge, there are no studies on the purifying capacity of nitric oxide by *C. sinensis* EOs.

The antimicrobial activity levels of the commercial orange oils obtained by cold-pressing (EOP) and cold-pressing followed by a hot distillation system (EOPD) were moderate and higher against *S. aureus* than *P. aeruginosa*. In concordance with our results, the *C. sinensis* EOs were more antimicrobial against pathogenic and food spoilage Gram-positive bacteria than Gram-negative bacteria [48]. It is well known that terpenoid and phenolic compounds have a wide range of biological activities, including antibacterial and antimicrobial activities. In particular, the pure main compounds limonene and myrcene from our sample have been reported as antibacterial agents against Gram-positive strains [49,50].

In addition, different *Citrus* EOs showed better antimicrobial effects against Gram-positive than Gram-negative bacteria [51]. Although limonene is the main constituent, the antimicrobial activity of orange essential oils was higher. This could be attributed mainly to the terpenes, and more specifically to a synergism between their components. Therefore, the antibacterial effects of these EOs are not uniform against different bacteria because they depend on the chemical composition; that is, the antimicrobial activity of the EOs depends on the presence of specific phytochemical components and their interactions, and these components vary according to the maturation stage of the plant. For example, when obtained from ripe fruit, the sweet orange essential oil is more effective against *P. aeruginosa* [52]. In another study, the behaviours of the compounds present in the essential oil of *C. sinensis* were compared according to the plant's maturation stage. It was observed that the limonene values did not vary, but the minority compounds were affected. Therefore, an essential oil's inhibitory activity results from a complex interaction between its different components, which can produce additive, synergistic, or antagonistic effects [53]. *Citrus* EOs such as grapefruit, bergamot, orange, lime, and lemon inhibit the growth of common foodborne and medically important bacterial pathogens, mostly having high minimal inhibitory concentration (MIC) values [51,54,55]. On the other hand, neither limonene nor *Citrus* EOs are bactericidal, even at high concentrations [56].

However, the orange EOs could be considered antipathogenic because they significantly inhibit biofilm production, virulence factors, and QS signals (AHLs). These results suggest that the inhibition effects displayed by EOP and EOPD did not relate exclusively to their action on the growth and that the QS mechanism was involved, coherently with previous studies on plant natural products [57,58]. Compounds with antipathogenic capacities instead of being involved in killing bacteria or stopping their growth act by controlling bacterial virulence factors such as the biofilm and elastase activity and prevent the development of resistant strains [59].

With respect to the biofilm formation, both essential oils interfered with its development in the same manner. The EOPD showed 18% greater inhibition of the *P. aeruginosa* ATCC biofilm than the EOP at 4 mg/mL ($p < 0.05$). It is important to note that the EOPD had a higher total content of monoterpenes and oxygenated monoterpenes (Table 1). Specifically, in relation to the linalool contents (0.55 vs. 0.83% in EOP and EOPD, respectively) and other compounds such as terpinen-4-ol (0.03 vs. 0.11), α-terpinol (0.08 vs. 0.20) and mentha-2,8-dien-1-ol were only found in EOPD. Concerning the minor sesquiterpene hydrocabons, δ-elemene and γ-muurolene were only found in EOPD.

It is important to note that the main compounds limonene (found in both EOs) and linalool have relevant antibiofilm and anti-QS properties against *P. aeruginosa* [29,59,60], which could explain the potential synergistic effect between them and the differential behaviours that were observed.

In agreement with the present results, previous studies have shown that *Citrus* EOs can be effective against bacterial biofilms. *Citrus limon* oils inhibited specific biofilm production and bacterial metabolic activities into biofilm in a dose-dependent manner for *P. aeruginosa* strains. Moreover, these EOs diminished 50% of the elastase activity at 0.1 mg/mL and decreased the pyocyanin biosynthesis. Additionally, another virulence factor, the swarming motility, was completely inhibited by 2 mg/mL. The results were correlated with the observed decrease (29–55%, 0.1–4 mg/mL) in QS signal synthesis [29]. *Citrus paradisi* (grapefruit) essential oils at low concentrations (0.1 mg/mL), mainly obtained by cold-pressing (EOP), were able to inhibit the biofilm establishment and the bacterial survival in the biofilm previously formed by *P. aeruginosa*. These EOs also reduced the *P. aeruginosa's* AHL production and elastase activity [59]. In addition, grapefruit EO has been shown to inhibit the formation of enterohemorrhagic *Escherichia coli* biofilms, while lemon EO inhibited both monomicrobial and mixed biofilms formed by *E. coli* and *Bacillus cereus* [61,62]. In the case of the mandarin EOs, they were not able to inhibit the *P. aeruginosa* growth at 4 mg/mL. However, they significantly inhibited the *P. aeruginosa* biofilm formation at 0.1 mg/mL, as well as the biofilm cell viability (41%), AHL production (33%), and elastase enzyme activity (75%) [24]. It was reported that many EOs could affect bacterial virulence through interference with QS. Rose, geranium, lavender, and rosemary EOs potently inhibited QS; eucalyptus and *Citrus* EOs moderately reduced violacein production; and the chamomile, orange, and juniper oils were ineffective [63]. Pekmezovic et al. [56] demonstrated that the QS inhibition in *P. aeruginosa* occurred through interference with AHL pathways. Several EOs have shown their ability to interfere with bacterial QS signalling and inhibit biofilm formation [61,62,64,65]. As QS inhibitors do not kill or inhibit bacterial growth, these agents have an advantage because they do not impose a selective pressure for resistance development compared to antibiotics [66]. Therefore, the fact that *C. sinensis* EOs have anti-QS activity is very significant, since it would allow the elimination of pathogens with resistance mechanisms. This is very important considering that they could be used in the food industry since they are safe, and many foods tolerate the presence of *Citrus* essential oils. Moreover, many of them are applied in industrial fields in various products, including cosmetics, drugs, foods, and beverages, due to their broad spectrum of biological activities, such as their antibacterial and antifungal activities [33].

Guo et al. [38] did not find a clear relationship when evaluating and comparing the chemical compositions and antimicrobial and antioxidant activities of essential oils from fourteen species of *Citrus*. Despite their promising antioxidant activities, the samples showed only marginal antimicrobial properties. In concordance with what was found in this work, both oils have poor antibacterial activity and significant antipathogenic and antioxidant capacity.

It is crucial to note that the free radical scavengers and antioxidants could also be linked to the promising attenuation of the quorum detection mechanism, because QS is activated by stress factors such as free radicals and oxidative agents.

5. Conclusions

The EOP and EOPD showed antioxidant activity by reducing metals, and particularly the EOP by also neutralizing free radicals. On the other hand, they partially affected the bacterial growth while strongly inhibiting the biofilm formation and viability of sessile bacteria living in a pre-existent biofilm (Gram-negative and Gram-positive). Moreover, the inhibition of AHL formation is reflected in the control of the production of other virulence factors such as elastase and pyocyanin. Therefore, they could represent natural and safe alternatives to extend the shelf life of food products by preventing oxidation and contamination by pathogens that spoil food, meaning the sweet orange EOs can be considered as an innovative dual strategy for food preservation.

Author Contributions: Conceptualization, M.E.A., E.C., M.R.A. and M.A.B.; methodology, M.C.L., M.E.A. and M.M.; software and formal analysis, M.C.L., M.E.A., M.M., E.C., M.R.A. and M.A.B., investigation.; writing—original draft preparation, M.E.A., E.C., M.R.A. and M.A.B.; writing—review and editing, M.E.A., E.C., M.R.A. and M.A.B.; visualization, M.E.A., E.C., M.R.A. and M.A.B.; supervision, M.E.A., E.C., M.R.A. and M.A.B.; project administration M.E.A. All authors have read and agreed to the published version of the manuscript.

Funding: This research received financial support from the SCAIT-UNT (PIUNT D26/638-1), the Agencia Nacional de Promoción Científica y Técnica ANPCyT (PICT 2018-02071 and -02514 and PICT 2021-0060), the Consejo Nacional de Investigaciones Científicas y Técnicas, CONICET (PIP 662, PUE 0021), and the Agencia Nacional de Laboratorios Públicos (ANLAP).

Institutional Review Board Statement: Not applicable.

Informed Consent Statement: Not applicable.

Data Availability Statement: The data are contained within the article.

Acknowledgments: The authors are grateful to the Central Service for Experimental Research (SCSIE) of the University of Valencia for providing the gas chromatography–mass spectrometry equipment and the Litoral Citrus Company for providing the samples.

Conflicts of Interest: The authors declare no conflict of interest.

References

1. Vazquez-Sanchez, D.; Cabo, M.L.; Rodriguez-Herrera, J.J. Antimicrobial activity of essential oils against *Staphylococcus aureus* biofilms. *Food Sci. Technol. Int.* **2014**, *21*, 559–570. [CrossRef] [PubMed]
2. Vazquez-Sanchez, D.; Antunes Galvao, J.; Rodrigues Mazine, M.; Micotti da Gloria, E.; Ferreira de Souza Vieira, T.M. Anti-biofilm efficacy of single and binary treatments based on plant essential oils against *Escherichia coli* persistent in food-processing facilities. *Food Sci. Technol. Int.* **2019**, *25*, 385–393. [CrossRef]
3. Randazzo, W.; Jimenez-Belenguer, A.; Settanni, L.; Perdones, A.; Moschetti, M.; Palazzolo, E.; Guarrasi, V.; Vargas, M.; Germanà, M.A.; Moschetti, G. Antilisterial effect of citrus essential oils and their performance in edible film formulations. *Food Control* **2016**, *59*, 750–758. [CrossRef]
4. Gismondi, A.; Di Marco, G.; Redi, E.L.; Ferrucci, L.; Cantonetti, M.; Canini, A. The antimicrobial activity of *Lavandula angustifolia* Mill. essential oil against *Staphylococcus* species in a hospital environment. *J. Herb. Med.* **2021**, *26*, 100426. [CrossRef]
5. Laird, K.; Armitage, D.; Phillips, C. Reduction of surface contamination and biofilms of *Enterococcus* sp. and *Staphylococcus aureus* using a citrus-based vapour. *J. Hosp. Infect.* **2012**, *80*, 61–66. [CrossRef] [PubMed]

6. Tian, J.; Huang, B.; Luo, X.; Zeng, H.; Ban, X.; He, J.; Wang, Y. The control of *Aspergillus flavus* with *Cinnamomum jensenianum* Hand.-Mazz essential oil and its potential use as a food preservative. *Food Chem.* **2012**, *130*, 520–527. [CrossRef]
7. Carson, C.F.; Hammer, K.A. Chemistry and bioactivity of essential oils. In *Lipids and Essential Oils as Antimicrobial Agents*; Thormar, H., Ed.; John Wiley & Sons, Ltd.: Chichester, UK, 2011; pp. 203–223.
8. Raut, J.S.; Karuppayil, S.M. A status review on the medicinal properties of essential oils. *Ind. Crops Prod.* **2014**, *62*, 250–264. [CrossRef]
9. Bakkali, F.; Averbeck, S.; Averbeck, D.; Idaomar, M. Biological effects of essential oils—A review. *Food Chem. Toxicol.* **2008**, *46*, 446–475.
10. Bora, H.; Kamle, M.; Mahato, D.K.; Tiwari, P.; Kumar, P. Citrus essential oils (CEOs) and their applications in food: An overview. *Plants* **2020**, *9*, 357.
11. Aydeniz Güneşer, B.; Demirel Zorba, N.N.; Yılmaz, E. Antimicrobial activity of cold pressed citrus seeds oils, some citrus flavonoids and phenolic acids. *Riv. Ital. Sostanze Grasse* **2018**, *95*, 119–131.
12. Kang, H.J.; Chawla, S.P.; Jo, C.; Kwon, J.H.; Byun, M.W. Studies on the development of functional powder from citrus peel. *Bioresour. Technol.* **2006**, *97*, 614–620. [CrossRef] [PubMed]
13. Agarwal, P.; Sebghatollahi, Z.; Kamal, M.; Dhyani, A.; Shrivastava, A.; Singh, K.K.; Sinha, M.; Mahato, N.; Mishra, A.K.; Baek, K.-H. Citrus essential oils in aromatherapy: Therapeutic effects and mechanisms. *Antioxidants* **2022**, *11*, 2374. [CrossRef] [PubMed]
14. Wu, Z.; Li, H.; Yang, Y.; Zhan, Y.; Tu, D. Variation in the components and antioxidant activity of *Citrus medica* L. var. *sarcodactylis* essential oils at different stages of maturity. *Ind. Crop. Prod.* **2013**, *46*, 311–316. [CrossRef]
15. Mitropoulou, G.; Fitsiou, E.; Spyridopoulou, K.; Tiptiri-Kourpeti, A.; Bardouki, H.; Vamvakias, M.; Panas, P.; Chlichlia, K.; Pappa, A.; Kourkoutas, Y. Citrus medica essential oil exhibits significant antimicrobial and antiproliferative activity. *LWT* **2017**, *84*, 344–352. [CrossRef]
16. Abebe, G.M. The role of bacterial biofilm in antibiotic resistance and food contamination. *Int. J. Microbiol.* **2020**, *2020*, 1705814.
17. Srey, S.; Jahid, I.K.; Ha, S.D. Biofilm formation in food industries: A food safety concern. *Food Control* **2013**, *31*, 572–585.
18. Sofos, J.N.; Geornaras, I. Overview of current meat hygiene and safety risks and summary of recent studies on biofilms, and control of *Escherichia coli* O157:H7 in nonintact, and *Listeria monocytogenes* in ready-to-eat, meat products. *Meat Sci.* **2010**, *86*, 2–4. [CrossRef]
19. Shikongo-Nambabi, M. Control of bacterial contamination during marine fish processing. *J. Biol.* **2011**, *3*, 1–17. [CrossRef]
20. Srivastava, S.; Bhargava, A. Biofilms and human health. *Biotechnol. Lett.* **2016**, *38*, 1–22. [CrossRef]
21. Ong, K.S.; Mawang, C.I.; Daniel-Jambun, D.; Lim, Y.Y.; Lee, S.M. Current anti-biofilm strategies and potential of antioxidants in biofilm control. *Expert Rev. Anti Infect. Ther.* **2018**, *16*, 855–864. [CrossRef]
22. Viola, C.M.; Torres-Carro, R.; Cartagena, E.; Isla, M.I.; Alberto, M.R.; Arena, M.E. Effect of wine wastes extracts on the viability and biofilm formation of *Pseudomonas aeruginosa* and *Staphylococcus aureus* strains. *Evid. Based Complement. Altern. Med.* **2018**, *4*, 9526878. [CrossRef] [PubMed]
23. Abdel Razek, M.M.M.; Moussa, A.Y.; El-Shanawany, M.A.; Singab, A.N.B. A new phenolic alkaloid from *Halocnemum strobilaceum* endophytes: Antimicrobial, antioxidant and biofilm inhibitory activities. *Chem. Biodivers.* **2020**, *17*, e2000496. [CrossRef] [PubMed]
24. Luciardi, M.C.; Blázquez, M.A.; Cartagena, E.; Bardón, A.; Arena, M.E. Mandarin essential oils inhibit quorum sensing and virulence factors of *Pseudomonas aeruginosa*. *LWT* **2016**, *68*, 373–380. [CrossRef]
25. Adams, R.P. *Identification of Essential Oil Components by Gas Chromatography/Mass Spectrometry*, 4th ed.; Allured Publishing Corporation: Carol Stream, IL, USA, 2007.
26. Zengin, G.; Aktumsek, A. Investigation of antioxidant potentials of solvent extracts from different anatomical parts of *Asphodeline anatolica* E. Tuzlaci: An endemic plant to Turkey. *Afr. J. Tradit. Complement. Altern. Med.* **2014**, *11*, 481–488. [CrossRef]
27. Re, R.; Pellegrini, N.; Proteggente, A.; Pannala, A.; Yang, M.; Rice-Evans, C. Antioxidant activity applying an improved ABTS radical cation decolorization assay. *Free Radic. Biol. Med.* **1999**, *26*, 1231–1237. [CrossRef]
28. Sadeer, N.; Montesano, D.; Albrizio, S.; Zengin, G.; Mahomoodally, M.F. The Versatility of antioxidant assays in food science and safety-chemistry, applications, strengths, and limitations. *Antioxidants* **2020**, *9*, 709. [CrossRef]
29. Luciardi, M.C.; Blázquez, M.A.; Alberto, M.R.; Cartagena, E.; Arena, M.E. Lemon oils attenuate the pathogenicity of *Pseudomonas aeruginosa* by quorum sensing inhibition. *Molecules* **2021**, *26*, 2863. [CrossRef]
30. Díaz, M.A.; González, S.N.; Alberto, M.R.; Arena, M.E. Human probiotic bacteria attenuate *Pseudomonas aeruginosa* biofilm and virulence by quorum-sensing inhibition. *Biofouling* **2020**, *36*, 597–609. [CrossRef]
31. Whiteley, M.; Lee, K.M.; Greenberg, E.P. Identification of genes controlled by quorum sensing in *Pseudomonas aeruginosa*. *Proc. Natl. Acad. Sci. USA* **1999**, *96*, 13904–13909. [CrossRef]
32. Miller, J.H. *Experiments in Molecular Genetics*; Cold Spring Harbor Laboratory Press: Woodbury, NY, USA, 1972; pp. 1–466.
33. Hosni, K.; Zahed, N.; Chrif, R.; Abid, I.; Medfei, W.; Kallel, M.; Ben Brahim, N.; Sebei, H. Composition of peel essential oils from four selected Tunisian *Citrus* species: Evidence for the genotypic influence. *Food Chem.* **2010**, *123*, 1098–1104. [CrossRef]
34. Raspo, M.A.; Vignola, M.B.; Andreatta, A.E.; Juliani, H.R. Antioxidant and antimicrobial activities of citrus essential oils from Argentina and the United States. *Food Biosci.* **2020**, *36*, 100651. [CrossRef]
35. Bhandari, D.P.; Poudel, D.K.; Satyal, P.; Khadayat, K.; Dhami, S.; Aryal, D.; Chaudhary, P.; Ghimire, A.; Parajuli, N. Volatile compounds and antioxidant and antimicrobial activities of selected citrus essential oils originated from Nepal. *Molecules* **2021**, *26*, 6683. [CrossRef] [PubMed]

36. Zhang, L.L.; Chen, Y.; Li, Z.J.; Fan, Z.G.; Li, X. Production, function, and applications of the sesquiterpenes valencene and nootkatone: A comprehensive review. *J. Agric. Food Chem.* **2022**, *71*, 121–142. [CrossRef] [PubMed]
37. González-Mas, M.C.; Rambla, J.L.; López-Gresa, M.P.; Blázquez, M.A.; Granell, A. Volatile compounds in citrus essential oils: A comprehensive review. *Front. Plant. Sci.* **2019**, *10*, 12. [CrossRef] [PubMed]
38. Guo, J.J.; Gao, Z.P.; Xia, J.L.; Ritenour, M.A.; Li, G.Y.; Shan, Y. Comparative analysis of chemical composition, antimicrobial and antioxidant activity of citrus essential oils from the main cultivated varieties in China. *LWT* **2018**, *97*, 825–839. [CrossRef]
39. Saini, R.K.; Ranjit, A.; Sharma, K.; Prasad, P.; Shang, X.; Gowda, K.G.M.; Keum, Y.-S. Bioactive compounds of citrus fruits: A review of composition and health benefits of carotenoids, flavonoids, limonoids, and terpenes. *Antioxidants* **2022**, *11*, 239. [CrossRef] [PubMed]
40. Frassinetti, S.; Caltavuturo, L.; Cini, M.; Della Croce, C.M.J.; Maserti, B.E. Antibacterial and antioxidant activity of essential oils from *Citrus* spp. *J. Essent. Oil Res.* **2011**, *23*, 27–31. [CrossRef]
41. Ruberto, G.; Baratta, M.T. Antioxidant activity of selected essential oil components in two lipid model systems. *Food Chem.* **2000**, *69*, 167–174. [CrossRef]
42. Miguel, M.G. Antioxidant and anti-inflammatory activities of essential oils: A short review. *Molecules* **2010**, *15*, 9252–9287. [CrossRef]
43. Bicas, J.L.; Neri-Numa, I.A.; Ruiz, A.L.; De Carvalho, J.E.; Pastore, G.M. Evaluation of the antioxidant and antiproliferative potential of bioflavors. *Food Chem. Toxicol.* **2011**, *49*, 1610–1615. [CrossRef]
44. Youcef-Ettoumi, K.; Zouambia, Y.; Moulai-Mostefa, N. Chemical composition, antimicrobial and antioxidant activities of Algerian *Citrus sinensis* essential oil extracted by hydrodistillation assisted by electromagnetic induction heating. *J. Food Sci. Technol.* **2021**, *58*, 3049–3055. [CrossRef] [PubMed]
45. Noshad, M.; Alizadeh Behbahani, B.; Jooyandeh, H.; Rahmati-Joneidabad, M.; Hemmati Kaykha, M.E.; Ghodsi Sheikhjan, M. Utilization of *Plantago major* seed mucilage containing *Citrus limon* essential oil as an edible coating to improve shelf-life of buffalo meat under refrigeration conditions. *Food Sci. Nutr.* **2021**, *19*, 1625–1639. [CrossRef] [PubMed]
46. Bonilla, J.; Sobral, P.J. Investigation of the physicochemical, antimicrobial and antioxidant properties of gelatin-chitosan edible film mixed with plant ethanolic extracts. *Food Biosci.* **2016**, *16*, 17–25. [CrossRef]
47. Burt, S. Essential oils: Their antibacterial properties and potential applications in foods—A review. *Int. J. Food Microbiol.* **2004**, *94*, 223–253. [CrossRef] [PubMed]
48. Settanni, L.; Palazzolo, E.; Guarrasi, V.; Aleo, A.; Mammina, C.; Moschetti, G.; Germanà, M.A. Inhibition of foodborne pathogen bacteria by essential oils extracted from citrus fruits cultivated in Sicily. *Food Control* **2012**, *26*, 326–330. [CrossRef]
49. Han, Y.; Sun, Z.; Chen, W. Antimicrobial susceptibility and antibacterial mechanism of limonene against *Listeria monocytogenes*. *Molecules* **2019**, *25*, 33. [CrossRef]
50. Połeć, K.; Broniatowski, M.; Wydro, P.; Hąc-Wydro, K. The impact of β-myrcene—The main component of the hop essential oil—On the lipid films. *J. Mol. Liq.* **2020**, *308*, 113028. [CrossRef]
51. Fisher, K.; Phillips, C.A. The effect of lemon, orange and bergamot essential oils and their components on the survival of *Campylobacter jejuni*, *Escherichia coli* O157, *Listeria monocytogenes*, *Bacillus cereus* and *Staphylococcus aureus* in vitro and in food systems. *J. Appl. Microbiol.* **2006**, *101*, 1232–1240. [CrossRef] [PubMed]
52. Bourgou, S.; Rahali, F.Z.; Ourghemmi, I.; Saïdani Tounsi, M. Changes of peel essential oil composition of four Tunisian citrus during fruit maturation. *Sci. World J.* **2012**, *2012*, 528593. [CrossRef]
53. Franco-Vega, A.; Ramírez-Corona, N.; Palou, E.; López-Malo, A. Estimation of mass transfer coefficients of the extraction process of essential oil from orange peel using microwave assisted extraction. *J. Food Eng.* **2016**, *170*, 136–143. [CrossRef]
54. Prabuseenivasan, S.; Jayakumar, M.; Ignacimuthu, S. In vitro antibacterial activity of some plant essential oils. *BMC Complement. Altern. Med.* **2006**, *6*, 39. [CrossRef] [PubMed]
55. Uysal, B.; Sozmen, F.; Aktas, O.; Oksal, B.S.; Kose, E.O. Essential oil composition and antibacterial activity of the grapefruit (*Citrus paradisi* L.) peel essential oils obtained by solvent-free microwave extraction: Comparison with hydrodistillation. *Int. J. Food Sci. Technol.* **2011**, *46*, 1455–1461. [CrossRef]
56. Pekmezovic, M.; Aleksic, I.; Barac, A.; Arsic-Arsenijevic, V.; Vasiljevic, B.; Nikodinovic-Runic, J.; Senerovic, L. Prevention of polymicrobial biofilms composed of *Pseudomonas aeruginosa* and pathogenic fungi by essential oils from selected *Citrus* species. *Pathog. Dis.* **2016**, *74*, ftw102. [CrossRef] [PubMed]
57. Molina, R.D.I.; Campos-Silva, R.; Díaz, M.A.; Macedo, A.J.; Blázquez, M.A.; Alberto, M.R.; Arena, M.E. Laurel extracts inhibit quorum sensing, virulence factors and biofilm of foodborne pathogens. *LWT* **2020**, *134*, 109899. [CrossRef]
58. Verni, M.C.; Orphèe, C.H.; González, S.N.; Bardón, A.; Arena, M.E.; Cartagena, E. *Flourensia fiebrigii* S.F. Blake in combination with *Lactobacillus paracasei* subsp. *paracasei* CE75. A novel anti-pathogenic and detoxifying strategy. *LWT* **2022**, *156*, 113023. [CrossRef]
59. Luciardi, M.C.; Blázquez, M.A.; Alberto, M.R.; Cartagena, E.; Arena, M.E. Grapefruit essential oils inhibit quorum sensing of *Pseudomonas aeruginosa*. *Food Sci. Technol. Int.* **2020**, *26*, 231–241. [CrossRef]
60. Lahiri, D.; Nag, M.; Dutta, B.; Dey, S.; Mukherjee, D.; Joshi, S.J.; Ray, R.R. Antibiofilm and anti-quorum sensing activities of eugenol and linalool from *Ocimum tenuiflorum* against *Pseudomonas aeruginosa* biofilm. *J. Appl. Microbiol.* **2021**, *131*, 2821–2837. [CrossRef]
61. Vikram, A.; Jesudhasan, P.R.; Jayaprakasha, G.K.; Pillai, B.S.; Patil, B.S. Grapefruit bioactive limonoids modulate *E. coli* O157:H7 TTSS and biofilm. *Int. J. Food Microbiol.* **2010**, *140*, 109–116. [CrossRef]

62. Kerekes, E.B.; Deák, É.; Takó, M.; Tserennadmid, R.; Petkovits, T.; Vágvölgyi, C.; Krisch, J. Anti-biofilm forming and antiquorum sensing activity of selected essential oils and their main components on food-related microorganisms. *J. Appl. Microbiol.* **2013**, *115*, 933–942. [CrossRef]
63. Szabó, M.A.; Varga, G.Z.; Hohmann, J.; Schelz, Z.; Szegedi, E.; Amaral, L.; Molnár, J. Inhibition of quorum sensing signals by essential oils. *Phytother. Res.* **2010**, *24*, 782–786. [CrossRef]
64. Kalia, M.; Yadav, V.K.; Singh, P.K.; Sharma, D.; Pandey, H.; Narvi, S.S.; Agarwal, V. Effect of cinnamon oil on quorum sensing-controlled virulence factors and biofilm formation in *Pseudomonas aeruginosa*. *PLoS ONE* **2015**, *10*, e0135495. [CrossRef] [PubMed]
65. Alves, S.; Duarte, A.; Sousa, S.; Domingues, F.C. Study of the major essential oil compounds of *Coriandrum sativum* against *Acinetobacter baumannii* and the effect of linalool on adhesion, biofilms, and quorum sensing. *Biofouling* **2016**, *32*, 155–165. [CrossRef] [PubMed]
66. Rasmussen, T.B.; Givskov, M. Quorum-sensing inhibitors as anti-pathogenic drugs. *Int. J. Med. Microbiol.* **2006**, *296*, 149–161. [CrossRef] [PubMed]

antioxidants

Article

Value-Added Compounds with Antimicrobial, Antioxidant, and Enzyme-Inhibitory Effects from Post-Distillation and Post-Supercritical CO_2 Extraction By-Products of Rosemary

Simon Vlad Luca [1], Gokhan Zengin [2,*], Kouadio Ibrahime Sinan [2], Izabela Korona-Glowniak [3], Mirjana Minceva [1], Krystyna Skalicka-Woźniak [4,*] and Adriana Trifan [5]

[1] Biothermodynamics, TUM School of Life Sciences, Technical University of Munich, 85354 Freising, Germany
[2] Physiology and Biochemistry Research Laboratory, Department of Biology, Science Faculty, Selcuk University, 42130 Konya, Turkey
[3] Department of Pharmaceutical Microbiology, Faculty of Pharmacy, Medical University of Lublin, 20-093 Lublin, Poland
[4] Department of Natural Products Chemistry, Medical University of Lublin, 20-093 Lublin, Poland
[5] Department of Pharmacognosy and Phytotherapy, Faculty of Pharmacy, "Grigore T. Popa" University of Medicine and Pharmacy Iasi, 700115 Iasi, Romania
* Correspondence: gokhanzengin@selcuk.edu.tr (G.Z.); kskalicka@pharmacognosy.org (K.S.-W.)

Citation: Luca, S.V.; Zengin, G.; Sinan, K.I.; Korona-Glowniak, I.; Minceva, M.; Skalicka-Woźniak, K.; Trifan, A. Value-Added Compounds with Antimicrobial, Antioxidant, and Enzyme-Inhibitory Effects from Post-Distillation and Post-Supercritical CO_2 Extraction By-Products of Rosemary. *Antioxidants* **2023**, *12*, 244. https://doi.org/10.3390/antiox12020244

Academic Editor: Filomena Nazzaro

Received: 24 December 2022
Revised: 14 January 2023
Accepted: 18 January 2023
Published: 21 January 2023

Abstract: Hydrodistillation is the main technique to obtain essential oils from rosemary for the aroma industry. However, this technique is wasteful, producing numerous by-products (residual water, spent materials) that are usually discarded in the environment. Supercritical CO_2 (SC-CO_2) extraction is considered an alternative greener technology for producing aroma compounds. However, there have been no discussions about the spent plant material leftover. Therefore, this work investigated the chemical profile (GC-MS, LC-HRMS/MS) and multi-biological activity (antimicrobial, antioxidant, enzyme inhibitory) of several raw rosemary materials (essential oil, SC-CO_2 extracts, solvent extracts) and by-products/waste materials (post-distillation residual water, spent plant material extracts, and post-supercritical CO_2 spent plant material extracts). More than 55 volatile organic compounds (e.g., pinene, eucalyptol, borneol, camphor, caryophyllene, etc.) were identified in the rosemary essential oil and SC-CO_2 extracts. The LC-HRMS/MS profiling of the solvent extracts revealed around 25 specialized metabolites (e.g., caffeic acid, rosmarinic acid, salvianolic acids, luteolin derivatives, rosmanol derivatives, carnosol derivatives, etc.). Minimum inhibitory concentrations of 15.6–62.5 mg/L were obtained for some rosemary extracts against *Micrococcus luteus*, *Bacilus cereus*, or *Staphylococcus aureus* MRSA. Evaluated in six different in vitro tests, the antioxidant potential revealed strong activity for the polyphenol-containing extracts. In contrast, the terpene-rich extracts were more potent in inhibiting various key enzymes (e.g., acetylcholinesterase, butyrylcholinesterase, tyrosinase, amylase, and glucosidase). The current work brings new insightful contributions to the continuously developing body of knowledge about the valorization of rosemary by-products as a low-cost source of high-added-value constituents in the food, pharmaceutical, and cosmeceutical industries.

Keywords: *Rosmarinus officinalis*; *Salvia rosmarinus*; aromatic herbs; wastes; residues; LC-HRMS/MS; essential oils

1. Introduction

Aromatic herbs/plants are important constituents of human nutrition, valued for their aroma, flavor, and color, as well as their preservative and nutraceutical properties. Moreover, these plants are acknowledged to contain a wide range of bioactive metabolites that make them promising drug lead candidates to treat numerous human ailments [1]. Among aromatic herbs, rosemary (*Salvia rosmarinus* Schleid., formerly *Rosmarinus officinalis* L.) has attracted particular attention over the years due to its culinary, cosmeceutical, and pharmaceutical uses [2]. Rosemary leaves have been used in traditional folk medicine to treat dysmenorrhea, muscle spasms, and renal colic [3]. Phytochemical studies have evidenced the abundance of volatile terpenes usually isolated as essential oils (EOs), such as α-pinene, eucalyptol, camphor, bornyl acetate, and eugenol. Further data have also indicated the presence of flavonoids (e.g., luteolin, hesperidin, diosmin, genkwanin), phenolic acids (e.g., rosmarinic acid, chlorogenic acid), diterpenes (e.g., carnosol, carnosic acid, rosmanol, rosmadial, rosmaquinones), and triterpenes (e.g., oleanolic acid, ursolic acid) in solvent extracts [4,5]. These constituents are assumed to be responsible for a plethora of bioactivities, such as antioxidant, anti-inflammatory, anti-ulcer, cardiovasculo-protective, neuroprotective, hepatoprotective, and anticancer effects [4–6]. Moreover, rosemary is also used in cosmetic formulations to treat ultraviolet oxidative damage, cellulite, alopecia, and aging [3].

The extraction of EOs represents one of the main reasons for the large-scale cultivation of rosemary. Steam and hydrodistillation remain the major technologies that satisfy industrial needs for rosemary EO due to their low cost, simplicity, and generation of high-quality oils [7]. However, various post-distillation by-products, including spent plant materials (solid residues), aqueous condensates (hydrolates, hydrosols), and residual waters (distillation waters or leachates), are generated in large amounts. Usually, these by-products are considered wastes and discarded in the environment without further processing [8]. Literature data on the recovery of post-distillation by-products from rosemary are scarce. For instance, Irakli et al. [2] developed a liquid chromatography method coupled with mass spectrometry to identify various phenolic compounds from a spent material extract of rosemary. Bouloumpasi et al. [9] explored the antioxidant and antibacterial properties of the material after steam distillation, whereas Yagoubi et al. [10] showed that rosemary distillation residues could reduce the lipid oxidation, increase the alpha-tocopherol content, and improve the fatty acid profile of lamb meat.

Supercritical CO_2 extraction (SC-CO_2) is an emerging technology alternative to steam or hydrodistillation that can provide several operational advantages. For instance, the preservation of thermosensitive terpenes is ensured as low temperatures and reduced extraction times are applied. SC-CO_2 is widely accepted by the food, cosmetic, and pharmaceutical industries as a green solvent since its complete removal at the end of the process can be achieved without additional energy consumption [11,12]. SC-CO_2 was briefly explored to extract volatile terpenes from rosemary [13–15]. Compared to steam and hydrodistillation, which generate multiple categories of by-products, SC-CO_2 extraction produces only the spent plant material as the primary residue. The utilization of post-SC-CO_2 by-products did not constitute the focus of previous works. Scrutiny of the literature retrieved one study that used the solid residues obtained after the SC-CO_2 extraction of *Melissa officinalis* L. for the further extraction of phenolic compounds [16].

This study aimed to provide a comparative assessment of the phytochemical profile and biological potential of raw (EO, SC-CO_2, and total extracts), post-distillation, and post-SC-CO_2 extracts. Thus, the EO and SC-CO_2 were profiled by gas chromatography coupled with mass spectrometry (GC-MS), whereas the remaining extracts were analyzed by liquid chromatography coupled with high-resolution tandem mass spectrometry (LC-HRMS/MS). A panel of Gram-positive and Gram-negative bacteria as well as yeast was used to evaluate the antimicrobial activity. The antioxidant activity was evaluated concerning the 1,1′-diphenyl-2-picrylhydrazyl (DPPH), 2,2′-azino-bis(3-ethylbenzothiazoline) 6-sulfonic acid (ABTS), cupric ion-reducing antioxidant capacity (CUPRAC), ferric ion-reducing antioxidant power (FRAP), metal chelating ability (MCA), and phosphomolybdenum (PBD), whereas the enzyme-inhibitory potential was assessed in acetylcholinesterase [17], butyrylcholinesterase (BChE), tyrosinase, amylase, and glucosidase tests.

2. Materials and Methods

2.1. Chemicals

Alkane standard solution (C8–C20, ~40 mg/L each, in hexane), 2′-azino-bis(3-ethylben zothiazoline-6-sulphonic acid) (ABTS), 5,5-dithio-bis(2-nitrobenzoic) acid, acarbose, acetonitrile, ammonium acetate, ammonium molybdate, amylase (EC. 3.2.1.1, from porcine pancreas), α-bisabolol, butyrylthiocholine chloride, caffeic acid, β-caryophyllene, cupric chloride, 1,1-diphenyl-2-picrylhydrazyl (DPPH), eel acetylcholinesterase (AChE, type: VI-S, EC 3.1.1.7), ethanol, ferric chloride, ferrous sulfate hexahydrate, ferrozine, Folin–Ciocalteu reagent, formic acid, galantamine, acetylthiocholine iodide, gallic acid, glucose, glucosidase (EC. 3.2.1.20, from *Saccharomyces cerevisiae*), α-humulene, horse serum butyrylcholinesterase (BChE, EC 3.1.1.8), hydrochloric acid, hydroxybenzoic acid, kojic acid, limonene, β-myrcene, Mueller–Hinton (MH) broth, rosmarinic acid, rutin, sodium carbonate, sodium hydroxide, sodium molybdate, sodium nitrate, 2,4,6-tris(2-pyridyl)-s-triazine, Trolox, ethylenediaminetetraacetate (EDTA), and tyrosinase (EC1.14.18.1, mushroom) were from Merck KGaA (Darmstadt, Germany). β-Caryophyllene oxide was from Thermo Scientific (Olching, Germany), α-humulene was from Biomol (Hamburg, Germany), and liquid CO_2 (≥99.7%) from Westfalen AG (Münster, Germany). Methanol was bought from VWR Chemicals (Ismaning, Germany). Acetonitrile and formic acid were from Avantor (Gliwice, Poland).

2.2. *Plant Material*

Dried leaves of rosemary were bought from a local market in Germany; the plant material was authenticated by one of the authors (A.T.). A voucher specimen (RO/220714) was deposited in Biothermodynamics, TUM School of Life Sciences, Technical University of Munich, Freising, Germany.

2.3. *Extraction*

2.3.1. Preparation of Essential Oil

The powdered dried rosemary leaves (50 g) were placed in a Clevenger-type apparatus with 500 mL of deionized water and distilled for 4 h. The obtained rosemary essential oil (**REO**) fraction was collected and dried over anhydrous sodium sulfate. The hydrodistillation procedure was performed in duplicate.

2.3.2. Preparation of SC-CO_2 Extracts

The SC-CO_2 extractions were performed on a Spe-ed SFE Zoran Extractor (Applied Separations, Allentown, PA, USA) which could operate at a maximum temperature and pressure of 180 °C and 690 bar, respectively. The extraction vessel was loaded with 40 g of powdered dried rosemary leaves, which were compressed to a bed length of 12.0 cm and a diameter of 3.0 cm. The vessel was sealed, placed in a thermostatic mantle, and connected to the extractor. The extraction was started after an equilibration static time of 10 min and performed for 30 min at a constant CO_2 flow (7 standard liter min^{-1}). All SC-CO_2 experiments were conducted at a pressure of 100 bar, whereas the temperature conditions were 40 °C, 50 °C, and 60 °C, yielding **RC1**, **RC2**, and **RC3**, respectively. Each experimental setup was performed in duplicate.

2.3.3. Preparation of Total, Spent, and Residual Water Extracts

At the end of the hydrodistillation process, the residual water was filtered and freeze-dried, affording the rosemary water extract (**RWE**). Totals of 10 g of the powdered dried rosemary leaves, spent plant material after hydrodistillation, and spent plant materials after the three SC-CO_2 extractions were separately extracted at room temperature in a Bandelin Sonorex Digitec ultrasound bath from BANDELIN Electronic GmbH & Co. KG (Berlin, Germany) with 3 × 100 mL methanol/water 75/25 (*v*/*v*) in three repeated ultrasound cycles (35 Hz), each 30 min long. All extractions were performed in duplicate. The following extracts were thus obtained: total extract (**RTE**), spent plant material extract (**RSE**), SC-CO_2 (100 bar, 40 °C) spent extract (**RSC1**), SC-CO_2 (100 bar, 50 °C) spent extract (**RSC2**), SC-CO_2 (100 bar, 60 °C) spent extract (**RSC3**). The extract yields are provided in Table 1.

Table 1. Extraction yields of rosemary extracts.

Extract	Code	Yield [g/100 g]
Extracts from raw materials		
Essential oil	REO	2.5 ± 0.1 *
SC-CO_2 (100 bar, 40 °C) extract	RC1	1.5 ± 0.2
SC-CO_2 (100 bar, 50 °C) extract	RC2	0.8 ± 0.2
SC-CO_2 (100 bar, 60 °C) extract	RC3	0.2 ± 0.0
Total extract	RTE	16.8 ± 3.1
Extracts from post-distillation materials		
Distillation (residual) water extract	RWE	25.6 ± 1.1
Spent plant material extract	RSE	9.2 ± 0.8
Extracts from post-SC-CO_2 materials		
SC-CO_2 (100 bar, 40 °C) spent extract	RSC1	17.1 ± 2.1
SC-CO_2 (100 bar, 50 °C) spent extract	RSC2	18.6 ± 0.8
SC-CO_2 (100 bar, 60 °C) spent extract	RSC3	20.9 ± 1.8

* mL oil/100 g plant material.

2.4. *Phytochemical Characterization*

2.4.1. GC-MS Analysis

A TRACE gas chromatograph [18] with an ISQ™ mass spectrometer (MS) from Thermo Fisher (Waltham, MA, USA) was used. The chromatographic separations were conducted on a Zebron™ ZB-5MS (30 m × 0.25 mm i.d., 0.25 µm film thickness) from Phenomenex (Torrance, CA, USA). Helium at a flow rate of 1.43 mL/min was the carrier gas. The inlet temperature was 250 °C, the split ratio was 50:1, and the injection volume was 1 µL. The oven temperature was held for 4 min at 60 °C; then it was increased to 280 °C at a rate of 10 °C/min and held for 5 min; finally, it was ramped to 300 °C at a rate of 10 °C/min and held for 10 min. The following MS settings were used: m/z: 50 to 350 amu; ionization energy: 70 eV; transfer line temperature: 320 °C; and source temperature: 230 °C. The linear retention indices were determined for each peak using a C8–C20 standard mixture of n-alkanes and compared with those of the literature. Furthermore, the recorded mass spectral information was compared with that from the NIST11 database. All measurements were performed in triplicate.

2.4.2. LC-HRMS/MS Analysis

An Agilent 1200 HPLC (Agilent Technologies, Palo Alto, CA, USA) with an accurate-mass quadrupole time-of-flight MS detector (G6530B) was used. The chromatographic separations were conducted on a Gemini C18 column (100 mm × 2 mm i.d., 3 µm) from Phenomenex (Torrance, CA, USA). The mobile phases comprised water (A) and acetonitrile (B), both acidified with 0.1% formic acid. The run started with 10% B and linearly increased to 60% B in 45 min at a flow rate of 0.2 mL/min; the injection volume was 10 µL. The following MS settings were used: m/z: 100–1700 amu; negative ionization mode; carrier gas (nitrogen) flow rate and temperature of 10 L/min and 275 °C, respectively; sheath gas (nitrogen) flow rate and temperature of 12 L/min and 325 °C, respectively; nebulizer pressure: 35 psi; capillary voltage: 4000 V; nozzle voltage: 1000 V; skimmer: 65 V; fragmentor: 140 V; and collision-induced dissociation: 30 V. The recorded mass spectral information was compared with that from databases and the literature.

2.4.3. Total Phenolic and Flavonoid Content

The total phenolic content (TPC) and total flavonoid content (TFC) were determined spectrophotometrically as described in [19]. Data were expressed as mg gallic acid equivalents (GAE)/g extract in TPC and mg rutin equivalents (RE)/g extract in TFC. All measurements were performed in triplicate.

2.5. Biological Activity Evaluation

2.5.1. Antimicrobial Assays

The microdilution method was used to determine the antimicrobial activity according to the European Committee on Antimicrobial Susceptibility Testing [20]. MH broth and MH broth with 2% glucose were used to grow the bacteria and yeasts, respectively. The following microbial strains were tested: *Bacillus subtilis* ATCC 6633, *Candida albicans* ATCC 2091, *Candida glabrata* ATCC 90030, *Candida parapsilosis* ATCC 22019, *Enterococcus faecalis* ATCC 29212, *Escherichia coli* ATCC 25922, *Klebsiella pneumoniae* ATCC 13883, *Micrococcus luteus* ATCC 10240, *Proteus mirabilis* ATCC 12453, *Pseudomonas aeruginosa* ATCC 90271, *Salmonella* Typhimurium ATCC 14028, *Staphylococcus aureus* ATCC 25923, *Staphylococcus aureus* ATCC BAA-1707 (methicillin-resistant strain), and *Staphylococcus epidermidis* ATCC 12228. The minimum inhibitory concentration (MIC) was calculated and reported for each sample and strain. All experiments were performed in triplicate.

2.5.2. Antioxidant Assays

The DPPH radical scavenging, ABTS radical scavenging, cupric ion reducing antioxidant capacity (CUPRAC), ferric ion reducing antioxidant power (FRAP), metal chelating activity (MCA), and phosphomolybdenum (PBD) were determined as presented in [19,21]. The activity data were expressed as mg Trolox equivalents (TE)/g extract in DPPH, ABTS, CUPRAC, and FRAP assays; mg EDTA equivalents (EDTAE)/g extract in the MCA assay; and mmol TE/g extract in the PBD assay.

2.5.3. Enzyme-Inhibitory Assays

AChE, BChE, tyrosinase, amylase, and glucosidase inhibition were determined as presented in [19,21]. The activity data were expressed as mg galanthamine equivalents (GALAE)/g extract in the AChE and BChE assays, mg kojic acid equivalents (KAE)/g extract in tyrosinase assay, and mmol acarbose equivalents (ACAE)/g extract in amylase and glucosidase assays.

2.6. Statistical and Data Processing

Data are presented as mean ± standard deviation of the respective number of replicates. One-way analysis of variance with Tukey's post-hoc test was conducted; $p < 0.05$ was considered statistically significant. The relationship between rosemary compounds vs. antimicrobial, antioxidant, and enzyme-inhibitory activities was assessed by calculating the Pearson correlation coefficient. Principal component analysis (PCA) and clustered image maps (CIM) were also performed, with the phytochemical data log transformed. The statistical analysis was done using R software v. 4.1.2 (R Foundation for Statistical Computing, Vienna, Austria).

3. Results and Discussion

3.1. GC-MS Characterization of Essential Oil and SC-CO_2 Rosemary Extracts

In this section, rosemary extracts rich in volatile compounds were obtained by hydrodistillation and SC-CO_2 extraction and characterized by GC-MS. The operating SC-CO_2 pressure (100 bar) and temperature range (40–60 °C) were selected based on previous systematic studies that presented a high recovery of rosemary volatiles under these conditions [11,15]. The EO yield was significantly higher than the SC-CO_2 yields (Table 1). This can be related to the different extraction mechanisms between the hydrodistillation and SC-CO_2 extraction. Hydrodistillation allows the recovery of only high-vapor-pressure (volatile, 'distillable') compounds, whereas SC-CO_2 extraction allows the recovery of compounds mostly based on their polarity and to a lower extent based on their vapor pressure. The high lipophilicity of the solvent (SC-CO_2) would allow high extraction rates of lipophilic compounds, including the low-polarity terpenes.

Within the three SC-CO_2 extracts, the yield decreased with the increase in temperature. This is in connection with the fact that temperature increments are known to reduce solvent density and negatively impact the solubility and extractability of compounds [22]. According to the GC-MS profiling (Table 2), the rosemary EO (**REO**) was characterized by 46 compounds, primarily monoterpenes (~97.2% of the total GC-MS peak area). The putative identity of the volatile compounds was established by comparing the linear retention indices with those of the literature data and the recorded mass spectra with those of NIST11 database. However, due to the lack of authentic standards, only a partial structural identification is possible with these resources.

Table 2. GC-MS profile (tentative annotation) of the essential oils obtained from thyme, oregano, and basil.

No.	Compound	LRI [a]	REO	RC1 (100 bar, 40 °C)	RC2 (100 bar, 50 °C)	RC3 (100 bar, 60 °C)
			% [b]			
1	Tricyclene	924	0.10 ± 0.01	–	–	–
2	*α*-Phellandrene	927	0.02 ± 0.00	–	–	–
3	*α*-Pinene	935	**13.27 ± 0.14**	1.07 ± 0.08	1.91 ± 0.07	–
4	Camphene	952	3.88 ± 0.03	–	0.55 ± 0.02	–
5	*β*-Pinene	980	2.59 ± 0.02	–	0.76 ± 0.02	–
6	*β*-Myrcene *	989	1.18 ± 0.01	–	0.48 ± 0.01	–
7	3-Thujene	1007	0.15 ± 0.00	–	–	–
8	*α*-Terpinene	1018	0.64 ± 0.00	1.85 ± 0.08	0.26 ± 0.01	–
9	*p*-Cymene	1027	2.72 ± 0.01	–	1.58 ± 0.03	–
10	Limonene *	1031	2.67 ± 0.04	0.36 ± 0.02	1.46 ± 0.03	–
11	Eucalyptol	1034	**41.68 ± 0.42**	**13.81 ± 0.89**	**29.75 ± 1.14**	2.71 ± 0.01
12	*γ*-Terpinene	1060	0.47 ± 0.01	–	0.30 ± 0.01	–
13	*cis*-*α*-Terpineol	1072		–	0.17 ± 0.01	–
14	*α*-Terpinolene	1087	0.20 ± 0.00	–	0.17 ± 0.01	–
15	*p*-Cymenene	1091	0.06 ± 0.01	–	–	–
16	Linalool	1099	1.38 ± 0.03	–	2.32 ± 0.01	2.94 ± 0.05
17	Fenchyl alcohol	1121	0.07 ± 0.01	–	0.10 ± 0.01	–
18	*cis*-*p*-Menth-2-en-1-ol	1127	0.04 ± 0.00	–	0.05 ± 0.01	–
19	*trans*-*p*-Menth-2-en-1-ol	1145	0.04 ± 0.00	–	–	–
20	Camphor	1150	**13.52 ± 0.06**	0.27 ± 0.04	**18.33 ± 0.27**	**18.75 ± 0.22**
21	*cis*-*α*-Terpineol	1159	0.04 ± 0.00	5.54 ± 0.38	0.03 ± 0.00	–
22	Pinocamphone	1163	0.17 ± 0.03	–	0.21 ± 0.00	–
23	*cis*-Terpin hydrate	1173	–	0.46 ± 0.03	–	2.57 ± 0.06
24	Borneol	1176	**4.47 ± 0.05**	3.54 ± 0.19	**6.74 ± 0.01**	**10.40 ± 0.16**
25	Isopinocamphone	1179	0.05 ± 0.01	–	0.06 ± 0.02	–
26	Terpinen-4-ol	1183	1.05 ± 0.03	0.41 ± 0.03	1.45 ± 0.02	2.18 ± 0.02
27	*p*-Cymen-8-ol	1188	0.03 ± 0.00	–	0.09 ± 0.01	–
28	*trans*-*α*-Terpineol	1197	**5.89 ± 0.09**	**5.13 ± 0.33**	**9.52 ± 0.04**	**20.41 ± 0.23**
29	Isoborneol	1198	0.03 ± 0.00	–		–
30	Verbenone	1202	0.20 ± 0.01	–	0.68 ± 0.01	1.01 ± 0.05
31	Thymol	1281	0.09 ± 0.01	0.55 ± 0.05	0.08 ± 0.01	–
32	Bornyl acetate	1286	0.31 ± 0.01		0.61 ± 0.02	0.96 ± 0.02
33	Carvacrol	1293	0.12 ± 0.04	0.33 ± 0.02	0.05 ± 0.01	–
34	Eugenol	1350	–	–	0.08 ± 0.01	–
35	Copaene	1380	0.05 ± 0.00	–	0.48 ± 0.03	0.70 ± 0.05
36	Methyl eugenol	1398	0.04 ± 0.01	–	0.12 ± 0.01	–
37	*β*-Caryophyllene *	1428	1.43 ± 0.06	3.24 ± 0.18	**9.80 ± 0.36**	**21.47 ± 0.37**
38	*γ*-Elemene	1447	0.04 ± 0.00	–	0.17 ± 0.02	–
39	*β*-Farnesene	1453	–	–	0.02 ± 0.00	–
40	*α*-Humulene *	1464	0.20 ± 0.01	–	1.33 ± 0.07	2.77 ± 0.03
41	*α*-Huaiene	1481	0.04 ± 0.01	–	0.43 ± 0.04	0.81 + 0.05
42	*trans*-Cadina-1(6),4-diene	1484	–	–	0.09 ± 0.02	–
43	*β*-Selinene	1498	–	–	0.14 ± 0.01	–
44	*α*-Selinene	1504	–	–	0.22 ± 0.02	–
45	*β*-Bisabolene	1511	–	–	0.36 ± 0.03	0.74 ± 0.01
46	*γ*-Cadinene	1520	0.02 ± 0.00	–	0.19 ± 0.01	0.65 ± 0.02

Table 2. *Cont.*

No.	Compound	LRI [a]	REO	RC1 (100 bar, 40 °C)	RC2 (100 bar, 50 °C)	RC3 (100 bar, 60 °C)
			% [b]			
47	*β*-Cadinene	1524	0.09 ± 0.01	–	0.88 ± 0.05	2.15 ± 0.04
48	*δ*-Cadinene	1528	–	–	0.23 ± 0.03	–
49	*β*-Caryophyllene oxide *	1593	0.26 ± 0.02	–	0.91 ± 0.06	3.26 ± 0.05
50	Aromadendrene oxide	1603	0.04 ± 0.01	–	0.07 ± 0.01	–
51	*trans*-(*E*)-*α*-Bisabolene epoxide	1620	0.04 ± 0.01	–	0.17 ± 0.02	–
52	*trans*-(*Z*)-*α*-Bisabolene epoxide	1644	0.06 ± 0.01	0.46 ± 0.00	1.05 ± 0.08	3.12 ± 0.13
53	*allo*-Aromadendrene oxide	1647	0.07 ± 0.01	0.53 ± 0.01	–	–
54	*β*-Eudesmol	1663	0.25 ± 0.02	2.56 ± 0.12	0.62 ± 0.06	1.20 ± 0.06
55	*allo*-Aromadendrene epoxide	1678	0.11 ± 0.01	1.58 ± 0.06	0.37 ± 0.05	–
56	*α*-Bisabolol *	1690	–	–	0.03 ± 0.01	–
	Hydrocarbon monoterpenes		*27.96 ± 0.14*	*3.56 ± 0.10*	*7.30 ± 0.18*	–
	Oxygenated monoterpenes		*69.23 ± 0.06*	*29.78 ± 1.88*	*70.62 ± 1.24*	*61.93 ± 0.65*
	Hydrocarbon sesquiterpenes		*1.86 ± 0.09*	*3.24 ± 0.18*	*14.34 ± 0.74*	*29.06 ± 0.65*
	Oxygenated sesquiterpenes		*0.83 ± 0.07*	*5.13 ± 0.17*	*3.30 ± 0.30*	*7.57 ± 0.18*
	Total identified		***99.89 ± 0.05***	***41.71 ± 2.18***	***95.56 ± 0.41***	***98.57 ± 0.28***

[a] Linear retention index on ZB-5MS column; [b] Expressed as the mean percentage area extracted from the GC-MS chromatograms of three repeated analyses; * standard injection: the major volatile compounds are in bold; sample codes as in Table 1.

The major volatile compounds (Figure 1) were represented by eucalyptol (41.7%), camphor (13.5%), and *α*-pinene (13.27%). Similarly, Ramzi et al. [23] documented eucalyptol (29.31%), camphor (24.7%), and *α*-pinene (12.8%) as the dominant terpenes in rosemary EO. In addition, Ouknin et al. [24] reported eucalyptol (27.6%), *α*-pinene (26.6%), verbenone (5.3%), camphene (4.5%), and camphor (4.3%) as the main compounds of rosemary EO.

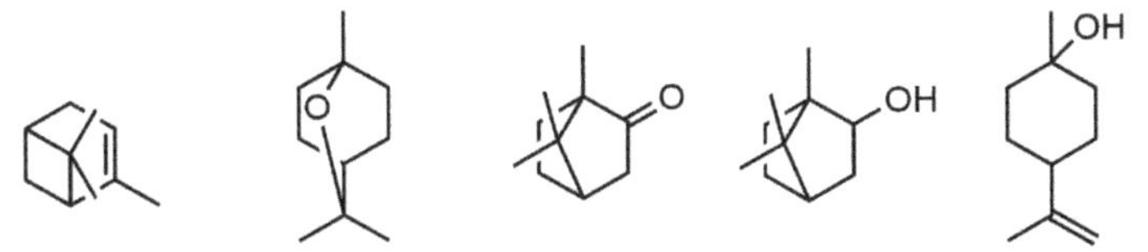

α-pinene eucalyptol camphor borneol *trans*-*α*-terpineol

Figure 1. Proposed chemical structures of the main volatile terpenes identified in rosemary essential oil.

Concerning the three SC-CO_2 extracts, significant differences in the profiles of the volatile compounds were noticed. The SC-CO_2 extraction at 100 bar and 40 °C (**RC1**) recovered the lowest number (17 compounds) and amount of terpenes (only 41.7% of the total GC-MS peak area). Eucalyptol remained the predominant compound in this extract, but its level was considerably low (13.8%). However, *α*-pinene (1.1%) and camphor (0.3%) were found in traces. The extract obtained at 100 bar and 50 °C (**RC2**) displayed a significantly high number (48 compounds) and concentration of volatile compounds (95.6% of the total GC-MS peak area). Eucalyptol (29.8%) and camphor (18.3%) were the major terpenes in **RC2**; in addition, the levels of trans-*α*-terpineol (9.5%) and caryophyllene (9.8%) were dramatically increased as compared to those of **REO** and **RC1**. Lastly, the third SC-CO_2 extraction conditions (100 bar and 60 °C) allowed the recovery of 24 terpenes (accounting for 98.6% of the total GC-MS peak area). Nonetheless, the concentration of eucalyptol decreased to 2.71%, whereas the concentration of camphor was kept high (at 18.8%). Furthermore, borneol (10.4%), trans-*α*-terpineol (20.4%), and caryophyllene (21.5%) reached their highest values in **RC3**.

Overall, it can be noticed that hydrodistillation was clearly more selective in recovering rosemary terpenes. However, the selectivity of the SC-CO_2 extraction toward volatiles increased considerably by increasing the temperature from 40 to 60 °C. Most likely, the SC-CO_2 process, especially at 100 bar and 40 °C, allowed the simultaneous extraction of non-volatile lipophilic compounds (e.g., waxes, fatty acids, lipophilic pigments, etc.). It is also worth emphasizing that the ratio of monoterpenes/sesquiterpenes significantly decreased from 36/1 in the **REO** to 4/1, 4.4/1, and 1.7/1 in **RC1, RC2,** and **RC3**, respectively. Even though various SC-CO_2 extraction conditions allow the efficient recovery of terpenes, the extraction yields, qualitative profile, and quantitative data of terpenes are significantly altered compared to those with hydrodistillation. Similar conclusions were also reported in previous works [12].

3.2. LC-HRMS/MS Characterization of the Total, Post-Distillation, and Post-SC-CO_2 Rosemary Extracts

In this section, various rosemary extracts, namely total (unspent plant material, **RTE**), post-hydrodistillation residual water (**RWE**), post-hydrodistillation spent material (**RSE**), and post-SC-CO_2 spent material (**RSC1, RSC2,** and **RSC3**) extracts, were obtained and characterized by LC-HRMS/MS. **RWE** was characterized by the highest extraction yield (25.6%); interestingly, the yield of **RSE** was significantly low (9.2%) (Table 1). The post-SC-CO_2 materials allowed extraction yields between 17.1% and 20.9%. The order is correlated with the SC-CO_2 extraction yields: the higher the SC-CO_2 extraction yields, the lower the post-SC-CO_2 extraction yields (Table 1). The LC-HRMS/MS profiling (Table 3) allowed the annotation of 25 specialized metabolites belonging to various phytochemical classes. The putative identity of the compounds was established by comparing the spectro-chromatographic data with those presented in the literature [2,25–28] and relevant databases (METLIN, KNApSacK, PubChem, NIST Chemistry WebBook). However, due to the lack of authentic standards, only a partial structural identification was possible with these resources.

Table 3. LC-HRMS/MS profile (tentative annotation) of extracts obtained from rosemary (raw materials, post-distillation materials, or post-supercritical CO_2 materials).

No	Compound	Class	T_R (min)	[M–H]⁻ (*m/z*)	MF	HRMS/MS (*m/z*)	Sample	Ref.
1	Quinic acid *	Organic acid	1.9	191.0599	$C_7H_{12}O_6$	173.0492, 127.0427, 111.0470	RTE, RWE, RSE, RSC1, RSC2, RSC3	[27]
2	Danshensu	Phenolic acid	5.4	197.0451	$C_9H_{10}O_5$	179.0355, 151.0408, 135.0455, 123.0452	RTE, RWE, RSE, RSC3	[2]
3	Hydroxybenzoic acid *	Phenolic acid	9.6	137.0243	$C_7H_6O_3$	108.0218	RWE, RSE	[25]
4	Caffeic acid *	Phenolic acid	13.9	179.0359	$C_9H_8O_4$	135.0450, 107.0503	RTE, RWE, RSE, RSC1, RSC2, RSC3	[28]
5	Gallocatechin *	Flavonoid	24.3	305.0773	$C_{15}H_{14}O_7$	225.118	RTE, RWE, RSE, RSC1, RSC2, RSC3	[27]
6	Rosmarinic acid *	Phenolic acid	26.7	359.0855	$C_{18}H_{16}O_8$	197.0491, 179.0380, 161.0272, 135.0475	RTE, RWE, RSE, RSC1, RSC2, RSC3	[28]
7	Luteolin-*O*-glucuronide	Flavonoid	28.3	461.0730	$C_{21}H_{18}O_{12}$	285.0471, 151.0064, 133.0320	RTE, RWE, RSE, RSC1, RSC2, RSC3	[27]
8	Salvianolic acid A	Phenolic acid	30.1	493.1190	$C_{26}H_{22}O_{10}$	313.0757, 295.0646, 197.0471, 185.0264	RTE, RWE, RSE, RSC1, RSC2, RSC3	[26]
9	Luteolin-*O*-acetylglucuronide I	Flavonoid	31.2	503.0837	$C_{27}H_{20}O_{10}$	285.0486, 133.0326	RTE, RWE, RSE, RSC1, RSC2, RSC3	[27]
10	Luteolin *O*-acetylglucuronide II	Flavonoid	31.9	503.0839	$C_{23}H_{20}O_{13}$	285.0372, 151.0023, 133.0283	RTE, RWE, RSE, RSC1, RSC2, RSC3	[27]
11	Cirsimaritin	Flavonoid	33.0	313.0710	$C_{17}H_{14}O_6$	161.0241, 151.0388, 133.0288	RSE	[27]

Table 3. *Cont.*

No	Compound	Class	T_R (min)	[M–H]⁻ (*m/z*)	MF	HRMS/MS (*m/z*)	Sample	Ref.
12	Ladanein	Flavonoid	34.0	313.0707	$C_{17}H_{14}O_6$	161.0239, 133.0293	RWE, RSE	[25]
13	Rosmanol	Diterpene	36.1	345.1694	$C_{20}H_{26}O_5$	301.1818, 283.1711	RTE, RWE, RSE, RSC1, RSC2, RSC3	[27]
14	Epirosmanol	Diterpene	37.4	345.1706	$C_{20}H_{26}O_5$	301.1802, 283.1706, 268.1467, 227.1078	RTE, RWE, RSE, RSC1, RSC2, RSC3	[27]
15	Isorosmanol	Diterpene	38.6	345.1707	$C_{20}H_{26}O_5$	301.1828, 283.1725, 268.1478, 227.1087	RTE, RWE, RSE, RSC1, RSC2, RSC3	[27]
16	Genkwanin	Flavonoid	40.5	283.0616	$C_{16}H_{12}O_5$	268.0367, 240.0416, 151.0030	RTE, RWE, RSE, RSC1, RSC2, RSC3	[27]
17	Epiisorosmanol	Diterpene	42.1	345.1730	$C_{20}H_{26}O_5$	301.1806, 285.1507	RTE, RSE, RSC1, RSC2, RSC3	[27]
18	Epirosmanol methyl ether	Diterpene	44.0	359.1865	$C_{21}H_{28}O_5$	315.1961, 300.1733, 283.1707	RTE, RSE, RSC1, RSC2, RSC3	[28]
19	Methoxyrosmanol	Diterpene	45.3	359.1883	$C_{21}H_{28}O_5$	315.1982, 300.1743, 283.1718	RTE, RSE, RSC1, RSC2, RSC3	[28]
20	Carnosol	Diterpene	46.0	329.1768	$C_{22}H_{26}O_4$	314.1506, 299.1286, 271.0977	RTE, RWE, RSE, RSC1, RSC2, RSC3	[27]
21	Rosmadial I	Diterpene	47.7	343.1549	$C_{20}H_{24}O_5$	315.1621, 299.1679, 287.1673	RTE, RSE, RSC1, RSC2, RSC3	[28]
22	Rosmadial II	Diterpene	48.5	343.1558	$C_{20}H_{24}O_5$	299.1665, 271.1716	RTE, RSE, RSC1, RSC2, RSC3	[28]
23	Carnosic acid I	Diterpene	50.1	331.1908	$C_{22}H_{28}O_4$	287.2095, 244.1529	RTE, RWE, RSE, RSC1, RSC2, RSC3	[27]
24	Carnosic acid methyl ester	Diterpene	51.5	345.2064	$C_{21}H_{30}O_4$	301.2225, 286.2012, 271.1777	RTE, RWE, RSE, RSC1, RSC2, RSC3	[27]
25	Carnosic acid II	Diterpene	52.1	331.1906	$C_{22}H_{28}O_4$	287.2033, 244.1523	RTE, RSE, RSC1, RSC2, RSC3	[27]

MF, molecular formula; * Confirmed by standard; sample codes as in Table 1.

Danshensu (**2**), hydroxybenzoic acid (**3**), caffeic acid (**4**), rosmarinic acid (**6**), and salvianolic acid A (**8**) were found as typical phenolic acids in the analyzed samples. In addition, seven flavonoids were spotted. They were putatively labeled as free aglycones: gallocatechin (**5**), cirsimaritin (**11**), ladanein (**12**), genkwanin (**16**), glycosylated flavonoids: luteolin-*O*-glucuronide (**7**), and two isomeric luteolin-*O*-acetylglucuronides (**9** and **10**). Besides ladanein, the other phenolic acids and flavonoids were previously documented in rosemary extracts [2,27,28]. Diterpenes constituted the representative class of phytochemicals (12 compounds). A few diterpenes were derivatives of rosmanol (**13**), such as epirosmanol (**14**), isorosmanol (**15**), epiisorosmanol (**17**), epirosmanol methyl ether (**18**), and methoxyrosmanol (**19**). The remaining diterpenes were either derivatives of carnosol (**20**), namely carnosic acid (**23** and **25**) and carnosic acid methyl ester (**24**), or two isomers of rosmadial (**21** and **22**). Previously, rosemary extracts were shown to be abundant in similar diterpenic compounds [2,27,28].

The by-product extracts that resulted after the hydrodistillation and SC-CO_2 extraction of rosemary can be regarded as rich sources of phytochemicals, especially phenolic compounds, such as phenolic acids, flavonoids, and diterpenes. Compared to the total (unspent material), no substantial qualitative differences were spotted in the spent material extracts. **RTE** and the three post-SC-CO_2 extracts showed a very similar metabolite profile. In the **RWE,** several non-polar diterpenes (e.g., **17**, **18**, **19**, **21**, **22**, and **25**) were not present, which could be linked to the high polarity of the solvent (water). However, **RSE** showed the highest abundance of compounds. Several hypotheses can be formulated. For instance, constituents found in small amounts in the original (unspent) plant materials could become more accessible to the solvent extraction that follows hydrodistillation. On the other hand, due to the long exposure of the plant material to boiling water, a cell permeation effect can be assumed, favoring the subsequent extraction of the metabolites. In addition, the harsh hydrodistillation conditions (high temperatures and long exposure times) can also lead to the formation of phenolic artifacts in the spent extracts.

To find more significant differences in the six extracts, a CIM analysis was next performed with the logarithmically transformed and scaled semi-quantitative data (peak area extracted from the base peak chromatograms of the LC-HRMS/MS analyses). As shown in Figure 2, the samples were distinguishable from each other, even if they seemed to form four clusters. Moreover, to describe the compounds characterizing each cluster, three blocks (I-III) were defined. In brief, **RWE** (cluster A) and **RSC3** (cluster B) contained low concentrations of compounds grouped in blocks I and III (Figure 2). Cirsimaritin, ladanein, and hydroxybenzoic acid were abundant in cluster C comprising **RSE**. In contrast, the samples of cluster D (**RSC1**, **RSC2**, and **RTE**) had low levels of the compounds mentioned above. In this cluster, **RTE** contained the highest concentration of caffeic acid and luteolin-*O*-acetylglucuronides.

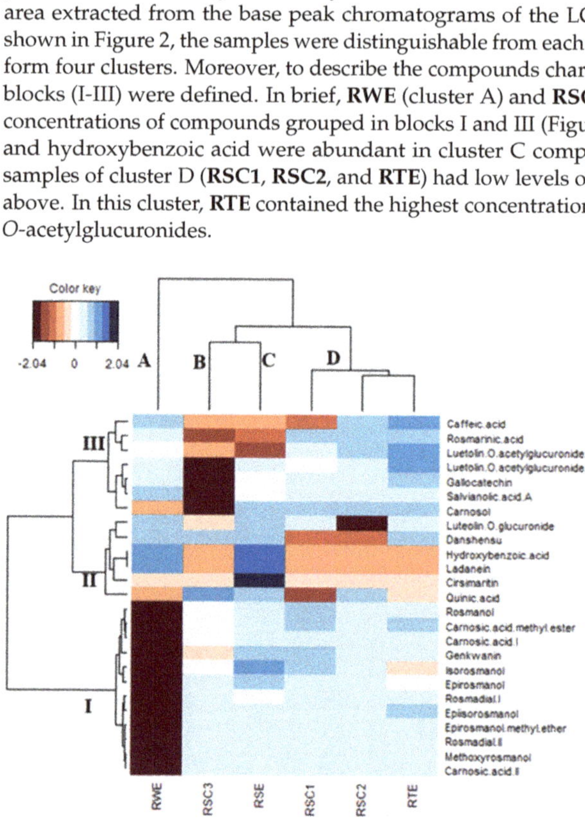

Figure 2. Global overview of the phytochemical differences among total, post-distillation, and post-SC-CO_2 rosemary extracts (Red color: low content. Blue color: high content).

3.3. *Total Phenolic and Flavonoid Content of the Total, Post-Distillation, and Post-SC-CO_2 Rosemary Extracts*

In this section, the TPC and TFC of the total (**RTE**), post-distillation (**RSE**, **RWE**), and post-SC-CO_2 (**RSC1-3**) rosemary extracts were determined. As can be seen from Table 4, the highest TPC was detected in **RWE** (108.10 mg GAE/g), followed by **RTE** (99.36 mg GAE/g), **RSC2** (98.02 mg GAE/g), and **RSC1** (97.68 mg GAE/g). **RSE** contained the lowest TPC. Regarding TFC, the highest content was recorded in **RSC2** (32.58 mg RE/g) and the lowest in **RSE** (19.86 mg RE/g). Altogether, **RWE** can be regarded as an extract with a very high amount of both phenolic and flavonoid compounds. Different results regarding the total bioactive content of rosemary extracts have been reported in the literature [29,30]. In a recent paper by Zeroual et al. [31], the TPC and TFC in rosemary extracts were dependent on extraction methods (Soxhlet and maceration) and solvents (hexane, ethyl acetate, methanol, and ethanol). In their study, the highest TPC (34.98 mg GAE/g) was lower than that of the current findings. In another study [32], sixty Jordanian plants were investigated, with the highest TPC recorded in the rosemary extract (101.339 mg GAE/g).

Table 4. Total phenolic and flavonoid content of rosemary extracts obtained from raw post-distillation, or post-SC-CO_2 materials.

Extracts	TPC (mg GAE/g)	TFC (mg RE/g)
RTE	99.36 ± 1.17^{c}	27.46 ± 0.19^{c}
RSE	57.68 ± 0.54^{d}	19.86 ± 0.13^{e}
RWE	108.10 ± 0.26^{a}	30.41 ± 0.17^{b}
RSC1	97.68 ± 5.95^{b}	27.99 ± 0.70^{c}
RSC2	98.02 ± 1.46^{b}	32.58 ± 0.11^{a}
RSC3	66.65 ± 4.11^{c}	21.63 ± 0.15^{d}

Values are reported as mean ± SD of three parallel measurements: TPC: Total phenolic content; TFC: Total flavonoid content; GAE: Gallic acid equivalent; RE: Rutin equivalent. Different letters indicate significant differences among the extracts from each species ($p < 0.05$); sample codes as in Table 1.

3.4. Post-Distillation and Post-SC-CO_2 Rosemary Extracts as Antimicrobials

The extensive use of antibiotics and the rapid emergence of multi-drug-resistant microbial strains represent severe issues for modern medicine. Numerous approaches are currently under evaluation, such as using novel plant-based antimicrobials with superior efficiency and safety profiles [33]. Various studies have repeatedly brought to attention the antimicrobial activity of rosemary EO and solvent extracts [34–39]. Thus, in this section, the activity of the ten rosemary raw and by-product extracts was evaluated by the micro-dilution method in a panel of 14 pathogenic strains. The criteria proposed by Kuete and Efferth [40] were used to categorize the observed activity into significant (MIC < 100 mg/L) and moderate-to-weak (MIC > 100 mg/L) activity. According to the results presented in Table 5, it was observed that **REO**, **RC2**, **RC3**, and **RWE** showed practically no relevant antimicrobial activity (MIC > 250 mg/L). In connection with the phytochemical composition (Tables 2 and 3), it can be assumed that the rosemary extracts rich in lipophilic compounds (the case of **REO**, **RC2**, and **RC3**) or hydrophilic compounds (the case of **RWE**) were inactive. Generally, the Gram-negative bacteria and yeasts were not inhibited by any extract. The most sensitive strains (MIC = 15.6 mg/L) were *S. aureus* after the treatment with **RSE** and **RSC1** and *M. luteus* after the treatment with **RC1**. With MIC values of 31.3 mg/L, **RC1**, **RTE**, **RSC2**, and **RSC3** also potently inhibited *S. aureus*. A similar effect was exhibited by **RC1** against *S. epidermidis*, **RSE** against *M. luteus* and *E. faecalis*, and **RSC1** against *M. luteus*. *S. aureus* MRSA was sensitive (MIC = 62.5 mg/L) to **RC1** and **RSE**, whereas *B. cereus* was inhibited to the same extent by **RC1**, **RTE**, and **RSE**.

Table 5. Antimicrobial activity of rosemary extracts obtained from raw, post-distillation or post-SC-CO_2.

Microorganism	REO	RC1	RC2	RC3	RTE	RSE	RWE	RSC1	RSC2	RSC3	Control
	MIC [mg/L]										
Gram-positive bacteria											**Vancomycin**
Staphylococcus aureus ATCC 25923	>1000	31.3	250	1000	31.3	15.6	>1000	15.6	31.3	31.3	0.98
Staphylococcus aureus ATCC BAA-1707 *	>1000	62.5	1000	>1000	125	62.5	>1000	125	125	125	0.98
Staphylococcus epidermidis ATCC 12228	>1000	31.3	250	1000	62.5	62.5	>1000	62.5	125	125	0.98
Micrococcus luteus ATCC 10240	>1000	15.6	250	250	62.5	31.3	250	31.3	62.5	62.5	0.12
Enterococcus faecalis ATCC 29212	>1000	125	>1000	1000	125	31.3	>1000	125	62.5	62.5	1.95
Bacillus cereus ATCC 10876	>1000	62.5	250	1000	62.5	62.5	>1000	125	125	125	0.98
Gram-negative bacteria											**Ciprofloxacin**
Salmonella Typhimurium ATCC 14028	>1000	>1000	>1000	>1000	1000	1000	>1000	1000	>1000	>1000	0.061
Escherichia coli ATCC 25922	>1000	>1000	>1000	>1000	1000	1000	>1000	1000	>1000	>1000	0.015
Proteus mirabilis ATCC 12453	>1000	>1000	>1000	>1000	250	250	1000	250	250	500	0.030
Klebsiella pneumoniae ATCC 13883	>1000	>1000	>1000	>1000	>1000	>1000	>1000	>1000	>1000	>1000	0.122
Pseudomonas aeruginosa ATCC 9027	>1000	>1000	>1000	>1000	1000	1000	>1000	1000	1000	1000	0.488
Yeasts											**Nystatin**
Candida glabrata ATCC 2091	1000	2000	2000	1000	2000	1000	>2000	1000	2000	2000	0.48
Candida albicans ATCC 102231	2000	>2000	1000	1000	1000	1000	>2000	1000	1000	1000	0.24
Candida parapsilosis ATCC 22019	1000	2000	500	500	500	250	1000	250	250	125	0.24

* Methicillin-resistant *Staphylococcus aureus* (MRSA) strain; MIC, minimum inhibitory concentration; sample codes as in Table 1.

The high MIC values for the rosemary EO agree with those of the literature [36,37]. For example, Hussain et al. [38] reported MIC values ranging from 300 mg/L to 1720 mg/L for rosemary EO against various Gram-positive and Gram-negative bacteria, whereas Ojeda-Sana et al. [39] documented values of 1000–2500 mg/L against *S. aureus*, *E. faecalis*, *E. coli*, and *K. pneumonia*. In contrast, various solvent extracts were more potent as antimicrobial agents. Amaral et al. [34] reported MIC values ranging from 16 to 256 mg/L for rosemary extracts obtained with ethyl acetate, dichloromethane, and ethanol against *S. aureus*, *S. epidermidis*, and *B. cereus*. Karadag et al. [35] showed MIC values between 78 and 156 mg/L against *S. aureus*, *E. faecalis*, and *H. pylori* for a hexane rosemary extract. In summary, it can be stated that the post-distillation and post-SC-CO_2 extracts are more efficient antimicrobial agents than the EO and SC-CO_2 extracts. In addition, some polyphenolic compounds' (e.g., epiirosmanol with *S. aureus* and *C. albicans*) volatile metabolites (e.g., camphor with *C. parapsilosis*) seemed to have been correlated to some extent with the antimicrobial activity (Figures 3 and 4).

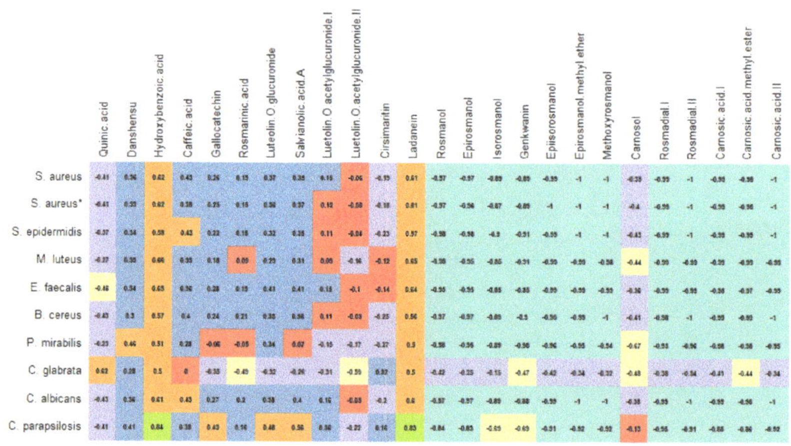

Figure 3. Correlation between polyphenols and antimicrobial activities of rosemary extracts.

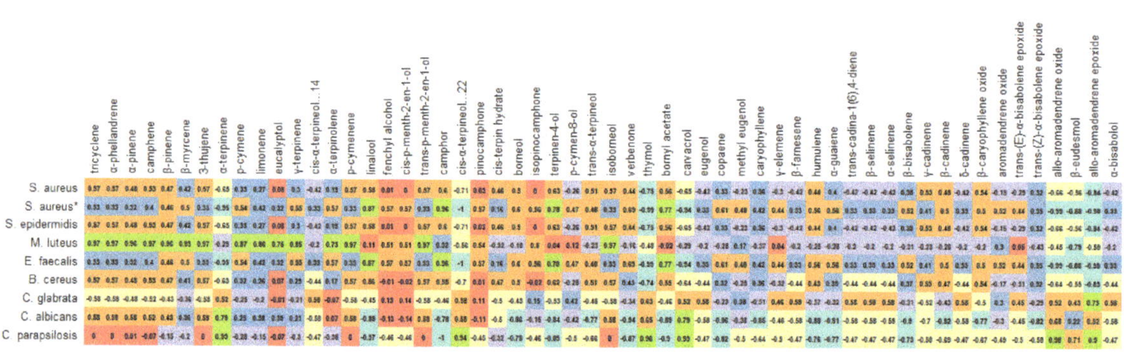

Figure 4. Correlation between volatile metabolites and antimicrobial activities of rosemary extracts.

3.5. Post-Distillation and Post-SC-CO_2 Rosemary Extracts as Antioxidants

Over the past decade, antioxidants have become increasingly popular in the treatment of oxidative-stress-related diseases, such as cardiovascular disease, diabetes, and cancer [41]. Thus, intensive efforts are carried out to identify new and safer sources of antioxidants. In this section, the antioxidant properties of rosemary extracts obtained from raw and by-product materials were investigated in six complementary assays, including radical quenching (ABTS and DPPH), reducing power (CUPRAC and FRAP), phosphomolybdenum, and metal chelating. The results are presented in Table 6. In the radical scavenging assays, the best ability was noted in **RSC2** (DPPH: 173.49 mg TE/g; ABTS: 202.19 mg TE/g), followed by **RWE** (DPPH: 164.08 mg TE/g; ABTS: 179.76 mg TE/g), and **RTE** (DPPH: 144.17 mg TE/g; ABTS: 155.03 mg TE/g). The weakest abilities for both radical quenching abilities were found in the samples obtained from the supercritical CO_2 extraction, and they can be ranked as **RC1** > **RC2** > **RC3**. The observed radical scavenging abilities were adversely affected by the increase in temperature during the supercritical CO_2 extraction procedure. In the spent extracts from the SC-CO_2 extractions, the radical scavenging ability decreased in the order **RSC2** > **RSC1** > **RSC3**, which corresponds to the level of total bioactive compounds. Additionally, when **REO** was compared to the SC-CO_2 extracts, **REO** demonstrated a higher radical scavenging ability than **RC3**. With values of 396.28 mg TE/g in CUPRAC and 205.38 mg TE/g in FRAP, **RWE** can be an excellent reducing agent compared to other samples. The reduction power of SC-CO_2 and post-SC-CO_2 extracts followed the same pattern as the radical scavenging activity. From these findings, it could be concluded that similar compounds could play a key role in the assays. As can be seen in Figure 3, some compounds (e.g., rosmarinic acid, luteolin, and caffeic acid) correlated strongly with radical scavenging and reducing abilities. Consistent with our approach, several researchers have already described these compounds as powerful antioxidants [18,42,43]. In addition, some volatile metabolites, such as thymol and carvacrol, could contribute significantly to the observed radical scavenging and reducing activities of **REO** and SC-CO_2 extracts.

Table 6. Antioxidant properties of rosemary extracts obtained from raw, post-distillation, or post-SC-CO_2 materials.

Extracts	DPPH (mg TE/g)	ABTS (mg TE/g)	CUPRAC (mg TE/g)	FRAP (mg TE/g)	MCA (mg EDTAE/g)	PBD (mg TE/g)
REO	3.70 ± 0.43 h	32.05 ± 0.12 f	33.49 ± 0.61 g	26.34 ± 1.00 g	na	18.08 ± 0.11 a
RC1	38.89 ± 1.66 f	55.61 ± 0.69 e	68.15 ± 3.73 f	47.04 ± 1.49 f	na	4.04 ± 0.16 b
RC2	10.63 ± 0.34 g	20.83 ± 0.17 f	33.07 ± 1.07 g	21.03 ± 0.41 gh	na	3.65 ± 0.32 bc
RC3	2.53 ± 0.61 h	6.30 ± 0.29 g	17.76 ± 0.39 h	11.24 ± 0.12 h	na	3.34 ± 0.35 cd
RTE	144.17 ± 1.93 c	155.03 ± 7.44 c	312.61 ± 6.76 b	197.87 ± 10.97 ab	1.60 ± 0.28 d	2.10 ± 0.02 g
RSE	48.45 ± 0.11 e	69.19 ± 0.02 e	166.92 ± 2.57 e	80.96 ± 1.92 e	na	1.41 ± 0.01 h
RWE	164.08 ± 4.50 b	179.76 ± 5.43 b	396.28 ± 7.48 a	205.38 ± 3.48 a	8.63 ± 0.45 a	2.33 ± 0.01 fg
RSC1	158.77 ± 4.48 b	157.31 ± 4.54 c	269.84 ± 6.85 c	173.25 ± 3.50 c	5.34 ± 0.70 b	2.91 ± 0.08 de
RSC2	173.49 ± 1.20 a	202.19 ± 10.73 a	317.00 ± 4.96 b	192.31 ± 5.22 b	3.13 ± 0.18 c	2.72 ± 0.19 ef
RSC3	94.77 ± 0.92 d	105.27 ± 0.69 d	182.52 ± 6.74 d	104.11 ± 4.03 d	1.56 ± 0.23 d	2.12 ± 0.14 g

Values are reported as mean ± SD of three parallel measurements. TE: Trolox equivalent. EDTAE: EDTA equivalent; na: not active; Different letters indicate significant differences among the extracts/essential oils from each species ($p < 0.05$); sample codes as in Table 1.

The highest metal chelating ability was observed in **RWE** with 8.63 mg EDTAE/g, followed by **RSC1** (5.34 mg EDTAE/g) and **RSC2** (3.13 mg EDTAE/g). Surprisingly, all non-polar samples, namely **REO** and SC-CO_2 extracts, showed no chelating effects. From Figure 5, only two compounds (rosmarinic acid and salvianolic acid A) moderately correlated with the chelating activity. In this sense, the observed ability can be explained by the presence of non-phenolic chelators, such as peptides or polysaccharides. In contrast to other assays, the highest value in the phosphomolybdenum assay was achieved by **REO** (18.08 mmol TE/g), followed by **RC1**, **RC2**, and **RC3**. The non-polar samples were more active than the polar samples. This fact was also observed in the correlation analysis. As shown in Figure 6, numerous volatile compounds were strongly associated with this propensity. These results are consistent with those of the literature that reported potent phosphomolybdenum properties for EOs [44,45]. Additionally, significant antioxidant properties of rosemary extracts, post-distillation, or essential oils have been reported in several studies [11,46,47].

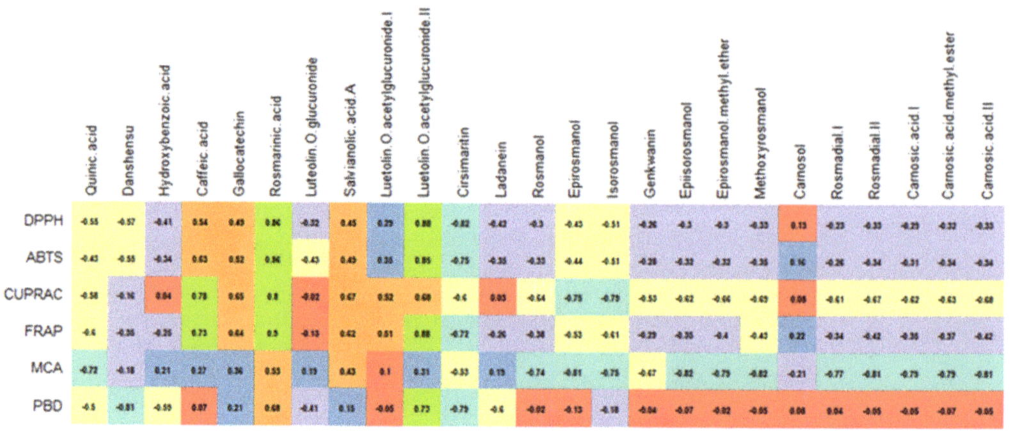

Figure 5. Correlation between polyphenols and antioxidant activities of rosemary extracts.

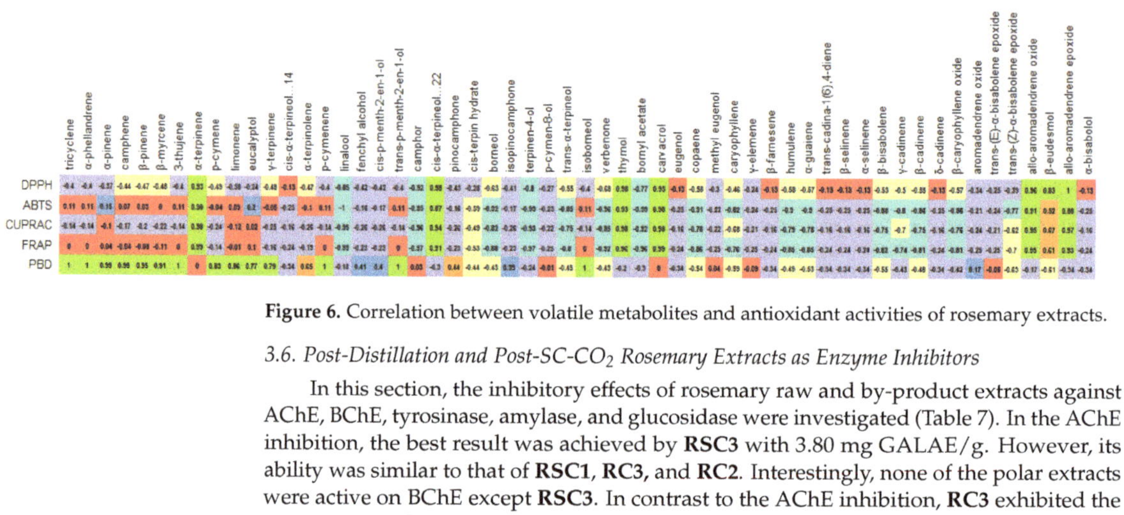

Figure 6. Correlation between volatile metabolites and antioxidant activities of rosemary extracts.

3.6. Post-Distillation and Post-SC-CO_2 Rosemary Extracts as Enzyme Inhibitors

In this section, the inhibitory effects of rosemary raw and by-product extracts against AChE, BChE, tyrosinase, amylase, and glucosidase were investigated (Table 7). In the AChE inhibition, the best result was achieved by **RSC3** with 3.80 mg GALAE/g. However, its ability was similar to that of **RSC1**, **RC3,** and **RC2**. Interestingly, none of the polar extracts were active on BChE except **RSC3**. In contrast to the AChE inhibition, **RC3** exhibited the most potent BChE inhibitory effect (3.01 mg GALAE/g). As shown in Figure 7, specific terpenoids, including linalool, terpinene-4-ol, and camphor, may be responsible for the anti-cholinesterase properties observed in the EO and SC-CO_2 extracts. In this sense, a good agreement with previous studies was found [48–50]. In addition, the cholinesterase-inhibiting effects of rosemary have been reported in several studies.

The highest tyrosinase inhibition was provided by **REO** with 59.23 mg KAE/g, followed by **RC1**, **RC2**, and **RSC1**. The anti-tyrosinase activity of post-SC-CO_2 extracts was less potent than their supercritical counterparts. This fact could be explained by some volatile compounds (α-pinene, β-pinene, *p*-cymene, etc.) and was confirmed as shown in Figure 7. The residual water, spent, and total extracts showed similar anti-tyrosinase abilities.

Table 7. Enzyme-inhibitory properties of rosemary extracts obtained from raw, post-distillation, or post-SC-CO_2 materials.

Extracts	AChE (mg GALAE/g)	BChE (mg GALAE/g)	Tyrosinase (mg KAE/g)	Amylase (mmol ACAE/g)	Glucosidase (mmol ACAE/g)
REO	3.05 ± 0.24 ^{b}	1.65 ± 0.19 ^{b}	59.23 ± 2.80 ^{a}	0.39 ± 0.03 ^{a}	na
RC1	na	1.76 ± 0.08 ^{b}	44.59 ± 0.60 ^{b}	0.33 ± 0.02 ^{b}	1.20 ± 0.03 ^{a}
RC2	3.53 ± 0.18 ^{a}	2.88 ± 0.35 ^{a}	42.23 ± 0.59 ^{bc}	0.27 ± 0.01 ^{cd}	1.17 ± 0.02 ^{a}
RC3	3.65 ± 0.14 ^{a}	3.01 ± 0.13 ^{a}	38.13 ± 0.53 ^{d}	0.24 ± 0.01 ^{d}	0.94 ± 0.01 ^{d}
RTE	1.36 ± 0.11 ^{c}	na	23.81 ± 0.47 ^{e}	0.08 ± 0.01 ^{e}	1.04 ±0.01 ^{bc}
RSE	1.12 ± 0.06 ^{c}	na	20.81 ± 0.07 ^{e}	0.07 ± 0.01 ^{e}	1.24± 0.02 ^{a}
RWE	1.12 ± 0.03 ^{c}	na	22.71 ± 0.31 ^{e}	0.06 ± 0.01 ^{e}	0.96 ± 0.01 ^{cd}
RSC1	3.79 ± 0.06 ^{a}	na	39.38 ± 2.06 ^{cd}	0.32 ± 0.01 ^{b}	1.08 ± 0.06 ^{b}
RSC2	3.09 ± 0.07 ^{b}	na	37.84 ± 0.46 ^{d}	0.29 ± 0.01 ^{bc}	0.78 ± 0.05 ^{e}
RSC3	3.80 ± 0.18 ^{a}	0.54 ± 0.04 ^{c}	37.10 ± 0.52 ^{d}	0.27 ± 0.01 ^{cd}	0.88 ± 0.03 ^{d}

Values are reported as mean ± SD of three parallel measurements. GALAE: Galanthamine equivalent; KAE: Kojic acid equivalent; na: not active; Different letters indicate significant differences among the extracts/essential oils from each species ($p < 0.05$); sample codes as in Table 1.

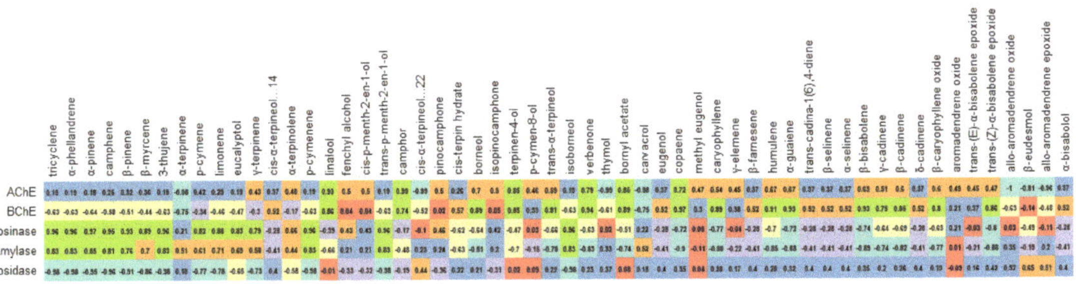

Figure 7. Correlation between volatile metabolites and enzyme-inhibitory activities of rosemary extracts.

REO achieved the most substantial inhibition value (0.39 mmol ACAE/g) in the case of amylase. Surprisingly, the same sample was not active on glucosidase; the highest glucosidase inhibitory activity was found in **RSE** (1.24 mmol ACAE/g), but the value was similar to **RC1** (1.20 mmol ACAE/g) and **RC2** (1.17 mmol ACAE/g). These results suggest terpenoids, such as α-pinene, β-pinene, or α-phellandrene, might be attributed to amylase inhibition. At the same time, some phenolics (including luteolin and ladanein) might also be the main players in the glucosidase-inhibitory capacity (Figure 8). Moreover, the mentioned compounds have been reported to have an inhibitory effect on the enzymes [51–53].

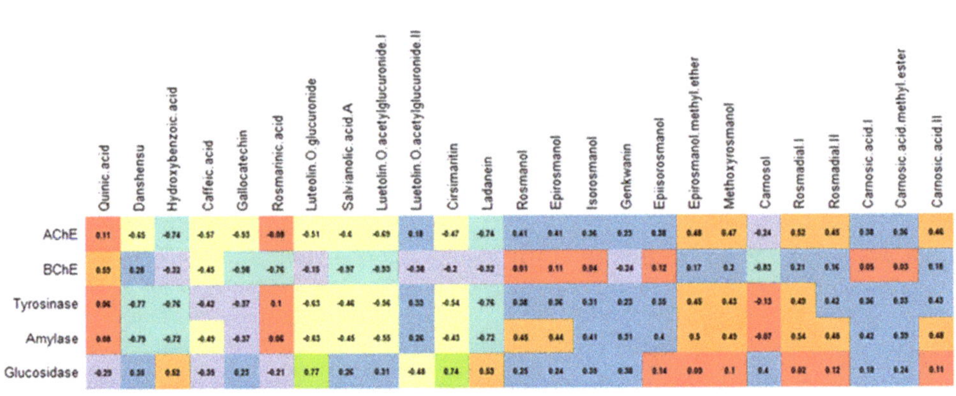

Figure 8. Correlation between polyphenols and enzyme-inhibitory activities of rosemary extracts.

3.7. Multivariate Analysis

The application of multivariate tools in biochemical sciences has been proven to be extremely convenient since it enables the cluster of different biological activities of different samples [54]. To establish the global overview of the similarities and differences between all the rosemary samples in terms of their bioactivities, two multivariate methods (PCA, CIM) were applied. Before the PCA analysis, the data were scaled to ensure the equal influence of all the bioactivities analyzed. In the CIM analysis, clusters were formed by the Ward method, and Euclidean distance was applied as a measure of diversity in the cluster analysis.

In Figure 9A, the first three dimensions represented practically 90% of the variance. The relationship of the three dimensions with the bioactivities is presented in Figure 9B. Due to its high percentage of explained variance (58.9%), the first dimension was linked to several bioactivities compared to the other two dimensions. Indeed, dimension 1 was positively correlated with DPPH, ABTS, CUPRAC, and FRAP and negatively correlated with BChE. The second dimension, which accounted for 21.4% of the variance, was positively bound to phosphomolybdenum, amylase, and tyrosinase and negatively bound to glucosidase. In the third dimension, only AChE showed a significant correlation.

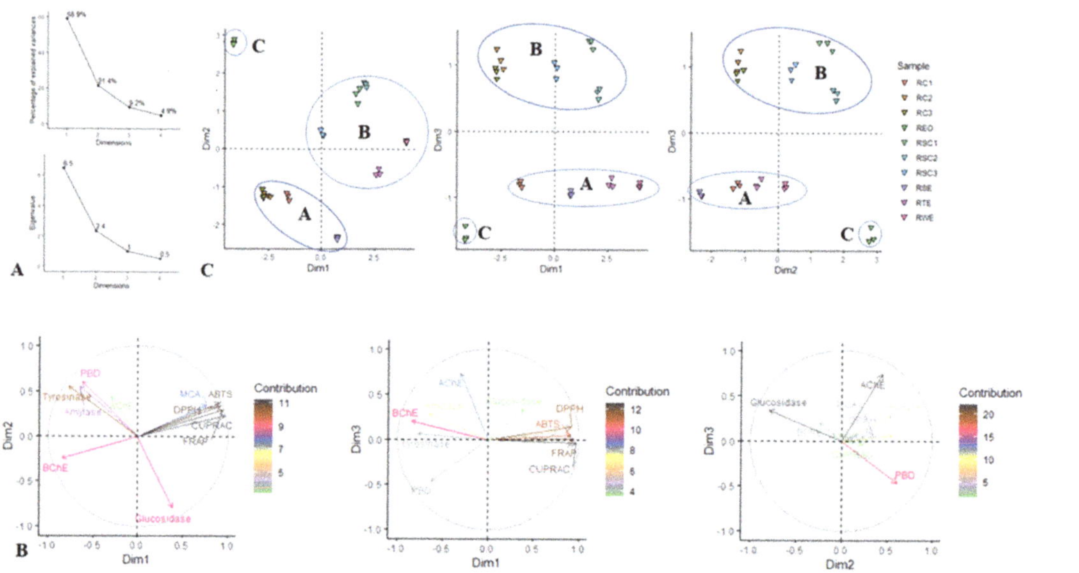

Figure 9. Principal component analysis. (**A**) Percentage of explained variance and eigenvalue. (**B**) Correlation circle showing the relationship of biological activities on each dimension of PCA. (**C**) Scatter plot showing the distribution of the samples in the factorial plan derived from the three retained dimensions.

Figure 9C shows the division of the tested samples based on the three dimensions. In each scatter plot, the samples were divided into three clusters. In each of these plots, the **REO** sample formed a standalone cluster. In addition, the samples forming clusters A and B in the first scatter plot differed from those constituting the same clusters in the remaining scatter plot. Thereby to evaluate the accuracy of the PCA classification, the clustering was adequately identified by the CIM analysis. Five clusters grouped into two large clusters were obtained (Figure 10). Cluster A comprised **REO**, which had remarkable phosphomolybdenum, anti-tyrosinase, and anti-amylase activities. Cluster B included **RC3** and **RC2**, which demonstrated a relatively high anti-BChE activity. Cluster C contained **RC1** and **RSE**. Clusters D (**RSC1**, **RSC2**, and **RSC3**) and E (**RTE** and **RWE**) were distinguished from the other clusters by their antioxidant activity. In short, the terpene-containing extracts (**REO, RC1-RC3**) showed better anti-enzymatic activity, while the remaining extracts showed good antioxidant activity.

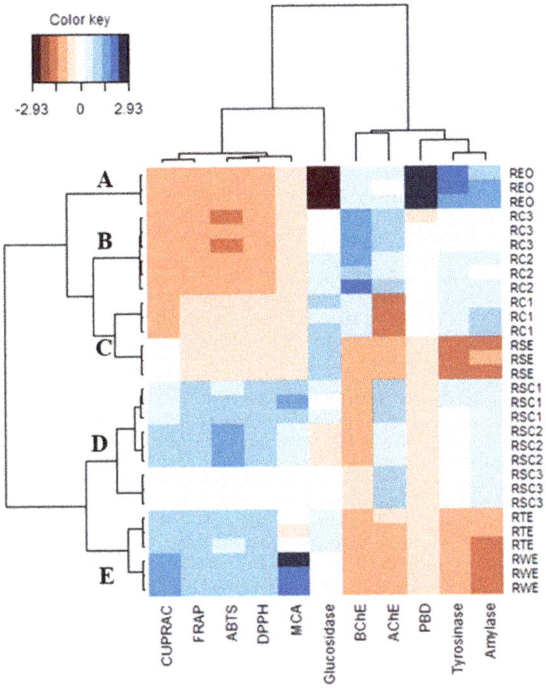

Figure 10. Clustered image map analysis on biological activities dataset. (Red color: low bioactivity. Blue color: high bioactivity).

4. Conclusions

In this work, rosemary extracts obtained from raw materials (essential oil, SC-CO_2, and total extracts) and post-distillation and post-SC-CO_2 materials were comparatively assessed for the first time from a phytochemical (GC-MS, LC-HRMS/MS) and multi-biological (antimicrobial, antioxidant, enzyme-inhibitory) approach. Overall, it can be concluded that the by-products can find uses beyond those of the terpene-rich extracts (EO, SC-CO_2) that are conventionally used as food preservatives (antioxidants) or aroma-active ingredients. The antimicrobial, antioxidant, and enzyme-inhibitory results could provide initial evidence for the health-promoting effects of the post-distillation and post-SC-CO_2 samples, which can constitute novel materials for the pharmaceutical, cosmetic, and nutraceutical industries. Furthermore, this can lead to finding new ways of exploiting the large amounts of waste produced worldwide by the rosemary essential oil industry, with beneficial environmental, technological, and economic advantages.

Author Contributions: Conceptualization, S.V.L. and A.T.; methodology, G.Z., K.I.S., I.K.-G. and S.V.L.; software, G.Z., K.I.S. and S.V.L.; validation, A.T. and K.S.-W.; formal analysis, S.V.L., G.Z., K.I.S. and A.T.; investigation A.T., G.Z., I.K.-G. and S.V.L.; resources, A.T., G.Z., K.S.-W. and S.V.L.; data curation, S.V.L.; writing—original draft preparation, S.V.L., G.Z., K.I.S. and A.T.; writing—review and editing, I.K.-G., M.M. and K.S.-W.; visualization, A.T. and S.V.L.; supervision, S.V.L. and A.T.; project administration, S.V.L.; funding acquisition, S.V.L. and M.M. All authors have read and agreed to the published version of the manuscript.

Funding: This research received funding from the Medical University of Lublin (Poland), DS 28.

Institutional Review Board Statement: Not applicable.

Informed Consent Statement: Not applicable.

Data Availability Statement: Data is contained within the article.

Conflicts of Interest: The authors declare no conflict of interest.

References

1. Trifan, A.; Zengin, G.; Brebu, M.; Skalicka-Woźniak, K.; Luca, S.V. Phytochemical Characterization and Evaluation of the Antioxidant and Anti-Enzymatic Activity of Five Common Spices: Focus on Their Essential Oils and Spent Material Extractives. *Plants* **2021**, *10*, 2692. [CrossRef]
2. Irakli, M.; Skendi, A.; Bouloumpasi, E.; Chatzopoulou, P.; Biliaderis, C.G. LC-MS Identification and Quantification of Phenolic Compounds in Solid Residues from the Essential Oil Industry. *Antioxidants* **2021**, *10*, 2016. [CrossRef] [PubMed]
3. De Macedo, L.M.; dos Santos, É.M.D.; Militão, L.; Tundisi, L.L.; Ataide, J.A.; Souto, E.B.; Mazzola, P.G. Rosemary (*Rosmarinus officinalis* L., syn *Salvia rosmarinus* Spenn.) and Its Topical Applications: A Review. *Plants* **2020**, *9*, 651. [CrossRef] [PubMed]
4. De Oliveira, J.R.; Camargo, S.E.A.; de Oliveira, L.D. *Rosmarinus officinalis* L. (rosemary) as therapeutic and prophylactic agent. *J. Biomed. Sci.* **2019**, *26*, 5. [CrossRef] [PubMed]
5. Nieto, G.; Ros, G.; Castillo, J. Antioxidant and Antimicrobial Properties of Rosemary (*Rosmarinus officinalis* L.): A Review. *Medicines* **2018**, *5*, 98. [CrossRef]
6. Ribeiro-Santos, R.; Carvalho-Costa, D.; Cavaleiro, C.; Costa, H.S.; Albuquerque, T.G.; Castilho, M.C.; Ramos, F.; Melo, N.R.; Sanches-Silva, A. A novel insight on an ancient aromatic plant: The rosemary (*Rosmarinus officinalis* L.). *Trends Food Sci. Technol.* **2015**, *45*, 355–368. [CrossRef]
7. Mahmoudi, H.; Marzouki, M.; M'Rabet, Y.; Mezni, M.; Ouazzou, A.A.; Hosni, K. Enzyme pretreatment improves the recovery of bioactive phytochemicals from sweet basil (*Ocimum basilicum* L.) leaves and their hydrodistilled residue by-products, and potentiates their biological activities. *Arab. J. Chem.* **2020**, *13*, 6451–6460. [CrossRef]
8. Peshev, D. Theoretical assessment of the use of nanofiltration for fractionation of waste aqueous fractions from the essential oil industry. *Bulg. Chem. Commun.* **2020**, *52*, 532.
9. Bouloumpasi, E.; Hatzikamari, M.; Lazaridou, A.; Chatzopoulou, P.; Biliaderis, C.G.; Irakli, M. Antibacterial and Antioxidant Properties of Oregano and Rosemary Essential Oil Distillation By-Products. *Biol. Life Sci. Forum* **2021**, *6*, 11020. [CrossRef]
10. Yagoubi, Y.; Joy, M.; Ripoll, G.; Mahouachi, M.; Bertolín, J.; Atti, N. Rosemary distillation residues reduce lipid oxidation, increase alpha-tocopherol content and improve fatty acid profile of lamb meat. *Meat Sci.* **2018**, *136*, 23–29. [CrossRef]
11. Conde-Hernández, L.A.; Espinosa-Victoria, J.R.; Trejo, A.; Guerrero-Beltrán, J.Á. CO_2-supercritical extraction, hydrodistillation and steam distillation of essential oil of rosemary (*Rosmarinus officinalis*). *J. Food Eng.* **2017**, *200*, 81–86. [CrossRef]
12. Luca, S.V.; Kittl, T.; Minceva, M. Supercritical CO_2 extraction of spices: A systematic study with focus on terpenes and piperamides from black pepper (*Piper nigrum* L.). *Food Chem.* **2023**, *406*, 135090. [CrossRef] [PubMed]
13. Reverchon, E.; Senatore, F. Isolation of rosemary oil: Comparison between hydrodistillation and supercritical CO_2 extraction. *Flavour Fragr. J.* **1992**, *7*, 227–230. [CrossRef]
14. Carvalho Jr, R.N.; Moura, L.S.; Rosa, P.T.; Meireles, M.A.A. Supercritical fluid extraction from rosemary (*Rosmarinus officinalis*): Kinetic data, extract's global yield, composition, and antioxidant activity. *J. Supecrit. Fluids* **2005**, *35*, 197–204. [CrossRef]
15. Bensebia, O.; Barth, D.; Dahmani, A. Supercritical CO_2 extraction of rosemary: Effect of extraction parameters and modelling. *J. Supercrit. Fluids* **2009**, *49*, 161–166. [CrossRef]
16. Ribeiro, M.; Bernardo-Gil, M.; Esquível, M. *Melissa officinalis* L.: Study of antioxidant activity in supercritical residues. *J. Supercrit. Fluids* **2001**, *21*, 51–60. [CrossRef]
17. Holdgate, G.A.; Meek, T.D.; Grimley, R.L. Mechanistic enzymology in drug discovery: A fresh perspective. *Nat. Rev. Drug Discov.* **2018**, *17*, 115–132. [CrossRef]
18. Tian, C.; Liu, X.; Chang, Y.; Wang, R.; Lv, T.; Cui, C.; Liu, M. Investigation of the anti-inflammatory and antioxidant activities of luteolin, kaempferol, apigenin and quercetin. *S. Afr. J. Bot.* **2021**, *137*, 257–264. [CrossRef]
19. Uysal, S.; Zengin, G.; Locatelli, M.; Bahadori, M.B.; Mocan, A.; Bellagamba, G.; De Luca, E.; Mollica, A.; Aktumsek, A. Cytotoxic and Enzyme Inhibitory Potential of Two Potentilla species (*P. speciosa* L. and *P. reptans* Willd.) and Their Chemical Composition. *Front. Pharmacol.* **2017**, *8*, 290. [CrossRef]
20. European Society of Clinical Microbiology and Infectious Diseases (ESCMID). Determination of minimum inhibitory concentrations (MICs) of antibacterial agents by broth dilution. *Clin. Microbiol. Infect.* **2003**, *9*, 9–15.
21. Grochowski, D.M.; Uysal, S.; Aktumsek, A.; Granica, S.; Zengin, G.; Ceylan, R.; Locatelli, M.; Tomczyk, M. In vitro enzyme inhibitory properties, antioxidant activities, and phytochemical profile of *Potentilla thuringiaca*. *Phytochem. Lett.* **2017**, *20*, 365–372. [CrossRef]
22. McHugh, M.; Krukonis, V. *Supercritical Fluid Extraction: Principles and Practice*; Elsevier: Amsterdam, The Netherlands, 2013.
23. Ramzi, A.; El Ouali Lalami, A.; Ez Zoubi, Y.; Assouguem, A.; Almeer, R.; Najda, A.; Ullah, R.; Ercisli, S.; Farah, A. Insecticidal effect of wild-grown *Mentha pulegium* and *Rosmarinus officinalis* essential oils and their main monoterpenes against *Culex pipiens* (Diptera: Culicidae). *Plants* **2022**, *11*, 1193. [CrossRef] [PubMed]

24. Ouknin, M.; Aghraz, A.; Chibane, M.; Boumezzourh, A.; Costa, J.; Majidi, L. Enzyme inhibitory, antioxidant activity and phytochemical analysis of essential oil from cultivated *Rosmarinus officinalis*. *J. Food Meas. Charact.* **2021**, *15*, 3782–3790. [CrossRef]
25. Gök, H.N.; Luca, S.V.; Ay, S.T.; Komsta, L.; Salmas, R.E.; Orhan, I.E.; Skalicka-Woźniak, K. Profiling the annual change of the neurobiological and antioxidant effects of five Origanum species in correlation with their phytochemical composition. *Food Chem.* **2021**, *368*, 130775. [CrossRef]
26. Darrag, H.M.; Almuhanna, H.T.; Hakami, E.H. Secondary metabolites in basil, bio-insecticide, inhibition effect, and in silico molecular docking against proteolytic enzymes of the red palm weevil (*Rhynchophorus ferrugineus*). *Plants* **2022**, *11*, 1087. [CrossRef]
27. Borrás-Linares, I.; Stojanović, Z.; Quirantes-Piné, R.; Arráez-Román, D.; Švarc-Gajić, J.; Fernández-Gutiérrez, A.; Segura-Carretero, A. *Rosmarinus officinalis* Leaves as a Natural Source of Bioactive Compounds. *Int. J. Mol. Sci.* **2014**, *15*, 20585. [CrossRef]
28. Kakouri, E.; Nikola, O.; Kanakis, C.; Hatziagapiou, K.; Lambrou, G.I.; Trigas, P.; Kanaka-Gantenbein, C.; Tarantilis, P.A. Cytotoxic Effect of *Rosmarinus officinalis* Extract on Glioblastoma and Rhabdomyosarcoma Cell Lines. *Molecules* **2022**, *27*, 6348. [CrossRef]
29. Afshar, M.; Najafian, S.; Radi, M. Seasonal variation on the major bioactive compounds: Total phenolic and flavonoids contents, and antioxidant activity of rosemary from shiraz. *Nat. Prod. Res.* **2022**, *36*, 4287–4292. [CrossRef]
30. Tzima, K.; Brunton, N.P.; Lyng, J.G.; Frontuto, D.; Rai, D.K. The effect of Pulsed Electric Field as a pre-treatment step in Ultrasound Assisted Extraction of phenolic compounds from fresh rosemary and thyme by-products. *Innov. Food Sci. Emerg. Technol.* **2021**, *69*, 102644. [CrossRef]
31. Zeroual, A.S.; Sakar, E.H.; Mahjoubi, F.; Chaouch, M.; Chaqroune, A.; Taleb, M. Effects of extraction technique and solvent on phytochemicals, antioxidant, and antimicrobial activities of cultivated and wild rosemary (*Rosmarinus officinalis* L.) from Taounate region. *Biointerface Res. Appl. Chem.* **2022**, *12*, 8441–8452.
32. Al-Sayyed, H.F.; Al-Kurd, R.A.; Mahmoud, I.F.; AbdelQader, S.M.; Sweidan, D.H.; Rizeq, L.T.; Arafat, T.A.; Mwalla, M.M. Developing a database for total phenolic content, total flavonoid content, and antioxidant activity of Jordanian crops. *Int. J. Food Prop.* **2022**, *25*, 1290–1301. [CrossRef]
33. Trifan, A.; Luca, S.V.; Greige-Gerges, H.; Miron, A.; Gille, E.; Aprotosoaie, A.C. Recent advances in tackling microbial multidrug resistance with essential oils: Combinatorial and nano-based strategies. *Crit. Rev. Microbiol.* **2020**, *46*, 338–357. [CrossRef] [PubMed]
34. Amaral, G.P.; Mizdal, C.R.; Stefanello, S.T.; Mendez, A.S.L.; Puntel, R.L.; de Campos, M.M.A.; Soares, F.A.A.; Fachinetto, R. Antibacterial and antioxidant effects of *Rosmarinus officinalis* L. extract and its fractions. *J. Tradit. Complement. Med.* **2019**, *9*, 383–392. [CrossRef] [PubMed]
35. Karadağ, A.E.; Demirci, B.; Çaşkurlu, A.; Okur, M.; Orak, D.; Sipahi, H.; Başer, K. In vitro antibacterial, antioxidant, anti-inflammatory and analgesic evaluation of *Rosmarinus officinalis* L. flower extract fractions. *S. Afr. J. Bot.* **2019**, *125*, 214–220. [CrossRef]
36. Wang, W.; Li, N.; Luo, M.; Zu, Y.; Efferth, T. Antibacterial Activity and Anticancer Activity of *Rosmarinus officinalis* L. Essential Oil Compared to That of Its Main Components. *Molecules* **2012**, *17*, 2704–2713. [CrossRef] [PubMed]
37. Okoh, O.O.; Sadimenko, A.P.; Afolayan, A.J. Comparative evaluation of the antibacterial activities of the essential oils of *Rosmarinus officinalis* L. obtained by hydrodistillation and solvent free microwave extraction methods. *Food Chem.* **2010**, *120*, 308–312. [CrossRef]
38. Hussain, A.I.; Anwar, F.; Chatha, S.A.S.; Jabbar, A.; Mahboob, S.; Nigam, P.S. Rosmarinus officinalis essential oil: Antiproliferative, antioxidant and antibacterial activities. *Br. J. Microbiol.* **2010**, *41*, 1070–1078. [CrossRef]
39. Ojeda-Sana, A.M.; van Baren, C.M.; Elechosa, M.A.; Juárez, M.A.; Moreno, S. New insights into antibacterial and antioxidant activities of rosemary essential oils and their main components. *Food Control* **2012**, *31*, 189–195. [CrossRef]
40. Kuete, V.; Efferth, T. Cameroonian Medicinal Plants: Pharmacology and Derived Natural Products. *Front. Pharmacol.* **2010**, *1*, 123. [CrossRef]
41. Demirci-Çekiç, S.; Özkan, G.; Avan, A.N.; Uzunboy, S.; Çapanoğlu, E.; Apak, R. Biomarkers of oxidative stress and antioxidant defense. *J. Pharm. Biomed. Anal.* **2022**, *209*, 114477. [CrossRef]
42. Adomako-Bonsu, A.G.; Chan, S.L.F.; Pratten, M.; Fry, J.R. Antioxidant activity of rosmarinic acid and its principal metabolites in chemical and cellular systems: Importance of physico-chemical characteristics. *Toxicol. In Vitro* **2017**, *40*, 248–255. [CrossRef] [PubMed]
43. Espíndola, K.M.M.; Ferreira, R.G.; Narvaez, L.E.M.; Rosario, A.C.R.S.; Da Silva, A.H.M.; Silva, A.G.B.; Vieira, A.P.O.; Monteiro, M.C. Chemical and Pharmacological Aspects of Caffeic Acid and Its Activity in Hepatocarcinoma. *Front. Oncol.* **2019**, *9*, 541. [CrossRef] [PubMed]
44. Pavlić, B.; Teslić, N.; Zengin, G.; Đurović, S.; Rakić, D.; Cvetanović, A.; Gunes, A.; Zeković, Z. Antioxidant and enzyme-inhibitory activity of peppermint extracts and essential oils obtained by conventional and emerging extraction techniques. *Food Chem.* **2021**, *338*, 127724. [CrossRef] [PubMed]
45. Rezzoug, M.; Bakchiche, B.; Gherib, A.; Roberta, A.; Guido, F.; Kilinçarslan, Ö.; Mammadov, R.; Bardaweel, S.K. Chemical composition and bioactivity of essential oils and Ethanolic extracts of *Ocimum basilicum* L. and *Thymus algeriensis* Boiss. & Reut. from the Algerian Saharan Atlas. *BMC Complement. Altern. Med.* **2019**, *19*, 146. [CrossRef]
46. Hcini, K.; Lozano-Pérez, A.A.; Luis Cenis, J.; Quílez, M.; José Jordán, M. Extraction and encapsulation of phenolic compounds of tunisian rosemary (*Rosmarinus officinalis* L.) extracts in silk fibroin nanoparticles. *Plants* **2021**, *10*, 2312. [CrossRef] [PubMed]

47. Ivanović, J.; Đilas, S.; Jadranin, M.; Vajs, V.; Babović, N.; Petrović, S.D.; Žižović, I. Supercritical carbon dioxide extraction of antioxidants from rosemary (*Rosmarinus officinalis* L.) and sage (*Salvia officinalis* L.). *J. Serb. Chem. Soc.* **2009**, *74*, 717–732. [CrossRef]
48. Amina, B.; Soumeya, B.; Salim, B.; Mahieddine, B.; Sakina, B.; Chawki, B.; Francesca, N.; Marzia, V.; Carmine, N.; Luigi, D.B. Chemical profiling, antioxidant, enzyme inhibitory and in silico modeling of *Rosmarinus officinalis* L. and *Artemisia herba alba* Asso. essential oils from Algeria. *S. Afr. J. Bot.* **2022**, *147*, 501–510. [CrossRef]
49. Farag, M.A.; Ezzat, S.M.; Salama, M.M.; Tadros, M.G.; Serya, R.A. Anti-acetylcholinesterase activity of essential oils and their major constituents from four *Ocimum* species. *Z. Naturfor. C* **2016**, *71*, 393–402. [CrossRef]
50. Mills, C.; Cleary, B.V.; Walsh, J.J.; Gilmer, J.F. Inhibition of acetylcholinesterase by Tea Tree oil. *J. Pharm. Pharmacol.* **2004**, *56*, 375–379. [CrossRef]
51. Jelenkovic, L.; Jovanovic, V.S.; Palic, I.; Mitic, V.; Radulovic, M. In Vitro Screening of α-Amylase Inhibition by Selected Terpenes from Essential Oils. *Trop. J. Pharm. Res.* **2014**, *13*, 1421. [CrossRef]
52. Oboh, G.; Ademosun, A.O.; Odubanjo, O.V.; Akinbola, I.A. Antioxidative Properties and Inhibition of Key Enzymes Relevant to Type-2 Diabetes and Hypertension by Essential Oils from Black Pepper. *Adv. Pharmacol. Sci.* **2013**, *2013*, 926047. [CrossRef] [PubMed]
53. Kashtoh, H.; Baek, K.-H. Recent Updates on Phytoconstituent Alpha-Glucosidase Inhibitors: An Approach towards the Treatment of Type Two Diabetes. *Plants* **2022**, *11*, 2722. [CrossRef] [PubMed]
54. Sinan, K.I.; Martinović, L.S.; Peršurić, Ž.; Pavelić, S.K.; Etienne, O.K.; Mahomoodally, M.F.; Sadeer, N.B.; Zengin, G. Novel insights into the biopharmaceutical potential, comparative phytochemical analysis and multivariate analysis of different extracts of shea butter tree-Vitellaria paradoxa CF Gaertn. *Proc. Bichem.* **2020**, *98*, 65–75. [CrossRef]

antioxidants

Article

First Report on the Chemical Composition, Antioxidant Capacity, and Preliminary Toxicity to *Artemia salina* L. of *Croton campinarensis* Secco, A. Rosário & PE Berry (Euphorbiaceae) Essential Oil, and In Silico Study

Leonardo Souza da Costa [1], Ângelo Antônio Barbosa de Moraes [1,2], Jorddy Neves Cruz [3], Suraj N. Mali [4], Lorena Queiroz Almeida [1], Lidiane Diniz do Nascimento [2], Oberdan Oliveira Ferreira [2], Everton Luiz Pompeu Varela [5], Sandro Percário [5], Mozaniel Santana de Oliveira [2,*] and Eloisa Helena de Aguiar Andrade [2]

1 School of Chemical Engineering, Institute of Technology, Universidade Federal do Pará, Belem 66075-110, Brazil
2 Adolpho Ducke Laboratory, Coordination of Botany, Emílio Goeldi Museum of Pará, Belem 66077-830, Brazil
3 Laboratory of Functional and Structural Biology, Institute of Biological Sciences, Universidade Federal do Pará, Belem 66075-110, Brazil
4 Department of Pharmacy, Government College of Pharmacy, Affiliated to Shivaji University, Kolhapur, Karad 415124, Maharashtra, India
5 Oxidative Stress Research Laboratory, Biological Sciences Institute, Universidade Federal do Pará, Belem 66075-110, Brazil
* Correspondence: mozanieloliveira@museu-goeldi.br

Citation: da Costa, L.S.; de Moraes, Â.A.B.; Cruz, J.N.; Mali, S.N.; Almeida, L.Q.; do Nascimento, L.D.; Ferreira, O.O.; Varela, E.L.P.; Percário, S.; de Oliveira, M.S.; et al. First Report on the Chemical Composition, Antioxidant Capacity, and Preliminary Toxicity to *Artemia salina* L. of *Croton campinarensis* Secco, A. Rosário & PE Berry (Euphorbiaceae) Essential Oil, and In Silico Study. *Antioxidants* **2022**, *11*, 2410. https://doi.org/10.3390/antiox11122410

Academic Editors: Delia Mirela Tit and Simona Bungau

Received: 3 November 2022
Accepted: 30 November 2022
Published: 6 December 2022

Abstract: *Croton campinarensis* Secco, A. Rosário & PE Berry is an aromatic species recently discovered in the Amazon region. This study first reports the chemical profile, antioxidant capacity, and preliminary toxicity to *A. salina* Leach of the essential oil (EO) of this species. The phytochemical profile of the essential oil was analyzed by gas chromatography (GC/MS) and (GC-FID). The antioxidant capacity of the EO was measured by its inhibition of $ABTS^{\bullet+}$ and $DPPH^{\bullet}$ radicals. Molecular modeling was used to evaluate the mode of interaction of the major compounds with acetylcholinesterase (AChE). The results indicate that the EO yield was 0.24%, and germacrene D (26.95%), bicyclogermacrene (17.08%), (*E*)-caryophyllene (17.06%), and δ-elemene (7.59%) were the major compounds of the EO sample. The EO showed a TEAC of 0.55 ± 0.04 mM·L^{-1} for the reduction of the $ABTS^{\bullet+}$ radical and 1.88 ± 0.08 mM·L^{-1} for the reduction of the $DPPH^{\bullet}$ radical. Regarding preliminary toxicity, the EO was classified as toxic in the bioassay with *A. salina* ($LC_{50} = 20.84 \pm 4.84$ μg·mL^{-1}). Through molecular docking, it was found that the majority of the EO components were able to interact with the binding pocket of AChE, a molecular target related to toxicity evaluated in *A. salina* models; the main interactions were van der Waals and π-alkyl interactions.

Keywords: Amazon; new species; natural products; bioactive compounds; molecular modeling

1. Introduction

Natural products, in particular volatile oils isolated from aromatic plants, have been the subject of several studies over the years. This may be related to their complex chemical compositions, which may include a series of classes of chemical compounds such as monoterpenes, sesquiterpenes, hydrocarbons, oxygenated mono- and sesquiterpenes, and phenylpropanoids [1–4]. In addition, this diverse chemical composition may be related to potential activities, such as antioxidant [5–7]. Amazonian flora, for example, include several aromatic plant species that produce EOs [8,9] that can serve as inputs for various sectors of the chemical, pharmaceutical, cosmetic, and food industries; these potential applications show promise in generating development in the region [4,10–12]. Among

Amazonian species that produce EOs, those belonging to the family Euphorbiaceae stand out, especially species of the genus *Croton* [13].

The *Croton* genus includes a variety of over 1200 species with a widespread global distribution in both the tropics and subtropics, especially in arid and semiarid zones [14]. In Brazil, 300 species are found, of which 230 are considered endemic. In the Amazon region, 61 species have been recorded [15]. *Croton* species have shrub characteristics, can resprout in rainy seasons, and grow widely, especially in deforested areas [16].

Several species of *Croton* have chemical compounds used for medicinal purposes, mainly as anti-inflammatory, antihypertensive, antifungal, antimicrobial, antidiabetic, antioxidant, antinociceptive, and antitumor agents [16]. In traditional medicine, the leaves of species of this genus are used to treat gastrointestinal disorders, rheumatism, migraine, diabetes, cholesterol level, inflammatory diseases, and bronchitis [17–19]. The EOs of species of the genus *Croton* are characterized by different chemical classes of compounds, with a predominance of terpenoids and phenylpropanoids [20]. Compounds such as limonene, (*E*)-caryophyllene, spathulenol, bicyclogermacrene, germacrene D, (*E*)-anathole, and estragol are common components of EOs of *Croton* species [21–26].

The EOs of *Croton* species have also been widely studied for the discovery of new natural antioxidants [27–31]. Antioxidant substances can inhibit free radicals [32,33], which may be responsible for the damage caused by oxidative stress related to various diseases, such as Alzheimer's disease, Parkinson's disease, cancer, and diabetes [34]. The use of natural antioxidant compounds rather than synthetic antioxidants is currently being widely explored, as the former pose lower human health risks, especially those related to high toxicity and the triggering of new diseases caused by synthetic products [35–37].

When evaluating the toxicity of EOs, preliminary tests are performed to ensure their safety for humans [38,39]. Preliminary toxicity tests on *A. salina* allow an initial response to the potential toxicity of an extract or isolated substance; this is due to the sensitivity similar to that of human cells that the microcrustacean presents. [40]. The EOs of some species of the genus *Croton* have moderate or high toxicity against *A. salina*, which is directly related to the high sesquiterpene content in their chemical composition [41–44].

This bioassay is also important due to the biochemical activity of the enzyme acetylcholinesterase (AChE), which mediates the larval mortality of *A. salina* when individuals come into contact with EO within 24 h; the number of deaths of microcrustaceans is counted, which indicates a potential biological activity of the essential oil [45]. This enzyme is also directly related to the behavior and physiology of *A. salina*, and its inhibition may cause deleterious effects in individuals [46].

Croton campinarensis Secco, A. Rosário & PE Berry is a recently discovered species recorded for the first time in 2012 [47]. The species is listed only in the state of Pará, located in the Brazilian Amazon [15]. Because it is a recently known species, there are no records in the literature related to the chemical composition and antioxidant and biological properties of its EO. As a result, the present study first reports the yield, chemical composition, antioxidant profile, and preliminary toxicity of the EO of dry leaves of *C. campinarensis*, aiming to contribute to the phytochemical knowledge of aromatic plants of the genus *Croton* from the Amazon region.

2. Materials and Methods

2.1. Collection and Processing of Botanical Material

Leaves of *C. campinarensis* were collected in the locality of Campina do Guajará, municipality of Bujaru (Latitude: 1°31′15″ S, Longitude: 48°2′37″ W), microregion of Castanhal, Pará State, Brazil, in July 2017. The sample was identified and deposited in the herbarium of the Museu Paraense Emílio Goeldi, Belém, Pará, with registration number MG167619.

2.2. Distillation of Essential Oil

The processed botanical material was subjected to hydrodistillation to obtain EO, using a modified Clevenger apparatus for 3 h. After distillation, the EO was centrifuged and dehydrated with anhydrous sodium sulfate (Na_2SO_4). Then, it was stored and preserved in a freezer. The EO yield was calculated on a moisture-free basis [45].

2.3. Analysis of the Chemical Composition of the Essential Oil

The analysis of the phytochemical profile of the essential oil of *C. campinarensis* was carried out following the same protocols of our research group, as well as the brand and model of the equipment [33,48–50]. Masses (GC-MS) and quantification were performed by gas chromatography with a flame ionization detector (GC-FID). The identifications in the GC/MS were performed based on the calculated retention index [51] and compared with the literature [52].

2.4. Determination of the Trolox Equivalent Antioxidant Capacity (TEAC) of the Essential Oil

2.4.1. DPPH Method

This method was performed to analyze the potential of EO of *C. campinarensis* to inhibit the 1,1-diphenyl-2-picrylhydrazyl (DPPH•) radical, a violet chromophore, resulting in the formation of the hydrogenated DPPH product, which is yellow or colorless [53]. Description of the method can be found in [54].

2.4.2. ABTS Method

This method was performed to analyze the potential of *C. campinarensis* EO to inhibit the 2,2-azino-bis (3-ethylbenzothiazoline-6-sulfonic acid) diammonium salt (ABTS; Sigma-Aldrich; A1888, São Paulo, Brazil) radical, according to modification by Re et al. [55] of the experimental method proposed by Miller et al. [56]. Description of the method can be found in [54].

2.5. Determination of Preliminary Toxicity against Artemia salina Leach

For toxicity tests on *A. salina*, the essential oil was prepared at concentrations ranging from 1–100 $\mu g \cdot mL^{-1}$, according to methods described by [45]. A total of ten *A. salina* larvae were added to each test flask with the aid of automatic micropipettes. In the control group and the positive group with lapachol, the same solvent was used for the samples and larvae, under the same conditions as the bioassay. The counting of *A. salina* was carried out after a period of 24 h at each concentration used and the IC50 was calculated, with the experiments being carried out in triplicate ($n = 3$).

2.6. Statistical Analysis

In the experimental tests, with the exception of the analysis of the chemical composition, the statistical Student's *t*-test was applied, and a significance level of 5% ($p \leq 0.05$) was considered.

2.7. Molecular Docking

For the molecular docking studies, data on the major compounds germacrene D (26.95%), bicyclogermacrene (17.98%), (*E*)-caryophyllene (17.60%), γ-terpinene (8.99%), and δ-elemene (7.59%) were obtained from the PubChem database (https://pubchem.ncbi.nlm.nih.gov/, accessed on 1 October 2022). Then, their structures were optimized with B3LYP/6-31G* using Gaussian 09 software (Gaussian, Inc., Wallingford, England) [57–60].

Molecular interactions were performed with the majority compounds and AChE in Molegro Virtual Docker (MVD) 5.5 software (Molexus IVS, Odder, Denmark) [61–64]; the structure of the protein used in the molecular modeling study can be obtained from the Protein Data Bank (https://www.rcsb.org/, accessed on 1 September 2022), using the ID code 4M0E [65]. The MolDock Score (GRID) was carried out as described by [45].

3. Results and Discussion

3.1. Yield and Chemical Composition of the Essential Oil

The EO yield for the studied sample was 0.24%, we can observe in Table 1. There are no records in the literature regarding the yield and volatile composition of the EO of *C. campinarensis*. However, the yield found in the present study is lower than that reported by Turiel et al. [66] for four *Croton* species from the Amazon (0.50% ± 1.10%). Another notable finding is the high concentration of sesquiterpene hydrocarbons found in both the present sample and the species analyzed by Turiel et al. [66], which had levels between 55.30% and 83.00%.

Table 1. Main chemical constituents and EO yield of *C. campinarensis*.

Yield (%)			0.24
RI_L	**RI_C**	**Chemical Constituents**	**Area (%)**
1014 [a]	1017	α-terpinene	0.31
1020 [a]	1024	*p*-cymene	0.49
1054 [a]	1059	γ-terpinene	8.99
1335 [a]	1339	δ-elemene	7.59
1374 [a]	1378	α-copaene	0.34
1387 [a]	1387	β-bourbonene	0.45
1389 [a]	1394	β-elemene	3.56
1409 [a]	1412	α-gurjunene	0.48
1417 [a]	1423	(*E*)-caryophyllene	17.60
1430 [a]	1431	β-copaene	1.37
1434 [a]	1434	γ-elemene	0.43
1432 [a]	1437	α-*trans*-bergamotene	1.03
1447 [b]	1446	isogermacrene D	0.60
1452 [a]	1456	α-humulene	2.49
1458 [a]	1463	*allo*-aromadrendene	0.38
1465 [a]	1465	*cis*-muurola-4(14).5-diene	0.21
1478 [a]	1479	γ-muurolene	0.39
1484 [a]	1484	germacrene D	26.95
1495 [a]	1492	γ-amorphene	0.39
1500 [a]	1500	bicyclogermacrene	17.98
1504 [a]	1506	cuparene	1.68
1514 [a]	1513	β-curcumene	0.28
1514 [a]	1516	(*Z*)-γ-bisabolene	1.01
1522 [a]	1525	δ-cadinene	1.38
1529 [a]	1533	(*E*)-γ-bisabolene	1.06
1533 [a]	1539	*trans*-cadina-1.4-diene	0.09
1537	1544	(*E*)-α-bisabolene	0.21
1577 [a]	1581	spathulenol	0.96
1582	1587 [a]	caryophyllene oxide	0.31
1638	1643 [a]	*epi*-α-cadinol	0.33
1644	1649 [a]	α-muurolol	0.21
1652	1655 [a]	α-cadinol	0.31
	monoterpene hydrocarbons		9.30
	oxygenated monoterpenes		0.00
	sesquiterpene hydrocarbons		87.95
	oxygenated sesquiterpenes		2.12
	other		0.49
	TOTAL		99.86

RI_C, retention index, RI_L: retention index from the literature: (a) = [52] and (b) = [67]. Relative percentage areas calculated based on the peak areas.

Regarding the chemical composition, we can observe in Table 1, 32 volatile constituents were identified, accounting for 99.86% of the total, and sesquiterpene hydrocarbons dominated (87.95%), mainly germacrene D (26.95%), bicyclogermacrene (17.08%),

(*E*)-caryophyllene (17.06%), and δ-elemene (7.59%). In addition, the monoterpene hydrocarbon γ-terpinene (8.99%) had a high content. Turiel et al. [66] reported that (*E*)-caryophyllene was the major component of the EOs of *C. campestris* (23.90%) and *C. eriocladus* (24.10%), bicyclogermacrene was the main volatile compound of the EO of *C. chaetocalyx* (13.90%), and spatulenol was the component with the highest content in *C. glandulosus* EO (19.70%). The authors also found germacrene D at high levels in the EOs of *C. campestris* (13.70%), *C. eriocladus* (9.70%), and *C. chaetocalyx* (17.90%); (*E*)-caryophyllene in the EOs of *C. eriocladus* (7.10%) and *C. glandulosus* (8.90%); bicyclogermacrene in the EO of *C. glandulosus* (9.60%); and δ-elemene in the EOs of *C. chaetocalyx* (13.50%) and *C. glandulosus* (8.00%). The authors also noted that γ-terpinene was identified only in *C. campestris* EO, with a low content (0.70%). These results indicate that the present sample of *C. campinarensis* EO has a chemical composition similar to those of other species of this genus present in the Amazon region.

Germacrene D, the major constituent of the sample, shows larvicidal activity against the mosquito *Aedes aegypti* [68]. EOs containing this compound as a major component also have anti-inflammatory and anti-AChE properties [69]. (*E*)-Caryophyllene also has potential anti-inflammatory activity and can be used to treat central nervous system diseases, cancer, and dental caries infections caused by etiological agents [70].

Franco et al. [33] reported that germacrene D and (*E*)-caryophyllene have antioxidant activity, and bicyclogermacrene has been associated with larvicidal and antiviral activity. Figueiredo et al. [71] reported that EOs with high levels of γ-terpinene have moderate activity against food pathogens. According to Dang et al. [72], δ-elemene has anticancer activity against HeLa cells.

Da Silva Júnior et al. [73] reported that germacrene D, bicyclogermacrene and δ-elemene are directly related to plant defense mechanisms and have higher concentrations in the rainy season in the Amazon region. According to the authors, this concentration trend may be directly related to a plant strategy for attracting pollinating agents, especially bees and flies, which are common in the rainy season.

3.2. Antioxidant Capacity and Preliminary Toxicity of the Essential Oil

The Table 2, below shows the TEAC of the EO of *C. campinarensis*, measured through the inhibition of $ABTS^{\bullet+}$ and $DPPH^{\bullet}$ radicals and its preliminary toxicity against *A. salina*.

Table 2. Trolox equivalent antioxidant capacity of *C. campinarensis* EO for inhibition of $ABTS^{\bullet+}$ and $DPPH^{\bullet}$ radicals and preliminary toxicity against *A. salina*.

TEAC		Preliminary Toxicity	
ABTS ($mM \cdot L^{-1}$)	DPPH ($mM \cdot L^{-1}$)	LC_{50} ($\mu g \cdot mL^{-1}$)	R^2
0.55 ± 0.04 [a]	1.88 ± 0.08 [b]	20.84 ± 4.84	0.85

Values are expressed as the mean and standard deviation (n = 3) of Trolox equivalent antioxidant capacity. Student's *t*-test was used to compare OE of *C. campinarensis* to the Trolox standard (1 $mM \cdot L^{-1}$). TEAC = Trolox equivalent antioxidant capacity. Different letters indicate that the samples are significantly different.

The TEAC of the EO to inhibit the $ABTS^{\bullet+}$ radical was 0.55 ± 0.04 $mM \cdot L^{-1}$ (p = 0.557). In the DPPH assay, the TEAC was 1.88 ± 0.08 $mM \cdot L^{-1}$ (p = 0.001). These results indicate that the $ABTS^{\bullet+}$ radical capture potential of the EO of *C. campinarensis* was lower than that presented by the Trolox standard (1 $mM \cdot L^{-1}$). On the other hand, for the inhibition of $DPPH^{\bullet}$ radicals, the TEAC of the EO was almost double that of the Trolox standard (1 $mM \cdot L^{-1}$). Regarding the preliminary toxicity against *A. salina*, the EO had an LC_{50} of 20.84 ± 4.84 $\mu g \cdot mL^{-1}$. According to Ramos et al. [74], EOs with LC_{50} values lower than 80 $\mu g \cdot mL^{-1}$ are classified as toxic.

There are no reports in the literature regarding the antioxidant capacity and preliminary toxicity of *C. campinarensis* EO. However, other *Croton* species do have literature data on these properties. Morais et al. [75] reported that the evaluation of antioxidant activity by the DPPH method showed that *C. campinarensis* EOs have moderate antioxidant activity.

According to the authors, the Croton EOs did not contain phenolic compounds, which is the main cause of their lower antioxidant activity than the phenolic compound thymol and the commercial antioxidant BHT. In addition, the authors identified high levels of oxygenated sesquiterpenes in the samples, especially caryophyllene oxide and spathulenol. Pino et al. [76] found that the EO of *Croton wagneri* from Ecuador had a moderate elimination effect in the DPPH and ferric reducing antioxidant power (FRAP) assays. According to the authors, *cis*-chrysanthenol (27.5%) and myrcene (19.2%) were the major components of the sample.

Do Vale et al. [77] analyzed the antioxidant capacity of the EO of *Croton piauhiensis*, characterized by compounds such as (*E*)-caryophyllene (21.58%), γ-terpinene (10.08%), and germacrene D (9.56%). The authors indicated that this EO showed high antioxidant capacity in the DPPH test, with higher results than the positive control (quercetin).

Regarding preliminary toxicity, Ribeiro et al. [78] evaluated the preliminary toxicity of the EO of *Croton rudolphianus* leaves by bioassays with *A. salina*. According to the authors, this EO exhibited high toxicity to microcrustaceans (LC_{50} = 68.33 $\mu g \cdot mL^{-1}$). In addition, the authors identified chemical constituents consistent with those found in the present sample, such as (*E*)-caryophyllene (17.33%), bicyclogermacrene (7.1%), and germacrene D (5.38%).

Andrade et al. [79] reported that the EO of *Croton zehntneri* leaves showed high toxicity against *A. salina*, with an LC_{50} of 4.54 $\mu g \cdot mL^{-1}$. However, the authors found that phenylpropanoid estragol was the major compound in the sample (84.70%). Lawal et al. [80] reported that the EO of *Croton gratissimus* showed toxicity against *A. salina*, with an LC_{50} of 8.52 $mg \cdot mL^{-1}$, corresponding to a classification of toxic. Regarding the chemical composition of the EO, the authors indicated that α-phellandrene (12.30%), β-phellandrene (10.70%), α-pinene (6.05%), and germacrene D (5.90%) were the major components.

Regarding the possible antioxidant capacity of the major components of the EO of *C. campinarensis*, EOs containing germacrene D, bicyclogermacrene, (*E*)-caryophyllene, γ-terpinene, and δ-elemene at high levels have shown relevant activity, scavenging $ABTS^{\bullet+}$ and $DPPH^{\bullet}$ radicals [81–85]. In addition, γ-terpinene can increase the protection of lipids and oxidizable substrates, in addition to prolonging the protective activity of the synthetic antioxidant α-tocopherol, making it a promising natural antioxidant for use in foods [86,87].

Regarding the toxicity of major compounds, Judzentienė et al. [88] reported that the EO of *Artemisia* vulgaris, containing high levels of germacrene D (10.60–30.50%), showed high toxicity against *A. salina*. The authors directly attributed the results to the presence of germacrene D in the sample. Machado et al. [89] reported that (*E*)-caryophyllene showed toxicity against *A. salina* only at high concentrations (3 mM). Schmitt et al. [90] showed that this sesquiterpene had no significant toxicity against rats.

Govindarajan et al. [91] stated that bicyclogermacrene showed high toxicity in mosquitoes of the species *Anopheles subpictus*, *Anopheles albopictus*, and *Culex tritaeniorhynchus*. De Oliveira et al. [92] found that the EO of *Lantana montevidensis* leaves, characterized by germacrene D (31.27%) and (*E*)-caryophyllene (28.15%), showed no fumigant toxicity in *Drosophila melanogaster* flies. EOs containing γ-terpinene and δ-elemene at high levels showed low toxicity against *A. salina* [93,94].

The biological properties of an EO may be associated with the major constituent of the sample and/or the synergistic and antagonistic effects exerted by all components present in the mixture [95]. Sesquiterpenes have higher toxicity than monoterpenes and phenylpropanoids [96]. Regarding free radical capture ability, sesquiterpenes generally have a lower antioxidant capacity than monoterpenes and phenylpropanoids [97]. These attributes may explain the antioxidant profile and preliminary toxicity exhibited by the EO of *C. campinarensis*.

3.3. Analysis of the Interactions of Major Compounds with AChE

In silico methods have been successfully used to evaluate the interaction between molecules of natural origin and molecular targets of pharmacological interest [98–101].

In this study, molecular docking was used to evaluate how the major compounds of *C. campinarensis* interact with the binding pocket of AChE, a molecular target related to toxicity that is investigated in *A. salina* models [45]. The energy values obtained for the interactions of the compounds with the target enzyme are summarized in Table 3.

Table 3. Moldock scores obtained from the docking protocol using MVD 5.5.

Molecule	MolDock Score	Rerank Score
Germacrene D	−01.107	−55.75
Bicyclogermacrene	−95.71	−71.41
(*E*)-caryophyllene	−103.70	−80.34
γ-terpinen	−49.42	−43.07
δ-elemene	−89.36	−70.84

The active site interactions are shown in Figure 1. Germacrene D established π-alkyl interactions with Tyr374, Tyr370, Phe371, Trp83, and Tyr71 (Figure 1A). The ligand bicyclogermacrene formed π-alkyl interactions with Tyr71, Tyr370, Tyr374, Trp472, and Trp83 (Figure 1B). For (*E*)-caryophyllene, the active site formed hydrophobic π-alkyl interactions with Trp472, Leu479, Trp83, Tyr71 Tyr374, and Tyr370 (Figure 1C). γ-Terpinene (8.99%) formed π-alkyl interactions with Tyr71, Tyr374, Trp83, Tyr370, Leu479, and Trp472 (Figure 1D). The binding of δ-elemene to the AChE binding pocket formed π-alkyl hydrophobic interactions with residues Tyr370, Tyr374, Tyr71, and Trp83 (Figure 1D).

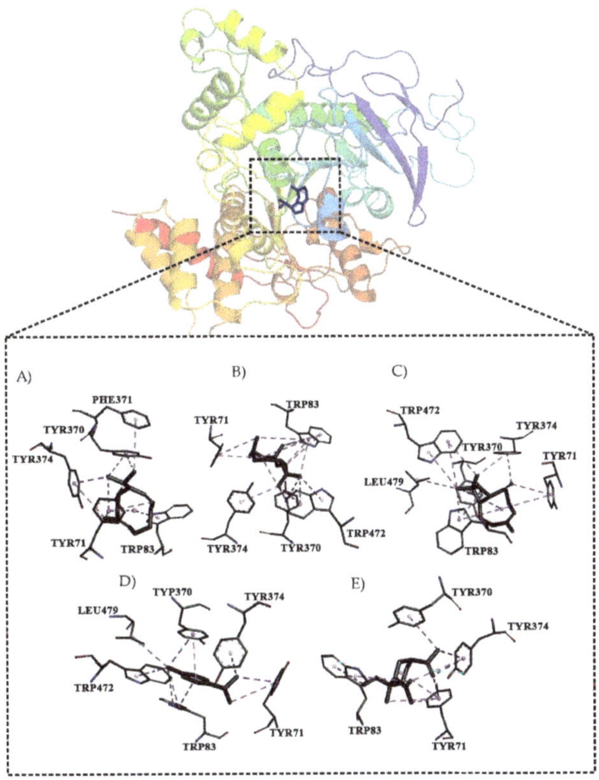

Figure 1. Molecular interactions in the binding pocket of AChE. Germacrene (**A**); bicyclogermacrene (**B**); (*E*)-caryophyllene (**C**); γ-terpinene (**D**); δ-elemene (**E**).

4. Conclusions

The yield of the *C. campinarensis* EO analyzed in this study was 0.24%. Terpenes characterized the aromatic profile of the EO, with a predominance of sesquiterpene hydrocarbons (87.95%), mainly germacrene D (26.95%), bicyclogermacrene (17.08%), (*E*)-caryophyllene (17.06%), and δ-elemene (7.59%). Regarding the antioxidant capacity of the EO, the evaluation of TEAC by the ABTS method showed that the EO has moderate antioxidant activity. However, the TEAC evaluation showed important inhibition of DPPH• radicals. The preliminary cytotoxicity test against *A. salina* indicated that the EO of *C. campinarensis* can be classified as toxic, with an LC_{50} of 20.84 ± 4.84 µg·mL^{-1}. The energy calculations in Table 1 showed that complex formation was favorable. Hydrophobic interactions dominated the interactions between the major compounds of *C. campinarensis* EO and AChE. These results may indicate that the chemical components with higher contents in the sample may be related to the high toxicity demonstrated by the EO against the microcrustacean *A. salina*. This study presents the first report on the chemical composition, antioxidant capacity, and preliminary toxicity of the EO of *C. campinarensis*, contributing to the knowledge of the phytochemistry of species of the genus *Croton* from the Amazon region.

Author Contributions: Conceptualization, L.S.d.C., O.O.F., Â.A.B.d.M., S.N.M. and J.N.C.; methodology, O.O.F., Â.A.B.d.M., L.Q.A. and E.L.P.V.; software, O.O.F. and Â.A.B.d.M., validation, S.P., M.S.d.O. and E.H.d.A.A.; formal analysis, S.P., M.S.d.O. and E.H.d.A.A.; investigation, L.S.d.C.; O.O.F. and Â.A.B.d.M.; resources, O.O.F. and Â.A.B.d.M.; data curation, O.O.F. and Â.A.B.d.M.; writing—original draft preparation, L.S.d.C., O.O.F., E.L.P.V. and Â.A.B.d.M.; writing—review and editing, O.O.F., Â.A.B.d.M. and S.N.M.; visualization, M.S.d.O. and E.H.d.A.A.; supervision, L.D.d.N., M.S.d.O. and E.H.d.A.A.; project administration, E.H.d.A.A. All authors have read and agreed to the published version of the manuscript.

Funding: Universidade Federal do Pará, Pró-Reitoria de Pesquisa e Pós-Graduação (PROPESP), Programa de Apoio à Publicação Qualificada—PAPQ, EDITAL 02/2022—PROPESP and the APC were funded by the Universidade Federal do Pará.

Institutional Review Board Statement: Not applicable.

Informed Consent Statement: Not applicable.

Data Availability Statement: The data are contained within this article.

Acknowledgments: The author M. S. de O. thanks PCI-MCTIC/MPEG as well as CNPq for the process number: [300983/2022-0].

Conflicts of Interest: The authors declare no conflict of interest.

References

1. Rout, S.; Tambe, S.; Deshmukh, R.K.; Mali, S.; Cruz, J.; Srivastav, P.P.; Amin, P.D.; Gaikwad, K.K.; de Aguiar Andrade, E.H.; de Oliveira, M.S. Recent Trends in the Application of Essential Oils: The next Generation of Food Preservation and Food Packaging. *Trends Food Sci. Technol.* **2022**, *129*, 421–439. [CrossRef]
2. de Oliveira, M.S. *Essential Oils*, 1st ed.; de Oliveira, M.S., Ed.; Springer International Publishing: Cham, Switzerland, 2022; ISBN 978-3-030-99475-4.
3. Cascaes, M.M.; Silva, S.G.; Cruz, J.N.; De Oliveira, S.; Oliveira, J.; Antonio, A.; Moraes, B.D.; Augusto, F.; Santana, K.; Diniz, L.; et al. First Report on the *Annona exsucca* DC. Essential Oil and *in silico* Identification of Potential Biological Targets of Its Major Compounds. *Nat. Prod. Res.* **2021**, *36*, 4009–4012. [CrossRef] [PubMed]
4. Bezerra, F.W.F.; de Oliveira, M.S.; Bezerra, P.N.; Cunha, V.M.B.; Silva, M.P.; da Costa, W.A.; Pinto, R.H.H.; Cordeiro, R.M.; da Cruz, J.N.; Chaves Neto, A.M.J.; et al. Extraction of Bioactive Compounds. In *Green Sustainable Process for Chemical and Environmental Engineering and Science*; Inamuddin, R.M., Boddula, R., Asiri, A.M., Eds.; Elsevier: Amisterdan, The Netherlands, 2020; pp. 149–167. ISBN 9780128173886.
5. Rodrigues, T.L.M.; Castro, G.L.S.; Viana, R.G.; Gurgel, E.S.C.; Silva, S.G.; de Oliveira, M.S.; de Aguiar Andrade, E.H. Physiological Performance and Chemical Compositions of the *Eryngium foetidum* L. (Apiaceae) Essential Oil Cultivated with Different Fertilizer Sources. *Nat. Prod. Res.* **2020**, *35*, 1–5. [CrossRef] [PubMed]
6. Gontijo, D.C.; do Nascimento, M.F.A.; Brandão, G.C.; de Oliveira, A.B. Phytochemistry and Antiplasmodial Activity of Xylopia Sericea Leaves. *Nat. Prod. Res.* **2019**, *34*, 3526–3530. [CrossRef] [PubMed]

7. Cruz, J.N.; de Oliveira, M.S.; de Aguiar Andrade, E.H.; Rodrigues Lima, R. Molecular Modeling Approaches Can Reveal the Molecular Interactions Established between a Biofilm and the Bioactive Compounds of the Essential Oil of *Piper divaricatum*. *Molecules* **2022**, *27*, 4199. [CrossRef]
8. Maia, O.G.S.; Andrade, L.H.A. Database of the Amazon Aromatic Plants and Their Essential Oils. *Quim. Nova* **2009**, *32*, 595–622. [CrossRef]
9. Cascaes, M.M.; Carneiro, O.D.S.; do Nascimento, L.D.; de Moraes, Â.A.B.; de Oliveira, M.S.; Cruz, J.N.; Guilhon, G.M.S.P.; Andrade, E.H.D.A. Essential Oils from Annonaceae Species from Brazil: A Systematic Review of Their Phytochemistry, and Biological Activities. *Int. J. Mol. Sci.* **2021**, *22*, 12140. [CrossRef]
10. Silva, S.G.; de Oliveira, M.S.; Cruz, J.N.; da Costa, W.A.; da Silva, S.H.M.; Barreto Maia, A.A.; de Sousa, R.L.; Carvalho Junior, R.N.; de Aguiar Andrade, E.H. Supercritical CO_2 Extraction to Obtain *Lippia thymoides* Mart. & Schauer (Verbenaceae) Essential Oil Rich in Thymol and Evaluation of Its Antimicrobial Activity. *J. Supercrit. Fluids* **2021**, *168*, 105064. [CrossRef]
11. Ferreira, O.O.; da Cruz, J.N.; Franco, C.D.J.P.; Silva, S.G.; da Costa, W.A.; de Oliveira, M.S.; Andrade, E.H.D.A. First Report on Yield and Chemical Composition of Essential Oil Extracted from *Myrcia eximia* DC (Myrtaceae) from the Brazilian Amazon. *Molecules* **2020**, *25*, 783. [CrossRef]
12. Da Cruz, E.D.N.S.; Peixoto, L.D.S.; da Costa, J.S.; Mourão, R.H.V.; do Nascimento, W.M.O.; Maia, J.G.S.; Setzer, W.N.; da Silva, J.K.; Figueiredo, P.L.B. Seasonal Variability of a Caryophyllane Chemotype Essential Oil of *Eugenia patrisii* Vahl Occurring in the Brazilian Amazon. *Molecules* **2022**, *27*, 2417. [CrossRef] [PubMed]
13. de Lima, E.J.S.P.; Alves, R.G.; D'elia, G.M.A.; da Anunciação, T.A.; Silva, V.R.; Santos, L.D.S.; Soares, M.B.P.; Cardozo, N.M.D.; Costa, E.V.; da Silva, F.M.A.; et al. Antitumor Effect of the Essential Oil from the Leaves of *Croton matourensis* Aubl. (Euphorbiaceae). *Molecules* **2018**, *23*, 2974. [CrossRef]
14. Cucho-Medrano, J.L.L.; Mendoza-Beingolea, S.W.; Fuertes-Ruitón, C.M.; Salazar-Salvatierra, M.E.; Herrera-Calderon, O. Chemical Profile of the Volatile Constituents and Antimicrobial Activity of the Essential Oils from *Croton adipatus*, *Croton thurifer*, and *Croton collinus*. *Antibiotics* **2021**, *10*, 1387. [CrossRef] [PubMed]
15. Caruzo, M.B.R.; Secco, R.S.; Medeiros, D.; Riina, R.; Torres, D.S.C.; Santos, R.F.D.; Pereira, A.P.N.; Rossine, Y.; Lima, L.R.; Muniz Filho, E.; et al. Croton in Flora e Funga Do Brasil. Available online: https://floradobrasil.jbrj.gov.br/FB17497 (accessed on 10 July 2022).
16. Guerra Júnior, J.I.; Ferreira, M.R.A.; de Oliveira, A.M.; Soares, L.A.L. Croton Sp.: A Review about Popular Uses, Biological Activities and Chemical Composition. *Res. Soc. Dev.* **2022**, *11*, e57311225306. [CrossRef]
17. Silva, P.M.D.S.; Fiaschitello, T.R.; de Queiroz, R.S.; Freeman, H.S.; da Costa, S.A.; Leo, P.; Montemor, A.F.; da Costa, S.M. Natural Dye from *Croton urucurana* Baill. Bark: Extraction, Physicochemical Characterization, Textile Dyeing and Color Fastness Properties. *Dye. Pigment.* **2020**, *173*, 107953. [CrossRef]
18. Ferreira, O.O.; Cruz, J.N.; de Moraes, Â.A.B.; de Jesus Pereira Franco, C.; Lima, R.R.; dos Anjos, T.O.; Siqueira, G.M.; do Nascimento, L.D.; Cascaes, M.M.; de Oliveira, M.S.; et al. Essential Oil of the Plants Growing in the Brazilian Amazon: Chemical Composition, Antioxidants, and Biological Applications. *Molecules* **2022**, *27*, 4373. [CrossRef] [PubMed]
19. Sá Firmino, N.C.; Alexandre, F.S.O.; de Vasconcelos, M.A.; Pinheiro, A.A.; Arruda, F.V.S.; Guedes, M.L.S.; Silveira, E.R.; Teixeira, E.H. Diterpenes Isolated from *Croton blanchetianus* Baill: Potential Compounds in Prevention and Control of the Oral Streptococci Biofilms. *Ind. Crops Prod.* **2019**, *131*, 371–377. [CrossRef]
20. Da Silva Brito, S.S.; Silva, F.; Malheiro, R.; Baptista, P.; Pereira, J.A. *Croton argyrophyllus* Kunth and *Croton heliotropiifolius* Kunth: Phytochemical Characterization and Bioactive Properties. *Ind. Crops Prod.* **2018**, *113*, 308–315. [CrossRef]
21. de Alencar Filho, J.M.T.; Araújo, L.D.C.; Oliveira, A.P.; Guimarães, A.L.; Pacheco, A.G.M.; Silva, F.S.; Cavalcanti, L.S.; Lucchese, A.M.; Almeida, J.R.G.D.S.; Araújo, E.C.D.C. Chemical Composition and Antibacterial Activity of Essential Oil from Leaves of *Croton heliotropiifolius* in Different Seasons of the Year. *Rev. Bras. Farmacogn.* **2017**, *27*, 440–444. [CrossRef]
22. Sousa, A.; Oliveira, G.; Fonseca, L.; Rocha, M.; Rai, M.; Santos, F.; de Lima, S. Antioxidant Properties of *Croton zehntneri* Pax et Hoffm. Essential Oil and Its Inclusion Complex with β-Cyclodextrin Prepared by Spray Drying. *J. Braz. Chem. Soc.* **2022**, *593*, 30–35. [CrossRef]
23. Coelho-de-Souza, A.N.; Rocha, M.V.A.P.; Oliveira, K.A.; Vasconcelos, Y.A.G.; Santos, E.C.; Silva-Alves, K.S.; Diniz, L.R.L.; Ferreira-da-Silva, F.W.; Oliveira, A.C.; Ponte, E.L.; et al. Volatile Oil of *Croton zehntneri* per Oral Sub-Acute Treatment Offers Small Toxicity: Perspective of Therapeutic Use. *Rev. Bras. Farmacogn.* **2019**, *29*, 228–233. [CrossRef]
24. de Souza, G.S.; Bonilla, O.H.; de Lucena, E.M.P.; Barbosa, Y.P. Rendimento e Composição Química Do Óleo Essencial de Três Espécies de Croton. *Cienc. Rural* **2017**, *47*, 1–8. [CrossRef]
25. Lima, C.C.; de Holanda-Angelin-Alves, C.M.; Pereira-Gonçalves, Á.; Kennedy-Feitosa, E.; Evangelista-Costa, E.; Bezerra, M.A.C.; Coelho-de-Souza, A.N.; Leal-Cardoso, J.H. Antispasmodic Effects of the Essential Oil of *Croton zehnteneri*, Anethole, and Estragole, on Tracheal Smooth Muscle. *Heliyon* **2020**, *6*, e05445. [CrossRef] [PubMed]
26. Rocha, R.R.; Matos, M.N.C.; Guerrero, J.A.P.; Cavalcante, R.M.B.; Melo, R.S.; Azevedo, Á.M.A.; Pereira, A.M.G.; Lopes, P.H.R.; Rodrigues, T.H.S.; Bandeira, P.N.; et al. Comparative Study of the Chemical Composition, Antibacterial Activity and Synergic Effects of the Essential Oils of *Croton tetradenius* Baill. And *C. pulegiodorus* Baill. Against *Staphylococcus aureus* Isolates. *Microb. Pathog.* **2021**, *156*, 104934. [CrossRef] [PubMed]

27. Almeida, J.; Souza, A.V.; Oliveira, A.P.; Santos, U.; Souza, M.; Bispo, L.; Turatti, Z.C.; Lopes, N. Chemical Composition of Essential Oils from *Croton conduplicatus* (Euphorbiaceae) in Two Different Seasons. *J. Essent. Oil Bear. Plants* **2014**, *17*, 1137–1145. [CrossRef]
28. Azevedo, M.M.B.; Chaves, F.C.M.; Almeida, C.A.; Bizzo, H.R.; Duarte, R.S.; Campos-Takaki, G.M.; Alviano, C.S.; Alviano, D.S. Antioxidant and Antimicrobial Activities of 7-Hydroxycalamenene-Rich Essential Oils from *Croton cajucara* Benth. *Molecules* **2013**, *18*, 1128–1137. [CrossRef] [PubMed]
29. Donati, M.; Mondin, A.; Chen, Z.; Miranda, F.M.; Do Nascimento, B.B.; Schirato, G.; Pastore, P.; Froldi, G. Radical Scavenging and Antimicrobial Activities of *Croton zehntneri*, *Pterodon emarginatus* and *Schinopsis brasiliensis* Essential Oils and Their Major Constituents: Estragole, Trans -Anethole, β-Caryophyllene and Myrcene. *Nat. Prod. Res.* **2015**, *29*, 939–946. [CrossRef]
30. Souto, E.B.; Severino, P.; Marques, C.; Andrade, L.N.; Durazzo, A.; Lucarini, M.; Atanasov, A.G.; El Maimouni, S.; Novellino, E.; Santini, A. *Croton argyrophyllus* Kunth Essential Oil-Loaded Solid Lipid Nanoparticles: Evaluation of Release Profile, Antioxidant Activity and Cytotoxicity in a Neuroblastoma Cell Line. *Sustainability* **2020**, *12*, 7697. [CrossRef]
31. Simionatto, E.; Bonani, V.F.L.; Morel, A.F.; Poppi, N.R.; Raposo Júnior, J.L.; Stuker, C.Z.; Peruzzo, G.M.; Peres, M.T.L.P.; Hess, S.C. Chemical Composition and Evaluation of Antibacterial and Antioxidant Activities of the Essential Oil of *Croton urucurana* Baillon (Euphorbiaceae) Stem Bark. *J. Braz. Chem. Soc.* **2007**, *18*, 879–885. [CrossRef]
32. Ramos da Silva, L.R.; Ferreira, O.O.; Cruz, J.N.; de Jesus Pereira Franco, C.; dos Anjos, T.O.; Cascaes, M.M.; Almeida da Costa, W.; Helena de Aguiar Andrade, E.; Santana de Oliveira, M. Lamiaceae Essential Oils, Phytochemical Profile, Antioxidant, and Biological Activities. *Evid. Based Complement. Altern. Med.* **2021**, *2021*, 6748052. [CrossRef]
33. Franco, C.D.J.P.; Ferreira, O.O.; de Moraes, Â.A.B.; Varela, E.L.P.; do Nascimento, L.D.; Percário, S.; de Oliveira, M.S.; Andrade, E.H.D.A. Chemical Composition and Antioxidant Activity of Essential Oils from *Eugenia patrisii* Vahl, *E. punicifolia* (Kunth) DC., and *Myrcia tomentosa* (Aubl.) DC., Leaf of Family Myrtaceae. *Molecules* **2021**, *26*, 3292. [CrossRef]
34. Diniz Do Nascimento, L.; De Moraes, A.A.B.; Da Costa, K.S.; Marcos, J.; Galúcio, P.; Taube, P.S.; Leal Costa, M.; Neves Cruz, J.; De Aguiar Andrade, E.H.; De Faria, L.J.G. Bioactive Natural Compounds and Antioxidant Activity of Essential Oils from Spice Plants: New Findings and Potential Applications. *Biomolecules* **2020**, *10*, 988. [CrossRef] [PubMed]
35. Mutlu-Ingok, A.; Devecioglu, D.; Dikmetas, D.N.; Karbancioglu-Guler, F.; Capanoglu, E. Antibacterial, Antifungal, Antimycotoxigenic, and Antioxidant Activities of Essential Oils: An Updated Review. *Molecules* **2020**, *25*, 4711. [CrossRef] [PubMed]
36. do Nascimento, L.D.; Silva, S.G.; Cascaes, M.M.; da Costa, K.S.; Figueiredo, P.L.B.; Costa, C.M.L.; Andrade, E.H.D.A.; de Faria, L.J.G. Drying Effects on Chemical Composition and Antioxidant Activity of Lippia Thymoides Essential Oil, a Natural Source of Thymol. *Molecules* **2021**, *26*, 2621. [CrossRef] [PubMed]
37. de Oliveira, M.S.; Silva, S.G.; da Cruz, J.N.; Ortiz, E.; da Costa, W.A.; Bezerra, F.W.F.; Cunha, V.M.B.; Cordeiro, R.M.; de Jesus Chaves Neto, A.M.; de Andrade, E.H.A.; et al. Supercritical CO_2 Application in Essential Oil Extraction. In *Industrial Applications of Green Solvents—Volume II.*; Inamuddin, R.M., Asiri, A.M., Eds.; Materials Research Foundations: Millersville, PA, USA, 2019; pp. 1–28.
38. Fuentes, C.; Fuentes, A.; Barat, J.M.; Ruiz, M.J. Relevant Essential Oil Components: A Minireview on Increasing Applications and Potential Toxicity. *Toxicol. Mech. Methods* **2021**, *31*, 559–565. [CrossRef]
39. da Silva Júnior, O.S.; de Jesus Pereira Franco, C.; de Moraes, A.A.B.; Cruz, J.N.; da Costa, K.S.; Diniz do Nascimento, L.; de Aguiar Andrade, E.H. In Silico Analyses of Toxicity of the Major Constituents of Essential Oils from Two *Ipomoea* L. Species. *Toxicon* **2021**, *195*, 111–118. [CrossRef]
40. Oliva, M.D.L.M.; Gallucci, N.; Zygadlo, J.A.; Demo, M.S. Cytotoxic Activity of Argentinean Essential Oils on *Artemia salina*. *Pharm. Biol.* **2007**, *45*, 259–262. [CrossRef]
41. Brasil, D.D.S.B.; Muller, A.H.; Guilhon, G.M.S.P.; Alves, C.N.; Andrade, E.H.A.; da Silva, J.K.R.; Maia, J.G.S. Essential Oil Composition of *Croton palanostigma* Klotzsch from North Brazil. *J. Braz. Chem. Soc.* **2009**, *20*, 1188–1192. [CrossRef]
42. Da Costa, J.G.M.; Rodrigues, F.F.G.; Angélico, E.C.; Pereira, C.K.B.; De Souza, E.O.; Caldas, G.F.R.; Silva, M.R.; Santos, N.K.A.; Mota, M.L.; Dos Santos, P.F. Chemical Composition and Evaluation of the Antibacterial Activity and Toxicity of the Essential Oil of *Croton zehntneri* (Variety Estragol). *Rev. Bras. Farmacogn.* **2008**, *18*, 583–586. [CrossRef]
43. De Lima, S.G.; Medeiros, L.B.P.; Cunha, C.N.L.C.; Da Silva, D.; De Andrade, N.C.; Moita Neto, J.M.; Lopes, J.A.D.; Steffen, R.A.; Araújo, B.Q.; De Reis, F.A.M. Chemical Composition of Essential Oils of *Croton hirtus* L'Her from Piauí (Brazil). *J. Essent. Oil Res.* **2012**, *24*, 371–376. [CrossRef]
44. Werka, J.S.; Boehme, A.K.; Setzer, W.N. Biological Activities of Essential Oils from Monteverde, Costa Rica. *Nat. Prod. Commun.* **2007**, *2*, 1934578X0700201204. [CrossRef]
45. Mesquita, K.D.S.M.; Feitosa, B.D.S.; Cruz, J.N.; Ferreira, O.O.; Franco, C.D.J.P.; Cascaes, M.M.; de Oliveira, M.S.; Andrade, E.H.D.A. Chemical Composition and Preliminary Toxicity Evaluation of the Essential Oil from *Peperomia circinnata* Link Var. circinnata. (Piperaceae) in Artemia salina Leach. *Molecules* **2021**, *26*, 7359. [CrossRef]
46. Baek, I.; Choi, H.J.; Rhee, J.S. Inhibitory Effects of Biocides on Hatching and Acetylcholinesterase Activity in the Brine Shrimp *Artemia salina*. *Toxicol. Environ. Health Sci.* **2015**, *7*, 303–308. [CrossRef]
47. Secco, R.D.S.; Rosário, A.S.D.; Berry, P.E. *Croton campinarensis* (Euphorbiaceae), a New Species from Eastern Amazonian Brazil. *Phytotaxa* **2012**, *49*, 1–5. [CrossRef]
48. Ferreira, O.O.; da Silva, S.H.M.; de Oliveira, M.S.; Andrade, E.H.D.A. Chemical Composition and Antifungal Activity of Myrcia Multiflora and Eugenia Florida Essential Oils. *Molecules* **2021**, *26*, 7259. [CrossRef] [PubMed]

49. de Oliveira, M.S.; da Cruz, J.N.; da Costa, W.A.; Silva, S.G.; Brito, M.D.P.; de Menezes, S.A.F.; de Jesus Chaves Neto, A.M.; de Aguiar Andrade, E.H.; de Carvalho Junior, R.N. Chemical Composition, Antimicrobial Properties of Siparuna Guianensis Essential Oil and a Molecular Docking and Dynamics Molecular Study of Its Major Chemical Constituent. *Molecules* **2020**, *25*, 3852. [CrossRef] [PubMed]
50. Ferreira, O.O.; Franco, C.D.J.P.; Varela, E.L.P.; Silva, S.G.; Cascaes, M.M.; Percário, S.; de Oliveira, M.S.; Andrade, E.H.D.A. Chemical Composition and Antioxidant Activity of Essential Oils from Leaves of Two Specimens of Eugenia Florida DC. *Molecules* **2021**, *26*, 5848. [CrossRef] [PubMed]
51. van Den Dool, H.; Kratz, P.D. A Generalization of the Retention Index System Including Linear Temperature Programmed Gas—Liquid Partition Chromatography. *J. Chromatogr. A* **1963**, *11*, 463–471. [CrossRef]
52. Adams, R.P. *Identification of Essential Oil Components by Gas Chromatography/Mass Spectrometry*, 4th ed.; Allured Publ.: Carol Stream, IL, USA, 2007; Volume 8, ISBN 978-1932633214.
53. Blois, M.S. Antioxidant Determinations by the Use of a Stable Free Radical. *Nature* **1958**, *181*, 1199–1200. [CrossRef]
54. de Moraes, Â.A.B.; Ferreira, O.O.; da Costa, L.S.; Almeida, L.Q.; Varela, E.L.P.; Cascaes, M.M.; de Jesus Pereira Franco, C.; Percário, S.; do Nascimento, L.D.; de Oliveira, M.S.; et al. Phytochemical Profile, Preliminary Toxicity, and Antioxidant Capacity of the Essential Oils of *Myrciaria floribunda* (H. West Ex Willd.) O. Berg. and *Myrcia sylvatica* (G. Mey) DC. (Myrtaceae). *Antioxidants* **2022**, *11*, 2076. [CrossRef]
55. Re, R.; Pellegrini, N.; Proteggente, A.; Pannala, A.; Yang, M.; Rice-Evans, C. Antioxidant Activity Applying an Improved ABTS Radical Cation Decolorization Assay. *Free Radic. Biol. Med.* **1999**, *26*, 1231–1237. [CrossRef]
56. Miller, N.J.; Rice-Evans, C.; Davies, M.J.; Gopinathan, V.; Milner, A. A Novel Method for Measuring Antioxidant Capacity and Its Application to Monitoring the Antioxidant Status in Premature Neonates. *Clin. Sci.* **1993**, *84*, 407–412. [CrossRef] [PubMed]
57. Neto, R.D.A.M.; Santos, C.B.R.; Henriques, S.V.C.; Machado, L.D.O.; Cruz, J.N.; da Silva, C.H.T.D.P.; Federico, L.B.; de Oliveira, E.H.C.; de Souza, M.P.C.; da Silva, P.N.B.; et al. Novel Chalcones Derivatives with Potential Antineoplastic Activity Investigated by Docking and Molecular Dynamics Simulations. *J. Biomol. Struct. Dyn.* **2022**, *40*, 2204–2216. [CrossRef] [PubMed]
58. Rego, C.M.A.; Francisco, A.F.; Boeno, C.N.; Paloschi, M.V.; Lopes, J.A.; Silva, M.D.S.; Santana, H.M.; Serrath, S.N.; Rodrigues, J.E.; Lemos, C.T.L.; et al. Inflammasome NLRP3 Activation Induced by Convulxin, a C-Type Lectin-like Isolated from *Crotalus durissus terrificus* Snake Venom. *Sci. Rep.* **2022**, *12*, 1–17. [CrossRef]
59. Almeida, V.M.; Dias, Ê.R.; Souza, B.C.; Cruz, J.N.; Santos, C.B.R.; Leite, F.H.A.; Queiroz, R.F.; Branco, A. Methoxylated Flavonols from Vellozia Dasypus Seub Ethyl Acetate Active Myeloperoxidase Extract: In Vitro and in Silico Assays. *J. Biomol. Struct. Dyn.* **2021**, *40*, 7574–7583. [CrossRef] [PubMed]
60. Lima, A.D.M.; Siqueira, A.S.; Möller, M.L.S.; Souza, R.C.D.; Cruz, J.N.; Lima, A.R.J.; da Silva, R.C.; Aguiar, D.C.F.; Junior, J.L.; Gonçalves, E.C. In Silico Improvement of the Cyanobacterial Lectin Microvirin and Mannose Interaction. *J. Biomol. Struct. Dyn.* **2020**, *40*, 1064–1073. [CrossRef] [PubMed]
61. Thomsen, R.; Christensen, M.H. MolDock: A New Technique for High-Accuracy Molecular Docking. *J. Med. Chem.* **2006**, *49*, 3315–3321. [CrossRef] [PubMed]
62. Leão, R.P.; Cruz, J.V.; da Costa, G.V.; Cruz, J.N.; Ferreira, E.F.B.; Silva, R.C.; de Lima, L.R.; Borges, R.S.; dos Santos, G.B.; Santos, C.B.R. Identification of New Rofecoxib-Based Cyclooxygenase-2 Inhibitors: A Bioinformatics Approach. *Pharmaceuticals* **2020**, *13*, 209. [CrossRef]
63. Mascarenhas, A.M.S.; de Almeida, R.B.M.; de Araujo Neto, M.F.; Mendes, G.O.; da Cruz, J.N.; dos Santos, C.B.R.; Botura, M.B.; Leite, F.H.A. Pharmacophore-Based Virtual Screening and Molecular Docking to Identify Promising Dual Inhibitors of Human Acetylcholinesterase and Butyrylcholinesterase. *J. Biomol. Struct. Dyn.* **2020**, *39*, 1–10. [CrossRef]
64. Santos, C.B.R.; Santos, K.L.B.; Cruz, J.N.; Leite, F.H.A.; Borges, R.S.; Taft, C.A.; Campos, J.M.; Silva, C.H.T.P. Molecular Modeling Approaches of Selective Adenosine Receptor Type 2A Agonists as Potential Anti-Inflammatory Drugs. *J. Biomol. Struct. Dyn.* **2021**, *39*, 3115–3127. [CrossRef] [PubMed]
65. Cheung, J.; Gary, E.N.; Shiomi, K.; Rosenberry, T.L. Structures of Human Acetylcholinesterase Bound to Dihydrotanshinone i and Territrem B Show Peripheral Site Flexibility. *ACS Med. Chem. Lett.* **2013**, *4*, 1091–1096. [CrossRef]
66. Turiel, N.A.; Ribeiro, A.F.; Carvalho, N.C.C.; Monteiro, O.S.; Lucas, F.C.A.; Carreira, L.M.M.; Andrade, E.H.A.; Maia, J.G.S. Variability in Essential Oil Composition of Croton Species with Occurrence in the Eastern Brazilian Amazon. *Rec. Nat. Prod.* **2016**, *10*, 380–384.
67. Stein, S.; Mirokhin, D.; Tchekhovskoi, D.; Mallard, G.; Mikaia, A.; Zaikin, V.; Sparkmanm, D. *Standard Reference Data Program of the National Institute of Standards and Technology*; National Institute of Standards and Technology: Gaithersburg, MD, USA, 2011; The NIST mass spectral search program for the nist/epa/nih mass spectra library.
68. Albuquerque, B.N.D.L.; Da Silva, M.F.R.; Da Silva, P.C.B.; De Lira Pimentel, C.S.; Lino Da Rocha, S.K.; De Aguiar, J.C.R.O.F.; Neto, A.C.A.; Paiva, P.M.G.; Gomes, M.G.M.; Da Silva-Júnior, E.F.; et al. Oviposition Deterrence, Larvicidal Activity and Docking of β-Germacrene-D-4-Ol Obtained from Leaves of *Piper corcovadensis* (Piperaceae) against *Aedes aegypti*. *Ind. Crops Prod.* **2022**, *182*, 114830. [CrossRef]
69. Formagio, A.S.N.; Vilegas, W.; Volobuff, C.R.F.; Kassuya, C.A.L.; Cardoso, C.A.L.; Pereira, Z.V.; Silva, R.M.M.F.; dos Santos Yamazaki, D.A.; de Freitas Gauze, G.; Manfron, J.; et al. Exploration of Essential Oil from Psychotria Poeppigiana as an Anti-Hyperalgesic and Anti-Acetylcholinesterase Agent: Chemical Composition, Biological Activity and Molecular Docking. *J. Ethnopharmacol.* **2022**, *296*, 115220. [CrossRef]

70. Francomano, F.; Caruso, A.; Barbarossa, A.; Fazio, A.; La Torre, C.; Ceramella, J.; Mallamaci, R.; Saturnino, C.; Iacopetta, D.; Sinicropi, M.S. β-Caryophyllene a Sesquiterpene with Countless. *Appl. Sci.* **2019**, *9*, 5420–5438. [CrossRef]
71. Reis, J.B.; Figueiredo, L.A.; Castorani, G.M.; Veiga, S.M.O.M. Avaliação Da Atividade Antimicrobiana Dos Óleos Essenciais Contra Patógenos Alimentares. *Braz. J. Health Rev.* **2020**, *3*, 342–363. [CrossRef]
72. Lu, J.J.; Dang, Y.Y.; Huang, M.; Xu, W.S.; Chen, X.P.; Wang, Y.T. Anti-Cancer Properties of Terpenoids Isolated from Rhizoma Curcumae—A Review. *J. Ethnopharmacol.* **2012**, *143*, 406–411. [CrossRef] [PubMed]
73. Da Silva Júnior, O.S.; Franco, C.D.J.P.; de Moraes, Â.A.B.; Pastore, M.; Cascaes, M.M.; do Nascimento, L.D.; de Oliveira, M.S.; Andrade, E.H.D.A. Chemical Variability of Volatile Concentrate from Two *Ipomoea* L. Species within a Seasonal Gradient. *Nat. Prod. Res.* **2022**, *36*, 1–8. [CrossRef] [PubMed]
74. Ramos, S.C.S.; De Oliveira, J.C.S.; Da Câmara, C.A.G.; Castelar, I.; Carvalho, A.F.F.U.; Lima-Filho, J.V. Antibacterial and Cytotoxic Properties of Some Plant Crude Extracts Used in Northeastern Folk Medicine. *Rev. Bras. Farmacogn.* **2009**, *19*, 376–381. [CrossRef]
75. Morais, S.M.; Cossolosso, D.S.; Silva, A.A.S.; de Moraes Filho, M.O.; Teixeira, M.J.; Campello, C.C.; Bonilla, O.H.; de Paula, V.F.; Vila-Nova, N.S. Essential Oils from Croton Species: Chemical Composition, in Vitro and in Silico Antileishmanial Evaluation, Antioxidant and Cytotoxicity Activities. *J. Braz. Chem. Soc.* **2019**, *30*, 2404–2412. [CrossRef]
76. Pino, J.A.; Terán-Portelles, E.C.; Hernández, I.; Rodeiro, I.; Fernández, M.D. Chemical Composition of the Essential Oil from *Croton wagneri* Müll. Arg. (Euphorbiaceae) Grown in Ecuador. *J. Essent. Oil Res.* **2018**, *30*, 347–352. [CrossRef]
77. Do Vale, J.P.C.; Vasconcelos, M.A.; Arruda, F.V.S.; Firmino, N.C.S.; Pereira, A.L.; Andrade, A.L.; Saker-Sampaio, S.; Sampaio, A.H.; Marinho, E.S.; Teixeira, A.M.R.; et al. Evaluation of Antimicrobial and Antioxidant Potential of Essential Oil from *Croton piauhiensis* Müll. Arg. *Curr. Microbiol.* **2021**, *78*, 1926–1938. [CrossRef] [PubMed]
78. Ribeiro, L.P.; Domingues, V.C.; Gonçalves, G.L.P.; Fernandes, J.B.; Glória, E.M.; Vendramim, J.D. Essential Oil from *Duguetia lanceolata* St.-Hil. (Annonaceae): Suppression of Spoilers of Stored-Grain. *Food Biosci.* **2020**, *36*, 100653. [CrossRef]
79. Andrade, T.C.B.; Lima, S.G.D.; Freitas, R.M.; Rocha, M.S.; Islam, T.; Silva, T.G.D.; Militão, G.C.G. Isolation, Characterization and Evaluation of Antimicrobial and Cytotoxic Activity of Estragole, Obtained from the Essential Oil of *Croton zehntneri* (Euphorbiaceae). *An. Da Acad. Bras. De Ciências* **2015**, *87*, 173–182. [CrossRef]
80. Lawal, O.A.; Ogunwande, I.A.; Osunsanmi, F.O.; Opoku, A.R.; Oyedeji, A.O. *Croton gratissimus* Leaf Essential Oil Composition, Antibacterial, Antiplatelet Aggregation, and Cytotoxic Activities. *J. Herbs Spices Med. Plants* **2017**, *23*, 77–87. [CrossRef]
81. Ascari, J.; de Oliveira, M.S.; Nunes, D.S.; Granato, D.; Scharf, D.R.; Simionatto, E.; Otuki, M.; Soley, B.; Heiden, G. Chemical Composition, Antioxidant and Anti-Inflammatory Activities of the Essential Oils from Male and Female Specimens of *Baccharis punctulata* (Asteraceae). *J. Ethnopharmacol.* **2019**, *234*, 1–7. [CrossRef]
82. Casiglia, S.; Bruno, M.; Bramucci, M.; Quassinti, L.; Lupidi, G.; Fiorini, D.; Maggi, F. *Kundmannia sicula* (L.) DC: A Rich Source of Germacrene D. *J. Essent. Oil Res.* **2017**, *29*, 437–442. [CrossRef]
83. Dahham, S.S.; Tabana, Y.M.; Iqbal, M.A.; Ahamed, M.B.K.; Ezzat, M.O.; Majid, A.S.A.; Majid, A.M.S.A. The Anticancer, Antioxidant and Antimicrobial Properties of the Sesquiterpene β-Caryophyllene from the Essential Oil of Aquilaria Crassna. *Molecules* **2015**, *20*, 11808–11829. [CrossRef]
84. de Castro, J.A.M.; Monteiro, O.S.; Coutinho, D.F.; Rodrigues, A.A.C.; da Silva, J.K.R.; Maia, J.G.S. Seasonal and Circadian Study of a Thymol/γ-Terpinene/p-Cymene Type Oil of *Ocimum gratissimum* L. And Its Antioxidant and Antifungal Effects. *J. Braz. Chem. Soc.* **2019**, *30*, 930–938. [CrossRef]
85. Zuccolotto, T.; Bressan, J.; Lourenço, A.V.F.; Bruginski, E.; Veiga, A.; Marinho, J.V.N.; Raeski, P.A.; Heiden, G.; Salvador, M.J.; Murakami, F.S.; et al. Chemical, Antioxidant, and Antimicrobial Evaluation of Essential Oils and an Anatomical Study of the Aerial Parts from *Baccharis* Species (Asteraceae). *Chem. Biodivers.* **2019**, *16*, e1800547. [CrossRef] [PubMed]
86. Guo, Y.; Liu, Z.; Hou, E.; Ma, N.; Fan, J.; Jin, C.Y.; Yang, R. Non-Food Bioactive Natural Forest Products as Insecticide Candidates: Preparation, Biological Evaluation and Molecular Docking Studies of Novel N-(1,3-Thiazol-2- Yl)Carboxamides Fused (+)-Nootkatone from Chamaecyparis Nootkatensis [D. Don] Spach. *Ind. Crops Prod.* **2020**, *156*, 112864. [CrossRef]
87. Mollica, F.; Gelabert, I.; Amorati, R. Synergic Antioxidant Effects of the Essential Oil Component γ-Terpinene on High-Temperature Oil Oxidation. *ACS Food Sci. Technol.* **2022**, *2*, 180–186. [CrossRef]
88. Judžentienė, A.; Būdienė, J. Mugwort (*Artemisia vulgaris* L.) Essential Oils Rich in Germacrene D, and Their Toxic Activity. *J. Essent. Oil Res.* **2021**, *33*, 256–264. [CrossRef]
89. Machado, K.D.C.; Paz, M.F.C.J.; de Oliveira Santos, J.V.; da Silva, F.C.C.; Tchekalarova, J.D.; Salehi, B.; Islam, M.T.; Setzer, W.N.; Sharifi-Rad, J.; de Castro e Sousa, J.M.; et al. Anxiety Therapeutic Interventions of β-Caryophyllene: A Laboratory-Based Study. *Nat. Prod. Commun.* **2020**, *15*, 1934578X20962229. [CrossRef]
90. Schmitt, D.; Levy, R.; Carroll, B. Toxicological Evaluation of β-Caryophyllene Oil: Subchronic Toxicity in Rats. *Int. J. Toxicol.* **2016**, *35*, 558–567. [CrossRef] [PubMed]
91. Govindarajan, M.; Rajeswary, M.; Hoti, S.L.; Bhattacharyya, A.; Benelli, G. Eugenol, α-Pinene and β-Caryophyllene from *Plectranthus barbatus* Essential Oil as Eco-Friendly Larvicides against Malaria, Dengue and Japanese Encephalitis Mosquito Vectors. *Parasitol. Res.* **2016**, *115*, 807–815. [CrossRef] [PubMed]
92. De Oliveira, M.R.C.; Barros, L.M.; Duarte, A.E.; De Lima Silva, M.G.; Da Silva, B.A.F.; Pereira Bezerra, A.O.B.; Oliveira Tintino, C.D.M.; De Oliveira, V.A.P.; Boligon, A.A.; Kamdem, J.P.; et al. Gc-Ms Chemical Characterization and in Vitro Evaluation of Antioxidant and Toxic Effects Using Drosophila Melanogaster Model of the Essential Oil of *Lantana montevidensis* (Spreng) Briq. *Medicina* **2019**, *55*, 194. [CrossRef]

93. Ojah, E.O.; Moronkola, D.O.; Petrelli, R.; Nzekoue, F.K. Chemical Composition of Ten Essential Oils from Calophyllum Inophyllum Linn and Their Toxicity Against *Artemia salina*. *Eur. J. Pharm. Med. Res.* **2019**, *6*, 185–194.
94. Fabri, N.T.; Gatto, L.J.; Furusho, A.S.; Garcia, M.J.B.; Marques, F.D.A.; Miguel, M.D.; Montrucchio, D.P.; Zanin, S.M.W.; Miguel, O.G.; Gaspari Dias, J.D.F. Composition, Antioxidant Properties, and Biological Activities of the Essential Oil Extracted from *Ocotea diospyrifolia* (Meisn.) Mez. *Braz. J. Pharm. Sci.* **2019**, *55*, 1–8. [CrossRef]
95. Cho, T.J.; Park, S.M.; Yu, H.; Seo, G.H.; Kim, H.W.; Kim, S.A.; Rhee, M.S. Recent Advances in the Application of Antibacterial Complexes Using Essential Oils. *Molecules* **2020**, *25*, 1752. [CrossRef]
96. Wojtunik-Kulesza, K.A. Toxicity of Selected Monoterpenes and Essential Oils Rich in These Compounds. *Molecules* **2022**, *27*, 1716. [CrossRef] [PubMed]
97. Iannone, M.; Ovidi, E.; Vitalini, S.; Laghezza Masci, V.; Iriti, M.; Tiezzi, A.; Garzoli, S.; Marianelli, A. From Hops to Craft Beers: Production Process, VOCs Profile Characterization, Total Polyphenol and Flavonoid Content Determination and Antioxidant Activity Evaluation. *Processes* **2022**, *10*, 517. [CrossRef]
98. dos Santos, K.L.B.; Cruz, J.N.; Silva, L.B.; Ramos, R.S.; Neto, M.F.A.; Lobato, C.C.; Ota, S.S.B.; Leite, F.H.A.; Borges, R.S.; da Silva, C.H.T.P.; et al. Identification of Novel Chemical Entities for Adenosine Receptor Type 2a Using Molecular Modeling Approaches. *Molecules* **2020**, *25*, 1245. [CrossRef]
99. Pinto, V.D.S.; Araújo, J.S.C.; Silva, R.C.; da Costa, G.V.; Cruz, J.N.; Neto, M.F.D.A.; Campos, J.M.; Santos, C.B.R.; Leite, F.H.A.; Junior, M.C.S. In Silico Study to Identify New Antituberculosis Molecules from Natural Sources by Hierarchical Virtual Screening and Molecular Dynamics Simulations. *Pharmaceuticals* **2019**, *12*, 36. [CrossRef] [PubMed]
100. Costa, E.B.; Silva, R.C.; Espejo-Román, J.M.; Neto, M.F.D.A.; Cruz, J.N.; Leite, F.H.A.; Silva, C.H.T.P.; Pinheiro, J.C.; Macêdo, W.J.C.; Santos, C.B.R. Chemometric Methods in Antimalarial Drug Design from 1,2,4,5-Tetraoxanes Analogues. *SAR QSAR Environ. Res.* **2020**, *31*, 677–695. [CrossRef]
101. Galucio, N.C.D.R.; Moysés, D.D.A.; Pina, J.R.S.; Marinho, P.S.B.; Gomes Júnior, P.C.; Cruz, J.N.; Vale, V.V.; Khayat, A.S.; Marinho, A.M.D.R. Antiproliferative, Genotoxic Activities and Quantification of Extracts and Cucurbitacin B Obtained from *Luffa operculata* (L.) Cogn. *Arab. J. Chem.* **2022**, *15*, 103589. [CrossRef]

antioxidants

MDPI

Review

Citrus Essential Oils in Aromatherapy: Therapeutic Effects and Mechanisms

Pooja Agarwal [1,†], Zahra Sebghatollahi [2,†], Mehnaz Kamal [3,†], Archana Dhyani [4,†], Alpana Shrivastava [5], Kiran Kumari Singh [6], Mukty Sinha [7], Neelima Mahato [8], Awdhesh Kumar Mishra [9,*] and Kwang-Hyun Baek [9,*]

1 Division of Chemistry, School of Basic and Applied Sciences, Galgotias University, Greater Noida 203 201, Uttar Pradesh, India
2 Department of Plant Breeding and Biotechnology, Faculty of Agricultural Sciences and Food Industries, Science and Research Branch, Islamic Azad University, Tehran 1477893855, Iran
3 Department of Pharmaceutical Chemistry, College of Pharmacy, Prince Sattam Bin Abdulaziz University, Al Kharj 11942, Saudi Arabia
4 Department of Applied Sciences, School of Engineering, University of Petroleum and Energy Studies, Dehradun 248 007, Uttarakhand, India
5 Department of Botany, Sri Shankar College, (A Constituent Unit of V.K.S.U., Ara), Rohtas, Sasaram 821 115, Bihar, India
6 Department of Geography, Central University of South Bihar, Gaya 151001, Bihar, India
7 Department of Medical Devices, National Institute of Pharmaceutical Education and Research, Ahmedabad 382 355, Gandhinagar, India
8 School of Chemical Engineering, Yeungnam University, Gyeongsan 38541, Gyeongsangbuk-do, Republic of Korea
9 Department of Biotechnology, Yeungnam University, Gyeongsan 38541, Gyeongsangbuk-do, Republic of Korea
* Correspondence: awdhesh@ynu.ac.kr (A.K.M.); khbaek@ynu.ac.kr (K.-H.B.); Tel.: +82-53-810-3029 (A.K.M.); Fax: +82-53-810-4769 (A.K.M.)
† These authors contributed equally to this work.

Citation: Agarwal, P.; Sebghatollahi, Z.; Kamal, M.; Dhyani, A.; Shrivastava, A.; Singh, K.K.; Sinha, M.; Mahato, N.; Mishra, A.K.; Baek, K.-H. Citrus Essential Oils in Aromatherapy: Therapeutic Effects and Mechanisms. *Antioxidants* **2022**, *11*, 2374. https://doi.org/10.3390/antiox11122374

Academic Editors: Delia Mirela Tit and Simona Bungau

Received: 25 October 2022
Accepted: 25 November 2022
Published: 30 November 2022

Abstract: Citrus is one of the main fruit crops cultivated in tropical and subtropical regions worldwide. Approximately half (40–47%) of the fruit mass is inedible and discarded as waste after processing, which causes pollution to the environment. Essential oils (EOs) are aromatic compounds found in significant quantities in oil sacs or oil glands present in the leaves, flowers, and fruit peels (mainly the flavedo part). Citrus EO is a complex mixture of ~400 compounds and has been found to be useful in aromatic infusions for personal health care, perfumes, pharmaceuticals, color enhancers in foods and beverages, and aromatherapy. The citrus EOs possess a pleasant scent, and impart relaxing, calming, mood-uplifting, and cheer-enhancing effects. In aromatherapy, it is applied either in message oils or in diffusion sprays for homes and vehicle sittings. The diffusion creates a fresh feeling and enhances relaxation from stress and anxiety and helps uplifting mood and boosting emotional and physical energy. This review presents a comprehensive outlook on the composition, properties, characterization, and mechanism of action of the citrus EOs in various health-related issues, with a focus on its antioxidant properties.

Keywords: citrus essential oils; aromatherapy; natural aromatic compounds; therapeutic effects of citrus EOs; characterization of citrus EOs

1. Introduction

Citrus is one of the world's most abundant fruits containing substantial amounts of beneficial secondary metabolites [1]. Among them, citrus essential oils (EOs) are important secondary metabolites; they are usually aromatic compounds found in oil glands present in the flowers, leaves, and fruit peels. However, most citrus EOs are extracted from fruit peels, viz., fruit rind, or flavedo (green part) and albedo (white part). These citrus EOs contain

85–99% volatile and 1–15% non-volatile components [2] and their content as well as chemical composition depend on species and extraction methods [3–5]. These volatile constituents contain large amounts of monoterpene hydrocarbons (70–95%) and *d*-limonene, a good source of antioxidants, in all the reported orange species [6].

Aromatherapy using citrus EOs has been practiced as a treatment method since ancient times. Aromatherapy is utilized to relieve many symptoms, such as body pain, nausea, vomiting, anxiety, depression, stress, insomnia, etc. [7]. Several scientific reports have been published regarding the use of EOs in the treatment of a number of medical issues, including hypertension, hypotension, cognitive dysfunction [8–12], physical and psychological stress, and exhaustion [13]. EOs are extracted from plants and used in a controlled manner in aromatherapy with few or no side effects. Currently, EOs are hugely popular as safe and natural agents with medicinal and therapeutic properties and have been approved by the US Food and Drug Administration (FDA).

Citrus EOs are fragrant volatile molecules, which upon inhalation can alter hemodynamic parameters or blood flow in the body by controlling circulation through the autonomous nervous system. Citrus EOs have also been investigated for their antimicrobial [14] and antioxidant activities [15,16]. Many citrus EOs, such as orange [17] and bitter orange [18,19] have shown anxiolytic, antidepressant, anticonvulsant, analgesic, and sedative effects and influence overall emotional behavior. Major components in the citrus EOs include bioactive compounds, such as monoterpenes and its derivatives, aldehydes, ketones, esters, alcohols, limonene, β-pinene, and γ-terpinene [20].

Global Production and Consumption of Citrus

Natural products are popular globally due to their nutritional value and little or no side effects. The demand for citrus EOs has been continuously increasing to produce higher quality nutraceuticals, food and beverages, bakery, natural preservatives for vegetables, meat and fish, pharmaceuticals, aromatherapy, perfumes, toiletries, and personal care, blending ingredients for herbal tea, cosmetic ingredients, and so on [21]. The major citrus-producing countries, climate sustainability, and annual production of the different citrus fruits in different geographical regions are shown in Figure S1 (Supplementary Materials).

The global market of citrus EOs in the year 2018 was 6.31 billion USD, which is predicted to grow at a rate of 6.5% by the year 2025. The market has been estimated to grow up to 9.43 billion by the year 2028 [22,23]. The market segregation of citrus essential oils and the global market for citrus EOs based on its major applications (in %; up to the year 2018) are shown in Figure S2 (Supplementary Materials). Global citrus oil market by application (by the year 2018) [22,24] and citrus EO market value forecast [25] are shown in Figure S3a,b (Supplementary Materials). The market of EOs obtained from citrus fruits (data year 2020) and its market value forecast for the decade are displayed in the world map in Figure 1. Many countries in the Asia-Pacific region have a high demand for citrus EOs because of their use in various food and beverages, cosmetic preparations, and therapeutics. Similarly, the demand is expected to increase in Europe and the US due to their higher usage in the food and beverage industry and the substantial use of these products in aromatherapies. In addition, citrus EOs are also becoming a preferred ingredient material in green repellents and pesticides against insects and pests [26].

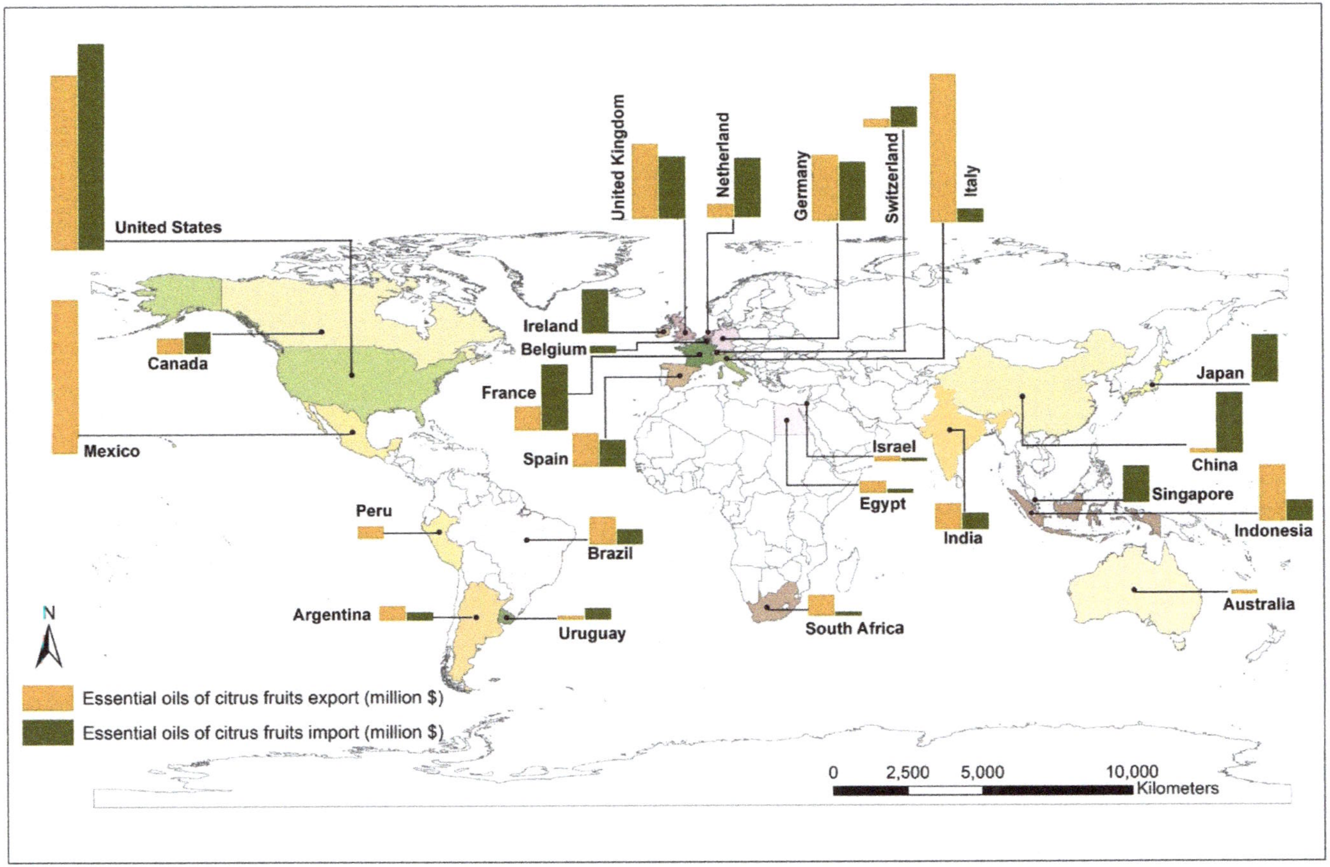

Figure 1. Market of essential oils obtained from citrus fruits, (2020) (Note: Bar graph was created for major countries involved in import and export; the map was created using ArcGIS 10.8.1 with UTM projection taking WGS84 datum) [24].

2. Extraction, Characterization, and Authentication Methods for Citrus EOs

EOs are present in the oil glands in the peel and cuticles of the citrus fruit peel or pericarp. The EOs are released when oil sacs are crushed during juice extraction or under pressure during cold pressing of the peel waste. The major component of the EOs is *d*-limonene [14,20]. Cold pressing has been a traditional method of extracting essential oils and the yield is a watery emulsion. The latter is centrifuged to recover the EOs [27]. Alternately, the extraction of EOs is also carried out using stream stripping and distillation methods. These methods have been found effective and efficient in removing oil components from oil-milled sludge. Modern methods include distillation techniques, e.g., Microwave Steam Distillation (MSD), Microwave Hydrodiffusion and Gravity (MHG), and Instant Controlled Pressure Drop Technique (DIC). Instant controlled pressure drop (DIC) technology facilitates the extraction of essential oil as well as the expansion of the plant matrix. This improves the extraction of the oil significantly. In this method, a high-steam pressure (~0.6–1.0 MPa) is applied constantly for a short time (~5–60 s) followed by an instantaneous drop in the pressure towards vacuum 5 kPa with a rate $\geq$0.5 MPa s^{-1}. This treatment results in the rapid expansion of the sample matrix, auto evaporation, and faster cooling, enabling the extraction of volatile compounds and EOs within 1–4 min. The EOs obtained by distillation have been observed to deteriorate easily and develop off-odor because of the instability of the terpene hydrocarbons, e.g., *d*-limonene [28]. Supercritical fluid extraction (SFE) is an emerging and inexpensive technique of extraction and isolation of EOs [29]. By this method, efficient and fast extraction can be done at ambient temperatures, without incorporating clean-up steps in absence of harmful organic solvents. Carbon dioxide (CO_2) is used in the SFE method because of its non-explosive, non-toxic nature along with the ease availability. CO_2 can be considered an ideal solvent and can be easily eliminated from extracted products [30,31].

MHG is a highly efficient method as it accelerates the extraction process many times over. In addition, it also enables the recovery of EOs without any changes in the oil composition. MSD techniques have an edge over MSD as it causes more rapid rupture of the cell wall of the plant material under strong microwaves which quickly releases the cell cytoplasm containing oils. The main extraction methods/techniques for obtaining citrus EOs are summarized in Table S1 (Supplementary Materials) [4].

The extraction process yields a matrix containing a mixture of phytochemicals which ought to undergo separation, purification, and isolation to obtain individual compounds. Citrus EO is a complex mixture of ~400 volatile and semi-volatile compounds. Column chromatography, high-speed counter current chromatography (HSCC), and high-performance liquid chromatography (HPLC) are generally employed involving solvent combinations, such as hexane:n-butanol, ethyl acetate:hexane, butanol:water, chloroform:methanol, etc., in the basic process of purification and separation of compounds. The different compounds are detected and quantitatively determined using a combination of modern instruments, viz., UV–visible, mass spectroscopy, and HPLC. GC and its extensions such as MDGC, enantioselective capillary gas chromatography (eCGC), ultra-high performance liquid chromatography, etc. are the most extensively employed for EO separation, identification, and quantitative characterization due to its volatility and complexity of most natural oils [32].

Gas chromatography (GC) is one of the most popular methods for the characterization of single-phase vapor samples and is suitable for samples with 2 (molecular hydrogen) to 1500 mass units. Almost all EOs fall within this mass range. A combination of GC and MDGC with other techniques, such as mass spectroscopy (MS) and Raman spectroscopy is employed to improve the efficiency of the separating power of chromatography and analyze more complex structures. Such coupled analyses improve data quality and exhibit good separation performance [33]. The use of two or more techniques detects adulteration in EOs extracted from the citrus peels and waste more accurately [34]. Researchers used simultaneous distillation and extraction (SDE)-GC-MS and MDGC-MS techniques to study and authenticate the enantiomeric ratios of chiral compounds present in the citrus EOs. These techniques helped in the identification of 67 volatile compounds, including limonene, γ-terpinene, and linalool, as the major compounds and sabinene, camphene, and β-phellandrene as minor and chiral aromatic components in lemon and lime. A combination of MDGC and GCC-IRMS is employed to determine the authenticity of the EOs extracted from neroli (Egyptian bitter orange flower) and lime [35]. A comparative analysis was performed for lime (*Citrus aurantifolia* Swingle and *Citrus latifolia* Tanaka)-based Eos following two different approaches using MDGC and gas chromatography–combustion-isotope ratio mass spectrometry (GC-C-IRMS). This study is the first to differentiate Eos extracted from Persian lime and key lime. A series of components were identified including limonene, geranial, β-caryophyllene, trans-α-bergamotene, α and β-pinene, and germacrene B, using GC–C-IRMS. MDGC determined the enantiomeric distribution of camphene, limonene, linalool, α-phellandrene, β-phellandrene, β-pinene, terpinen-4-ol, α-terpineol, sabinene, and α-thujene in lime oils. Such hyphenated techniques are also used successfully in the investigation of citrus oil-based flavored drinks. Italian alcoholic syrup was examined by comparing carbon isotope ratios to identify the presence of genuine cold-pressed peel oils. For this purpose, solid phase microextraction was performed, followed by GC with IRMS. GC was used to determine the enantiomeric distribution of the selected volatile chiral samples, whereas qualitative analyses of the samples were performed by mass spectrometry. The results were confirmed using enantioselective gas chromatography [36].

Ultra-high-performance liquid chromatography–time-of-flight–mass spectrometry (UHPLC–TOF–MS) profiling and ^{1}H nuclear magnetic resonance (NMR) near-infrared spectroscopy are employed for the fingerprinting of lemon oil [37]. Metabolite variations have been also investigated in lemon oil samples. Such analysis has high demand in the fragrance and flavor industries for terpenoids, citropten, bergamottin, furocoumarins, flavonoids, and fatty acids. Characterization based on quantitative analysis of substances present in EOs is an important process in essential oil-based industries. The different methods/techniques of characterization/authentication of citrus EOs have been summarized in Table S2 (Supplementary Materials).

3. Components of Citrus EOs

Citrus species are rich in various EOs, with many chemical components of interest for aromatherapy. Several ingredients used in pharmaceuticals and cosmetics are procured from citrus EOs [38–41]. Around 400 compounds, which cover 85–99% of the total oil fraction, have volatile and semi-volatile properties [38,39,42,43]. Citrus EOs can be grouped into five major classes: hydrocarbon monoterpenes, oxygenated monoterpenes, hydrocarbon sesquiterpenes, and oxygenated sesquiterpenes. The major component of citrus Eos is limonene, which can be found in quantities ranging from 32% to 98% [44]. Hydrocarbons, aliphatic aldehydes, and oxygen-containing mono- and sesquiterpenes are the next most significant classes of compounds present in citrus EOs which show antioxidant properties. Several terpenes are present as their functionalized derivatives, which are volatile compounds, and flavonoids, diterpenoids, sterols, coumarins, and fatty acids are some of the non-volatile compounds [45]. Several studies have reported the chemical composition of EOs derived from the citrus flower, leaf, and peel. The composition of citrus EOs varies with citrus species, origin, climatic and geographical conditions, ripening, method of extraction, etc. [14]. The molecular structures of the volatile and non-volatile compounds present in citrus EOs are displayed in Figure S4 (Supplementary Materials). The composition of the aromatherapeutic components present in the EOs of common citrus species [3,14] are summarized in Figure S5 (Supplementary Materials).

4. Aromatherapy: Mechanisms

4.1. Evolution of Aromatherapy

Stress conditions alter the respiration process, and an altered respiration activates the limbic system (amygdala, hippocampus, and hypothalamus) in the brain and induces psycho-physiological responses. The latter can alter the emotional responses. This is how respiration relates to emotional behavior and brain functions. Furthermore, pulmonary diseases have been observed to affect brain-cell growth, reduce oxygen supply in the body and brain causing neurophysiological and neurobehavioral disorders, namely anxiety and depression. Moreover, the systemic circulation carrying blood with insufficient oxygen supply also transports lung-induced inflammation mediators. The latter induces adaptive responses in the brain and the body. Applications of EOs have been observed to impart neuroprotective and anti-aging effects, and relief from respiratory congestion, pain, insomnia, anxiety, depression, stress, and other psychological and physiological disorders mostly due to its antioxidant properties [46]. When inhaled, the EOs can stimulate the olfactory, respiratory, and gastrointestinal systems, and the EOs release endorphins to initiate a feeling of well-being and an analgesic effect [7]. Citrus EOs have been reported to be safe and effective for treating insomnia. Moreover, these can decrease the side effects of drugs and sleep illnesses owing to their short- or long-term usage [47]. The EOs have gained attention in clinical and scientific research because they are harmless and do not have any side effects [46]. There are three ways by which EOs can reach and act directly on the respiratory, circulatory, and central nervous systems, viz., (i) inhalation through the respiratory tract; (ii) oral intake in the form of capsules, drops, or food; and (iii) topical absorption through the skin [48].

4.2. Mechanism

4.2.1. Inhalation

A human can differentiate more than 10,000 types of aromas. Humans possess ~400 functional gene coding for olfactory sensory neurons (OSNs). Each receptor expresses a specific type of odorant reception [49,50]. Inhalation is the fastest and most effective way to induce responses in the central nervous system within a few seconds. The inhalation of the EO molecules delivers active volatile compounds to the brain and the circulatory system via (a) the olfactory lobe and (b) the respiratory system, respectively. The olfactory system begins with the nasal cavity which leads to the olfactory lobe located close to the brain. The olfactory lobe is connected to several brain areas, e.g., the hypothalamus and hippocampus. The volatile molecules in the citrus EOs enter through the nasal cavity, pass through the olfactory lobe, activate the sensory neurons present in the olfactory mucosa, and the axons of the sensory neuron cells ultimately deliver the signals to the central nervous system (CNS) [51–53]. The 'activation' is the initiation of electrical signals (by the odorant molecules) in the olfactory lobe. The signal is transmitted from the olfactory lobe to the olfactory cortex. The stimuli modulate specific physiological responses involving mood and behavioral actions (emotion and cognition), hormone production, regulation of body temperature, digestive reactions, memory, stress responses, sedation, sex stimulation, blood pressure, heart rate, etc. [54,55]. It has been observed that if the sense of smell is lost in patients with anxiety and depression, inhaled volatile molecules enter the lungs through the circulatory system via gas exchange and trigger the healing process. Another pathway of the EOs post inhalation is through the respiratory system involving gaseous exchange via diffusion into the blood circulation in the alveoli [49]. The EOs action toward brain functioning has been explained to take place via three basic mechanisms: (a) Activation of nasal olfactory chemoreceptors, (b) direct absorption of the EO active molecules into the neuronal pathway, (c) absorption of EO active molecules in the alveolar blood circulation.

The pathways followed by citrus EOs are illustrated in Figure 2.

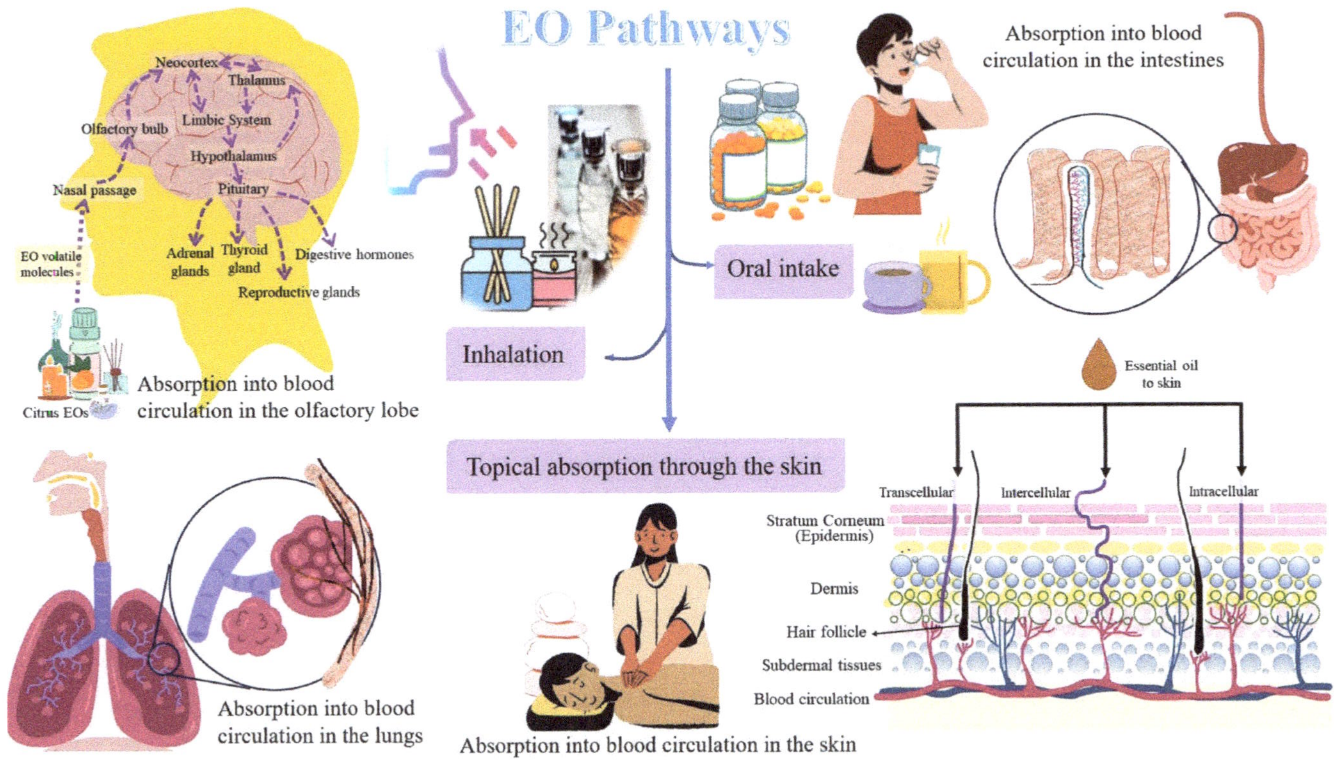

Figure 2. Pathways followed by citrus EOs for aromatherapy.

(a) Activation of nasal olfactory chemoreceptors

This involves the activation of nasal olfactory chemoreceptors and the consequent effects of the olfactory signals on the respective brain segments. The EO, upon inhalation, travels through the interiors of the nasal passage where the endothelium in the inner lining is thin and located close to the brain. Therefore, the EO molecules readily enter the local circulation and the brain. A particular odorant can activate a single or a set of OSN receptors and generate an electrophysiological signal for transmission into the brain. This is how different odors can be identified and differentiated. The olfactory epithelial layer is facilitated by different types of OSNs. A smell is identified by the activation of nasal olfactory chemoreceptors. The odorant molecules approach the olfactory epithelium and bind with the dendrite receptors present in the OSNs. This generates an electrophysiological signal via induction of an action potential. The axons of the OSNs extend and converge into the corresponding glomerulus cell. The latter is associated with a specific mitral and tufted cell. The signals are transmitted via dendrites of the glomerulus through connected mitral and tufted cells and eventually to the pyramidal neurons present in the olfactory cortex. In the cortex region, the transmitted electrophysical signals further stimulate the target regions in the brain [56,57]. The olfactory cortex of the brain is divided into other smaller regions, namely the piriform cortex, olfactory tubercle, and entorhinal cortex. Each of these regions project information to the amygdala (regulates aggression, eating, drinking, sexual behavior), hippocampus (regulates emotion, learning, memory, odor memory), and hypothalamus (regulates blood glucose levels, salt, blood pressure, and hormones) or 'limbic system'. The olfactory signals directly transmits into the cortex and responses to the stimuli are expressed in terms of odor, memory, emotions, and endocrine functions [58].

(b) Direct absorption of the EO active molecules into the neuronal pathway

The small and volatile molecules present in the EOs can be transported either by extracellular or by intracellular transport mechanisms. In the intracellular mechanism, the EO active molecules directly pass through the neuronal pathway in the olfactory lobe and transmitted to the brain. The molecules bind with the olfactory receptor surface of the neurons and initiate receptor-mediated endocytosis (cells take in substances present outside the cell body by engulfing them in a vesicle which reopens inside the cell and the substance becomes a part of the cytoplasm). The molecules absorbed in the OSN are transported to the olfactory bulb along the axons by endosomes. In the extracellular transport mechanism, the EO active molecules pass through the paracellular cleft between the OSN and supporting cells and enter the lamina propria (connective tissues) through movement in the fluid. From lamina propria, the EO active molecules are further transported to perineural space along the axons and eventually arrive at the brain parenchyma. Finally, the EO active molecules enter across the blood–brain barrier and blood–cerebrospinal fluid barrier to spread into different regions in the brain. The EO active molecules now interact with the neurotransmitter receptors, namely transient receptor potential (TRP) channel proteins, glutamate, and γ-amino-butyric acid (GABA), 5-hydroxytryptamine (5-HT), and dopamine (DA), and produce anxiolytic and antidepressant effects [58].

(c) Absorption of EO active molecules in the alveolar blood circulation

The EO vapor molecules, upon inhalation, travel to the lungs and induce an immediate and easing impact on breathing difficulties that appear during cold and congestion. The EO active molecules present in the inhaled vapor pass through the respiratory tract, enter the lungs, and reach the alveolar sacs where gaseous exchange between the cells of the alveoli and blood cells in the capillaries take place. Simultaneously, some molecules are also absorbed by the inner mucous linings of the respiratory tract, bronchi, and bronchioles. Deep breathing tends to increase the quantity of any EO components absorbed into the body by this route. EO active molecules enter the neuronal pathway, and absorption of the EO active molecules in the alveolar blood circulation is illustrated in Figure 3.

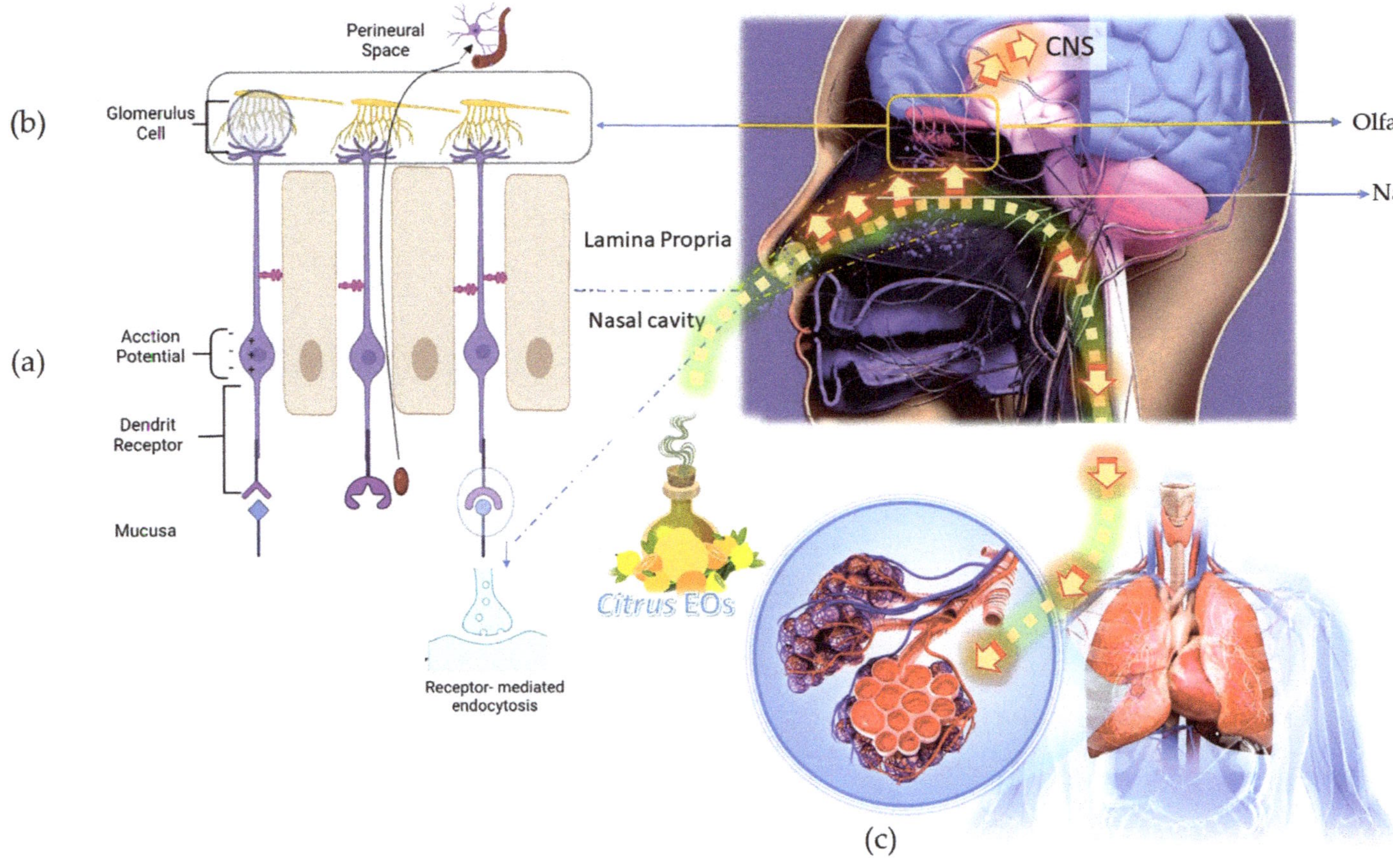

Figure 3. Inhalation of the citrus EOs and response delivery to the central nervous system through the olfactory lobe and respiratory and circulatory system. (**a**) Activation of nasal olfactory chemoreceptors (**b**) Direct absorption of the EO active molecules into the neuronal pathway (**c**) Absorption of EO active molecules in the alveolar blood circulation.

The soluble molecules present in the EO vapor carried with the inhaled air can cross the air–blood barrier. A majority of the EO components are lipophilic and hydrophobic in nature (lipid soluble terpene family). Lipophilic EO components can cross the blood–brain barrier and transport to the CNS [58] The EO action in aromatherapy through the inhalation process towards the brain functioning has been explained to take place via three basic mechanisms, viz., activation of nasal olfactory chemoreceptors and direct absorption of the active molecules. Aromatherapy is known to improve mood and certain mild symptoms of stress-related disorders, such as anxiety, depression, loss of appetite, loss of concentration, and chronic pain. The benefits of aromatherapy have been established by both the physiological and psychological effects upon inhalation of volatile EO components. The EO active components act via the limbic system, namely the hippocampal, the hypothalamus, and the pyriform cortex.

4.2.2. Oral Intake

Citrus and its juice have been a major medicinal recipe for abdominal problems since ancient times in tropical and subtropical countries besides its use in foods, bakeries, and confectionaries. The lime fruits have been used for making anti-odorant agents due to the fragrance and freshness effects of their aromatic volatiles. Bergamot essential oils are utilized in pharmaceutical industries to absorb unpleasant odors of medicinal products and add antiseptic and antibacterial properties. In addition, the juice is added to drinking water, alcoholic, and non-alcoholic beverages to enhance flavor and antioxidants. The characteristic flavor of citrus oils is mainly due to the presence of certain components, namely linalool, citral, and linalyl acetate [59]. However, limonene and pinene present in the EO composition have not been much favored. Moreover, they are relatively unstable compounds and decompose when exposed to heat and light and they are removed from the oil to enhance the life of the products [59,60]. The roots of the lime tree have been used as a febrifuge and antipyretic in traditional medicine. The bark of the lemon tree is often boiled in water to obtain a decoction and taken as a remedy for gonorrhea and related disorders. In many tribal populations, the roots of the plant are dried and chewed for headache and vermifuge effects in the stomach and the intestines. The citrus EO components have several benefits when taken orally due to their antiviral, antiseptic, antimicrobial, astringent, restorative, stimulant, and antioxidant properties [12,46,48,61–65].

Bergamot EO possesses a bitter aromatic taste and a characteristic pleasant aroma. It is a popular pharmacopoeias in many countries. It has been also reported for its hypolipemic and hypoglycemics activities, anti-inflammatory, and anti-cancer properties [66–70]. In folk medicine in many countries, it has been popularly used for fever and parasitic diseases. Due to its significant antimicrobial properties, it has been found useful in treating infections in the mouth, skin, respiratory and urinary tract, gonococcal infections, leucorrhea, vaginal pruritus, tonsillitis, and sore throats [71]. BEO and vapors have been observed to exhibit resistance against common food-borne pathogens. The EO component linalool is reported to be the most effective antibacterial component [72]. BEO has also been reported for its antibacterial and anti-fungal activities against several microbial strains, such as *Campylobacter jejuni, Escherichia coli* O157, *Listeriamono cytogenes, Bacillus cereus, Staphylococcus aureus,* dermatophytes, and *Candida species*-induced infections [73–75]. BEO-incorporated chitosan-based films with concentrations, *viz.*, 0.5, 1, 2, and 3% *w/w* have been reported to exhibit a significant dose-dependent inhibitory effect against *Penicillium italicum* [76].

Bergamottine (5-geranoxypsoralen), an important component in the Eos, is a natural furanocoumarin. It can be extracted from the pulp of pomelos and grapefruits and the peel and pulp of bergamot oranges. It has been found to decrease the electrocardiographic changes significantly during experimental studies in guinea pigs. The latter is typical of coronary arterial spasms and cardiac arrhythmias provoked by pitressin. Bergamottine is also found to increase the dose of ouabain required to induce ventricular premature beats, ventricular tachyarrhythmias, and death. The experimental studied suggest that bergamottine possesses potential anti-anginal and antiarrhythmic properties [77]. In another experimental model of rat angioplasty, a pretreatment with a volatile fraction of bergamot EO in a dose-dependent manner has been observed to reduce the neointima proliferation, together with the free radical formation and LOX-1 expression. Lectin-like oxy LDL receptor-1(LOX-1) is known to be involved in smooth muscle cell proliferation and neo-intima formation occurring in injured blood vessels [66]. Furthermore, the bergamot EO has also been observed to induce vasorelaxation of the mouse aorta by activating K^+ channels and inhibiting Ca^{2+} influx [78]. The latter differentially modulates intracellular Ca^{2+} levels in vascular endothelial and smooth muscle cells [79]. These research findings altogether indicate that bergamot EO possesses potential activity as a vasodilator agent in cardiovascular diseases. Citrus EOs in oral administration has been observed to be beneficial in treating anxiety [80].

The citrus EOs undergo significant biotransformation after being absorbed in the digestive system which has been observed to alter their effects on health. When ingested orally, the EOs enter the digestive system and its components begin a wide range of actions. Primarily the monoterpenoids, namely *d*-limonene, carvone, *cis*- and *trans*- carveol (CAR), perillyl alcohol (POH), and geraniol have been observed to alleviate the carcinogenesis of exogenous substances. Other EO components, such as linalool and citral along with carvone and geraniol have been found to impart antimicrobial activities in the digestive system. The antimicrobial properties of the citrus EOs are attributed to the presence of abundant amounts of limonene and flavonoids in their composition [81]. Liver CYPs (Cytochrome P450) transform limonene into a variety of products. The CYPs act on various types of substrates or target molecules, and it has been observed that more than one P450 can act on the same type of substrate which produces multiple products from the same substrate. In human beings, the biotransformation of limonene occurs via four pathways, namely oxidation of endo- and exocyclic double bonds, oxidation of methyl side chain, and allylic oxidation of C6-ring [82]. The oxidation of the exocyclic double bond present in the limonene molecule produces Limonene (LMN)-8,9-OH, whereas the other three pathways produce perillyl alcohol (POH), perillic acid (PA), and *cis*- and *trans*- carveol (CAR).

The biotransformation of α-pinene, the second major component of citrus EOs produces myrtenol, *cis*- and *trans*- verbenol. In addition, carene is transformed into caren-10-ol, caren-10-carboxylic acid and caren-3,4-diol [82]. Biotransformation of the citrus EOs alters its bioavailability. For example, the major component in the citrus Eos, limonene, is readily absorbed into the blood from the digestive tract. It is reported that the d-limonene (labeled with radioactive substance) is absorbed in the liver in 1.0 h with a peak concentration of 45.1 dpm (disintegration per min)/mg. Within the next 1.0 h, the peak concentration of the labeled d-limonene in adrenal glands and kidney was found to be 77.3 and 21.8 dpm/mg, respectively [83]. The biotransformation of limonene is a rapid process and the concentration of limonene, and its metabolites become undetectable within 24 h of ingestion (oral intake). The products of biotransformation of limonene (in citrus EOs) are excreted from the body via urine (~60%), feces, and breath after oral consumption [83].

The products of the citrus EOs post biotransformation exhibit certain health-promoting effects. Perillyl alcohol has been observed to reduce the incidence and diversity of colonic invasive adenocarcinoma in rats (induced by injecting methoxymethane (or azoxymethane (AOM) carcinogen). Furthermore, perillyl alcohol has been found to be more effective compared with limonene in terms of chemoprotection against malignant cancer [84]. The metabolism of *d*-limonene and α-pinene in the liver, and absorption of citrus EO components into the circulatory system through the intestinal villi is shown in Figure 4. The mechanisms of gastroprotection, anti-cancer, anti-tumor, anti-inflammation, anti-microbial, and lipolytic actions of citrus EO components are summarized in Figure 5.

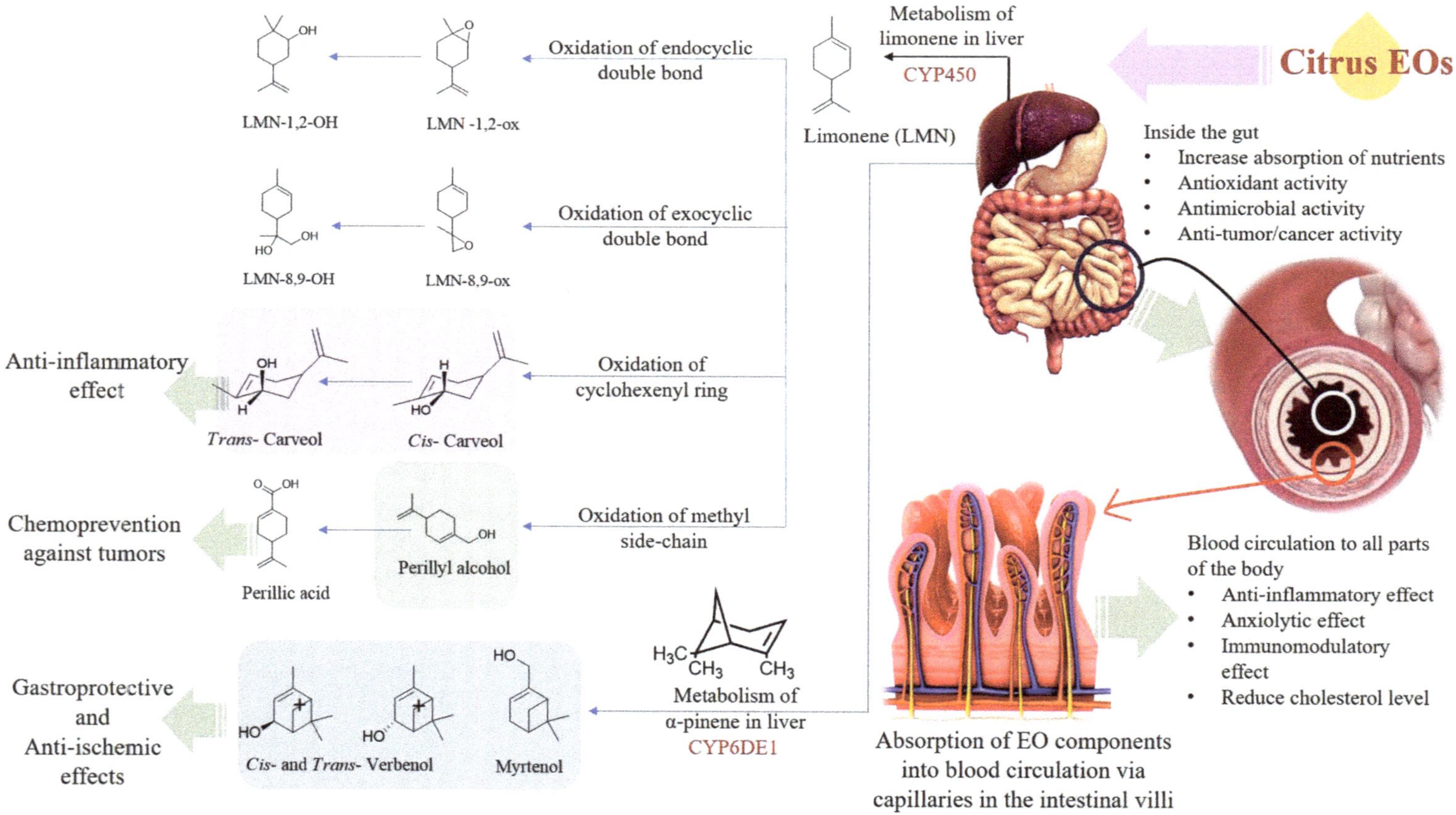

Figure 4. Metabolism of *d*-limonene and *α*-pinene in the liver, absorption of citrus EO components into the circulatory system through the intestinal villi.

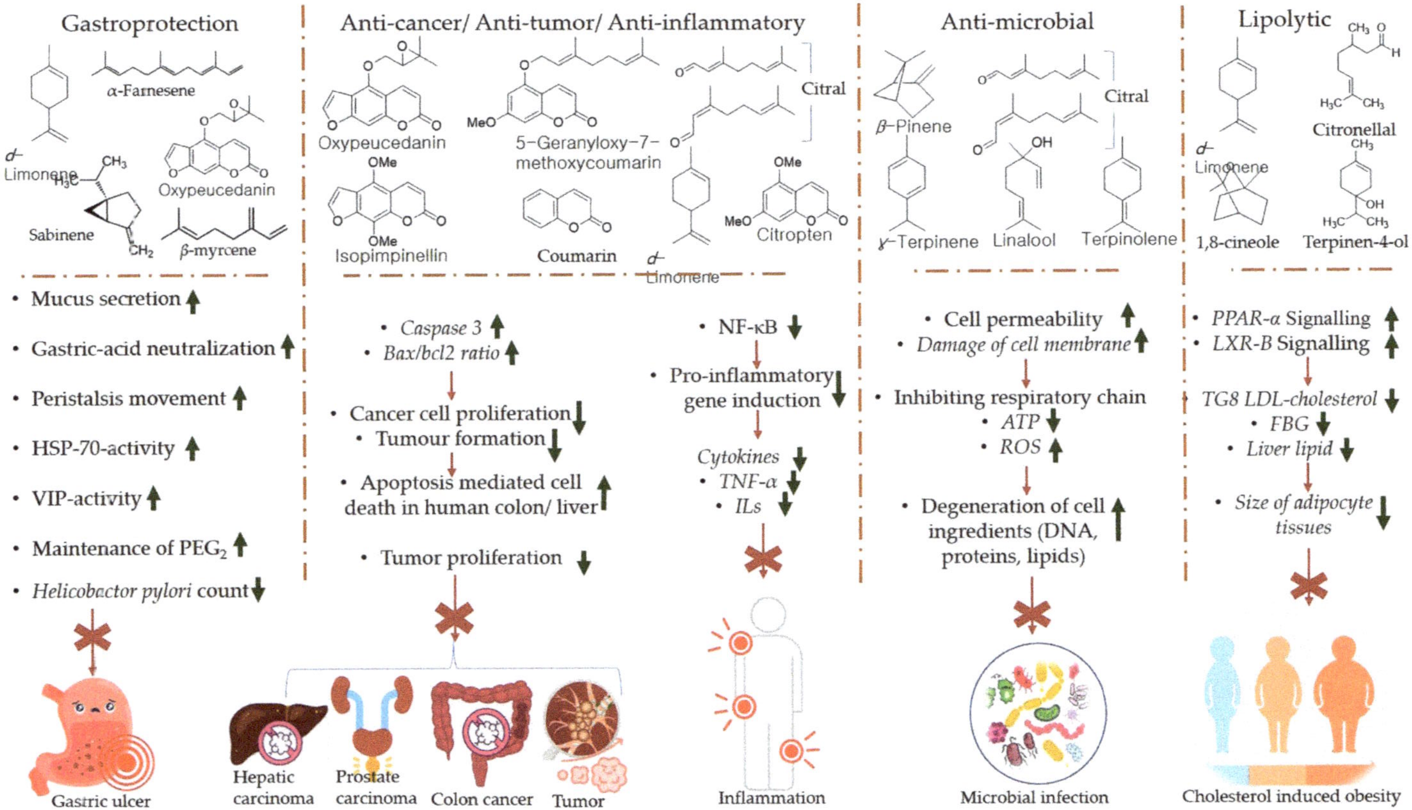

Figure 5. Mechanisms of gastroprotection, anti-cancer, anti-tumor, anti-inflammation, anti-microbial, and lipolytic actions of citrus EO components. Caspase (key apoptosis-inducing protein) [85–92]. Abbreviations; PPAR-α (Peroxisome proliferator-activated receptor alpha), bcl 2 (B-cell lymphoma protein 2), Bax (bcl 2-associated X), NF-KB (Nuclear factor-κB), LXR-β (Liver X receptor beta), TG8 LDL (Triglycerides 8 Low-density lipoprotein), FBG (fasting blood glucose), ROS (Reactive Oxygen Species), TNF-α (tumor necrosis factor alpha), ILs (Interleukins), ATP (Adenosine triphosphate).

The mechanism of chemoprotection by monoterpenes has been explained via several hypotheses, viz., G1block, induction of cell apoptosis or cell death, aggravation of stressed condition inside endoplasmic reticulum, and alteration in mevalonate metabolism pathway. Perillyl alcohol is believed to mainly block the modification of Ras oncoproteins; inhibit farnesyl-protein transferase (FPTase) and geranylgeranyl protein transferases (GGPTases), whereas other metabolites of limonene biotransformation, viz., cis- and trans-carveol (CAR) induce anti-inflammatory activity by suppressing the generation of superoxide dismutase (SOD) and nitric oxide and NF-κB signaling pathway. Furthermore, myrtenol and *cis*- and *trans*-verbenol (products of α-pinene biotransformation) have been observed to induce gastroprotective and anti-ischemic effects [82,93].

4.2.3. Applications on Skin

Skin is the largest organ of the human body. Cell cytoplasm contains 90% of its composition as water, and therefore the skin acts as a protective barrier to resist water loss. However, the skin is semi-permeable to water and water-soluble substances. The barrier protection is attributed to the stratum corneum (the epidermis). It is an outer tough, durable keratinized layer with a thickness up to 20 layers of dead cells, and self-repairing. Beneath the epidermis is the dermis, a complex structure comprising lymph, blood vessels and capillaries, nerves, sweat and oil glands, hair follicles, collagen, elastin, fibroblasts, mast cells, and so on. Due to the lipids present in all cell membranes, the penetration of molecules through the dermis is relatively easier. The fundamental physicochemical properties of the external molecules which decide the rate and quantity of external molecules to penetrate the skin are the molecular weight of the molecule, its spatial structure and arrangement of functional groups, polarity, optical activity, liposolubility, coefficients of diffusion, dissociation, and so on. Due to the presence of lipids in the stratum corneum, the liposoluble compounds in the EOs make their way into the inner layers of the skin and finally reach into the blood stream.

Once the EO components penetrate the epidermis and enter the dermis, they are absorbed into the blood circulation and carried to every cell in the body. The hydrophilic and lipophilic molecules present in the citrus Eos can penetrate the skin through sweat gland openings, hair follicles, and sebaceous glands, respectively. The EO molecules progress through the passage between cells, i.e., fatty cement of the skin layers as well as through the cells themselves by intervening through the cell's membrane made of phospholipids. The skin epidermis thickness is uneven in different parts of the body. For example, the skin epidermis of the forehead and scalp is relatively thinner and contains a large number of oil glands. Therefore, lipophilic molecules penetrate readily through the partial barrier and enter the blood stream.

The lipophilic EO components are lipid soluble and tend to accumulate in lipid-rich areas of the body to form reservoir(s) and possibly to be sequestrated [94,95]. The EO constituent molecule reservoirs present in the outer layers of the epidermis and subcutaneous fat are retained in the fat for some time and do no disperse to the adjacent layers of the skin because of poor blood supply in this region [94]. Furthermore, the enzymes present in the skin participate in regulating (activation/inactivation) the natural chemicals present in the body, such as hormones, steroids, and inflammatory mediators as well as externally applied chemicals, such as medicines/drugs and EO components. In addition, the enzymes also participate in the metabolism of EO components which may result in the change of molecular structure of the original compound. The latter changes the effect on the body. The skin enzymatic activities vary differently in different age group individuals which define skin elasticity, dehydration, damage, broken, pigmentation, inflammation, diseases, and so on [94].

Bergamot EOs (BEO) have been a part of homemade ointments, soaps, toiletries, bodywash, shampoos, anti-dandruff and hair care products, masks and cleansers, candles, and massage oils (a mixture of oils) employed for skin disinfection [96], as an astringent [97], antiseptic or aid for healing minor wounds [98], insect bites, sunburn, aromatherapy massage, and cosmetics [99]. BEO aromatherapy massage as a complementary therapy to the patients suffering from cancer has been observed to induce relaxation from symptoms of clinical anxiety and depression for up to two weeks [100]. Furthermore, aromatherapy involving BEO has been observed to help in improving mood, and symptoms of mild stress and facilitating sleep induction [101]. The commonly used EOs from citrus in aromatherapy (in the form of aroma-sticks) in clinical studies are lemon (*Citrus limon* (L.) *Osbeck*) [102], bergamot (*Citrus bergamia*) [103] and orange sweet (*Citrus sinensis* (L.) *Osbeck*) [104] along with EOs from other herbs, such as lavender (*Lavandula angustifolia Mill.*), frankincense (*Boswellia carterii*), and peppermint (*Mentha piperita*) [105,106].

BEO is reported to exhibit anti-inflammatory activity while conducting a carrageenan-induced rat paw edema test. The optimal response for the anti-inflammatory activity was observed with a 0.10 mL/kg dosage injected intraperitoneally while the median effective dose was 0.079 mL/kg [73]. The absorbed EO and its components penetrated the skin and can be detected in exhaled air of the breath within 20–60 min. For example, the times taken to detect citrus EO components in exhaled breath post penetration into the skin are 1,8-cineole and α-pinene (20 min); linalyl acetate, geranyl acetate (between 20 and 40 min), bergamot, and lemon oils (40 and 60 min, respectively), and geraniol and citral (up to 2 h) [94,107].

Lemon EO has antioxidant properties, i.e., fighting free radicals which cause premature aging of the skin, and is therefore a popular ingredient in skincare products. The antibacterial property of lemon EOs is attributed to its components, viz., citric acid, limonene, and pinene. This makes it a suitable component in formulating cleansers, body washes, and soaps, as it helps in removing bacteria, and other microbes from pores of the skin of acne-prone oily skin. Furthermore, lemon EOs also possess astringent properties which are effective in closing the pores in the skin and preventing the blockages from being inflamed. For a typical formulation to be utilized in topical applications, citrus EOs in a blend with other EOs, such as lavender and chamomile EOs, are employed for calming skin inflammation and reducing redness. In skin lotions and ointments, the citrus EOs are mixed with a carrier oil, such as jojoba oil or olive oil to dilute the potency of the oil for applications at sensitive areas such as the face, neck, and chest. Some citrus EOs, e.g., EOs from bergamot, lemon, and grapefruit exhibit phototoxic effects (e.g., skin-irritation, damage) upon exposure to sunlight/UV rays in the Sun's radiation owing to furanocoumarins, especially 5-MOP (5-methoxypsoralen or bergapten) present in the EO composition. Removal of psoralen (the parent compound in a family of naturally occurring linear furanocoumarins) from the citrus EOs-containing formulations has been found to eliminate the possibility of phototoxicity [108].

The volatile constituents in the EOs penetrate the skin through deeper layers of the skin via different mechanisms of action, viz., interaction with the highly ordered intercellular lipid structure in stratum corneum (SC), and interaction with intercellular proteins resulting in conformational changes, and the latter increases permeability of the skin [109]. The penetration of the EO components also forms a pathway for different drugs (hydrophobic and hydrophilic), and vitamins in the topical formulation to enter lower skin layers [109–112]. Furthermore, the EO components are rapidly metabolized, not accumulated in the skin and the body, and rapidly excreted after application to the skin; therefore, regarded as safe penetration enhancers [109]. The absorption of citrus EOs to the deeper layers of the skin, molecular structures of the skin penetration enhancers (frequently employed in topical lotions/ointments for facilitating transdermal drug delivery), and molecules participating in anti-inflammatory, anti-microbial, and anti-carcinoma activities are displayed in Figure 6.

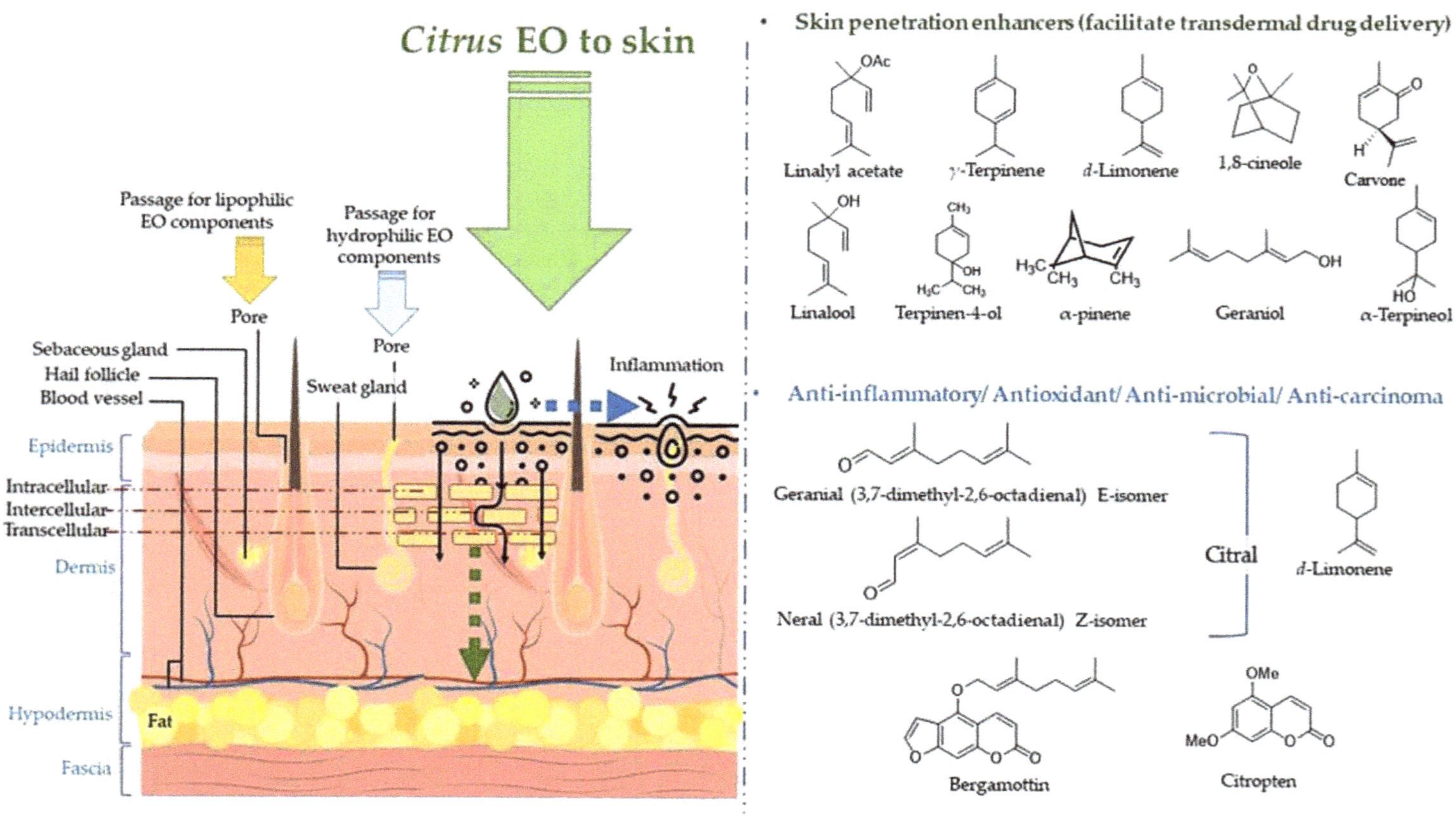

Figure 6. Absorption of citrus EOs to the deeper layers of the skin, molecular structures of the skin penetration enhancers (frequently employed in topical lotions/ointments for facilitating transdermal drug delivery), and molecules participating in anti-inflammatory, anti-microbial, and anti-carcinoma activities.

Terpenes in citrus EOs are proven as promising nontoxic, non-irritating penetration enhancers for both hydrophilic and lipophilic drugs [113,114]. Some of the well-known penetration enhancer molecules present in the citrus EOs are *d*-limonene, α-pinene, α-terpineol, carvone, and 1,8-cineole. The terpenes exhibit a significantly efficient fluidizing effect on the lipid bilayer structure. Limonene has been found to induce change in the barrier structure of the skin in the presence of ethanol and facilitate the permeation of EO components and drug molecules utilizing its affinity with alcohol. In addition, sesquiterpenes also have been found to increase the penetrability of the skin by altering the structure of intercellular lipid bilayer and formation of a complex with the drug molecule [109,111].

5. Aromatherapy Using Citrus EOs for Health and Treatment of Diseases

5.1. Oxidative Stress

Free radicals, such as reactive oxygen species (ROS), and reactive nitrogen species (RNS) are produced during cellular aerobic respiration in mitochondria (endogenous). ROS are also produced when skin is exposed to ultraviolet (UV) light (UV-A; 320–400 nm, and UV-B; 290–320 nm) and this is known as the exogenous origin of free radicals. In addition to the ROS, superoxide anion radical ($^{*}O_2^{\bullet -}$), hydrogen peroxide (H_2O_2), hydroxyl radical ($^{*}OH$), singlet oxygen ($^{*}O_2$), lipid peroxides (LOOH), and their radicals (LOO*) are also formed which participate in the process of skin aging, phototoxicity, induction of inflammation, and inflammation-induced malignant tumors [115–119]. The free radicals attack and degenerate structural molecules, such as collagen; and functional biomolecules, such as RNA and DNA, fatty acids, proteins, and other essential molecules. This gives rise to several complications which result in aging, inflammation, cancer, Alzheimer's disease, Parkinson's disease, diabetes, atherosclerosis, liver disease, etc. Oxidative stress is one of the main reasons behind allergic and inflammatory skin diseases, e.g., atopic dermatitis, urticaria, and psoriasis. Furthermore, microbial infections, e.g., that are caused by *S. aureus*, may worsen the damaged and lesioned skin by the production of ROS [120].

Aerobic respiration at the cellular level takes place in the mitochondria. The latter is a double-walled organelle (in eukaryotic cells) which carries out aerobic respiration and produces adenosine triphosphate (ATP). ATP is the utilizable form of the chemical energy consumed by the cell in its various functions. In diseased conditions, such as Alzheimer's disease, dementia, or aging, the mitochondria undergo a dysfunctional stage during which oxidizing free radicals are generated in excessive amounts which eventually leads to oxidative stress and oxidative damages to essential molecules in the cell and ultimately pathological abnormalities. Beta-amyloid (Aβ) is an initiator of reactive oxygen species (ROS) and reactive nitrogen species (RNS). The free radicals attack and damage the essential molecules present in the cell including membrane lipids, and cellular organelles and generate mitochondrial toxins, such as hydroxynonenal (HNE) and malondialdehyde. When the membrane bound ion selective ATPase is damaged because of oxidative stress, it stimulates the NMDA receptors, membrane attack complex (MAC), and ion-specific Aβ pore formation. As a result, an influx of calcium ions increases and consequently cytosolic and mitochondrial calcium load. In the next stage, cellular amyloid targets essential enzymes, namely cytochrome-C oxidase, α-ketoglutarate dehydrogenase, pyruvate dehydrogenase, and manganese superoxide dismutase (MnSOD). This causes damage to the mitochondrial DNA and ultimately fragmentation of the structure. Aβ stimulates stress-induced protein kinases-p38, c-jun N-terminal kinase (JNK), and tumor suppressor protein (P^{53}) leading to apoptosis or cellular damage.

In natural and healthy physiological conditions, the free radicals generated are neutralized to non-radical forms under the action of certain enzymes, e.g., catalase (CAT) and hydroxy peroxidase. In acute and chronic cases or low immunity, the production of free radicals becomes radically high. To elaborate on this, products of lipid peroxidation stimulate phosphorylation and aggregation of tau proteins. The latter inhibits complex-I in a cell under oxidative stress, and excessive quantities of ROS and RNS are produced at complexes I and III. In the final stage, mitochondrial membrane potential (MMP) drops, and the permeability-transition pores (ψm) opens. The latter results in the activation of caspases and cellular damage. Ultimately, the reactive species (ROS and RNS) readily initiate oxidative degradation of somatic and brain cells (neural, microglial, and cerebrovascular cells). In such conditions, supplementary administration of free radical scavengers is recommended [58,121].

The citrus EOs possess antioxidative properties due to the ability of the component molecules to donate a hydrogen atom, or an electron to the free radicals which can delocalize the unpaired electrons (in conjugated/aromatic structure), thus neutralizing the free radicals and protecting the biological molecules from being damaged by oxidation or oxidative stress. The EO components also interfere with lipid metabolism in animal tissues by upregulating the activities of antioxidative enzymes, such as superoxide dismutase, catalase, and glutathione peroxidase. This results in the inhibition of the formation of reactive oxygen species and oxidation of polyunsaturated fatty acids which give rise to off-flavors in the food materials [122,123]. Inhalation of Citrus EOs can increase the amount of GSH and cause a reduction in lipid peroxidation in the brain, and it helps prevent DNA cleavage and cell apoptosis by scavenging free radicals (ROS) via antioxidant effects. Inhalation of EOs augments the level of antioxidant enzymes involved in the immune system, e.g., superoxide dismutase (SOD), glutathione peroxidase, and catalase (CAT). It has been found that terpenes present in the citrus EOs can reduce inflammation symptoms by decreasing/inhibiting the release of pro-inflammatory cytokines, such as NF-κB (nuclear transcription factor-kappa B), IL-1β (interleukin-1β), and TNF-α (tumor necrosis factor-alpha) [124].

In addition to monoterpene hydrocarbons, limonene can also inhibit the production of pro-inflammatory cytokines in lipopolysaccharide (LPS)-induced inflammation symptoms, and the production of ROS in H_2O_2-induced oxidative stress and wound healing. EOs obtained from bergamot and sweet orange have been found to heal acne vulgaris caused by excessive secretion of androgen by reducing the growth rate of as well as secretion from, sebaceous glands. This activates the inhibition of triglyceride (TG) accumulation and the release of inflammatory cytokines in the sebaceous glands. This results in apoptosis in sebaceous glands leading to a decrease of T/E_2 ratio. The EOs act to lower the IL-1α levels in sebaceous glands which help in improving acne lesions by alleviating inflammatory responses [121,125,126]. Another study investigating limonene's anti-inflammatory response on human eosinophilic leukemia HL-60 clone 15 cells revealed interesting results. Hirota et al. [127] reported that a low concentration of limonene (7.34 mmol/L) can inhibit ROS production for eotaxin-stimulated HL-60 clone 15 cells.

A higher limonene concentration of 14.68 mmol/L was found to diminish diesel exhaust particles (DEP)-induced MCP-1 production significantly, indicating that the antioxidant activity of limonene can help restrict monocyte infiltration into the lungs and prevent migration of eosinophil, protecting asthmatic lungs and prevent damage from DEPs in the lung. Furthermore, NF-κB formation was also diminished upon the addition of proteasome inhibitor MG132. The limonene can inhibit DEP induced p38 MAPK signaling pathway and inhibit eotaxin-induced chemotaxis by eosinophils [127]. Citrus EO components exhibit antioxidative activities against the oxidation of linoleic acid. In addition, antioxidant activities have also been reported against in vitro oxidation of human low-density lipoprotein induced by Cu^{2+}, and 2, 2′-azobis (2-aminopropane) hydrochloride [128]. The antioxidant properties of citrus EOs are attributed to the presence of phenolic compounds in their composition. Monoterpene hydrocarbons (limonene, thujene), and oxygenated monoterpenes (monoterpenes with different functional groups, such as phenols, alcohols, aldehydes, ethers, esters, and ketones) contribute significantly to the antioxidant properties of the citrus EOs [129]. The events and consequences of oxidative stress in a somatic and nerve cell, and the therapeutic effects of citrus EO aromatherapy are displayed in Figures 7–9.

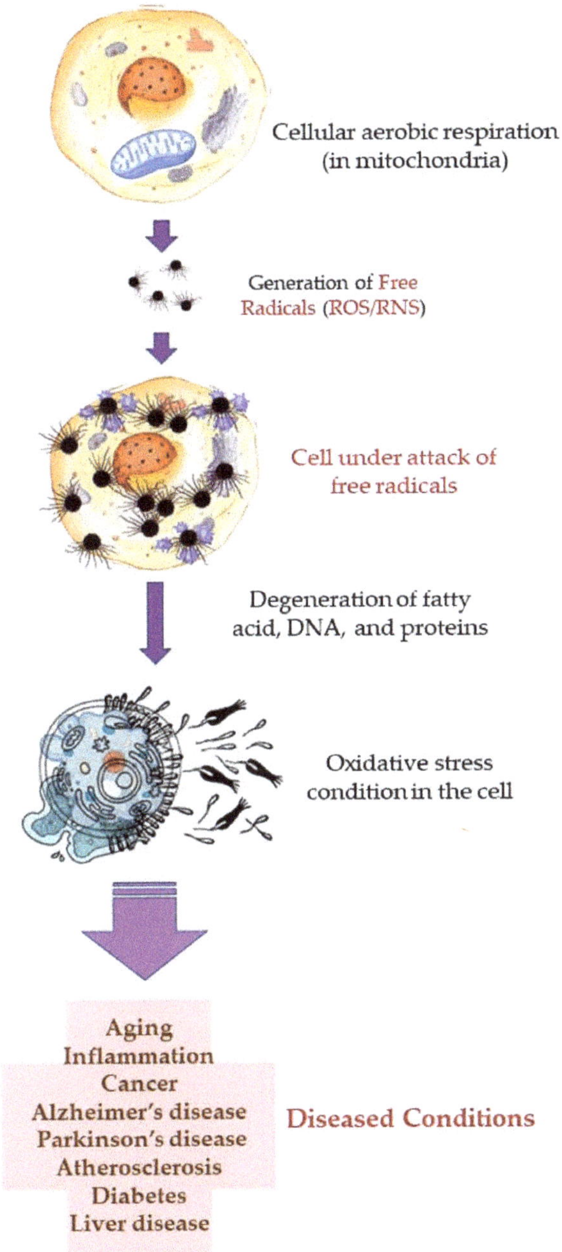

Figure 7. Oxidative stress in the cell: Events and consequences-I: Somatic cell.

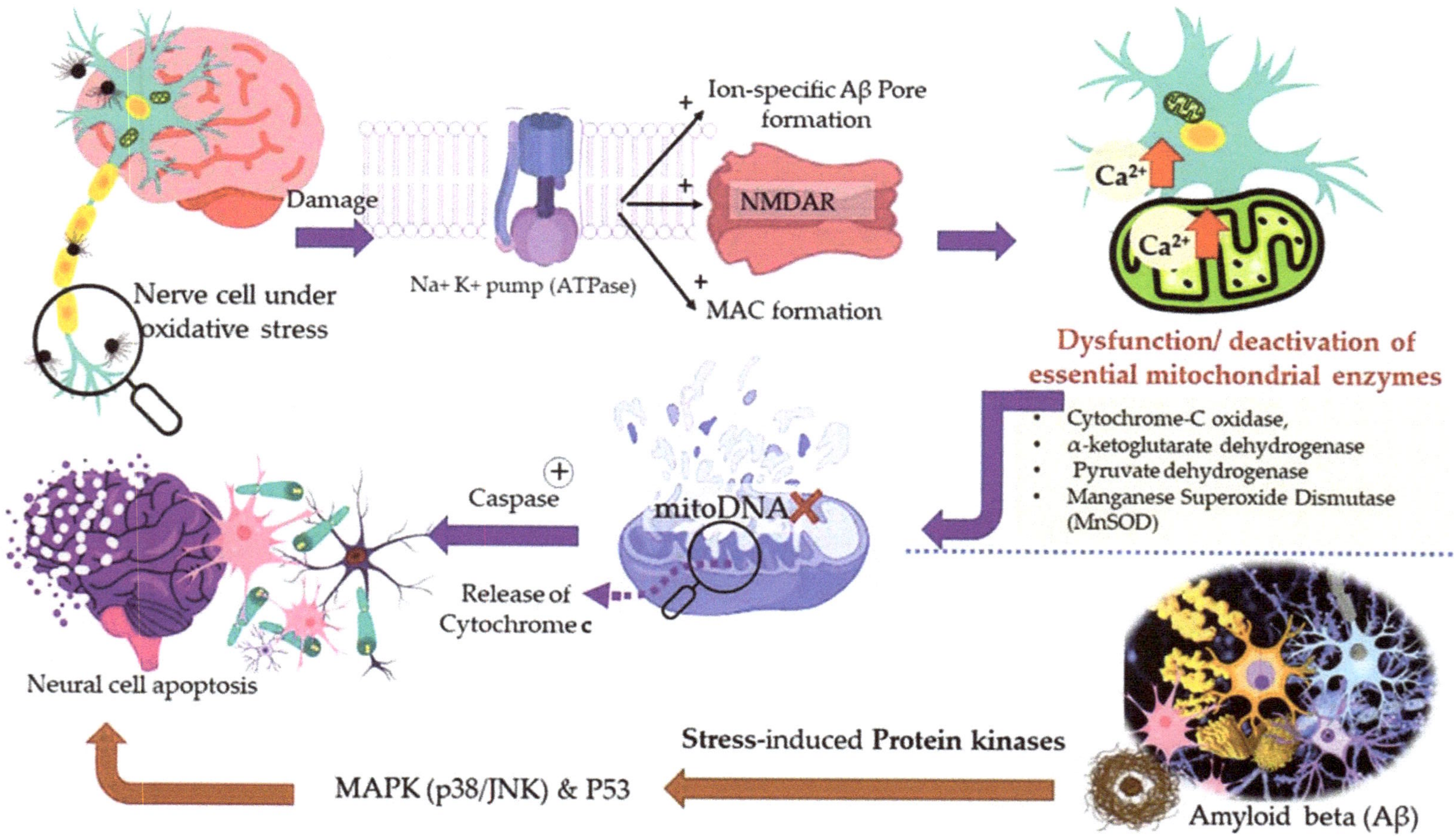

Figure 8. Oxidative stress in the cell: Events and consequences-II: Nerve cell.

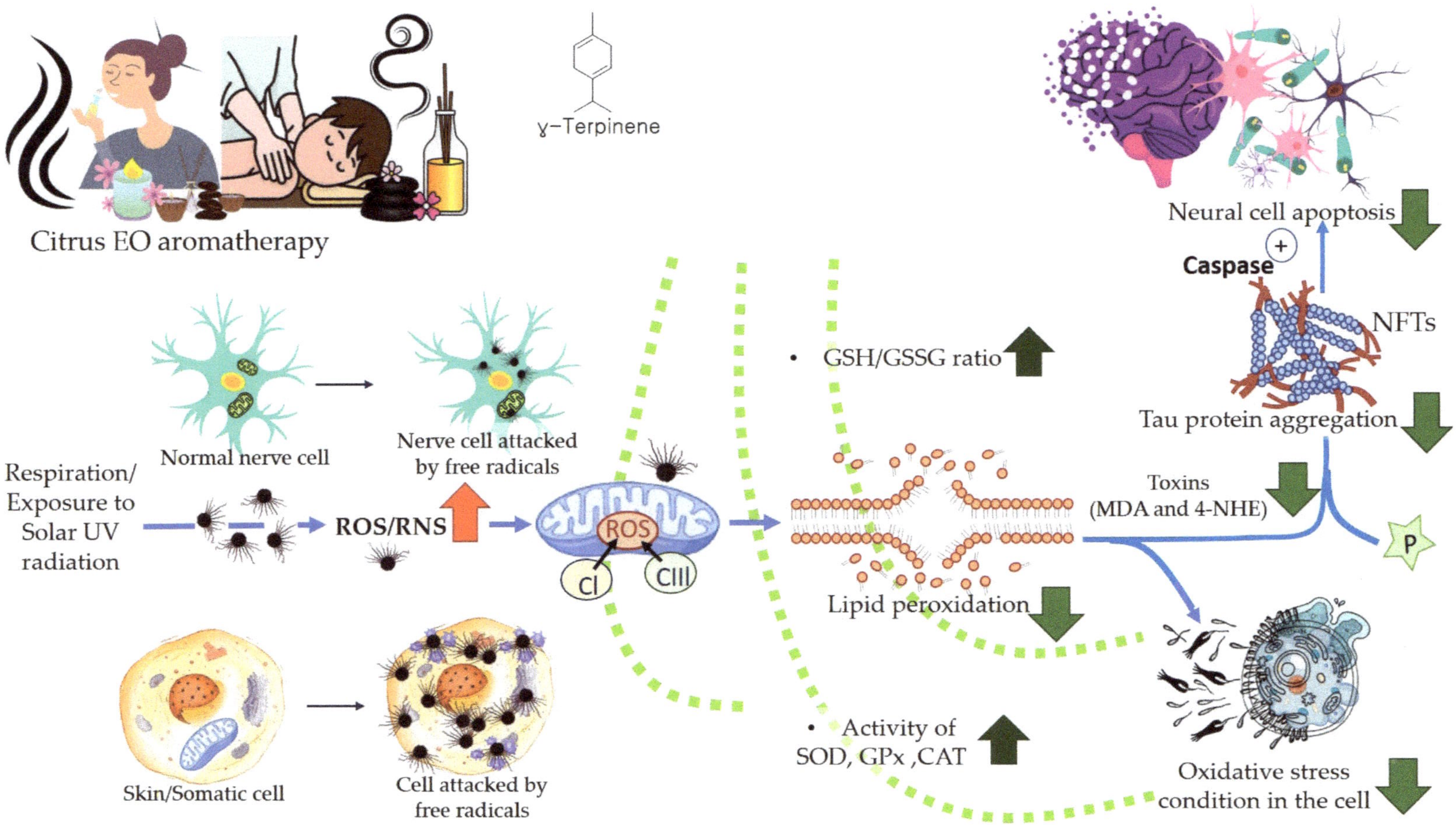

Figure 9. Therapeutic effect of citrus EOs aromatherapy.

Thujene, a monoterpene, has been reported to exhibit good antioxidant activity due to its ability to quench singlet oxygen efficiently [130]. The alcohol compounds, e.g., carveol and perillyl alcohol; ketones, e.g., carvone and aldehydes, perillyl aldehyde; esters, e.g., citronellyl acetate, geranyl acetate, neryl acetate exhibit good antioxidant activities. Among the compounds, γ-terpinene, geranial, R-(+) limonene, and β-pinene have been reported for possessing the highest antioxidant capacities [131–133].

5.2. Stress-Related Disorders/Mood Disorders

Stress-related disorders or mood disorders have become very common in everyday life. Mood disorders include several psychiatric illnesses which significantly (sometimes severely) impact the mood related function of an individual (patient). The disorders are characterized by cognitive deficits such as impaired learning, loss of memory, and inability to focus/concentrate. Sudden, significant, and persistent changes in emotions or state of mind, sadness, anxiety, depression, sleep disorders, and insomnia are symptoms associated with chronic stress or trauma. Mood disorders originate from physiological, psychological disturbances, organic damage, nerve injury, side effects of medications, chronic stress, etc. Depression is characterized by a combination of symptoms associated with traumatic emotions (sadness and anhedonia), cognition deficit, and somatic symptoms (change in appetite, such as over/under eating), sleep disorders, insomnia, melancholy, hopeless, despair, detachment from daily life/routine activities, fatigue, and even suicidal tendencies. Anxiety is mainly caused by physiological and psychological disturbances, e.g., emotional, behavioral, environmental, somatic, and social elements. When any of these elements invoke unpleasant situation or sensations, fret, phobias, disquietedness, or restlessness, the human mind enters a stressed condition or anxiety. Prolonged stressed conditions lead to a stage when the person faces the onset of anxiety symptoms, such as unusual panic situations characterized by hypertension, sweating, palpitation, chest pain, migraine, papillary dilation, shortness of breath, and so on [134,135]. According to a WHO report, more than 260 million people are suffering from depression with varying levels and approximately 800,000 people die by committing suicide every year [136]. Furthermore, more than 50 million people are known to be suffering from dementia/Alzheimer's disease which is projected to rise in number up to 82 to 152 million by the years 2030 and 2050, respectively. A stressed or diseased person finds it difficult to perform his/her daily life and respond approximately to the problems, challenges, or important events on time. Moreover, the disease further progresses with loss of memory. In the pathological aspect, the diseased person is diagnosed by the presence of amyloid plaques, neurofibrillary tangles, and loss of neural transmission in the brain [137,138]. Insomnia patients have common symptoms of depression and anxiety, and no single medication is known to cure this condition accurately. Insomnia is also characterized by acute sleep disorder. Prolonged disturbance in sleep patterns may result in high blood pressure, cardiovascular diseases, and severe risks of acute mental illnesses [139–141].

Bergamot oil has been found to reduce blood pressure and heart rate and help induce sleep and relief from restlessness. EOs extracted from sweet orange and lavender EO have been observed to improve sleep quality and provide relief from tiredness in hemodialysis patients [142]. Takeda et al. carried out a study on inhalation aromatherapy in elderly dementia patients by applying EO drop on towels covering their pillows during their sleep time. The researchers recorded a better sleep latency and improved total sleep time and effectiveness of sleep among the treated people [143]. The aromatic EO molecules enter limbic system in the brain via nasal passages and stimulate GABA receptors in the hypothalamus. The overall process induces and maintains restful sleep [144]. Citrus EO (with 95% citral in the composition) has been observed to induce a pleasant mood in people suffering from sadness [145]. The molecular pathways involved in the pathophysiology of depression include the hypothalamic–pituitary–adrenal axis, sympathetic nervous system, monoamine neurotransmission system (e.g., serotonergic (5-HT), dopaminergic (DA), and GABAergic pathways), cyclic adenosine monophosphate (c-AMP) response element-binding (CREB) protein signaling pathway [58,146–152]. According to neurotropic hypothesis, depression is associated with a deficit of neurotropic factors caused by prolonged exposure to stress which results in loss of neural plasticity [153]. Brain-derived neurotropic factors (BDNF), a protein in the brain produced by BDNF gene, and neurotrophins, a class of growth factors, promote the growth of the neurons and maintain adequate neural plasticity. During depression, the level of BDNF in the serum decreases. Therefore, deficiency of neurogenesis or production of new neurons in the brain hippocampus is a major reason behind depression. EO-based aromatherapy involving EOs of lavender, lemon, and bergamot has been reported to prevent negative symptoms of depression, such as deficiency of neurogenesis, suppressed dendritic growth of immature neurons, and low serum BDNF levels in the brain hippocampus [154–157]. In a clinical study involving patients diagnosed with stress- and depression-related symptoms, such as attention deficit and hyperactivity disorder, four weeks employing EO-based aromatherapy resulted in a decrease in the level of anxiety and depression and a simultaneous increase in blood plasma BDNF levels in the brain hippocampal tissues [157]. Moreover, regarding neurogenic and enhancing neurotropic factors in the human brain, citrus EOs have also been observed to participate in the regulation of the neuroendocrine system. Depression and anxiety disorder release the stress hormone cortisol. Aromatherapy involving lavender EO has been observed to downregulate the release of stress hormones and a decrease in salivary and serum cortisol levels was recorded [48,158]. In addition, bergamot EO and grapefruit seed EO have also been reported to induce lowering of cortisol levels in the blood, thereby lower stress related symptoms. There have also been recorded improved coronary flow velocity and enhancement in relaxation. Bergamot EOs have been observed to cause an alteration of HPA axis and attenuate the rise of corticosterone levels in the blood [159]. Lemon EOs have been recorded to produce antidepressant effects in terms of accelerated turnover of dopamine in the brain hippocampal region establishing therapeutic effects of EOs in healing the patients from depression and related symptoms [58].

Anshen EO, a mixture of EOs from lavender, sweet orange, and sandalwood, has been observed to have anxiolytic, antidepressant, sedative, and hypnotic effects. Researchers have performed sleep latency and sleep duration experiments, where they compared diazepam—generally used to treat insomnia—with anshen EOs [160]. Mouse brain responses were analyzed using ELISA test to detect changes in 5-HT and GABA levels. The results showed a significant decrease in impulsive activities and reduced sleep potential. An increase in the levels of 5-HT and GABA was observed in the mouse brain. Anxiolytic effects of BEO (1.0, 2.5, and 5.0% w/w) were studied by administering it to rats subjected to anxiety-related behaviors, the elevated plus-maze, and the hole-board tests, and then measuring the stress-induced levels of plasma corticosterone in comparison with the effects of diazepam. BEO (2.5%) and diazepam exhibited anxiolytic-like effects and attenuated the corticosterone response to acute stress [159]. After perfusion into the hippocampus via the dialysis probe (having volumetric flow rate 20 μL/min), BEO produced a dose-dependent and Ca^{2+}-independent increase of extra cellular aspartate, glycine, taurine, GABA, and glutamate [161]. Inhalation of orange EO for 90 s has been observed to cause a significant decrease in oxyhemoglobin concentration in the right prefrontal cortex of the brain which increases comfortable, relaxed, and natural feelings [104]. Osbeck EO from *Citrus sinensis* Osbeck is found to exert antidepressant effects, being suitable to treat minor stress. The effects of Osbeck EO inhalation on CUMS (Chronic Unpredictable Mild Stress) mice were found to tackle depression along with decreased body weight, interest, movement, and dyslipidemia. Limonene is not metabolized in the brain immediately after inhalation. An in-depth study revealed that limonene is significantly effective as an antidepressant and shows healing progress in the neuroendocrine, neurotrophic, and monoaminergic systems [17].

Moradi et al. [162] conducted a study on patients who underwent coronary angiography. The patients were divided into two intervention groups, each comprising 40 patients. Patients of the test group inhaled EO from *Citrus aurantium* for 15–20 min about 60 min before the procedure. In the control group, distilled water was used instead of EO. Following *Citrus aurantium* EO inhalation, noticeable responses were observed. Vital signs of anxiety such as pulse rate, systolic blood pressure (SBP), and diastolic blood pressure (DBP) were significantly decreased after the intervention [162]. Li et al. [163] compared the effects of an essential oil mixture (EOM) (a mixture of *Citrus sinensis* L., *Mentha piperita* L., *Syzygium aromaticum* L. and *Rosmarinus officinalis* L.), with peppermint EO on physical exhaustion in two rat groups. After swimming, the two rat groups were maintained in an environment of EOM and peppermint EO, respectively. Various body parameters were studied after three continuous days of nebulization. Blood lactic acid (BLA) and malondialdehyde (MDA) levels were found to decrease in both groups. An improved duration of fatigue and increased superoxide dismutase (SOD) activity were observed in both groups. The results observed in the EOM group were noticeable, such as an increase in blood glucose and a reduction of blood urea nitrogen (BUN) and glutathione peroxidase (GSH-PX). This study determined that exercise-induced fatigue can be effectively relieved by inhalation of EOs [163]. Another study was performed on Swiss male mice to observe the neurotransmission contribution of nitric oxide when *C. sinensis* EO was used for its anxiolytic effects. To perform this study, mice were placed in an environment of *C. sinensis* for inhalation of EOs at different concentrations. Nitric oxide was used as a precursor to observe the mediation behavior of the nitrergic system, and it was found to play a significant role in the anxiolytic effect of *C. sinensis.* Bergamot essential oil (BEO), obtained from the fruit of *Citrus bergamia*, is used in aromatherapy as a pain reliever, improves sleep disorders, and reduces anxiety. BEO can induce neurotransmission which is associated with its anxiolytic-relaxant effects. Anxiolytic effects are shown to be the result of the collaborative action of BEO and the 5-hydroxytryptamine (5-HT) 1A along with the involvement of multiple and complex mechanisms [19].

5.3. Diseased Conditions

5.3.1. Neurogenic Inflammation

Neurogenic inflammation is inflammation in neurons caused by the release of pro-inflammatory mediators, namely Substance P, calcitonin gene-related peptide (CGRP), neurokinin A (NKA), and endothelin-3 (ET-3). The release of pro-inflammatory mediators in the neurons is stimulated by the activation of ion channels (transient receptor potential ion channel-1 or TRPA-1) in response to harmful/unpleasant environmental stimuli. Acute neurogenic inflammation is caused by the activation of TRPA-1 channels induced by LPS. Following the release of inflammation causing neuropeptides is the release of histamine from the mast cells present in the vicinity of the affected neurons. The latter stimulates release of Substance P and calcitonin gene-related peptide, thereby establishing a bidirectional link between histamine and neuropeptide in the causation of neurogenic inflammation. Approximately 25% of migraine cases lead to temporary dysfunction of the central nervous system associated with visual field disturbances, sensitivity to light/sound, nausea, and/or vomiting [164].

Terpenes and terpene derivatives have been investigated for anti-inflammatory bioactivities. In this regard, limonene, α-pinene, β-caryophyllene, and β-myrcene have been most preferred for migraine cases [165]. Alpha-pinene (α-pinene) present in citrus EOs has been found to reduce NF-κB/p65 nucleus of LPS-stimulated THP-1 cells and increase the cytoplasmic concentration of Iκ-Bα protein. Alpha-pinene (α-pinene) also significantly decreases the levels of IL-6, TNF-α, and NO, as well as the expression of iNOS and Cox-2 induced by LPS. An in vitro study on *d*-limonene activity revealed an increase in IL-10/IL-2 ratio, consequently enhancing IL-10 levels. The latter is a cytokine synthesis inhibitory factor and inhibits proinflammatory Th1 cytokine production (IL-2) [166]. Furthermore, *d*-limonene epoxide has been observed to prevent the release of inflammatory mediators, inhibit vascular permeability, reduce migration of neutrophils, and display systematic and peripheral analgesic effects towards the brain's opioid system (associated with regulating pain, reward, and addictive behavior) [167]. The pathophysiological mechanism of migraine induced by 5-HT and neuroprotective mechanisms of α-pinene in migraine are displayed in Figures 10 and 11, respectively.

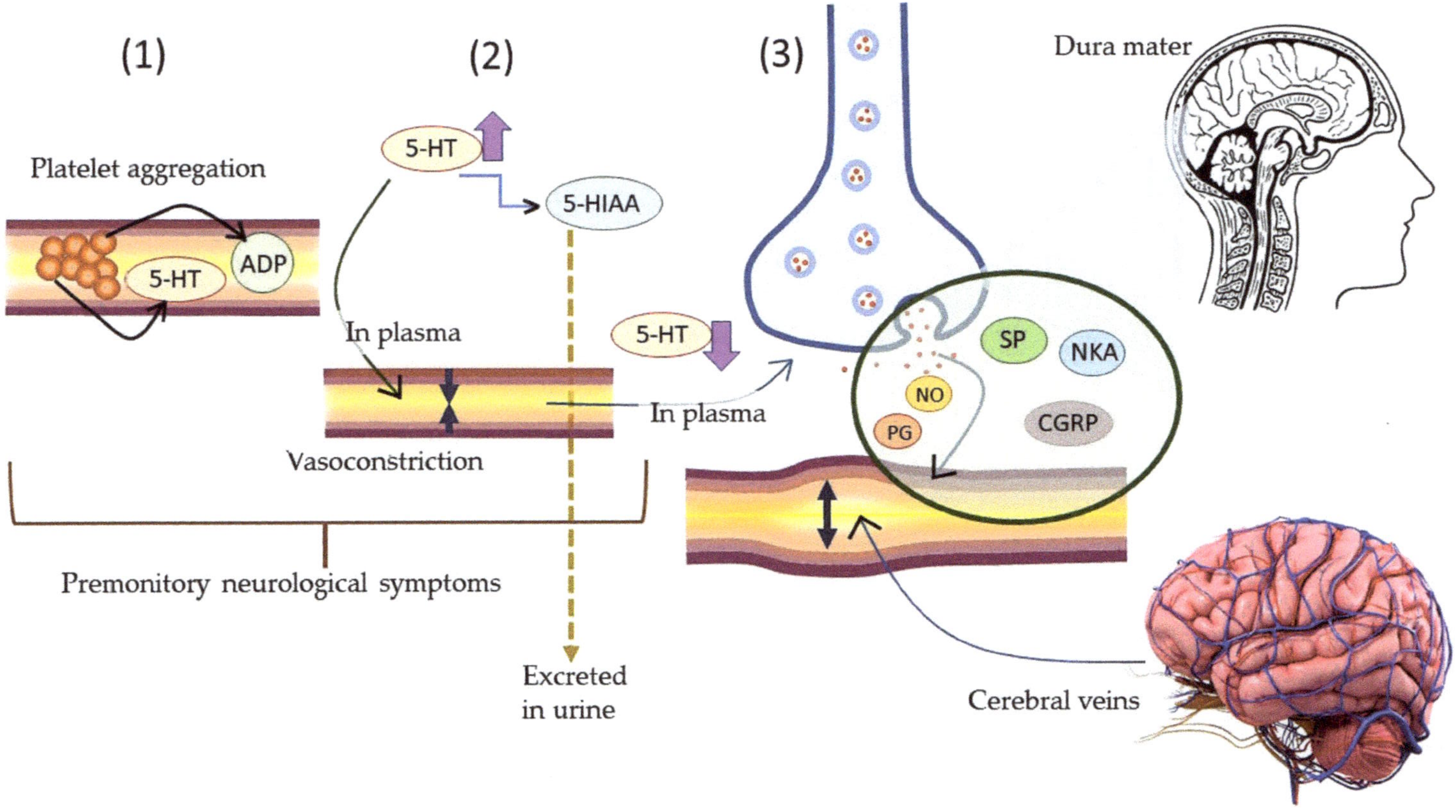

Figure 10. The pathophysiological mechanism of migraine induced by 5-HT. (1) Platelet aggregation trigger release of 5-HT and ADP in blood plasma. (2) High level of plasma 5-HT causes reversible vasoconstriction followed by conversion of 5-HT to its metabolite 5-HIAA. The latter is excreted in urine. (3) Low level of plasma 5-HT stimulates perivascular neurons to release neuropeptides (NO, PG, SP, NKA, CGRP) causing vasodilation of cerebral veins. The latter leads to migraine symptoms.

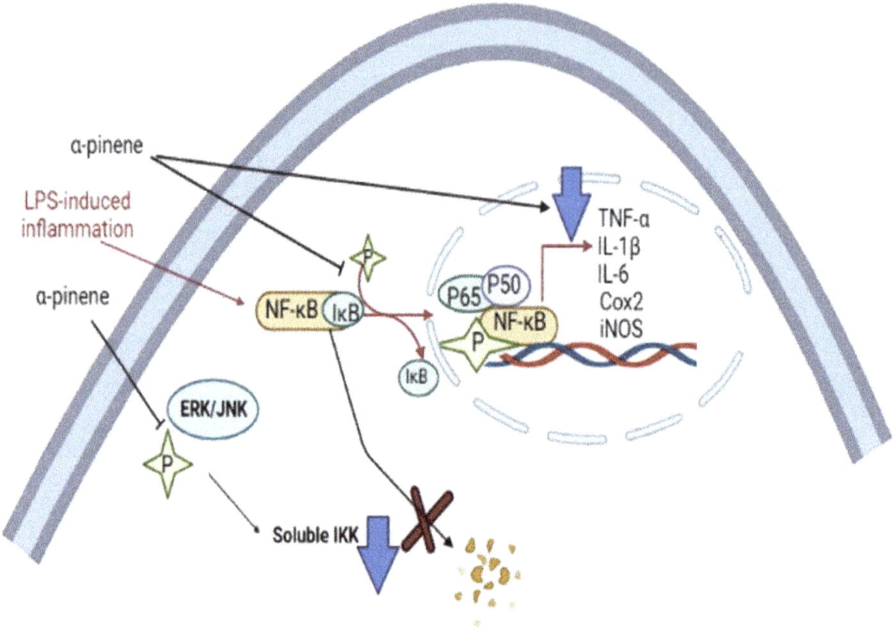

Figure 11. Neuroprotective mechanisms of α-pinene in migraine [168]. The α-pinene can reduce LPS-induced inflammation in macrophages. α-pinene can block phosphorylation of MAPKs (ERK/JNK) in macrophages and reduce the level of active (soluble IKK). This can prevent degradation of the NF-kB/IkB complex. Also, α-pinene can hinder NF-kB phosphorylation and formation of the P65/p50/NF-kB complex that leads to its nuclear translocation and induction of inflammatory genes to generate cytokines. Abbreviations; TNF-α (tumor necrosis factor alpha), IL-1β (Interleukin-1β), IL-6 (Interleukin), Cox-2 (Cyclooxygenase-2), Inos (Inducible nitric oxide synthase).

Neurogenic inflammation further causes conditions for the pathogenesis of several other neurogenic diseases, namely multiple sclerosis, migraine, psoriasis, asthma, vasomotor rhinitis, and so on. In migraine, the stimulation of the trigeminal nerve takes place which releases neuropeptides, such as Substance P, nitric oxide, 5-HT, vasoactive intestinal polypeptide, neurokinin A, and CGRP which eventually results in "sterile neurogenic inflammation". The release of Substance P stimulates the production of several other pro-inflammatory cytokines, namely interleukins (IL-1, IL-6), and TNF-alpha (TNF-α). Migraine is characterized by a strong headache accompanied with nausea, vomiting, and sensitivity to light which may persist up to 72 h or longer. The phases in migraine can be explained to take place in four stages, *viz.*, (a) prodrome: this stage persists for a few hours to few days and is characterized by irritability, depression, yawning, nausea, fatigue, muscle stiffness, difficulty in concentration and sleep; (b) aura: this persists for 5 to 60 min and is characterized by visual disturbances, temporary loss of sight, numbness in hands and feet, and tingling sensations in the body; (c) headache; this persists for 4 to 72 h and is characterized by throbbing pain, sensitivity to light, noise, odors, nausea, vomiting, giddiness, insomnia, neck and body pain and stiffness, and burning; and (d) postdrome: this is characterized by an inability to concentrate, fatigue, and lack of comprehension.

5.3.2. Dementia, Alzheimer's Disease (AD), and Parkinson's Disease (PD)

Alzheimer's disease is an age-related neurodegenerative disorder characterized by gradual memory loss and dementia. It also shows cognitive dysfunctions and turbulent behavioral patterns. At a physio-chemical level, it is diagnosed by scarcity in cholinergic neurotransmission in the cranial (brain) nerves, cognitive dysfunction, behavioral turbulence, gradual memory loss, accumulation of amyloid plaques (amyloid-β, Aβ) and neurofibrillary tangles (NFTs) in the specific brain areas, reduced glutathione (GSH) content in the hippocampus, mitochondrial dysfunction in the cells, and excess production of free radicals leading to oxidative stress [169]. The cholinesterase (ChEs) enzyme hydrolyses acetylcholine (Ach) into choline and acetate and the concentration of Ach neurotransmitter molecules in the brain drops resulting in the termination of neurotransmission. Acetylcholine is involved in the key function of learning and memory. In addition, monoamines, *viz.*, dopamine and serotonin (5HT), released in the brain are also attributed to learning and memory. A decrease in the dopamine amount in the brain, and consequently, functional degradation of dopamine receptors has been identified as one of the common causes of Parkinson's disease and Alzheimer's disease [170]. For symptomatic management of AD, inhibitors of acetylcholinesterase (AChE) and butyrylcholinesterase (BChE) enzymes responsible for the degradation of the essential neurotransmitter acetylcholine (ACh) are considered for the development of anti-AD drugs. The choline esterase inhibitors reversibly bind to the active sites of acetylcholinesterase (AChE)/butyrylcholinesterase (BChE) enzymes. As a result, the hydrolytic degradation of ACh neurotransmitter molecules into choline and acetate is inhibited. Consequently, the concentration of ACh increases at the synaptic gaps in cholinergic neurons in the hippocampus cerebral cortex and some parts of the new striatum. Other neurodegenerative pathological conditions in patients suffering from AD include an increase in monoamine oxidase (MAO) activity and lipid oxidation induced by Fe^{2+} ions. The increase of MAO deactivates neuroactive amines, such as serotonin, dopamine, and norepinephrine, and enhances the production of free radicals (or ROS) in the patient's brain [171]. Fe^{2+} ions have the ability to cross the blood–brain barrier which induces lipid oxidation via Fenton's reaction. This leads to an abundance of polyunsaturated fatty acids in the brain tissues and causes vulnerability to free radical attacks. The latter causes the formation of radical species, e.g., MDA which participates in neurodegeneration. As a remedy, if an antioxidant mechanism stops or inhibits the lipid peroxidation products (MDA), it is possible to deplete the concentration of free Fe^{2+} ions in the cytosol. Consequently, the level of oxidative stress decreases in the brain as well as in the entire body [172–177].

Most of the drugs employed in the treatment for AD are synthesized chemically and have been observed to cause side effects, e.g., nausea or vomiting, hepatotoxicity, dyspepsia, myalgia, dizziness, anorexia, and so on. EOs have been observed to interact with a range of neurotransmitter pathways, namely noradrenergic (related to norepinephrine), 5-HTergic (related to serotonin), GABAergic (related to γ-aminobutyric acid), DAergic or dopaminergic (related to dopamine), etc. Furthermore, the specific compounds present in the EOs participate in specific action mechanisms, e.g., benzyl benzoate activates 5-HTergic and dopaminergic pathways and consequently exerts anxiolytic and anti-depressant effects [178]. Linalool and *β*-pinene interact with GABAergic pathway to produce similar effects. In this direction, other EO components, namely limonene benzyl alcohol has also been found to produce anxiolytic and anti-depressant effects. EOs can inhibit enzymes linked with hydrolysis of neurotransmitters, such as monoamine oxidase (MAO). Moreover, EOs possess antioxidative properties and can penetrate the blood–brain barrier. In this direction, Ademosun et al. carried out carried out inhibition assays of AChE and BChE, MAO, and lipid peroxidation [173]. The pathophysiological targets in diseased conditions of dementia, Alzheimer's, and Parkinson's are summarized in Figure 12. The Mechanism of action of citrus EOs to inhibit acetylcholinesterase (AChE), thereby increasing levels and duration of acetylcholine in the brain and assisting with cognition (learning and memory retention) is shown in Figure 13. The syntheses of different neurotransmitter molecules in the brain, namely GABA, dopamine, and serotonin, and the mechanism of neurotransmission are shown in Figure 14. The neurotransmission pathways in GABAergic, DAergic (dopaminergic), and 5-HTergic (serotoninergic) neurons and citrus EO components that activate neurotransmission and exhibit anti-proliferative effects on human neuroblastoma cell growth are shown in Figure 15.

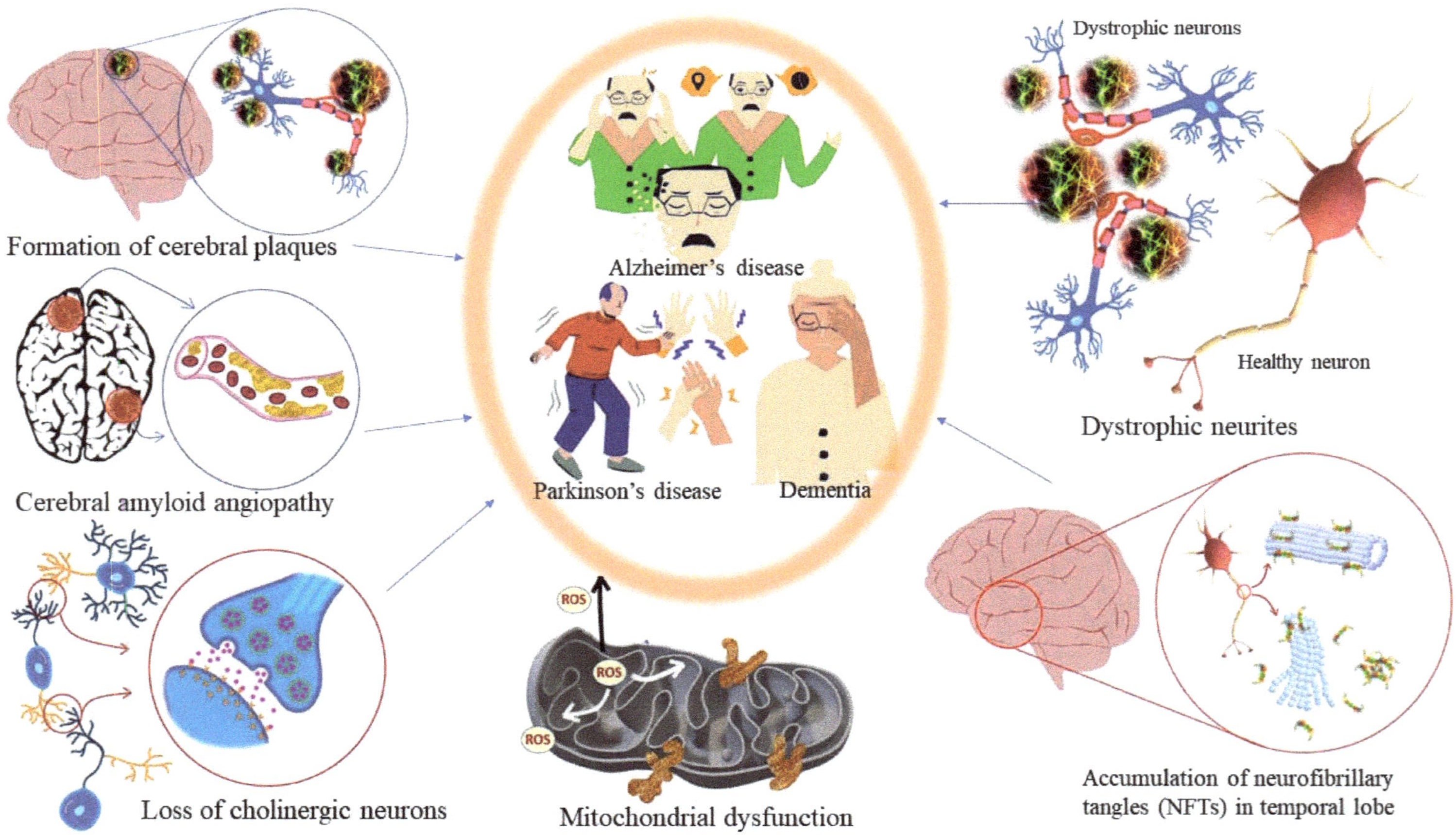

Figure 12. Pathological targets in diseased conditions of dementia, Alzheimer's, and Parkinson's.

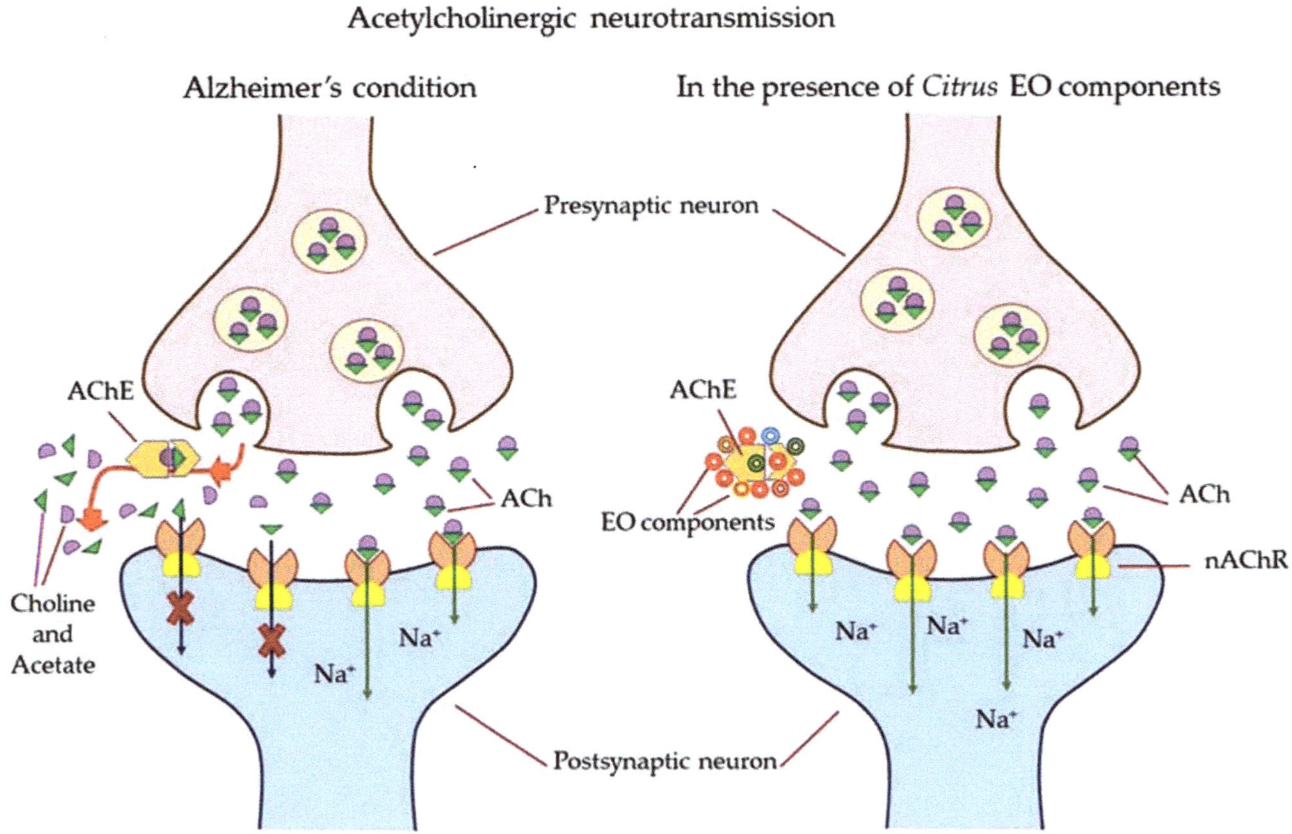

Figure 13. Mechanism of action of citrus EOs to inhibit acetylcholinesterase (AChE), thereby increasing levels and duration of acetylcholine in the brain and assisting with cognition (learning and memory retention). Abbreviation; ACh—acetylcholine; AChE—acetylcholinesterase; nAChr—nicotinic acetylcholine receptors; EOs—Citrus essential oil components.

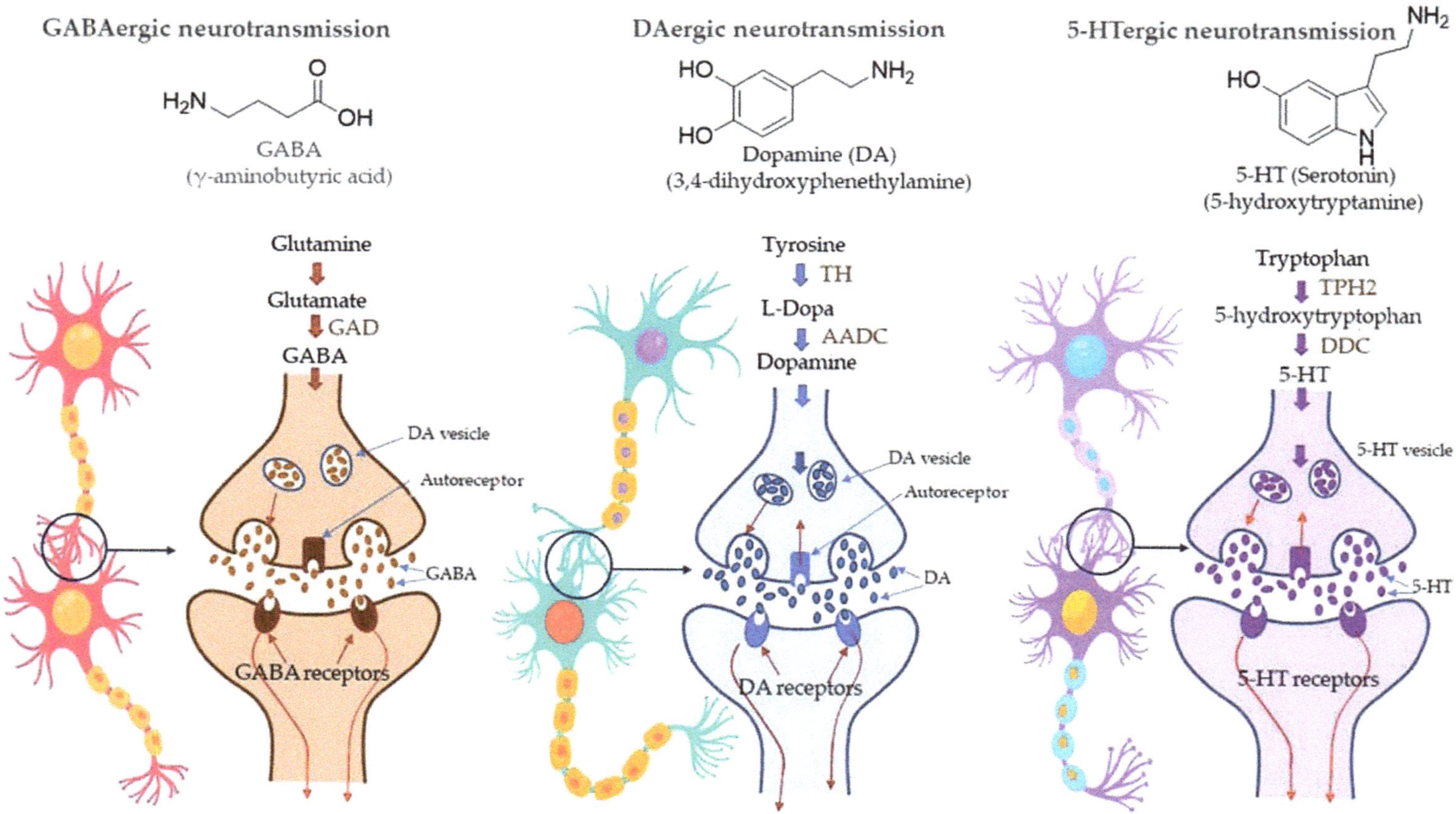

Figure 14. Syntheses of neurotransmitter molecules, viz., GABA (γ- Aminobutyric acid), dopamine, and serotonin (also called as 5-HT) and the mechanism of neurotransmission. AADC also known as DDC. Abbreviations; GAD (glutamate decarboxylase), TH (Tyrosine hydroxylase), AADC (aromatic amino acid decarboxylase), DDC (DOPA decarboxylase), TPH2 (s tryptophan hydroxylase 2).

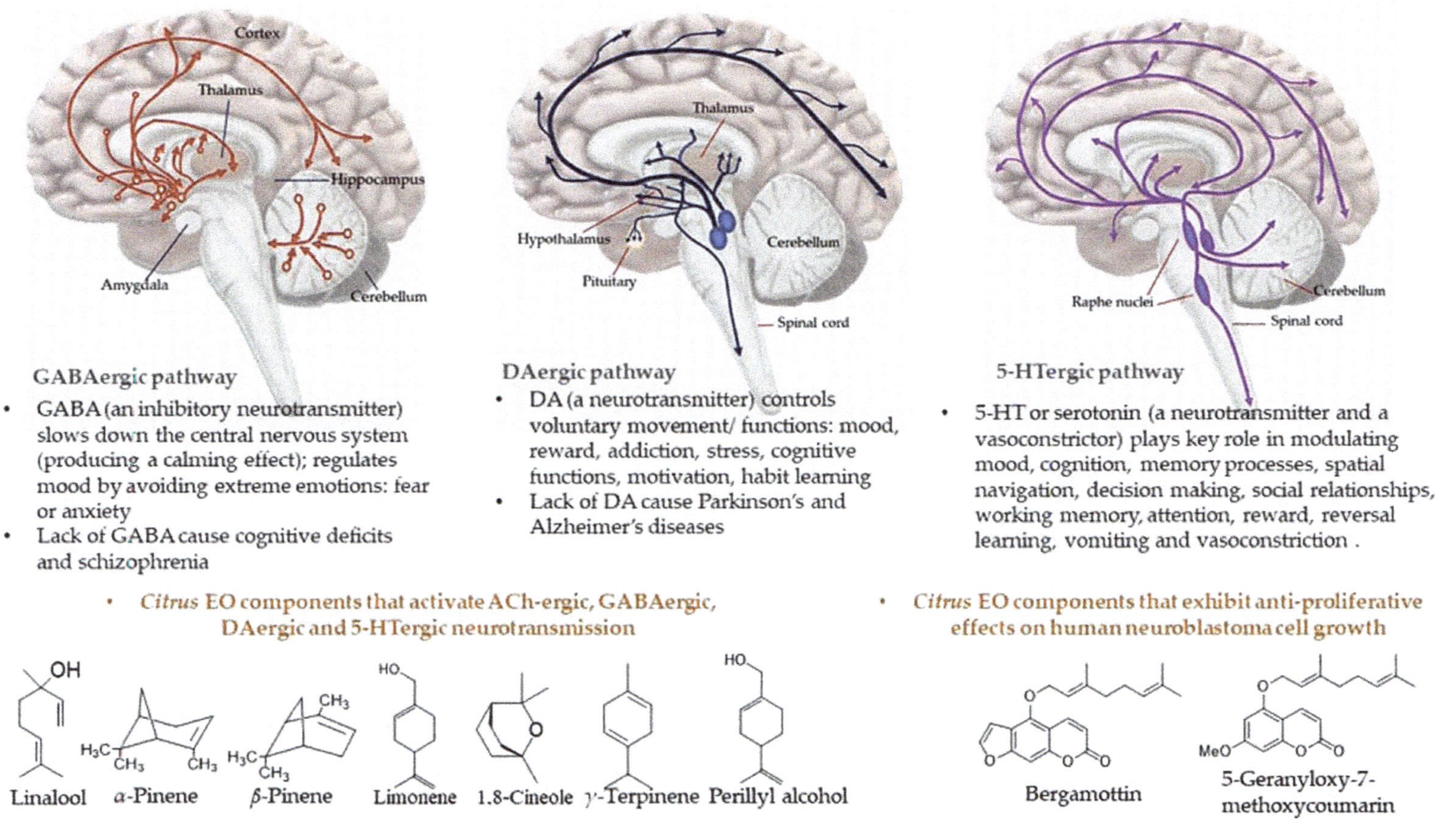

Figure 15. Neurotransmission pathways in GABAergic, DAergic, and 5-HTergic neurons and *Citrus* EO components that activate neurotransmission and exhibit anti-proliferative effects on human neuroblastoma cell growth.

The EO has been observed to inhibit AChE, BChE, and MAO in a dose-dependent manner. However, the EOs extracted from peels exhibited a significantly higher inhibition towards AChE compared with the EOs extracted from the seeds. On the other hand, the EOs from seeds exhibited a higher inhibition towards MAO activity compared with the peel EOs. Furthermore, the EOs also exhibited a decreasing effect on malondialdehyde (MDA) production which is present inside the brain homogenates. MAO activity is a crucial determinant in deactivating main neurotransmitters, such as serotonin, dopamine in the brain cells. This affects the overall behavior and mood of the patients suffering from Alzheimer's disease. Zhou et al. [179] carried out passive avoidance test (PA) and open field habituation test (OFT) employing lemon EO components, *viz.*, s-limonene and its derivatives-perillyl alcohol to investigate the effect of EOs on memory in rats. The rats were fed with s-limonene (100 mg/kg), s-perillyl alcohol (50 mg/kg) in their diets and scopolamine (1 mg/kg) was injected subcutaneously 30 min before the training test [179]. The lemon EO components showed a strong ability to improve learning and memory impaired by scopolamine in rats. BEO has been reported to exhibit antiproliferative activities in terms of inhibition against the survival and proliferation of SH-SY5Y neuroblastoma cells by activating multiple pathways resulting into necrosis and apoptotic cell death [69,180,181]. A summary of the studies on the application of citrus EOs in aromatherapy is presented in Table 1.

Table 1. Pharmacological behavior of citrus EOs in aromatherapy.

Citrus Type	Particulars	In Vitro/In Vivo/Animal Model	Activity	References
Bergamot orange (*C. bergamia*) essential oil (CBEO)	Antioxidant behavior	In vivo model obtained from mouse hearts	Increase in transcription of genes involved in antioxidant responses Having lower IC_{50} $O_2^{\bullet -}$ value in scavenging activity test than ascorbic acid and higher FRAP activity	[182]
	Mood disorder	BEO aromatherapy in alleviating depressive mood in postpartum women	Significantly improve the depressive mood Sleep quality was not significantly different	[103]
	Diseased condition	Acclimatization of the rats was performed	Relieve symptoms of stress-induced anxiety No overlapping between BEO and benzodiazepines behavioral effects Integrated effect on both 5-HT and GABA-A receptors	[183]
	Neuropharmacological studies	The elevated plus-maze and the hole-board tests were performed to study of BEO on rats	Usefulness in neuroprotection Chronic pain control Management of stress, anxiety, and anxiety-related conditions	[184]
	Antinociceptive effect	Effect of inhalation BEO on formalin-induced nociceptive response in mice.	Inhalation of BEO exerted antinociceptive activity. reduces formalin-induced licking/biting behavior. chronic pain relief in a stepwise therapeutic manner	[161]
Sweet orange or navel orange (*C. sinensis* L.) essential oil (CSEO)	Antioxidant behavior	Evaluated against the ROS-generating compound	Activity in DPPH assay was in a range of 6–23% for *C. sinensis* Decreased apoptosis in HaCat cells stimulated with H2O2. The levels of intracellular superoxide ion found to be lower	[185]

Table 1. *Cont.*

Citrus Type	Particulars	In Vitro/In Vivo/Animal Model	Activity	References
	Mood disorder	Aromatherapy during dental treatment	Lower degree of anxiety and a more cheerful attitude. To reduce salivary cortisol and pulse rate	[104]
	Physiological and psychological effect	Measurements were performed in a chamber with an artificial climate with 20 females	Significant decrease in oxyhemoglobin concentration in the right prefrontal cortex of the brain. Increases comfortable, relaxed, and natural feelings.	[104]
	Anxiolytic effect	Forty (40) male volunteers were allocated for the inhalation	Decreases the symptoms of anxiety Improves the mood	[186]
	Unpredictable mild stress	Randomized three-arm controlled trial	Significantly improved depression-like behaviors in CUMS mice by lowering sucrose preference, body weight, curiosity, and mobility Reducing immobility time and dyslipidemia	[17]
Bitter orange (*C. aurantium*) essential oil (CAEO)	Antioxidant behavior	DPPH scavenging test	contribute to the prevention of oxidation as antioxidants and free radical scavengers	[187]
	Mood disorder/anxiolytic effect	Collection of medullary material in patients with chronic myeloid leukemia (CML)	Anxiolytic effect and reduces the signs and symptoms associated with anxiety Decrease in the SBP and DBP	[188]
	Diseased condition/premenstrual syndrome (PMS)	Inhalation of 0.5 percent CAEO during the luteal phase of the menstrual cycle	Improved the symptoms of PMS Effective as a new and complementary therapeutic method for the emotions PMS symptoms in female.	[18]
	Sedative and hypnotic effects	Spielberger's State-Trait Anxiety Inventory (STAI) was used after giving bitter orange flower powder capsule to post-menopausal women	Inhaling the CAEO greatly reduced anxiety	[189]
	Reduces pain	Study was a randomized clinical trial conducted with 126 eligible primiparous patients	Controls the enzymes in prostaglandins and reduces pain; controls the contractions caused by oxytocin and prostaglandins and exert anti-uterine pain effects	[190]
Lemon (*C. limon*) essential oil (CLEO)	Antioxidant behavior	DPPH radical scavenging assay	Lemon peel EO showed 55.09% inhibition of DPPH considerable antioxidant properties both in vitro and barley soup as food model	[191]
	Mood disorder/Anxiety	Thirty-nine sophomore nursing students (35 female and 4 males)	Positive effect on cognitive test anxiety	[192]

Table 1. *Cont.*

Citrus Type	Particulars	In Vitro/In Vivo/Animal Model	Activity	References
	Diseased condition/anxiolytic-like effect	Swiss mice model	Induce an anxiolytic behavior in mice no toxicity in vitro	[193]
	Treatment of dysmenorrhea	Population of this study amounted to 185	Psychological and physical benefits	[194]
	Effect on nausea among pregnant women	Control trial on 90 pregnant women	Effective in reducing pregnancy nausea and vomiting	[195]
Mandarin (*C. reticulata*) essential oil (CREO)	Antioxidant behavior	DPPH), 3-(N-morpholino) propane sulfonic acid (ABTS)	Exhibited moderate radical scavenging activity	[196]
	Mood disorder/mood and as a relaxing hypnotic agent	Frontal and parietal skulls of male Wistar rats implanted with electrodes for electroencephalographic (EEG)	CREO reduces REM sleep latency and enhanced the overall time and number of REM sleep episodes	[197]
	Anti-proliferative	Protective effects on bleomycin (BLM)-induced lung fibrosis in rats	Preventive effects on BLM-induced pulmonary fibrosis in rats Anti-proliferative effect against human embryonic lung fibroblasts	[198]
Kaffir lime (*C. hystrix*) essential oil (CHEO)	Antioxidant behavior	DPPH free radical scavenging assay	Potential antioxidant activity	[199]
	Stimulating effect	Forty healthy volunteers participated in the experiments	Reducing depression and stress in humans more alert, attentive, cheerful attitude	[200]
Yuzu (*C. junos*) essential oil (CJEO)	Antioxidant behavior	DPPH free radical scavenging test	Mature yuzu contains higher amounts of vitamin C and phenolics than other citrus fruits Significant dietary source of antioxidants	[201]
	Mood disorder	Inhaled administration (i.h.) of EOCJ for 90 min on mouse	Increased locomotor activity The anxiolytic-like effect	[202]
	Autonomic nervous system (ANS)	Study on seventeen women with subjective premenstrual symptoms	Therapeutic effects of yuzu fragrance on premenstrual symptoms (PMS) Can reduce premenstrual emotional symptoms Increased parasympathetic activity	[203]
	Physiological effect	Effect of 10-min inhalation of the yuzu scent on 21 women	Reduced heart rate (HR) and enhanced high-frequency power of heart rate variability (HRV), exhibiting parasympathetic nervous system activation, alleviation of negative emotional stress	[204]

Table 1. *Cont.*

Citrus Type	Particulars	In Vitro/In Vivo/Animal Model	Activity	References
	Human psychology	32 healthy participants enrolled in the study (16 men and 16 women, aged 20–24 years)	Oxyhemoglobin concentration in the prefrontal cortex increased Task performance improved after inhaling yuzu essential oil	[203]
Neroli (*C. aurantium*) essential oil (CAEO)	Antioxidant behavior	DPPH test	Prevention of oxidation as antioxidants and free radical scavengers. Essential oils in the old leaves had the maximum antioxidant activity	[43]
	Diseased condition/neurological disorder	Study on scopolamine-induced learning and memory deficit in rats	Repairing effects on memory and behavioral disorders Treatment of AD, insomnia, anxiety, and epilepsy	[205]
	Mood disorder/anxiolytic Effect	Study on patients with chronic myeloid leukemia (CML)	Diastolic pressure decreases Exhibits an anxiolytic effect and reduces the signs and symptoms associated with anxiety in patients with CML	[188]
	Antiseizure and anticonvulsant effect	Assessed in pentylenetetrazole (PTZ)-induced in mice	Anticonvulsant activity which supports the ethnomedicinal claims of the use of the plant in the management of seizure	[206]
	Effect on anxiety and perceived pain in women during labor	Study on 88 women during labor	Used as an alternative tool to relieve anxiety and perceived pain in women during all stages of labor	[207]

6. Summary

Citrus EOs are economical, eco-friendly, and natural alternatives to the synthetic compounds used in aromatherapy. Citrus-based EOs are mainly obtained from the leaves, flowers, and peels of young and ripened fruits, indirectly emphasizing waste management to save the environment from pollution and prevent contamination of the underground water table. Citrus EOs from waste peels used in aromatherapy help in relieving stress and stress-related disorders/diseases. The majorly occurring components present in citrus EOs and their therapeutic effects in aromatherapy have been summarized pictorially as below (Figure 16).

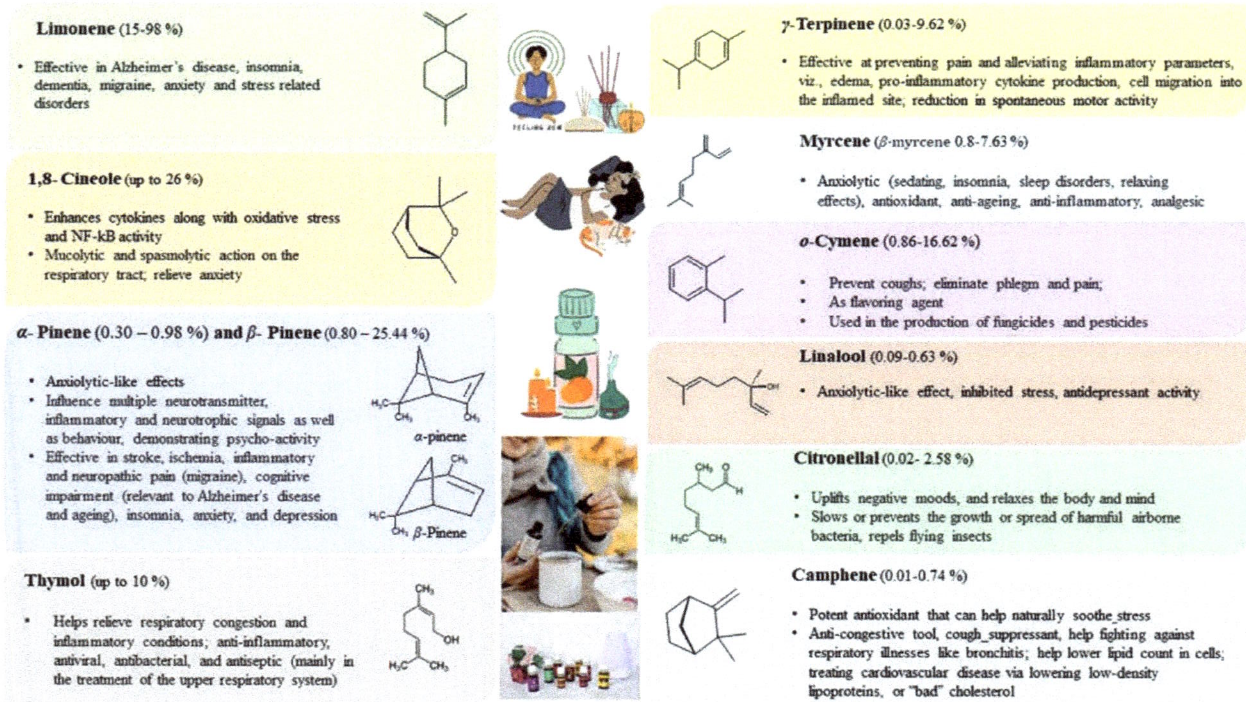

Figure 16. Therapeutic effects of majorly occurring component in citrus essential oil [202,208–211].

Supplementary Materials: The following supporting information can be downloaded at: https://www.mdpi.com/article/10.3390/antiox11122374/s1, Figure S1: Climate sustainability and the annual production of citrus fruits in different geographical regions across the globe; Figure S2: Market segmentation of citrus essential oils; Figure S3: (a) Global citrus oil market by application, by the year 2018, (b) Citrus essential oil market value forecast (Citrus Oil Market by Product Type, 2022); Figure S4: Molecular structures of the volatile and non-volatile components present in Citrus EOs; Figure S5: Composition of EOs in different Citrus varieties; Table S1: Methods/Techniques of extracting Citrus essential oils; Table S2: Methods/techniques of characterization/authentication of Citrus essential oils. References [3,4,14,21,22,24,25,34–37,42,170,212–219] are cited in Supplementary Materials.

Author Contributions: P.A.: conceptualization, writing original draft; Z.S.: designing the schematic diagrams and creating figures; M.K.: conceptualization, writing original draft; A.D.: writing original draft; A.S.: writing—reviewing and editing; K.K.S.: maps and graphical content; M.S.: writing—reviewing; N.M.: content collection, reconstructing text and figures, and editing; A.K.M.: reviewing and editing and resources; K.-H.B.: reviewing, editing, and supervision. All authors have read and agreed to the published version of the manuscript.

Funding: This research was funded by the Rural Development Administration, Republic of Korea, grant number PJ0157260.

Acknowledgments: This work was supported by the Cooperative Research Program for Agriculture Science and Technology Development (project no. PJ015726), Republic of Korea.

Conflicts of Interest: The authors declare that there is no conflict of interest.

References

1. Zayed, A.; Badawy, M.T.; Farag, M.A. Valorization and extraction optimization of Citrus seeds for food and functional food applications. *Food Chem.* **2021**, *355*, 129609. [CrossRef] [PubMed]
2. Fisher, K.; Phillips, C. Potential antimicrobial uses of essential oils in food: Is citrus the answer? *Trends Food Sci. Technol.* **2008**, *19*, 156–164. [CrossRef]
3. Mahato, N.; Sharma, K.; Koteswararao, R.; Sinha, M.; Baral, E.R.; Cho, M.H. Citrus essential oils: Extraction, authentication and application in food preservation. *Crit. Rev. Food Sci. Nutr.* **2019**, *59*, 611–625. [CrossRef] [PubMed]
4. Mahato, N.; Sinha, M.; Sharma, K.; Koteswararao, R.; Cho, M.H. Modern extraction and purification techniques for obtaining high purity food-grade bioactive compounds and value-added co-products from citrus wastes. *Foods* **2019**, *8*, 523. [CrossRef] [PubMed]
5. Ferhat, M.-A.; Boukhatem, M.N.; Hazzit, M.; Meklati, B.Y.; Chemat, F. Cold pressing, hydrodistillation and microwave dry distillation of Citrus essential oil from Algeria: A comparative study. *Electron. J. Biol. S* **2016**, *1*, 30–41.
6. Boughendjioua, H.; Boughendjioua, Z. Chemical composition and biological activity of essential oil of mandarin (*Citrus reticulata*) cultivated in Algeria. *Int. J. Pharm. Sci. Rev. Res.* **2017**, *44*, 179–184.
7. Farrar, A.J.; Farrar, F.C. Clinical Aromatherapy. *Nurs. Clin. N. Am.* **2020**, *55*, 489–504. [CrossRef]
8. Goepfert, M.; Liebl, P.; Herth, N.; Ciarlo, G.; Buentzel, J.; Huebner, J. Aroma oil therapy in palliative care: A pilot study with physiological parameters in conscious as well as unconscious patients. *J. Cancer Res. Clin. Oncol.* **2017**, *143*, 2123–2129. [CrossRef]
9. Fernández, L.F.; Palomino, O.M.; Frutos, G. Effectiveness of Rosmarinus officinalis essential oil as antihypotensive agent in primary hypotensive patients and its influence on health-related quality of life. *J. Ethnopharmacol.* **2014**, *151*, 509–516. [CrossRef]
10. Choi, S.Y.; Kang, P.; Lee, H.S.; Seol, G.H. Effects of Inhalation of Essential Oil of Citrus aurantium L. var. amara on Menopausal Symptoms, Stress, and Estrogen in Postmenopausal Women: A Randomized Controlled Trial. *Evid. Based. Complement. Alternat. Med.* **2014**, *2014*, 796518. [CrossRef]
11. Doweidar, H.; El-Damrawi, G.; El-Stohy, S. Structure and properties of CdO–B_2O_3 and CdO–MnO–B_2O_3 glasses; Criteria of getting the fraction of four coordinated boron atoms from infrared spectra. *Phys. B Condens. Matter* **2017**, *525*, 137–143. [CrossRef]
12. Jimbo, D.; Kimura, Y.; Taniguchi, M.; Inoue, M.; Urakami, K. Effect of aromatherapy on patients with Alzheimer's disease. *Psychogeriatrics* **2009**, *9*, 173–179. [CrossRef]
13. Matsubara, E.; Tsunetsugu, Y.; Ohira, T.; Sugiyama, M. Essential oil of Japanese cedar (*Cryptomeria japonica*) wood increases salivary dehydroepiandrosterone sulfate levels after monotonous work. *Int. J. Environ. Res. Public Health* **2017**, *14*, 97. [CrossRef]
14. Dosoky, N.S.; Setzer, W.N. Biological activities and safety of citrus spp. Essential oils. *Int. J. Mol. Sci.* **2018**, *19*, 1966. [CrossRef]
15. Lin, X.; Cao, S.; Sun, J.; Lu, D.; Zhong, B.; Chun, J. The chemical compositions, and antibacterial and antioxidant activities of four types of Citrus essential oils. *Molecules* **2021**, *26*, 3412. [CrossRef]
16. Badalamenti, N.; Bruno, M.; Schicchi, R.; Geraci, A.; Leporini, M.; Gervasi, L.; Tundis, R.; Loizzo, M.R. Chemical compositions and antioxidant activities of essential oils, and their combinations, obtained from flavedo by-product of seven cultivars of Sicilian *Citrus aurantium* L. *Molecules* **2022**, *27*, 1580. [CrossRef]

17. Zhang, L.L.; Yang, Z.Y.; Fan, G.; Ren, J.N.; Yin, K.J.; Pan, S.Y. Antidepressant-like Effect of *Citrus sinensis* (L.) Osbeck Essential Oil and Its Main Component Limonene on Mice. *J. Agric. Food Chem.* **2019**, *67*, 13817–13828. [CrossRef]
18. Heydari, N.; Abootalebi, M.; Jamalimoghadam, N.; Kasraeian, M.; Emamghoreishi, M.; Akbarzadeh, M. Investigation of the effect of aromatherapy with *Citrus aurantium* blossom essential oil on premenstrual syndrome in university students: A clinical trial study. *Complement. Ther. Clin. Pract.* **2018**, *32*, 1–5. [CrossRef]
19. Rombolà, L.; Scuteri, D.; Adornetto, A.; Straface, M.; Sakurada, T.; Sakurada, S.; Mizoguchi, H.; Corasaniti, M.T.; Bagetta, G.; Tonin, P.; et al. Anxiolytic-like effects of Bergamot Essential Oil are insensitive to Flumazenil in Rats. *Evid.-Based Complement. Altern. Med.* **2019**, *2019*, 2156873. [CrossRef]
20. Bora, H.; Kamle, M.; Mahato, D.K.; Tiwari, P.; Kumar, P. Citrus essential oils (CEOs) and their applications in food: An overview. *Plants* **2020**, *9*, 357. [CrossRef]
21. Dima, C.; Dima, S. Essential oils in foods: Extraction, stabilization, and toxicity. *Curr. Opin. Food Sci.* **2015**, *5*, 29–35. [CrossRef]
22. Global Citrus Oil Market: By Origin: Organic, Conventional; By Oil Type: Orange, Lemon, Lime, and Bergamot Oil, Others; By Application: Food and Beverage, Cosmetic and Personal Care, Pharmaceuticals, Others; Regional Analysis; Historical Market and Forecast (2018–2028); Competitive Landscape; Industry Events and Developments; 2022. Available online: https://www.expertmarketresearch.com/reports/citrus-oil-market (accessed on 22 October 2022).
23. Shahbandeh, M. World Production of Citrus Fruits in 2020, by Region (in Thousand Metric Tons). 2022. Available online: https://www.statista.com/statistics/264002/production-of-citrus-fruits-worldwide-by-region/ (accessed on 22 October 2022).
24. Citrus Oil Market by Product Types (Orange Oil, Lemon Lime Oil, Others), by Application (Food & Beverages, Personal Care, Cosmetics, Aromatherapy & Others), by Extraction Method, by Application, by Sales Channel & By Region. Available online: https://www.factmr.com/report/279/citrus-oil-market (accessed on 22 October 2022).
25. Global And United States Citrus Oil Market Report & Forecast 2022–2028. 2022. Available online: https://www.industryresearch.biz/global-and-united-states-citrus-oil-market-20679913 (accessed on 22 October 2022).
26. Simoes, A.J.G.; Hidalgo, C.A. The economic complexity observatory: An analytical tool for understanding the dynamics of economic development. In Proceedings of the Workshops at the Twenty-Fifth AAAI Conference on Artificial Intelligence, San Francisco, CA, USA, 7–8 August 2011; pp. 39–42.
27. Bousbia, N.; Vian, M.A.; Ferhat, M.A.; Meklati, B.Y.; Chemat, F. A new process for extraction of essential oil from Citrus peels: Microwave hydrodiffusion and gravity. *J. Food Eng.* **2009**, *90*, 409–413. [CrossRef]
28. Yamauchi, Y.; Saito, M. Fractionation of lemon-peel oil by semi-preparative supercritical fluid chromatography. *J. Chromatogr. A* **1990**, *505*, 237–246. [CrossRef]
29. Aresta, A.; Cotugno, P.; De Vietro, N.; Massari, F.; Zambonin, C. Determination of polyphenols and vitamins in wine-making by-products by supercritical fluid extraction (SFE). *Anal. Lett.* **2020**, *53*, 2585–2595. [CrossRef]
30. Yousefi, M.; Rahimi-Nasrabadi, M.; Pourmortazavi, S.M.; Wysokowski, M.; Jesionowski, T.; Ehrlich, H.; Mirsadeghi, S. Supercritical fluid extraction of essential oils. *TrAC—Trends Anal. Chem.* **2019**, *118*, 182–193. [CrossRef]
31. Rajput, S.; Kaur, S.; Panesar, P.S.; Thakur, A. Supercritical fluid extraction of essential oils from Citrus reticulata peels: Optimization and characterization studies. *Biomass Convers. Biorefinery* **2022**, 1–10. [CrossRef]
32. Rubiolo, P.; Sgorbini, B.; Liberto, E.; Cordero, C.; Bicchi, C. Essential oils and volatiles: Sample preparation and analysis. A review. *Flavour Fragr. J.* **2010**, *25*, 282–290. [CrossRef]
33. Brinkman, U.A.T. Hyphenation: Hype and Fascination. *J. Chromatogr.* **1999**, *856*, 535.
34. Hong, J.H.; Khan, N.; Jamila, N.; Hong, Y.S.; Nho, E.Y.; Choi, J.Y.; Lee, C.M.; Kim, K.S. Determination of volatile flavour profiles of *Citrus* spp. Fruits by SDE-GC–MS and enantiomeric composition of chiral compounds by MDGC–MS. *Phytochem. Anal.* **2017**, *28*, 392–403. [CrossRef]
35. Bonaccorsi, I.; Sciarrone, D.; Schipilliti, L.; Dugo, P.; Mondello, L.; Dugo, G. Multidimensional enantio gas chromtography/mass spectrometry and gas chromatography-combustion-isotopic ratio mass spectrometry for the authenticity assessment of Lime essential oils (*C. aurantifolia* Swingle and *C. latifolia* Tanaka). *J. Chromatogr. A* **2012**, *1226*, 87–95. [CrossRef]
36. Schipilliti, L.; Bonaccorsi, I.; Cotroneo, A.; Dugo, P.; Mondello, L. Evaluation of gas chromatography-combustion-isotope ratio mass spectrometry (GC-C-IRMS) for the quality assessment of *Citrus liqueurs. J. Agric. Food Chem.* **2013**, *61*, 1661–1670. [CrossRef]
37. Marti, G.; Boccard, J.; Mehl, F.; Debrus, B.; Marcourt, L.; Merle, P.; Delort, E.; Baroux, L.; Sommer, H.; Rudaz, S.; et al. Comprehensive profiling and marker identification in non-volatile citrus oil residues by mass spectrometry and nuclear magnetic resonance. *Food Chem.* **2014**, *150*, 235–245. [CrossRef]
38. González-Mas, M.C.; Rambla, J.L.; López-Gresa, M.P.; Amparo Blázquez, M.; Granell, A. Volatile compounds in citrus essential oils: A comprehensive review. *Front. Plant Sci.* **2019**, *10*, 12. [CrossRef]
39. Tranchida, P.Q.; Bonaccorsi, I.; Dugo, P.; Mondello, L.; Dugo, G. Analysis of Citrus essential oils: State of the art and future perspectives. A review. *Flavour Fragr. J.* **2012**, *27*, 98–123. [CrossRef]
40. Palazzolo, E.; Armando Laudicina, V.; Antonietta Germanà, M. Current and potential use of Citrus Essential Oils. *Curr. Org. Chem.* **2013**, *17*, 3042–3049. [CrossRef]
41. Md Othman, S.; Hassan, M.; Nahar, L.; Basar, N.; Jamil, S.; Sarker, S. Essential Oils from the Malaysian Citrus (Rutaceae) Medicinal Plants. *Medicines* **2016**, *3*, 13. [CrossRef]
42. Bonaccorsi, I.; Sciarrone, D.; Cotroneo, A.; Mondello, L.; Dugo, P.; Dugo, G. Enantiomeric distribution of key volatile components in Citrus essential oils. *Rev. Bras. Farmacogn.* **2011**, *21*, 841–849. [CrossRef]

43. Sarrou, E.; Chatzopoulou, P.; Dimassi-Theriou, K.; Therios, I. Volatile constituents and antioxidant activity of peel, flowers and leaf oils of Citrus aurantium L. growing in Greece. *Molecules* **2013**, *18*, 10639–10647. [CrossRef]
44. Svoboda, K.P.; Greenaway, R.I. Lemon scented plants. *Int. J. Aromather.* **2003**, *13*, 23–32. [CrossRef]
45. Modzelewska, A.; Sur, S.; Kumar, S.K.; Khan, S.R. Sesquiterpenes: Natural products that decrease cancer growth. *Curr. Med. Chem.—Anti-Cancer Agents* **2005**, *5*, 477–499. [CrossRef]
46. Perry, N.; Perry, E. Aromatherapy in the management of psychiatric disorders. *CNS Drugs* **2006**, *20*, 257–280. [CrossRef]
47. Lillehei, A.S.; Halcon, L.L. A systematic review of the effect of inhaled essential oils on sleep. *J. Altern. Complement. Med.* **2014**, *20*, 441–451. [CrossRef]
48. Shiina, Y.; Funabashi, N.; Lee, K.; Toyoda, T.; Sekine, T.; Honjo, S.; Hasegawa, R.; Kawata, T.; Wakatsuki, Y.; Hayashi, S.; et al. Relaxation effects of lavender aromatherapy improve coronary flow velocity reserve in healthy men evaluated by transthoracic Doppler echocardiography. *Int. J. Cardiol.* **2008**, *129*, 193–197. [CrossRef]
49. Kotlik, B.A.; Petrova, L.M.; Red'ko, A.I.; Elu, K. Sources of equipment measurement error in stimulation electromyography. *Fiziol. Cheloveka* **1986**, *12*, 691–694.
50. Zimmerman, W.B. Sleep mentation and auditory awakening thresholds. *Psychophysiology* **1970**, *6*, 540–549. [CrossRef] [PubMed]
51. Momjian, S.; Tyrand, R.; Landis, B.N.; Boëx, C. Intraoperative monitoring of olfactory function: A feasibility study. *J. Neurosurg.* **2019**, *132*, 1659–1664. [CrossRef] [PubMed]
52. Marciniak-Firadza, R. The Sense of Smell in Logopaedic Theory and Practice. *Logop. Lodz.* **2021**, *5*, 123–145. [CrossRef]
53. Naidich, T.P.; Som, P.M. The Olfactory System: Part III: Central processing of olfaction and central olfactory dysfunction. *Neurographics* **2018**, *8*, 298–316. [CrossRef]
54. Loos, H.M.; Schreiner, L.; Karacan, B. A systematic review of physiological responses to odours with a focus on current methods used in event-related study designs. *Int. J. Psychophysiol.* **2020**, *158*, 143–157. [CrossRef]
55. Liao, W.-C.; Yao, R.-A.; Chen, L.-Y.; Renn, T.-Y.; Klimenkov, I.V.; Sudakov, N.P.; Mai, F.-D.; Chen, Y.-T.; Chang, H.-M. Olfactory stimulation successfully modulates the neurochemical, biochemical and behavioral phenotypes of the visceral pain. *Molecules* **2022**, *27*, 7659. [CrossRef]
56. Llorens, J. The physiology of taste and smell: How and why we sense flavors. *Water Sci. Technol.* **2004**, *49*, 1–10. [CrossRef]
57. Lee, A.J.; Ling, R.S. Improved cementing techniques. *Instr. Course Lect.* **1981**, *30*, 407–413.
58. Fung, T.K.H.; Lau, B.W.M.; Ngai, S.P.C.; Tsang, H.W.H. Therapeutic effect and mechanisms of essential oils in mood disorders: Interaction between the nervous and respiratory systems. *Int. J. Mol. Sci.* **2021**, *22*, 4844. [CrossRef]
59. Fang, T.; Goto, M.; Sasaki, M.; Hirose, T. Combination of supercritical CO_2 and vacuum distillation for the fractionation of bergamot oil. *J. Agric. Food Chem.* **2004**, *52*, 5162–5167. [CrossRef]
60. Reverchon, E.; Iacuzio, G. Supercritical desorption of bergamot peel oil from silica gel—Experiments and mathematical modelling. *Chem. Eng. Sci.* **1997**, *52*, 3553–3559. [CrossRef]
61. Adokoh, C.K.; Asante, D.B.; Acheampong, D.O.; Kotsuchibashi, Y.; Armah, F.A.; Sirikyi, I.H.; Kimura, K.; Gmakame, E.; Abdul-Rauf, S. Chemical profile and in vivo toxicity evaluation of unripe Citrus aurantifolia essential oil. *Toxicol. Rep.* **2019**, *6*, 692–702. [CrossRef]
62. Smith, C.A.; Collins, C.T.; Crowther, C.A. Aromatherapy for pain management in labour. *Cochrane Database Syst. Rev.* **2011**, *6*, CD009215. [CrossRef]
63. George, A.; Sanjay, M.R.; Srisuk, R.; Parameswaranpillai, J.; Siengchin, S. A comprehensive review on chemical properties and applications of biopolymers and their composites. *Int. J. Biol. Macromol.* **2020**, *154*, 329–338. [CrossRef]
64. Lai, T.K.T.; Cheung, M.C.; Lo, C.K.; Ng, K.L.; Fung, Y.H.; Tong, M.; Yau, C.C. Effectiveness of aroma massage on advanced cancer patients with constipation: A pilot study. *Complement. Ther. Clin. Pract.* **2011**, *17*, 37–43. [CrossRef]
65. Steps, S.B. An Introductory Guide to 1000′s of Uses for Essential Oils. Available online: http://www.sustainablebabysteps.com/uses-for-essential-oils.html (accessed on 14 October 2022).
66. Mollace, V.; Sacco, I.; Janda, E.; Malara, C.; Ventrice, D.; Colica, C.; Visalli, V.; Muscoli, S.; Ragusa, S.; Muscoli, C. Hypolipemic and hypoglycaemic activity of bergamot polyphenols: From animal models to human studies. *Fitoterapia* **2011**, *82*, 309–316. [CrossRef]
67. Impellizzeri, D.; Bruschetta, G.; Di Paola, R.; Ahmad, A.; Campolo, M.; Cuzzocrea, S.; Esposito, E.; Navarra, M. The anti-inflammatory and antioxidant effects of bergamot juice extract (BJe) in an experimental model of inflammatory bowel disease. *Clin. Nutr.* **2015**, *34*, 1146–1154. [CrossRef] [PubMed]
68. Delle Monache, S.; Sanità, P.; Trapasso, E.; Ursino, M.R.; Dugo, P.; Russo, M.; Ferlazzo, N.; Calapai, G.; Angelucci, A.; Navarra, M. Mechanisms underlying the anti-tumoral effects of *Citrus bergamia* juice. *PLoS ONE* **2013**, *8*, e61484. [CrossRef] [PubMed]
69. Navarra, M.; Ursino, M.R.; Ferlazzo, N.; Russo, M.; Schumacher, U.; Valentiner, U. Effect of *Citrus bergamia* juice on human neuroblastoma cells *in vitro* and in metastatic xenograft models. *Fitoterapia* **2014**, *95*, 83–92. [CrossRef] [PubMed]
70. Visalli, G.; Ferlazzo, N.; Cirmi, S.; Campiglia, P.; Gangemi, S.; Di Pietro, A.; Calapai, G.; Navarra, M. Bergamot juice extract inhibits proliferation by inducing apoptosis in human colon cancer cells. *Anticancer. Agents Med. Chem.* **2014**, *14*, 1402–1413. [CrossRef] [PubMed]
71. Pendino, G.M. Il bergamotto in terapia medica: Attualità e prospettive. *Essenze Deriv. Agrum.* **1998**, *68*, 57–62.
72. Fisher, K.; Phillips, C.A. The effect of Lemon, Orange and Bergamot essential oils and their components on the survival of *Campylobacter jejuni*, *Escherichia coli* O_{157}, *Listeria monocytogenes*, *Bacillus cereus* and *Staphylococcus aureus* in vitro and in food systems. *J. Appl. Microbiol.* **2006**, *101*, 1232–1240. [CrossRef]

73. Karaca, M.; Özbek, H.; Him, A.; Tütüncü, M.; Akkan, H.A.; Kaplanoğlu, V. Investigation of anti-inflammatory activity of bergamot oil. *Eur. J. Gen. Med.* **2007**, *4*, 176–179.
74. Sanguinetti, M.; Posteraro, B.; Romano, L.; Battaglia, F.; Lopizzo, T.; De Carolis, E.; Fadda, G. In vitro activity of *Citrus bergamia* (bergamot) oil against clinical isolates of dermatophytes. *J. Antimicrob. Chemother.* **2007**, *59*, 305–308. [CrossRef]
75. Romano, L.; Battaglia, F.; Masucci, L.; Sanguinetti, M.; Posteraro, B.; Plotti, G.; Zanetti, S.; Fadda, G. *In vitro* activity of bergamot natural essence and furocoumarin-free and distilled extracts, and their associations with boric acid, against clinical yeast isolates. *J. Antimicrob. Chemother.* **2005**, *55*, 110–114. [CrossRef]
76. Sánchez-González, L.; Cháfer, M.; Chiralt, A.; González-Martínez, C. Physical properties of edible chitosan films containing bergamot essential oil and their inhibitory action on *Penicillium italicum*. *Carbohydr. Polym.* **2010**, *82*, 277–283. [CrossRef]
77. Occhiuto, F.; Circosta, C. Antianginal and antiarrhythmic effects of bergamottine, a furocoumarin isolated from Bergamot oil. *Phyther. Res.* **1996**, *10*, 491–496. [CrossRef]
78. Kang, P.; Suh, S.H.; Min, S.S.; Seol, G.H. The essential oil of *Citrus bergamia* Risso induces vasorelaxation of the mouse aorta by activating K+ channels and inhibiting Ca^{2+} influx. *J. Pharm. Pharmacol.* **2013**, *65*, 745–749. [CrossRef]
79. You, T.; Arsenis, N.C.; Disanzo, B.L.; LaMonte, M.J. Effects of exercise training on chronic inflammation in obesity. *Sports Med.* **2013**, *43*, 243–256. [CrossRef]
80. Mannucci, C.; Navarra, M.; Calapai, F.; Squeri, R.; Gangemi, S.; Calapai, G. Clinical Pharmacology of *Citrus bergamia:* A Systematic Review. *Phyther. Res.* **2017**, *31*, 27–39. [CrossRef]
81. Karn, A.; Zhao, C.; Yang, F.; Cui, J.; Gao, Z.; Wang, F.; Xiao, H.; Zheng, J. In-vivo biotransformation of citrus functional components and their effects on health. *Crit. Rev. Food Sci. Nutr.* **2021**, *61*, 756–776. [CrossRef]
82. Schmidt, L.; Göen, T. Human metabolism of α-pinene and metabolite kinetics after oral administration. *Arch. Toxicol.* **2017**, *91*, 677–687. [CrossRef]
83. Igimi, H.; Nishimura, M.; Kodama, R.; Ide, H. Studies on the metabolism of d-limonene (p-mentha-1, 8-diene): I. the absorption, distribution and excretion of d-limonene in rats. *Xenobiotica* **1974**, *4*, 77–84. [CrossRef]
84. Chen, T.C.; Da Fonseca, C.O.; Schönthal, A.H. Preclinical development and clinical use of perillyl alcohol for chemoprevention and cancer therapy. *Am. J. Cancer Res.* **2015**, *5*, 1580.
85. Jayaprakasha, G.K.; Murthy, K.N.C.; Uckoo, R.M.; Patil, B.S. Chemical composition of volatile oil from *Citrus limettioides* and their inhibition of colon cancer cell proliferation. *Ind. Crop. Prod.* **2013**, *45*, 200–207. [CrossRef]
86. Elegbede, J.A.; Maltzman, T.H.; Verma, A.K.; Tanner, M.A.; Elson, C.E.; Gould, M.N. Mouse skin tumor promoting activity of orange peel oil and d-limonene: A re-evaluation. *Carcinogenesis* **1986**, *7*, 2047–2049. [CrossRef]
87. Guo, P.; Li, P.; Zhang, X.; Liu, N.; Wang, J.; Yang, S.; Yu, L.; Zhang, W. The effectiveness of aromatherapy on preoperative anxiety in adults: A systematic review and meta-analysis of randomized controlled trials. *Int. J. Nurs. Stud.* **2020**, *111*, 103747. [CrossRef] [PubMed]
88. Patil, J.R.; Jayaprakasha, G.K.; Murthy, K.N.C.; Tichy, S.E.; Chetti, M.B.; Patil, B.S. Apoptosis-mediated proliferation inhibition of human colon cancer cells by volatile principles of *Citrus aurantifolia*. *Food Chem.* **2009**, *114*, 1351–1358. [CrossRef]
89. Bardon, S.; Foussard, V.; Fournel, S.; Loubat, A. Monoterpenes inhibit proliferation of human colon cancer cells by modulating cell cycle-related protein expression. *Cancer Lett.* **2002**, *181*, 187–194. [CrossRef] [PubMed]
90. Karlson, J.; Borg-Karlson, A.-K.; Unelius, R.; Shoshan, M.C.; Wilking, N.; Ringborg, U.; Linder, S. Inhibition of tumor cell growth by monoterpenes in vitro: Evidence of a Ras-independent mechanism of action. *Anticancer. Drugs* **1996**, *7*, 422–429. [CrossRef] [PubMed]
91. Feng, K.; Zhu, X.; Liu, G.; Kan, Q.; Chen, T.; Chen, Y.; Cao, Y. Dietary citrus peel essential oil ameliorates hypercholesterolemia and hepatic steatosis by modulating lipid and cholesterol homeostasis. *Food Funct.* **2020**, *11*, 7217–7230. [CrossRef]
92. Zhang, Y.; Gong, J.; Yu, H.; Guo, Q.; Defelice, C.; Hernandez, M.; Yin, Y.; Wang, Q. Alginate-whey protein dry powder optimized for target delivery of essential oils to the intestine of chickens. *Poult. Sci.* **2014**, *93*, 2514–2525. [CrossRef]
93. Marques, F.M.; Figueira, M.M.; Schmitt, E.F.P.; Kondratyuk, T.P.; Endringer, D.C.; Scherer, R.; Fronza, M. In vitro anti-inflammatory activity of terpenes via suppression of superoxide and nitric oxide generation and the NF-κB signalling pathway. *Inflammopharmacology* **2019**, *27*, 281–289. [CrossRef]
94. Price, L. How essential oils enter the body. In *Aromatherapy for Health Professionals*, 4th ed.; Price, S., Price, L., Eds.; Churchill Livingstone/Elsevier: Edinburgh, Scotland, 2012; pp. 125–139. ISBN 9780702035647.
95. Buchbauer, G. Molecular interaction: Biological effects and modes of action of essential oils. *Int. J. Aromather.* **1993**, *5*, 11–14.
96. Parker, R.A.; Gabriel, K.T.; Graham, K.D.; Butts, B.K.; Cornelison, C.T. Antifungal activity of select essential oils against *Candida auris* and Their interactions with antifungal drugs. *Pathogens* **2022**, *11*, 821. [CrossRef]
97. Shutes, J.; Galper, A. *The Aromatherapy Companion: A Portable Guide to Blending Essential Oils and Crafting Remedies for Body, Mind, and Spirit*; Shutes, J., Galper, A., Eds.; Fair Winds Press: Beverly, MA, USA, 2022; ISBN 9780760377925.
98. Scuteri, D.; Watanabe, C.; Sakurada, S.; Hamamura, K.; Sakurada, T.; Tonin, P.; Bagetta, G.; Corasaniti, M.T. Pharmacotechnological Advances for clinical translation of essential oils for the treatment of pain and agitation in severe dementia. *Processes* **2022**, *10*, 1340. [CrossRef]
99. Mohammed, M.; Badeggi, U.M.; Mathew, J.T.; Umar, T.M. Therapeutic benefits of selected plant herbs with essential oils. *Lapai J. Sci. Technol.* **2021**, *7*, 158–168.

100. Wilkinson, S.M.; Love, S.B.; Westcombe, A.M.; Gambles, M.A.; Burgess, C.C.; Cargill, A.; Young, T.; Maher, E.J.; Ramirez, A.J. Effectiveness of aromatherapy massage in the management of anxiety and depression in patients with cancer: A multicenter randomized controlled trial. *J. Clin. Oncol.* **2007**, *25*, 532–539. [CrossRef]
101. Halcón, L.L. Aromatherapy: Therapeutic applications of plant essential oils. *Minn. Med.* **2002**, *85*, 42–46.
102. Viana, M.D.M.; Cardoso, R.M.; Silva, N.K.G.T.; Falcão, M.A.P.; Vieira, A.C.S.; Alexandre-Moreira, M.S.; Campesatto, E.A. Anxiolytic-like effect of *Citrus limon* (L.) Burm f. essential oil inhalation on mice. *Rev. Bras. Plantas Med.* **2016**, *18*, 96–104. [CrossRef]
103. Chen, M.; Chen, Y.; Lee, H. The effect of bergamot essential oil aromatherapy on improving depressive mood and sleep quality in postpartum women: A randomized controlled trial. *J. Nurs. Res.* **2022**, *30*, e201. [CrossRef]
104. Igarashi, M.; Ikei, H.; Song, C.; Miyazaki, Y. Effects of olfactory stimulation with rose and orange oil on prefrontal cortex activity. *Complement. Ther. Med.* **2014**, *22*, 1027–1031. [CrossRef]
105. Keyhanmehr, A.S.; Kolouri, S.; Heydarirad, G.; Mofid, B.; Mosavat, S.H. Aromatherapy for the management of cancer complications: A narrative review. *Complement. Ther. Clin. Pract.* **2018**, *31*, 175–180. [CrossRef]
106. Dyer, J.; Cleary, L.; Ragsdale-Lowe, M.; McNeill, S.; Osland, C. The use of aromasticks at a cancer centre: A retrospective audit. *Complement. Ther. Clin. Pract.* **2014**, *20*, 203–206. [CrossRef]
107. Kliszcz, A.; Danel, A.; Puła, J.; Barabasz-Krasny, B.; Możdżeń, K. Fleeting beauty—The world of plant fragrances and their application. *Molecules* **2021**, *26*, 2473. [CrossRef]
108. Sebghatollahi, Z.; Ghanadian, M.; Agarwal, P.; Ghaheh, H.S.; Mahato, N.; Yogesh, R.; Hejazi, S.H. Citrus flavonoids: Biological activities, implementation in skin health, and topical applications: A Review. *ACS Food Sci. Technol.* **2022**, *2*, 1417–1432. [CrossRef]
109. Herman, A.; Herman, A.P. Essential oils and their constituents as skin penetration enhancer for transdermal drug delivery: A review. *J. Pharm. Pharmacol.* **2015**, *67*, 473–485. [CrossRef] [PubMed]
110. Valgimigli, L.; Gabbanini, S.; Berlini, E.; Lucchi, E.; Beltramini, C.; Bertarelli, Y.L. Lemon (*Citrus limon*, Burm.f.) essential oil enhances the trans-epidermal release of lipid-(A, E) and water-(B_6, C) soluble vitamins from topical emulsions in reconstructed human epidermis. *Int. J. Cosmet. Sci.* **2012**, *34*, 347–356. [CrossRef] [PubMed]
111. Jiang, Q.; Wu, Y.; Zhang, H.; Liu, P.; Yao, J.; Yao, P.; Chen, J.; Duan, J. Development of essential oils as skin permeation enhancers: Penetration enhancement effect and mechanism of action. *Pharm. Biol.* **2017**, *55*, 1592–1600. [CrossRef] [PubMed]
112. Parhi, R.; Suresh, P.; Mondal, S.; Kumar, P.M. Novel Penetration Enhancers for Skin Applications: A Review. *Curr. Drug Deliv.* **2012**, *9*, 219–230. [CrossRef] [PubMed]
113. Furuishi, T.; Kato, Y.; Fukami, T.; Suzuki, T.; Endo, T.; Nagase, H.; Ueda, H.; Tomono, K. Effect of terpenes on the skin permeation of lomerizine dihydrochloride. *J. Pharm. Pharm. Sci.* **2013**, *16*, 551–563. [CrossRef] [PubMed]
114. Moghadam, S.H.; Saliaj, E.; Wettig, S.D.; Dong, C.; Ivanova, M.V.; Huzil, J.T.; Foldvari, M. Effect of chemical permeation enhancers on stratum corneum barrier lipid organizational structure and interferon alpha permeability. *Mol. Pharm.* **2013**, *10*, 2248–2260. [CrossRef]
115. Bech-Thomsen, N.; Wulf, H.C. Carcinogenic and melanogenic effects of a filtered metal halide UVA source and a tubular fluorescent UVA tanning source with or without additional solar-simulated UV radiation in hairless mice. *Photochem. Photobiol.* **1995**, *62*, 773–779. [CrossRef]
116. Kligman, A.M. Early destructive effect of sunlight on human skin. *Jama* **1969**, *210*, 2377–2380. [CrossRef]
117. Oikarinen, A.; Karvonen, J.; Uitto, J.; Hannuksela, M. Connective tissue alterations in skin exposed to natural and therapeutic UV-radiation. *Photodermatology* **1985**, *2*, 15–26.
118. Sakurai, H.; Yasui, H.; Yamada, Y.; Nishimura, H.; Shigemoto, M. Detection of reactive oxygen species in the skin of live mice and rats exposed to UVA light: A research review on chemiluminescence and trials for UVA protection. *Photochem. Photobiol. Sci.* **2005**, *4*, 715–720. [CrossRef]
119. Watson, R.E.B.; Griffiths, C.E.M. Pathogenic aspects of cutaneous photoaging. *J. Cosmet. Dermatol.* **2005**, *4*, 230–236. [CrossRef]
120. Okayama, Y. Oxidative stress in allergic and inflammatory skin diseases. *Curr. Drug Targets Inflamm. Allergy* **2005**, *4*, 517–519. [CrossRef]
121. Ishfaq, M.; Akhtar, B.; Muhammad, F.; Sharif, A.; Akhtar, M.F.; Hamid, I.; Sohail, K.; Muhammad, H. Antioxidant and Wound healing potential of essential oil from *Citrus reticulata* peel and its chemical characterization. *Curr. Pharm. Biotechnol.* **2020**, *22*, 1114–1121. [CrossRef]
122. Simitzis, P.E. Enrichment of animal diets with essential oils—A great perspective on improving animal performance and quality characteristics of the derived products. *Medicines* **2017**, *4*, 35. [CrossRef]
123. Giannenas, I.; Bonos, E.; Christaki, E.; Florou-Paneri, P. Oregano: A feed additive with functional properties. In *Therapeutic Foods*; Holban, A.M., Grumezescu, A.M., Eds.; Academic Press: San Diego, CA, USA, 2018; pp. 179–208. ISBN 9780128115176.
124. Del Prado-Audelo, M.L.; Cortés, H.; Caballero-Floran, I.H., González-Torres, M.; Escutia-Guadarrama, L.; Bernal-Chávez, S.A.; Giraldo-Gomez, D.M.; Magaña, J.J.; Leyva-Gómez, G. Therapeutic applications of terpenes on inflammatory diseases. *Front. Pharmacol.* **2021**, *12*, 2114. [CrossRef]
125. Maurya, A.K.; Mohanty, S.; Pal, A.; Chanotiya, C.S.; Bawankule, D.U. The essential oil from *Citrus limetta* Risso peels alleviates skin inflammation: In-vitro and in-vivo study. *J. Ethnopharmacol.* **2018**, *212*, 86–94. [CrossRef]
126. Li, Z.H.; Cai, M.; Liu, Y.S.; Sun, P.L.; Luo, S.L. Antibacterial activity and mechanisms of essential oil from *Citrus medica* L. var. sarcodactylis. *Molecules* **2019**, *24*, 1577. [CrossRef]

127. Hirota, R.; Roger, N.N.; Nakamura, H.; Song, H.S.; Sawamura, M.; Suganuma, N. Anti-inflammatory effects of limonene from yuzu (*Citrus junos* tanaka) essential oil on eosinophils. *J. Food Sci.* **2010**, *75*, H87–H92. [CrossRef]
128. Bertuzzi, G.; Tirillini, B.; Angelini, P.; Venanzoni, R. Antioxidative Action of Citrus limonum Essential Oil on Skin. *European J. Med. Plants* **2013**, *3*, 1–9. [CrossRef]
129. Ruberto, G.; Baratta, M.T. Antioxidant activity of selected essential oil components in two lipid model systems. *Food Chem.* **2000**, *69*, 167–174. [CrossRef]
130. El Bouamri, M.; Gorrichon, J.P.; Braun, A.M.; Oliveros, E. The reactivity of citronellol and α-thujene with singlet oxygen. rate constants of chemical reaction and physical quenching. *Photochem. Photobiol.* **1991**, *54*, 619–623. [CrossRef]
131. Ambrosio, C.M.S.; Diaz-Arenas, G.L.; Agudelo, L.P.A.; Stashenko, E.; Contreras-Castillo, C.J.; Da Gloria, E.M. Chemical composition and antibacterial and antioxidant activity of a citrus essential oil and its fractions. *Molecules* **2021**, *26*, 2888. [CrossRef]
132. Song, H.-S.; Ukeda, H.; Sawamura, M. Antioxidative activities of citrus peel essential oils and their components against linoleic acid oxidation. *Food Sci. Technol. Res.* **2001**, *7*, 50–56. [CrossRef]
133. Takahashi, Y.; Inaba, N.; Kuwahara, S.; Kuki, W. Antioxidative effect of citrus essential oil components on human low-density lipoprotein in vitro. *Biosci. Biotechnol. Biochem.* **2003**, *67*, 195–197. [CrossRef] [PubMed]
134. Xue, L.; Liu, G.; Parfitt, J.; Liu, X.; Van Herpen, E.; Stenmarck, Å.; O'Connor, C.; Östergren, K.; Cheng, S. Missing Food, Missing Data? A Critical Review of Global Food Losses and Food Waste Data. *Environ. Sci. Technol.* **2017**, *51*, 6618–6633. [CrossRef]
135. Borkovec, T.D.; Ruscio, A.M. Psychotherapy for generalized anxiety disorder. *J. Clin. Psychiatry* **2001**, *62*, 37–45.
136. WHO. Depression. Available online: https://www.who.int/news-room/fact-sheets/detail/depression (accessed on 14 October 2022).
137. Association, A.S. What Is Alzheimer's Disease? Available online: https://www.alz.org/alzheimers-dementia/what-is-alzheimers (accessed on 14 October 2022).
138. Alzheimer's Disease Fact Sheet. National Institute on Aging. Available online: https://www.nia.nih.gov/health/alzheimers-disease-fact-sheet (accessed on 14 October 2022).
139. Santamaria, J.; Iranzo, A. Sleep disorders matter in neurology. *Lancet Neurol.* **2014**, *13*, 18–20. [CrossRef]
140. Freeman, D.; Sheaves, B.; Waite, F.; Harvey, A.G.; Harrison, P.J. Sleep disturbance and psychiatric disorders. *Lancet Psychiatry* **2020**, *7*, 628–637. [CrossRef]
141. Momen, N.C.; Plana-Ripoll, O.; Agerbo, E.; Benros, M.E.; Børglum, A.D.; Christensen, M.K.; Dalsgaard, S.; Degenhardt, L.; de Jonge, P.; Debost, J.-C.P.G. Association between mental disorders and subsequent medical conditions. *N. Engl. J. Med.* **2020**, *382*, 1721–1731. [CrossRef]
142. Muz, G.; Taşcı, S. Effect of aromatherapy via inhalation on the sleep quality and fatigue level in people undergoing hemodialysis. *Appl. Nurs. Res.* **2017**, *37*, 28–35. [CrossRef]
143. Takeda, A.; Watanuki, E.; Koyama, S. Effects of inhalation aromatherapy on symptoms of sleep disturbance in the elderly with dementia. *Evid.-Based Complement. Altern. Med.* **2017**, *2017*, 1902807. [CrossRef]
144. Tang, Y.; Gong, M.; Qin, X.; Su, H.; Wang, Z.; Dong, H. The therapeutic effect of aromatherapy on insomnia: A meta-analysis. *J. Affect. Disord.* **2021**, *288*, 1–9. [CrossRef]
145. Pause, B.M.; Miranda, A.; Göder, R.; Aldenhoff, J.B.; Ferstl, R. Reduced olfactory performance in patients with major depression. *J. Psychiatr. Res.* **2001**, *35*, 271–277. [CrossRef]
146. Guillemain, J.; Rousseau, A.; Delaveau, P. Neurodepressive effects of the essential oil of Lavandula angustifolia Mill. *Ann. Pharm. Fr.* **1989**, *47*, 337–343.
147. Tafet, G.E.; Idoyaga-Vargas, V.P.; Abulafia, D.P.; Calandria, J.M.; Roffman, S.S.; Chiovetta, A.; Shinitzky, M. Correlation between cortisol level and serotonin uptake in patients with chronic stress and depression. *Cogn. Affect. Behav. Neurosci.* **2001**, *1*, 388–393. [CrossRef]
148. Raison, C.L.; Capuron, L.; Miller, A.H. Cytokines sing the blues: Inflammation and the pathogenesis of depression. *Trends Immunol.* **2006**, *27*, 24–31. [CrossRef]
149. Chung, S.; Son, G.H.; Kim, K. Circadian rhythm of adrenal glucocorticoid: Its regulation and clinical implications. *Biochim. Biophys. Acta (BBA)-Mol. Basis Dis.* **2011**, *1812*, 581–591. [CrossRef]
150. Slavich, G.M.; Irwin, M.R. From stress to inflammation and major depressive disorder: A social signal transduction theory of depression. *Psychol. Bull.* **2014**, *140*, 774. [CrossRef]
151. Chen, J.; Jiang, Q.D.; Chai, Y.P.; Zhang, H.; Peng, P.; Yang, X.X. Natural terpenes as penetration enhancers for transdermal drug delivery. *Molecules* **2016**, *21*, 1709. [CrossRef]
152. Xiong, M.; Li, Y.; Tang, P.; Zhang, Y.; Cao, M.; Ni, J.; Xing, M. Effectiveness of aromatherapy massage and inhalation on symptoms of depression in Chinese community-dwelling older adults. *J. Altern. Complement. Med.* **2018**, *24*, 717–724. [CrossRef]
153. Shimizu, E.; Hashimoto, K.; Watanabe, H.; Komatsu, N.; Okamura, N.; Koike, K.; Shinoda, N.; Nakazato, M.; Kumakiri, C.; Okada, S. Serum brain-derived neurotrophic factor (BDNF) levels in schizophrenia are indistinguishable from controls. *Neurosci. Lett.* **2003**, *351*, 111–114. [CrossRef]
154. Sánchez-Vidaña, D.I.; Po, K.K.-T.; Fung, T.K.-H.; Chow, J.K.-W.; Lau, W.K.-W.; So, P.-K.; Lau, B.W.-M.; Tsang, H.W.-H. Lavender essential oil ameliorates depression-like behavior and increases neurogenesis and dendritic complexity in rats. *Neurosci. Lett.* **2019**, *701*, 180–192. [CrossRef]

155. Ruan, L.; Lau, B.W.-M.; Wang, J.; Huang, L.; ZhuGe, Q.; Wang, B.; Jin, K.; So, K.-F. Neurogenesis in neurological and psychiatric diseases and brain injury: From bench to bedside. *Prog. Neurobiol.* **2014**, *115*, 116–137. [CrossRef] [PubMed]
156. Malberg, J.E.; Eisch, A.J.; Nestler, E.J.; Duman, R.S. Chronic antidepressant treatment increases neurogenesis in adult rat hippocampus. *J. Neurosci.* **2000**, *20*, 9104–9110. [CrossRef] [PubMed]
157. Ayuob, N.N. Evaluation of the antidepressant-like effect of musk in an animal model of depression: How it works. *Anat. Sci. Int.* **2017**, *92*, 539–553. [CrossRef] [PubMed]
158. Atsumi, T.; Tonosaki, K. Smelling lavender and rosemary increases free radical scavenging activity and decreases cortisol level in saliva. *Psychiatry Res.* **2007**, *150*, 89–96. [CrossRef]
159. Saiyudthong, S.; Marsden, C.A. Acute effects of bergamot oil on anxiety-related behaviour and corticosterone level in rats. *Phyther. Res.* **2011**, *25*, 858–862. [CrossRef]
160. Zhong, Y.; Zheng, Q.; Hu, P.; Huang, X.; Yang, M.; Ren, G.; Du, Q.; Luo, J.; Zhang, K.; Li, J.; et al. Sedative and hypnotic effects of compound Anshen essential oil inhalation for insomnia. *BMC Complement. Altern. Med.* **2019**, *19*, 306. [CrossRef]
161. Rombolà, L.; Tridico, L.; Scuteri, D.; Sakurada, T.; Sakurada, S.; Mizoguchi, H.; Avato, P.; Corasaniti, M.T.; Bagetta, G.; Morrone, L.A. Bergamot Essential Oil Attenuates Anxiety-Like Behaviour in Rats. *Molecules* **2017**, *22*, 614. [CrossRef]
162. Moradi, K.; Ashtarian, H.; Danzima, N.Y.; Saeedi, H.; Bijan, B.; Akbari, F.; Mohammadi, M.M. Essential Oil from Citrus aurantium Alleviates Anxiety of Patients Undergoing Coronary Angiography: A Single-Blind, Randomized Controlled Trial. *Chin. J. Integr. Med.* **2021**, *27*, 177–182. [CrossRef]
163. Li, Z.; Wu, F.; Shao, H.; Zhang, Y.; Fan, A.; Li, F. Does the Fragrance of Essential Oils Alleviate the Fatigue Induced by Exercise? A Biochemical Indicator Test in Rats. *Evid.-Based Complement. Altern. Med.* **2017**, *2017*, 5027372. [CrossRef]
164. Tanaka, M.; Szab, Á.; Spekker, E.; Tanaka, M.; Szabó, Á.; Vécsei, L. Neurogenic inflammation: The participant in migraine and recent advancements in translational research. *Biomedicines* **2021**, *10*, 76.
165. Cutillas, A.-B.; Carrasco, A.; Martinez-Gutierrez, R.; Tomas, V.; Tudela, J. Rosmarinus officinalis L. essential oils from Spain: Composition, antioxidant capacity, lipoxygenase and acetylcholinesterase inhibitory capacities, and antimicrobial activities. *Plant Biosyst. Int. J. Deal. All Asp. Plant Biol.* **2018**, *152*, 1282–1292. [CrossRef]
166. Ku, C.-M.; Lin, J.-Y. Anti-inflammatory effects of 27 selected terpenoid compounds tested through modulating Th_1/Th_2 cytokine secretion profiles using murine primary splenocytes. *Food Chem.* **2013**, *141*, 1104–1113. [CrossRef]
167. De Almeida, A.A.C.; Silva, R.O.; Nicolau, L.A.D.; de Brito, T.V.; de Sousa, D.P.; Barbosa, A.L.d.R.; de Freitas, R.M.; Lopes, L.d.S.; Medeiros, J.-V.R.; Ferreira, P.M.P. Physio-pharmacological investigations about the anti-inflammatory and antinociceptive efficacy of (+)-limonene epoxide. *Inflammation* **2017**, *40*, 511–522. [CrossRef]
168. Tang, M.Z.; Wang, Z.F.; Shi, Y.L. Involvement of cytochrome c release and caspase activation in toosendanin-induced PC_{12} cell apoptosis. *Toxicology* **2004**, *201*, 31–38. [CrossRef]
169. Jackson, H.; Anzures-Cabrera, J.; Taylor, K.I.; Pagano, G. Hoehn and Yahr stage and striatal Dat-SPECT uptake are predictors of Parkinson's disease motor progression. *Front. Neurosci.* **2021**, *15*, 765765.
170. Zhou, X.; Hai-Yan, G.; Tun-Hai, X.; Tian, S. Physicochemical evaluation and essential oil composition analysis of Hyssopus cuspidatus Boriss from Xinjiang, China. *Pharmacogn. Mag.* **2010**, *6*, 278–281.
171. Liu, J.L.; Fan, Y.G.; Yang, Z.S.; Wang, Z.Y.; Guo, C. Iron and Alzheimer's disease: From pathogenesis to therapeutic implications. *Front. Neurosci.* **2018**, *12*, 632. [CrossRef]
172. Swaminathan, S.; Fonseca, V.A.; Alam, M.G.; Shah, S.V. The role of iron in diabetes and its complications. *Diabetes Care* **2007**, *30*, 1926–1933. [CrossRef]
173. Ademosun, A.O.; Oboh, G.; Olupona, A.J.; Oyeleye, S.I.; Adewuni, T.M.; Nwanna, E.E. Comparative study of chemical composition, in vitro inhibition of cholinergic and monoaminergic enzymes, and antioxidant potentials of essential oil from peels and seeds of sweet orange (*Citrus sinensis* L. Osbeck) Fruits. *J. Food Biochem.* **2016**, *40*, 53–60. [CrossRef]
174. Oboh, G.; Puntel, R.L.; Rocha, J.B.T. Hot pepper (*Capsicum annuum*, Tepin and Capsicum chinese, Habanero) prevents Fe^{2+}-induced lipid peroxidation in brain–in vitro. *Food Chem.* **2007**, *102*, 178–185. [CrossRef]
175. Oboh, G.; Olasehinde, T.A.; Ademosun, A.O. Essential oil from lemon peels inhibit key enzymes linked to neurodegenerative conditions and pro-oxidant induced lipid peroxidation. *J. Oleo Sci.* **2014**, *63*, 373–381. [CrossRef] [PubMed]
176. Chaiyana, W.; Okonogi, S. Inhibition of cholinesterase by essential oil from food plant. *Phytomedicine* **2012**, *19*, 836–839. [CrossRef] [PubMed]
177. Uttara, B.; Singh, A.V.; Zamboni, P.; Mahajan, R. Oxidative stress and neurodegenerative diseases: A review of upstream and downstream antioxidant therapeutic options. *Curr. Neuropharmacol.* **2009**, *7*, 65–74. [CrossRef] [PubMed]
178. Zhang, N.; Zhang, L.; Feng, L.; Yao, L. The anxiolytic effect of essential oil of Cananga odorata exposure on mice and determination of its major active constituents. *Phytomedicine* **2016**, *23*, 1727–1734. [CrossRef] [PubMed]
179. Zhou, W.; Fukumoto, S.; Yokogoshi, H. Components of lemon essential oil attenuate dementia induced by scopolamine. *Nutr. Neurosci.* **2009**, *12*, 57–64. [CrossRef]
180. Navarra, M.; Mannucci, C.; Delbò, M.; Calapai, G. Citrus bergamia essential oil: From basic research to clinical application. *Front. Pharmacol.* **2015**, *6*, 36. [CrossRef]
181. Celia, C.; Trapasso, E.; Locatelli, M.; Navarra, M.; Ventura, C.A.; Wolfram, J.; Carafa, M.; Morittu, V.M.; Britti, D.; Di Marzio, L.; et al. Anticancer activity of liposomal Bergamot essential oil (BEO) on human neuroblastoma cells. *Colloids Surf. B Biointerfaces* **2013**, *112*, 548–553. [CrossRef]

182. Da Pozzo, E.; De Leo, M.; Faraone, I.; Milella, L.; Cavallini, C.; Piragine, E.; Testai, L.; Calderone, V.; Pistelli, L.; Braca, A.; et al. Antioxidant and Antisenescence Effects of Bergamot Juice. *Oxid. Med. Cell. Longev.* **2018**, *2018*, 9395804. [CrossRef]
183. Valussi, M.; Donelli, D.; Firenzuoli, F. Bergamot Oil: Botany, Production, Pharmacology. *Encyclopedia* **2021**, *1*, 16. [CrossRef]
184. Scuteri, D.; Rombolà, L.; Tridico, L.; Mizoguchi, H.; Watanabe, C.; Sakurada, T.; Sakurada, S.; Corasaniti, M.T.; Bagetta, G.; Morrone, L.A. Neuropharmacological Properties of the Essential Oil of Bergamot for the Clinical Management of Pain-Related BPSDs. *Curr. Med. Chem.* **2019**, *26*, 3764–3774. [CrossRef]
185. Toscano-Garibay, J.D.; Arriaga-Alba, M.; Sánchez-Navarrete, J.; Mendoza-García, M.; Flores-Estrada, J.J.; Moreno-Eutimio, M.A.; Espinosa-Aguirre, J.J.; González-Ávila, M.; Ruiz-Pérez, N.J. Antimutagenic and antioxidant activity of the essential oils of Citrus sinensis and Citrus latifolia. *Sci. Rep.* **2017**, *7*, 11479. [CrossRef]
186. Goes, T.C.; Teixeira-silva, F. Effect of Sweet Orange Aroma on Experimental Anxiety in Humans. *J. Altern. Complement. Med.* **2012**, *18*, 798–804. [CrossRef]
187. Hsouna, A.B.; Hamdi, N.; Halima, N.B.; Abdelkafi, S. Characterization of essential oil from Citrus aurantium L. flowers: Antimicrobial and antioxidant activities. *J. Oleo Sci.* **2013**, *62*, 763–772. [CrossRef]
188. Pimenta, F.C.F.; Alves, M.F.; Pimenta, M.B.F.; Melo, S.A.L.; De Almeida, A.A.F.; Leite, J.R.; Pordeus, L.C.D.M.; Diniz, M.D.F.F.M.; Almeida, R.N. De Anxiolytic Effect of *Citrus aurantium* L. on Patients with Chronic Myeloid Leukemia. *Phytother. Res.* **2016**, *30*, 613–617. [CrossRef]
189. Farshbaf-Khalili, A.; Kamalifard, M.; Namadian, M. Comparison of the effect of lavender and bitter orange on anxiety in postmenopausal women: A triple-blind, randomized, controlled clinical trial. *Complement. Ther. Clin. Pract.* **2018**, *31*, 132–138. [CrossRef]
190. Namazi, M.; Amir Ali Akbari, S.; Mojab, F.; Talebi, A.; Alavi Majd, H.; Jannesari, S. Aromatherapy with *Citrus aurantium* oil and anxiety during the first stage of labor. *Iran. Red Crescent Med. J.* **2014**, *16*, e18371. [CrossRef]
191. Moosavy, M.H.; Hassanzadeh, P.; Mohammadzadeh, E.; Mahmoudi, R.; Khatibi, S.A.; Mardani, K. Antioxidant and antimicrobial activities of essential oil of Lemon (*Citrus limon*) peel in vitro and in a food model. *J. Food Qual. Hazards Control* **2017**, *4*, 42–48.
192. Johnson, C.E. Effect of aromatherapy on cognitive test anxiety among nursing students. *Altern. Complement. Ther.* **2014**, *20*, 84–87. [CrossRef]
193. Viana, M.D.M.; Silva Neto, G.J.D.; Lima, A.A.D.; Leite, A.B.; Souza, I.T.; Santana, A.E.G.; Campesatto, E.A.; Moreira, M.S.A. *Citrus limon* (L.) Burm f. Essential oil has anxiolytic and sedative properties by modulating GABAA-receptors. *Braz. Arch. Biol. Technol.* **2020**, *63*, e20200206. [CrossRef]
194. Khotimah, H.; Subagio, S.U. The Application of Acupressure Phytopharmaceutical Using Lemon Essential Oil Aromatherapy to Overcome Dysmenorrhea in Adolescents. *Faletehan Health J.* **2021**, *8*, 187–193. [CrossRef]
195. Kustriyanti, D.; Putri, A.A. The effect of ginger and lemon aromatherapy on nausea and vomiting among pregnant women. *J. Keperawatan Soedirman* **2019**, *14*, 15–22. [CrossRef]
196. Yi, F.; Jin, R.; Sun, J.; Ma, B.; Bao, X. Evaluation of mechanical-pressed essential oil from Nanfeng mandarin (*Citrus reticulata* Blanco cv. Kinokuni) as a food preservative based on antimicrobial and antioxidant activities. *Lwt* **2018**, *95*, 346–353. [CrossRef]
197. Kwangjai, J.; Cheaha, D.; Manor, R.; Sa-ih, N.; Samerphob, N.; Issuriya, A.; Wattanapiromsakul, C.; Kumarnsit, E. Modification of brain waves and sleep parameters by *Citrus reticulata* Blanco. cv. Sai-Nam-Phueng essential oil. *Biomed. J.* **2021**, *44*, 727–738. [CrossRef]
198. Zhou, X.M.; Zhao, Y.; He, C.C.; Li, J.X. Preventive effects of *Citrus reticulata* essential oil on bleomycin-induced pulmonary fibrosis in rats and the mechanism. *J. Chin. Integr. Med.* **2012**, *10*, 200–209. [CrossRef]
199. Ali, M.; Akhter, R.; Narjish, S.N.; Shahriar, M.; Bhuiyan, M.A. Studies of preliminary phytochemical screening, membrane stabilizing activity, thrombolytic activity and in-vitro antioxidant activity of leaf extract of *Citrus hystrix*. *Int. J. Pharm. Sci. Res.* **2015**, *6*, 2367.
200. Hongratanaworakit, T.; Buchbauer, G. Chemical composition and stimulating effect of *Citrus hystrix* oil on humans. *Flavour Fragr. J.* **2007**, *22*, 443–449. [CrossRef]
201. Yoo, K.M.; Lee, K.W.; Park, J.B.; Lee, H.J.; Hwang, I.K. Variation in major antioxidants and total antioxidant activity of Yuzu (*Citrus junos* Sieb ex Tanaka) during maturation and between cultivars. *J. Agric. Food Chem.* **2004**, *52*, 5907–5913. [CrossRef]
202. Satou, T.; Matsuura, M.; Takahashi, M.; Umezu, T.; Hayashi, S.; Sadamoto, K.; Koike, K. Anxiolytic-like effect of essential oil extracted from Abies sachalinensis. *Flavour Fragr. J.* **2011**, *26*, 416–420. [CrossRef]
203. Ohata, M.; Zhou, L.; Ando, S.; Kaneko, S. Application of integrative physiological approach to evaluate human Application of integrative physiological approach to evaluate human physiological responses to the inhalation of essential oils of Japanese Citrus fruits. *Biosci. Biotechnol. Biochem.* **2021**, *86*, 109–116. [CrossRef]
204. Matsumoto, T.; Kimura, T.; Hayashi, T. Aromatic effects of a Japanese citrus fruit-yuzu (*Citrus junos* Sieb. ex Tanaka)-on psychoemotional states and autonomic nervous system activity during the menstrual cycle: A single-blind randomized controlled crossover study. *Biopsychosoc. Med.* **2016**, *10*, 1–11. [CrossRef]
205. Rahnama, S.; Rabiei, Z.; Alibabaei, Z.; Mokhtari, S.; Rafieian-Kopaei, M.; Deris, F. Anti-amnesic activity of *Citrus aurantium* flowers extract against scopolamine-induced memory impairments in rats. *Neurol. Sci.* **2015**, *36*, 553–560. [CrossRef]
206. Azanchi, T.; Shafaroodi, H.; Asgarpanah, J. Anticonvulsant activity of *Citrus aurantium* blossom essential oil (neroli): Involvment of the GABAergic system. *Nat. Prod. Commun.* **2014**, *9*, 1615–1618.

207. Scandurra, C.; Mezzalira, S.; Cutillo, S.; Zapparella, R.; Statti, G.; Maldonato, N.M.; Locci, M.; Bochicchio, V. The effectiveness of neroli essential oil in relieving anxiety and perceived pain in women during labor: A Randomized controlled trial. *Healthcare* **2022**, *10*, 366. [CrossRef]
208. Satou, T.; Miyahara, N.; Murakami, S.; Hayashi, S.; Koike, K. Differences in the effects of essential oil from Citrus junos and (+)-limonene on emotional behavior in mice. *J. Essent. Oil Res.* **2012**, *24*, 493–500. [CrossRef]
209. Eddin, L.B.; Jha, N.K.; Meeran, M.F.; Kesari, K.K.; Beiram, R.; Ojha, S. Neuroprotective potential of limonene and limonene containing natural products. *Molecules* **2021**, *26*, 4535. [CrossRef]
210. Ali, B.; Al-Wabel, N.A.; Shams, S.; Ahamad, A.; Khan, S.A.; Anwar, F. Essential oils used in aromatherapy: A systemic review. *Asian Pac. J. Trop. Biomed.* **2015**, *5*, 601–611. [CrossRef]
211. Kobayashi, Y.; Takemoto, H.; Fu, Z.; Shimizu, E.; Kinjo, Y. Enhancement of pentobarbital-induced sleep by the vaporized essential oil of citrus keraji var. Kabuchii and its characteristic component, γ-Terpinene. *Nat. Prod. Commun.* **2016**, *11*, 1175–1178. [CrossRef]
212. Blanch, G.P.; Nicholson, G.J. Determination of the Enantiomeric Composition of Limonene and Limonene-1,2-epoxide in Lemon Peel by Multidimensional Gas Chromatography with Flame-Ionization Detection and Selected Ion Monitoring Mass Spectrometry. *J. Chromatogr. Sci.* **1998**, *36*, 37–43.
213. Flores, G.; Blanch, G.P.; Ruiz del Castillo, M.L.; Herraiz, M. Enantiomeric composition studies in Lavandula species using supercritical fluids. *J. Sep. Sci.* **2005**, *28*, 2333–2338.
214. Mosandl, A. Authenticity assessment: A permanent challenge in food flavor and essential oil analysis. *J. Chromatogr. Sci.* **2004**, *42*, 440–449.
215. Nhu-Trang, T.T.; Casabianca, H.; Grenier-Loustalot, M.F. Authenticity control of essential oils containing citronellal and citral by chiral and stable-isotope gas-chromatographic analysis. *Anal. Bioanal. Chem.* **2006**, *386*, 2141–2152.
216. Hör, K.; Ruff, C.; Weckerle, B.; König, T.; Schreier, P. Flavor authenticity studies by 2H/1H ratio determination using on-line gas chromatography pyrolysis isotope ratio mass spectrometry. *J. Agric. Food Chem.* **2001**, *49*, 21–25.
217. Faulhaber, S.; Hener, U.; Mosandl, A. GC/IRMS Analysis of Mandarin Essential Oils. 2. δ13CPDB Values of Characteristic Flavor Components. *J. Agric. Food Chem.* **1997**, *45*, 4719–4725.
218. König, W.A.; Hochmuth, D.H. Enantioselective gas chromatography in flavor and fragrance analysis: Strategies for the identification of known and unknown plant volatiles. *J. Chromatogr. Sci.* **2004**, *42*, 423–439.
219. Schulz, H.; Schrader, B.; Quilitzsch, R.; Steuer, B. Quantitative analysis of various citrus oils by ATR/FT-IR and NIR-FT Raman spectroscopy. *Appl. Spectrosc.* **2002**, *56*, 117–124.

antioxidants

MDPI

Article

Phytochemical Profile, Preliminary Toxicity, and Antioxidant Capacity of the Essential Oils of *Myrciaria floribunda* (H. West ex Willd.) O. Berg. and *Myrcia sylvatica* (G. Mey) DC. (Myrtaceae)

Ângelo Antônio Barbosa de Moraes [1,2], Oberdan Oliveira Ferreira [2,3], Leonardo Souza da Costa [1], Lorena Queiroz Almeida [1], Everton Luiz Pompeu Varela [3,4], Márcia Moraes Cascaes [5], Celeste de Jesus Pereira Franco [2], Sandro Percário [3,4], Lidiane Diniz do Nascimento [2], Mozaniel Santana de Oliveira [2,*] and Eloisa Helena de Aguiar Andrade [1,2,3,5]

1 Faculdade de Engenharia Química, Instituto de Tecnologia, Universidade Federal do Pará, Rua Augusto Corrêa, 01, Guamá, Belém 66075-900, Brazil
2 Laboratório Adolpho Ducke, Coordenação de Botânica, Museu Paraense Emílio Goeldi, Av. Perimetral, 1901, Terra Firme, Belém 66077-830, Brazil
3 Programa de Pós-Graduação em Biodiversidade e Biotecnología—Rede Bionorte, Universidade Federal do Pará, Rua Augusto Corrêa, 01, Guamá, Belém 66075-900, Brazil
4 Laboratório de Pesquisas em Estresse Oxidativo, Universidade Federal do Pará, Rua Augusto Corrêa 01, Guamá, Belém 66075-900, Brazil
5 Programa de Pós-graduação em Química, Universidade Federal do Pará, Rua Augusto Corrêa, 01, Guamá, Belém 66075-900, Brazil
* Correspondence: mozanieloliveira@museu-goeldi.br

Citation: de Moraes, Â.A.B.; Ferreira, O.O.; da Costa, L.S.; Almeida, L.Q.; Varela, E.L.P.; Cascaes, M.M.; de Jesus Pereira Franco, C.; Percário, S.; Nascimento, L.D.d.; de Oliveira, M.S.; et al. Phytochemical Profile, Preliminary Toxicity, and Antioxidant Capacity of the Essential Oils of *Myrciaria floribunda* (H. West ex Willd.) O. Berg. and *Myrcia sylvatica* (G. Mey) DC. (Myrtaceae). *Antioxidants* **2022**, *11*, 2076. https://doi.org/10.3390/antiox11102076

Academic Editors: Delia Mirela Tit and Simona Bungau

Received: 14 September 2022
Accepted: 18 October 2022
Published: 21 October 2022

Abstract: The essential oils (EOs) of *Myrciaria floribunda* (Mflo) and *Myrcia sylvatica* (Msyl) (Myrtaceae) were obtained by hydrodistillation. The analysis of volatile constituents was performed by GC/MS. Preliminary toxicity was assessed on *Artemia salina* Leach. The antioxidant capacity was measured by the $ABTS^{\bullet+}$ and $DPPH^{\bullet}$ radical inhibitory activities. The results indicate that the Mflo EO had the highest yield (1.02%), and its chemical profile was characterized by high levels of hydrocarbon (65.83%) and oxygenated (25.74%) monoterpenes, especially 1,8-cineole (23.30%), terpinolene (22.23%) and α-phellandrene (22.19%). Regarding the Msyl EO, only hydrocarbon (51.60%) and oxygenated (46.52%) sesquiterpenes were identified in the sample, with (*Z*)-α-trans-bergamotene (24.57%), α-sinensal (13.44%), and (*Z*)-α-bisabolene (8.33%) at higher levels. The EO of Mflo exhibited moderate toxicity against *A. salina* (LC_{50} = 82.96 ± 5.20 μg.mL^{-1}), while the EO of Msyl was classified as highly toxic (LC_{50} = 2.74 ± 0.50 μg.mL^{-1}). In addition, relative to Trolox, the EOs of Mflo and Msyl showed significant inhibitory effects ($p < 0.0001$) against the $DPPH^{\bullet}$ radical. This study contributes to the expansion of chemical and biological knowledge on the EOs of Myrtaceae species from the Amazon region.

Keywords: Amazonian natural products; *Artemia salina*; bioactive compounds; 1,8-cineole; (*Z*)-α-trans-bergamotene

1. Introduction

The Amazonian flora is widely studied from chemical and biological perspectives due to the existence of species used in traditional medicine for the treatment of various endemic diseases in an effort to relate the practical uses of these species with the chemical composition of their natural products [1,2]. The essential oils (EOs) of aromatic members of the Amazonian flora are extensively studied by the scientific community due to their high-value-added properties; these organisms have also aroused industrial and economic interest due to their strong prospects for wealth generation and development in the region [3,4]. Among the families of Amazonian species that produce EOs, the Myrtaceae family stands out as one of the most important of the Neotropics, with great medicinal interest [5,6].

The family Myrtaceae is represented in Brazil by 29 genera and 1193 species, of which 266 occur in the Amazon [7]. Myrtaceae species are also economically important because many of them are edible and are sources of natural products with pharmacological potential [8–14]. The species of the family Myrtaceae are also widely known for their antioxidant potential, especially species of the genera *Eugenia*, *Syzygium*, *Myrciaria*, and *Eucalyptus* [15–22]. The volatile compounds present in the EOs of species of this family can eliminate or inhibit the effects of free radicals of oxidizing substances, attenuating the effects of oxidative stress and reducing the possibility of occurrence of degenerative diseases, such as Alzheimer's disease, Parkinson's disease, cancer, diabetes, and sclerosis [23–25]. The use of natural products as antioxidant substances is being increasingly studied today because they are environmentally friendly, cause no damage to the environment, and ensure greater safety of human health [26,27].

To determine possible biological activities, it is necessary to perform preliminary toxicity tests to evaluate the possible risks to human health [28,29]. Thus, the bioassay against the microcrustacean *Artemia salina* Leach is used to determine the level of cytotoxicity of natural products [30,31]. This test also allows the investigation of highly toxic EOs for possible uses as inputs in the manufacture of pest repellents and related products, especially in the agroindustry [32,33].

Myrciaria floribunda (H. West ex Willd.) O. Berg is one of the species of Myrtaceae found in the Amazon and is popularly known as "camboim" [34,35]. Its fruit is consumed in the form of gelatin or used as flavoring in the distilled beverage industry [36]. *Myrcia sylvatica* (G. Mey) DC. is a shrub popularly known as "cumatê-Folha-miúda", "myrtle", or "broom" that is used in folk medicine for the treatment of dysentery and intestinal diseases; like another species of Myrtaceae known as "pedra-ume-caá", it also has potential for use against diabetes [37–39]. Despite the use of these species by the traditional communities of the Amazon, there are still few studies on the chemical profile and biological and antioxidant potentials of the EOs of these two species. Thus, the present study aimed to evaluate the phytochemical profile, antioxidant potential, and preliminary toxicity of the EOs of *M. floribunda* and *M. sylvatica* species.

2. Materials and Methods

2.1. Collection and Processing of Botanical Material

Shoots of *M. floribunda* and *M. sylvatica* were collected at Campina do Guarujá in the city of Bujaru, Lower Tocantins microregion, in the state of Pará, Brazil (01°57′36″ South latitude and 48°11′51″ longitude), in July 2017, following conventional botanical procedures. After collection, the material was dried in a convection oven at 35 °C for 5 days, ground, homogenized, weighed, and subjected to hydrodistillation to obtain the EO. The exsiccates, botanical identification, and registration were incorporated into the Collection of Aromatic Plants of the João Murça Herbarium collection of the Emílio Goeldi Paraense Museum, Belém, Pará, with the following registration numbers: *M. floribunda* MG237492 and *M. sylvatica* MG237516.

2.2. Production and Yield of Essential Oils

The EOs were obtained by hydrodistillation using a modified Clevenger apparatus for 3 h. After the end of distillation, the EOs were centrifuged for 5 min at 3000 rpm, dehydrated with anhydrous sodium sulfate (Na_2SO_4), and again centrifuged under the same conditions. Then, they were stored in amber glass ampoules and kept in a refrigerated environment at a constant temperature of 5 °C. The percentage of water in the studied samples was determined using an infrared moisture analyzer. The EO yield was calculated using the relationship between mass, oil, and moisture, as established by Santos et al. [40].

2.3. Identification of Chemical Components by GC/MS

The chemical composition of the EOs was analyzed in the Adolpho Ducke laboratory of the Goeldi Museum (LAD/MPEG, Belém, Brazil) by gas chromatography coupled to mass spectrometry (GC/MS) in a SHIMADZU QP Plus-2010 system equipped with a DB-5MS silica capillary column (30 m × 0.25 mm; 0.25 m film thickness) under the following operating conditions: carrier gas: helium, linear velocity of 36.5 cm.s^{-1}; injection mode: splitless (2 μL of oil in 0.5 mL of hexane); injector temperature: 250 °C, temperature program: 60–250 °C with a gradient of 3 °C.min^{-1}; temperature of the ion source and other parts: 220 °C.

The quadrupole scan rate was 39 to 500 daltons.s^{-1}. Ionization was performed in electron impact mode at 70 eV. The identification of volatile components was based on the linear retention index (RI) and the fragmentation patterns in the mass spectra by comparison with standard samples in the NIST-11 and FFNSC-2 databases and from the literature [41]. The RIs were obtained using the homologous series of *n*-alkanes (C_8-C_{40}) from Sigma–Aldrich (San Luis, AZ, USA).

2.4. *Preliminary Toxicity Bioassay with Artemia Salina*

The preliminary toxicity of the EOs was tested against larvae of the microcrustacean *A. salina*. The *A. salina* cysts were incubated at room temperature (27–30 °C) under artificial lighting in an aquarium with artificial salt water: 46 g of NaCl, 22 g of $MgCl_2.6H_2O$, 8 g of Na_2SO_4, 2.6 g of $CaCl_2.6H_2O$, and 1.4 g of KCl in 2.0 L of distilled water. The pH was adjusted to 8.0–9.0 using Na_2CO_3.

Twenty-four hours after hatching, EO solutions were prepared in triplicate at concentrations of 100, 50, 25, 10, 5, and 1 μg.mL^{-1} using brine water as a vehicle and 5% dimethyl sulfoxide (DMSO) as a diluent. Ten larvae of *A. salina* in the meta-nauplius stage were placed in each tube. After 24 h, the mortality rate of the larvae was quantified, and the mean lethal concentration (LC_{50}) was calculated using the Probitos statistical method, according to the methodology adapted from Góes et al. [42].

2.5. *Tests of the Antioxidant Capacity of Essential Oils*

2.5.1. ABTS Assay

The radical inhibition activity (AIR) of 2,2′-azino-bis (3-ethylbenzothiazoline-6-sulfonic acid) diammonium salt ($ABTS^{•+}$) was analyzed according to the initial principles proposed by Miller et al. [42] with the reaction conditions modified by Re et al. [43]. The method is based on the ability of substances to eliminate the cationic $ABTS^{•+}$ radical, a blue-green chromophore with maximum absorption at 734 nm, resulting in the formation of the stable, colorless ABTS product.

Initially, ABTS (7 mM.L^{-1}; Sigma–Aldrich; A1888; São Paulo/SP, Brazil) and potassium persulfate (140 mM.L^{-1}; $K_2O_8S_2$; Sigma–Aldrich; 216224; São Paulo/SP, Brazil) were mixed and left in the dark for 16 h to form the $ABTS^{•+}$ radical (2.45 mM.L^{-1}). Then, the radical was diluted with phosphate-buffered saline until reaching an absorbance of 0.700 ± 0.020 in an 800XI spectrophotometer (Femto; São Paulo/SP, Brazil) at 734 nm. Then, 30 μL of sample or standard was added to the solution (in triplicate), and after 5 min, the final absorbance was read. In addition, Trolox® (1 mM.L^{-1}; Sigma–Aldrich; 23881-3; São Paulo/SP, Brazil) was used as a standard antioxidant. We calculated the inhibition activity according to the following equation:

$$AIR\ (\%) = [(A_{control} - A_{sample})/A_{control} \times 100] \quad (1)$$

where $A_{control}$ represents the absorbance of the $ABTS^{•+}$ radical (2.5 mM.L^{-1}), and A_{sample} represents the absorbance of the sample.

2.5.2. DPPH Assay

The AIR of 2,2-diphenyl-1-picrylhydrazyl ($DPPH^•$) was determined according to the method proposed by Blois [44] with modifications. This assay assesses the total antioxidant capacity of a substance to eliminate the radical $DPPH^•$ (Sigma–Aldrich; D9132;

São Paulo/SP, Brazil), a violet chromophore with absorption at 517 nm, resulting in the formation of the hydrogenated DPPH product, which is yellow or colorless.

First, DPPH• solution (0.1 mM.L^{-1}) was prepared from the reaction between DPPH (394.32 g.mol^{-1}; Sigma–Aldrich; A1888; São Paulo/SP, Brazil) and ethyl alcohol (PA; C_2H_6O; Sigma–Aldrich; 216224; São Paulo/SP, Brazil). Then, the absorbance of the DPPH• solution was read in an 800XI spectrophotometer (Femto; São Paulo/SP, Brazil) at 517 nm. Next, 50 µL of the sample or standard (Trolox; triplicate) were mixed in 950 µL of DPPH• solution and placed in a water bath at 30 °C for 30 min. Trolox was also used as a standard antioxidant. We calculated the AIR as described in the ABTS assay. Please refer to Supplemental Material S1, for better understanding of antioxidant methods.

2.6. Statistical Analysis

For analysis of the preliminary toxicity and ABTS•+ and DPPH• AIR of the EOs of *M. floribunda* and *M. sylvatica*, analysis of variance (ANOVA (Analysis of Variance),) was applied. Significant differences were compared between groups using Tukey's post hoc analysis. In all tests, a significance level of 5% ($p \leq 0.05$) was considered.

3. Results and Discussion

3.1. Analysis of the Yield and Chemical Composition of Essential Oils

The following Table 1 shows the results for the yield and chemical composition of the EOs of the two species of Myrtaceae. In total, 26 volatile constituents were identified for *M. floribunda* and 30 for *M. sylvatica*.

Table 1. Chemical composition of the essential oils of the two species of Myrtaceae.

	Species		***M. floribunda***	***M. sylvatica***
	Yield		**1.02%**	**0.22%**
IR_L	**IR_C**	**Chemical Component**	**Area (%)**	**Area (%)**
932	935	α-pinene	3.30	
988	991	Myrcene	2.67	
1002	1008	α-phellandrene	22.19	
1014	1018	α-terpinene	1.53	
1026	1030	*o*-cymene	7.04	
1024	1033	Limonene	1.00	
1026	1036	1,8-cineole	23.30	
1054	1059	γ-terpinene	5.87	
1086	1091	terpinolene	22.23	
1095	1100	linalool	0.81	
1174	1178	terpinen-4-ol	1.63	
1186	1191	α-terpineol	2.45	
1374	1375	α-copaene	0.57	0.50
1389	1387	β-elemene		0.52
1411	1410	α-*cis*-bergamotene		0.41
1417	1419	(*E*)-caryophyllene	2.21	4.82
1430	1425	β-copaene		0.16
1434	1429	γ-elemene		2.89
1432	1431	α-*trans*-bergamotene		7.06
1440	1451	(*E*)-β-farnesene		1.79
1457	1456	β-santalene		2.44
1452	1456	α-humulene	0.16	
1484	1477	germacrene D		1.30
1480	1481	β-*trans*-bergamotene		5.07
1489	1488	β-selinene	0.17	
1496	1491	valencene	0.04	

Table 1. *Cont.*

	Species		*M. floribunda*	*M. sylvatica*
	Yield		1.02%	0.22%
IR_L	IR_C	Chemical Component	Area (%)	Area (%)
1500	1493	bicyclogermacrene		3.34
1496	1497	viridiflorene	0.51	
1506	1498	(Z)-α-bisabolene		8.33
1500	1502	α-muurolene	0.15	
1505	1504	β-bisabolene		4.27
1514	1506	β-curcumene		0.56
1502	1509	*trans*-β-guaiene	0.11	
1514	1511	(Z)-γ-bisabolene		0.56
1521	1519	β-sesquiphellandrene		3.94
1522	1525	δ-cadinene	0.70	
1529	1527	(E)-γ-bisabolene		0.92
1537	1537	(E)-α-bisabolene		0.76
1559	1554	germacrene B		1.96
1561	1564	(E)-nerolidol	0.61	
1564	1573	davanone B	0.04	
1577	1574	spathulenol		1.46
1586	1581	thujopsan-2-α-ol		1.73
1592	1595	viridiflorol	0.19	
1595	1597	cubeban-11-ol	0.06	0.56
1640	1645	*epi*-α-muurolol		1.74
1652	1651	α-cadinol		1.25
1670	1666	*epi*-β-bisabolol		0.57
1674	1675	(Z)-α-santalol		0.39
1690	1694	(Z)-α-trans-bergamotene	0.11	24.57
1755	1748	α-sinensal		13.44
1806	1802	nootkatone		0.81
		Hydrocarbon monoterpenes	65.83	0.00
		Oxygenated monoterpenes	25.74	0.00
		Hydrocarbon sesquiterpenes	4.62	51.60
		Oxygenated sesquiterpenes	1.01	46.52
		TOTAL	99.65	98.12

RI_L, retention index in the literature [41]; RI_C, retention index calculated from a homologous series of *n*-alkanes (C_8-C_{40}) in a DB5-MS column. Relative area (%) calculated based on the peak areas.

The EO of *M. floribunda* predominantly contained hydrocarbon monoterpenes (65.82%) and oxygenated monoterpenes (25.74%). Tietbhol et al. [45] identified high levels of hydrocarbon sesquiterpenes (53.50%), oxygenated monoterpenes (16.70%), and oxygenated sesquiterpenes (16.50%) in the aromatic profile of *M. floribunda*. These results indicate a difference in chemical composition between the current sample and the sample reported in the literature. Regarding the EO of *M. sylvatica*, only hydrocarbon sesquiterpenes (51.60%) and oxygenated sesquiterpenes (46.52%) were observed in the sample. Raposo et al. [46] also identified high levels of hydrocarbon (28.00–63.90%) and oxygenated (15.30–51.40%) sesquiterpenes in the EO of the species.

3.1.1. Yield and Chemical Composition of the Essential Oil of *Myrciaria floribunda*

The yield of the EO of *M. floribunda* in the present study was 1.02%. According to Tietbohl et al. [45], the yield of the EO of fresh leaves of *M. floribunda* originating in Rio de Janeiro, Brazil, was 0.37%. Barbosa et al. [47] reported a yield of the EO of *M. floribunda* fruits of 0.60%. These comparisons indicate that the yield of the present sample is higher than that recorded in the literature. Regarding the phytochemical profile of the EO of *M. floribunda*, the oxygenated monoterpene 1,8-cineole, also known as eucalyptol, was the component with the highest content (23.0%), followed by the hydrocarbon monoterpenes terpinolene (22.23%), α-phellandrene (22.19%), *o*-cimene (7.04%), and γ-terpinene (5.87%).

Tietbohl et al. [45] evaluated the chemical composition of the EO of the fresh leaves of a specimen of *M. floribunda* from the Restinga de Jurubatiba National Park, Rio de Janeiro, Brazil. According to the authors, 1,8-cineole (10.40%), β-selinene (8.40%), and α-selinene (7.40%) were the volatile constituents with the highest contents. The 1,8-cineole content found in the present study was twice that recorded by those authors. In addition, α- and β-selinene were not present in the chemical composition of the sample analyzed in this study.

Tietbohl et al. [45] emphasized that the EO of *M. floribunda* significantly increased mortality and interrupted the metamorphosis of the parasite *Trypanosoma cruzi*, indicating that the secondary metabolites present in the chemical composition of this natural product are promising for use as bioinsecticides and for the environmentally appropriate control of vectors of Chagas disease. Tietbohl et al. [48] reported that the EO of another specimen found in Rio de Janeiro, Brazil, was characterized by high levels of monoterpenes (53.90%), among which 1,8-cineole was the major component (38.40%), with a concentration higher than that indicated in the present study.

Barbosa et al. [47] analyzed the chemical composition of the EO of *M. floribunda* fruits collected in the Brazilian state of Pernambuco. According to the authors, the hydrocarbon sesquiterpenes δ-cadinene (26.84%) and γ-cadinene (15.69%) were the major components of the sample. De Oliveira et al. [49] reported β-(*Z*)-*o*-cymene (50.80%), 1,8-cineole (3.14%), γ-terpinene (2.51%), and (*E*)-caryophyllene (1.16%) as the constituents with the highest contents in the EO of lyophilized isolates of fruit of *M. floribunda* from the Restinga de Maricá in Rio de Janeiro, Brazil.

These results indicate that the phytochemical profiles of the EOs of the fruits and leaves are different. According to Ferreira et al. [50], species from different geographic locations have different chemotypes [50]. This statement may explain the differences between the chemical composition of the present sample and those of samples reported in the literature.

The major constituent of the sample (1,8-cineole) has potential as an anti-inflammatory drug for the treatment of cancer [51]. In addition, the encapsulation of 1,8-cineole in nanofibers can prolong fungal activities against *Candida* species given that the isolated compound is not a strong fungal agent [52–54]. 1,8-Cineole also has potential as an antimicrobial, repellent, and anticancer agent and can be used in the treatment of respiratory diseases, including COVID-19 [55–63].

This compound also has several industrial applications, mainly in the production of pharmaceuticals and as a flavoring of foods and toothpastes [64–67]. Recent studies also point to the use of this compound as a biosolvent and anti-corrosion agent, as it is an ecological alternative to synthetic products [68–70].

Terpinolene has larvicidal, insecticidal, antifungal, antibacterial, antiproliferative, cytoprotective, antiviral, antimicrobial, and antibacterial activities, with activity against dangerous multidrug-resistant bacteria in the treatment of industrial wastewater [71–74].

The compound α-phellandrene also has biological activities described in the literature, including potential antifungal properties, and could potentially be used as an endosomal gel for the treatment of gout [75,76]. Regarding *o*-cymene, recent studies point to possible antiviral effects against COVID-19 due to its anti-inflammatory and anti-influenza activities and as an antifungal agent [77–79]. There are few reports of the properties of γ-terpinene in the literature. However, according to Reis et al. [80], oils containing this compound as a major component or at high levels have moderate antimicrobial activity against food pathogens.

3.1.2. Yield and Chemical Composition of *Myrcia sylvatica* Essential Oil

The yield obtained for the EO of *M. sylvatica* was 0.22%. According to Raposo et al. [46] The EO content of the leaves of a sample from Santarém, Pará, Brazil, ranged from 0.90 to 1.60%, reaching the highest content in the month of July (Amazonian dry season) and the lowest content in the month of January (rainy season). In addition, Raposo et al. [46]

analyzed the yield of the EO of the fruit of the species and found a value of 1.70%. Saccol et al. [81] found an analyzed yield of 1.10% for EO from the dry leaves of a specimen from Santarém, Pará, Brazil. Rosa et al. [82] found a yield of 0.50% for the EO of *M. sylvatica*. These results indicate that the EO content of the present sample is lower than the results in the literature.

The EO of *M. sylvatica* analyzed in the present study showed the oxygenated sesquiterpene (Z)-α-trans-bergamotene as the volatile component with the highest content (24.57%), followed by the oxygenated sesquiterpene α-sinensal (13.44%) and hydrocarbon sesquiterpenes (Z)-α-bisabolene (8.33%), α-*trans*-bisabolene (7.06%), and β-*trans*-bisabolene (5.07%). Raposo et al. [46] evaluated the phytochemical profile of the EO of the leaves of a specimen of *M. sylvatica* collected in Santarém, Pará, Brazil. According to the authors, the hydrocarbon sesquiterpenes β-selinene (6.2–10.5%), cadalene (4.7–8.2%), α-calacorene (1.5–6.0%), δ-cadinene (0.0–6.0%), and *trans*-calamenene (3.5–6.5%) and the oxygenated sesquiterpene 1-*epi*-cubenol (5.9–9.8%) and muskatone (2.7–6.2%) characterized the chemical profile of the sample.

Raposo et al. [46] also evaluated the chemical composition of the EO of the fruit of *M. sylvatica* in the fertile period and found δ-cadinene (11.3%), β-selinene (6.0%), and 1-*epi*-cubenol (5.1%) as the volatile constituents with the highest contents. Saccol et al. [81] studied the volatile composition of the EO of the aerial parts of *M. sylvatica* from Santarém, Pará, Brazil. According to the authors, the hydrocarbon sesquiterpenes *ar*-curcumene (8.09%), β-selinene (6.48%), cadalene (6.24%), α-calamenene (5.89%), and (Z)-calamenene (5.54%) and the oxygenated sesquiterpene 1-*epi*-cubenol (7.41%) were the major components of the sample. The authors also found that this EO attenuated molecular and biochemical effects and physiological changes in *Rhamdia quelen* under different stress events.

Saccol et al. [83] analyzed the chemical composition of the EO of the aerial parts of *M. sylvatica* collected in Santarém, Pará, Brazil. According to the authors, α-selinene (16.08%), calamenene (11.68%), and α-calacorene (11.47%) were the volatile constituents with the highest contents. The authors also noted that this EO attenuates stress induced in the freshwater fish *Sparus aurata*. Saccol et al. [84] identified β-selinene (9.96%), cadalene (9.36%), α-calacorene (9.17%), and (Z)-calamene (8.17%) as major compounds of the EO of a specimen from Santarém, Pará, Brazil. Rosa et al. [82] identified (*E*)-caryophyllene (45.88%) and 14-hydroxy-(Z)-caryophyllene (10.15%) as the main components of the EO of a sample from Carolina, Maranhão, Brazil. According to Cascaes et al. [85], *M. sylvatica* has great genetic variability, which is directly responsible for the different chemotypes presented by the species.

Regarding the properties and applications of the major components, EOs containing high levels of (Z)-α-trans-bergamotene may have antibacterial and antifungal activities [86,87]. This compound is also used in industry as a flavoring agent and can be found in the aromas of cereals and other derivatives [88]. α-Sinensal is responsible for the sapodilla aroma and is also used in the food, cosmetics, and perfumery industries as a flavoring because it has an intense orange aroma accompanied by distinct floral notes [89,90]. Yi et al. [91] observed that α-sinensal showed antimicrobial activity and inhibited Gram-negative *Staphylococcus aureus* and *Bacillus subtilis*.

Lancaster et al. [92] showed that (Z)-α-bisabolene is a potential candidate insecticide against *Murgantia histrionica* (harlequin bug) and *Phyllotreta striolata* (flea beetle). There are few reports in the literature about the biological properties of the α- and β-*trans*-bergamotene isomers. However, Moraes et al. [93] stated that the two isomers are defensive components of insects common in Brazil. The potential of the two compounds and other derivatives as insecticides was also explored by Cribb et al. [94] for management of the southern green stink bug (*Nezara viridula*).

3.2. Preliminary Toxicity

In the *A. salina* bioassay, no mortality was observed for the control, showing that the use of 5% DMSO as a diluent is feasible [95]. The average lethal concentration (LC_{50}) values of the EOs were calculated by fitting a logarithmic curve to the number of dead individuals and extracting the equation in Probitos. The table below shows the mortality results and the LC_{50} concentrations, with their respective coefficients of determination (R^2) as well as the preliminary cytotoxicity classification of the EOs (Table 2).

Table 2. Preliminary cytotoxicity of essential oils.

Species	Concentration (μg.mL^{-1})	Mortality (%)	LC_{50} (μg.mL^{-1})	R^2	Classification (Ramos et al. [96])
M. floribunda	100	56.67	82.96 ± 5.20 ^{b}	0.76	Moderate toxicity
	50	50.00			
	25	33.33			
	10	0.00			
	5	0.00			
	1	0.00			
M. sylvatica	100	96.67	2.74 ± 0.50 ^{a}	0.88	High toxicity
	50	93.33			
	25	86.67			
	10	70.00			
	5	56.67			
	1	40.00			

Values are expressed as the mean and standard deviation (n = 3) of the preliminary toxicity. Results with different lowercase letters (b or a) demonstrate that they are statistically different from each other.

The *A. salina* bioassay is used to efficiently evaluate the cytotoxicity of natural products aquatic environments; it is simple and fast and has low requirements [97]. The test allows the screening of a large number of toxic substances because the microcrustacean larvae are quite sensitive to various chemical constituents, and in some cases, toxicity to *A. salina* coincides with toxicity to mammalian cells [98,99].

Previous studies show that results of preliminary toxicity bioassay with *A. salina* vary according to the type of sample studied (essential oil or extract) and its chemical composition. For the brine shrimp bioassay performed with the essential oils obtained from *Conobea scoparioides* fresh and dried leaves, it was verified a CL_{50} equal to 7.8 ± 0.3 μg.mL^{-1} and 7.5 ± 0.3 μg mL^{-1}, respectively. According to Ramos et al. [96], an EO is classified as toxic when LC_{50} is below 80 μg.mL^{-1}, moderately toxic when LC_{50} is between 80 and 250 μg.mL^{-1}, and nontoxic or slightly toxic when LC_{50} is higher than 250 μg.mL^{-1}.

3.2.1. Toxicity of the Essential oil of *Myrciaria floribunda*

The EO of the dry leaves of *M. floribunda* had a mean LC_{50} of 82.96 ± 5.20 μg.mL^{-1}. This result indicates that the EO of *M. floribunda* is moderately toxic according to the classification of Ramos et al. [96]. Barbosa et al. [47] reported that the EO of *M. floribunda* showed high inhibitory activity of the enzyme acetylcholinesterase, the main target of *A. salina* and the compound responsible for the activity of the microcrustacean.

According to Barbosa et al. [100], the EO of *M. floribunda* showed moderate cytotoxic potential in mammalian host cells. However, the chemical composition reported by those authors differed significantly from that of the present sample because it had high levels of sesquiterpenes. According to Tietbohl et al. [48], the EO of *M. floribunda*, composed mainly of 1,8-cineole (38.40%), showed acute toxicity against *Oncopeltus fasciatus* and *Dysdercus peruvianus*.

According to Caldas et al. [101] and Izham et al. [102], 1,8-cineole, the major component of the present sample, showed no indication of toxicity in tests performed with mice. Izham et al. [102] also highlighted that 1,8-cineole had a cytotoxic effect against breast cancer cells without harming the health of the mice subjected to the test. Elmhalli et al. [103] showed that the EOs of *Salvadora persica* and *Rosmarinus officinalis*, which contain 1,8-cineole as a major constituent, have moderate toxicity against nymphs of *Ixodes ricinus*.

Bhowal and Gopal [104] stated that 1,8-cineole has no reported negative effects in animal experiments and that the toxicity reported thus far in animal experiments appeared only after the application of high doses. According to Galan et al. [105], the amount of 1,8-cineole used commercially does not produce harmful effects to human health, and the compound has toxic effects only when administered at high doses.

Ribeiro et al. [106] reported that terpilonene showed low acute toxicity by residual contact against the mite *Tetranychus urticae* and was less efficient than other monoterpenes. However, Agus [107] stated that terpinolene is among the most toxic monoterpenes along with α-terpineol and linalool. According to Scherf et al. [108], terpinolene was toxic to *Drosophila melanogaster*, with an LC_{50} of 34.60 $\mu L.L^{-1}$ at 12 h of exposure.

Scherer et al. [109] found that terpinolene and α-phellandrene showed no cytotoxic effect against L929 fibroblasts or RAW 267.7 macrophages. Martínez et al. [110] stated that α-phellandrene exhibited high toxicity against larvae of the insect *Tenebrio molitor*. Abdelgaleila and El-Sabrout [111] indicated that the EO of *Schinus molle*, consisting of 29.78% α-phellandrene, was toxic to *Culex pipiens* mosquito larvae. Other compounds present in lower concentrations in the EO of *M. floribunda* may have toxic effects, such as γ-terpinene and γ-cymene, which were likely responsible for the toxic effect of the EO of *Eucalyptus camaldulensis* against several cancer cells [112].

The biological activities of EOs may be related to the presence of certain constituents at a higher content and to the synergistic and/or antagonistic effects exerted by all substances present in the samples [113–118]. Considering that 1,8-cineole has low toxicity, and the other compounds present at high levels have moderate or high toxicity, the combined effects of the volatile constituents of the EO of *M. floribunda* may explain its moderate toxicity. In addition, monoterpenes have low toxicity when compared to sesquiterpenes and phenylpropanoids [119–121].

3.2.2. Toxicity of the Essential Oil of *Myrcia sylvatica*

Regarding the EO of *M. sylvatica*, the LC_{50} was 2.74 $\mu g.mL^{-1}$, indicating that the natural product showed very pronounced toxicity against *A. salina* and that the oil of the species is highly toxic. This EO consists only of sesquiterpenes, which generally have high toxicity [122,123]. Many sesquiterpenes have shown promising potential for use as raw materials or additives in bioinsecticides, natural repellents, and biopesticides due to their high toxicities [124–126].

Rosa et al. [82] indicated that the EO of a specimen of *M. sylvatica* from Carolina, Maranhão, Brazil, showed toxicity to *A. salina*, with an LC_{50} of 79.44 $\mu g.mL^{-1}$. The authors emphasized that the EO is composed mainly of sesquiterpenes, primarily (*E*)-caryophyllene and 14-hydroxy-(*Z*)-caryophyllene. Toxic effects of other species of *Myrcia* with high levels of sesquiterpenes have been described in the literature. Scalvenzi et al. [127] reported that the EO of *M. splendens* showed cytotoxic activity against the human tumor cell lines MCF-7 (breast) and A549 (lung) and the nontumor cell line HaCaT (human keratinocytes). Melo et al. [128] stated that the EO of *M. lundiana* showed toxicity against the insect *Acromyrmex balzani*.

According to Alves et al. [129], the major constituent of the EO of *M. sylvatica* is present in high content in the EO of *Cordia verbenacea*, which inhibited 25.9% of cowpea weevils at a concentration of 0.40 $\mu L.cm^{-3}$. Powers et al. [87] indicated that the EOs of *Santalum album* and *S. paniculatum*, containing (Z)-α-trans-bergamotene as one of their major components, showed toxicity against Hep-G2 tumor cells (liver cancer). Regarding α-sinensal, there are no reports in the literature regarding its toxicity.

Fernandes et al. [130] found that the EOs of the leaves and stems of *Eremanthus erythropappus*, which contain (Z)-α-bisabolene at high levels, showed marked toxicity in mice by the MTT assay. Mahdavi et al. [131] showed that the EO of the *Zingiber officinale* rhizomes showed substantial toxicity against the potato moth (*Phthorimaea operculella*). According to the authors, this EO contains (Z)-α-bisabolene as one of its major components.

According to Fernandez et al. [132] α-*trans*-bergamotene was the major component of the EO of the fruits of *Garcinia gardneriana* and showed toxicity against *Aedes aegyptus* mosquito larvae. Vinturelle et al. [133] reported that the EO of *Copaifera officinalis* contained α-*trans*-bergamotene as one of the compounds with the highest content and caused 84.60%

mortality in engorged females of the tick *Rhipicephalus microplus*. The authors attributed this toxicity to the presence of sesquiterpenes, including α-*trans*-bergamotene, in the chemical profile of the EO. Matsuda et al. [134] reported that the compound β-*trans*-bergamotene has moderate toxicity against cancer cells and has not yet shown satisfactory results for possible use as a drug.

The possible toxic effects of the five major components of the EO of *M. sylvatica* and the synergistic effects of the other sesquiterpenes present in the sample may have contributed to the high toxicity of the EO against the microcrustacean *A. salina* [135]. Thus, further studies should be conducted to evaluate the possible harmful effects of this EO and its major components on mammalian cells and human health in addition to its possible use as an additive for the production of larvicides, pesticides, and insecticides.

3.3. Antioxidant Capacity of Essential Oils

The antioxidant capacity of the EOs of *M. floribunda* and *M. sylvatica* was determined from the $ABTS^{\bullet+}$ and $DPPH^{\bullet}$ AIR and by comparison with Trolox (1 mM/L), a water-soluble synthetic analog of vitamin E and a potent antioxidant.

According to Ferreira et al. [136], in the ABTS assay, the reaction between ABTS and $K_2O_8S_2$ generates the $ABTS^{\bullet+}$ radical, which is then reduced again to ABTS in the presence of antioxidant compounds; the degree and time scale of this reaction are dependent on the capacity antioxidant concentration, concentration of antioxidants present in the Eos, and duration of the reaction between the compounds. According to the authors, in the DPPH assay, when the antioxidant power of the EO is high, the color of the solution changes from purple to yellowish over time due to the donation of a hydrogen atom or the transfer of electrons from substances in the EO to the $DPPH^{\bullet}$ radical, which then transforms into a stable diamagnetic molecule. In the present results, the RIAs of the EOs were proportional to their TEAC values (Figure 1).

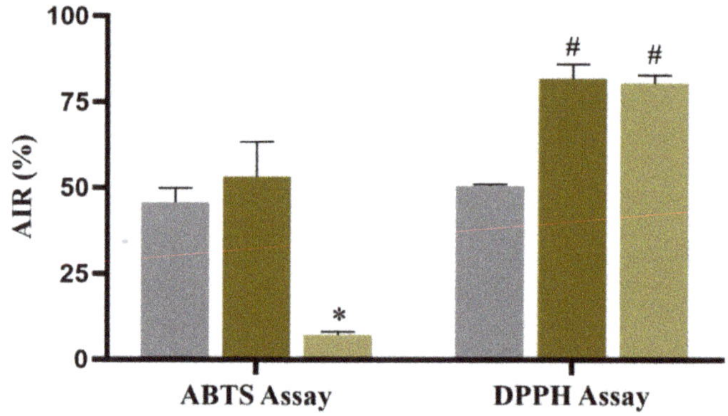

Figure 1. Percentage $ABTS^{\bullet+}$ and $DPPH^{\bullet}$ radical inhibition activity (AIR) of the essential oils of *Myrciaria floribunda* and *Myrcia sylvatica*. The results are expressed as the mean and standard deviation (n = 3). * p = 0.0001 versus Trolox and *Myrciaria floribunda*; # p = 0.0001 versus Trolox. ■ = *Myrciaria floribunda;* ■ = *Myrcia sylvatica;* ■ = Trolox.

3.3.1. Antioxidant Capacity of the Essential Oil of *Myrciaria floribunda*

The results (Figure 1) showed that the EO of *M. floribunda* had an $ABTS^{\bullet+}$ AIR of 53.27 ± 8.27%, indicating that this EO had an antioxidant potential (p = 0.45) similar to that of Trolox (45.74 ± 4.16%). In the DPPH assay, the AIR of the EO was 81.91 ± 3.46%, indicating that the effect of the *M. floribunda* EO was superior to that of Trolox ($p < 0.0001$), with 50.53 ± 0.52%. These results indicate that the EO of *M. floribunda* has excellent antioxidant activity. This activity can be attributed to the chemical composition of the EO, mainly due to the presence of high levels of monoterpenes with antioxidant potential [137].

According to De Oliveira et al. [49], the EO of lyophilized fruits of *M. floribunda* showed high antioxidant potential toward the free radicals $ABTS^{\bullet+}$ and $DPPH^{\bullet}$. According to the authors, the antioxidant capacity of $ABTS^{\bullet+}$ free radicals was 550.14 μmol Trolox.g^{-1}. In the $DPPH^{\bullet}$ assay, the EC_{50} was 85.68 g·g^{-1}. Moreover, the EO was characterized mainly by hydrocarbon sesquiterpenes, and its 1,8-cineole content was almost eight times lower than that recorded in the present study.

There are few reported on the antioxidant potential of the EOs of other species of *Myrciaria*. However, Da Costa et al. [138] stated that *M. dubia* fruits are a natural source of antioxidants due to their substantial contents of ascorbic acid and phenolic compounds. In addition, the antioxidant potential of the components of *M. floribunda* and other species of *Myrciaria* are well-known, and further studies are needed to verify the antioxidant effects of their volatile components [139,140].

1,8-Cineole, the major component of the EOs in this study, is volatile and has been reported to have antioxidant potential [141]. EOs containing this oxygenated monoterpene as the major constituent have shown high capture power of the $DPPH^{\bullet}$ and $ABTS^{\bullet+}$ radicals, as demonstrated by Boukhatem et al. [142]. According to the authors, the EO of *Eucalyptus globulus* is composed of more than 94% 1,8-cineole and showed good results for the inhibition of $DPPH^{\bullet}$ radicals and a metal chelating activity superior to those of the synthetic antioxidants gallic acid and ascorbic acid. Limam et al. [143] also indicated that the EOs of *Eucalyptus* species with high levels of 1,8-cineole (>48.00%) exhibited a good ability to inhibit $DPPH^{\bullet}$ radicals.

Terpilonene is a constituent with important antioxidant activities highlighted in the literature [144]. Aydin, Turkez, and Taşdemir [145] stated that terpinolene showed excellent antioxidant capacity and great potential to inhibit oxidative stress. In addition, the authors showed that the compound's antioxidant properties make it a strong candidate for anti-cancer treatment, but further studies are needed to corroborate these results. According to Osanloo, Ghanbariasad, and Taghinizhad [146], the EO of *Anethum graveolens* is composed mainly of α-phellandrene (26.75%). According to the authors, the natural product showed a good ability to capture $DPPH^{\bullet}$ radicals at concentrations above 160 μg.mL^{-1}.

Other compounds at lower concentrations in the EO of *M. floribunda*, such as *o*-cymene, γ-terpinene, α-terpineol, myrcene, and α-pinene, also have antioxidant potential [147–151]. The synergistic and/or antagonistic effects of the constituents of this EO may explain its high capture of free radicals considering that the volatile components in the sample are known for their excellent antioxidant properties [152,153].

3.3.2. Antioxidant Capacity of *Myrcia sylvatica* Essential Oil

According to the results (Figure 1), the EO of *M. sylvatica* exhibited an $ABTS^{\bullet+}$ AIR of 7.20 ± 0.72%, showing a lower effect than the EO of *M. floribunda* ($p < 0, 0001$) and the standard ($p < 0.0001$). On the other hand, in the DPPH assay, the AIR of the EO of *M. sylvatica* was 80.55 ± 2.00%; this effect was similar to that of *M. floribunda* oil ($p = 0.99$) and higher than that of Trolox ($p < 0.0001$). These data showed that the EO of *M. sylvatica* had different AIR effects. We believe that this difference in effects can be attributed mainly to the oxygenated sesquiterpenes present in the EO of *M. sylvatica*, which are highly reactive molecules due to their oxygenation content. In addition, the $ABTS^{\bullet+}$ radical exhibits steric hindrance around its nitrogen-centered atom, which hinders reactions with other highly reactive molecules, such as oxygenated sesquiterpenes.

Saccol et al. [84] evaluated the antioxidant effects of *M. sylvatica* EO on the sedation process of the tambaqui (*Colossoma macropomum*). According to the authors, the total reactive antioxidant potential in the brain and gills of anesthetized tambaqui was higher than that of the control. Franco et al. [154] analyzed the antioxidant capacity of the EO of *M. tomentosa*. According to the authors, the EO showed a high percent inhibition of free radicals $DPPH^{\bullet}$ (53.60 ± 0.15%) and $ABTS^{\bullet+}$ (213.00 ± 0.91). The authors also emphasized that this EO is characterized by high levels of hydrocarbon (56.74–75.82%) and oxygenated (15.09–16.83%)

sesquiterpenes, mainly γ-elemene, germacrene D, (*E*)-caryophyllene, spathulenol, and α-zingiberene.

Gatto et al. [155] stated that the EO of *M. hatschbachii* had a low capacity to inhibit DPPH• free radicals (9.14 ± 0.33%). The authors also showed that the EO is composed mainly of hydrocarbon (47.81%) and oxygenated (30.05%) sesquiterpenes, mainly *trans*-calamenene (19.10%), (*E*)-caryophyllene (10.96%), and spathulenol (5.03%). Scalvenzi et al. stated that the EO of *M. splendens* showed an antioxidant potential six times higher than that of vitamin E for the inhibition of DPPH• free radicals. According to the authors, the EO is characterized by the presence of *trans*-nerolidol (67.81%) and α-bisabolol (17.51%).

Regarding the antioxidant properties of the major component of the EO of *M. sylvatica*, there are few reports in the literature. However, Mohankumar et al. [156] stated that (Z)-α-trans-bergamotene at high concentrations in the EO of *Santalum album* may be responsible for the antioxidant and oxidative-stress-modulation activities of the EO, possibly by direct elimination of free radicals and activation of the antioxidant defense system in vitro and in vivo. M. Yi et al. [91] showed that in addition to (Z)-α-trans-bergamotene, α-sinensal was one of the main factors responsible for the $ABTS^{\bullet+}$ free radical capture ability of the EO of *Citrus reticulata*. Saroglou et al. [157] demonstrated that the high concentration of sesquiterpenes, including (Z)-α-bisabolene, in the EO of *Teucrium royleanum* was responsible for the DPPH• RIA.

There are no reports on the possible antioxidant potentials of α-*trans*-bergamotene and β-*trans*-bergamotene in the literature. In general, sesquiterpenes have a lower antioxidant profile than monoterpenes [158,159]. In addition, oxygenated compounds have a greater capacity for free radical scavenging and reduced deleterious effects of lipid peroxidation, especially alcohols, phenols, and enols, due to the presence of hydroxyl groups [160]. These properties may explain the antioxidant behaviors of the EOs of *M. sylvatica* and *M. floribunda* against the $ABTS^{\bullet+}$ and DPPH• radicals.

4. Conclusions

The present study indicated that the EO of *M. floribunda* is characterized by high levels of hydrocarbon and oxygenated monoterpenes, predominantly 1,8-cineole, terpinolene, and α-phellandrene. Regarding the EO of *M. sylvatica*, only hydrocarbon and oxygenated sesquiterpenes were identified in the sample, among which (Z)-α-trans-bergamotene, α-sinensal, and (Z)-α-bisabolene were the volatile constituents with the highest contents. The preliminary toxicity results indicated that the EO of *M. floribunda* exhibited moderate toxicity against *A. salina*, while the EO of *M. sylvatica* showed high toxicity. In addition, the EO of *M. floribunda* exhibited a greater capacity to inhibit the DPPH• radical. This study contributes to the knowledge of the aromatic profile and antioxidant and biological properties of species of the family Myrtaceae from the Amazon region.

Supplementary Materials: The following supporting information can be downloaded at: https://www.mdpi.com/article/10.3390/antiox11102076/s1, Supplementary Material S1, Detailed preparation method of the potential antioxidant activity.

Author Contributions: Conceptualization, Â.A.B.d.M.; methodology, Â.A.B.d.M., C.d.J.P.F., L.S.d.C., L.Q.A., L.D.d.N., O.O.F., M.M.C., E.L.P.V. and M.S.d.O.; software, Â.A.B.d.M.; formal analysis, L.D.d.N., M.S.d.O. and E.H.d.A.A.; research, Â.A.B.d.M.; writing—preparation of the original draft, S.P. and Â.A.B.d.M.; writing—revision and editing, O.O.F., E.L.P.V., C.d.J.P.F., L.D.d.N., M.M.C., M.S.d.O. and E.H.d.A.A.; visualization, L.D.d.N., M.S.d.O. and E.H.d.A.A.; supervision, L.D.d.N. and E.H.d.A.A.; project administration, E.H.d.A.A. All authors have read and agreed to the published version of the manuscript.

Funding: This research received no external funding.

Institutional Review Board Statement: Not applicable.

Informed Consent Statement: Not applicable.

Data Availability Statement: Not applicable.

Acknowledgments: Â.A.B.d.M., L.S.d.C. and L.Q.A. thank CNPq for the scientific initiative scholarship. M.S.d.O. thanks PCI-MCTI/MPEG, Process number: 300983/2022-0. The authors would like to thank the Federal University of Pará—Propesp/PAPQ—Support Program for Qualified Publication Edital 02/2022 – PAPQ/PROPESP.

Conflicts of Interest: The authors declare no conflict of interest.

References

1. Cascaes, M.M.; Carneiro, O.D.S.; Nascimento, L.D.D.; de Moraes, Â.A.B.; de Oliveira, M.S.; Cruz, J.N.; Guilhon, G.M.S.P.; Andrade, E.H.D.A. Essential oils from annonaceae species from brazil: A systematic review of their phytochemistry, and biological activities. *Int. J. Mol. Sci.* **2021**, *22*, 12140. [CrossRef] [PubMed]
2. Ferreira, O.O.; Cruz, J.N.; de Moraes, Â.A.B.; de Jesus Pereira Franco, C.; Lima, R.R.; Anjos, T.O.D.; Siqueira, G.M.; Nascimento, L.D.D.; Cascaes, M.M.; de Oliveira, M.S.; et al. Essential oil of the plants growing in the brazilian amazon: Chemical composition, antioxidants, and biological applications. *Molecules* **2022**, *27*, 4373. [CrossRef] [PubMed]
3. Maia, J.G.S.; Andrade, E.H.A. Database of the amazon aromatic plants and their essential oils. *Quim. Nova* **2009**, *32*, 595–622. [CrossRef]
4. Nascimento, L.D.D.; Almeida, L.Q.; Sousa, E.M.P.D.; Costa, C.M.L.; Costa, K.S.D.; Andrade, E.H.D.A.; Faria, L.J.G.D. Microwave-assisted extraction: An alternative to extract piper aduncum essential oil. *Brazilian J. Dev.* **2020**, *6*, 40619–40638. [CrossRef]
5. García, Y.M.; Lemos, E.E.P.D.; Augusti, R.; Melo, J.O.F. Optimization of extraction and identification of volatile compounds from myrciaria floribunda. *Rev. Ciênc. Agron.* **2021**, *52*, e20207199. [CrossRef]
6. Ferreira, O.O.; Cruz, J.N.D.; Nascimento, L.D.D.; Cascaes, M.M.; Pereira, S.F.M.; Franco, C.D.J.P.; Anjos, T.O.D.; Silva, M.V.C.; Oliveira, M.S.D.; Andrade, E.H.D.A. Antimicrobial property of essential oils from myrtaceae species. In *Recent Progress in Medicinal Plants: Antimicrobial Resistance and Bioactive Natural Product*; Panda, S.K., Dubey, D., Govin, J.N., Eds.; Studium Press LLC: Houston, TX, USA, 2020; Volume 51, pp. 192–214. ISBN 1626991154.
7. Proença, C.E.B.; Amorim, B.S.; Antonicelli, M.C.; Bünger, M.; Burton, G.P.; Caldas, D.K.D.; Costa, I.R.; Faria, J.E.Q.; Fernandes, T.; Gaem, P.H.; et al. Myrtaceae in Flora Do Brasil 2020. Available online: https://floradobrasil2020.jbrj.gov.br/reflora/floradobrasil/FB600341 (accessed on 14 October 2022).
8. Batiha, G.E.S.; Alkazmi, L.M.; Wasef, L.G.; Beshbishy, A.M.; Nadwa, E.H.; Rashwan, E.K. Syzygium Aromaticum l. (Myrtaceae): Traditional uses, bioactive chemical constituents, pharmacological and toxicological activities. *Biomolecules* **2020**, *10*, 202. [CrossRef]
9. Cilingir-Kaya, O.T.; Bihter Gurler, E. Therapeutic potential of essential oil of melaleuca quinquenervia (myrtaceae) in a rat model of ethanolinduced peptic ulcer. *Trop. J. Pharm. Res.* **2021**, *20*, 981–986. [CrossRef]
10. de Paulo Farias, D.; Neri-Numa, I.A.; de Araújo, F.F.; Pastore, G.M. A critical review of some fruit trees from the myrtaceae family as promising sources for food applications with functional claims. *Food Chem.* **2020**, *306*, 125630. [CrossRef]
11. Filomeno, C.A.; Almeida Barbosa, L.C.; Teixeira, R.R.; Pinheiro, A.L.; de Sá Farias, E.; Ferreira, J.S.; Picanço, M.C. Chemical diversity of essential oils of myrtaceae species and their insecticidal activity against rhyzopertha dominica. *Crop Prot.* **2020**, *137*, 105309. [CrossRef]
12. Ferreira, O.O.; da Silva, S.H.M.; de Oliveira, M.S.; Andrade, E.H.D.A. Chemical composition and antifungal activity of myrcia multiflora and eugenia florida essential oils. *Molecules* **2021**, *26*, 7259. [CrossRef]
13. Oliveira, M.S.D. *Essential Oils—Applications and Trends in Food Science and Technology*, 1st ed.; de Oliveira, M.S., Ed.; Springer International Publishing: Cham, Switzerland, 2022; ISBN 978-3-030-99475-4.
14. Bezerra, F.W.F.; de Oliveira, M.S.; Bezerra, P.N.; Cunha, V.M.B.; Silva, M.P.; da Costa, W.A.; Pinto, R.H.H.; Cordeiro, R.M.; da Cruz, J.N.; Chaves Neto, A.M.J.; et al. Extraction of bioactive compounds. In *Green Sustainable Process for Chemical and Environmental Engineering and Science*; Inamuddin, Isloor, A.M., Eds.; Elsevier: Amsterdam, The Netherlands, 2020; pp. 149–167. ISBN 9780128173886.
15. Figueiredo, P.L.B.; Pinto, L.C.; da Costa, J.S.; da Silva, A.R.C.; Mourão, R.H.V.; Montenegro, R.C.; da Silva, J.K.R.; Maia, J.G.S. Composition, antioxidant capacity and cytotoxic activity of eugenia uniflora L. chemotype-oils from the amazon. *J. Ethnopharmacol.* **2019**, *232*, 30–38. [CrossRef] [PubMed]
16. da Costa, J.S.; Barroso, A.S.; Mourão, R.H.V.; da Silva, J.K.R.; Maia, J.G.S.; Figueiredo, P.L.B. Seasonal and antioxidant evaluation of essential oil from eugenia uniflora L., curzerene-rich, thermally produced in situ. *Biomolecules* **2020**, *10*, 328. [CrossRef] [PubMed]
17. da Costa, J.S.; da Cruz, E.D.N.S.; Setzer, W.N.; da Silva, J.K.D.R.; Maia, J.G.S.; Figueiredo, P.L.B. Essentials oils from brazilian eugenia and syzygium species and their biological activities. *Biomolecules* **2020**, *10*, 1155. [CrossRef] [PubMed]
18. Cascaes, M.M.; Guilhon, G.M.S.P.; Zoghbi, M.D.G.; Andrade, E.H.A.; Santos, L.S.; Kelly, R.; da Silva, J.; Trovatti Uetanabaro, A.P.; Araújo, I.S. Flavonoids, antioxidant potential and antimicrobial activity of myrcia rufipila mcvaugh leaves (myrtaceae). *Nat. Prod. Res.* **2019**, *35*, 1717–1721. [CrossRef] [PubMed]
19. Ribeiro, A.R.C.; Cordeiro, M.L.D.S.; Silva, L.M.P.; Cadavid, C.O.M.; Caland, R.B.D.O.; Fernandes-Negreiros, M.M.; Queiroz, M.F.; Barbosa, J.D.S.; Aragão, C.F.S.; Zucolotto, S.M.; et al. *Myrciaria tenella* (DC.) O. berg (myrtaceae) leaves as a source of antioxidant compounds. *Antioxidants* **2019**, *8*, 310. [CrossRef] [PubMed]

20. Usman, L.A.; Oguntoye, O.S.; Ismaeel, R.O. Effect of seasonal variation on chemical composition, antidiabetic and antioxidant potentials of leaf essential oil of *Eucalyptus globulus* L. *J. Essent. Oil Bear. Plants* **2020**, *23*, 1314–1323. [CrossRef]
21. Wang, W.-H.; Tyan, Y.-C.; Chen, Z.-S.; Lin, C.-G.; Yang, M.-H.; Yuan, S.-S.; Tsai, W.-C. Evaluation of the antioxidant activity and antiproliferative effect of the jaboticaba (*Myrciaria cauliflora*) seed extracts in oral carcinoma cells. *BioMed Res. Int.* **2014**, *2014*, 185946. [CrossRef]
22. de Sá, L.Z.C.M.; Castro, P.F.S.; Lino, F.M.A.; Bernardes, M.J.C.; Viegas, J.C.J.; Dinis, T.C.P.; Santana, M.J.; Romao, W.; Vaz, B.G.; Lião, L.M.; et al. Antioxidant potential and vasodilatory activity of fermented beverages of jabuticaba berry (*Myrciaria jaboticaba*). *J. Funct. Foods* **2014**, *8*, 169–179. [CrossRef]
23. Sihoglu Tepe, A.; Ozaslan, M. Anti-alzheimer, anti-diabetic, skin-whitening, and antioxidant activities of the essential oil of cinnamomum zeylanicum. *Ind. Crops Prod.* **2020**, *145*, 112069. [CrossRef]
24. Nascimento, L.D.D.; Silva, S.G.; Cascaes, M.M.; Costa, K.S.D.; Figueiredo, P.L.B.; Costa, C.M.L.; Andrade, E.H.D.A.; de Faria, L.J.G. Drying effects on chemical composition and antioxidant activity of *Lippia thymoides* essential oil, a natural source of thymol. *Molecules* **2021**, *26*, 2621. [CrossRef]
25. Diniz Do Nascimento, L.; Antônio Barbosa De Moraes, A.; Santana Da Costa, K.; Marcos, J.; Galúcio, P.; Taube, P.S.; Leal Costa, M.; Neves Cruz, J.; Helena De Aguiar Andrade, E.; Guerreiro De Faria, L.J. Bioactive natural compounds and antioxidant activity of essential oils from spice plants: New findings and potential applications. *Biomolecules* **2020**, *10*, 988. [CrossRef] [PubMed]
26. De Souza Farias, S.A.; Da Costa, K.S.; Martins, J.B.L. Analysis of conformational, structural, magnetic, and electronic properties related to antioxidant activity: Revisiting flavan, anthocyanidin, flavanone, flavonol, isoflavone, flavone, and flavan-3-Ol. *ACS Omega* **2021**, *6*, 8908–8918. [CrossRef] [PubMed]
27. Figueiredo, P.L.B.; Silva, S.G.; Nascimento, L.D.; Ramos, A.R.; Setzer, W.N.; Silva, J.K.R.D.; Andrade, E.H.A. Seasonal study of methyleugenol chemotype of ocimum campechianum essential oil and its fungicidal and antioxidant activities. *NPC Nat. Prod. Commun.* **2018**, *13*, 1055–1058. [CrossRef]
28. Mesquita, K.D.S.M.; Feitosa, B.D.S.; Cruz, J.N.; Ferreira, O.O.; Franco, C.D.J.P.; Cascaes, M.M.; Oliveira, M.S.D.; Andrade, E.H.D.A. Chemical composition and preliminary toxicity evaluation of the essential oil from peperomia circinnata link var. circinnata. (piperaceae) in artemia salina leach. *Molecules* **2021**, *26*, 7359. [CrossRef]
29. Nadeem, H.R.; Akhtar, S.; Sestili, P.; Ismail, T.; Neugart, S.; Qamar, M.; Esatbeyoglu, T. Toxicity, antioxidant activity, and phytochemicals of basil (*Ocimum basilicum* L.) leaves cultivated in southern punjab, pakistan. *Foods* **2022**, *11*, 1239. [CrossRef]
30. Farias, A.L.F.; Rodrigues, A.B.L.; Martins, R.L.; Rabelo, É.D.M.; Farias, C.W.F.; de Almeida, S.S.M.D.S. Chemical characterization, antioxidant, cytotoxic and microbiological activities of the essential oil of leaf of tithonia diversifolia (hemsl) A. gray (asteraceae). *Pharmaceuticals* **2019**, *12*, 34. [CrossRef]
31. Betim, F.C.M.; Oliveira, C.F.D.; Souza, A.M.D.; Szabo, E.M.; Zanin, S.M.W.; Miguel, O.G.; Miguel, M.D.; Dias, J.D.F.G. Ocotea nutans (nees) mez (lauraceae): Chemical composition, antioxidant capacity and biological properties of essential oil. *Brazilian J. Pharm. Sci.* **2019**, *55*, 1–10. [CrossRef]
32. Cossolin, J.F.S.; Pereira, M.J.B.; Martínez, L.C.; Turchen, L.M.; Fiaz, M.; Bozdoğan, H.; Serrão, J.E. Cytotoxicity of piper aduncum (piperaceae) essential oil in brown stink bug euschistus heros (heteroptera: Pentatomidae). *Ecotoxicology* **2019**, *28*, 763–770. [CrossRef]
33. Lee, M.Y. Essential oils as repellents against arthropods. *Biomol. Res. Int.* **2018**, *2018*, 6860271. [CrossRef]
34. Tietbohl, L.A.C.; Oliveira, A.P.; Esteves, R.S. Antiproliferative activity in tumor cell lines, antioxidant capacity and total phenolic, flavonoid and tannin contents of *Myrciaria floribunda*. *An. Acad. Bras. Ciências* **2017**, *89*, 1111–1120. [CrossRef]
35. de Azevedo, M.M.L.; Cascaes, M.M.; Guilhon, G.M.S.P.; Andrade, E.H.A.; Zoghbi, M.D.G.B.; da Silva, J.K.R.; Santos, L.S.; da Silva, S.H.M. Lupane triterpenoids, antioxidant potential and antimicrobial activity of *Myrciaria floribunda* (H. West ex Willd.) O. Berg. *Nat. Prod. Res.* **2019**, *33*, 506–515. [CrossRef] [PubMed]
36. García, Y.M.; Ramos, A.L.C.C.; de Paula, A.C.C.F.F.; Nascimento, M.H.D.; Augusti, R.; de Araújo, R.L.B.; de Lemos, E.E.P.; Melo, J.O.F. Chemical Physical characterization and profile of fruit volatile compounds from different accesses of *Myrciaria floribunda* (H. West Ex Wild.) O. Berg through polyacrylate fiber. *Molecules* **2021**, *26*, 5281. [CrossRef] [PubMed]
37. Jerônimo, L.B.; da Costa, J.S.; Pinto, L.C.; Montenegro, R.C.; Setzer, W.N.; Mourão, R.H.V.; da Silva, J.K.R.; Maia, J.G.S.; Figueiredo, P.L.B. Antioxidant and cytotoxic activities of myrtaceae essential oils rich in terpenoids from brazil. *Nat. Prod. Commun.* **2021**, *16*, 1934578X21996156. [CrossRef]
38. Silva, F.K.S.; Rosário, A.S.; Secco, R.S.; Zoghbi, M.G.B. Levantamento das espécies conhecidas como pedra-ume-caá (myrtaceae), com ênfase nas comercializadas na cidade de belém, pará, brasil. *Biota Amaz.* **2015**, *5*, 7–15. [CrossRef]
39. Oliveira, E.S.C.; Pontes, F.L.D.; Acho, L.D.R.; do Rosário, A.S.; da Silva, B.J.P.; Bezerra, J.D.A.; Campos, F.R.; Lima, E.S.; Machado, M.B. QNMR quantification of phenolic compounds in dry extract of myrcia multiflora leaves and its antioxidant, anti-age, and enzymatic inhibition activities. *J. Pharm. Biomed. Anal.* **2021**, *201*, 114109. [CrossRef]
40. Santos, A.S.; Alves, S.D.M.; Figueiredo, F.J.C.; Neto, O.G.D.R. *Descrição de Sistema e de Métodos de Extração de Óleos Essenciais e Determinação de Umidade de Biomassa em Laboratório*; Ministério da Agricultura, Pecuária e Abastecimento: Belo Horizonte, Brazil, 2004; pp. 1–6.
41. Adams, R.P. *Identification of Essential Oil Components by Gas Chromatograpy/Mass Spectrometry*, 4th ed.; Allured Publishing Corporation: Carol Stream, IL, USA, 2007; pp. 804–806.

42. Góes, D.F.; Araújo, J.M.; Oliveira, N.R.D.S.; Lima, R.Q.D. Atividade toxica do óleo essencial de piper duckei (piperaceae) sobre microcrustáceo artemia salina / toxic activity of essential oil of piper duckei (piperaceae) on microcrustacean artemia salina. *Braz. J. Dev.* **2020**, *6*, 96278–96284. [CrossRef]
43. Re, R.; Pellegrini, N.; Proteggente, A.; Pannala, A.; Yang, M.; Rice-Evans, C. Antioxidant activity applying an improved abts radical cation decolorization assay. *Free Radic. Biol. Med.* **1999**, *26*, 1231–1237. [CrossRef]
44. BLOIS, M.S. Antioxidant determinations by the use of a stable free radical. *Nature* **1958**, *181*, 1199–1200. [CrossRef]
45. Tietbohl, L.A.C.; Mello, C.B.; Silva, L.R.; Dolabella, I.B.; Franco, T.C.; Enríquez, J.J.S.; Santos, M.G.; Fernandes, C.P.; Machado, F.P.; Mexas, R.; et al. Green insecticide against chagas disease: Effects of essential oil from *Myrciaria floribunda* (myrtaceae) on the development of *Rhodnius prolixus* nymphs. *J. Essent. Oil Res.* **2020**, *32*, 1–11. [CrossRef]
46. Raposo, J.D.A.; Figueiredo, P.L.B.; Santana, R.L.; da Silva Junior, A.Q.; Suemitsu, C.; da Silva, R.; Mourão, R.H.V.; Maia, J.G.S. Seasonal and circadian study of the essential oil of myrcia sylvatica (g. mey) dc., a valuable aromatic species occurring in the lower amazon river region. *Biochem. Syst. Ecol.* **2018**, *79*, 21–29. [CrossRef]
47. Barbosa, D.C.; Holandada Silva, V.N.; de Assis, C.R.D.; de Oliveira Farias de Aguiar, J.C.R.; DoNascimento, P.H.; da Silva, W.V.; do Amaral Ferraz Navarro, D.M.; Silva, M.V.D.; de Menezes Lima, V.L.; dos Santos Correia, M.T. Chemical composition and acetylcholinesterase inhibitory potential, in silico, of *Myrciaria floribunda* (H. West Ex Willd.) O. Berg fruit peel essential oil. *Ind. Crops Prod.* **2020**, *151*, 112372. [CrossRef]
48. Tietbohl, L.A.C.; Barbosa, T.; Fernandes, C.P.; Santos, M.G.; Machado, F.P.; Santos, K.T.; Mello, C.B.; Araújo, H.P.; Gonzalez, M.S.; Feder, D.; et al. Laboratory evaluation of the effects of essential oil of myrciaria floribunda leaves on the development of dysdercus peruvianus and oncopeltus fasciatus. *Rev. Bras. Farmacogn.* **2014**, *24*, 316–321. [CrossRef]
49. de Oliveira, L.M.; Porte, A.; de Oliveira Godoy, R.L.; da Costa Souza, M.; Pacheco, S.; de Araujo Santiago, M.C.P.; Gouvêa, A.C.M.S.; da Silva de Mattos do Nascimento, L.; Borguini, R.G. Chemical characterization of *Myrciaria floribunda* (H. West Ex willd) fruit. *Food Chem.* **2018**, *248*, 247–252. [CrossRef] [PubMed]
50. Ferreira, O.O.; Neves da Cruz, J.; de Jesus Pereira Franco, C.; Silva, S.G.; da Costa, W.A.; de Oliveira, M.S.; de Aguiar Andrade, E.H. First report on yield and chemical composition of essential oil extracted from myrcia eximia dc (myrtaceae) from the brazilian amazon. *Molecules* **2020**, *25*, 783. [CrossRef] [PubMed]
51. Boukhatem, M.N.; Sudha, T.; Darwish, N.H.E.; Chader, H.; Belkadi, A.; Rajabi, M.; Houche, A.; Benkebailli, F.; Oudjida, F.; Mousa, S.A. A New Eucalyptol-rich lavender (*Lavandula stoechas* L.) essential oil: Emerging potential for therapy against inflammation and cancer. *Molecules* **2020**, *25*, 3671. [CrossRef]
52. Ivanov, M.; Kannan, A.; Stojković, D.S.; Glamočlija, J.; Calhelha, R.C.; Ferreira, I.C.F.R.; Sanglard, D.; Soković, M. Camphor and eucalyptol—anticandidal spectrum, antivirulence effect, efflux pumps interference and cytotoxicity. *Int. J. Mol. Sci.* **2021**, *22*, 483. [CrossRef]
53. Saracino, I.M.; Foschi, C.; Pavoni, M.; Spigarelli, R.; Valerii, M.C.; Spisni, E. Antifungal activity of natural compounds vs. candida spp.: A mixture of cinnamaldehyde and eugenol shows promising in vitro results. *Antibiotics* **2022**, *11*, 73. [CrossRef]
54. Mishra, P.; Gupta, P.; Srivastava, A.K.; Poluri, K.M.; Prasad, R. Eucalyptol/β-cyclodextrin inclusion complex loaded gellan/pva nanofibers as antifungal drug delivery system. *Int. J. Pharm.* **2021**, *609*, 121163. [CrossRef]
55. Li, X.W.; Zhang, Z.J.; Hafeez, M.; Huang, J.; Zhang, J.M.; Wang, L.K.; Lu, Y. Bin rosmarinus officinialis L. (lamiales: Lamiaceae), a promising repellent plant for thrips management. *J. Econ. Entomol.* **2021**, *114*, 131–141. [CrossRef]
56. Moo, C.L.; Osman, M.A.; Yang, S.K.; Yap, W.S.; Ismail, S.; Lim, S.H.E.; Chong, C.M.; Lai, K.S. Antimicrobial activity and mode of action of 1,8-cineol against carbapenemase-producing klebsiella pneumoniae. *Sci. Rep.* **2021**, *11*, 20824. [CrossRef]
57. Reyes, E.I.M.; Farias, E.S.; Silva, E.M.P.; Filomeno, C.A.; Plata, M.A.B.; Picanço, M.C.; Barbosa, L.C.A. Eucalyptus resinifera essential oils have fumigant and repellent action against hypothenemus hampei. *Crop Prot.* **2019**, *116*, 49–55. [CrossRef]
58. Rodenak-Kladniew, B.; Castro, M.A.; Crespo, R.; Galle, M.; García de Bravo, M. Anti-cancer mechanisms of linalool and 1,8-cineole in non-small cell lung cancer A549 cells. *Heliyon* **2020**, *6*, e05639. [CrossRef] [PubMed]
59. Sampath, S.; Veeramani, V.; Krishnakumar, G.S.; Sivalingam, U.; Madurai, S.L.; Chellan, R. Evaluation of in vitro anticancer activity of 1,8-cineole–containing n-hexane extract of callistemon citrinus (curtis) skeels plant and its apoptotic potential. *Biomed. Pharmacother.* **2017**, *93*, 296–307. [CrossRef] [PubMed]
60. Sheikh, Z.; Amani, A.; Basseri, H.R.; Kazemi, S.H.M.; Sedaghat, M.M.; Azam, K.; Azizi, M.; Amirmohammadi, F. Repellent Efficacy of eucalyptus globulus and syzygium aromaticum essential oils against malaria vector, anopheles stephensi (diptera: Culicidae). *Iran. J. Public Health* **2021**, *50*, 1668–1677. [CrossRef] [PubMed]
61. Panikar, S.; Shoba, G.; Arun, M.; Sahayarayan, J.J.; Usha Raja Nanthini, A.; Chinnathambi, A.; Alharbi, S.A.; Nasif, O.; Kim, H.J. Essential oils as an effective alternative for the treatment of covid-19: Molecular interaction analysis of protease (mpro) with pharmacokinetics and toxicological properties. *J. Infect. Public Health* **2021**, *14*, 601–610. [CrossRef] [PubMed]
62. Valussi, M.; Antonelli, M.; Donelli, D.; Firenzuoli, F. Appropriate use of essential oils and their components in the management of upper respiratory tract symptoms in patients with COVID-19. *J. Herb. Med.* **2021**, *28*, 100451. [CrossRef] [PubMed]
63. Sharma, A.D.; Kaur, I. Eucalyptol (1,8 cineole) from eucalyptus essential oil a potential inhibitor of COVID-19 corona virus infection by molecular docking studies. *Preprints* **2020**, *3*, 55. [CrossRef]
64. Foudah, A.I.; Alam, P.; Alam, A.; Salkini, M.A.; Alqarni, M.H.; Yusufoglu, H.S. Development of high-performance thin-layer chromatography (hptlc) validated method for simultaneous quantification of eucalyptol and α-pinene in lamiaceae plants. *J. Pharm. Res. Int.* **2019**, *31*, 1–11. [CrossRef]

65. Griggs, J.; Almohanna, H.; Ahmed, A.; Ren, S.; Tosti, A. "Fresh Breath" on toothpaste: Peppermint as cause of cheilitis. *Dermatitis* **2019**, *30*, 74–75. [CrossRef]
66. Hodgson, A.; Cochran, J. Vacuum ultraviolet spectroscopy as a new tool for gc analysis of terpenes in flavors and fragrances. *J. AOAC Int.* **2019**, *102*, 655–658. [CrossRef]
67. Gogoi, R.; Begum, T.; Sarma, N.; Kumar Pandey, S.; Lal, M. Chemical composition of callistemon citrinus (curtis) skeels aerial part essential oil and its pharmacological applications, neurodegenerative inhibitory, and genotoxic efficiencies. *J. Food Biochem.* **2021**, *45*, e13767. [CrossRef] [PubMed]
68. Campos, J.F.; Ferreira, V.; Berteina-Raboin, S. Eucalyptol: A bio-based solvent for the synthesis of o,s,n-heterocycles. application to hiyama coupling, cyanation, and multicomponent reactions. *Catalysts* **2021**, *11*, 222. [CrossRef]
69. Campos, J.F.; Berteina-Raboin, S. Eucalyptol, an all-purpose product. *Catalysts* **2022**, *12*, 48. [CrossRef]
70. Nabah, R.; Lgaz, H.; Zarrok, H.; Larouj, M.; Benhiba, F.; Ourrak, K.; Cherkaoui, M.; Zarrouk, A.; Touir, R.; Oudda, H. Anti-Corrosion properties of niaouli essential oil for tinplate in 3 %NaCl medium. *J. Mater. Environ. Sci.* **2017**, *8*, 3730–3739.
71. Menezes, I.O.; Scherf, J.R.; Martins, A.O.B.P.B.; Ramos, A.G.B.; Quintans, J.D.S.S.; Coutinho, H.D.M.; Ribeiro-Filho, J.; de Menezes, I.R.A. Biological properties of terpinolene evidenced by in silico, in vitro and in vivo studies: A systematic review. *Phytomedicine* **2021**, *93*, 153768. [CrossRef]
72. Pavela, R. Acute toxicity and synergistic and antagonistic effects of the aromatic compounds of some essential oils against culex quinquefasciatus say larvae. *Parasitol. Res.* **2015**, *114*, 3835–3853. [CrossRef]
73. Pontin, M.; Bottini, R.; Burba, J.L.; Piccoli, P. Allium sativum produces terpenes with fungistatic properties in response to infection with sclerotium cepivorum. *Phytochemistry* **2015**, *115*, 152–160. [CrossRef]
74. Mukherjee, A.; Ahn, Y.H. Terpinolene as an enhancer for ultrasonic disinfection of multi-drug-resistant bacteria in hospital wastewater. *Environ. Sci. Pollut. Res.* **2022**, *29*, 34500–34514. [CrossRef]
75. Zhang, J.H.; Sun, H.L.; Chen, S.Y.; Zeng, L.I.; Wang, T. tao Anti-fungal activity, mechanism studies on α-phellandrene and nonanal against penicillium cyclopium. *Bot. Stud.* **2017**, *58*, 13. [CrossRef]
76. Soba, S.V.; Babu, M.; Panonnummal, R. Ethosomal Gel formulation of alpha phellandrene for the transdermal delivery in gout. *Adv. Pharm. Bull.* **2021**, *11*, 137–149. [CrossRef]
77. Gavanji, S.; Sayedipour, S.S.; Larki, B.; Bakhtari, A. Antiviral activity of some plant oils against herpes simplex virus type 1 in vero cell culture. *J. Acute Med.* **2015**, *5*, 62–68. [CrossRef]
78. Marchese, A.; Arciola, C.; Barbieri, R.; Silva, A.; Nabavi, S.; Tsetegho Sokeng, A.; Izadi, M.; Jafari, N.; Suntar, I.; Daglia, M.; et al. Update on monoterpenes as antimicrobial agents: A particular focus on p-cymene. *Materials* **2017**, *10*, 947. [CrossRef]
79. Garzoli, S.; Božović, M.; Baldisserotto, A.; Sabatino, M.; Cesa, S.; Pepi, F.; Vicentini, C.B.; Manfredini, S.; Ragno, R. Essential oil extraction, chemical analysis and anti-candida activity of foeniculum vulgare miller–new approaches. *Nat. Prod. Res.* **2018**, *32*, 1254–1259. [CrossRef] [PubMed]
80. Reis, J.B.; Figueiredo, L.A.; Castorani, G.M.; Veiga, S.M.O.M. Avaliação da atividade antimicrobiana dos óleos essenciais contra patógenos alimentares. *Braz. J. Health Rev.* **2020**, *3*, 342–363. [CrossRef]
81. Saccol, E.M.H.; Jerez-Cepa, I.; Ourique, G.M.; Pês, T.S.; Gressler, L.T.; Mourão, R.H.V.; Martínez-Rodríguez, G.; Mancera, J.M.; Baldisserotto, B.; Pavanato, M.A.; et al. Myrcia sylvatica essential oil mitigates molecular, biochemical and physiological alterations in rhamdia quelen under different stress events associated to transport. *Res. Vet. Sci.* **2018**, *117*, 150–160. [CrossRef]
82. Rosa, C.S.; Veras, K.S.; Silva, P.R.; Lopes Neto, J.J.; Cardoso, H.L.M.; Alves, L.P.L.; Brito, M.C.A.; Amaral, F.M.M.; Maia, J.G.S.; Monteiro, O.S.; et al. Composição química e toxicidade frente *Aedes aegypti* L. e Artemia salina Leach do óleo essencial das folhas de *Myrcia sylvatica* (G. Mey.) DC. *Rev. Bras. Plantas Med.* **2016**, *18*, 19–26. [CrossRef]
83. Saccol, E.M.H.; Parrado-Sanabria, Y.A.; Gagliardi, L.; Jerez-Cepa, I.; Mourão, R.H.V.; Heinzmann, B.M.; Baldisserotto, B.; Pavanato, M.A.; Mancera, J.M.; Martos-Sitcha, J.A. Myrcia sylvatica essential oil in the diet of gilthead sea bream (sparus aurata L.) attenuates the stress response induced by high stocking density. *Aquac. Nutr.* **2018**, *24*, 1381–1392. [CrossRef]
84. Saccol, E.M.H.; Toni, C.; Pês, T.S.; Ourique, G.M.; Gressler, L.T.; Silva, L.V.F.; Mourão, R.H.V.; Oliveira, R.B.; Baldisserotto, B.; Pavanato, M.A. Anaesthetic and Antioxidant effects of *Myrcia sylvatica* (G. Mey.) DC. and *Curcuma longa* L. essential oils on tambaqui (*Colossoma macropomum*). *Aquac. Res.* **2017**, *48*, 2012–2031. [CrossRef]
85. Cascaes, M.M.; Guilhon, G.M.S.P.; de Aguiar Andrade, E.H.; das Graças Bichara Zoghbi, M.; da Silva Santos, L. Constituents and pharmacological activities of *Myrcia* (Myrtaceae): A review of an aromatic and medicinal group of plants. *Int. J. Mol. Sci.* **2015**, *16*, 23881–23904. [CrossRef]
86. Ashrafi, B.; Beyranvand, F.; Ashouri, F.; Rashidipour, M.; Marzban, A.; Kheirandish, F.; Veiskarami, S.; Ramak, P.; Shahrokhi, S. Characterization of phytochemical composition and bioactivity assessment of pseudotrachydium kotschyi essential oils. *Med. Chem. Res.* **2020**, *29*, 1676–1688. [CrossRef]
87. Powers, C.N.; Osier, J.L.; McFeeters, R.L.; Brazell, C.B.; Olsen, E.L.; Moriarity, D.M.; Satyal, P.; Setzer, W.N. Antifungal and cytotoxic activities of sixty commercially-available essential oils. *Molecules* **2018**, *23*, 1549. [CrossRef] [PubMed]
88. Turgumbayeva, A.; Ustenova, G.; Datkhayev, U.; Rahimov, K.; Abramavicius, S.; Tunaityte, A.; Zhakipbekov, K.; Kozhanova, K.; Tulemissov, S.; Ustenova, O.; et al. Safflower (Carthamus Tinctorius L.) a potential source of drugs against Cryptococcal infections, malaria and Leishmaniasis. *Phyton* **2020**, *89*, 137–146. [CrossRef]
89. Lasekan, O.; Yap, S.P. Characterization of the aroma compounds in fresh and dried sapodilla (Manikara Zapota, L.) by the application of aroma extract dilution analysis. *CYTA—J. Food* **2018**, *16*, 801–806. [CrossRef]

90. Silvestre, W.P.; Agostini, F.; Muniz, L.A.R.; Pauletti, G.F. Fractionating of green mandarin (Citrus deliciosa Tenore) essential oil by vacuum fractional distillation. *J. Food Eng.* **2016**, *178*, 90–94. [CrossRef]
91. Yi, F.; Jin, R.; Sun, J.; Ma, B.; Bao, X. Evaluation of mechanical-pressed essential oil from Nanfeng mandarin (Citrus reticulata Blanco cv. Kinokuni) as a food preservative based on antimicrobial and antioxidant activities. *LWT* **2018**, *95*, 346–353. [CrossRef]
92. Lancaster, J.; Lehner, B.; Khrimian, A.; Muchlinski, A.; Luck, K.; Köllner, T.G.; Weber, D.C.; Gundersen-Rindal, D.E.; Tholl, D. An IDS-type sesquiterpene synthase produces the pheromone precursor (Z)-α-Bisabolene in Nezara viridula. *J. Chem. Ecol.* **2019**, *45*, 187–197. [CrossRef]
93. Moraes, M.C.B.; Pareja, M.; Laumann, R.A.; Borges, M. The chemical volatiles (semiochemicals) produced by neotropical stink bugs (Hemiptera: Pentatomidae). *Neotrop. Entomol.* **2008**, *37*, 489–505. [CrossRef]
94. Cribb, B.W.; Siriwardana, K.N.; Walter, G.H. Unicellular pheromone glands of the pentatomid bugnezara viridula (heteroptera: Insecta): Ultrastructure, classification, and proposed function. *J. Morphol.* **2006**, *267*, 831–840. [CrossRef]
95. Chanda, S.; Baravalia, Y. Brine shrimp cytotoxicity of caesalpinia pulcherrima aerial parts, antimicrobial activity and characterisation of isolated active fractions. *Nat. Prod. Res.* **2011**, *25*, 1955–1964. [CrossRef]
96. Ramos, S.C.S.; de Oliveira, J.C.S.; da Câmara, C.A.G.; Castelar, I.; Carvalho, A.F.F.U.; Lima-Filho, J.V. Antibacterial and cytotoxic properties of some plant crude extracts used in northeastern folk medicine. *Rev. Bras. Farmacogn.* **2009**, *19*, 376–381. [CrossRef]
97. Ribeiro, I.A.T.A.; Sá, J.L.F.; Lima, M.V.; Veras, S.T.S.; Aguiar, J.C.R.O.F.; Aires, A.L.; Albuquerque, M.C.P.A.; da Silva, M.V.; Melo, A.M.M.A.; Navarro, D.M.A.F.; et al. Toxic effect of croton rudolphianus leaf essential oil against biomphalaria glabrata, schistosoma mansoni cercariae and artemia salina. *Acta Trop.* **2021**, *223*, 106102. [CrossRef] [PubMed]
98. Santana de Oliveira, M.; Pereira da Silva, V.M.; Cantão Freitas, L.; Gomes Silva, S.; Nevez Cruz, J.; Aguiar Andrade, E.H. Extraction yield, chemical composition, preliminary toxicity of bignonia nocturna (bignoniaceae) essential oil and in silico evaluation of the interaction. *Chem. Biodivers.* **2021**, *18*, cbdv.202000982. [CrossRef] [PubMed]
99. do Nascimento, J.C.; David, J.M.; Barbosa, L.C.; de Paula, V.F.; Demuner, A.J.; David, J.P.; Conserva, L.M.; Ferreira, J.C.; Guimarães, E.F. Larvicidal activities and chemical composition of essential oils from *Piper klotzschianum* (Kunth) C. DC.(Piperaceae). *Pest Manag. Sci.* **2013**, *69*, 1267–1271. [CrossRef]
100. Barbosa, D.C.D.S.; Holanda, V.N.; Ghosh, A.; Maia, R.T.; da Silva, W.V.; Lima, V.L.D.M.; da Silva, M.V.; dos Santos Correia, M.T.; de Figueiredo, R.C.B.Q. Leishmanicidal and cytotoxic activity of essential oil from the fruit peel of *Myrciaria floribunda* (H. West ex Willd.) O. Berg: Molecular docking and molecular dynamics simulations of its major constituent onto Leishmania enzyme targets. *J. Biomol. Struct. Dyn.* **2021**, *39*, 1–16. [CrossRef]
101. Caldas, G.F.R.; Limeira, M.M.F.; Araújo, A.V.; Albuquerque, G.S.; Silva-Neto, J.D.C.; Silva, T.G.D.; Costa-Silva, J.H.; Menezes, I.R.A.D.; Costa, J.G.M.D.; Wanderley, A.G. Repeated-doses and reproductive toxicity studies of the monoterpene 1, 8-cineole (eucalyptol) in Wistar rats. *Food Chem. Toxicol.* **2016**, *97*, 297–306. [CrossRef] [PubMed]
102. Izham, M.N.M.; Hussin, Y.; Rahim, N.F.C.; Aziz, M.N.M.; Yeap, S.K.; Rahman, H.S.; Masarudin, M.J.; Mohamad, N.E.; Abdullah, R.; Alitheen, N.B. Physicochemical characterization, cytotoxic effect and toxicity evaluation of nanostructured lipid carrier loaded with eucalyptol. *BMC Complement. Med. Ther.* **2021**, *21*, 254. [CrossRef]
103. Elmhalli, F.; Garboui, S.S.; Borg-Karlson, A.K.; Mozūraitis, R.; Baldauf, S.L.; Grandi, G. The repellency and toxicity effects of essential oils from the libyan plants salvadora persica and rosmarinus officinalis against nymphs of ixodes ricinus. *Exp. Appl. Acarol.* **2019**, *77*, 585–599. [CrossRef]
104. Bhowal, M.; Gopal, M. Eucalyptol: Safety and pharmacological profile. *RGUHS J. Pharm Sci* **2016**, *5*, 125–131. [CrossRef]
105. Galan, D.M.; Ezeudu, N.E.; Garcia, J.; Geronimo, C.A.; Berry, N.M.; Malcolm, B.J. Eucalyptol (1, 8-cineole): An underutilized ally in respiratory disorders? *J. Essent. Oil Res.* **2020**, *32*, 103–110. [CrossRef]
106. Ribeiro, N.C.; da Camara, C.A.G.; Melo, J.P.R.; de Moraes, M.M. Acaricidal properties of essential oils from agro-industrial waste products from citric fruit against tetranychus urticae. *J. Appl. Entomol.* **2019**, *143*, 731–743. [CrossRef]
107. Agus, H.H. Terpene toxicity and oxidative stress. In *Toxicology: Oxidative Stress and Dietary Antioxidants*; Patel, V.B., Preedy, V.R., Eds.; Academic Press: Cambridge, UK, 2021; pp. 33–42.
108. Scherf, J.R.; Barbosa dos Santos, C.R.; Sampaio de Freitas, T.; Rocha, J.E.; Macêdo, N.S.; Mascarenhas Lima, J.N.; Melo Coutinho, H.D.; Bezerra da Cunha, F.A. Effect of terpinolene against the resistant Staphylococcus aureus strain, carrier of the efflux pump QacC and β-lactamase gene, and its toxicity in the Drosophila melanogaster model. *Microb. Pathog.* **2020**, *149*, 104528. [CrossRef] [PubMed]
109. de Christo Scherer, M.M.; Marques, F.M.; Figueira, M.M.; Peisino, M.C.O.; Schmitt, E.F.P.; Kondratyuk, T.P.; Endringer, D.C.; Scherer, R.; Fronza, M. Wound healing activity of terpinolene and α-phellandrene by attenuating inflammation and oxidative stress in vitro. *J. Tissue Viability* **2019**, *28*, 94–99. [CrossRef] [PubMed]
110. Martínez, L.C.; Plata-Rueda, A.; Colares, H.C.; Campos, J.M.; Dos Santos, M.H.; Fernandes, F.L.; Serrão, J.E.; Zanuncio, J.C. Toxic effects of two essential oils and their constituents on the mealworm beetle, tenebrio molitor. *Bull. Entomol. Res.* **2018**, *108*, 716–725. [CrossRef]
111. Abdelgaleil, S.A.M.; El-Sabrout, A.M. Composition, toxicity and developmental potential of three essential oils on the west nile virus mosquito, *Culex pipiens* L. *Int. J. Pest Manag.* **2020**, *66*, 1–9. [CrossRef]
112. Mubarak, E.E.; Ali, L.Z.; Ahmed, I.F.A.; Ahmed, A.B.A.; Taha, R.M. Essential oil compositions and cytotoxicity from various organs of eucalyptus camaldulensis. *Int. J. Agric. Biol.* **2015**, *17*, 320–326.

113. Andrade-Ochoa, S.; Sánchez-Aldana, D.; Chacón-Vargas, K.F.; Rivera-Chavira, B.E.; Sánchez-Torres, L.E.; Camacho, A.D.; Nogueda-Torres, B.; Nevárez-Moorillón, G.V. Oviposition deterrent and larvicidal and pupaecidal activity of seven essential oils and their major components against Culex quinquefasciatus Say (Diptera: Culicidae): Synergism–antagonism effects. *Insects* **2018**, *9*, 25. [CrossRef]
114. Kwiatkowski, P.; Łopusiewicz, Ł.; Pruss, A.; Kostek, M.; Sienkiewicz, M.; Bonikowski, R.; Wojciechowska-Koszko, I.; Dołęgowska, B. Antibacterial activity of selected essential oil compounds alone and in combination with β-lactam antibiotics against mrsa strains. *Int. J. Mol. Sci.* **2020**, *21*, 7106. [CrossRef]
115. Feng, Y.X.; Wang, Y.; Geng, Z.F.; Zhang, D.; Almaz, B.; Du, S.S. Contact toxicity and repellent efficacy of valerianaceae spp. to three stored-product insects and synergistic interactions between two major compounds camphene and bornyl acetate. *Ecotoxicol. Environ. Saf.* **2020**, *190*, 110106. [CrossRef]
116. Cascaes, M.M.; Silva, S.G.; Cruz, J.N.; Oliveira, S.D.; Oliveira, J.; Antonio, A.; Moraes, B.D.; Augusto, F.; Santana, K.; Diniz, L.; et al. First report on the annona exsucca dc. essential oil and in silico identification of potential biological targets of its major compounds. *Nat. Prod. Res.* **2021**, *36*, 4009–4012. [CrossRef]
117. Souza da Silva Júnior, O.; de Jesus Pereira Franco, C.; Barbosa de Moraes, A.A.; Cruz, J.N.; Santana da Costa, K.; Diniz do Nascimento, L.; Helena de Aguiar Andrade, E. In silico analyses of toxicity of the major constituents of essential oils from two Ipomoea L. species. *Toxicon* **2021**, *195*, 111–118. [CrossRef]
118. Da Silva Júnior, O.S.; Franco, C.D.J.P.; Moraes, Â.A.B.D.; Pastore, M.; Cascaes, M.M.; do Nascimento, L.D.; Oliveira, M.S.D.; Andrade, E.H.D.A. Chemical variability of volatile concentrate from two Ipomoea L. species within a seasonal gradient. *Nat. Prod. Res.* **2022**, *36*, 1–8. [CrossRef] [PubMed]
119. Brari, J.; Thakur, D.R. Fumigant toxicity and cytotoxicity evaluation of monoterpenes against four stored products pests. *Int. J. Dev. Res.* **2015**, *5*, 5661–5667.
120. Santos, P.L.; Matos, J.P.S.C.F.; Picot, L.; Almeida, J.R.G.S.; Quintans, J.S.S.; Quintans-Júnior, L.J. *Citronellol, A Monoterpene Alcohol with Promising Pharmacological Activities—A Systematic Review*; Elsevier Ltd.: Amsterdam, The Netherlands, 2019; Volume 123, ISBN 5579210566.
121. Wojtunik-Kulesza, K.A. Toxicity of selected monoterpenes and essential oils rich in these compounds. *Molecules* **2022**, *27*, 1716. [CrossRef] [PubMed]
122. Amina, M.; Alam, P.; Parvez, M.K.; Al-Musayeib, N.M.; Al-Hwaity, S.A.; Al-Rashidi, N.S.; Al-Dosari, M.S. Isolation and validated hptlc analysis of four cytotoxic compounds, including a new sesquiterpene from aerial parts of plectranthus cylindraceus. *Nat. Prod. Res.* **2018**, *32*, 804–809. [CrossRef] [PubMed]
123. Pang, X.; Almaz, B.; Qi, X.J.; Wang, Y.; Feng, Y.X.; Geng, Z.F.; Xi, C.; Du, S.S. Bioactivity of essential oil from atalantia buxifolia leaves and its major sesquiterpenes against three stored-product insects. *J. Essent. Oil-Bear. Plants* **2020**, *23*, 38–50. [CrossRef]
124. Barreto, I.C.; de Almeida, A.S.; Sena Filho, J.G. Taxonomic insights and its type cyclization correlation of volatile sesquiterpenes in vitex species and potential source insecticidal compounds: A review. *Molecules* **2021**, *26*, 6405. [CrossRef] [PubMed]
125. Hussain, A.; Rizwan-Ul-Haq, M.; AlJabr, A.M.; Al-Ayedh, H. Lethality of sesquiterpenes reprogramming red palm weevil detoxification mechanism for natural novel biopesticide development. *Molecules* **2019**, *24*, 1648. [CrossRef]
126. Wang, F.; Park, Y.L.; Gutensohn, M. Glandular trichome-derived sesquiterpenes of wild tomato accessions (solanum habrochaites) affect aphid performance and feeding behavior. *Phytochemistry* **2020**, *180*, 112532. [CrossRef]
127. Scalvenzi, L.; Grandini, A.; Spagnoletti, A.; Tacchini, M.; Neill, D.; Ballesteros, J.; Sacchetti, G.; Guerrini, A. *Myrcia splendens* (Sw.) DC.(syn. *M. fallax* (Rich.) DC.)(Myrtaceae) essential oil from Amazonian Ecuador: A chemical characterization and bioactivity profile. *Molecules* **2017**, *22*, 1163. [CrossRef]
128. Melo, C.R.; Blank, A.F.; Oliveira, B.M.S.; Santos, A.C.C.; Cristaldo, P.F.; Araújo, A.P.A.; Bacci, L. Formicidal activity of essential oils of myrcia lundiana chemotypes on acromyrmex balzani. *Crop Prot.* **2021**, *139*, 105343. [CrossRef]
129. Alves, M.S.; Santos, D.P.; Silva, L.C.P.; Pontes, E.G.; Souza, M.A.A. Essential oils composition and toxicity tested by fumigation against Callosobruchus maculatus (Coleoptera: Bruchidae) pest of stored cowpea. *Rev. Virtual Quim.* **2015**, *7*, 2387–2399. [CrossRef]
130. Fernandes, C.C.; Andrade, P.M.D.; Santos, T.C.L.D.; Santiago, M.B.; Crotti, A.E.M.; Martins, C.H.G.; Magalhães, L.G.; Miranda, M.L.D. In vitro evaluation of anticaries, antimycobacterial, antileishmanial and cytotoxic activities of essential oils from eremanthus erythropappus and of α-bisabolol, their major sesquiterpene. *Aust. J. Crop Sci.* **2020**, *14*, 236–243. [CrossRef]
131. Mahdavi, V.; Rafiee-Dastjerdi, H.; Asadi, A.; Razmjou, J.; Achachlouei, B.F. Synthesis of Zingiber officinale essential oil-loaded nanofiber and its evaluation on the potato tuber moth, Phthorimaea operculella (Lepidoptera: Gelechiidae). *J. Crop Prot.* **2018**, *7*, 39–49.
132. Fernandez, C.M.M.; Lorenzetti, F.B.; Kleinubing, S.A.; de Andrade, J.P.P.; Bortolucci, W.D.C.; Gonçalves, J.E.; Júnior, R.P.; Cortez, D.A.G.; Gazim, Z.C.; Filho, B.P.D. Chemical composition and insecticidal activity of Garcinia gardneriana (Planchon & Triana) Zappi (Clusiaceae) essential oil. *Bol. Latinoam. y del Caribe de Plantas Med. y Aromát.* **2021**, *20*, 503–514. [CrossRef]
133. Vinturelle, R.; Mattos, C.; Meloni, J.; Lamberti, H.D.; Nogueira, J.; da Silva Vaz Júnior, I.; Rocha, L.; Lione, V.; Folly, E. Evaluation of essential oils as an ecological alternative in the search for control Rhipicephalus microplus (Acari: Ixodidae). *Vet. Parasitol. Reg. Stud. Rep.* **2021**, *23*, 100523. [CrossRef]
134. Matsuda, S.; Tsunematsu, Y.; Matsushita, T.; Ogata, Y.; Hachiya, S.; Kishimoto, S.; Miyoshi, N.; Watanabe, K. Toward engineered biosynthesis of drugs in human cells. *ChemBioChem* **2022**, *23*, 2022. [CrossRef]

135. Ambrož, M.; Boušová, I.; Skarka, A.; Hanušová, V.; Králová, V.; Matoušková, P.; Szotáková, B.; Skálová, L. The influence of sesquiterpenes from myrica rubra on the antiproliferative and pro-oxidative effects of doxorubicin and its accumulation in cancer cells. *Molecules* **2015**, *20*, 15343–15358. [CrossRef]
136. Ferreira, O.O.; Franco, C.D.J.P.; Varela, E.L.P.; Silva, S.G.; Cascaes, M.M.; Percário, S.; de Oliveira, M.S.; Andrade, E.H.D.A. Chemical Composition and Antioxidant Activity of Essential Oils from Leaves of Two Specimens of *Eugenia florida* DC. *Molecules* **2021**, *26*, 5848. [CrossRef]
137. Ola, O.S.; Sofolahan, T.A. A monoterpene antioxidant, linalool, mitigates benzene-induced oxidative toxicities on hematology and liver of male rats. *Egypt. J. Basic Appl. Sci.* **2021**, *8*, 39–53. [CrossRef]
138. da Costa, J.S.; Andrade, W.M.S.; de Figueiredo, R.O.; Santos, P.V.L.; Freitas, J.J.D.S.; Setzer, W.N.; da Silva, J.K.R.; Maia, J.G.S.; Figueiredo, P.L.B. Chemical composition and variability of the volatile components of myrciaria species growing in the amazon region. *Molecules* **2022**, *27*, 2234. [CrossRef]
139. Plaza, M.; Batista, Â.G.; Cazarin, C.B.B.; Sandahl, M.; Turner, C.; Östman, E.; Maróstica Júnior, M.R. Characterization of antioxidant polyphenols from myrciaria jaboticaba peel and their effects on glucose metabolism and antioxidant status: A pilot clinical study. *Food Chem.* **2016**, *211*, 185–197. [CrossRef] [PubMed]
140. Santos, I.B.D.S.; Santos dos Santos, B.; Oliveira, J.R.S.D.; Costa, W.K.; Zagmignan, A.; da Silva, L.C.N.; Ferreira, M.R.A.; Lermen, V.L.; Lermen, M.S.B.D.S.; da Silva, A.G.; et al. Antioxidant action and in vivo anti-inflammatory and antinociceptive activities of myrciaria floribunda fruit peels: Possible involvement of opioidergic system. *Adv. Pharmacol. Pharm. Sci.* **2020**, *2020*, 1258707. [CrossRef] [PubMed]
141. Xu, G.; Guo, J.; Sun, C. Eucalyptol ameliorates early brain injury after subarachnoid haemorrhage via antioxidant and anti-inflammatory effects in a rat model. *Pharm. Biol.* **2021**, *59*, 114–120. [CrossRef] [PubMed]
142. Boukhatem, M.N.; Boumaiza, A.; Nada, H.G.; Rajabi, M.; Mousa, S.A. *Eucalyptus globulus* essential oil as a natural food preservative: Antioxidant, antibacterial and antifungal properties in vitro and in a real food matrix (orangina fruit juice). *Appl. Sci.* **2020**, *10*, 5581. [CrossRef]
143. Limam, H.; Ben Jemaa, M.; Tammar, S.; Ksibi, N.; Khammassi, S.; Jallouli, S.; Del Re, G.; Msaada, K. Variation in chemical profile of leaves essential oils from thirteen tunisian eucalyptus species and evaluation of their antioxidant and antibacterial properties. *Ind. Crops Prod.* **2020**, *158*, 112964. [CrossRef]
144. Chambre, D.R.; Moisa, C.; Lupitu, A.; Copolovici, L.; Pop, G.; Copolovici, D.M. Chemical composition, antioxidant capacity, and thermal behavior of satureja hortensis essential oil. *Sci. Rep.* **2020**, *10*, 21322. [CrossRef]
145. Aydin, E.; Türkez, H.; Taşdemir, Ş. Anticancer and antioxidant properties of terpinolene in rat brain cells. *Arch. Ind. Hyg. Toxicol.* **2013**, *64*, 415–424. [CrossRef]
146. Osanloo, M.; Ghanbariasad, A.; Taghinezhad, A. Antioxidant and anticancer activities of *Anethum graveolens* L., *Citrus limon* (L.) Osbeck and Zingiber officinale Roscoe essential oils. *Tradit. Integr. Med.* **2021**, *6*, 333–347. [CrossRef]
147. De Castro, J.A.M.; Monteiro, O.S.; Coutinho, D.F.; Rodrigues, A.A.C.; Da Silva, J.K.R.; Maia, J.G.S. Seasonal and circadian study of a thymol/γ-terpinene/p-cymene type oil of *Ocimum gratissimum* L. and its antioxidant and antifungal effects. *J. Braz. Chem. Soc.* **2018**, *30*, 930–938. [CrossRef]
148. Julaeha, E.; Dewi, K.S.; Nurzaman, M.; Wahyudi, T.; Herlina, T. Chemical compositions and antioxidant activities of indonesian citrus essential oils and their elucidation using principal component analysis. *Preprints* **2020**, *11*, 86. [CrossRef]
149. Xanthis, V.; Fitsiou, E.; Voulgaridou, G.P.; Bogadakis, A.; Chlichlia, K.; Galanis, A.; Pappa, A. Antioxidant and Cytoprotective Potential of the Essential Oil *Pistacia lentiscus* var. *chia* and Its Major Components Myrcene and α-Pinene. *Antioxidants* **2021**, *10*, 127. [CrossRef] [PubMed]
150. Aydin, E.; Türkez, H.; Geyikoğlu, F. Antioxidative, anticancer and genotoxic properties of α-pinene on N2a neuroblastoma cells. *Biologia* **2013**, *68*, 1004–1009. [CrossRef]
151. Türkez, H.; Aydın, E. In vitro assessment of cytogenetic and oxidative effects of α-pinene. *Toxicol. Ind. Health* **2016**, *32*, 168–176. [CrossRef] [PubMed]
152. Ciesla, L.M.; Wojtunik-Kulesza, K.A.; Oniszczuk, A.; Waksmundzka-Hajnos, M. Antioxidant synergism and antagonism between selected monoterpenes using the 2,2-diphenyl-1-picrylhydrazyl method. *Flavour Fragr. J.* **2016**, *31*, 412–419. [CrossRef]
153. Olszowy-Tomczyk, M. Synergistic, antagonistic and additive antioxidant effects in the binary mixtures. *Phytochem. Rev.* **2020**, *19*, 63–103. [CrossRef]
154. Franco, C.D.J.P.; Ferreira, O.O.; Antônio Barbosa de Moraes, Â.; Varela, E.L.P.; Nascimento, L.D.D.; Percário, S.; de Oliveira, M.S.; Andrade, E.H.D.A. Chemical composition and antioxidant activity of essential oils from Eugenia patrisii vahl, E. punicifolia (Kunth) dc., and Myrcia tomentosa (Aubl.) dc., leaf of family Myrtaceae. *Molecules* **2021**, *26*, 3292. [CrossRef]
155. Gatto, L.J.; Fabri, N.T.; Souza, A.M.D.; Fonseca, N.S.T.D.; Furusho, A.D.S.; Miguel, O.G.; Dias, J.D.F.G.; Zanin, S.M.W.; Miguel, M.D. Chemical composition, phytotoxic potential, biological activities and antioxidant properties of myrcia hatschbachii D. legrand essential oil. *Braz. J. Pharm. Sci.* **2020**, *56*, 1–9. [CrossRef]
156. Mohankumar, A.; Kalaiselvi, D.; Levenson, C.; Shanmugam, G.; Thiruppathi, G.; Nivitha, S.; Sundararaj, P. Antioxidant and stress modulatory efficacy of essential oil extracted from plantation-grown *Santalum album* L. *Ind. Crops Prod.* **2019**, *140*, 111623. [CrossRef]
157. Saroglou, V.; Arfan, M.; Shabir, A.; Hadjipavlou-Litina, D.; Skaltsa, H. Composition and antioxidant activity of the essential oil of Teucrium royleanum Wall. ex Benth growing in Pakistan. *Flavour Fragr. J.* **2007**, *22*, 154–157. [CrossRef]

158. Iannone, M.; Ovidi, E.; Vitalini, S.; Laghezza Masci, V.; Iriti, M.; Tiezzi, A.; Garzoli, S.; Marianelli, A. From hops to craft beers: Production process, vocs profile characterization, total polyphenol and flavonoid content determination and antioxidant activity evaluation. *Processes* **2022**, *10*, 517. [CrossRef]
159. Zárybnický, T.; Boušová, I.; Ambrož, M.; Skálová, L. Hepatotoxicity of monoterpenes and sesquiterpenes. *Arch. Toxicol.* **2018**, *92*, 1–17. [CrossRef] [PubMed]
160. Pisoschi, A.M.; Pop, A. The role of antioxidants in the chemistry of oxidative stress: A review. *Eur. J. Med. Chem.* **2015**, *97*, 55–74. [CrossRef] [PubMed]

antioxidants

MDPI

Article

Phytochemical Profile, In Vitro Bioactivity Evaluation, In Silico Molecular Docking and ADMET Study of Essential Oils of Three *Vitex* Species Grown in Tarai Region of Uttarakhand

Himani Karakoti [1], Sonu Kumar Mahawer [1], Monika Tewari [1], Ravendra Kumar [1,*], Om Prakash [1], Mozaniel Santana de Oliveira [2,*] and Dharmendra Singh Rawat [3]

1 Department of Chemistry, College of Basic Sciences and Humanities, G.B. Pant University of Agriculture and Technology, Pantnagar 263145, U.S. Nagar, Uttarakhand, India

2 Campus de Pesquisa-Museu Paraense Emílio Goeldi-Botany Coordination, Av. Perimetral, 1901-Terra Firme, Belém 66077-830, PA, Brazil

3 Department of Biological Sciences, College of Basic Sciences and Humanities, G.B. Pant University of Agriculture and Technology, Pantnagar 263145, U.S. Nagar, Uttarakhand, India

* Correspondence: ravendra.kumar@gbpuat-cbsh.ac.in (R.K.); mozanieloliveira@museu-goeldi.br (M.S.d.O.)

Abstract: A comparative study of volatiles, antioxidant activity, phytotoxic activity, as well as in silico molecular docking and ADMET study, was conducted for essential oils from three *Vitex* species, viz., *V. agnus-castus*, *V. negundo*, and *V. trifolia*. Essential oils (OEs) extracted by hydrodistillation were subjected to compositional analysis using GC-MS. A total number of 37, 45, and 43 components were identified in *V. agnus-castus*, *V. negundo*, and *V. trifolia*, respectively. The antioxidant activity of EOs, assessed using different radical-scavenging (DPPH, H_2O_2 and NO), reducing power, and metal chelating assays, were found to be significant as compared with those of the standards. The phytotoxic potential of the EOs was performed in the receptor species *Raphanus raphanistrum* (wild radish) and the EOs showed different levels of intensity of seed germination inhibition and root and shoot length inhibition. The molecular docking study was conducted to screen the antioxidant and phytotoxic activity of the major and potent compounds against human protein target, peroxiredoxin 5, and 4-hydroxyphenylpyruvate dioxygenase protein (HPPD). Results showed good binding affinities and attributed the strongest inhibitory activity to 13-*epi*-manoyl oxide for both the target proteins.

Keywords: natural products; bioactive compounds; antioxidant; phytotoxic; molecular modeling; virtual ligand screening

Citation: Karakoti, H.; Mahawer, S.K.; Tewari, M.; Kumar, R.; Prakash, O.; de Oliveira, M.S.; Rawat, D.S. Phytochemical Profile, In Vitro Bioactivity Evaluation, In Silico Molecular Docking and ADMET Study of Essential Oils of Three *Vitex* Species Grown in Tarai Region of Uttarakhand. *Antioxidants* **2022**, *11*, 1911. https://doi.org/10.3390/antiox11101911

Academic Editors: Delia Mirela Tit and Simona Bungau

Received: 29 August 2022
Accepted: 23 September 2022
Published: 27 September 2022

1. Introduction

The plant genus *Vitex* (family: *Verbenaceae*) consists of 250 accepted species and has a wide distribution all over the world, ranging from shrubs to trees in the tropical, subtropical regions, and temperate zones [1]. The members of this genus have been widely used in folk medicine and are greatly valued as medicinal plants in several Asian countries, including India, Pakistan, Nepal, China, Sri Lanka, and Bangladesh [2]. The leaves, seeds, flowers, and the whole aerial part of different species of *Vitex* genus have several external and internal uses. The most popular uses of these plants are in the curing of asthma, ophthalmodynia, headaches, coughs, premenopausal syndrome, etc., but various other uses have also been reported [2]. For instance, *V. agnus-castus* fruits are being used in the treatment of menstrual disorders (amenorrhea, dysmenorrhea), and in other female conditions like premenstrual dysphoric disorder, infertility, disrupted lactation, acne, breast pain, menopause, and inflammatory conditions [3,4]. *V. negundo* species is used as tonic, vermifuge, lactagogue, and is also used to treat catarrhal fever, eye diseases, inflammation, skin ulcers, rheumatoid arthritis, and bronchitis [5]. *V. trifolia* is used as a sedative for headaches, as an anti-inflammatory agent, and for the cure of the common cold. The plant is also used for the treatment of cancers in Chinese folk medicine [6,7]. Phytochemical studies revealed that the members of the *Vitex* genus are rich sources of bioactive compounds, including essential oils (terpenoids), flavonoids, glycosides, phenolic acids, ecdysteroids, etc. [7]. Available literature advocated that the essential oils (EOs) and other bioactive compounds from *Vitex* species reported a number of biological activities, such as antioxidant, antibacterial, estrogenic, cytotoxic, antifeedant, antifungal, antidiabetic, enzyme inhibitory, antiproliferative, antipyretic, antimalarial, antinociceptive, and phytotoxic [7–13].

Modern pharmacological therapies for oxidative stress caused by free radicals are effective, but are costly, and are associated with several undesirable side effects such as carcinogenic and teratogenic effects [14,15]. In respect to the chemical antioxidants, the search for alternative sources, based on natural, plant-based origins, is increasing nowadays, as they are supposed to be safer. Essential oils from a wide number of plant sources have been tested and have been reported to have excellent antioxidant properties [16–19]. Similarly, in recent years, research regarding the pesticidal properties of plant-based products (botanicals) has been gradually increasing as they are safer to the environment, easily degradable, and have low toxicity as compared to the chemical pesticides [20]. Moreover, in previous studies, the essential oil obtained from *V. negundo* and *V. agnus-castus* has been reported to cause potent phytotoxic activity [21,22], however, there is no report on the phytotoxic potential of *V. trifolia*. EOs are excellent options among natural sources, as they are sources of highly phytotoxic allelochemicals that affect the growth and development of plants and unwanted plants (weeds) [23].

Studies highlighted *Vitex* species as a source of antioxidant and phytotoxic agent due to their essential oil composition [7,21,22]. The efficiency and activity of the essential oil greatly depends on its chemical constituents, which depends on the genotypes of the plant and on the environmental, climatic, and agronomic conditions [24]. The essential oils of *V. agnus castus* and *V. trifolia*, from the Tarai region of Uttarakhand, have not been screened for their phytochemical analysis. Therefore, briefly, the present study aims (i) to check the chemical diversity among the essential oil composition of *Vitex agnus-castus*, *Vitex negundo*, and *Vitex trifolia* (from Tarai region of Uttarakhand, India); (ii) to evaluate the in vitro antioxidant and phytotoxic (herbicidal) activities; (iii) to carry out the in silico studies about the inhibitory effect of major volatiles of the essential oils on the crystal structures of some proteins; (iv) to perform the ADMET prediction of major compounds identified in essential oils under investigation.

2. Materials and Methods

2.1. Collection of Plant Material

The fresh leaves of different plant species of the *Vitex* were collected from Pantnagar (28°58′12″ N, 79°24′36″ E), Tarai region of Uttarakhand, India. The plant specimens were identified by one of the authors (D.S. Rawat), a taxonomist. The voucher specimen for different identified *Vitex* species viz., *Vitex agnus-castus*, *Vitex negundo* L., and *Vitex trifolia* L., with voucher numbers GBPUH-1439, GBPUH-1438, and GBPUH-1440, were deposited at the herbarium of Department of Biological Sciences, for future references.

2.2. Extraction of Essential Oil

Fresh leaves of different *Vitex* species were subjected to hydrodistillation for 4 h to isolate essential oils using a Clevenger-type apparatus, and the isolated essential oils were designated as VAO, VNO, and VTO for *Vitex agnus-castus*, *Vitex negundo*, and *Vitex trifolia*, respectively. The obtained essential oils were dried over anhydrous sodium sulphate (Na_2SO_4) in order to remove any trace of water and then stored in amber color glass vials at a low temperature (4 °C in refrigerator) for further uses. The oil yield (v/w) was recorded as 0.9% (0.45 mL/100 gm dry matter), 0.8% (0.4 mL/100 gm dry matter), and 0.6% (0.3 mL/100 gm dry matter) for VAO, VNO, and VTO, respectively.

2.3. Chemical Composition Analysis

To check the chemical diversity in tested *Vitex* species, the essential oils were analyzed by GC-MS (Shimadzu QP 2010 plus) with GCMS-QP 2010 Ultra DB-5 and GCMS-QP 2010 Ultra Rtx-5MS column (30 m × 0.25 mm i.d., 0.25 μm). The following experimental conditions used helium as the carrier gas (flow rate = 1.21 mL/min, split ratio = 10.0). Oven temperature was programmed at 50–280 °C with a temperature gradient of 3 °C/min up to 210 °C (isotherm for 2 min), then 6 °C/min up to 280 °C. Identification of essential oil components was done by comparing their relative retention index (RI) values with mass spectra NIST (NIST version 2.1) and WILEY (7th edition) libraries, and by matching the fragmentation pattern of the mass spectral data with those reported in the literature [25,26].

2.4. Antioxidant Activity

Different in vitro tests were performed to evaluate the antioxidant activity of the essential oils, and the results were presented as mean ± SD of triplicate.

2.4.1. DPPH Radical Scavenging Assay

Previously proposed methods have been followed to perform the assay [27,28]. In brief, different concentrations of VAO, VNO, and VTO (10 μL/mL–50 μL/mL) were added to 5 mL of freshly prepared methanolic solution of DPPH (0.004%), solution was kept for incubation under dark for half an hour, and further, the absorbance was taken in triplicates at 517 nm in a UV spectrophotometer (Thermo Fisher Scientific, Evolution-201, Waltham, MA, USA) against a blank. The standard antioxidant used was BHT, in the same concentrations as the tested essential oils (10 μL/mL–50 μL/mL). The % inhibition of DPPH free radical of the oils and standard was calculated by using the following formula:

$$\%\ \text{DPPH radical scavenging activity} = \frac{(A_o - A_t)}{A_o} \times 100$$

where A_o and A_t are the absorbance values of control and test essential oils, respectively. Percent inhibition was plotted against concentrations, and the equation for the line was used to obtain the IC_{50} (half-maximal inhibitory concentration) values.

2.4.2. Hydrogen Peroxide (H_2O_2) Radical Scavenging Activity

H_2O_2 radical scavenging activity of tested samples was performed as per the prescribed protocol reported earlier [29,30]. Here, 0.6 mL of H_2O_2 solution (40 mM) prepared in phosphate buffer (0.1 M; pH 7.4) was added to 0.4 mL methanolic solution of different concentrations of essential oils and the standard (10–50 μL/mL). The above solution was incubated at room temperature for 10 min. Further, the absorbance was taken at 230 nm against the blank, i.e., methanol. Here, *L*-ascorbic acid (10–50 μL/mL) was taken as a positive control. The percentage scavenging of H_2O_2 was calculated by using the following formula:

$$\%\ H_2O_2\ \text{radical scavenging activity} = \frac{(A_o - A_t)}{A_o} \times 100$$

where A_o and A_t are the absorbance values of control and test essential oils, respectively. Percent inhibition was plotted against concentrations and the equation for the line was used to obtain the IC_{50} values.

2.4.3. Nitric Oxide Radical Scavenging Activity

The nitric oxide (NO) radical scavenging activity of the tested essential oils was determined by the method as described earlier [31]. Briefly, 2 mL of sodium nitroprusside (10 mM) prepared in phosphate buffer saline (0.5 mM, pH 7.4) was added to different concentrations of essential oils and the standard (10–50 μL/mL) separately, and incubated at 25 °C for 150 min. Further, 0.5 mL of griess reagent containing 1.0 mL sulphanilic acid reagent was added to 0.5 mL of each incubated solution. The mixture was again incubated for 30 min at room temperature and the absorbance was taken at 540 nm. *L*-ascorbic acid was taken as the standard antioxidant. The percentage scavenging of NO was calculated by using the following formula:

$$\%\ \text{NO radical scavenging activity} = \frac{(A_o - A_t)}{A_o} \times 100$$

where A_o and A_t are the absorbance values of control and test essential oils, respectively. Percent inhibition was plotted against concentrations and the equation for the line was used to obtain the IC_{50} values.

2.4.4. Reducing Power Assay

The reducing power assay of different essential oils was determined by the method developed earlier [32]. Different concentrations of tested samples (essential oils and the standard (10–50 μL/mL)) were added to 2.5 mL of phosphate buffer (200 mM, pH = 6.6). Further, 2.5 mL of 1% potassium ferricyanide, $K_3[FeCN_6]$, was added to the above solution. The solution was incubated for 20 min at 50 °C and then 2.5 mL of trichloroacetic acid was added to the incubated solution, followed by centrifugation at 650 rpm for 10 min. 5 mL of distilled water and 1 mL of 0.1% ferric chloride were added to the upper layer (1 mL). The absorbance of the final solution was taken at 700 nm, and gallic acid (10–50 μL/mL) was taken as a positive control. The percentage reducing power was calculated by using the following formula:

$$\%\ \text{Reducing power activity} = \frac{(A_o - A_t)}{A_o} \times 100$$

where A_o and A_t are the absorbance values of control and test essential oils, respectively. RP_{50} values were calculated using regression equations for the percent inhibition plotted against concentrations.

2.4.5. Fe^{2+} Metal Chelating Activity

The Fe^{2+} metal-chelation activity of VAO, VNO, and VTO was measured as per the prescribed and developed protocol [33]. Different concentration of oils (10–50 μL/mL), as well as the standard, were mixed with 0.1 mL of $FeCl_2{\cdot}4H_2O$ (2 mM) and 0.2 mL of (5 Mm) ferrozine separately. Further, methanol (4.7 mL) was added to the solution, making the final volume 5 mL. The solution was shaken and was incubated for 30 min at 25 °C, and the absorbance was taken at 562 nm using spectrophotometer (Thermo Fisher Scientific, Evolution-201, USA). Na_2-EDTA (10–50 μL/mL) was used as a standard antioxidant. The ability of the samples to chelate ferrous ion was calculated using the following formula:

$$\%\ Fe^{2+}\ \text{metal-chelation activity} = \frac{(A_o - A_t)}{A_o} \times 100$$

where A_o and A_t are the absorbance values of control and test essential oils, respectively. IC_{50} values were obtained using regression equations for the plots of percent inhibition against concentrations.

2.5. Herbicidal (Phytotoxic) Activity

The herbicidal activity on the receptor plant, *Raphanus raphanistrum*, was carried out with essential oils of *Vitex* species. Different parameters were used, such as inhibition of seed germination, inhibition of shoot, and root length growth, using the method reported earlier [34,35]. For the experiment, radish seeds were obtained from the VRC (Vegetable Research Centre), G.B.P.U.A. & T. Pantnagar, Uttarakhand, India.

2.5.1. Seed Germination Inhibition

For evaluating the seed germination inhibition, different concentrations of essential oils (50–200 μL/mL) were prepared in Tween-20 (1%) solution of distilled water. In order to break dormancy, radish seeds were surface sterilized in 5% hypochlorite solution for 15 min. Ten sterilized seeds of radish were placed in each petri plates, which were lined with sheets of qualitative filter papers. Further, 2 mL of various concentrations of the tested sample (50–200 μL/mL) were applied onto the plates and the seeds were allowed to germinate at controlled condition of 25 ± 1 °C and a photoperiod of 12 h in an incubator. Seeds with a root length of 2 mm were considered germinated. Distilled water was taken as the control while pendimethalin (50–200 μL/mL) was used as a standard herbicide, and the bioassay was performed in triplicate. After 120 h, the numbers of germinated seeds in each petri dish were counted, followed by the calculation of percent seed germination inhibition values using the following formula:

$$\text{Inhibition of seed germination (\% Inhibition)} = 100 \times (1 - Gt/Gc)$$

where Gt = no. of seeds germinates in treatment, Gc = No. of seeds germinate in control.

2.5.2. Inhibition of Shoot and Root Elongation

Assessment of shoot and root elongation were performed at controlled condition of 25 °C for a photoperiod of 24 h. Each Petri dish received 2.0 mL of the test solution, and two pre-germinated seeds were placed in each petri plate. The EOs were tested at the same concentrations as the germination bioassay. At the end of the 120 h of incubation, the length of the shoot and root were measured. Distilled water was taken as the controlled treatment while pendimethalin (50–200 μL/mL) was used as a standard herbicide, and the bioassays were performed in triplicate. The formulae used for determining the inhibition of shoot and root growth were as follows:

$$\text{Inhibition of hypocotyl (shoot length) growth (\% Inhibition)} = 100 \times (1 - Ct/Cc)$$

where, Ct = shoot length growth in treatment, Cc = shoot length growth in control.

$$\text{Inhibition of radicle (root length) growth (\% Inhibition)} = 100 \times (1 - Rt/Rc)$$

where, Rt = root length growth in treatment, Rc = root length growth in control.

2.6. Molecular Docking Studies

Virtual ligand screening is an in silico method used to dock small molecules (ligand) to macromolecule (protein) to discover potent compounds that have the necessary biological effect [36]. The molecular docking study of the selected volatiles from VAO, VNO, and VTO was carried out on 4-hydroxyphenylpyruvate dioxygenase (HPPD) receptors, as this protein has been reported as a molecular target for compounds with post-emergence herbicidal activity [37,38], and the second protein taken was human peroxiredoxin 5, which has a broader activity against reactive oxygen species [39]. The three-dimensional (3D) structures of the HPPD and human peroxiredoxin 5 proteins were obtained from the RCSB ProteinData Bank with PDB ID: 6J63 and 1HD2, respectively. The 3D structures of the selected proteins converted into PDB formats by deleting the water molecules, HETATOMS, and adding polar hydrogens using Biovia Discovery Studio-2021 Client. The compounds from the essential oils for docking studies were selected based on their higher percentage contents and their concerned structures were obtained from the PUBCHEM database (https://pubchem.ncbi.nlm.nih.gov/, accessed on 12 August 2022) in the SDF (structure data file) format. The selected compounds were 1,8-cineole (CID:2758), sabinene (CID:18818), α-pinene (CID:6654), α-terpinyl acetate (CID:111037), β-farnesene (CID:5281517), viridiflorol (CID:11996452), β-caryophyllene (CID:5281515), β-iraldiene (CID: 5375218), terpine-4-ol (CID:2724161), 5-(1-isopropenyl-4,5-dimethylbicyclo[4.3.0]nonan-5-yl)-3-methyl-2-pentenyl acetate (CID:5375240), 13-*epi*-manoyl acetate (CID:18529657), α-phellandrene (CID:7460), and caryophyllene oxide (CID:1742210). Structures of the ligands in their SDF format were then imported into PyRx Software using an open babel tool embedded in PyRx software. Energy minimization (optimization) was performed by adding charges and optimizing the universal force field. Further, the ligands were converted into AutoDock Ligand format (PDBQT). To find out the binding affinity and to know the various ligand–receptor interactions responsible for the antioxidant and phytotoxic activity, the molecular docking of the selected major constituents was performed using PyRx with Vina Wizard tool. The protein and multiple ligands to be docked were selected in the PyRx software using the Vina Wizard Control. The "Run Vina" control was selected to initiate the docking process. The results were observed by selecting the "Analyze Vina" tool and exported as CSV files [36]. Biovia Discovery Studio-2021 Client was used for the visualization of 2D and 3D interactions of docking poses.

2.7. In Silico ADMET Study

The structures of the selected compounds from the essential oils were drawn using ChemDraw Ultra 8.0 for the pharmacokinetics (absorption, distribution, metabolism, and excretion (ADME)) studies. The legends were converted into the SMILES format and then the drug-like and pharmacokinetic properties of the selected compounds were predicted using ADME tool by a SwissADME online server (http://www.swissadme.ch/, accessed on 12 August 2022), as per the developed protocol [40]. ProTox-II webserver (http://tox.charite.de/protox_II, accessed on 12 August 2022) was used to study the toxicity profile. It calculates the prediction based on different parameters such as organ toxicity (hepatotoxicity), oral toxicity, and toxicological endpoints (cytotoxicity, mutagenicity, carcinotoxicity, and immunotoxicity).

2.8. Statistical Analysis

Two and one factor Analysis of variance (ANOVA), followed by the Tukey test, was performed using RStudio (Version 2021.09.2) developed by RStudio team, PBC, Boston, MA and OriginPro, Version 2022b developed by OriginLab, Northampton, MA, USA Student trial version software, respectively, to analyze the significant difference among the treatment means. The p value < 0.05 was considered to show the significant difference. All the data in the experiment were reported as mean $\pm$ SD (standard deviation). To define the variability in different essential oils based on chemical composition, Chemometric Analysis was performed based on the heatmap clustering using heatmapper, free web server available (http://www.heatmapper.ca, accessed on 12 August 2022) developed at University of Alberta, Canada [41]. We performed Principal Component Analysis (PCA) on chemical composition for the three *Vitex* species under investigation to identify the most significant features in the dataset and Pearson's correlation test to analyze the correlation among the chemical compounds of essential oils, and their biological activities were performed using OriginPro, Version 2022b.

3. Results and Discussion

3.1. Chemical Composition

A total of 37, 45, and 43 components were detected in VAO (0.1–25.0%), VNO (0.1–19.4%), and VTO (0.1–16.2%), respectively. A total of 22 components were found to be common in all three EOs, which were as follows: α-thujene, α-pinene, sabinene, β-pinene, myrcene, 1,8-cineole, γ-terpinene, *p*-cymene, linalool, *trans*-sabinenehydrate, *cis*-*p*-menth-2-en-1-ol, terpinen-4-ol, α-terpineol, dihydroedulan II, β-caryophyllene, α-humulene, β-iraldeine, β-caryophyllene oxide, α-muurolol, drimenol, and manool. However, they varied in their relative percentage. As summarized in Table 1, 1,8-Cineole (25.0%), sabinene (13.3%), α-pinene (8.2%), and α-terpinyl acetate (5.5%) were the dominant compounds in *V. agnus-castus* oil; sabinene (19.4%), viridiflorol (17.8%), β-caryophyllene (7.5%), and β-iraldiene (6.4%) were dominant in *V. negundo* oil, while β-caryophyllene (16.2%), 5-

(1-isopropenyl-4,5-dimethylbicyclo[4.3.0]nonan-5-yl)-3-methyl-2-pentenol acetate (11.7%), 13-epi-manoyl oxide (5.6%), and caryophyllene oxide (4.6%) were the abundant compounds in *V. trifolia* oil. In terms of chemical class composition, VAO was dominated by oxygenated monoterpene (40.6%), followed by monoterpene hydrocarbons (31.2%) and others. The only diterpenoid present in VAO was manool (0.5%). On the other hand, VNO was mainly dominated by monoterpene hydrocarbons (29.4%), followed by oxygenated sesquiterpenes (24.8%) and oxygenated monoterpenoids (11.3%). The most abundant class found in VTO was sesquiterpene hydrocarbon (21.9%) followed by oxygenated sesquiterpene (15.8%) and oxygenated diterpenes (13.8%). For more details on chromatograms, and chemical composition mass spectra, please refer to Supplemental Material S1.

Table 1. Comparative chemical composition of essential oil of *Vitex* species.

S. No.	Compound Name	Molecular Formula	R.I.	% Composition		
				VAO	VNO	VTO
1	α-Thujene (MH)	$C_{10}H_{16}$	930	1.4	0.2	0.1
2	α-Pinene (MH)	$C_{10}H_{16}$	939	8.2	2.6	1.6
3	Sabinene (MH)	$C_{10}H_{16}$	975	13.3	19.4	2.0
4	β-Pinene (MH)	$C_{10}H_{16}$	979	1.2	0.4	0.3
5	Oct-1-en-3-ol	$C_8H_{16}O$	979	-	0.5	4.0
6	Myrcene (MH)	$C_{10}H_{16}$	990	3.1	0.7	0.1
7	α-Phellandrene (MH)	$C_{10}H_{16}$	1002	-	-	4.2
8	α-Terpinene (MH)	$C_{10}H_{16}$	1017	-	2.3	-
9	β-Phellandrene (MH)	$C_{10}H_{16}$	1029	-	1.0	-
10	1,8-Cineole (OM)	$C_{10}H_{18}O$	1031	25.0	1.2	2.1
11	β-Ocimene (MH)	$C_{10}H_{16}$	1044	1.5	-	-
12	γ-Terpinene (MH)	$C_{10}H_{16}$	1059	0.3	0.8	3.3
13	Linalool oxide (OM)	$C_{10}H_{18}O_2$	1086	-	0.4	-
14	α-Terpinolene (MH)	$C_{10}H_{16}$	1088	-	1.2	-
15	*p*-Cymene (MH)	$C_{10}H_{12}$	1091	2.2	0.6	0.2
16	Linalool (OM)	$C_{10}H_{18}O$	1096	0.8	0.6	2.2
17	*trans*-Sabinenehydrate (OM)	$C_{10}H_{18}O$	1098	0.2	0.3	-
18	Isoamyl isovalerate (Fatty Acid Ester)	$C_{10}H_{20}O_2$	1103	-	0.1	-
19	*cis-p*-menth-2-en-1-ol (OM)	$C_{10}H_{18}O$	1121	0.2	0.2	0.1
20	δ-Terpineol (OM)	$C_{10}H_{18}O$	1166	0.5	-	-
21	Terpinen-4-ol (OM)	$C_{10}H_{18}O$	1177	1.9	5.4	1.8
22	Cryptone (OM)	$C_9H_{14}O$	1185	0.1	-	-
23	α-Terpineol (OM)	$C_{10}H_{18}O$	1188	2.5	1.4	0.3
24	Myrtenol (OM)	$C_{10}H_{16}O$	1195	-	0.1	-
25	*cis*-Piperitol (OM)	$C_{10}H_{18}O$	1196	-	0.1	-
26	γ-Terpineol (OM)	$C_{10}H_{18}O$	1199	0.1	-	-
27	*trans*-Piperitol (OM)	$C_{10}H_{18}O$	1208	-	-	0.8
28	β-Citronellol (OM)	$C_{10}H_{20}O$	1225	1.4	-	0.5
29	*cis*-Verbenyl acetate (OM)	$C_{12}H_{18}O_2$	1282	1.1	-	-
30	Dihydroedulan II	$C_{13}H_{22}O_2$	1284	0.1	0.1	0.5
31	Theaspirane A (OM)	$C_{13}H_{22}O_2$	1290	-	1.5	0.2
32	α-Terpinyl acetate (OM)	$C_{12}H_{20}O_2$	1349	5.5	-	0.5
33	β-Citronellyl acetate (OM)	$C_{12}H_{22}O_2$	1352	1.3	-	-
34	β-Damascenone (OM)	$C_{13}H_{18}O$	1384	-	0.1	-
35	β-Bourbonene (SH)	$C_{15}H_{24}$	1388	-	-	0.7
36	β-Elemene (SH)	$C_{15}H_{24}$	1390	-	-	0.8
37	β-Caryophyllene (SH)	$C_{15}H_{24}$	1419	3.7	7.5	16.2
38	Methyl-isoeugenol (Phenylpropanoid)	$C_{11}H_{14}O_2$	1453	-	-	3.1
39	α-Humulene (SH)	$C_{15}H_{24}$	1454	0.2	0.4	1.4
40	β-Farnesene (SH)	$C_{15}H_{24}$	1456	4.5	0.6	-
41	β-Selinene (SH)	$C_{15}H_{24}$	1490	-	0.3	-
42	*epi*-Cubebol (OS)	$C_{15}H_{26}O$	1494	-	-	2.3

Table 1. *Cont.*

S. No.	Compound Name	Molecular Formula	R.I.	% Composition		
				VAO	VNO	VTO
43	Bicyclogermacrene (SH)	$C_{15}H_{24}$	1500	0.3	-	-
44	α-Muurolene (SH)	$C_{15}H_{24}$	1500	-	-	1.5
45	γ-Cadinene (SH)	$C_{15}H_{24}$	1513	-	-	0.8
46	δ-Cadinene (SH)	$C_{15}H_{24}$	1523	-	-	0.5
47	Hedycaryol (OS)	$C_{15}H_{26}O$	1548	-	0.5	-
48	β-Iraldeine (ionone)	$C_{14}H_{22}O$	1557	3.8	6.4	2.0
49	Nerolidol (OS)	$C_{15}H_{26}O$	1563	-	-	2.2
50	Spathulenol (OS)	$C_{15}H_{24}O$	1578	1.4	-	-
51	β-caryophyllene oxide (OS)	$C_{15}H_{24}O$	1583	1.9	1.3	4.6
52	Globulol (OS)	$C_{15}H_{26}O$	1590	-	0.2	1.4
53	Viridiflorol (OS)	$C_{15}H_{26}O$	1592	-	17.8	-
54	Ledol (OS)	$C_{15}H_{26}O$	1602	0.4	1.0	-
55	Humulene epoxide II (OS)	$C_{15}H_{24}O$	1608	-	-	1.2
56	Humulane-1,6-dien-3-ol (OS)	$C_{15}H_{26}O$	1619	2.4	2.3	-
57	*epi*-α-Cadinol (OS)	$C_{15}H_{26}O$	1640	2.1	-	-
58	α-Muurolol (OS)	$C_{15}H_{26}O$	1646	0.1	0.2	1.7
59	β-eudesmol (OS)	$C_{15}H_{26}O$	1650	0.1	1.1	-
60	Pogostol (OS)	$C_{15}H_{26}O$	1653	-	-	1.6
61	Drimenol (OS)	$C_{15}H_{26}O$	1767	0.3	0.4	0.8
62	Flourensadiol (OS)	$C_{15}H_{26}O_2$	1870	0.1	-	-
63	Cubitene (DT)	$C_{20}H_{32}$	1878	-	1.3	-
64	Phytol (OD)	$C_{20}H_{40}O$	1943	-	1.2	1.3
65	13-epi-manoyl oxide (OD)	$C_{20}H3_4O$	2002	-	-	5.6
66	Manool (OD)	$C_{20}H3_4O$	2057	0.5	1.9	1.7
67	Sclareolide (OD)	$C_{16}H_{26}O_2$	2066	-	0.2	-
68	Sclareol (OD)	$C_{20}H_{36}O_2$	2223	-	1.7	1.0
69	5-(1-isopropenyl-4,5-dimethylbicyclo[4.3.0]nonan-5-yl)-3-methyl-2-pentenol acetate	$C_{22}H_{36}O_2$	2265	-	5.2	11.7
70	Larixol (OD)	$C_{20}H_{34}O_2$	2266	-	0.7	-
71	Verticiol (OD)	$C_{20}H_{34}O$	2273	-	-	1.4
72	16-oxo-cleroda-3,13(14)-(e)-dien-15-oic acid (OD)		-	-	-	2.8
	Monoterpene hydrocarbon (MH)			31.2	29.4	11.8
	Oxygenated monoterpene (OM)			40.6	11.3	8.5
	Sesquiterpene hydrocarbon (SH)			8.7	8.8	21.9
	Oxygenated sesquiterpene (OS)			8.8	24.8	15.8
	Diterpene hydrocarbon (DT)			-	1.3	-
	Oxygenated diterpene (OD)			0.5	5.7	13.8
	Other than terpenoids			3.9	12.3	21.3
	Total			93.7	93.6	93.1

VAO = *V. agnus-castus;* VNO = *V. negundo;* VTO = *V. trifolia;* RI value = Retention index value on a DB-5MS column in reference Adams, 2007 [22], or on NIST webbook [23]. MH = Monoterpene hydrocarbon; OM = Oxygenated monoterpene; SH = Sesquiterpene hydrocarbons; OS = Oxygenated sesquiterpene; DT = Diterpenene hydrocarbon; OD = Oxygenated diterpene.

Rezaei et al. [42] evaluated the effects of different irrigation regimes on the essential oil composition of *V. agnus-castus* under three shading levels, collected from Isfahan, Iran, and reported α-pinene (16.4–60.7%), β terpinyl acetate (15.4–40.2%), caryophyllene (2.7–9.0%), and camphene (0.09–6%) as the main constituents. The compounds 1,8-cineole, sabinene, myrcene, α-thujene. α-terpineol, β-farnesene, spathulenol, β-caryophyllene oxide, humulane-1,6-dien-3-ol, and manool were not detected in any of the samples studied in this work, although these are present in noticeable amounts in the current study. However, in other studies, 1,8-cineole and sabinene were reported as the main EO constituents of *V. agnus castus* leaves [43–45]. These findings are generally consistent with those of the current investigation. The oxygenated sesquiterpenes identified in VAO, such as spathulenol, ledol, and *epi*-α-cadinol, have previously been identified in the EO of *V. agnus castus* leaves collected in Ogliastra, Sardinia, Italy [46], however, other sesquiterpenoids such as β-eudesmol, drimenol, and flourensadiol were not detected previously in *V. agnus castus* leaves essential oil. VAO also lacks compounds like limonene, viridiflorol, and globulol that are often present in most of the previous reports [45–48]. Thus, the study demonstrates different chemo-variants of *V. agnus castus* both qualitatively and quantitatively.

Previous researchers have also studied the EOs of *Vitex negundo* under investigation herein. For instance, the major compounds detected in VNO, sabinene (19.4%), viridiflorol (17.8%) and β-caryophyllene (7.5%), were also found to be present in the hydrodistilled *Vitex negundo* leaves EO in variable amounts [22,49,50]. 5-(1-Isopropenyl-4,5-dimethylbicyclo[4.3.0]nonan-5-yl)-3-methyl-2-pentenol acetate (5.2%), another major compound detected in VNO, was also found to be present in leaves essential oil of *V. negundo* in notable amounts [51]. The chemical composition of EO of *Vitex negundo* extracted during the spring season from the same location (Pantnagar) revealed the presence of over 33 compounds, in which the major compounds detected were viridiflorol (23.8%), sabinene (11.2%), unidentified diterpene M^+ = 272 (11.0%), and caryophyllene (6.7%) [50]. The composition was lacking the compounds, α-thujene, β-pinene, α-terpinene, β-phellandrene, γ-terpinene, linalool oxide, *p*-cymene, *trans*-sabinenehydrate, theaspirane A, β-iraldeine, β-caryophyllene oxide, ledol, humulane-1,6-dien-3-ol, and the diterpenes cubetene, phytol, manool, and sclereol, however, in the present study, these compounds are detected in noticeable amounts. Thus, the composition might vary as a result of the harvesting season. However, in another study of Indian origin, α-copaene (25.3%), β-elemene (19.2%), and camphene (21.1%) were reported as the predominant compounds in leaf essential oil of *V. negundo* [52]. Khokra et al. [53] reported ethyl-9-hexadecenoate (28.5%), δ-guaiene (18.0%), and caryophyllene oxide (10.2%) as the major components in leaf essential oil of *V. negundo*. On the other hand, the leaf essential oil of *V. negundo* from Chinese origin revealed δ-guaiene (50.0%) and β-caryophyllene (38.0%) as the major constituents [54]. Both qualitative and quantitative variations in essential oils of *V. negundo* from different geographic regions might be due to the different geographical and climatic conditions.

Thomas et al. [55], investigated the essential oil of *V. trifolia* and obtained caryophyllene (38.36%) and 1,8-cineole (25.72%) as the predominant compounds. However, in the present study, the amount of 1,8-cineole is only 2.1% in VTO. β-caryophyllene is also identified as the major constituent of *V. trifolia* oil by several other reports [56–58], which is in agreement with the present study. Arpiwi et al. [59] detected five components in *V. trifolia* essential oil in which *cis*-ocimene (44.57%), α-thujene (25.63%) and cyclopentene,3-isopropenyl-5,5-dimethyl (18.19%) were identified as the major constituents. However, in the present study such compounds were not detected, and the amount detected for α-thujene was also negligible (0.2%). The second major compound detected in VTO, 5-(1-isopropenyl-4,5-dimethylbicyclo[4.3.0]nonan-5-yl)-3-methyl-2-pentenolacetate (11.7%), has also been found in other *Vitex* species such as *V. agnus castus* and *V. negundo* [51,60]. The noticeable diterpenes identified in VTO, 13-epi-manoyl oxide (5.6%), and 16-oxo-cleroda-3,13(14)-(e)-dien-15-oic acid (2.8%) are also being detected for the first time in *V. trifolia* oil. These differences in the essential oil constituents might be due to internal and external factors and their interactions.

In addition, the compounds identified in the tested essential oils have potent biological applications. 1,8-cineole is used in cosmetic products and as a flavoring agent because of its pleasant aroma and taste. The compound has several other properties: insecticidal, antioxidant, and anti-inflammatory [61]. Viridiflorol has prominent use as an anti-inflammatory, antioxidant, and anti-tuberculosis agent [62]. Sabinene has antimicrobial, anti-inflammatory, and antioxidant properties described in literature [63]. Further, the diterpene, 13-*epi*-manoyl oxide, has cytotoxic antibacterial and antifungal activities [64].

3.2. Chemometric Analysis

The main chemical components common for all the essential oils of tested species (α-thujene, α-pinene, sabinene, β-pinene, myrcene, 1,8-cineole, γ-terpinene, *p*-cymene, linalool, cis-p-menth-2-en-1-ol, terpinen-4-ol, α-terpineol, dihydroedulan II, β-caryophyllene, α-humulene, β-iraldeine, β-caryophyllene oxide, α-muurolol, drimenol, and manool) were compared with hierarchical cluster analysis with Euclidean distance as the similarity index. The heat map clustering diagram is depicted in Figure 1. 1,8-Cineole, sabinene, and β-caryophyllene form separate clusters with different values compared to the rest of the analyzed common constituents. Based on Euclidean distance in the heat map clustering, all the tested species are clearly divided into two main clusters on the basis of their common chemical constituents. VNO and VTO are clustered in one cluster, whereas VAO is in the separate cluster.

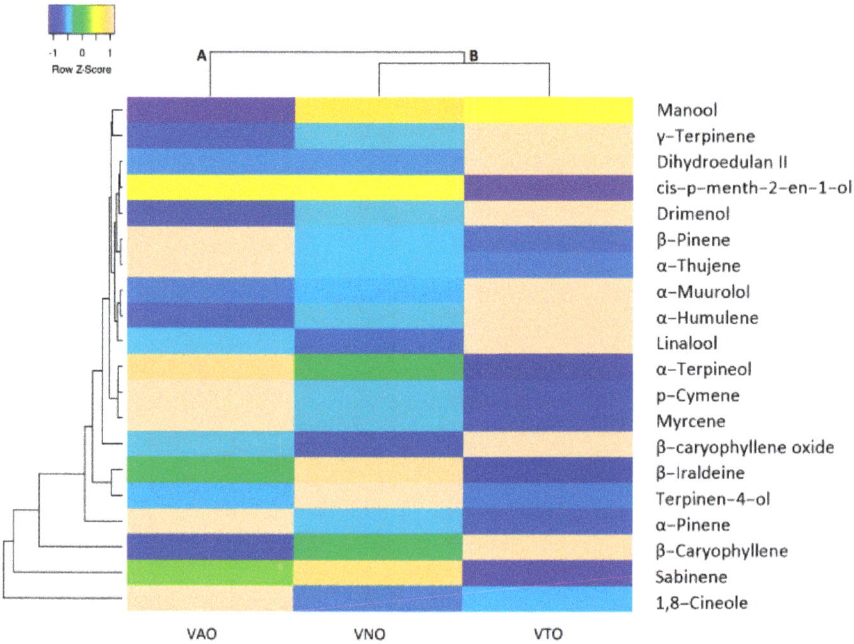

Figure 1. The heatmap analysis of the common essential oil constituents and tested species (The distribution of trait (common essential oil components) was identified by colors, where yellow color showed the maximum value of the trait, and blue color represented the minimum value).

3.3. Principal Component Analysis

Principal component analysis (PCA) is one of the greatest multivariate statistical techniques used to identify a dataset's most important features. To assess the chemical profiling changes caused by interspecies as well as altitudinal influences, distinct essential oils can be used in PCA pattern recognition. The PCA approach determined that the cumulative contribution rate of variance of the first two principal components (PC1 and PC2) could account for 81.2% of the variance information for changes in chemical composition. In order to define the compositional variations in the essential oils, PC1 and PC2 were used. PC1 was favorably linked with terpinen-4-ol, β-iraldeine, β-caryophyllene, viridiflorol, and sabinene, and contributed 48.7% of the total variance. However, PC2 makes up 32.5% of the variation and has a strong positive correlation with α-pinene and 1,8-cineole (Figure 2).

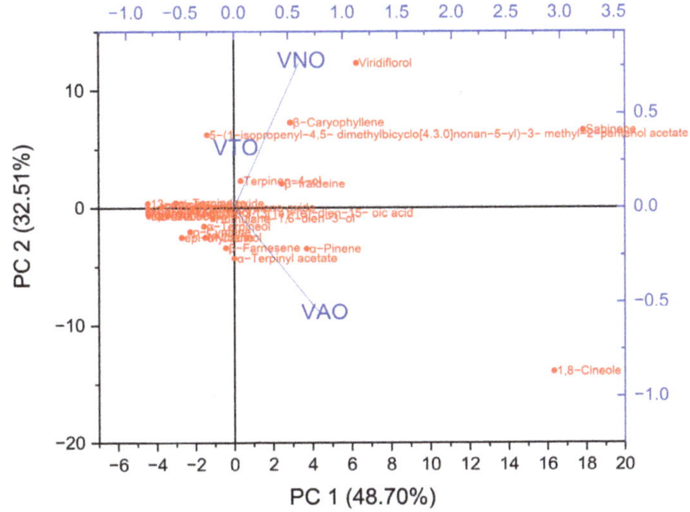

Figure 2. Principal Component Analysis of tested essential oil's chemical constituents.

3.4. *Antioxidant Activity*

The antioxidant activity was determined by using different chemical-based methodologies. Figure 3A–E depict the antioxidant activity of tested essential oil in terms of percent inhibition. Results revealed that all the antioxidant activities were in a concentration-dependent manner. The percent inhibition of free radicals (DPPH, H_2O_2, NO), reducing power, and metal chelation increased, with increasing concentration from 10 μL/mL to 50 μL/mL. Further, the percent inhibition by the tested essential oils and the standards for different antioxidant assays were plotted against concentrations, and the equation for the line was used to obtain the IC_{50} (half-maximal inhibitory concentration) values.

Figure 4A–E represent the antioxidant activity of tested essential oils in terms of their IC_{50} values. In the DPPH assay, the reduction of the stable radical DPPH (violet) to the yellow-colored DPPH-H is employed to measure the potential of an antioxidant molecule to act as a donor of hydrogen atoms or electrons. Figure 4A shows that VNO reduced DPPH with an IC_{50} value of 23.16 ± 0.5 μL/mL, which is close to the standard antioxidant taken for the assay, BHT (18.84 ± 0.6 μL/mL). VAO and VTO displayed moderate and weak antioxidant activity, with IC_{50} 25.39 ± 0.0 μL/mL and 32.49 ± 0.5 μL/mL, respectively. H_2O_2 can cross the biological membrane, and as a result it can damage the human body by forming reactive OH· radicals following Fenton reaction [65]. In H_2O_2 radical scavenging assay, VAO (IC_{50} = 24.49 ± 0.1 μL/mL) displayed good scavenging activity when compared to the standard, ascorbic acid (28.33 ± 0.5 μL/mL), followed by VNO (32.38 ± 0.5 μL/mL) and VTO (34.30 ± 0.5 μL/mL). The extent of nitrite scavenging by the samples was compared with ascorbic acid and showed IC_{50} values as: ascorbic acid (24.49 ± 0.1 μL/mL) > VNO (27.58 ± 0.1 μL/mL) > VTO (32.27 ± 0.1 μL/mL) > VAO (32.95 ± 0.5 μL/mL). The reducing power of a compound is related to its ability to transfer electrons, which indicates its significant antioxidant potential. As shown in Figure 4D, VNO displayed good reducing capability (RP_{50} = 19.05 ± 0.6 μL/mL) that is very close and lower than that of the standard gallic acid (20.22 ± 0.4 μL/mL). The order of RP_{50} values for different samples is in the order: VNO (19.05 ± 0.6 μL/mL) > gallic acid (20.22 ± 0.4 μL/mL) > VAO (20.97 ± 0.5 μL/mL) > VTO (22.74 ± 0.7 μL/mL). In auto-oxidation reactions, metal ion is a powerful catalyst as it can inhibit the generation of oxygen radicals. The IC_{50} values of different samples and standards towards their antioxidant potentiality in terms of chelating ability were observed as: Na_2-EDTA (IC_{50} = 26.23 ± 0.26 μL/mL) > VTO (IC_{50} = 29.77 ± 0.2 μL/mL) > VNO (IC_{50} = 31.18 ± 0.2 μL/mL) > VAO (IC_{50} = 36.60 ± 0.1 μL/mL).

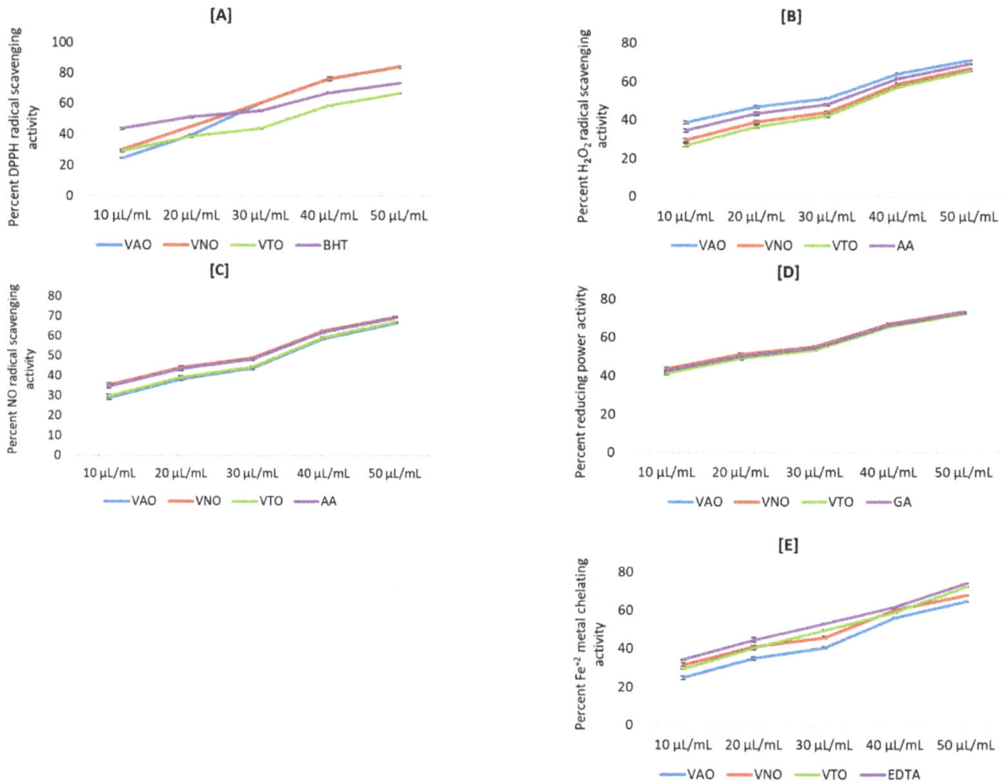

Figure 3. (**A–E**) Antioxidant activity for essential oils of *Vitex* species; (**A**) Percent DPPH radical scavenging activity; (**B**) Percent H_2O_2 scavenging activity; (**C**) Percent NO radical scavenging activity; (**D**) Percent reducing power activity; (**E**) Percent Fe^{2+} metal chelating activity.

Such high antioxidant activity of VNO for the DPPH and NO radical scavenging is likely due to high amount of sabinene as well as other constituents of VNO such as β-caryophyllene, terpinen-4-ol, 1,8-cineole, which already possess antioxidant potential via different parameters [66–68]. Additionally, Kazemi [69] showed that sabinene exhibited potent NO-scavenging effect and inhibited the expression of inducible NO synthase. Similar results were observed in previous studies in antioxidant activity of *V. negundo* essential oil in which the major component was sabinene [49]. In H_2O_2 radical scavenging assay, VAO showed good scavenging activity, which may be due to the presence of 1,8-cineole, sabinene, and β-caryophyllene as the major constituents [69,70]. In earlier reports, essential oil and extracts of aerial parts of *V. agnus castus* have been tested for antioxidant activity as having a high amount of 1,8-cineole and β-caryophyllene in their composition, and the samples showed good antioxidant activity [64,71,72]. Since essential oils are complexed mixtures of number of compounds, their whole biological activity is hard to explain. Therefore, research on the antioxidant activity of essential oils typically indicates that other minor chemical constituents that may interact synergistically or antagonistically to produce an additive and effective system against free radicals may also be responsible for the antioxidant activity [68,73].

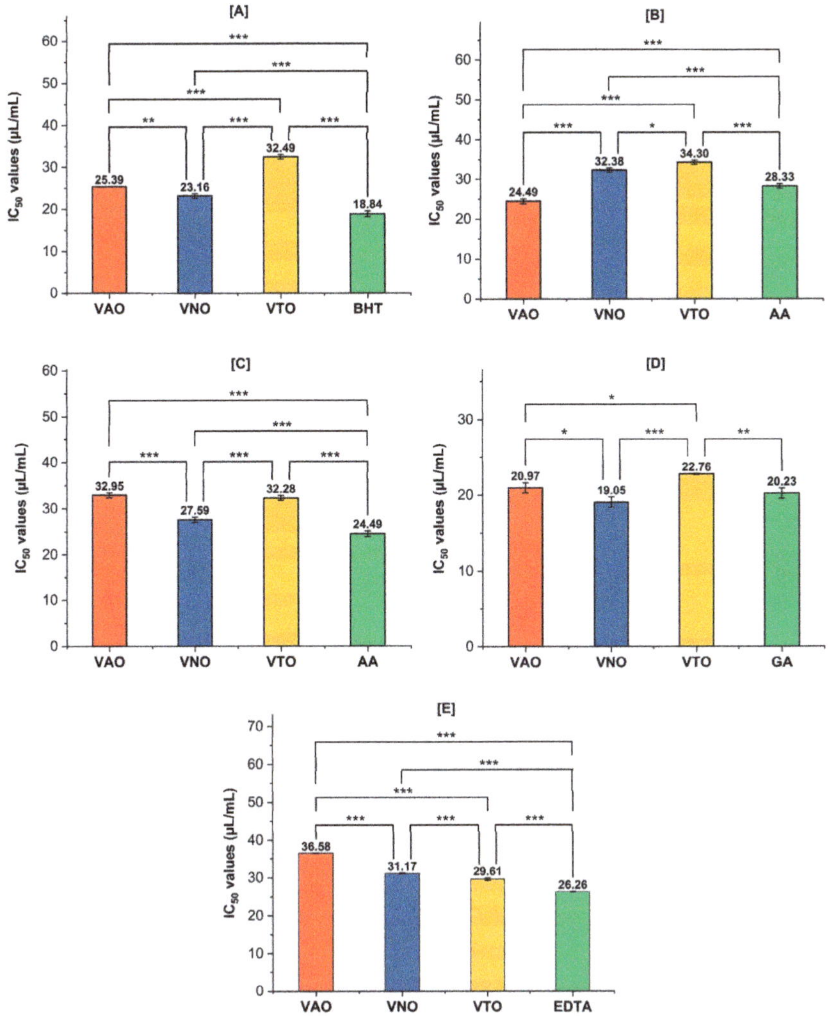

Figure 4. (**A–E**) Antioxidant activity in terms of IC_{50} values (µL/mL) for VAO, VNO, and VTO, (**A**) DPPH radical scavenging, (**B**) H_2O_2 radical scavenging, (**C**) NO radical scavenging, (**D**) reducing power activity, (**E**) Metal chelating activity. Statistically significant differences were examined using one-way ANOVA and Tukey posthoc tests. *** $p < 0.001$, ** $p < 0.005$, * $p < 0.05$ above columns indicate significant differences between treated groups. Values are mean ± SD, $n = 3$.

3.5. Herbicidal (Phytotoxic) Activity

The tested samples demonstrated notable phytotoxic activity against seed germination and seedling growth of the wild radish (*R. raphanistrum*) in a concentration-dependent manner. At the highest concentration (100 µL/mL), VAO showed inhibition of seed germination, root growth, and shoot growth of *R. raphanistrum* by 66.67%, 96.66%, and 89.09%, respectively, VNO showed inhibition values of 90.0%, 89.39%, and 97.57%, respectively, while VTO showed inhibition values of 100%, 99.39%, and 92.12%, respectively (Tables 2–4). Based on the IC_{50} values, VAO showed IC_{50} values of 82.89, 19.468, and 37.95 µL/mL regrading seed germination, root growth, and shoot growth, respectively. For VNO, the IC_{50} values were 50.13, 47.06, and 16.75 µL/mL, respectively. For VTO, the IC_{50} values were 29.5, 9.33, and 27.13 µL/mL, respectively (Tables 2–4).

Table 2. Mean % inhibition and IC_{50} values for seed germination inhibition by tested essential oils.

Samples	% Inhibition of Seed Germination					IC_{50} Values (μL/mL) in Triplicates			Mean IC_{50} Values (μL/mL) ± SD
	20 μL/mL	40 μL/mL	60 μL/mL	80 μL/mL	100 μL/mL	I	II	III	
VAO	3.33 ± 5.77 ^{h}	3.33 ± 5.77 ^{h}	40 ± 0.00 ^{f}	43.33 ± 5.77 ^{ef}	66.66 ± 5.77 ^{cd}	78.94	81.17	88.57	82.89 ± 5.04
VNO	23.33 ± 5.77 ^{g}	43.33 ± 5.77 ^{ef}	56.66 ± 5.77 ^{de}	76.66 ± 5.77 ^{bc}	90.00 ± 0.00 ^{ab}	51.11	55.294	44.00	50.13 ± 5.7
VTO	33.33 ± 5.77 ^{fg}	60.00 ± 0.00 ^{d}	90.00 ± 0.00 ^{ab}	93.33 ± 5.77 ^{a}	100.00 ± 0.00 ^{a}	31.764	25.00	31.764	29.50 ± 3.9
Pendimethalin	100.00 ± 0.00	100.00 ± 0.00	100.00 ± 0.00	100.00 ± 0.00	100.00 ± 0.00				

VAO = *V. agnus-castus*; VNO = *V. negundo*; VTO = *V. trifolia*; SD = standard deviation; According to Tukey's test ($p < 0.05$), mean values that are followed by the same letter inside a column are not statistically different from one another.

Table 3. Mean % inhibition and IC_{50} values for root length inhibition by tested essential oils.

Samples	% Inhibition of Root Length					IC_{50} Values (μL/mL) in Triplicates			Mean IC_{50} Values (μL/mL) ± SD
	20 μL/mL	40 μL/mL	60 μL/mL	80 μL/mL	100 μL/mL	I	II	III	
VAO	45.15 ± 1.0 ^{g}	66.36 ± 0.9 ^{d}	80.90 ± 0.9 ^{c}	89.69 ± 0.5 ^{b}	96.66 ± 0.5 ^{a}	18.963	19.161	20.28	19.468 ± 0.7
VNO	32.27 ± 1.6 ^{h}	45.45 ± 0.9 ^{g}	58.48 ± 1.0 ^{e}	68.78 ± 2.2 ^{d}	89.39 ± 2.6 ^{b}	46.529	47.00	47.67	47.06 ± 0.5
VTO	52.42 ± 2.2 ^{f}	70.00 ± 1.8 ^{d}	83.03 ± 1.3 ^{c}	92.12 ± 1.04 ^{b}	99.39 ± 1.04 ^{a}	9.766	9.766	8.479	9.337 ± 0.7
Pendimethalin	100.00 ± 0.00	100.00 ± 0.00	100.00 ± 0.00	100.00 ± 0.00	100.00 ± 0.00				

VAO = *V. agnus-castus*; VNO = *V. negundo*; VTO = *V. trifolia*; SD = standard deviation. According to Tukey's test ($p < 0.05$), mean values that are followed by the same letter inside a column are not statistically different from one another.

Table 4. Mean % inhibition and IC_{50} values for shoot length inhibition by tested essential oils.

Samples	% Inhibition of Shoot Length					IC_{50} Values (μL/mL) in Triplicates			Mean IC_{50} Values (μL/mL) ± SD
	20 μL/mL	40 μL/mL	60 μL/mL	80 μL/mL	100 μL/mL	I	II	III	
VAO	37.57 ± 0.5 ^{k}	52.72 ± 1.8 ^{i}	63.48 ± 0.6 ^{g}	77.57 ± 0.5 ^{e}	89.09 ± 0.9 ^{bc}	37.52	37.58	38.75	37.95 ± 0.6
VNO	51.21 ± 1.3 ^{i}	64.24 ± 2.7 ^{g}	74.54 ± 0.9 ^{e}	83.78 ± 0.2 ^{d}	97.57 ± 0.5 ^{a}	17.43	17.42	15.41	16.75 ± 1.2
VTO	45.15 ± 0.5 ^{j}	58.48 ± 1.0 ^{h}	68.03 ± 0.6 ^{f}	85.90 ± 1.2 ^{cd}	92.12 ± 0.5 ^{b}	28.65	26.417	26.33	27.13 ± 1.3
Pendimethalin	100.00 ± 0.01	100.00 ± 0.01	100.00 ± 0.0	100.00 ± 0.00	100.00 ± 0.00				

VAO = *V. agnus-castus*; VNO = *V. negundo*; VTO = *V. trifolia*; SD = standard deviation. According to Tukey's test ($p < 0.05$), mean values that are followed by the same letter inside a column are not statistically different from one another.

The phytotoxic potential of EOs from various *Vitex* species such as *V. agnus castus*, *V. negundo V. simplicifolia* has also been reported previously in other plants and weeds [15]. However, there is no study reported on phytotoxic potential of *V. trifolia*. Based on the present study, it was evident that VTO was more effective against *R. raphanistrum* than VNO and VAO. The suppressing effect of VTO on *R. raphanistrum* could be due to high amounts of β-caryophyllene (16.2%) and the synergetic effect of β-caryophyllene with other major and minor compounds present in the oil. In previous reports, β-caryophyllene was found to be responsible for the inhibition of germination and seedling growth of several plant species such as *Brassica campestris*, *Raphanus sativus*, *Vigna radiata*, and *Solanum lycopersicum* [22]. VNO also showed good inhibition values for seed germination and shoot growth, while VAO showed better inhibition value for root growth. The inhibition effect of samples could be due to the presence of phytotoxic compounds such as β-caryophyllene, 1,8-cineole, and sabinene, which are the main components in essential oil possessing phytotoxic activity [22,74]. In addition, 1,8-cineole was reported to interfere with the normal growth *Nicotiana tabacum* by blocking the DNA synthesis in their cell nuclei and organelles in root apical meristem cells [75]. Studies have also demonstrated that the terpenoids in EOs have a phytotoxic effects on plants, resulting in morphological and physiological alterations in the cells that impair plant growth [76].

3.6. Correlation of Essential Oil Components and Biological Activities

Pearson's correlation coefficient of major essential oil constituents (>2.00%) and antioxidant and herbicidal activities of *Vitex* species revealed that 5-(1-isopropenyl-4,5-dimethylbicyclo[4.3.0]nonan-5-yl)-3-methyl-2-pentenol acetate and β-caryophyllene strong positive correlation with DPPH radical scavenging activity and Fe^{2+} metal chelating activity, whereas γ-terpinene was also found positively correlated with Fe^{2+} metal chelating activity. Dahham et al. [77] also reported significant DPPH scavenging activity of β-caryophyllene. α-terpinyl acetate, 1,8-cineole and *epi*-α-cadinol, α-pinene, and β-farnesene showed moderate correlation with H_2O_2 radical scavenging activity. Terpinen-4-ol, α-terpinene, and viridiflorol showed moderate correlation with NO radical scavenging activity, as well as reducing power activity. The terpinen-4-ol was also reported to induce relaxation and was unlikely to be mediated by induction of NO release in rabbit duodenum relaxation in rabbit duodenum [78]. In terms of herbicidal activity of tested essential oils, 5-(1-isopropenyl-4,5-dimethylbicyclo[4.3.0]nonan-5-yl)-3-methyl-2-pentenol acetate and β-caryophyllene showed strong positive correlation with seed germination inhibition, whereas linalool and β-caryophyllene have moderated correlation with root length inhibition and terpinen-4-ol, α-terpinene, and viridiflorol were found to have strong positive correlation with shoot length inhibition. High β-caryophyllene-containing plants were also reported to have phytotoxic effects on weed species [79]. The correlation coefficient results were also supported by in vitro activities in presentation investigation, as well as previously reported studies. Pearson's correlation coefficient representation of essential oil constituents with their biological activities is demonstrated in Figure 5.

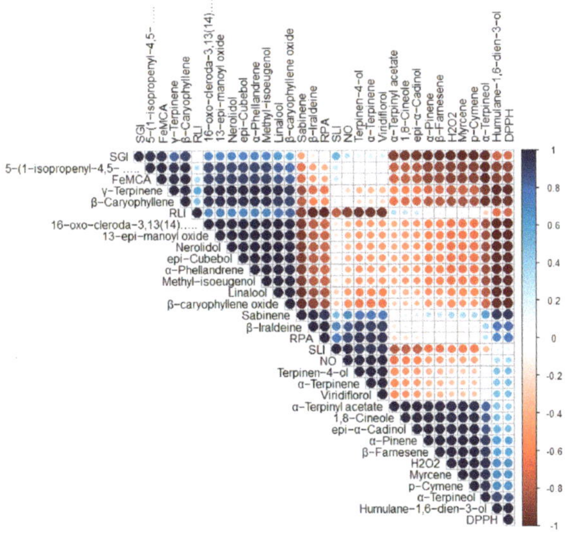

Figure 5. Correlation among chemical components of essential oils and biological activities of *Vitex* species (here, DPPH = percent inhibition of DPPH radical scavenging activity at 50 μL/mL; H_2O_2 = percent inhibition of H_2O_2 radical scavenging activity at 50 μL/mL; NO = percent inhibition of NO radical scavenging activity at 50 μL/mL; RPA = percent inhibition of reducing power activity at 50 μL/mL; FeMCA = percent inhibition of Fe^{2+} metal chelating activity at 50 μL/mL; SGI = percent inhibition of seed germination at 100 μL/mL; RLI = percent inhibition of root length at 100 μL/mL; SLI = percent inhibition of shoot length at 100 μL/mL.

3.7. Molecular Docking

From in vitro studies, it was found that essential oils have potent antioxidant and phytotoxic activity. We also examined whether the major phytoconstituents from VAO, VNO, and VTO physically bind with antioxidant protein (human peroxiredoxin 5, PDB: 1HD2) and 4-hydroxyphenylpyruvate dioxygenase (HPPD, PDB: 6J63) receptors. The tested essential oils displayed good inhibition of the free radicals, for which the enzyme human peroxiredoxin 5 was selected, as it has broader activity against the reactive oxygen species (ROS) and is mostly involved in the stress protection mechanism [80,81]. The reason for selecting HPPD is that it is known to be the target protein for compounds with post-emergence herbicidal activity. In our results, the tested essential oils were found to have good post-emergence herbicidal activity against the receptor species, for which HPPD was selected as a target enzyme [18,38]. Among all selected phytocompounds, 13-*epi*-manoyl oxide demonstrated the best binding affinity with human peroxiredoxin 5 (−6.2 kcal mol) and HPPD (−8.7 kcal/mol). By introspecting the multiple dock poses, the best docked pose was selected as having the lowest binding energy. The best docked pose of 13-*epi*-manoyl oxide exhibited 2 pi-alkyl interaction, 1 pi-sigma interaction, and other Van der Waal interactions with 6J63 containing amino acid residues such as Phe A:424, Phe A:419, and Phe A:381, as represented in Figure 6B. Similarly, the best docked pose of 13-*epi*-manoyl oxide exhibited alkyl interaction with 1HD2 containing amino acid Ala A:90, Arg A:86, and exhibited Van der Waal interaction. For comparison purposes, a docking study of Nitisinone (CID:115355) was also performed with HPPD. Nitisinone (2-[2-nitro-4-(trifluoromethyl)benzoyl]cyclohexane-1,3-dione, (NTBC)) is a known inhibitor of HPPD. The docking study of ascorbic acid (CID:54670067), a known antioxidant, was performed with 1HD2. The binding energy for NTBC complexed with 6J63 was −8.9 kcal/mol, which is very close to that of 13-*epi*-manoyl oxide (−8.7 kcal/mol). On the other hand, binding energy of ascorbic acid complexed with 1HD2 came out to be −5.7 kcal/mol, which was higher than most of the compounds such as 13-*epi*-manoyl oxide (−6.2 kcal/mol), caryophyllene oxide (−6.1 kcal/mol), 5-(1-isopropenyl-4,5-dimethylbicyclo [4.3.0]nonan-5-yl)-3-methyl-2-pentenol acetate (−6.1 kcal/mol), β-caryophyllene (−6.0 kcal/mol), and viridiflorol (−5.9 kcal/mol), as shown in Figure 7. The lower values of binding free energy demonstrate more significant interaction between the receptor and the ligand. Our results were consistent with previous in silico studies reported by Alminderej et al. [73], where a phenylpropanoid-rich *Piper cubeba* EO gave similar results in terms of a proposed in vitro antioxidant activity by targeting human periredoxin 5. In this study, the compounds viridiflorol and caryophyllene oxide showed significant interaction with 1HD2 receptor as in the present study. In a recent study, focusing on the phytotoxic potential of *Calycolpus goetheanus* EO, it was found that the major components of the specimen, 1,8-cineole and β-caryophyllene interacted favorably with the HPPD protein [18]. These results are in general agreement with those obtained in the present study.

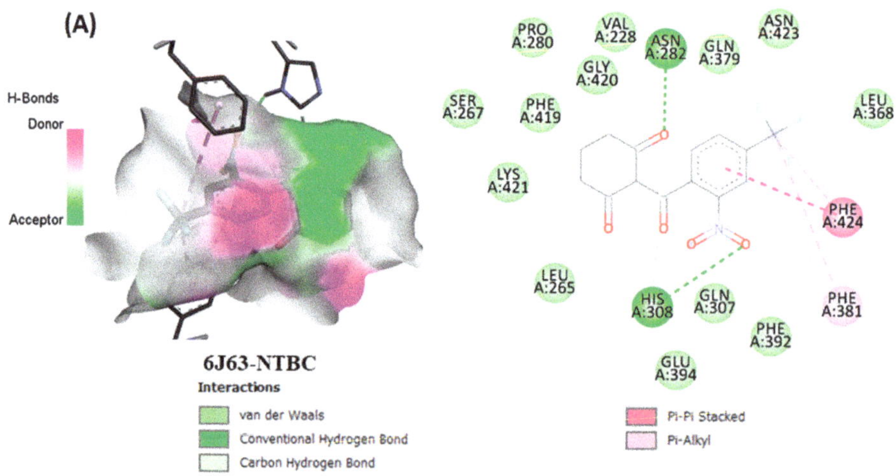

Figure 6. *Cont.*

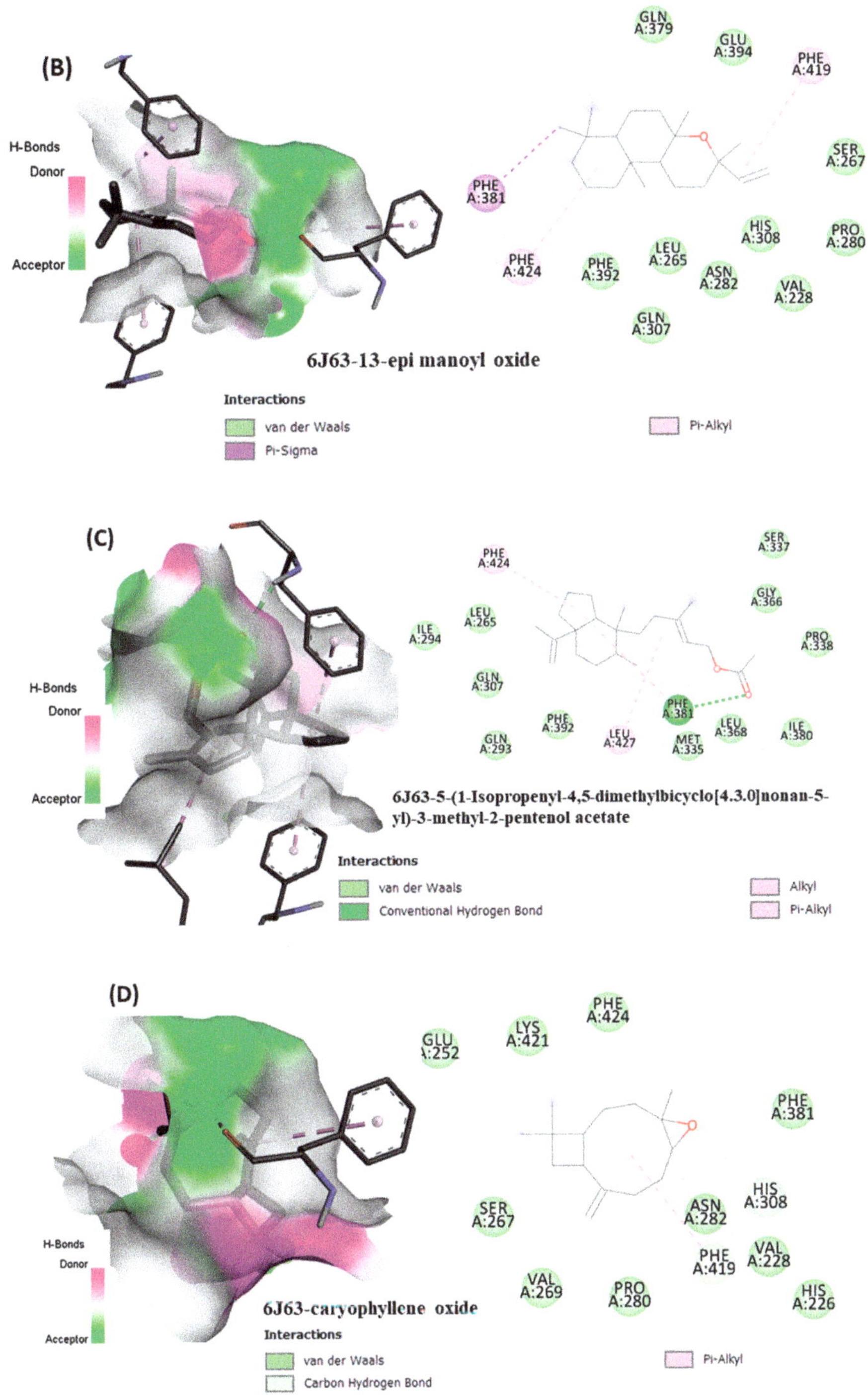

Figure 6. *Cont.*

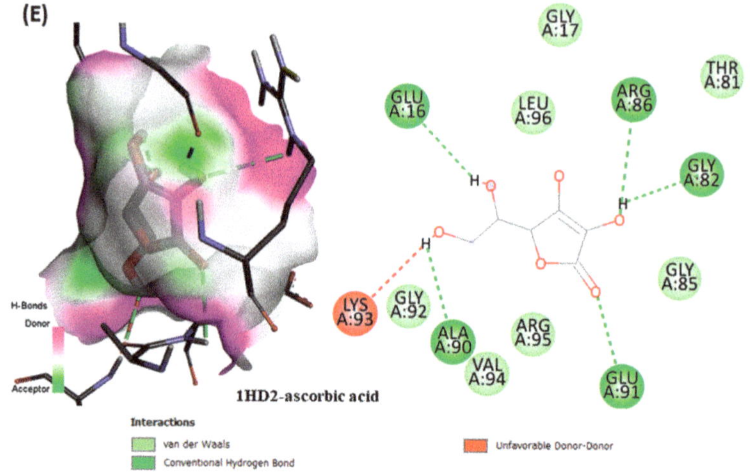

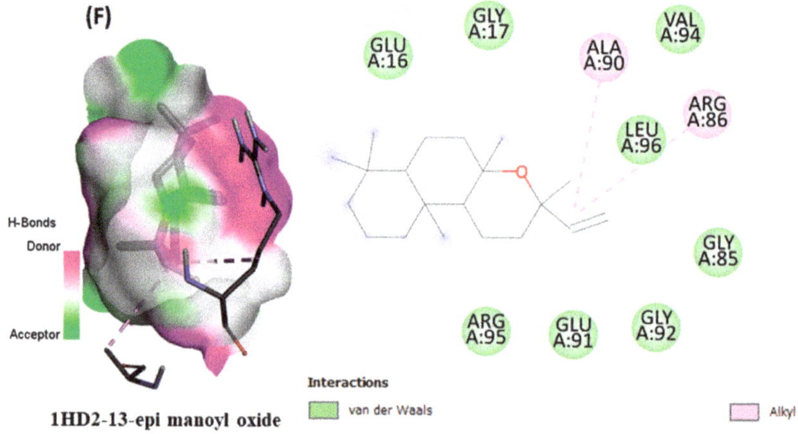

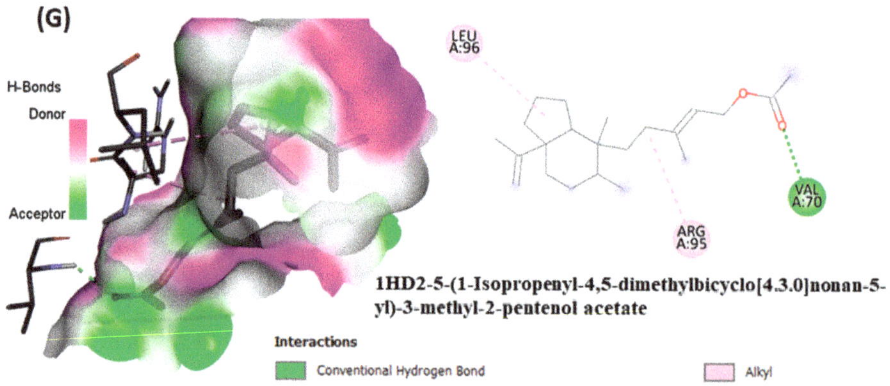

Figure 6. *Cont.*

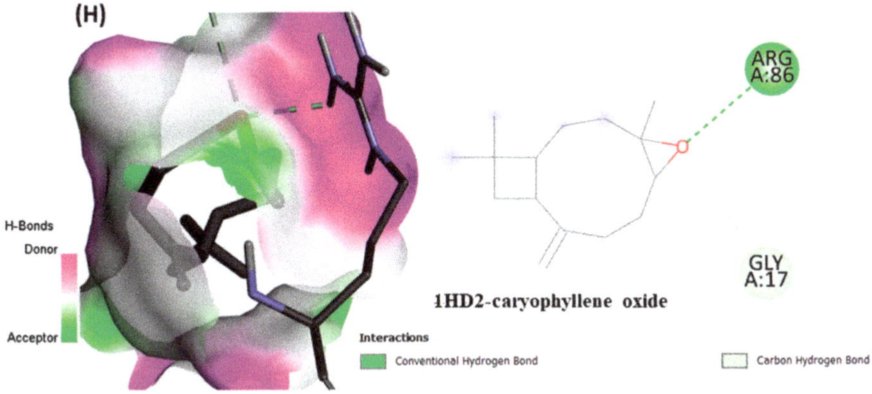

Figure 6. (**A–H**) Docked conformations of molecules in the binding cavity of HPPD (PDB: 6J63) and human periredoxin 5 (PDB: 1HD2) with least binding energies. The complex established are (**A**) 6J63-NTBC, (**B**) 6J63-13-*epi*-manoyl oxide, (**C**) 6J63-5-(1-Isopropenyl-4,5-dimethylbicyclo [4.3.0]nonan-5-yl)-3-methyl-2-pentenol acetate, (**D**) 6J63-caryophyllene oxide, (**E**) 1HD2-ascorbic acid; (**F**) 1HD2-13-*epi*-manoyl oxide, (**G**) 1HD2-5-(1-Isopropenyl-4,5-dimethylbicyclo [4.3.0]nonan-5-yl)-3-methyl-2-pentenol acetate, (**H**) 1HD2-caryophyllene oxide.

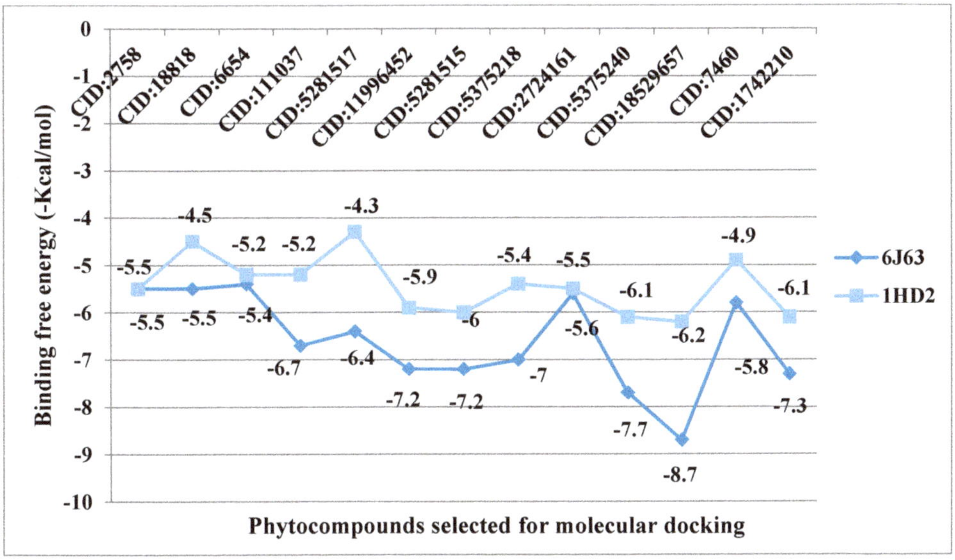

Figure 7. Binding energy (−kcal/mol) of selected phytocompounds from VAO, VNO, and VTO, complexed with 6J63 and 1HD2.

The listed binding energies of the volatiles docked with human peroxiredoxin 5 and HPPD (Figure 7) were found to be in the range −6.2 to −4.3 kcal/mol and −8.7 to −5.4 kcal/mol, respectively. Based on the study, it was observed that the major constituents interacted favorably with the receptors—most of which are the Van der Waal interactions. The analysis of ligand recognition reveals that the compounds can be good antioxidant and phytotoxic agents. Figure 6A–H shows the interaction of selected volatiles with the receptors (6J63 and 1HD2) having the least binding energies (higher docking scores), along with their 2D interaction with amino acid residues.

3.8. ADMET Analysis

The forecasting of ADME (absorption, distribution, metabolism, and excretion) properties of the selected compounds, including their pharmacokinetic and drug-like properties, have been estimated using SwissADME online server (http://www.swissadme.ch/, accessed on 12 August 2022). The collective laws of Lipinski's [82], Egan's [83], and Veber's [84], which determine the properties of a drug, were followed. According to the rule that the compound should not violate more than 1 Lipinski rule, molecular weight (MW) < 500, topological surface area (TPSA) < 140, number of H-bond acceptors (nOHA) $\leq$ 5, number of H-bond donors (nOHD) $\leq$ 5, water partition coefficient (WLOGP) $\leq$ 5.88, number of rotatable bonds (nRB) $\leq$ 10. Based on the current findings, 12 out of 13 compounds selected followed the Lipinski's, Egan's, and Verber's rule, indicating the good drug-like properties of the compounds. The bioavailability score was found to be 0.55 for all the compounds selected, indicating higher bioactivity of the molecule. The compounds share TPSA values less than 30 Å^2, indicating good brain penetration and good lipophilicity behavior, with the consensus Log Po/w coming in the range 2.60–5.14 (Table 5). There was no P-glycoprotein (P-gp) substrate found, suggesting the good intestinal absorption of compounds. Except sabinene, α-pinene, β-farnesene, β-caryophyllene, 13-*epi*-manoyl acetate, and α-phellandrene, all compounds showed high gastrointestinal absorption. The compounds that were predicted to not cross the blood–brain barrier (BBB) were β-farnesene, β-caryophyllene, 5-(1-isopropenyl-4,5-dimethylbicyclo[4.3.0]nonan-5-yl)-3-methyl-2-pentenol acetate, and 13-*epi*-manoyl oxide.

Some of the compounds interacted mainly with two isoenzymes of the cytochrome (CYP) family, namely CYP2C19 and CYP2C9, suggesting their efficiency while having minimal toxicity. Drug-like properties and GI absorption of selected compounds from VAO, VNO, and VTO were also represented by the boiled-egg prediction (Figure 8) and bioavailability radar graph (Figure 9). The compounds present in the yellow zone in the boiled-egg graph can permeate through the blood–brain barrier (BBB), and the pink area of the bioavailability radar graphs shows the drug-likeness of the compounds.

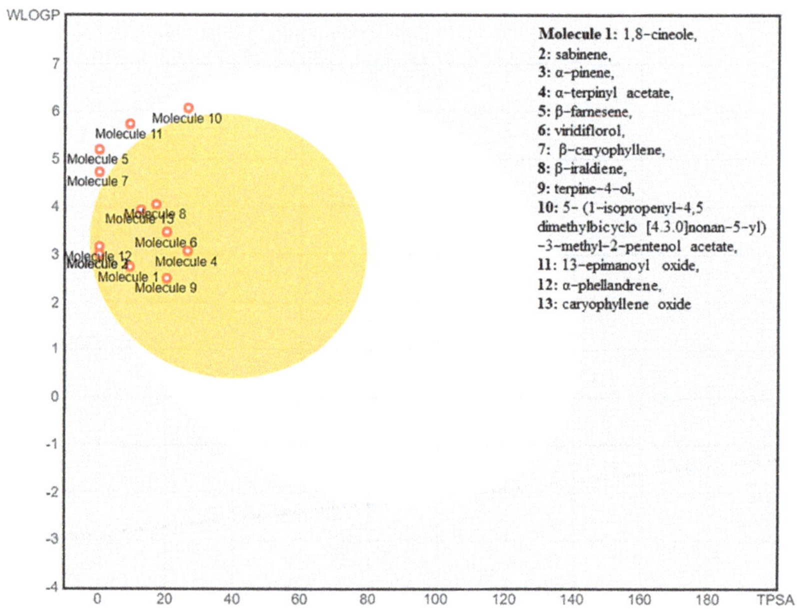

Figure 8. Boiled-egg graph of the selected phytoconstituents.

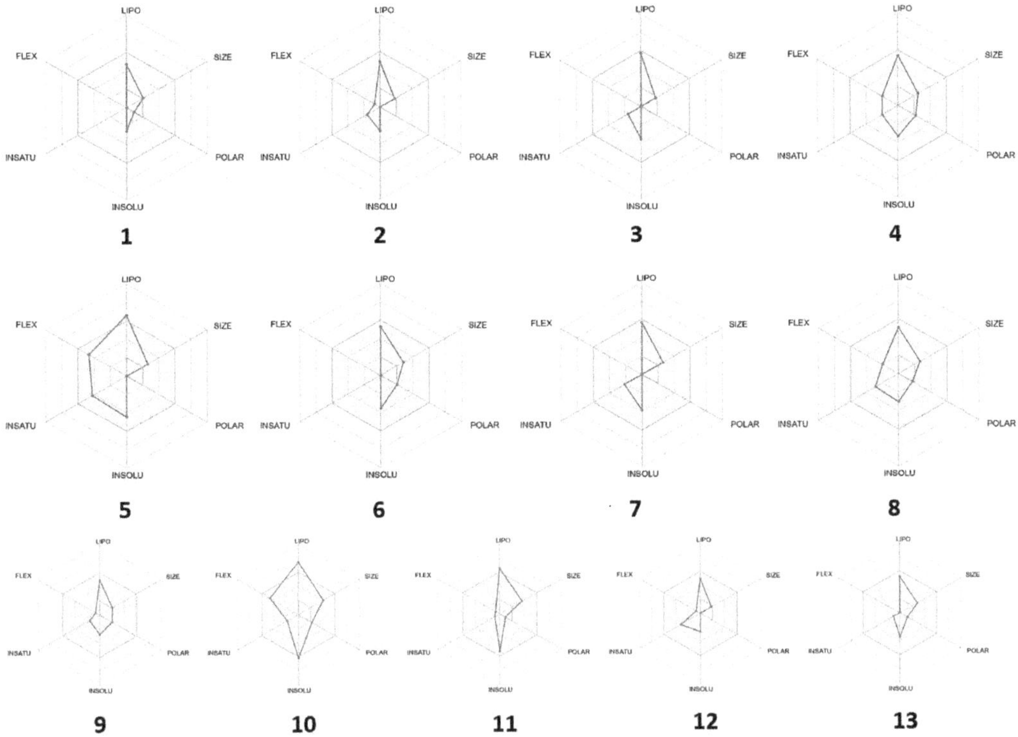

Figure 9. Bioavailability radar of selected phytoconstituents (pink area showed the drug likeness properties of selected compounds) 1: 1,8-cineole, 2: sabinene, 3: α-pinene, 4: α-terpinyl acetate, 5: β-farnesene, 6: viridiflorol, 7: β-caryophyllene, 8: β-iraldiene, 9: terpine-4-ol, 10: 5-(1-isopropenyl-4,5-dimethylbicyclo[4.3.0]nonan-5-yl)-3-methyl-2-pentenol acetate, 11: 13-epimanoyl oxide, 12: α-phellandrene, 13: caryophyllene oxide.

The toxicity parameters of selected phytocompounds were predicted using web server ProTox II (Table 6). All the selected compounds were predicted not to be hepatotoxic, carcinogenic, cytotoxic, immunotoxic, and mutagenic, except α-terpinyl acetate (hepatotoxic), 5-(1-isopropenyl-4,5-dimethylbic clo[4.3.0]nonan-5-yl)-3-methyl-2-pentenol acetate (carcinogenic), β-iraldiene, and caryophyllene oxide (immunotoxic). The LD_{50} values were also calculated to ensure the safety of the selected compounds as shown in Table 6. The compounds with LD_{50} > 2000 mg/kg, suggesting their safety for biological administration and as potential drugs.

Table 5. In silico ADMET analysis of major constituents of VAO, VNO, and VTO.

Entry	1	2	3	4	5	6	7	8	9	10	11	12	13
TPSA * (Å^2)	9.23	0.00	0.00	26.30	0.00	20.23	0.00	17.07	20.23	26.30	9.23	0.00	12.53
Consensus * Log Po/w	2.67	3.25	3.44	3.04	4.97	3.42	4.24	3.56	2.60	4.25	5.14	2.97	3.68
Mol wt (g/mol)	154.25	136.23	136.23	196.29	204.35	222.37	204.35	206.32	154.25	332.52	290.48	136.23	220.35
nRB	0	1	0	3	7	0	0	3	1	7	1	1	0
nOHA	1	0	0	2	0	1	0	1	1	2	1	0	1
nOND	0	0	0	0	0	1	0	0	1	0	0	0	0
WLOGP	2.74	3.00	3.00	3.07	5.20	3.47	4.73	4.05	3.50	6.07	5.74	3.16	3.94
Water solubility	Soluble	Soluble	Soluble	Soluble	Soluble	Soluble	Soluble	Soluble	Soluble	Moderately	Moderately	Soluble	Soluble
GI absorption **	High	Low	Low	High	Low	High	Low	High	High	High	Low	Low	High
BBB permeant **	Yes	Yes	Yes	Yes	No	Yes	No	Yes	Yes	No	No	Yes	Yes
P-gp substrate **	No	No	No	No	No	No	No	No	No	No	No	No	No
CYP1A2 inhibitor **	No	No	No	No	Yes	No	No	No	No	No	No	No	No
CYP2C19 inhibitor **	No	No	No	No	No	Yes	Yes	No	No	Yes	Yes	No	Yes
CYP2C9 inhibitor **	No	No	Yes	Yes	Yes	No	Yes	No	No	Yes	Yes	No	Yes
CYP2D6 inhibitor **	No	No	No	No	No	No	No	No	No	No	No	No	No
CYP3A4 inhibitor **	No	No	No	No	No	No	No	No	No	No	No	No	No
Log K_p (cm/s) (Skin permeation)	−5.30	−4.94	−3.95	−4.69	−3.27	−5.00	−4.44	−5.16	−4.93	−2.97	−3.86	−4.85	−5.12
Lipinski ***	Yes	Yes	Yes	Yes	Yes	Yes	Yes	Yes	Yes	Yes	Yes	Yes	Yes
Lipinski violation	0	1	1	0	1	0	1	0	0	1	1	0	0
Bioavailability score ***	0.55	0.55	0.55	0.55	0.55	0.55	0.55	0.55	0.55	0.55	0.55	0.55	0.55

VAO: *V. agnus-castus* essential oil, VNO: *V. negundo* essential oil, VTO: *V. trifolia* essential oil, ADMET: absorption, distribution, metabolism, excretion and toxicity, Lipophilicity *, Pharmacokinetics **, Drug Likeliness ***, TPSA: topological polar surface area, nRB: no. of rotable bonds, nOHA: no. of H-bond acceptor, nOHD: no. of H-bond donor, WLOGP: water partition coefficient, GI absorption: gastrointestinal absorption, BBB: blood–brain barrier, P-gp: permeability glycoprotein, CYP: cytochrome P450, Entry 1: 1,8-cineole, 2: sabinene, 3: α-pinene, 4: α-terpinyl acetate, 5: β-farnesene, 6: viridiflorol, 7: β-caryophyllene, 8: β-iraldiene, 9: terpine-4-ol, 10: 5-(1-isopropenyl-4,5-dimethylbicyclo [4.3.0]nonan-5-yl)-3-methyl-2-pentenol acetate, 11: 13-epimanoyl oxide, 12: α-phellandrene, 13: caryophyllene oxide.

Table 6. Toxicological properties of selected compounds from VAO, VNO, and VTO.

Compounds	Hepatotoxicity		Carcinogenicity		Cytotoxicity		Immunotoxicity		Mutagenicity		Predicted LD_{50} (mg/kg)	Toxicity Class
	Pr	Pb	Pr	Pb	Pr	Pb	Pr	Pb	Pr	Pb		
1,8-Cineole	NH	0.86	NC	0.68	NCy	0.75	NI	0.99	NM	0.96	2480	V
Sabinene	NH	0.81	NC	0.59	NCy	0.71	NI	0.51	NM	0.82	5000	V
α-Pinene	NH	0.86	NC	0.60	NCy	0.75	NI	0.99	NM	0.93	3700	V

Table 6. *Cont.*

Compounds	Hepatotoxicity		Carcinogenicity		Cytotoxicity		Immunotoxicity		Mutagenicity		Predicted LD_{50} (mg/kg)	Toxicity Class
	Pr	Pb	Pr	Pb	Pr	Pb	Pr	Pb	Pr	Pb		
α-Terpinyl acetate	H	0.53	NC	0.66	NCy	0.80	NI	0.97	NM	0.94	4800	V
β-Farnesene	NH	0.79	NC	0.73	NCy	0.81	NI	0.99	NM	0.98	5000	V
Viridiflorol	NH	0.77	NC	0.69	NCy	0.89	NI	0.87	NM	0.75	2000	IV
β-Caryophyllene	NH	0.80	NC	0.70	NCy	0.75	I	0.54	NM	0.95	5300	V
β-Iraldiene	NH	0.68	NC	0.79	NCy	0.78	NI	0.97	NM	0.93	4590	V
Terpine-4-ol	NH	0.80	NC	0.72	NCy	0.88	NI	0.99	NM	0.83	1016	IV
5-(1-isopropenyl-4,5-dimethylbicyclo[4.3.0]nonan-5-yl)-3-methyl-2-pentenol acetate	NH	0.68	C	0.58	NCy	0.76	NI	0.89	NM	0.87	5000	V
13-epi manoyl oxide	NH	0.86	NC	0.69	NCy	0.75	NI	0.71	NM	0.91	4300	V
α-Phellandrene	NH	0.83	NC	0.52	NCy	0.80	NI	0.88	NM	0.92	5700	VI
Caryophyllene oxide	NH	0.80	NC	0.57	NCy	0.79	I	0.83	NM	0.88	5000	V

Pr: Prediction, Pb: Probability, NH: Nonhepatotoxic, NC: Noncarcinogenic, NCy: Noncytotoxic, NI: Nonimmunotoxic, NM: Nonmutagenic, H: Hepatotoxic, I: Immunotoxic, C: Carcinogenic, Toxicity class: (Class I: fatal if swallowed ($LD_{50} \leq 5$), Class II: fatal if swallowed ($5 < LD_{50} \leq 50$), Class III: toxic if swallowed ($50 < LD_{50} \leq 300$), Class IV: harmful if swallowed ($300 < LD_{50} \leq 2000$), Class V: may be harmful if swallowed ($2000 < LD_{50} \leq 5000$), Class VI: non-toxic ($LD_{50} > 5000$)).

4. Conclusions

In this study, the chemical diversity among the EOs obtained from three *Vitex* species from Tarai region, India, was revealed and analyzed. The chemical profile of EOs was characterized by high content of terpenoids. Moreover, the in vitro antioxidant and phytotoxic activities of the EOs were investigated to check the biological potentials of the plant-derived products of these *Vitex* species. All the tested EOs showed moderate to good antioxidant and phytotoxic potentials as assessed with different assays. The molecular docking study suggested that the compounds from the EOs can be good antioxidant and phytotoxic agents by the analysis of ligand interaction with the proteins. The ADMET analysis revealed the safety of most of the major compounds in the EOs. Overall, this study unveiled some interesting biological activities of these EOs, especially as natural antioxidants and phytotoxic agents, which justifies the use of the plant species in traditional medicine, as well as in the crop protection field. However, the in vivo study is necessary to investigate and assess the potency and safety of these EOs and their active components.

Supplementary Materials: The following supporting information can be downloaded at: https://www.mdpi.com/article/10.3390/antiox11101911/s1, Supplementary Material S1, ion-chromatograms, and mass spectra of the major compounds can be observed.

Author Contributions: Conceptualization, H.K., S.K.M. and R.K.; methodology, H.K., S.K.M., M.T. and R.K.; software, H.K. and S.K.M.; validation, H.K., S.K.M. and R.K.; formal analysis, O.P. and D.S.R.; investigation, R.K. and M.S.d.O.; resources, R.K., O.P. and D.S.R.; writing-original draft preparation, H.K. and S.K.M.; data curation, S.K.M.; writing-review and editing, H.K., S.K.M. and R.K.; supervision, R.K.; funding. M.S.d.O.; funding acquisition, M.S.d.O. All authors have read and agreed to the published version of the manuscript.

Funding: This research received no external funding.

Institutional Review Board Statement: Not applicable.

Informed Consent Statement: Not applicable.

Data Availability Statement: Data are contained within the article.

Acknowledgments: Authors are grateful to G.B. Pant University of Agriculture and Technology, Pantnagar for providing all necessary research facilities and AIRF (Advance Instrumentation Research Facility), JNU, New Delhi, India for GC-MS analysis. The author Dr Mozaniel de Oliveira., thanks PCI-MCTIC/MPEG, as well as CNPq for the process number: [300983/2022-0].

Conflicts of Interest: The authors declare no conflict of interest.

References

1. Ganapaty, S.; Vidyadhar, K. Phytoconstituents and biological activities of *Vitex*—A review. *J. Nat. Rem.* **2005**, *5*, 75–95. [CrossRef]
2. Gill, B.S.; Mehra, R.; Kumar, S. *Vitex negundo* and its medicinal value. *Mol. Biol. Rep.* **2018**, *45*, 2925–2934. [CrossRef] [PubMed]
3. Dugoua, J.J.; Seely, D.; Perri, D.; Koren, G.; Mills, E. Safety and efficacy of chastetree (*Vitex agnus-castus*) during pregnancy and lactation. *Can. J. Clin. Pharmacol.* **2008**, *15*, e74–e79.
4. Costa, M.R.; Ribeiro, C.G.; Santos-Filho, S.D.; Neves, R.F.; Fonseca, A.S.; Bernardo-Filho, M. An aqueous extract of *Vitex agnus-castus* alters the labeling of blood constituents with Technetium-99m. *Braz. Arch. Biol. Technol.* **2007**, *50*, 183–188. [CrossRef]
5. Chaturvedi, G.; Singh, R. Experimental studies on the antiarthritic effect of certain indigenous drugs. *Indian J. Med. Res.* **1965**, *53*, 71.
6. Li, W.X.; Cui, C.B.; Cai, B.; Wang, H.Y.; Yao, X.S. Flavonoids from *Vitex trifolia* L. inhibit cell cycle progression at G2/M phase and induce apoptosis in mammalian cancer cells. *J. Asian Nat. Prod. Res.* **2005**, *7*, 615–626. [CrossRef]
7. Rani, A.; Sharma, A. The genus *Vitex*: A review. *Pharmacogn. Rev.* **2013**, *7*, 188. [CrossRef]
8. Koirala, N.; Dhakal, C.; Munankarmi, N.N.; Ali, S.W.; Hameed, A.; Martins, N.; Salehi, B. *Vitex negundo* Linn.: Phytochemical composition, nutritional analysis, and antioxidant and antimicrobial activity. *Cell. Mol. Biol.* **2020**, *66*, 1–7. [CrossRef]
9. Senatore, F.; Napolitano, F.; Ozcan, M. Chemical composition and antibacterial activity of essential oil from fruits of *Vitex agnus-castus* L. (Verbenaceae) growing in Turkey. *J. Essent. Oil-Bear. Plants* **2003**, *6*, 185–190. [CrossRef]
10. Jarry, H.; Spengler, B.; Porzel, A.; Schmidt, J.; Wuttke, W.; Christoffel, V. Evidence for estrogen receptor beta-selective activity of *Vitex agnus-castus* and isolated flavones. *Planta Med.* **2003**, *69*, 945–947. [CrossRef]
11. Sathiamoorthy, B.; Gupta, P.; Kumar, M.; Chaturvedi, A.K.; Shukla, P.K.; Maurya, R. New antifungal flavonoid glycoside from *Vitex negundo*. *Bioorg. Med. Chem. Lett.* **2007**, *17*, 239–242. [CrossRef] [PubMed]

12. Sundaram, R.; Naresh, R.; Shanthi, P.; Sachdanandam, P. Antihyperglycemic effect of iridoid glucoside, isolated from the leaves of *Vitex negundo* in streptozotocin-induced diabetic rats with special reference to glycoprotein components. *Phytomedicine* **2012**, *19*, 211–216. [CrossRef] [PubMed]
13. Xin, H.; Hu, Y.; Zhang, Q.; Huang, B.; Zheng, H.H.; Qin, L. Study on chemical constituents in fruits of *Vitex trifolia*. *Acad. J. Sec. Mil. Med. Univ.* **2006**, *27*, 1038–1040.
14. Devcich, D.A.; Pedersen, I.K.; Petrie, K.J. You eat what you are: Modern health worries and the acceptance of natural and synthetic additives in functional foods. *Appetite* **2007**, *48*, 333–337. [CrossRef] [PubMed]
15. Botterweck, A.A.M.; Verhagen, H.; Goldbohm, R.A.; Kleinjans, J.; Van den Brandt, P.A. Intake of butylated hydroxyanisole and butylated hydroxytoluene and stomach cancer risk: Results from analysis in the Netherlands cohort study. *Food Chem. Toxicol.* **2000**, *38*, 599–605. [CrossRef]
16. Himani, S.K.M.; Arya, S.; Kumar, R.; Prakash, O. Essential Oil: Source of Antioxidants and Role in Food Preservation. In *Essential Oils: Applications and Trends in Food Science and Technology*; Springer: Berlin/Heidelberg, Germany, 2022; pp. 173–189. [CrossRef]
17. Bezerra, F.W.F.; de Oliveira, M.S.; Bezerra, P.N.; Cunha, V.M.B.; Silva, M.P.; da Costa, W.A.; Pinto, R.H.H.; Cordeiro, R.M.; da Cruz, J.N.; Chaves Neto, A.M.J.; et al. Extraction of Bioactive Compounds. In *Green Sustainable Process for Chemical and Environmental Engineering and Science*; Inamuddin, A.M., Asiri, A., Isloor, A., Eds.; Elsevier: Amsterdam, The Netherlands, 2020; pp. 149–167. ISBN 9780128173886.
18. De Franco, C.J.P.; Ferreira, O.O.; Antônio Barbosa de Moraes, Â.; Varela, E.L.P.; do Nascimento, L.D.; Percário, S.; de Oliveira, M.S.; de Andrade, E.H.A. Chemical Composition and Antioxidant Activity of Essential Oils from Eugenia Patrisii Vahl, E. Punicifolia (Kunth) DC., and Myrcia Tomentosa (Aubl.) DC., Leaf of Family Myrtaceae. *Molecules* **2021**, *26*, 3292. [CrossRef]
19. de Santana Oliveira, M. (Ed.) *Essential Oils-Applications and Trends in Food Science and Technology*, 1st ed.; Springer International Publishing: Berlin/Heidelberg, Germany, 2022; ISBN 978-3-030-99475-4. [CrossRef]
20. Lengai, G.M.; Muthomi, J.W.; Mbega, E.R. Phytochemical activity and role of botanical pesticides in pest management for sustainable agricultural crop production. *Sci. Afr.* **2020**, *7*, e00239. [CrossRef]
21. Ulukanli, Z.; Çenet, M.; Öztürk, B.; Bozok, F.; Karabörklü, S.; Demirci, S.C. Chemical characterization, phytotoxic, antimicrobial and insecticidal activities of *Vitex agnus-castus'* essential oil from East Mediterranean region. *J. Esset. Oil Bear. Plants* **2015**, *18*, 1500–1507. [CrossRef]
22. Issa, M.; Chandel, S.; Singh, H.P.; Batish, D.R.; Kohli, R.K.; Yadav, S.S.; Kumari, A. Appraisal of phytotoxic, cytotoxic and genotoxic potential of essential oil of a medicinal plant *Vitex negundo*. *Ind. Crops Prod.* **2020**, *145*, 112083. [CrossRef]
23. Werrie, P.Y.; Durenne, B.; Delaplace, P.; Fauconnier, M.L. Phytotoxicity of Essential Oils: Opportunities and Constraints for the Development of Biopesticides. A Review. *Foods* **2020**, *9*, 1291. [CrossRef]
24. Mejri, J.; Abderrabba, M.; Mejri, M. Chemical composition of the essential oil of *Ruta chalepensis* L: Influence of drying, hydro-distillation duration and plant parts. *Ind. Crops Prod.* **2010**, *32*, 671–673. [CrossRef]
25. Adams, R.P. *Identification of Essential Oil Components by Gas Chromatography/Mass Spectrometry*; Allured Publishing Corporation: Carol Stream, IL, USA, 2007; pp. 544–545.
26. Acree, W., Jr.; Chickos, J.S. Phase Transition Enthalpy Measurements of Organic and Organometallic Compounds. In *NIST Chemistry WebBook, NIST Standard Reference Database Number 69*; Linstrom, P.J., Mallard, W.G., Eds.; National Institute of Standards and Technology: Gaithersburg, MD, USA, 2022. [CrossRef]
27. Blois, M.S. Antioxidant determinations by the use of a stable free radical. *Nature* **1958**, *181*, 1199–1200. [CrossRef]
28. Diniz do Nascimento, L.; Moraes, A.A.B.D.; Costa, K.S.D.; Pereira Galúcio, J.M.; Taube, P.S.; Costa, C.M.L.; Faria, L.J.G.D. Bioactive natural compounds and antioxidant activity of essential oils from spice plants: New findings and potential applications. *Biomolecules* **2020**, *10*, 988. [CrossRef] [PubMed]
29. Moller, A.C.; Parpa, C.; Said, B.; Werner, E.; Flores, S.; Villena, J.; Madrid, A. Antioxidant and anti-proliferative activity of essential oil and main components from leaves of *Aloysia polystachya* harvested in Central Chile. *Molecules* **2021**, *26*, 131. [CrossRef]
30. Joshi, M.; Kumar, R.; Prakash, O.; Pant, A.K.; Rawat, D.S. Chemical composition and biological activities of *Nepeta hindostana* (Roth) Haines, *Nepeta graciliflora* Benth. and *Nepeta cataria* L. from India. *J. Med. Herb.* **2021**, *12*, 35–46.
31. Makhija, I.K.; Ram, H.A.; Shreedhara, C.S.; Kumar, S.V.; Devkar, R. In vitro antioxidant studies of *Sitopaladi churna*, a polyherbal ayurvedic formulation. *Free Radic. Antioxid.* **2011**, *1*, 37–41. [CrossRef]
32. Gedikoğlu, A.; Sökmen, M.; Çivit, A. Evaluation of *Thymus vulgaris* and *Thymbra spicata* essential oils and plant extracts for chemical composition, antioxidant, and antimicrobial properties. *Food Sci. Nutr.* **2019**, *7*, 1704–1714. [CrossRef]
33. Decker, E.A.; Welch, B. Role of Ferritin as a Lipid Oxidation Catalyst in Muscle. *Food. J. Agric. Food Chem.* **1990**, *38*, 674–677. [CrossRef]
34. de Oliveira, M.S.; da Costa, W.A.; Pereira, D.S.; Botelho, J.R.S.; de Alencar Menezes, T.O.; de Aguiar Andrade, E.H.; da Silva, S.H.M.; da Silva Sousa Filho, A.P.; de Carvalho, R.N. Chemical composition and phytotoxic activity of Clove (*Syzygium aromaticum*) essential oil obtained with supercritical CO_2. *J. Supercrit. Fluids* **2016**, *118*, 185–193. [CrossRef]
35. Arya, S.; Kumar, R.; Prakash, O.; Kumar, S.; Mahawer, S.K.; Chamoli, S.; de Oliveira, M.S. Chemical Composition and Biological Activities of *Hedychium coccineum* Buch.-Ham. ex Sm. Essential Oils from Kumaun Hills of Uttarakhand. *Molecules* **2022**, *27*, 4833. [CrossRef]
36. Dallakyan, S.; Olson, A.J. Small-molecule library screening by docking with PyRx. In *Chemical Biology*; Humana Press: New York, NY, USA, 2015; pp. 243–250.

37. Song, H.M.; Zhao, L.X.; Zhang, S.Q.; Ye, T.; Fu, Y.; Ye, F. Design, Synthesis, Structure–Activity Relationship, Molecular Docking, and Herbicidal Evaluation of 2-Cinnamoyl-3-Hydroxycyclohex-2-en-1-one Derivatives as Novel 4-Hydroxyphenylpyruvate Dioxygenase Inhibitors. *J. Agric. Food Chem.* **2021**, *69*, 12621–12633. [CrossRef]
38. Wang, H.; Wang, L.; Zhang, X.; Bai, S.; Jin, T.; Liu, W.; Wang, J. Unravelling Phytotoxicity and Mode of Action of Tripyrasulfone, a Novel Herbicide. *J. Agric. Food Chem.* **2021**, *69*, 7168–7177. [CrossRef] [PubMed]
39. Alam, A.; Jawaid, T.; Alam, P. In vitro antioxidant and anti-inflammatory activities of green cardamom essential oil and in silico molecular docking of its major bioactives. *J. Taibah Univ. Sci.* **2021**, *15*, 757–768. [CrossRef]
40. Daina, A.; Michielin, O.; Zoete, V. SwissADME: A free web tool to evaluate pharmacokinetics, drug-likeness and medicinal chemistry friendliness of small molecules. *Sci. Rep.* **2017**, *7*, 42717. [CrossRef] [PubMed]
41. Babicki, S.; Arndt, D.; Marcu, A.; Liang, Y.; Grant, J.R.; Maciejewski, A.; Wishart, D.S. Heatmapper: Web-enabled heat mapping for all. *Nucleic Acids Res.* **2016**, *44*, W147–W153. [CrossRef] [PubMed]
42. Rezaei, M.; Razmjoo, J.; Ehtemam, M.H.; Karimmojeni, H.; Zahedi, M. The interaction between shade and drought affects essential oil quantity and quality of *Vitex agnus-castus* L. leaves and seeds. *Ind. Crops Prod.* **2019**, *137*, 460–467. [CrossRef]
43. Senatore, F.; Della Porta, G.; Reverchon, E. Constituents of *Vitex agnus-castus* L. Essential Oil. *Flavour Fragr. J.* **1996**, *11*, 179–182. [CrossRef]
44. Novak, J.; Draxler, L.; Göhler, I.; Franz, C.M. Essential oil composition of *Vitex agnus-castus*-comparison of accessions and different plant organs. *Flavour Fragr. J.* **2005**, *20*, 186–192. [CrossRef]
45. Stojković, D.; Soković, M.; Glamočlija, J.; Džamić, A.; Ćirić, A.; Ristić, M.; Grubišić, D. Chemical composition and antimicrobial activity of *Vitex agnus-castus* L. fruits and leaves essential oils. *Food Chem.* **2011**, *128*, 1017–1022. [CrossRef]
46. Marongiu, B.; Piras, A.; Porcedda, S.; Falconieri, D.; Gonçalves, M.J.; Salgueiro, L.; Maxia, A.; Lai, R. Extraction, separation and isolation of volatiles from *Vitex agnus-castus* L. (Verbenaceae) wild species of Sardinia, Italy, by supercritical CO_2. *Nat. Prod. Res.* **2010**, *24*, 569–579. [CrossRef]
47. Kustrak, D.; Kuftinec, J.; Blazević, N. Composition of the Essential Oil of *Vitex agnus-castus* L. *J. Essent. Oil Res.* **1994**, *6*, 341–344. [CrossRef]
48. Maleš, Ž.; Blažević, N.; Antolić, A. The essential oil composition of *Vitex agnus-castus f. rosea* leaves and flowers. *Planta Med.* **1998**, *64*, 286–287. [CrossRef] [PubMed]
49. Huang, H.C.; Chang, T.Y.; Chang, L.Z.; Wang, H.F.; Yih, K.H.; Hsieh, W.Y.; Chang, T.M. Inhibition of melanogenesis versus antioxidant properties of essential oil extracted from leaves of *Vitex negundo* Linn and chemical composition analysis by GC-MS. *Molecules* **2012**, *17*, 3902–3916. [CrossRef] [PubMed]
50. Padalia, R.C.; Verma, R.S.; Chauhan, A.; Chanotiya, C.S.; Thul, S. Phytochemical diversity in essential oil of *Vitex negundo* L. populations from India. *Rec. Nat. Prod.* **2016**, *10*, 452–464.
51. Raju, S.; Dharani, J.; Ravi, S. Essential oil composition of *Vitex negundo*, acetylcholine esterase inhibition activity and molecular docking studies against bacterial proteins. *J. Pharm. Sci. Res.* **2017**, *9*, 1677–1681.
52. Singh, P.; Mishra, G.; Jha, K.K.; Garg, V.K.; Khosa, R.L. Chemical composition and antimicrobial activity of essential oil of leaves of *Vitex negundo* Linn. (Verbenaceae). *Int. J. Chem. Tech. Res.* **2010**, *2*, 1686–1690.
53. Khokra, S.L.; Prakash, O.; Jain, S.; Aneja, K.R.; Dhingra, Y. Essential oil composition and antibacterial studies of *Vitex negundo* Linn. extracts. *Indian J. Pharm. Sci.* **2008**, *70*, 522. [CrossRef]
54. Chen, Y.; Liu, K.H.; Chen, H.B.; Liao, Y.L. Year-round yield of essential oil of *Vitex negundo* and its chemical constituent analysis. *Planta Med.* **2012**, *78*, PF84. [CrossRef]
55. Thomas, R.P.; Ramachandran, A.; Paul, J.; Mohan, M. Essential oils studies of the genus *Vitex* L. (Verbenaceae). *Int. J. Adv. Res.* **2019**, *7*, 568–574. [CrossRef]
56. Suksamrarn, A.; Werawattanametin, K.; Brophy, J.J. Variation of essential oil constituents in *Vitex trifolia* species. *Flavour Fragr. J.* **1991**, *6*, 97–99. [CrossRef]
57. Boveiri Dehsheikh, A.; Mahmoodi Sourestani, M.; Boveiri Dehsheikh, P.; Vitalini, S.; Iriti, M.; Mottaghipisheh, J. A comparative study of essential oil constituents and phenolic compounds of Arabian Lilac (*Vitex trifolia* var. Purpurea): An evidence of season effects. *Foods* **2019**, *8*, 52. [CrossRef] [PubMed]
58. Musa, N.; Banerjee, S.; Maspalma, G.A.; Usman, L.U.; Hussaini, B. Assessment of the phytochemical, antioxidant and larvicidal activity of essential oil extracted from Simple leaf Chastetree (*Vitex trifolia*) leaves obtained from Ganye Local Government, Adamawa State-Nigeria. *Mater. Today Proc.* **2022**, *49*, 3435–3438. [CrossRef]
59. Arpiwi, N.L.; Muksin, I.K.; Kriswiyanti, E. Essential oils from *Vitex trifolia* as an effective repellent for *Aedes aegypti*. *Biodivers. J. Biol. Divers.* **2020**, *21*, 4536–4544. [CrossRef]
60. Al Saka, F.; Daghestani, M.; Karabet, F. Composition and antioxidant activity of *Vitex agnus-castus* L. and *Rosmarinus officinalis*, L. leaves essential oils cultivated in Syria. *SM Anal. Bioanal. Tech.* **2017**, *2*, 1010. [CrossRef]
61. Cai, Z.M.; Peng, J.Q.; Chen, Y.; Tao, L.; Zhang, Y.Y.; Fu, L.Y.; Shen, X.C. 1, 8-Cineole: A review of source, biological activities, and application. *J. Asian Nat. Prod. Res.* **2021**, *23*, 938–954. [CrossRef]
62. Trevizan, L.N.F.; do Nascimento, K.F.; Santos, J.A.; Kassuya, C.A.L.; Cardoso, C.A.L.; do Carmo Vieira, M.; Formagio, A.S.N. Anti-inflammatory, antioxidant and anti-*Mycobacterium tuberculosis* activity of viridiflorol: The major constituent of *Allophylus edulis* (A. St.-Hil., A. Juss. & Cambess.) Radlk. *J. Ethnopharmacol.* **2016**, *192*, 510–515.

63. Cao, Y.; Zhang, H.; Liu, H.; Liu, W.; Zhang, R.; Xian, M.; Liu, H. Biosynthesis and production of sabinene: Current state and perspectives. *Appl. Microbiol. Biotechnol.* **2018**, *102*, 1535–1544. [CrossRef]
64. Angelopoulou, D.; Demetzos, C.; Dimas, C.; Perdetzoglou, D.; Loukis, A. Essential oils and hexane extracts from leaves and fruits of *Cistus monspeliensis*. Cytotoxic activity of ent-13-epi-manoyl oxide and its isomers. *Planta Med.* **2021**, *67*, 168–171. [CrossRef]
65. Lu, C.; Li, H.; Li, C.; Chen, B.; Shen, Y. Chemical composition and radical scavenging activity of *Amygdalus pedunculata* Pall leaves' essential oil. *Food Chem. Toxicol.* **2018**, *119*, 368–374. [CrossRef]
66. Donati, M.; Mondin, A.; Chen, Z.; Miranda, F.M.; do Nascimento Jr, B.B.; Schirato, G.; Froldi, G. Radical scavenging and antimicrobial activities of *Croton zehntneri*, *Pterodon emarginatus* and *Schinopsis brasiliensis* essential oils and their major constituents: Estragole, trans-anethole, β-caryophyllene and myrcene. *Nat. Prod. Res.* **2015**, *29*, 939–946. [CrossRef]
67. Souza, C.F.; Baldissera, M.D.; Silva, L.D.L.; Geihs, M.A.; Baldisserotto, B. Is monoterpene terpinen-4-ol the compound responsible for the anesthetic and antioxidant activity of *Melaleuca alternifolia* essential oil (tea tree oil) in silver catfish. *Aquaculture* **2018**, *486*, 217–223. [CrossRef]
68. Ciftci, O.; Ozdemir, I.; Tanyildizi, S.; Yildiz, S.; Oguzturk, H. Antioxidative effects of curcumin, β-myrcene and 1, 8-cineole against 2, 3, 7, 8-tetrachlorodibenzo-p-dioxin-induced oxidative stress in rats liver. *Toxicol. Ind. Health* **2011**, *27*, 447–453. [CrossRef] [PubMed]
69. Kazemi, M. Phenolic profile, antioxidant capacity and anti-inflammatory activity of *Anethum graveolens* L. essential oil. *Nat. Prod. Res.* **2015**, *29*, 551–553. [CrossRef] [PubMed]
70. Aftab, A.; Rita, S.M. Antioxidant activity of essential oil of three cultivars of *Amomum subulatum* and standardization of high performance thin layer chromatography (HPTLC) method for the estimation of 1,8-cineole. *Afr. J. Biotechnol.* **2018**, *17*, 1129–1137. [CrossRef]
71. Sağlam, H.; Pabuçcuoğlu, A.; Kıvçak, B. Antioxidant activity of *Vitex agnus-castus* L. extracts. *Phytother. Res.* **2007**, *21*, 1059–1060. [CrossRef]
72. Sarikurkcu, C.; Arisoy, K.; Tepe, B.; Cakir, A.; Abali, G.; Mete, E. Studies on the antioxidant activity of essential oil and different solvent extracts of *Vitex agnus castus* L. fruits from Turkey. *Food Chem. Toxicol.* **2009**, *47*, 2479–2483. [CrossRef]
73. Asdadi, A.; Hamdouch, A.; Oukacha, A.; Moutaj, R.; Gharby, S.; Harhar, H.; Hassani, L.I. Study on chemical analysis, antioxidant and in vitro antifungal activities of essential oil from wild *Vitex agnus-castus* L. seeds growing in area of Argan Tree of Morocco against clinical strains of Candida responsible for nosocomial infections. *J. Mycol. Med.* **2015**, *25*, e118–e127. [CrossRef]
74. Zhou, S.; Wei, C.; Zhang, C.; Han, C.; Kuchkarova, N.; Shao, H. Chemical composition, phytotoxic, antimicrobial and insecticidal activity of the essential oils of *Dracocephalum integrifolium*. *Toxins* **2019**, *11*, 598. [CrossRef]
75. Yoshimura, H.; Sawai, Y.; Tamotsu, S.; Sakai, A. 1, 8-cineole inhibits both proliferation and elongation of BY-2 cultured tobacco cells. *J. Chem. Ecol.* **2011**, *37*, 320–328. [CrossRef]
76. Wagner, L.S.; Sequin, C.J.; Foti, N.; Campos-Soldini, M.P. Insecticidal, fungicidal, phytotoxic activity and chemical composition of *Lavandula dentata* essential oil. *Biocatal. Agric. Biotechnol.* **2021**, *35*, 102092. [CrossRef]
77. Dahham, S.S.; Tabana, Y.M.; Iqbal, M.A.; Ahamed, M.B.; Ezzat, M.O.; Majid, A.S.; Majid, A.M. The anticancer, antioxidant and antimicrobial properties of the sesquiterpene β-caryophyllene from the essential oil of *Aquilaria crassna*. *Molecules* **2015**, *20*, 11808–11829. [CrossRef] [PubMed]
78. Nascimento, N.R.; Leal-Cardoso, J.H.; Lessa, L.M.; Roriz-Filho, J.S.; Cunha, K.M.; Fonteles, M.C. Terpinen-4-ol: Mechanisms of relaxation on rabbit duodenum. *J. Pharm. Pharmacol.* **2005**, *57*, 467–474. [CrossRef] [PubMed]
79. Dias, J.D.F.G.; Miguel, O.G.; Miguel, M.D. Composition of essential oil and allelopathic activity of aromatic water of *Aster lanceolatus* Willd:(Asteraceae). *Braz. J. Pharm. Sci.* **2009**, *45*, 469–474. [CrossRef]
80. Alminderej, F.; Bakari, S.; Almundarij, T.I.; Snoussi, M.; Aouadi, K.; Kadri, A. Antioxidant activities of a new chemotype of *Piper cubeba* L. fruit essential oil (Methyleugenol/Eugenol): In Silico molecular docking and ADMET studies. *Plants* **2020**, *9*, 1534. [CrossRef] [PubMed]
81. Aouadi, K.; Hajlaoui, H.; Arraouadi, S.; Ghannay, S.; Snoussi, M.; Kadri, A. HPLC/MS Phytochemical Profiling with Antioxidant Activities of *Echium humile* Desf. Extracts: ADMET Prediction and Computational Study Targeting Human Peroxiredoxin 5 Receptor. *Agronomy* **2021**, *11*, 2165. [CrossRef]
82. Lipinski, C.A.L.F. Poor aqueous solubility—an industry wide problem in drug discovery. *Am. Pharm Rev.* **2002**, *5*, 82–85.
83. Egan, W.J. Predicting ADME properties in drug discovery. In *Drug Design: Structure-and Ligand-Based Approaches*; Merz, K.M., Ringe, D., Jr., Reynolds, C.H., Eds.; Cambridge University Press: Cambridge, UK, 2010; pp. 165–173.
84. Pollastri, M.P. Overview on the Rule of Five. *Curr. Protoc. Pharmacol.* **2010**, *49*, 9–12. [CrossRef]

antioxidants

MDPI

Article

Phytochemical Profile, Antioxidant Potential and Toxicity Evaluation of the Essential Oils from *Duguetia* and *Xylopia* Species (*Annonaceae*) from the Brazilian Amazon

Márcia Moraes Cascaes [1], Ângelo Antônio Barbosa De Moraes [2], Jorddy Neves Cruz [3], Celeste de Jesus Pereira Franco [2], Renan Campos E Silva [1], Lidiane Diniz do Nascimento [2], Oberdan Oliveira Ferreira [4], Tainá Oliveira dos Anjos [5], Mozaniel Santana de Oliveira [2,*], Giselle Maria Skelding Pinheiro Guilhon [1] and Eloisa Helena de Aguiar Andrade [1,2,4]

1 Programa de Pós-Graduação em Química, Universidade Federal do Pará, Rua Augusto Corrêa S/N, Guamá, Belém 66075-900, PA, Brazil
2 Laboratório Adolpho Ducke-Coordenação de Botânica, Museu Paraense Emílio Goeldi, Av. Perimetral, 1901, Terra Firme, Belém 66077-830, PA, Brazil
3 Laboratory of Functional and Structural Biology, Institute of Biological Sciences, Universidade Federal do Pará, Rua Augusto Corrêa S/N, Guamá, Belém 66075-900, PA, Brazil
4 Programa de Pós-Graduação em Biodiversidade e Biotecnologia-Rede Bionorte, Universidade Federal do Pará, Rua Augusto Corrêa S/N, Guamá, Belém 66075-900, PA, Brazil
5 Programa de Pós-Graduação em Ciências Biológicas-Botânica Tropical, Museu Paraense Emilio Goeldi/Universidade Federal Rural da Amazônia, Av. Perimetral, 1901, Terra Firme, Belém 66077-830, PA, Brazil
* Correspondence: mozaniel.oliveira@yahoo.com.br or mozanieloliveira@museu-goeldi.br; Tel.: +55-91-988647823

Citation: Cascaes, M.M.; De Moraes, Â.A.B.; Cruz, J.N.; Franco, C.d.J.P.; E Silva, R.C.; Nascimento, L.D.d.; Ferreira, O.O.; Anjos, T.O.d.; de Oliveira, M.S.; Guilhon, G.M.S.P.; et al. Phytochemical Profile, Antioxidant Potential and Toxicity Evaluation of the Essential Oils from *Duguetia* and *Xylopia* Species (*Annonaceae*) from the Brazilian Amazon. *Antioxidants* **2022**, *11*, 1709. https://doi.org/10.3390/antiox11091709

Academic Editor: Stanley Omaye

Received: 28 July 2022
Accepted: 27 August 2022
Published: 30 August 2022

Abstract: The essential oils (EOs) of Duguetia echinophora, D. riparia, Xylopia emarginata and X. frutescens (Annonaceae) were obtained by hydrodistillation and the chemical composition was analyzed by GC-MS. An antioxidant assay using the ABTS and DPPH radicals scavenging method and cytotoxic assays against Artemia salina were also performed. We evaluated the interaction of the major compounds of the most toxic EO (X. emarginata) with the binding pocket of the enzyme Acetylcholinesterase, a molecular target related to toxicity in models of Artemia salina. The chemical composition of the EO of D. echinophora was characterized by β-phellandrene (39.12%), sabinene (17.08%) and terpinolene (11.17%). Spathulenol (22.22%), caryophyllene oxide (12.21%), humulene epoxide II (11.86%) and allo-aromadendrene epoxide (10.20%) were the major constituents of the EO from D. riparia. Spathulenol (5.65%) and caryophyllene oxide (5.63%) were the major compounds of the EO from X. emarginata. The EO of X. frutescens was characterized by α-pinene (20.84%) and byciclogermacrene (7.85%). The results of the radical scavenger DPPH assays ranged from 15.87 to 69.38% and the highest percentage of inhibition was observed for the EO of X. emarginata, while for ABTS radical scavenging, the antioxidant capacity of EOs varied from 14.61 to 63.67%, and the highest percentage of inhibition was observed for the EO of X. frutescens. The EOs obtained from D. echinophora, X. emarginata and X. frutescens showed high toxicity, while the EO of D. riparia was non-toxic. Because the EO of X. emarginata is the most toxic, we evaluated how its major constituents were able to interact with the Acetylcholinesterase enzyme. The docking results show that the compounds are able to bind to the binding pocket through non-covalent interactions with the residues of the binding pocket. The species X. emarginata and X. frutescens are the most promising sources of antioxidant compounds; in addition, the results obtained for preliminary cytotoxicity of the EOs of these species may also indicate a potential biological activity.

Keywords: natural products; Amazon; medicinal plant; volatile compounds; bioactive compounds

1. Introduction

Essential oils (EOs) are complex mixtures of substances formed in the secondary metabolism of plants [1,2], and the substances present in EOs are intended to protect plants against pests, herbivores, fungi and bacteria [3]. Among these substances, sesquiterpenes, monoterpenes, aldehydes, alcohols, esters, and ketones stand out [4–8]. In aromatic species belonging to the *Annonaceae* family, compounds belonging to the class of mono and sesquiterpenoids have been identified as predominant [9,10]. Due to the strong demand for pure natural ingredients in various fields, EOs have been widely used all over the world for various applications in industrials sectors, such as food, pharmaceuticals and cosmetics production [11].

The antioxidant activity of EOs is a property of great interest because the EOs may preserve foods, cosmetics, perfumes and other products from the toxic effects of oxidants. Moreover, the ability of EOs to scavenge free radicals may play an important role in prevention of some diseases such as brain dysfunction, cancer, heart disease and immune system decline. Increasing evidence has suggested that these diseases may result from cellular damage caused by free radicals [12–14]. Furthermore, the *Artemia salina* Leach assay is a preliminary toxicity test that screens a large number of biosynthesized compounds from plant secondary metabolism and can quickly indicate the potential biological activity of EOs [15]. In general, authors report that the molecular target in toxicity tests with *A. salina* is acetylcholinesterase, so it is important to investigate the interaction mechanisms using in silico studies [16,17].

Annonaceae has numerous species that produce EOs. This family consists of 2106 species and more than 130 genera concentrated in the tropics. Around 900 species are neotropical, 450 are Afrotropical and the other species are Indomalayan [18]. In the Amazon region it is estimated that there are approximately 268 species [19]. The biological activities described for the EOs of these species include antioxidant [20–22] and cytotoxic activities [23]. Considering the large number of species of *Annonaceae* occurring in the Amazon region, there are still few studies investigating the chemical composition and the biological activities of the EOs of these species. In this paper, the chemical composition and the antioxidant and cytotoxic properties of the EOs obtained from the *Annonaceae* species collected in the State of Pará-Brazil (*Duguetia echinophora* R.E.Fr., *D. riparia* Huber, *Xylopia emarginata* Mart. and *X. frutescens* Aubl) were evaluated. We also studied the interaction of the major compounds of the most toxic EO with the binding pocket of the enzyme Acetylcholinesterase.

It is worth mentioning that there is still no literature available on the biological properties of the EOs from *D. echinophora, D. riparia* or *X. emarginata* nor on the chemical composition of the EO of the species *D. echinophora*. The chemical composition of EOs from *D. riparia, X. emarginata* and *X. frutescens* has been evaluated and is characterized by mono and sesquiterpenes [24–26]. The EO from *X. frutescens* showed interesting anticancer [26] and repellent activities [27]. In folk medicine, this species is known in Brazil as "embira", "semente-de-embira", "embira-vermelha" and "pau carne", and is widely used to treat flu, digestive problems, rheumatism, halitosis, tooth decay and as a bladder stimulant [26,28,29].

The present work provides new information related to the antioxidant potential of EOs from the species *D. echinophora, D. riparia, X. emarginata* and *X. frutescens* for use in areas such as food conservation. In addition, we investigate preliminary toxicity that provides important information related to the application of these EOs in potential biological activities.

2. Materials and Methods

2.1. Botanical Material

The leaves of *Annonaceae* species were collected in the municipality of Magalhães Barata (State of Pará, Amazon region, Brazil) in March 2018 (00°47′51.6″ S; 047°33′38.4″ W). The samples were identified by Jorge Oliveira, a parataxonomist from the Museu Paraense Emílio Goeldi (MPEG), Belém, Pará, Brazil. The voucher specimens were deposited at the Herbarium of MPEG under the registration codes MG-237446 for *D. riparia*, MG-237477 for *D. echinophora*, MG-237444 for *X. frutescens* and MG-237449 for *X. emarginata*.

2.2. Preparation of Botanical Material and Extraction of Essential Oils

The leaves of *Annonaceae* species were dried in an air-circulation oven for five days at 35 °C and then crushed in a knife mill (Tecnal, model TE-631/3, Piracicaba, São Paulo, Brazil). The moisture content was analyzed using a moisture analyzer (Marte, model ID50, São Paulo, Brazil). The EOs were extracted from the leaves of *Annonaceae* species by hydrodistillation in a glass modified Clevenger-type apparatus [30,31], using 150 g of plant material for each experiment. Hydrodistillations were carried out for 3 h at 100 °C. The obtained EOs were dried over anhydrous sodium sulfate and stored in a freezer at −10 °C. The yields of EOs (%) were calculated based on plant dry weight and expressed in mL/100 g of dried material.

2.3. Analysis of Chemical Profile of Essential Oil

The phytochemical profiles of the EOs were analyzed using chromatography/mass spectrometry (GC/MS) using a Shimadzu QP Plus 2010 GC-MS (Kyoto, Japan) following protocols reported earlier by our research group [32,33]. The retention index was calculated for all volatile constituents using a homologous series of *n*-alkanes (C_8-C_{40}, Sigma-Aldrich, St. Louis, MO, USA) according Van den Dool and Kratz [34], and the compounds were identified by comparing their mass spectrum and retention index with the data from the libraries [35].

2.4. ABTS•+ Radical Scavenging Assay

The ABTS•+ assay was performed according to the methodology adapted from Miller et al. [36], and modified by Re et al. [37]. ABTS•+ was prepared using 7 mM ABTS•+ and 140 mM of potassium persulfate incubated at room temperature without light for 16 h. The solution was then diluted with phosphate-buffered saline until it reached an absorbance of 0.700 ± 0.02 at 734 nm. To measure the antioxidant capacity, 2.97 mL of the ABTS•+ solution was transferred to the cuvette, and the absorbance at 734 nm was determined using a Biospectro SP 22 spectrophotometer. Then, 0.03 mL of the sample was added to the cuvette containing the ABTS•+ radical, and after 5 min, the second reading was performed. The data were expressed as percent inhibition.

2.5. DPPH• Radical Scavenging Assay

The test was carried out according to the method proposed by Blois et al. [38]. To measure the antioxidant capacity, initially, the absorbance of DPPH• 0.1 mM diluted in ethanol was determined. Subsequently, 0.6 mL of DPPH• solution, 0.35 mL of distilled water, and 0.05 mL of the sample were mixed and placed in a water bath at 37 °C for 30 min. Thereafter, the absorbances were determined in a spectrophotometer at 517 nm. The data were expressed as percent inhibition.

2.6. Preliminary Toxicity

The toxicity of the essential oils was tested against larvae of the microcrustacean *Artemia salina* leach (brine shrimp). The eggs of *A. salina* (25 mg) were incubated at room temperature (27–30 °C) in an aquarium with artificial salt water composed of a mixture of 46 g of NaCl, 22 g of $MgCl_2.6H_2O$, 8 g of Na_2SO_4, 2.6 g of $CaCl_2.6H_2O$, and 1.4 g of KCl in 2.0 L of distilled water. The pH was adjusted to the 8.0–9.0 range using Na_2CO_3 to avoid the risk of larvae death by lowering the pH during incubation. After 24 h of egg hatching, oil solutions were prepared at concentrations of 100, 50, 25, 10, 5 and 1 $\mu g{\cdot}mL^{-1}$ using brine as vehicle and 5% dimethyl sulfoxide as diluent. Ten larvae of *A. salina* were placed in each tube containing the solution, and the mortality rate of the larvae after 24 h was calculated. The mean lethal concentration (LC_{50}) was estimated using the Probitos statistical method. All the experiments were performed in triplicate using same protocols as described by Rebelo et al. [39].

2.7. In silico analysis

To carry out the in silico study, the molecules spathulenol and caryophyllene oxide (the major constituents present in the EO of *Xylopia emarginata*) were constructed using GaussView 5.5 software [40,41]. Their molecular structures were optimized with B3LYP/6-31G* [42,43] with Gaussian 09 [44]. We used the molecular method to evaluate the compounds interaction mode with Acetylcholinesterase (AChE). For this we used the Molegro Virtual Docker (MVD) 5.5 [45–47], and the crystal structure used as a molecular target can be found in the Protein Data Bank using the ID: 4M0E [48]. The MolDock Score (GRID) scoring function was used with a Grid resolution of 0.30 Å and 5 Å radius encompassing the entire connection cavity. The MolDock SE algorithm was used with the following parameter settings: number of runs equal to 10, maximum of 1500 interactions, and maximum population size equal to 50. The maximum evaluation of 300 steps with a neighbor distance factor equal to 1 and energy threshold equal to 100 was used during the molecular docking simulation.

2.8. Multivariate Analysis

The multivariate analysis was performed using the Minitab 17® software (free version number 17, Minitab Inc., State College, PA, USA). The chemical constituents of the EOs from the leaves of *D. echinophora*, *D. riparia*, *X. emarginata* and *X. frutescens*, (≥3%), were set as the experimental variables, thus forming a matrix of 4 (samples) × 23 (variables) according to the literature [15,32,33].

3. Results and discussion

3.1. Chemical Composition

The EOs yields from the leaves of the Annonaceae species were 1.76, 0.08, 0.27 and 1.50% for *D. echinophora*, *D. riparia*, *X. emarginata* and *X. frutescens*, respectively. The yield found in this study for the EO of *D. riparia* was close to those found in studies with other species of the *Duguetia* genus (0.1–0.6%) [24]. The EOs yields found for the *Xylopia* species were also very close to those found in others studies [25,26]. The yields and EOs compositions of the species are shown in Table 1.

Table 1. Yield and chemical compositions of the *Annonaceae* species essential oils.

				DE	DR	XE	XF
		Essential Oil Yield (%)		1.76	0.08	0.27	1.50
R_T	**RI_L**	**RI_C**	Constituents (%)				
5.19	801	798	Hexanal			0.95	
6.95	844	845	Hex-(3*E*)-enol			1.35	
7.25	863	857	Hexanol			0.85	
8.30	924	925	α-Thujene				4.89
11.17	932	932	α-Pinene	4.14	1.31	3.14	20.84
11.23	946	948	Camphene			2.72	
11.98	969	974	Sabinene	17.08			
12.32	974	974	β-Pinene			2.01	25.95
12.57	988	991	Myrcene	3.61			
13.09	1002	1002	α-Phellandrene	1.27			1.73
13.56	1008	1011	δ-3-Carene	0.95			
13.89	1014	1016	α-Terpinene				0.85
14.02	1020	1022	*p*-Cymene	0.65		0.54	0.44
14.32	1024	1027	Limonene				3.00
14.79	1025	1029	β-Phellandrene	39.12			2.60
14.90	1026	1031	1,8-Cineole			3.36	1.00
14.99	1032	1043	(Z)-β-Ocimene				0.44

Table 1. *Cont.*

				DE	DR	XE	XF
		Essential Oil Yield (%)		1.76	0.08	0.27	1.50
R_T	RI_L	RI_C	Constituents (%)				
15.02	1054	1055	γ-Terpinene	1.00			1.40
15.13	1065	1068	*cis*-Hydrate sabinene				0.17
15.74	1086	1084	Terpinolene	11.17			0.39
15.87	1095	1099	Linalool			0.30	1.74
16.07	1112	1118	*trans*-Thujone				0.09
16.15	1114	1119	*endo*-Fenchol			0.33	
16.48	1118	1123	*cis*-*p*-Ment-2-en-1-ol				0.08
16.89	1122	1126	α-Campholenal			0.38	0.14
17.32	1135	1140	*trans*-Pinocarveol			4.46	0.36
17.54	1137	1149	*cis*-Verbenol			0.49	0.15
17.76	1154	1156	Sabina ketone			0.27	
17.94	1160	1162	Pinocarvone			2.35	0.16
17.98	1166	1168	*p*-Mentha-1,5-dien-8-ol			1.26	0.11
18.03	1167	1169	Umbellulone				0.04
18.09	1174	1180	Terpinen-4-ol	1.16			1.06
18.51	1179	1186	*p*-Cymen-8-ol	3.36		0.71	
18.74	1186	1194	α -Terpineol				0.97
18.82	1195	1196	Myrtenal			3.24	
18.91	1204	1207	Verbenone			1.62	0.1
19.13	1215	1218	*trans*-Carveol			0.33	
19.22	1239	1243	Carvone			0.23	
19.38	1249	1248	Geraniol			0.39	
19.53	1335	1335	δ-Elemene			2.32	4.41
19.68	1345	1345	α-Cubebene			0.74	0.08
19.95	1373	1367	α-Ylangene			1.35	
20.37	1374	1368	Isoledene				0.02
20.90	1374	1374	α-Copaene	0.25		1.07	0.25
21.56	1379	1378	Geranyl acetate				1.38
22.02	1387	1381	β-Bourbonene			0.93	
22.95	1389	1389	β-Elemene	0.74	0.49	3.10	0.54
23.68	1409	1405	α-Gurjunene			0.11	0.06
23.82	1417	1422	(*E*)-Caryophyllene	2.98	1.56	0.93	0.03
24.17	1419	1416	β-Ylangene				0.72
25.04	1434	1429	γ-Elemene		0.19	0.39	
25.26	1439	1439	Aromadendrene			0.75	0.39
26.13	1442	1442	6,9-Guaiadiene			0.06	
26.58	1451	1450	*trans*-Muurola-3,5-diene			0.38	
26.81	1452	1452	α-Humulene	0.73	1.40	0.35	0.10
27.05	1458	1456	*allo*-Aromadendrene	0.11			
27.19	1464	1465	(*E*)-9-*epi*-caryophyllene				0.26
27.81	1471	1470	Dauca-5,8-diene			0.29	
27.98	1478	1484	γ -Muurolene			3.06	
28.10	1484	1492	Germacrene D	1.24	1.34	1.08	3.26
28.18	1489	1487	β-Selinene			1.61	
28.29	1493	1494	*epi*-Cubebol			0.91	
28.33	1495	1490	γ-Amorphene			0.67	
28.52	1496	1489	Viridiflorene				0.56
28.61	1500	1497	Bicyclogermacrene	0.21			7.85
28.91	1500	1498	α-Muurolene			0.95	
29.03	1513	1513	γ-Cadinene			2.67	0.13
29.17	1514	1513	Cubebol		0.68		0.05
29.57	1522	1520	δ-Cadinene			1.61	0.38
29.98	1528	1520	*cis*-Calamenene	2.06	4.01	0.48	

Table 1. *Cont.*

				DE	DR	XE	XF
		Essential Oil Yield (%)		1.76	0.08	0.27	1.50
R_T	RI_L	RI_C	Constituents (%)				
30.06	1533	1531	*trans*-Cadina-1,4-diene			0.16	0.01
30.24	1533	1534	10-*epi*-Cubebol				0.09
30.39	1537	1536	α-Cadinene			0.24	
30.58	1539	1540	α-Copaen-11-ol				0.04
30.67	1544	1540	α-Calacorene			1.47	
30.83	1548	1548	Elemol				0.06
31.57	1564	1561	β-Calacorene			0.65	
31.97	1577	1579	Spathulenol	1.87	22.22	5.65	2.18
32.28	1582	1583	Caryophyllene oxide	2.49	12.21	5.63	0.18
32.51	1590	1589	Globulol				1.10
32.62	1592	1593	Viridiflorol		0.61		0.54
32.76	1595	1594	Cubeban-11-ol				0.23
32.85	1596	1596	Fokienol		2.48		
32.92	1600	1604	Rosifoliol				0.23
33.09	1602	1601	Ledol				0.10
33.25	1608	1609	Humulene Epoxide II	0.21	11.86	1.41	
33.81	1630	1630	Muurola-4,10(14)-dien-1-β-ol			4.70	
34.18	1638	1643	*epi*-α-Cadinol				0.09
34.36	1639	1657	*Allo*-Aromadendrene Epoxide		10.20		0.02
34.45	1639	1661	Caryophylla-4(12),8(13)-dien-5-α-ol		1.36		
34.51	1640	1664	*epi*-α-Muurolol				0.12
34.71	1644	1669	α-Muurolol			0.83	
34.84	1645	1672	Cubenol		2.57	0.68	
34.89	1648	1678	*cis*-Guaia-3,9-dien-11-ol				0.54
34.93	1652	1681	α-Cadinol		3.45		0.30
35.49	1668	1684	*trans*-Calamenen-10-ol			0.30	
35.62	1668	1692	14-Hydroxy-9-*epi*-(*E*)-caryophyllene			1.00	
35.94	1676	1694	Mustakone		3.36		
36.28	1685	1695	Germacra-4(15),5,10(14)-trien-1-α-ol		0.76	1.40	0.04
39.41	1767	1768	14-oxi-α-Muurolene			0.48	
59.61	2400	2408	Tetracosane				0.02
62.38	2500	2512	Pentacosane				0.02
		Monoterpenes hydrocarbon		78.99	1.80	8.41	62.53
		Oxygenated monoterpenes		4.52	0	19.45	6.17
		Sesquiterpenes hydrocarbon		8.32	8.50	27.42	19.05
		Oxygenated sesquiterpenes		4.57	71.76	22.99	5.03
		Others class		-	-	3.42	2.30
		Total		96.4	82.06	81.69	95.08

RT: Retention Time; RI_C = Calculated retention index; RI_L = Literature retention index; DE: *Duguetia echinophora*; DR: *Duguetia riparia*; XE: *Xylopia emarginata*; XF: *Xylopia frutescens*.

The chemical compositions of the EOs of *D. echinophora*, *D. riparia*, *X. emarginata* and *X. frutescens* were characterized by GC-MS, and a total of 22, 19, 59 and 62 components were identified, representing 96.40, 82.06, 81.69 and 95.08% of the total EOs for each species, respectively. The hydrocarbon monoterpenes compounds represented the most abundant class in the EOs of *D. echinophora* (78.99%) and *X. frutescens* (62.53%), and the oxygenated sesquiterpenes class characterized the EO of *D. riparia* (71.76%). The compounds β-Phellandrene (39.12%), sabinene (17.08%) and terpinolene (11.17%) were dominant in the *D. echinophora* EO, while spathulenol (22.22%), caryophyllene oxide (12.21%), humulene epoxide II (11.86%) and *allo*-aromadendrene epoxide (10.20%) were the major constituents of the *D. riparia* EO. The EO of *X. emarginata* was characterized by spathulenol (5.65%) and caryophyllene oxide (5.63%), and *X. frutescens* EO was characterized by α-pinene (20.84%) and byciclogermacrene (7.85%). Ion chromatograms are available in the Supplementary Material.

According to data previous published, the chemical compositions of the EOs of *Annonaceae* species occurring in Brazil are predominantly characterized by substances belonging to the class of mono and sesquiterpenes, and among these compounds, the most abundant are β-elemene, α-pinene, β-pinene limonene, bicyclogermacrene, (E)-caryophyllene, caryophyllene oxide, spathulenol, and germacrene D, [9].

Previous reports have investigated the chemical composition of the EOs from the *Annonaceae* species described in this work (*D. riparia*, *X. emarginata* and *X. frutescens*). The leaves and fine stems EO of *D. riparia*, also collected in State of Pará-Brazil, showed spathulenol (46.5%), caryophyllene oxide (28.9%) and α-pinene (6.1%) as their main compounds [24], and quantitative differences were observed for the constituents spathulenol and caryophyllene oxide in relation to the *D. riparia* EO described in the present work. The EO from the leaves of *X. emarginata*, collected in Caxiuanã National Forest, Melgaço, State of Pará-Brazil, showed a high percentage of sesquiterpene spathulenol (73.0%) [11], whereas in the present work, this constituent was obtained at a low percentage (5.65%) [25]. The EO from the leaves of *X. frutescens*, collected in Municipality of Capela, Sergipe State, Brazil, had as its major compounds (E)-caryophyllene (31.48%), bicyclogermacrene (15.13%), germacrene D (9.66%), δ-cadinene (5.44%), viridiflorene (5.09%) and α-copaene (4.35%) [26], while the EO from the leaves of the specimen collected in the city of Itabaiana, Sergipe-Brazil, had as its major constituents bicyclogermacrene (23.23%), (E)-caryophyllene (17.24%), β-elemene (6.35%) and (E)-β-ocimene (5.23%) [27].

The chemical composition of EOs can be strongly influenced by several factors, including season, climate, geography, age, genotype, organ, development periods, collection place and even extraction method, etc. [49–51]. Figueiredo and collaborators evaluated the influence of seasonal variation on the EO of *Eugenia patrisii* Vahl (*Myrtaceae*) and verified a potential correlation between the content of the main constituents of the essential oil and climatic parameters (temperature, insolation and humidity rate) [52]. The EOs of *Flos Chrysanthemi indici*, an important medicinal and aromatic plant in China, were obtained by different extraction techniques, hydrodistillation (HD), steam distillation (SD), solvent-free microwave extraction (SFME) and supercritical fluid extraction (SFE), and the authors found that the EO yield, chemical composition and bioactivities varied according to the extraction method used [53]. Some *Annonaceae* species have shown qualitative and quantitative variability in their EO compositions according to different collection sites. The EOs from the leaves of *Annona vepretorum* Mart. collected in the State of Sergipe, Brazil, showed bicyclogermacrene, spathulenol and α-phellandrene as the major constituents [54], while another specimen collected in the State of Pernambuco, Brazil, showed α-pinene, limonene, spathulenol and caryophyllene oxide as the compounds with higher percentage [55]. The compounds α-selinene, aristolochene, (*E*)-caryophyllene and (*E*)-calamenene were identified as the major constituents of EO from leaves of a specimen of *Duguetia lanceolata* collected in the state of Minas Gerais, Brazil [23], while another specimen collected in the State of São Paulo, Brazil, had as its main constituents of the EO the compounds *trans*-muurola-4(14),5-diene, β-bisabolene, 3,4,5-trimethoxy-styrene and 2,4,5-trimethoxy-styrene [56].

3.2. Multivariate Analyses

Figures 1 and 2 show the correlations between the classes of compounds identified in the different samples according to the multivariate analyses, principal component analysis (PCA) and hierarchical cluster analysis (HCA), respectively. PC1 and PC2 represent the principal components (PC), which contained 39.0% and 32.0% of the variables, respectively, and accounted for 71.0% of the variance in the analyzed data. In the HCA analysis, tree groups were observed that show the similarity between the identified classes. Group I, including the samples of EOs from *D. echinophora* and *X. frutescens* showed a similarity of 10.67% (Figure 2) and comprised the compounds β-phellandrene, *p*-cymen-8-ol, bicyclogermacrene, terpinolene, α-pinene, sabinene, myrcene, limonene, β-pinene and α-thujene (Figure 1). Groups II and III contained only one sample each and comprised β-elemene, 1,8-cineol, muurola-4,10(14)-dien-1-β-ol, trans-pinocarveol, myrtenal and γ-muurolene

(*X. emarginata* EO) and cis-calamenene, α-cadinol, mustakone, *allo*-aromadendrene epoxide, humulene epoxide II, spathulenol and caryophyllene oxide (*D. riparia* EO), with similarities of 7.18% and 0%, respectively (Figure 2).

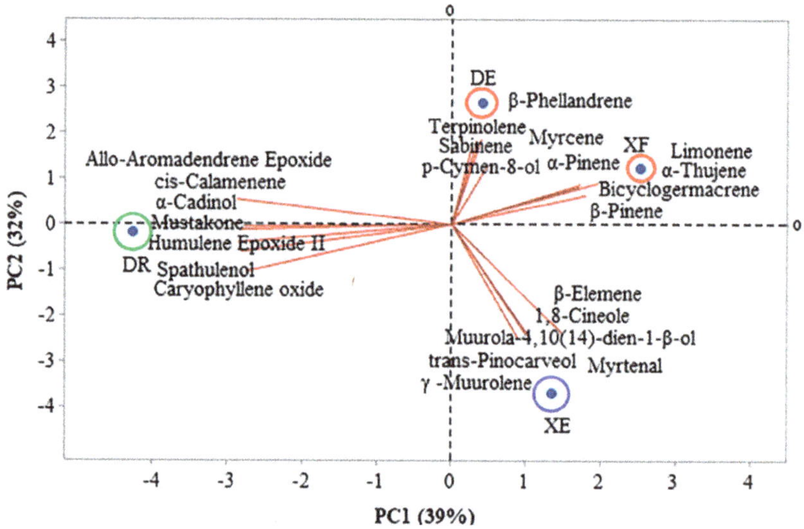

Figure 1. Biplot (principal component analysis) for the chemical analysis of the essential oils from *Annonaceae* species. DE: *Duguetia echinophora*; DR: *Duguetia riparia*; XE: *Xylopia emarginata*; XF: *Xylopia frutescens*.

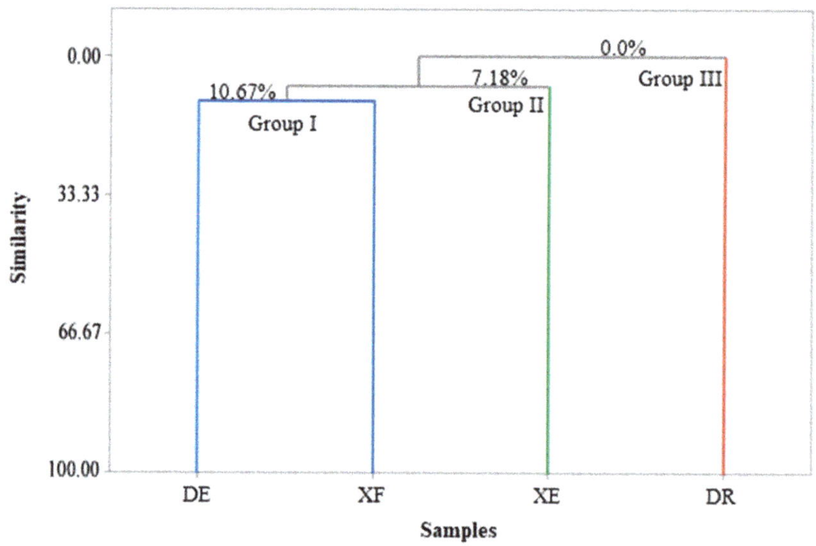

Figure 2. Dendrogram presenting the relational similarity of the chemical composition of the essential oils from *Annonaceae* species. DE: *Duguetia echinophora*; DR: *Duguetia riparia*; XE: *Xylopia emarginata*; XF: *Xylopia frutescens*.

3.3. Antioxidant Capacity

The antioxidant potential of the EOs from *Annonaceae* species was evaluated based on their ability to scavenge stable free DPPH• (2,2-diphenyl-1-picrylhydrazyl) and ABTS•+ (2,2′-Azino-bis (3-ethylbenzothiazoline-6-sulfonic acid) radicals; the results are shown in Figure 3. The results of the DPPH assays ranged from 15.87 to 69.38% and the highest percentage of inhibition was observed for the EO of *X. emarginata*, characterized by spathulenol (5.65%) and caryophyllene oxide (5.63%). For ABTS radical scavenging, the antioxidant capacity of EOs ranged from 14.61 to 63.67%. The species *X. frutescens* showed the higher antioxidant capacity by the ABTS•+ assay. This may be due to the presence of α-Pinene (20.84%) and β-Pinene (25.95%), the major components present in this EO. Possibly, the antioxidant activity of the *X. emarginata* EO can also be attributed to its main components which are described as antioxidants [57,58]. The high free radical scavenging effect of this sample may be related to the fact that the combination of the numerous organic chemical constituents present in EOs have a synergistic effect, increasing the biological activity or, conversely, an antagonistic effect [59]. In addition, bioactive compounds belonging to the monoterpenoid class have antioxidant activity, as reported in the literature [60].

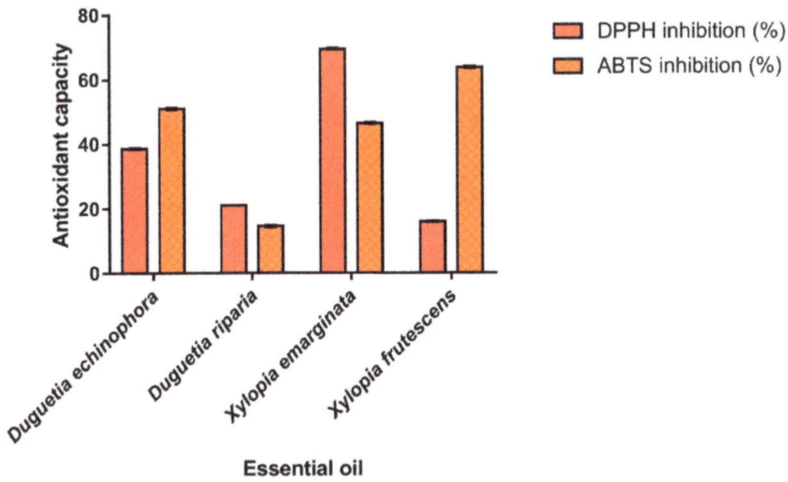

Figure 3. ABTS•+ and DPPH• radical scavenging assay and Trolox equivalent antioxidant capacity of essential oils. Values are expressed as mean and standard deviation (n = 3) of Trolox equivalent antioxidant capacity.

Other studies investigating EOs of the *Duguetia* and *Xylopia* genera identified antioxidant effects. The EO of *Xylopia sericea* A. St.-Hil. showed significant antioxidant activity using DPPH (IC_{50} 49.1 μg·mL^{-1}), β-carotene/linoleic acid bleaching (IC_{50} 6.9 μg·mL^{-1}), TAC (IC_{50} 78.2 μg·mL^{-1}) and TBARS (IC_{50} 80.0 μg·mL^{-1}) methods [20]. The EO of *Duguetia lanceolata* St. Hil. branches showed antioxidant effects using a DPPH assay (EC_{50} 159.4 μg·mL^{-1}), Fe^{+3} reduction (EC_{50} 187.8 μg·mL^{-1}) and inhibition of lipid peroxidation (41.5%); the authors suggest that caryophyllene oxide is one of the active compounds found in this EO [21].

3.4. Cytotoxic Activity of Essential Oils

The toxicity of the EOs from *Annonaceae* species was measured in terms of LC_{50} (lethal concentration) with two negative control groups (control 1:10 nauplii and artificial sea-water with DMSO 0.1%; control 2: 10 nauplii and artificial seawater) and one positive control ($K_2Cr_2O_7$, 50 μg·mL^{-1}). The values are shown in Table 2. Values of LC50 < 80 μg·mL^{-1} are considered highly toxic [15,61,62]; values of LC_{50} within the range 80 to 250 μg·mL^{-1} are moderately toxic; and LC_{50}> 250 μg·mL^{-1} are considered as low toxicity or non-toxic [63].

The EOs of *D. echinophora*, *X. emarginata* and *X. frutescens* showed high toxicity, whereas the EO of *D. riparia* showed low toxicity or was non-toxic. The major compounds from the EOs of *X. emarginata* (spathulenol and caryophyllene oxide) [64,65], *D. echinophora* (β-phellandrene and terpinolene) [66,67] and *X. frutescens* (α-pinene and byciclogermacrene) [68,69] showed cytotoxic effects and these results indicate that the cytotoxic potential observed for the EOs tested may be related to the presence of these secondary metabolites.

Table 2. LC_{50} concentrations of the essential oils using *Artemia salina* assay.

Essential Oil	LC_{50} ($\mu g \cdot mL^{-1}$)
Duguetia echinophora	28.00 ± 0.30
Duguetia riparia	310.80 ± 0.70
Xylopia emarginata	26.72 ± 0.17
Xylopia frutescens	54.36 ± 0.20
Positive control ($K_2Cr_2O_7$)	50.00 ± 0.00

Values are expressed as mean and standard deviation (n = 3).

Toxicity tests in *A. Salina* performed with *Coriandrum sativum* L. (*Apiaceae*) showed an LC_{50} value of 23 $\mu g \cdot mL^{-1}$ [70], which is similar to those obtained in the present work for *D. echinophora* and X. *emarginata* EOs. Oliva and coworkers evaluated toxicity of the EOs from some medicinal plants, and the results showed a decreasing activity in the brine assay of *Aloysia polystachia* (*Verbenaceae*) (LC_{50} 6459 $\mu g \cdot mL^{-1}$), *Aloysia triphylla (Verbenaceae)* (LC_{50} 1279 $\mu g \cdot mL^{-1}$), *Minthostachys verticillata* (*Myrtaceae*) (LC_{50} 1848 $\mu g \cdot mL^{-1}$), and *Schinus poligamus* (*Anacardiaceae*) (LC_{50} 1179 $\mu g \cdot mL^{-1}$), that were considered non-toxic [71], Other authors have also reported the toxicity of essential oils from a variety of plants [17,72,73].

The essential oils of *Duguetia* species have been studied by using the *A. salina* bioassay. The EOs from the leaves, underground heartwood and underground stem bark of *Duguetia furfuracea* (A. St. -Hil.) Saff. showed potent activity against *A. salina* larvae (LC50 6.01, 7.79 and 9.98 $\mu g \cdot mL^{-1}$, respectively) and the leaf EO from *D. lanceolata* also showed potent activity against the same larvae (LC_{50} 0.89 $\mu g \cdot mL^{-1}$) [23]. In another study, the EOs of *D. lanceolata* showed toxicity against *A. salina* with LC_{50} values of 49.0 $\mu g \cdot mL^{-1}$ (2 h of hydrodistillation extraction) and 60.7 $\mu g \cdot mL^{-1}$ (4h of hydrodistillation extraction) [74].

3.5. Molecular Docking

Molecular modeling approaches have been used to investigate how natural compounds interact with molecular targets of pharmacological interest [75–78]. One of the tools used has been molecular docking, which can provide insights into how these compounds interact with the binding pocket of proteins. Here, we use this approach to assess how the major compounds of the EO from *X. emarginata* interact with the AChE active site, as this target is closely related to the toxicity mechanism observed in the *A. salina* assays [79,80]. Spathulenol formed hydrophobic interactions with various residues such as Ser293, Phe297, Trp286, Tyr72, Tyr341, and Phe338. A hydrogen bond was established with Ser293. Caryophyllene oxide established pi-alkyl hydrophobic interactions with Trp286, Tyr341 and Tyr337 (Figure 4). The interaction between spathulenol and caryophyllene with the active site of AChE has already been described [58] and this could be the likely mechanism responsible for the cytotoxicity of the EO from *X. emarginata*.

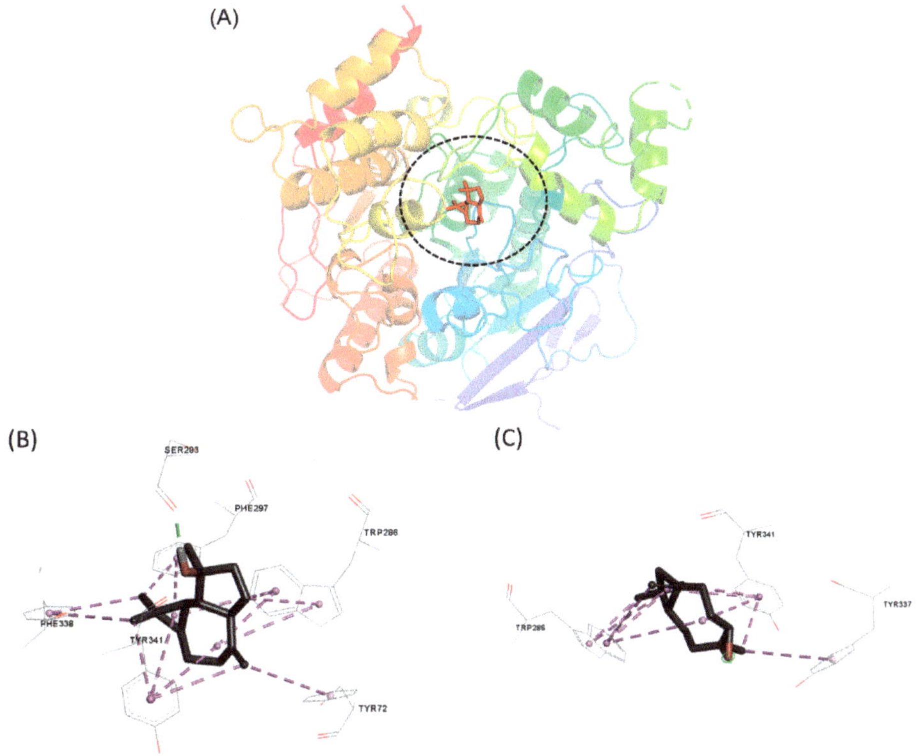

Figure 4. (**A**) Binding pocket of interaction of compounds with AChE. Molecular interactions established by (**B**) spathulenol and (**C**) caryophyllene oxide with the AChE active site.

4. Conclusions

The present study presents new insights concerning the chemical composition, antioxidant activity and preliminary toxicity of some *Annonaceae* EOs. Essential oils obtained from *D. echinophora, X. emarginata* and *X. frutescens* showed high toxicity, compared with EO obtained from *D. riparia*, which showed low toxicity or was non-toxic. The cytotoxicity test against *A. salina* can be considered as a good preliminary assessment of bioactive compounds, and may indicate a potential biological activity. The docking results elucidated the interaction mode of the major compounds of *X. emarginata* EO, spathulenol and caryophyllene, with the active site of the enzyme Acetylcholinesterase. The greatest capacities to scavenge DPPH and ABTS radicals were found in the essential oils of *X. emarginata* and *X. frutescens*, respectively, and the main constituents of the EO of this species may play the main role in the observed antioxidant capacity; however, the impact of less abundant constituents also should be considered.

Supplementary Materials: The following supporting information can be downloaded at: https://www.mdpi.com/article/10.3390/antiox11091709/s1, Figure S1: Ions-chromatogram relating to the chemical profile of essential oils from different species of *Annonaceae Xylopia*.

Author Contributions: Conceptualization, M.M.C., M.S.d.O. and L.D.d.N.; methodology, M.M.C., O.O.F., L.D.d.N., Â.A.B.D.M.; R.C.E.S.; J.N.C., T.O.d.A. and C.d.J.P.F.; writing—original draft preparation, M.M.C.; writing—review and editing, L.D.d.N., G.M.S.P.G. and M.S.d.O.; visualization, G.M.S.P.G. and E.H.d.A.A.; supervision, G.M.S.P.G. and E.H.d.A.A.; project administration, E.H.d.A.A. All authors have read and agreed to the published version of the manuscript.

Funding: This study was financed in part by the Coordenação de Aperfeiçoamento de Pessoal de Nível Superior-Brazil (CAPES)-Finance Code 001. Universidade Federal do Pará. Pró-Reitoria de Pesquisa e Pós-Graduação (PROPESP), Programa de Apoio à Publicação Qualificada-PAPQ, EDITAL 02/2022–PROPESP. The APC was funded by Universidade Federal do Pará.

Institutional Review Board Statement: Not applicable.

Informed Consent Statement: Not applicable.

Data Availability Statement: The data are contained within the article and supplementary materials.

Acknowledgments: The author M.M.C. thanks CAPES for the Ph.D. scholarship process number: [88887.497476/2020-00]. Â.A.B.D.M. thanks CNPq for the scientific initiation scholarship. The author M.S.d.O., thanks PCI-MCTIC/MPEG, as well as CNPq for the process number: [302050/2021-3]. The authors would like to thank the Universidade Federal do Pará.

Conflicts of Interest: The authors declare no conflict of interest.

References

1. Rehman, R.; Hanif, M.A.; Mushtaq, Z.; Al-Sadi, A.M. Biosynthesis of essential oils in aromatic plants: A review. *Food Rev. Int.* **2016**, *32*, 117–160. [CrossRef]
2. Ferreira, O.O.; Cruz, J.N.; de Moraes, Â.A.; de Jesus Pereira Franco, C.; Lima, R.R.; Anjos, T.O.; Siqueira, G.M.; Nascimento, L.D.; Cascaes, M.M.; de Oliveira, M.S.; et al. Essential Oil of the Plants Growing in the Brazilian Amazon: Chemical Composition, Antioxidants, and Biological Applications. *Molecules* **2022**, *27*, 4373. [CrossRef] [PubMed]
3. Perricone, M.; Arace, E.; Corbo, M.R.; Sinigaglia, M.; Bevilacqua, A. Bioactivity of essential oils: A review on their interaction with food components. *Front. Microbiol.* **2015**, *6*, 76. [CrossRef] [PubMed]
4. Tohidi, B.; Rahimmalek, M.; Trindade, H. Industrial Crops & Products Review on essential oil, extracts composition, molecular and phytochemical properties of Thymus species in Iran. *Ind. Crop. Prod.* **2019**, *134*, 89–99. [CrossRef]
5. Santana de Oliveira, M.; da Cruz, J.N.; Almeida da Costa, W.; Silva, S.G.; Brito, M.d.P.; de Menezes, S.A.F.; de Jesus Chaves Neto, A.M.; de Aguiar Andrade, E.H.; de Carvalho Junior, R.N. Chemical Composition, Antimicrobial Properties of Siparuna guianensis Essential Oil and a Molecular Docking and Dynamics Molecular Study of its Major Chemical Constituent. *Molecules* **2020**, *25*, 3852. [CrossRef]
6. Ferreira, O.O.; Neves da Cruz, J.; de Jesus Pereira Franco, C.; Silva, S.G.; da Costa, W.A.; de Oliveira, M.S.; de Aguiar Andrade, E.H. First report on yield and chemical composition of essential oil extracted from myrcia eximia DC (Myrtaceae) from the Brazilian Amazon. *Molecules* **2020**, *25*, 783. [CrossRef] [PubMed]
7. Silva, S.G.; de Oliveira, M.S.; Cruz, J.N.; da Costa, W.A.; da Silva, S.H.M.; Barreto Maia, A.A.; de Sousa, R.L.; Carvalho Junior, R.N.; de Aguiar Andrade, E.H. Supercritical CO_2 extraction to obtain Lippia thymoides Mart. & Schauer (Verbenaceae) essential oil rich in thymol and evaluation of its antimicrobial activity. *J. Supercrit. Fluids* **2021**, *168*, 105064. [CrossRef]
8. Carvalho, R.N.d.; Oliveira, M.S.d.M.S.d.; Silva, S.G.; Cruz, J.N.d.; Ortiz, E.; Costa, W.A.d.; Bezerra, F.W.F.; Cunha, V.M.B.; Cordeiro, R.M.; Neto, A.M.d.J.C.; et al. Supercritical CO_2 Application in Essential Oil Extraction. In *Materials Research Foundations*; Inamuddin, R.M., Asiri, A.M., Eds.; Materials Research Foundations: Millersville PA, USA, 2019; Volume 54, pp. 1–28.
9. Cascaes, M.M.; Dos, O.; Carneiro, S.; Diniz Do Nascimento, L.; Antônio Barbosa De Moraes, Â.; Santana De Oliveira, M.; Neves Cruz, J.; Skelding, G.M.; Guilhon, P.; Helena De Aguiar Andrade, E.; et al. Essential Oils from Annonaceae Species from Brazil: A Systematic Review of Their Phytochemistry, and Biological Activities. *Int. J. Mol. Sci.* **2021**, *22*, 12140. [CrossRef]
10. Bezerra, F.W.F.W.F.; De Oliveira, M.S.S.; Bezerra, P.N.N.; Cunha, V.M.B.M.B.; Silva, M.P.P.; da Costa, W.A.A.; Pinto, R.H.H.H.H.; Cordeiro, R.M.M.; Da Cruz, J.N.N.; Chaves Neto, A.M.J.M.J.; et al. Extraction of bioactive compounds. In *Green Sustainable Process for Chemical and Environmental Engineering and Science*; Elsevier: Amsterdam, The Netherlands, 2020; pp. 149–167; ISBN 9780128173886.
11. Sharmeen, J.B.; Mahomoodally, F.M.; Zengin, G.; Maggi, F. Essential Oils as Natural Sources of Fragrance Compounds for Cosmetics and Cosmeceuticals. *Molecules* **2021**, *26*, 666. [CrossRef]
12. Miguel, M.G. Antioxidant and anti-inflammatory activities of essential oils: A short review. *Molecules* **2010**, *15*, 9252–9287. [CrossRef]
13. Do Nascimento, L.D.; de Moraes, A.A.B.; da Costa, K.S.; Galúcio, J.M.P.; Taube, P.S.; Costa, C.M.L.; Cruz, J.N.; Andrade, E.H.d.A.; de Faria, L.J.G. Bioactive natural compounds and antioxidant activity of essential oils from spice plants: New findings and potential applications. *Biomolecules* **2020**, *10*, 988. [CrossRef] [PubMed]
14. Ramos Da Silva, L.R.; Ferreira, O.O.; Cruz, J.N.; De Jesus Pereira Franco, C.; Oliveira Dos Anjos, T.; Cascaes, M.M.; Almeida Da Costa, W.; Helena De Aguiar Andrade, E.; Santana De Oliveira, M. Lamiaceae Essential Oils, Phytochemical Profile, Antioxidant, and Biological Activities. *Evid.-Based Complement. Altern. Med.* **2021**, *2021*, 6748052. [CrossRef] [PubMed]
15. Mesquita, K.d.S.M.; Feitosa, B.d.S.; Cruz, J.N.; Ferreira, O.O.; Franco, C.d.J.P.; Cascaes, M.M.; Oliveira, M.S.d.; Andrade, E.H.d.A. Chemical Composition and Preliminary Toxicity Evaluation of the Essential Oil from Peperomia circinnata Link var. circinnata. (Piperaceae) in Artemia salina Leach. *Molecules* **2021**, *26*, 7359. [CrossRef] [PubMed]

16. Radulović, N.S.; Mladenović, M.Z.; Randjelovic, P.J.; Stojanović, N.M.; Dekić, M.S.; Blagojević, P.D. Toxic essential oils. Part IV: The essential oil of Achillea falcata L. as a source of biologically/pharmacologically active trans-sabinyl esters. *Food Chem. Toxicol.* **2015**, *80*, 114–129. [CrossRef]
17. Santana de Oliveira, M.; Pereira da Silva, V.M.; Cantão Freitas, L.; Gomes Silva, S.; Nevez Cruz, J.; Aguiar Andrade, E.H. Extraction Yield, Chemical Composition, Preliminary Toxicity of Bignonia nocturna (Bignoniaceae) Essential Oil and in Silico Evaluation of the Interaction. *Chem. Biodivers.* **2021**, *18*, cbdv.202000982. [CrossRef]
18. Tamokou, J.d.D.; Mbaveng, A.T.; Kuete, V. Antimicrobial Activities of African Medicinal Spices and Vegetables. In *Medicinal Spices and Vegetables from Africa: Therapeutic Potential Against Metabolic, Inflammatory, Infectious and Systemic Diseases*; Elsevier Inc.: Amsterdam, The Netherlands, 2017; pp. 207–237; ISBN 9780128094419.
19. Maas, P.; Lobão, A.; Rainer, H. Annonaceae Juss. Available online: http://floradobrasil.jbrj.gov.br (accessed on 28 July 2022).
20. Mendes, R.d.F.; Pinto, N.d.C.C.; da Silva, J.M.; da Silva, J.B.; Hermisdorf, R.C.d.S.; Fabri, R.L.; Chedier, L.M.; Scio, E. The essential oil from the fruits of the Brazilian spice Xylopia sericea A. St.-Hil. presents expressive in-vitro antibacterial and antioxidant activity. *J. Pharm. Pharmacol.* **2017**, *69*, 341–348. [CrossRef]
21. de Sousa, O.V.; Del-Vechio-Vieira, G.; Santos, B.C.S.; Yamamoto, C.H.; Araújo, A.L.S.d.; de Araújo, A.d.A.; Pinto, M.A.d.; Rodarte, M.P.; Alves, M.S. In-vivo and vitro bioactivities of the essential oil of Duguetia lanceolata branches. *Afr. J. Pharm. Pharmacol.* **2016**, *10*, 298–310. [CrossRef]
22. Costa, E.V.; Dutra, L.M.; Ramos De Jesus, H.C.; De Lima Nogueira, P.C.; De Souza Moraes, V.R.; Salvador, M.J.; De Holanda Cavalcanti, S.C.; La Corte Dos Santos, R.; Do Nacimento Prata, A.P. Chemical composition and antioxidant, antimicrobial, and larvicidal activities of the essential oils of Annona salzmannii and A. pickelii (Annonaceae). *Nat. Prod. Commun.* **2011**, *6*, 907–912. [CrossRef]
23. Maia, D.S.; Lopes, C.F.; Saldanha, A.A.; Silva, N.L.; Sartori, Â.L.B.; Carollo, C.A.; Sobral, M.G.; Alves, S.N.; Silva, D.B.; de Siqueira, J.M. Larvicidal effect from different Annonaceae species on Culex quinquefasciatus. *Environ. Sci. Pollut. Res.* **2020**, *27*, 36983–36993. [CrossRef]
24. Maia, J.G.S.; Andrade, E.H.A.; Carreira, L.M.M.; Oliveira, J. Essential oil composition from duguetia species (annonaceae). *J. Essent. Oil Res.* **2006**, *18*, 60–63. [CrossRef]
25. Maia, J.G.S.; Andrade, E.H.A.; Carla, A.; Silva, M.; Oliveira, J.; Carreira, L.M.M.; Araújo, J.S. Leaf volatile oils from four Brazilian Xylopia species. *Flavour Fragr. J.* **2005**, *20*, 474–477. [CrossRef]
26. Ferraz, R.P.C.; Cardoso, G.M.B.; Da Silva, T.B.; Fontes, J.E.D.N.; Prata, A.P.D.N.; Carvalho, A.A.; Moraes, M.O.; Pessoa, C.; Costa, E.V.; Bezerra, D.P. Antitumour properties of the leaf essential oil of Xylopia frutescens Aubl. (Annonaceae). *Food Chem.* **2013**, *141*, 542–547. [CrossRef]
27. Nascimento, A.M.D.; Maia, T.D.S.; Soares, T.E.S.; Menezes, L.R.A.; Scher, R.; Costa, E.V.; Cavalcanti, S.C.H.; La Corte, R. Repellency and Larvicidal Activity of Essential oils from Xylopia laevigata, Xylopia frutescens, Lippia pedunculosa, and Their Individual Compounds against Aedes aegypti Linnaeus. *Neotrop. Entomol.* **2017**, *46*, 223–230. [CrossRef] [PubMed]
28. Alcântara, J.M.; De Lucena, J.M.V.M.; Facanali, R.; Marques, M.O.M.; Da Paz Lima, M. Chemical composition and bactericidal activity of the essential oils of four species of annonaceae growing in brazilian amazon. *Nat. Prod. Commun.* **2017**, *12*, 619–622. [CrossRef] [PubMed]
29. Agra, M.D.F.; Freitas, P.F.D.; Barbosa-filho, J.M. Divulgação Synopsis of the plants known as medicinal and poisonous in Northeast of Brazil. *Brazilian J. Pharmacogn.* **2007**, *17*, 114–140. [CrossRef]
30. Cascaes, M.M.; Silva, S.G.; Cruz, J.N.; Santana de Oliveira, M.; Oliveira, J.; Moraes, A.A.B.d.; Costa, F.A.M.d.; da Costa, K.S.; Diniz do Nascimento, L.; Helena de Aguiar Andrade, E. First report on the Annona exsucca DC. Essential oil and in silico identification of potential biological targets of its major compounds. *Nat. Prod. Res.* **2021**, *35*, 4009–4012. [CrossRef] [PubMed]
31. Do Nascimento, L.D.; Silva, S.G.; Cascaes, M.M.; da Costa, K.S.; Figueiredo, P.L.B.; Costa, C.M.L.; Andrade, E.H.d.A.; de Faria, L.J.G. Drying effects on chemical composition and antioxidant activity of lippia thymoides essential oil, a natural source of thymol. *Molecules* **2021**, *26*, 2621. [CrossRef]
32. Franco, C.d.J.P.; Ferreira, O.O.; Antônio Barbosa de Moraes, Â.; Varela, E.L.P.; Do Nascimento, L.D.; Percário, S.; de Oliveira, M.S.; Andrade, E.H.D.A. Chemical composition and antioxidant activity of essential oils from eugenia patrisii vahl, e. Punicifolia (kunth) dc., and myrcia tomentosa (aubl.) dc., leaf of family myrtaceae. *Molecules* **2021**, *26*, 3292. [CrossRef]
33. Ferreira, O.O.; da Silva, S.H.M.; de Oliveira, M.S.; Andrade, E.H.d.A. Chemical Composition and Antifungal Activity of Myrcia multiflora and Eugenia florida Essential Oils. *Molecules* **2021**, *26*, 7259. [CrossRef]
34. van Den Dool, H.; Kratz, P.D. A generalization of the retention index system including linear temperature programmed gas—liquid partition chromatography. *J. Chromatogr. A* **1963**, *11*, 463–471. [CrossRef]
35. Adams, R.P. *Identification of Essential Oil Components by Gas Chromatography/Mass Spectrometry*, 4th ed.; Allured Publ.: Carol Stream, IL, USA, 2007; Volume 8, ISBN 978-1932633214.
36. Miller, N.J.; Rice-Evans, C.; Davies, M.J.; Gopinathan, V.; Milner, A. A novel method for measuring antioxidant capacity and its application to monitoring the antioxidant status in premature neonates. *Clin. Sci.* **1993**, *84*, 407–412. [CrossRef] [PubMed]
37. Re, R.; Pellegrini, N.; Proteggente, A.; Pannala, A.; Yang, M.; Rice-Evans, C. Antioxidant activity applying an improved ABTS radical cation decolorization assay. *Free Radic. Biol. Med.* **1999**, *26*, 1231–1237. [CrossRef]
38. Blois, M.S. Antioxidant determinations by the use of a stable free radical. *Nature* **1958**, *181*, 1199–1200. [CrossRef]

39. Rebelo, M.M.; Da Silva, J.K.R.; Andrade, E.H.A.; Maia, J.G.S. Antioxidant capacity and biological activity of essential oil and methanol extract of Hyptis crenata Pohl ex Benth. *Rev. Bras. Farmacogn.* **2009**, *19*, 230–235. [CrossRef]
40. Dennington, R.; Keith, T.; Millam, J. *GaussView, version 5*; Semichem Inc.: Shawnee Mission, KS, USA, 2009.
41. Almeida, V.M.; Dias, Ê.R.; Souza, B.C.; Cruz, J.N.; Santos, C.B.R.; Leite, F.H.A.; Queiroz, R.F.; Branco, A. Methoxylated flavonols from Vellozia dasypus Seub ethyl acetate active myeloperoxidase extract: In vitro and in silico assays. *J. Biomol. Struct. Dyn.* **2021**, 1–10. [CrossRef]
42. Becke, A.D. Density-functional thermochemistry. III. The role of exact exchange. *J. Chem. Phys.* **1993**, *98*, 5648–5652. [CrossRef]
43. dos Santos, K.L.B.; Cruz, J.N.; Silva, L.B.; Ramos, R.S.; Neto, M.F.A.; Lobato, C.C.; Ota, S.S.B.; Leite, F.H.A.; Borges, R.S.; da Silva, C.H.T.P.; et al. Identification of novel chemical entities for adenosine receptor type 2a using molecular modeling approaches. *Molecules* **2020**, *25*, 1245. [CrossRef]
44. Frisch, M.J.; Trucks, G.W.; Schlegel, H.B.; Scuseria, G.E.; Robb, M.A.; Cheeseman, J.R.; Barone, V.; Mennucci, B.; Petersson, G.A.; Nakatsuji, H.; et al. Gaussian 09. 2009. Available online: https://gaussian.com/g09citation/ (accessed on 28 July 2022).
45. Thomsen, R.; Christensen, M.H. MolDock: A new technique for high-accuracy molecular docking. *J. Med. Chem.* **2006**, *49*, 3315–3321. [CrossRef]
46. Pinto, V.d.S.; Araújo, J.S.C.; Silva, R.C.; da Costa, G.V.; Cruz, J.N.; Neto, M.F.D.A.; Campos, J.M.; Santos, C.B.R.; Leite, F.H.A.; Junior, M.C.S. In silico study to identify new antituberculosis molecules from natural sources by hierarchical virtual screening and molecular dynamics simulations. *Pharmaceuticals* **2019**, *12*, 36. [CrossRef]
47. Castro, A.L.G.; Cruz, J.N.; Sodré, D.F.; Correa-Barbosa, J.; Azonsivo, R.; de Oliveira, M.S.; de Sousa Siqueira, J.E.; da Rocha Galucio, N.C.; de Oliveira Bahia, M.; Burbano, R.M.R.; et al. Evaluation of the genotoxicity and mutagenicity of isoeleutherin and eleutherin isolated from Eleutherine plicata herb. using bioassays and in silico approaches. *Arab. J. Chem.* **2021**, *14*, 103084. [CrossRef]
48. Cheung, J.; Gary, E.N.; Shiomi, K.; Rosenberry, T.L. Structures of human acetylcholinesterase bound to dihydrotanshinone i and territrem B show peripheral site flexibility. *ACS Med. Chem. Lett.* **2013**, *4*, 1091–1096. [CrossRef] [PubMed]
49. Liao, Z.; Huang, Q.; Cheng, Q.; Khan, S.; Yu, X. Seasonal variation in chemical compositions of essential oils extracted from lavandin flowers in the yun-gui plateau of china. *Molecules* **2021**, *26*, 5639. [CrossRef] [PubMed]
50. Cantor, M.; Vlas, N.; Szekely, Z.; Jucan, D.; Zaharia, A. The influence of distillation time and the flowering phenophase on quantity and quality of the essential oil of Lavandula angustifolia cv. *'Codreanca.' Rom. Biotechnol. Lett.* **2018**, *23*, 14146–14152.
51. Méndez-Tovar, I.; Novak, J.; Sponza, S.; Herrero, B.; Asensio-S-Manzanera, M.C. Variability in essential oil composition of wild populations of Labiatae species collected in Spain. *Ind. Crops Prod.* **2016**, *79*, 18–28. [CrossRef]
52. Nazar, E.D.; Peixoto, L.D.S.; Costa, J.S.; Helena, R.; Mour, V.; Maria, W.; Nascimento, O.; Maia, G.S.; Setzer, W.N.; Kelly, J.; et al. Seasonal Variability of a Caryophyllane Chemotype Essential Oil of Eugenia patrisii Vahl Occurring in the Brazilian Amazon. *Molecules* **2022**, *27*, 2417.
53. Jing, C.L.; Huang, R.H.; Su, Y.; Li, Y.Q.; Zhang, C.S. Variation in chemical composition and biological activities of flos chrysanthemi indici essential oil under different extraction methods. *Biomolecules* **2019**, *9*, 518. [CrossRef]
54. De Souza Araújo, C.; De Oliveira, A.P.; Lima, R.N.; Alves, P.B.; Diniz, T.C.; Da Silva Almeida, J.R.G. Chemical constituents and antioxidant activity of the essential oil from leaves of Annona vepretorum Mart. (Annonaceae). *Pharmacogn. Mag.* **2015**, *11*, 615–618. [CrossRef]
55. Meira, C.S.; Guimarães, E.T.; MacEdo, T.S.; Da Silva, T.B.; Menezes, L.R.A.; Costa, E.V.; Soares, M.B.P. Chemical composition of essential oils from Annona vepretorum Mart. and *Annona squamosa* L. (Annonaceae) leaves and their antimalarial and trypanocidal activities. *J. Essent. Oil Res.* **2015**, *27*, 160–168. [CrossRef]
56. Ribeiro, L.P.; Domingues, V.C.; Gonçalves, G.L.P.; Fernandes, J.B.; Glória, E.M.; Vendramim, J.D. Essential oil from *Duguetia lanceolata* St.-Hil. (Annonaceae): Suppression of spoilers of stored-grain. *Food Biosci.* **2020**, *36*, 100653. [CrossRef]
57. do Nascimento, K.F.; Moreira, F.M.F.; Alencar Santos, J.; Kassuya, C.A.L.; Croda, J.H.R.; Cardoso, C.A.L.; Vieira, M.d.C.; Góis Ruiz, A.L.T.; Ann Foglio, M.; de Carvalho, J.E.; et al. Antioxidant, anti-inflammatory, antiproliferative and antimycobacterial activities of the essential oil of *Psidium guineense* Sw. and spathulenol. *J. Ethnopharmacol.* **2018**, *210*, 351–358. [CrossRef] [PubMed]
58. Karakaya, S.; Yilmaz, S.V.; Özdemir, Ö.; Koca, M.; Pınar, N.M.; Demirci, B.; Yıldırım, K.; Sytar, O.; Turkez, H.; Baser, K.H.C. A caryophyllene oxide and other potential anticholinesterase and anticancer agent in Salvia verticillata subsp. amasiaca (Freyn & Bornm.) Bornm. (Lamiaceae). *J. Essent. Oil Res.* **2020**, *32*, 512–525. [CrossRef]
59. Caesar, L.K.; Cech, N.B. Synergy and antagonism in natural product extracts: When 1 + 1 does not equal 2. *Nat. Prod. Rep.* **2019**, *36*, 869–888. [CrossRef] [PubMed]
60. Rao, M.J.; Wu, S.; Duan, M.; Wang, L. Antioxidant metabolites in primitive, wild, and cultivated citrus and their role in stress tolerance. *Molecules* **2021**, *26*, 5801. [CrossRef] [PubMed]
61. Bezerra, J.W.A.; Costa, A.R.; da Silva, M.A.P.; Rocha, M.I.; Boligon, A.A.; da Rocha, J.B.T.; Barros, L.M.; Kamdem, J.P. Chemical composition and toxicological evaluation of *Hyptis suaveolens* (L.) Poiteau (LAMIACEAE) in *Drosophila melanogaster* and *Artemia salina*. *S. Afr. J. Bot.* **2017**, *113*, 437–442. [CrossRef]
62. Braguini, W.L.; Alves, B.B.; Pires, N.V. Toxicity assessment of *Lavandula officinalis* extracts in Brine Shrimp (*Artemia salina*). *Toxicol. Mech. Methods* **2019**, *29*, 411–420. [CrossRef]
63. Ramos, S.C.S.; De Oliveira, J.C.S.; Da Câmara, C.A.G.; Castelar, I.; Carvalho, A.F.F.U.; Lima-Filho, J.V. Antibacterial and cytotoxic properties of some plant crude extracts used in Northeastern folk medicine. *Rev. Bras. Farmacogn.* **2009**, *19*, 376–381. [CrossRef]

64. Mirzaei, H.H.; Firuzi, O.; Baldwin, I.T.; Jassbi, A.R. Cytotoxic activities of different iranian solanaceae and lamiaceae plants and bioassay-guided study of an active extract from salvia lachnocalyx. *Nat. Prod. Commun.* **2017**, *12*, 1563–1566. [CrossRef]
65. Fidyt, K.; Fiedorowicz, A.; Strządała, L.; Szumny, A. Β-Caryophyllene and Β-Caryophyllene Oxide—Natural Compounds of Anticancer and Analgesic Properties. *Cancer Med.* **2016**, *5*, 3007–3017. [CrossRef]
66. Fraternale, D.; Ricci, D.; Calcabrini, C.; Guescini, M.; Martinelli, C.; Sestili, P. Cytotoxic activity of essential oils of aerial parts and ripe fruits of echinophora spinosa (apiaceae). *Nat. Prod. Commun.* **2013**, *8*, 1645–1649. [CrossRef]
67. Aydin, E.; Türkez, H.; Taşdemir, Ş. Anticancer and antioxidant properties of terpinolene in rat brain cells. *Arh. Hig. Rada Toksikol.* **2013**, *64*, 415–424. [CrossRef]
68. Salehi, B.; Upadhyay, S.; Orhan, I.E.; Jugran, A.K.; Baghalpour, N.; Cho, W.C.; Sharifi-rad, J. Therapeutic Potential of α- and β-Pinene_A Miracle Gift of Nature. *Biomolecules* **2019**, *9*, 738. [CrossRef]
69. Grecco, S.D.S.; Martins, E.G.A.; Girola, N.; De Figueiredo, C.R.; Matsuo, A.L.; Soares, M.G.; Bertoldo, B.D.C.; Sartorelli, P.; Lago, J.H.G. Chemical composition and in vitro cytotoxic effects of the essential oil from Nectandra leucantha leaves. *Pharm. Biol.* **2015**, *53*, 133–137. [CrossRef]
70. Soares, B.V.; Morais, S.M.; Dos Santos Fontenelle, R.O.; Queiroz, V.A.; Vila-Nova, N.S.; Pereira, C.M.C.; Brito, E.S.; Neto, M.A.S.; Brito, E.H.S.; Cavalcante, C.S.P.; et al. Antifungal activity, toxicity and chemical composition of the essential oil of coriandrum sativum L. Fruits. *Molecules* **2012**, *17*, 8439–8448. [CrossRef]
71. Oliva, M.D.L.M.; Gallucci, N.; Zygadlo, J.A.; Demo, M.S. Cytotoxic activity of Argentinean essential oils on Artemia salina. *Pharm. Biol.* **2007**, *45*, 259–262. [CrossRef]
72. Sharififar, F.; Assadipour, A.; Moshafi, M. Bioassay Screening of the Essential Oil and Various Extracts of Nigella sativa L. Seeds Using Brine Shrimp Toxicity Assay. *Herb. Med.* **2017**, *2*, 1–6. [CrossRef]
73. Do Nascimento, J.C.; David, J.M.; Barbosa, L.C.; De Paula, V.F.; Demuner, A.J.; David, J.P.; Conserva, L.M.; Ferreira, J.C.; Guimarães, E.F. Larvicidal activities and chemical composition of essential oils from *Piper klotzschianum* (Kunth) C. DC. (Piperaceae). *Pest Manag. Sci.* **2013**, *69*, 1267–1271. [CrossRef] [PubMed]
74. Sousa, O.V.; Del-Vechio-Vieira, G.; Alves, M.S.; Araújo, A.A.L.; Pinto, M.A.O.; Amaral, M.P.H.; Rodarte, M.P.; Kaplan, M.A.C. Chemical composition and biological activities of the essential oils from *Duguetia lanceolata* St. Hil. barks. *Molecules* **2012**, *17*, 11056–11066. [CrossRef]
75. Galucio, N.C.d.R.; Moysés, D.d.A.; Pina, J.R.S.; Marinho, P.S.B.; Gomes Júnior, P.C.; Cruz, J.N.; Vale, V.V.; Khayat, A.S.; Marinho, A.M.d.R. Antiproliferative, genotoxic activities and quantification of extracts and cucurbitacin *B obtained* from *Luffa operculata* (L.) Cogn. *Arab. J. Chem.* **2022**, *15*, 103589. [CrossRef]
76. Rego, C.M.A.; Francisco, A.F.; Boeno, C.N.; Paloschi, M.V.; Lopes, J.A.; Silva, M.D.S.; Santana, H.M.; Serrath, S.N.; Rodrigues, J.E.; Lemos, C.T.L.; et al. Inflammasome NLRP3 activation induced by Convulxin, a C-type lectin-like isolated from *Crotalus durissus terrificus* snake venom. *Sci. Rep.* **2022**, *12*, 4706. [CrossRef]
77. Costa, E.B.B.; Silva, R.C.C.; Espejo-Román, J.M.M.; Neto, M.F.d.A.F.d.A.; Cruz, J.N.N.; Leite, F.H.A.H.A.; Silva, C.H.T.P.H.T.P.; Pinheiro, J.C.C.; Macêdo, W.J.C.J.C.; Santos, C.B.R.B.R. Chemometric methods in antimalarial drug design from 1,2,4,5-tetraoxanes analogues. *SAR QSAR Environ. Res.* **2020**, *31*, 677–695. [CrossRef]
78. Neto, R.d.A.M.M.; Santos, C.B.R.R.; Henriques, S.V.C.C.; Machado, L.d.O.; Cruz, J.N.; da Silva, C.H.T.d.P.T.d.P.; Federico, L.B.; Oliveira, E.H.C.d.C.d.; de Souza, M.P.C.C.; da Silva, P.N.B.B.; et al. Novel chalcones derivatives with potential antineoplastic activity investigated by docking and molecular dynamics simulations. *J. Biomol. Struct. Dyn.* **2020**, *40*, 2204–2216. [CrossRef] [PubMed]
79. da Silva Júnior, O.S.; Franco, C.d.J.P.; de Moraes, A.A.B.; Cruz, J.N.; da Costa, K.S.; do Nascimento, L.D.; Andrade, E.H.d.A. In silico analyses of toxicity of the major constituents of essential oils from two *Ipomoea* L. species. *Toxicon* **2021**, *195*, 111–118. [CrossRef] [PubMed]
80. Baek, I.; Choi, H.J.; Rhee, J.S. Inhibitory effects of biocides on hatching and acetylcholinesterase activity in the brine shrimp Artemia salina. *Toxicol. Environ. Health Sci.* **2015**, *7*, 303–308. [CrossRef]

antioxidants

MDPI

Article

Bridging the Chemical Profile and Biomedical Effects of *Scutellaria edelbergii* Essential Oils

Muddaser Shah [1,2], Shabana Bibi [3,4], Zul Kamal [5], Jamal Nasser Al-Sabahi [6], Tanveer Alam [2], Obaid Ullah [2,7], Waheed Murad [1,*], Najeeb Ur Rehman [2,*] and Ahmed Al-Harrasi [2,*]

1 Department of Botany, Abdul Wali Khan University Mardan, Mardan 23200, Pakistan
2 Natural and Medical Sciences Research Center, University of Nizwa, Birkat Al-Mauz, P.O. Box 33, Nizwa 616, Oman
3 Department of Biosciences, Shifa Tameer-e-Millat University, Islamabad 44000, Pakistan
4 Yunnan Herbal Laboratory, College of Ecology and Environmental Sciences, Yunnan University, Kunming 650091, China
5 Department of Pharmacy, Shaheed Benazir Bhutto University, Upper Dir 18000, Pakistan
6 Central Instrument Laboratory, College of Agriculture and Marine Sciences, Sultan Qaboos University, Muscat 123, Oman
7 Department of Chemistry, University of Chakdara, Chakdara 18800, Pakistan
* Correspondence: waheedmurad@awkum.edu.pk (W.M.); najeeb@unizwa.edu.om (N.U.R.); aharrasi@unizwa.edu.om (A.A.-H.)

Abstract: The present study explored chemical constituents of *Scutellaria edelbergii* essential oils (SEEO) for the first time, extracted through hydro-distillation, and screened them against the microbes and free radicals scavenging effect, pain-relieving, and anti-inflammatory potential employing standard techniques. The SEEO ingredients were noticed via Gas Chromatography-Mass-Spectrometry (GC-MS) analysis and presented fifty-two bioactive compounds contributed (89.52%) with dominant volatile constituent; 3-oxomanoyl oxide (10.09%), 24-norursa-3,12-diene (8.05%), and methyl 7-abieten-18-oate (7.02%). The MTT assay via 96 well-plate and agar-well diffusion techniques against various microbes was determined for minimum inhibitory concentration (MIC), minimum bactericidal concentration (MBC), IC_{50}, and zone of inhibitions (ZOIs). The SEEO indicated considerable antimicrobial significance against tested bacterial strains viz. *Escherichia coli*, *Pseudomonas aeruginosa*, *Klebsiella pneumoniae*, and *Enterococcus faecalis* and the fungal strains *Fusarium oxysporum* and *Candida albicans*. The free radicals scavenging potential was noticed to be significant in 1,1-Diphenyl-2-picryl-hydrazyl (DPPH) as compared to 2,2′-azino-bis-3-ethylbenzotiazolin-6-sulfonic acid (ABTS) assays with $IC_{50} = 125.0 \pm 0.19$ µg/mL and $IC_{50} = 153.0 \pm 0.31$ µg/mL correspondingly; similarly, the antioxidant standard in the DPPH assay was found efficient as compared to ABTS assay. The SEEO also offered an appreciable analgesic significance and presented 54.71% in comparison with standard aspirin, 64.49% reduction in writhes, and an anti-inflammatory potential of 64.13%, as compared to the standard diclofenac sodium inhibition of 71.72%. The SEEO contain bioactive volatile ingredients with antimicrobial, free radical scavenging, pain, and inflammation relieving potentials. Computational analysis validated the anti-inflammatory potential of selected hit "methyl 7-abieten-18-oate" as a COX-2 enzyme inhibitor. Docking results were very good in terms of docked score (−7.8704 kcal/mol) and binding interactions with the functional residues; furthermore, MD simulation for 100 ns has presented a correlation with docking results with minor fluctuations. In silico, ADMET characteristics supported that methyl 7-abieten-18-oate could be recommended for further investigations in clinical tests and could prove its medicinal status as an anti-inflammatory drug.

Keywords: *S. edelbergii*; essential oils; GC-MS analysis; in vitro and in vivo biological activities

Citation: Shah, M.; Bibi, S.; Kamal, Z.; Al-Sabahi, J.N.; Alam, T.; Ullah, O.; Murad, W.; Rehman, N.U.; Al-Harrasi, A. Bridging the Chemical Profile and Biomedical Effects of *Scutellaria edelbergii* Essential Oils. *Antioxidants* **2022**, *11*, 1723. https://doi.org/10.3390/antiox11091723

Academic Editor: Stanley Omaye

Received: 27 July 2022
Accepted: 24 August 2022
Published: 30 August 2022

1. Introduction

Medicinal herbs are the most prevalent form of conventional therapies used for maintaining human health, in addition to inhibiting or curing physical and mental diseases [1,2].

Plants produce diverse chemical ingredients, such as essential oils (EOs), which have multiple properties; for example, they can resist microbes [3], scavenge free radicals [4], cure inflammation [5], and relieve pain [6]. The EOs are complex concoctions with an aroma accumulated in the special plant cells and are obtained from the aromatic plants by hydro wave, microwave, or steam distillation.

They are highly efficacious and are economically important due to lower toxic effects as compared to the available commercial drugs. Thus, the EOs have been reported and tested as alternative remedies and are particularly important antimicrobial agents [7]. The widespread and irrational use of the currently used antibiotics has resulted in microbial mutation and antimicrobial resistance, thus lowering their therapeutic efficacy. This, in turn, increases the demand for plant-based constituents as alternatively safe and effective antimicrobial resources [8,9]. The EOs offered promising potential against the microbes as well as presented the significant capability to scavenge the free radicals with fewer side effects [10].

Oxidation is a biochemical process involving the transfer of an electron from a rich to a deficient unit. The deficient electron molecule is named an oxidizing agent [11]. The agents required for neutralizing the effects of the oxidizing agents are termed antioxidant agents, thus protecting our body from cellular damage because of oxidation effects [12]. The damaging effects of reactive oxygen species (ROS) are well-adjusted by the antioxidant activity of plant-based natural products [13].

These properties are interconnected and lead toward the control of inflammatory objectives. Hence, this chemical constituent proved itself as an essential mediator to incite or maintain the inflammatory condition and balance the human body by scavenging free radicals [10]. It is also reported that anti-oxidation, inflammation, and pains are interconnected objects that affect other mechanisms in the body; this is because oxidative stress can influence a variety of transcriptional factors and hence could the differential expression of important genes and gene elements affecting inflammatory mechanism [14]. Pain is a distasteful sensation triggered through internal or external stimuli, and it can also influence the sensory experience related to tissue damage [15]. The sensation of pain and its response to analgesics have an integrated relationship that involves many biochemical channels that are controlled by significant hereditary factors and may modify the pain stimuli or hinder the response to analgesics [12]. Moreover, there is a terrific deal of inter-individual variability in pain perception as well as the dosage required to relieve pain, and EOs have the potential to cope with and overcome pain [16]. Inflammation is the human body's defensive mechanism to combat chemical, physical, or biological hostility. Thus, it is the basic need to annihilate the detrimental agent and diminish its toxic effect by reducing its propagation [12]. Furthermore, the inappropriate usage of EOs obtained from plants belonging to the family Rutaceae might cause adversative properties in human beings, irritation of the skin, inflammation, headache, and nausea. Vigilance is usually needed if EOs is to be taken internally or used on foodstuff for the reason of the probable cancer-causing consequences among them [17]. With the use of a non-recommended dosage, EOs might cause functional harm to organs such as the stomach and liver in animals and, almost certainly, in humans [18].

The currently used anti-inflammatory and antioxidant drugs contain steroidal or non-steroidal anti-inflammatory constituents, and prolonged use is associated with adverse effects such as gastrointestinal intolerance, depression of bone marrow, and retention of salt and water [19]. Natural remedies, on the other hand, are an alternative source with a higher therapeutic efficacy rate and fewer side effects. Hence, the essential oils bearing herbs and their products can overcome inflammatory-related complications [20]. Lamiaceae is a cosmopolitan family and contains around 252 genera and 6700 taxa [21]. Mostly the species of Lamiaceae contain terpenes and many other bioactive constituents, predominantly occurring in the epidermal glands of flowers, leaves, and stems [22]. The genus *Scutellaria* L. comprises 360 plant species, distributed, and practiced as healing agents in conventional medicines in China, Korea, North America, and Pakistan [1,2,23]. Currently, more than 300 bioactive ingredients, including terpenes, flavonoids, flavones, glycosides, and terpenoids, have been isolated from *Scutellaria* [23,24], which have multiple health benefits; thus, *Scutellaria* species are well-known for their diverse promising medicinal purposes and offered significance resistance to the microbes, also act as an antioxidant, anti-inflammatory, and analgesic agent [25]. *Scutellaria edelbergii* Rech. F. (Lamiaceae) is a perennial herb, spreading through its hard woody rootstock, with slender stems and a procumbent or softly ascending, round-quadrangular, much branched ovate or acute margins leaves. The flowers are observed subtended, with a small calyx with purple scutellum, become puffy during fruit, yellow petals or occasionally blue-violet, lower lip darker, spreading erect or erect, densely glandular pilose. The flowering period is April–July, found between 1660 and 2200 m [26].

S. edelbergii is locally named panra and is used to cure inflammation and relieve pain, make green tea (kawa), and purify the blood. It is distributed over North America, Europe, East Asia, and Pakistan. However, in Pakistan, *S. edelbergii* is found in the mountain trails of Swat, Kalam, and Chitral [1], with moist loamy soil habitats. The EOs composition of the plant is influenced by climatic, topographic, edaphic, harvesting time, quality, and availability of water. The EOs of the genus *Scutellaria* mainly contain α-pinene, β-pinene, thymol, myrcene, linalool, sabinene, and γ-eudesmole, which are well known for their biomedical significance as reported by [27–30]. The crude extract and subfractions of the selected plant *S. edelbergii*, as well as its n-hexane extracted fatty acid esters, were found as antimicrobial, antioxidant, antidiabetic, analgesic, and anti-inflammatory agents [31]. However, currently, no scientific data are available for the therapeutic applications of *S. edelbergii* essential oil. Thus, the present exploration was designed to describe the SEEO constituents and screened them for their antimicrobial, antioxidant, analgesic, and anti-inflammatory effects and further validated using a computational approach.

2. Materials and Methods

2.1. Plant Material

The fresh *S. edelbergii* plant species were gathered from mountain tails in the Kalam region (Ushu and Mataltan), District Swat, at random intervals (April–June 2019), during the flowering season. After collection, *S. edelbergii* plant species were shifted to the research laboratory and placed in the open air to dry. The air-dried whole plant (8.7 kg) was pressed and preserved for identification and placed at the herbarium at the Department of Botany, Abdul Wali Khan University, Mardan (AWKUM/Herb/2234).

2.2. Essential Oils Extraction

The understudy plant species were chopped using an electric blender after they were completely dried. The obtained powder (2.0 kg) was weighed, and the essential oils (1.6 g) were extracted using a Clevenger device via hydro-distillation for 6 h, three times, until no more essential oils could be extracted. The EOs (0.08%) were measured after being collected off the top of the hydrosol in a glass bottle and were passed through anhydrous sodium sulfate to remove the moisture and placed in the refrigerator until further use [32].

2.3. GC-MS Analysis and Compounds Identification

The bioactive ingredients in the SEEO were determined using gas chromatography-mass spectrometry (GC-MS) analysis using a Perkin-Elmer-Clarus (PEC) 600 GC device, coupled with Rtx5MS, with a capillary column (30 m × 0.25 µm) at a maintained temperature of 260 °C attached with PEC 600 MS. The ultra-pure helium (99.99%) was used as a carrier gas at a constant flow rate of 1.0 mL/min. The injection, transfer line, and ion source temperatures were 260 °C, 270 °C, and 280 °C, respectively. The ionizing energy was noted as 70 eV. The electron multiplier voltage was operated from auto-tune. The full-scan mass spectra were obtained at 45–550 a.m.u. scan range. The tested sample at 1 µL quantity was loaded with a specified split ratio of 10:1. The oven temperature was maintained at 60 °C for a minute, while the temperature from 4 °C/min up to 260 °C was maintained for

4 min. The process was completed in 50 min [31]. The chemical constituents were identified through their corresponding chromatograms peaks obtained for each oil via GC-MS analysis in terms of retention indices (RI) compared with standards and the spectral mass data of each chromatogram through the National Institute of Standard Technology NIST-14 (2011 Ver. 2.3) [33] and were further authenticated by using the available literature [32].

2.4. Antimicrobial Screening

The EOs were examined for their antimicrobial screenings from low to high dosages. OD_{600nm} through the 96-well-plate method was used to determine MIC, whereas its agar-plate method was used to calculate MBC, as ZOIs were carried out by employing the agar-well diffusion technique [31]. The clinical isolates microbial strains were identified and authenticated by Chairman Dr. Hazir Rahman, Department of Microbiology AWKUM, Mardan.

2.4.1. Antibacterial Assay

Fresh bacterial strains inoculum of *K. pneumoniae*, *P. aeruginosa*, *E. coli*, and *E. faecalis* from a single colony were transferred to a sterile nutrient broth media and kept overnight at 37 °C/220 rpm. The MIC at OD_{600nm} (~0.5 McFarland standards) of *S. edelbergii* essential oil were determined for 1, 5, 10, 15, 25, 50, 100, 150, 200, and 500 µL two-fold serial dilution in sterile broth concentrations. Similarly, 20 µL of respective bacterial inoculum (1.5×10^6 CFU) were added to 100 µL of each concentration in triplicate wells and incubated for 18-24 h at 37 °C ± 0.5. Then, 50 µL of 3-(4,5-dimethylthiazol-2-yl)-2,5-diphenyltetrazolium bromide (0.2 mg/mL) MTT were added, and the plate was incubated at 37 °C for 45 min, along with an appropriate sterile broth, run as negative control while the bacterial suspension was comprised as a positive control. The absorbance was measured at 570 nm, and IC_{50} = OD of positive control-OD of test sample/OD of positive control x100 were calculated. Similarly, MBC for all these tested samples was confirmed by the agar plate spreading method (100 µL from MIC results of all concentrations were coated on TSA plates and cultured overnight, zero growth shows MBC. For scientific validity and authenticity, the entire data were taken in triplicates and listed as (Mean ± SEM).

The above available bacterial strains at a concentration of 1.5×10^8 CFU/mL of the bacterial cell density (BCD) were spread over the solidified media using a wire loop. In these Petri dishes, four wells at the same distance of size 3 mm were made. The essential oils at 50 µg/mL and 100 µg/mL was injected through micropipette into the 1st and 2nd wells, while the negative control (DMSO) and levofloxacin as standard was employed in the 3rd and 4th well, respectively, and were incubated at ±37 °C for 24 h and the obtained ZOI was measured in mm.

2.4.2. Antifungal Assay

The same procedures as those used in the antibacterial assay were used for antifungal activity assays on the SEEO. The *C. albicans* and *F. oxysporum* strains were sub-cultured on a fresh potato dextrose agar (PDA) plate for 24 h before antifungal assays. The same dose and same concentrations were used for SEEO to determine MIC, minimal fungicidal concentrations (MFC), and ZOIs [31]. The DMSO was run as negative control, while clotrimazole was used as a positive control accordingly. Then, during the incubation, the zone of inhibition was measured in mm. For scientific validity and authenticity, the entire data were taken in triplicates and listed as (Mean ± SEM).

2.5. Antioxidant Activity

2.5.1. DPPH Assay

The SEEO were screened for their free radicals scavenging effect using the DPPH assay [31,34] with a slight variation. The DPPH 3 mg was dissolved in 100 mL distilled methanol. The homogenized solution was placed in the dark for 30 min to generate free radicals for investigating the antioxidant activity of the EO. The samples were tested at different concentrations of 1000, 500, 250, 125, and 62.5 µg/mL. Next, 2 mL from each sample was mixed with 2 mL of the already prepared DPPH solution and placed for incubation in the dark for 25 min. The absorbance of the test samples was then determined at 517 nm using UV/Vis spectrophotometry. Ascorbic acid was employed as standard. The antioxidant potential of the test samples was determined using the following Equation (1).

$$\% \text{ Scavenging activity} = \text{A} - \text{B}/\text{A} \times 100 \quad (1)$$

where (A) absorbance of the control and (B) is the absorbance of the tested samples (standard and essential oil).

2.5.2. ABTS Assay

The free radicals scavenging significance of the tested samples was conducted using ABTS assay. About 383 mg of ABTS and 66.2 mg of $K_2S_2O_8$ were separately dissolved in 100 mL analytical grade methanol and then combined. Next, 2 mL from the ABTS solution was incubated with 2 mL of test samples for 25 min using similar concentrations as described in the earlier DPPH assay. Furthermore, the absorbance of the EOs and ascorbic acid was determined at 746 nm using UV/V is spectrophotometry. The free radical scavenging effect was estimated using Equation (1).

2.6. *Approval of Experimental Animals*

Healthy Swiss mice of (24–30 g) were obtained from Veterinary Research Institute (VRI), Peshawar, and were accommodated in cages in AWKUM animal house under a controlled temperature of 20 °C for 6.5 weeks. The required materials (rodent pellets, foodstuff, and water) were given under cleaned conditions following ARRIVE guidelines.

2.7. *Analgesic Activity*

The EOs of the understudy plant was tested for relieving pain using an acetic acid-induced writhing assay as stated by [31] with slight modification. Moreover, swiss albino mice were used and divided into 5 groups ($n = 6$). The EOs, control, and standard were injected into the mice through intraperitoneal muscle with a sterilized syringe. All the mice groups were pretreated with 1 mL acetic acid (0.7%) at a concentration of 5 mL/kg body weight (BW) and then after 45 min. Next, the experimental animal of group 1 indicated as normal control, was treated with 1 mL of normal saline, while the animals in Group 2 were infused with 1 mL aspirin as a standard. Furthermore, the SEEO was administrated to the remaining swiss albino mice in groups 3, 4, and 5 at doses of 25, 50, and 100 mg/kg BW doses, respectively.

Writhes numbers were counted for determining the analgesic effect of the tested samples in comparison with normal saline and standard for 10 min. The results obtained were expressed in % inhibition using Equation (2).

$$\% \text{ Inhibition} = \text{A} - \frac{\text{B}}{\text{A}} \times 100 \quad (2)$$

(A) is the writhes inducer (acetic acid); while (B) is the tested samples significance (EO, standard, and control).

However, in the anti-inflammatory assay, (A) in the equation indicates the paw edema induced through carrageenan.

2.8. *Anti-Inflammatory Activity*

The efficacy of SEEO to treat inflammation was tested using carrageenan-induced paw edema in Swiss albino mice, as stated by [31], with slight modification. The mice were grouped as mentioned in the analgesic activity: The inflammation was induced by injecting 1 mL (1%) carrageenan into all six groups of the swiss albino mice [35]. After 30 min, the 1st group was given 1 mL of normal saline, while group 2 was given 1 mL of diclofenac Na (50 mg/kg) using a sterile syringe. The SEEO at 25, 50, and 100 mg/kg body weight doses were injected into groups 3, 4, and 5, respectively, following the safety measures. The anti-inflammatory activity of tested samples was observed by measuring the paw diameter of the experimental animals right after each 1st, 2nd, and 3rd hour, respectively. In addition, the resulting data were expressed as % inhibition and calculated using Equation (2).

2.9. Computational Analysis

2.9.1. Construction of Chemical Compounds Database

The literature is available to highlight the importance of identifying promising anti-inflammatory medicines and already available synthetic drugs on the market that are not so healthy because of certain adverse side effects [36]. Computer-assisted screening applications highly support the identification of novel drugs for different diseases [37,38]. An integrated computer-assisted scheme was employed using the database of 52 compounds to identify novel drugs to combat anti-inflammatory diseases. The structure of each compound was drawn by using the ChemDraw software [39], and information on each structure was crosschecked from the PubChem database [40] to reduce the chance of ambiguity and saved in SDF format for further analysis.

2.9.2. Selection of Target Protein

It is highly significant to understand the disease mechanism and then select the appropriate protein structure to initiate the drug design pipeline, and it could explain the important parameters necessary to clarify the action of bound ligands; these ligands or chemicals are the drugs that could selectively inhibit the activity of the Cyclooxygenase-2 (COX-2) enzyme for inflammatory disease [41,42]. Therefore, COX-2 protein (PDB ID: 5KIR) was selected to execute a protein-ligand docking experiment in this study [41,43].

2.9.3. Molecular Docking and Interactions Investigation

Protein-ligand docking is an appropriate technique to understand the protein-ligand bounded conformation and explains the molecular mechanism of small drug-like entities in cellular pathways [44]. The GC-MS-based fifty-two identified compounds that were used for molecular docking and a three-dimensional structure of the COX-2 enzyme (PBD ID: 5KIR) in PDB format were imported to the MOE software [45]. Heteroatoms, 3D protonation, and water molecules, along with the default ligand attached to the target protein, were removed to prepare the protein for the docking procedure. An active site was identified in the selected protein (5KIR) based on the previous literature [41], and structural optimization was performed by following parameters, such as the addition of hydrogen atoms and energy minimization with the Amber14 force field method was applied with chiral constraints and geometrical parameter. By using the surfaces and maps panel module, the transparency of the front and the back surface was adjusted and resulted in the information of significant residues in the selected substrate-binding pocket of 5KIR protein in native conformation [41]. MOE software creates a database of 52 compounds identified from experimental studies to perform molecular docking simulations and saves them with MDB extension for further analysis. Top-ranked poses were subjected to refinement and calculation of binding free energies (ΔG), which is evaluated by scoring function (GBVI/WSA dg) as described by Aldeghi et al. [46]. A reliable scoring scheme that results in the docking score of the correct binding poses was established by the number of molecular interactions (hydrogen, Pi, and Van der Waals interactions) documented in the literature of Ahmad et al. [47] and generated by Discovery Studio [43]; MOE database of the docked complexes was visualized carefully for understanding the mode of binding interactions of COX-2 inhibitors bound in the selected site of the target protein.

2.9.4. Molecular Dynamics Simulations

Molecular dynamic (MD) simulations are a popular and interesting technique to understand the selected docked complex at a further atomic level, as reported by Bibi et al. [38]. For this study, the best-docked complex was selected for the MD simulation analysis at

100 ns; hence the conformational stability of the ligand bound in the vicinity of the active binding pocket of COX-2 protein in the system was analyzed for practical applications [48]. The study was divided into three major phases of MD simulation application. Initially, the parameter files were fixed, then subsequently moved towards the pre-processing, and finally performed the simulations as documented in the data of Ahmad et al. [47]. The antechamber module is important for preparing the files in AMBER20 software, as stated by Lee et al. [48]. Complex libraries and different MD simulations were fixed using ligand and protein information, and with the help of the Leap module, it was solvated at 12 Å, and appropriate measurement was accomplished. Molecular interacting residues were resolute with the help of the force field (ff14SB) [49]. The requirement of charge neutralization was fulfilled by the addition of Na+ ions to the system. In the system pre-processing, the selected binding energies were retrieved after the optimization of the system many times, hydrogen atoms were abated for the 500 steps, and the solvation energy was calculated for the 1000 steps by keeping the average limit of 200 kcal/mol-$Å^2$, energy minimization was followed by the system carbon alpha atoms, complete set for once again until 1000 steps with the pragmatic scale of 5 kcal/mol-$Å^2$, and similarly, for non-heavy atoms; 300 steps of minimization was performed with the scale of 100 kcal/mol-$Å^2$. Hence, the system was heated for 300 K by NVT ensemble explained with Langevin dynamics and SHAKE algorithm as stated by Krautler et al. [50] and restrict the number of hydrogen bonds (HBs). The equilibration was attained by 100-ps. MD-simulation system was assisted with pressure using NPT ensemble parameters and fixed the system carbon alpha atoms. MD-simulation was subsequently performed using the time scale of 2-fs until 100 ns. By using CPPTRAJ, inter-and intra-molecular interactions were noted, keeping the cut-off range of distance at 8.0 Å [51]. For molecular and behavioral explanations, different trajectories were generated to explain the stability of the system with the help of the Visual molecular dynamic (VMD) tool, as earlier described by Humphrey et al. [52].

2.9.5. Binding Free Energy (BFE) Estimation

Amber 20 was used for the estimation of binding interaction and the solvation-free energies for the COX-2 enzyme. Enzyme/protein-ligand complex was subjected to MM-PBSA calculation for absolute BFE estimation, which is the sum of gas-phase and solvation-free energies during MD simulation for 100 ns, as stated by Daina et al. [53], and its corresponding MM-GBSA calculations were performed with aims to develop the distinction between the bounded and detached presentation of solvated conformations of the selected potential target COX-2 as described by Sander et al. [54] Mathematically, the BFE can be analyzed by the Following Equation (3).

$$\Delta G_{\text{bind,solv}} = \Delta G_{\text{bind,vacuum}} + \Delta G_{\text{solv}},\ \text{target protien} - \text{ligand} - (\Delta G_{\text{ solv}},\ \text{ligand} + \Delta G_{\text{ solv}},\ \text{target protein}) \quad (3)$$

For all three conditions of the MM system, the solvation energy referred to as the transfer of molecules from the gas phase to solvent was assessed by resolving any one Poisson Boltzmann (PB) or Generalized Born (GB) equation. Therefore, it contributes to the electrostatic role of the solvation phase. Similarly, it permits the calculation of empirical terms for hydrophobic assistances, as presented in Equation (4).

$$\Delta G_{\text{solv}} = G_{\text{electrostatic},\ \epsilon=80} - G_{\text{electrostatic},\ \epsilon=1} + \Delta G_{\text{hydrophobic}} \quad (4)$$

The estimation of the average interaction energy among the ligand and protein gives to delta-$\Delta G_{\text{ vacuum}}$ (Equation (5)).

$$\Delta G_{\text{vacuum}} = \Delta E_{\text{molecular mechanics}} - T \cdot \Delta S \quad (5)$$

2.9.6. In Silico Pharmacokinetic/ADMET Profile Calculations

Based on docking results, the best compound was used for the calculation of the ADMET (absorption, distribution, metabolism, excretion, and toxicity) profile, and it is a significant criterion for drug-like screening of chemical compounds, as stated by

Khan et al. [55]. For ADMET profile estimation, SwissADME, as stated by Daina et al. [53], and Data-Warrior tools as reported by Sander et al. [54], were used.

2.10. Statistical Analysis

The data of the current study were taken in triplets and estimated using one-way analysis of variance (ANOVA), followed by Bonferroni's test at significance level $p = 0.05$ represented (*) and 0.01 denoted as (**) using two-way ANOVA. While the Sidak's multiple comparisons test [p = (ns > 0.9999, **** <0.0001)], for statistical authenticity. However, for antioxidant significance, a nonlinear regression graph was marked among % inhibition and concentration of the tested samples, and the IC_{50} was estimated via the GraphPad prism 9 programs for windows (GraphPad software, San Diego, CA, USA, 2020) by applying the below equation.

Y = 100/1 + (^HillSlope)
The equation can be explained as:
1 = Denote the concentration of the inhibitor.
Indicates the inhibitor's reaction.
HillSlope indicates the steepness of the curve.

3. Result and Discussion

The current study was undertaken to evaluate the chemical ingredients and determine the microbial, free radicals scavenging, antinociceptive, and analgesic effects of EOs of *S. edelbergii* to scientifically validate their multiple therapeutic applications. The chemical constituents in plants are screened through various chromatographic techniques, including column chromatography, high-performance liquid chromatography, and gas chromatography-mass spectrometry.

3.1. GC-MS Analysis

The essential oils serve as alternative remedies aromatherapy contributed to their valuable capacities, the EOs are extracted from medicinal plants via hydro-distillation, microwave oven, and steam distillation, and the active ingredients are highlighted in GC-MS analysis. These bioactive chemical ingredients serve as a basis for the pharmaceutical industries to use the active chemical ingredients for their diverse biomedical applications as available drugs become ineffective over time. Plants are categorized as medicinal due to the presence of chemical constituents that are affected by various topographic, climatic, and numerous other factors that alter the composition of the bioactive ingredients. Fifty-two compounds were identified in the essential oil of the understudy plant *S. edelbergii* via GC-MS analysis, which contributed to around 89.52% (Table 1). The dominant compounds among the screened constituents were 3-Oxomanoyl oxide, followed by 24-Norursa-3,12-diene, methyl 7-abieten-18-oate, and β-eudesmol with 10.09%, 8.05%, 7.02%, and 6.39% respectively (Figure 1). These constituents were reported in some species, as described by Bekana et al. [56], and in the genus *Scutellaria*, earlier reported by Kurkcuoglu et al. [57] Furthermore, the essential oil of the under-study plant contains α-pinene (0.08%), myrcene (0.03%), and heptadecane (0.14%) in fewer amounts as compared to the same constituents, which were earlier reported from *S. diffusa* and *S. heterophylla* documented by Cicek et al. [58] at the quantity of 0.2%, 0.5%, and 0.2%, respectively. However, caryophyllene was observed in higher amounts (7.4%), and caryophyllene oxide (6.8%) from various species of the genus *Scutellaria* reflected in the literature of Formisano et al. [59], while the species of the same genus *S. edelbergii* depicted the same constituents in low amount with the amount of 2.30% and 3.94%, respectively. In addition to that, some species of the same genus *Scutellaria* possess the same constituents as reported in our studied plant as stated by Mamadalieva et al. [24].

Table 1. GC-MS detected EOs of *S. edelbergii*.

C. No	Name of the Compounds	Rt	Contents (%)	$RI_{Rep.}$	$RI_{calc.}$
1	α-Pinene	7.83	0.08	931	942
2	β-Pinene	8.91	0.02	943	982
3	β-Myrcene	9.37	0.03	958	999
4	α-Pellandrene	9.75	0.04	964	1013
5	Sabinene	10.43	0.10	969	1021
6	psi-Limonene	10.48	0.02	992	998
7	(+)-4-Carene	11.24	0.03	1017	1026
8	Linalool	12.31	0.02	1081	1104
9	Nonanal	12.46	0.14	1080	1110
10	α-Campholenal	13.10	0.08	1102	1134
11	L-Pinocarveol	13.47	0.15	1143	1147
12	cis-Verbenol	13.61	0.09	1131	1153
13	α-Phellandren-8-ol	14.19	0.09	1148	1174
14	Terpinen-4-ol	14.48	0.24	1160	1185
15	α-Terpineol	14.82	0.10	1172	1198
16	Myrtenal	15.01	0.13	1175	1205
17	Cuminal	16.11	0.08	1214	1248
18	Carvone	16.21	0.08	1229	1251
19	Bornyl acetate	17.26	0.59	1273	1292
20	Phenol,2-methyl-5-(1methylethyl)-	17.36	0.60	1278	1296
21	α-Terpinyl acetate	18.79	0.62	1322	1354
22	(-)-β-Bourbonene	19.71	0.46	1386	1392
23	β-Elemene	19.84	2.09	1398	1401
24	Caryophyllene	20.54	2.30	1421	1428
25	Humulene	21.32	0.40	1454	1462
26	gamma-muurolene	21.79	0.46	1471	1483
27	β-eudesmene	22.07	4.10	1478	1494
28	α-Selinene	22.26	1.83	1500	1503
29	Cadina-1(10), 4-diene	22.83	1.90	1514	1529
30	Elemol	23.38	0.93	1535	1554
31	Caryophyllene oxide	24.20	3.94	1575	1591
32	γ-Eudesmole	24.793	0.54	1627	1619
33	tau-Cadinol	25.35	0.60	1628	1646
34	β-Eudesmol	25.58	6.39	1644	1658
35	tau-Muurolol	25.63	4.43	1628	1660
36	Ar-Turmerone	25.77	0.19	1638	1667
37	α-Phellandrene, dimer	28.33	1.52	1801	1811
38	Linalyl phenylacetate	29.08	1.51	1945	1953
39	Thunbergen	29.70	1.74	1934	1942
40	p-Camphorene	30.47	2.47	1977	1986
41	Geranyl.alpha-terpinene	31.10	1.72	1962	1973
42	Thunbergol	31.83	0.50	2073	2082
43	Verticiol	32.38	0.96	2106	2118
44	Linoleic acid, methyl ester	33.46	4.12	2071	2080
45	3-Oxomanoyl oxide	33.96	10.09	2133	2140
46	Methyl pimar-8-en-18-oate	36.97	1.40	2231	2297
47	Methyl 7-abieten-18-oate	37.25	7.02	2164	2178
48	Methyl dehydroabietate	37.53	2.79	2293	2335
49	Methyl abietate	38.20	1.36	2339	2379
50	24-Norursa-3,9(11),12-triene	46.50	3.70	3042	3057
51	24-Norursa-3,12-diene	47.21	8.05	3105	3062
52	24-Norursa-3,12-dien-11-one	50.09	6.68	3351	3308
	Identified compounds		89.52		

Identified compounds: Elution order on HP-5MS column. $RI_{rep.}$: Retention Index from the database (NIST, 2011). C. No = Compound number; $RI_{calc.}$ = Retention Index calculated.

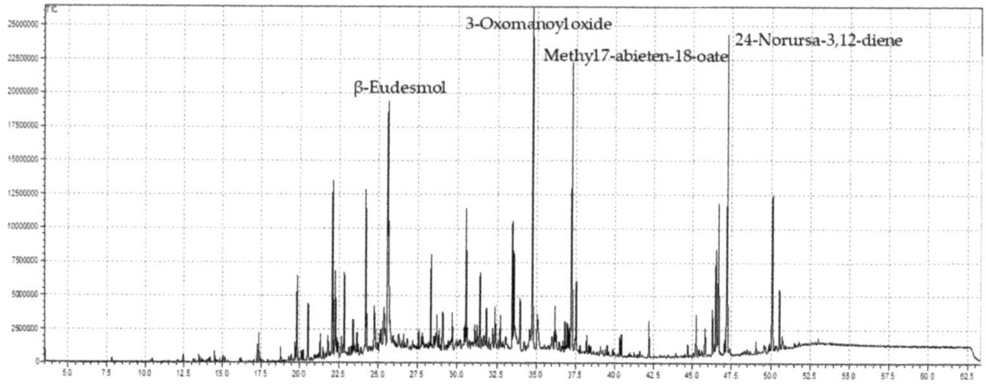

Figure 1. GC chromatogram of SEEO.

These chemical ingredients as serve as antioxidants, anti-cancer, and antimicrobial agents, as reported by Hussain et al. [60] and Mahmood et al. [61] These constituents have antimicrobial activities, as stated by Utegenova et al. [27], as well as the capacity to scavenge free radicals, cure inflammation, and relieve pain, as reported by Surendran et al. [28]. The EOs present in the selected plant *S. edelbergii* were also present in some plants belonging to the same genus *Scutellaria* as reflected in the studies reported by Kasaian et al. [62], Yilmaz et al. [63], and lawson et al. [64].

3.2. Antibacterial Capacities

The SEEO were tested against the human pathogenic bacterial strains, *K. pneumonia, P. aeruginosa, E. coli,* and *E. faecalis,* for MIC, MBS, optical density (OD_{600nm}), and inhibitory concentrations fifty (IC_{50}) for 1/2MIC, MIC, and 2MIC concentrations displayed in Figure 2A,B, respectively. The zone of inhibitions was also determined at low to high doses, as well as compared with the standard (Levofloxacin) and negative control (DMSO) (Figure 3A). For SEEO at MIC, the IC_{50} for *K. pneumonia* was 84.60%, *P. aeruginosa* was 87.14%, *E. coli* was 87.96%, and *E. faecalis was* 83.70%. The current findings of SEEO exhibited considerable ability against the tested Gram-negative and Gram-positive bacterial strains. In addition, appreciable resistance of 15.8 $\pm$ 0.03 and 21.2 $\pm$ 0.02 mm was exhibited against the *E. faecalis* from low to high doses, as compared to levofloxacin 18.1 $\pm$ 0.01 and 23.7 $\pm$ 0.02; the *E. coli* and *P. aeruginosa* were also susceptible to SEEO. The essential oil was known for its antimicrobial significance, as stated by Seow et al. [65]. The substantial antibacterial activity of the SEEO might probably be due to caryophyllene [66], γ-eudesmole [67], linoleic acid, methyl ester [61], methyl stearate as stated by Adnan et al. [68], methyl pimar-8-en-18-oate [69], methyl-7-abieten-18-oate [70], 24-norursa-3,9(11),12-triene,24-norursa-3,12-diene [60], α-pinene and β-pinene [71], α-pellandrene [72], sabinene [27], psi-limonene [73], (+)-4-carene antibacterial resistance [74], α-campholenal, and L-pinocarveol can resist microbes [16], which was reported in the understudy plant (Table 1), represented the analysis depicts all these active fractions and compounds, which possess these antibacterial activities. However, Gram-positive strains are more susceptible to the tested sample as compared to Gram-negative bacterial strains, which agrees with the findings of Kasaian et al. [62], Skaltsa et al. [75], and Skaltsa et al. [76] in some species of *Scutellaria.* In addition that our findings were not agreed with the data reported by Bogdan et al. [77] for the essential oils of *Lavandula angustifolia* as the constituents in the plant might be influenced by phenological behaviors as stated by Moisa et al. [78] for *Thymus valgaris.*

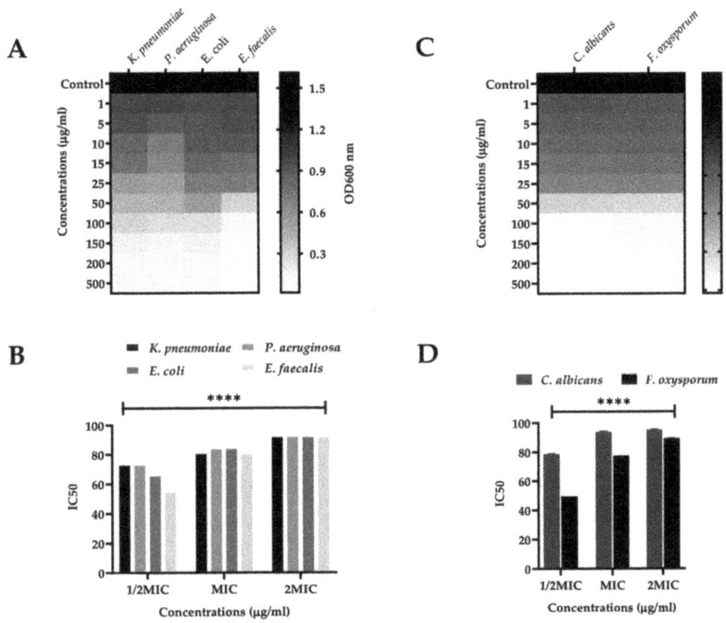

Figure 2. (**A**,**B**) antibacterial OD_{600nm} and IC_{50} of *Klebsiella pneumoniae, Pseudomonas aeruginosa, Escherichia coli,* and *Enterococcus faecalis* against various concentrations of essential oil of *S. edelbergii;* (**C**,**D**) Antifungal OD_{600nm} and IC_{50} of the potential of essential oils of *S. edelbergii* against *Candida albicans* and *Fusarium oxysporum* against various concentrations. Data were taken in triplicate ($n = 3$) and analyzed through two-way ANOVA, via Tukey's multiple comparison test, p = **** < 00.0001).

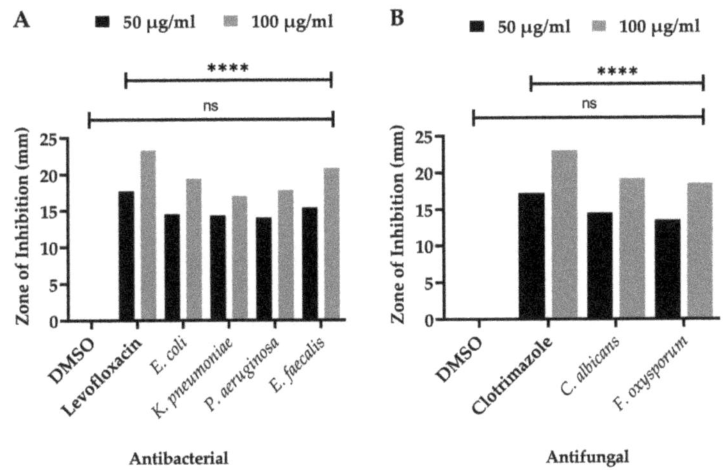

Figure 3. Zone of inhibition's (**A**) antibacterial and (**B**) antifungal potential zone of inhibitions of essential oils of *S. edelbergii*; DMSO = negative control; levofloxacin and clotrimazole = positive control. Data were taken in triplicate ($n = 3$) and analyzed through two-way ANOVA, via Sidak's multiple comparison test, ns = >0.9999, p = **** < 0.0001).

3.3. Antifungal Significance

Similarly, the SEEO also produced a substantial effect against the tested fungal strains *C. albicans* and *F. oxysporum* in a dose-dependent manner. The MIC and MFC of the respective fungal strains, OD_{600nm} and IC_{50}, are shown in Figure 2C and 2D, respectively. IC_{50} of SEEO at MIC was 93.93% and 77.54%, respectively, for *C. albicans* and *F. oxysporum*. Similarly, the maximum zone of inhibition (Figure 3B). It was observed against the SEEO displayed 14.3 ± 0.33 and 19.6 ± 0.05 mm resistance against *C. albicans* as compared with clotrimazole 17.6 ± 0.11 and 23.4 ± 0.4, respectively, from low to high doses. The antifungal effect is attributed to the existence of the bioactive compounds present in the SEEO, such as α-phellandrene [79], sabinene [27], α-phellandrene, dimer [79], thunbergen [80], thunbergol [81], and methyl dehydroabietic [82], which were stated in our GC-MS analysis (Table 1). The previous literature revealed that essential oils have a promising ability to overcome the complications caused by the fungus, as described by Singh [83]. Our findings were consistent with the research conducted by Yu et al. [84] and Zhu et al. [85] on the species of the same genus S. *barbata* and *S. strigillosa*, respectively. In addition, our outcomes also agreed with Zahra et al. [23] for the *S. multicaulis* and *S. bornmuelleri*. The current findings equated with the reported literature, using the same method and dose, as well due to similar genus. However, our data do not support the literature presented by Ullah et al. [86] for the essential oils of *Ochradenus arabicus* essential oils various parts due to the variation among the plant family and habitat, which influence the quality and quantity of the chemical ingredients.

3.4. Antioxidant Significance

The free radicals scavenging activity of different concentrations of SEEO was determined via DPPH and ABTS assays. The SEEO showed significant antioxidant activity due to the presence of diverse bioactive ingredients. The EOs of *S. edelbergii* presented a substantial mechanism to neutralize the free radicals in the DPPH assay with IC_{50} = 125.0 ± 0.19 μg/mL, while in the ABTS assay, the activity of the EOs was observed with an IC_{50} = 153.0 ± 0.31μg/mL). Furthermore, the antioxidant effect of the EOs of the studied plant was assessed to the ascorbic using the same concentrations and exhibited an IC_{50} = 70.19 ± 0.16 μg/mL and IC_{50} = 90.70 ± 0.32 μg/mL for the DPPH and ABTS bioassays, respectively (Figure 4). Our results strongly comply with those stated by Mamadalieva et al. [24] for some species of *Scutellaria*. The significance of scavenging the free radicals described by Lawson et al. [64] by the plant species belonging to the same genus *Scutellaria* consented to our data. However, different compounds were identified through GC-MS analysis in the same plant, *S. edelbergii* n-hexane crude oil reported by Shah et al. [31], as compared to the essential oil extracted through hydro-distillation. Furthermore, the essential oils extracted through hydro-distillation were observed to be efficient in neutralizing the free radicals as compared to the n-hexane-extracted crude oils of the same plant; hence, the mode of extraction also affects the quality and quantity of the compounds.

Our findings about the details of essential oils as a source of aromatherapy also conform to the data described by Mot. et al. [87] for *Salvia officinalis* essential oils and essential oils of *Ochradenus arabicus*, as stated by Ullah et al. [86] This highlights the idea that each molecule has a unique chemical structure that determines characteristic biochemical, physiological, pharmacological, toxic, etc. properties, and highlights this strong interdependence as reflected in the literature described by Glevitzky et al.'s [88] statistical analysis of the relationship between antioxidant activity and the structure of flavonoid compounds. This might be attributed to the difference in the chemical constituents among the essential and crude oils of the same plant. The main constituent responsible for scavenging the free radical is β-myrcene, linalool, α-campholenal, cuminal, and (-)-β-bourbonene, as revealed by Surendran et al. [28], Kamatou and Viljoen [89], Zhang et al. [79], and Ghasemi et al. [90]. It also complied with the literature by Khalilov et al. [91], with mainly flavonoids and phenols acting as promising antioxidant agents.

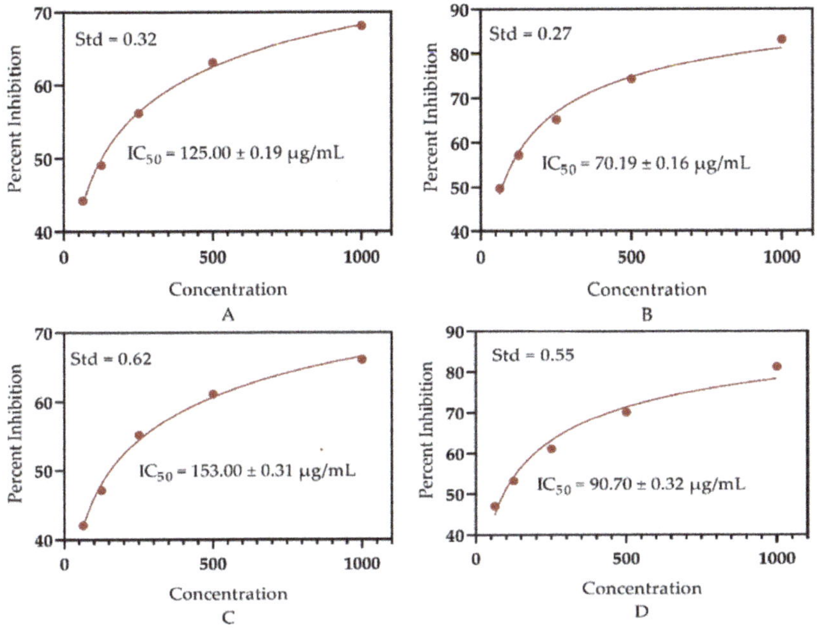

Figure 4. (**A**) antioxidant potential of the *S. edelbergii* essential via DPPH assay, (**B**) antioxidant activity of ascorbic acid at DPPH assay, (**C**) antioxidant activity of SEEO via ABTS assay, (**D**) free radicals scavenging effect of ascorbic acid via ABTS assay, (Std) standard deviation.

3.5. Analgesic Potential

The analgesic effect of SEEO was determined using Swiss albino mice at low to high doses (25, 50, and 100 mg/kg) and showed significant pain-relieving activity when tested against pain induced by acetic acid, as shown in Table 2. The tested sample at a dose of 25 mg/kg depicted 33.33% and exhibited significant inhibition of 43.11% and 54.71% at a concentration of (50 and 100 mg/kg). Moreover, the normal saline has no effect, and the standard (aspirin) presented 64.49% in writhes reduction tested in the Swiss albino mice. The current findings were consistent with the literature presented by Uritu et al. [92] for the *Scutellaria* species, as well as by Mondal et al. [35], which described the analgesic significance in the leaves of *Eucalyptus camaldulensis* and with the study presented by Chen et al. [93], Sarmento-Neto et al. [15], and Mishra et al. [94], in which the EOs were extracted using the same hydro-distillation methods. By applying different oils extraction methods, the same plant yields different types of compounds; thus, the essential oils extracted via hydro-distillation were much more effective in curing pain as compared to the crude oils obtained through column chromatography of the n-hexane fraction as a part of the continuous studies earlier documented by Shah et al. [31] The promising analgesic activity was depicted by the SEEO due to the presence of bioactive ingredients such as β-myrcene previously described by Surendran et al. [28]; γ-eudesmole, as reflected in the literature of Aati et al. [29]; tau-muurolol, documented by Hameed et al. [95]; and some other bioactive compounds with the capacity to reduce pain. Furthermore, our data also matched with the findings of Liang et al. [96]; the root EOs of *Illicium lanceolatum and B. persicum* EOs, as described by Hajhashemi et al. [97], presented the significant potential to cure pain. Moreover, *S. rufinervis* is an aromatic plant comprising Eos, which also has aromatherapeutic significance, especially in relieving pain, as documented by Santos et al. [98].

Table 2. Analgesic significance of essential oils of *S. edelbergii*.

Treatment	Dose Conc.	No. of Writhes Mean ± SEM	% Reduction in Writhes after 45 min
Acetic acid	1 mL	27.6 ± 0.03	
Normal saline	1 mL	27.3 ± 0.05	-
Aspirin	1 mL	9.8 ± 0.02	64.49
SEEO	25 (mg/kg)	18.4 ± 0.02 **	33.33
	50	15.7 ± 0.03 **	43.11
	100	12.5 ± 0.05 **	54.71

(Acetic acid) Inducer, (Normal saline) negative control, (Aspirin) standard, (SEEO) *S. edelbergii* essential oil, $p = 0.01$ denoted as (**).

3.6. Anti-Inflammatory Capabilities

The SEEO were tested to cure the inflammation induced by carrageenan in the experimental animals, as displayed in Table 3. The SEEO demonstrated considerable capability with 53.10% inhibition at the dose of 25 mg/kg, whereas the tested samples at the concentrations of 50 mg/kg and 100 mg/kg displayed 57.93% and 64.13% inhibition, as compared to Diclofenac Sodium, which depicted 71.72% inhibition. Moreover, the normal saline does not affect the paw diameter of the swiss albino mice caused by carrageenan. The use of essential oil for remedial practices is documented in the various literature and specially to reduce paw edema, as stated by Colares et al. [99] The potential exhibited by the SEEO to cure inflammation is also supported by the literature given by Mogosan et al. [100] in some *Mentha* species and further validated by Boukhatem et al. [101] in some other species of the same family Lamiaceae. Moreover, the SEEO extracted through hydro-distillation offered a significant ability to cure inflammation when equated with the crude oils extracted from the same plant *S. edelbergii* using column chromatography of the n-hexane fraction described by Shah et al. [31] However, the available literature stated that the bioactive compounds responsible for curing inflammation are β-myrcene as conveyed by Surendran et al. [28], sabinene described by Zhang et al. [79], linalool illustrated by [89], myrtenal documented by Dragomanova et al. [102], and many other potent chemical ingredients with the capability to cure the inflammation and act as an anti-inflammatory agent. Our data were consistent with the finding noticed for *Jatropha curcas* described by Adesosun et al. [103] However, our data are not equated with the reported literature by Apel et al. [104] for *Myrciaria tenella* and *Calycorectes sellowianus* essential oils; they presented high significance as compared to our studied plant.

Table 3. Anti-inflammatory significance of *S. edelbergii* EOs.

		Changes in Paw Diameter in Swiss Albino Mice (Mean ± SEM)				
Samples Used	Dose Conc.	1 h	2 h	3 h	Av. Paw Diameter	% Inhibition
Carrageenan	1 mL	1.21 ± 0.03	1.42 ± 0.03	1.74 ± 0.05	1.45 ± 0.03	
NS	1 mL	1.18 ± 0.02	1.40 ± 0.04	1.71 ± 0.02	1.43 ± 0.03	-
Standard	50 (mg/kg)	0.49 ± 0.02	0.42 ± 0.03	0.33 ± 0.01	0.41 ± 0.02	71.72
SEEO	25	0.73 ± 0.04	0.68 ± 0.02	0.63 ± 0.03	0.68 ± 0.03 *	53.10
	50	0.67 ± 0.01	0.62 ± 0.03	0.56 ± 0.02	0.61 ± 0.04 *	57.93
	100	0.58 ± 0.06	0.53 ± 0.04	0.46 ± 0.02	0.52 ± 0.03 *	64.13

(NS) normal saline, (standard) diclofenac Na, (SEEO) *S. edelbergii* essential oils, $p = 0.05$ denoted as (*).

3.7. Molecular Docking and Interactions Analysis

The structural information of the COX-2 enzyme (PDB ID: 5KIR) was used. The prepared biomolecule structure without any bounded ligand and considerable active binding residues pocket for the protein-ligand interaction analysis was performed by Sander et al. [54]. A database of 52 bioactive compounds was identified through GC-MS and subjected to MOE with MDB extension and performed protein-ligand docking simulations by the Dock module of MOE software [54]. Phytochemicals with the best binding poses and molecular interactions with the significant amino acids intricate the mechanism of inhibition of COX-2 enzyme to manage inflammatory disease. By the evaluation of the selected active binding site residues, targeted docking was applied. Dock score and RMSD values calculated for the database of 52 compounds with COX-2 protein by MOE software are enlisted in Table 4. While the selected highest dock scored four compounds were demonstrated with the best binding poses, noteworthy binding interactions with active site residues of COX-2 enzyme within the range of 4.5 Å are presented in Figure 5, and a summary of docking results is enlisted in Table 4.

Table 4. Dock score and RMSD values were calculated for 52 compounds with COX-2 protein.

Numbering	Score	RMSD	Numbering	Score	RMSD
1	−5.323	1.2178	**27**	−6.234	0.651
2	−5.392	1.44	**28**	−6.198	3.501
3	−5.421	1.0412	**29**	−6.144	2.123
4	−5.422	1.042	**30**	−6.543	1.577
5	−5.364	1.4674	**31**	−5.922	2.437
6	−5.319	1.3645	**32**	−6.477	0.716
7	−5.311	0.6295	**33**	−6.054	1.429
8	−5.691	1.9427	**34**	−6.032	1.647
9	−5.758	1.0724	**35**	−6.503	1.481
10	−5.550	1.1354	**36**	−6.248	1.524
11	−5.677	1.1992	**37**	−7.130	2.209
12	−5.566	0.8491	**38**	−7.043	2.447
13	−5.490	0.5436	**39**	−6.715	0.929
14	−5.564	2.2633	**40**	−6.711	1.964
15	−5.536	0.4417	**41**	−7.949	0.639
16	−5.565	0.4645	**42**	−6.640	2.445
17	−5.160	1.2301	**43**	−6.237	2.991
18	−5.529	0.7864	**44**	−7.626	1.845
19	−6.139	2.7774	**45**	−6.808	1.376
20	−5.323	0.8127	**46**	−6.731	1.114
21	−6.572	1.2823	**47**	−7.870	1.920
22	−6.329	2.1625	**48**	−6.802	2.784
23	−6.006	1.2073	**49**	−7.422	2.663
24	−5.9901	1.3014	**50**	−6.3748	4.1046
25	−5.9146	2.6874	**51**	−6.315	2.3077
26	−6.3482	0.9651	**52**	−6.6796	3.2421

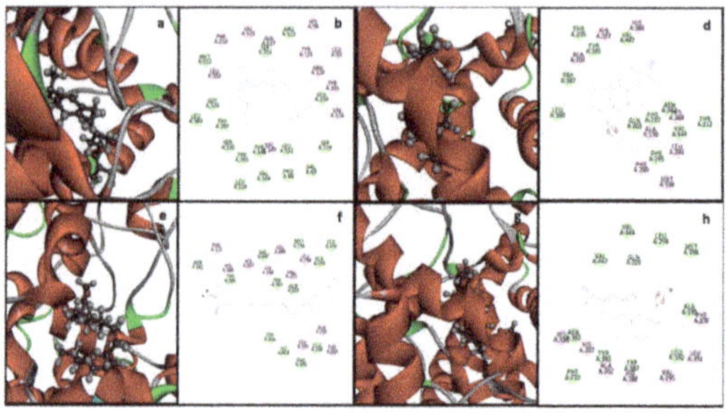

Figure 5. Four compounds (((**41**) **a**,**b**), ((**47**) **c**,**d**), ((**44**) **e**,**f**), and ((**49**) **g**,**h**)) with the highest dock score represent the best binding poses within the active binding site of Cyclooxygenase-2 (COX-2) protein (PBD ID: 5KIR). Three-dimensional view of the best-bounded conformation of docked ligands (**a**,**c**,**e**,**g**) and a two-dimensional plot showing significant protein-ligand binding interactions (green colored circles present Van der Waals interactions and carbon-hydrogen bond, while purple-colored circles present alkyl and Pi-alkyl interactions) (**b**,**d**,**f**,**g**).

All four selected potential hits presented the Van der Waals, carbon-hydrogen, alkyl, and Pi-alkyl bonding, and most of the selected active site is hydrophobic (Table 5). The dock score range is from −7.9497 to −7.4221 kcal/mol and shows that these hits were bounded in the best conformation within the target COX-2 protein selected active site (Figure 5).

Table 5. Summary of molecular docking results of top 4 dock score compounds with the COX-2 target protein.

Numbering	Dock Score (kcal/mol)	Functional Residues	Binding Interactions
41	−7.9497	PRO86 (A), VAL89 (A), HIS90 (A), LEU93 (A), TYR115 (A), VAL116 (A), SER119 (A), ARG120 (A), VAL344 (A), TYR348 (A), LEU352 (A), SER353 (A), TYR355 (A), LEU359 (A), LEU384 (A), TYR385 (A), TRP387 (A), ARG513 (A), PHE518 (A), MET522 (A), VAL523 (A), VAL523 (A), GLY526 (A), ALA527 (A), SER530 (A), LEU531 (A), LEU534 (A).	Van der Waals, Carbon-hydrogen bond, Alkyl, Pi-Alkyl
47	−7.8704	MET196 (A), ALA199 (A), PHE200 (A), ALA202 (A), GLN203 (A), THR206 (A), HIS207 (A), PHE210 (A), THR212 (A), ASN382 (A), TYR385 (A), HIS386 (A), TRP387 (A), HIS388 (A), LEU390 (A), LEU391 (A), PHE395 (A), VAL444 (A), VAL447 (A).	Van der Waals, Carbon-hydrogen bond, Alkyl, Pi-Alkyl
44	−7.6261	MET196 (A), ALA199 (A), PHE200 (A), ALA202 (A), GLN203 (A), HIS207 (A), PHE210 (A), LEU294 (A), VAL295 (A), LEU298 (A), LEU390 (A), LEU291 (A), ASN382 (A), TYR385 (A), HIS386 (A), TRP387 (A), HIS388 (A), LEU390 (A), PHE395 (A), TYR404 (A), ILE408 (A), VAL444 (A), VAL447 (A).	Van der Waals, Carbon-hydrogen bond, Alkyl, Pi-Alkyl
49	−7.4221	MET196 (A), ALA199 (A), PHE200 (A), ALA202 (A), GLN203 (A), HIS207 (A), PHE210 (A), VAL295 (A), LEU298 (A), ASN382 (A), TYR385 (A), HIS386 (A), TRP387 (A), HIS388 (A), LEU390 (A), LEU391(A), VAL444 (A), VAL447 (A).	Van der Waals, Carbon-hydrogen bond, Alkyl, Pi-Alkyl

3.7.1. Molecular Dynamic Simulation Analysis

Based on the most stable and best-scored conformation, the compound (**47**) bounded with COX-2 enzyme complex was subjected to MD simulation analysis for 100 ns and retrieved very good results. The superimposed complex at 0 ns and 100 ns is demonstrated in Figure 6. The calculated RMSD value of the superimposed complex is 1.157 Å, and it is a very good value means there is minor fluctuation observed during the MD simulation system at 100 ns; hence, to differentiate the superimposed structures, it is demonstrated in different colors, such as protein at 0 ns is presented in Cyan and ligand in yellow color, while protein at 100 ns is presented in Sienna and ligand in coral color. Critical structural changes are shown in the zoomed view in Figure 6.

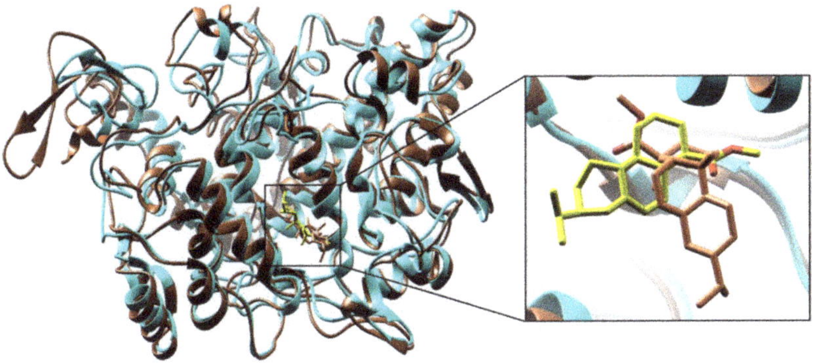

Figure 6. Superimposition of the compound (**47**) bounded with Cyclooxygenase-2 (COX-2) at 0 ns and 100 ns. The calculated RMSD value is 1.157 Å; protein at 0 ns is presented in Cyan and ligand in yellow color, while protein at 100 ns is presented in Sienna and ligand in coral color. Critical structural changes are shown in zoomed view.

During the MD simulation run, 2D plots were generated to explain the fluctuating behavior of the docked complex at different time frames during MD simulation productions, as explained in many previous studies for the identification of COX-2 inhibitors, as described by Razzaghi et al. [105] These plots are important for the MD simulation's statistical analysis; hence it could be significant to decode the backbone stability and flexibility of the residues during the different time frames of MD simulations as carried out by Razzaghi et al. [105] RMSD was calculated as the small atoms convergence from a reference state (Figure 7A), while the RMSF of a protein explained the protein's dynamic nature that contributed to the system's overall versatility and residual mobility from its mean position (Figure 7B). The average RMSD value was noted as 2.722 Å, and the average RMSF value was noted as 1.351 Å for the COX-2 inhibitor, Methyl 7-abieten-18-oate complex; the results perceived that those key amino acids residues were correlated in the binding interactions at 0 ns and 100 ns, showing maximum correlation with the docking results, and minor fluctuations of the COX-2 macromolecules residues were observed from the initial state during MD simulation runs during 0–100 ns simulation period. The radius of gyration (Rg) was evaluated to confirm the distribution of protein elements and define the equilibrium conformation of the system during MD simulations; it is expected that the high and low values of radius of gyration describe the magnitude of the system in terms of molecules tight packing scheme as reflected in the literature of Lobanov et al. [106]. The estimated average Rg of the system was 24.338 Å (Figure 7C). Beta-factor (BF) of the protein complex is a very complicated and challenging aspect; it explains the thermal residual deviation throughout the MD simulation run (Figure 7D) and seems effectively correlated with RMSF, and hereafter approves the stability of the system. BF and RMSF are complements of each other and explain the overall simulation system stability along with the aspects of elements flexibility of protein as stated by Raniolo and Limongelli [107].The average BF of the system analyzed for COX-2 inhibitor and enzyme complex is 79.941 Å.

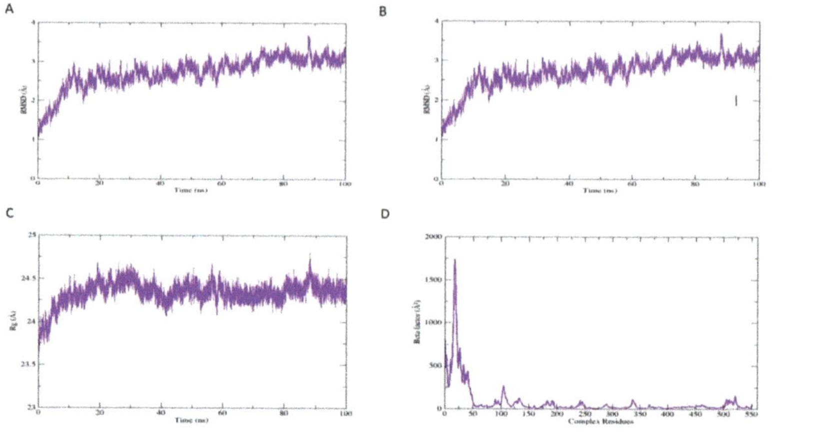

Figure 7. Root means square deviation (RMSD) of COX-2 protein and Methyl 7-abieten-18-oate complex (**A**) in-unit Angstrom (Å) is shown at Y-axis, while the variation in bonded conformation through time in nanoseconds (ns) is shown at X-axis. Root Mean Square Fluctuation (RMSF) of protein COX-2 (**B**) in-unit Angstrom (Å) is shown at the Y-axis, while the X-axis shows the index of the residue during 0–100 ns. The radius of gyration (Rg) of protein and COX-2 inhibitor complex (**C**) is shown at Y-axis in-unit Angstrom (Å), while X-axis shows the variation in bonded conformation throughout simulations. A beta factor of protein COX-2 (**D**) is shown at Y-axis in-unit Angstrom (Å), while X-axis shows the index of the residue until 100 ns.

The frequency of the HBs plays an important function in the overall stability of the protein-ligand complex and increases the capacity of HBs, which upsurges the binding attributes of the biomolecule Alkorta et al. [108]. These HBs were obtained through by VMD hydrogen bond plugin. The estimated maximum number of HBs between COX-2 residues and inhibitor atom was 9, and the average number of HBs was 1.779, respectively, during MD analysis until 0–100 ns (Figure 8A). Solvent accessible surface area (SASA) is a very expedient examination in which changes in the accessibility of the COX-2 protein to solvent were determined. The stability of each complex was perceived during the simulation, and the average value of SASA was calculated at 145.520 nm^2 (Figure 8B).

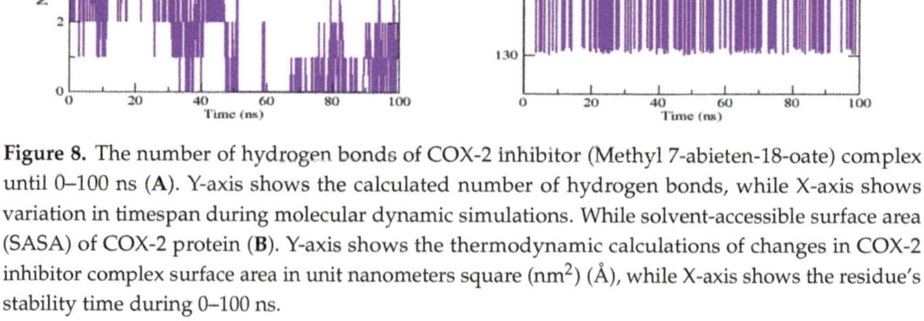

Figure 8. The number of hydrogen bonds of COX-2 inhibitor (Methyl 7-abieten-18-oate) complex until 0–100 ns (**A**). Y-axis shows the calculated number of hydrogen bonds, while X-axis shows variation in timespan during molecular dynamic simulations. While solvent-accessible surface area (SASA) of COX-2 protein (**B**). Y-axis shows the thermodynamic calculations of changes in COX-2 inhibitor complex surface area in unit nanometers square (nm^2) (Å), while X-axis shows the residue's stability time during 0–100 ns.

3.7.2. Binding Free Energies (BFE) Calculations (MM-GBSA/MM-PBSA)

In molecular mechanics (MM), PBSA and GBSA energy model generation are the acceptable and necessary techniques used in the biomolecular investigation of BFE calculations of COX_2 and inhibitor complex and support the analysis of protein folding and stability in drug design and discovery Ismail et al. [109]. The MM energy of the selected complex (ΔTOTAL) followed by generalized surface area Born (MM/GBSA) outlines the well-organized estimations deprived of any loss in terms of accuracy and Poisson-Boltzmann (MM/PBSA) benchmark parameters (Genheden and Ryde, 2015), could be significant to expose the promising macromolecule COX-2—methyl 7-abieten-18-oate ligand complex in pure water. The total energy estimated for the MM/GBSA and MM/PBSA model of the Methyl 7-abieten-18-oate complex is −50.9841 kcal and −42.4442 kcal/mol, respectively. For molecular mechanics' energy parameters, more influence was observed from the gas-phase energy (ΔG gas) when assessed to extremely irrelevant influences from the solvation energy (ΔG solv). In the MM/GBSA model of Methyl 7-abieten-18-oate-complex, the ΔG gas energy was estimated as −92.2470 kcal/mol; however, in the MM/PBSA model with minor difference energy parameters, estimation was recorded as −92.2092 kcal/mol. For the Methyl 7-abieten-18-oate-complex, the ΔG solv energy for the MM/GBSA model was projected as 40.6321 kcal/mol, though in the case of MM/PBSA, 50.4040 kcal/mol was noticed. For Methyl 7-abieten-18-oate-complex, the electrostatic forces play a role in the stability of the complex, and the system estimated by the MM forces field in PBSA is in the range of total premeditated energy calculated as −32.3777 kcal/mol for Methyl 7-abieten-18-oate-complex; for GBSA models, it is −32.1854 kcal/mol. Similarly, the Van der Waals forces are also estimated from MM associated with the system stability as −60.8915 kcal/mol and −60.2943 kcal/mol. For the Methyl 7-abieten-18-oate-complex complex, the electrostatic energy impact (EGB and EPB) to the ΔG solv was observed, and the principal constraint following towards the out-of-the-range calculations in MM/GBSA solv energy. The surface area energy highlighted as ESURF computed in the MM/GBSA model is −6.0429 kcal/mol. In the MM/PBSA model, the two important ENPOLAR and EDISPER, the repulsive and attractive free energies presented as −4.6961 kcal/mol and Zero kcal/mol. For Methyl 7-abieten-18-oate-complex complex, the electrostatic energy impact, EGB, and EPB to the ΔG solv were counted as non-favorable results in MM/GBSA solv energy. The individual total bounded conformation presenting the free energy for the selected COX-2 macromolecule as receptor and selected phytochemicals complex as a ligand (47) are explained in Table 6.

Table 6. Binding free energies of the COX-2 protein and selected potential bioactive compound complex.

MM/GBSA Model				MM/PBSA Model			
Energy (E) Component	**Average Values**	**Standard Deviation Values**	**Standard Error of Mean Values**	**Energy (E) Components**	**Average Values**	**Standard Deviation Values**	**Standard Error of Mean Values**
Van der Waals	−60.8915	2.3527	0.2469	Van der Waals	−60.2943	4.3561	0.4356
EEL	−32.3777	5.9753	0.7345	EEL	−32.1854	5.9826	0.7432
EGB	46.3402	4.5936	0.5086	EPB	55.1001	5.1322	0.4423
ESURF	−6.0429	0.1611	0.0163	ENPOLAR	−4.6961	0.1102	0.0110
				EDISPER	0	0	0
ΔG gas	−92.2470	6.5432	0.6949	ΔG gas	−92.2092	6.3912	0.6391
ΔG solv	40.6321	4.6615	0.4661	ΔG solv	50.4040	5.0137	0.5014
Δtotal	−50.9841	4.1324	0.4319	ΔTOTAL	−42.4442	5.2303	0.5140

3.7.3. In Silico Pharmacokinetic/ADMET Profile Calculations

Many studies explained the importance of pharmacokinetic/ADMET profile estimation as carried out by Chandrasekaran et al. [110] for the screening of databases to identify potential drug-like and lead-like compounds that could be better tolerable in the discovery and development of novel drug candidates for the management of inflammatory diseases, as reflected in the literature by Sudha et al. [111], ADMET properties are calculated by SwissADME, as described by Daina et al. [53], and Data-warrior tools, as performed by Sander et al. [54] Physicochemical properties of the selected compound, such as molecular weight, partition coefficient/lipophilic parameters (logP values), hydrogen bond acceptor, hydrogen bond donor, total polar surface area, molar refractivity, and rotatable bond are important drug-like characteristics calculated for the selected compound as documented by Chandrasekaran et al. [110]. In the initial phase, drug discovery protocols promote the calculation of drug-likeness, water-solubility, pharmacokinetics, and toxicity estimations, along with the medicinal chemistry perspective, as mentioned for selected compounds in Table 7. The enlisted chemical descriptors of the selected compound are in the acceptable range without any violation, as described by Lipinski [112] and Veber et al. [113] theory of drug-likeness. Lipophilicity and water solubility classes also presented very good outcomes. Gastrointestinal drug absorption as stated by Kimura and Higaki, [114] and blood-brain barrier permeability [115] are also in the range of acceptable pharmacokinetic parameters. Compound **47** showed CYP2C19 and CYP2C9 inhibitory potential and presented nonsubstrate characteristics for not P-glycoprotein. Prediction of Log Kp value (skin permeation), as reported by Alonso et al. [116], is good for selected compound −4.15 cm/s, PAINS alert and Brenk alert, supported by the medicinal chemistry parameter evaluation, as reported by Bibi et al. [117]. Our results enlisted the minor violations as Brenk alert; one isolated alkene entity is required to be optimized. It is suggested to improve before moving a drug to the next phase of development. Synthetically compound **47** is highly accessible with a score of 4.69, and in silico toxicity estimations with four major aspects of mutagenicity, tumorigenicity, irritant, and reproductive effects are completed in an acceptable range. Therefore, this selected lead compound has presented very good ADMET results, as given in Table 7.

Table 7. Summary of in silico ADMET profile estimated for selected one compound.

Descriptors	Methyl 7-Abieten-18-Oate (47)
Formula	$C_{21}H_{34}O_2$
Molecular weight	318.49 g/mol
Number of rotatable bonds	3
Number of hydrogen bond acceptors	2
Number of hydrogen bond donors	0
Molar refractivity	97.01
Total polar surface area	26.30 Å^2
Lipophilicity (Log P)	4.71
Water Solubility (Log S)	−4.46
Solubility Class	Moderately soluble
Drug-pharmacokinetic parameters	
Gastrointestinal absorption	Yes
Blood-brain barrier permeability	Yes
P-glycoprotein substrate	No
CYP1A2 inhibitor	No
CYP2C19 inhibitor	Yes
CYP2C9 inhibitor	Yes
CYP2D6 inhibitor	No
CYP3A4 inhibitor	No
Log Kp (skin permeation)	−4.15 cm/s
Drug-like characteristics	
Lipinski rule	Acceptable
Veber rule	Acceptable
Drug-likeness	Yes
Drug-likeness score	0.55
Parameters of medicinal chemistry	
PAINS alert	None
Brenk alert	1 alert: isolated alkene
Synthetic accessibility score	4.69
Toxicity	None

4. Conclusions

Due to their diverse health-promoting benefits, Physico-chemical, biological properties, and versatile chemical structures, essential oils are currently gaining considerable scientific attention as alternative effective and safe therapeutic candidates for a variety of disorders. The SEEO are reported for the first time for in vitro and in vivo pharmacological activities, and they showed significant antimicrobial activity, the ability to scavenge free radicals, a significant effect on pain, and the potential to cure inflammation when compared to their standards. The results indicate that the essential oils of *S. edelbergii* contain a bioactive chemical constituent, Methyl 7-abieten-18-oate (**47**), which has the potential to act as a possible candidate molecule against microbes, as an antioxidant, and as an effective pain reliever and anti-inflammatory. Computational analysis has further validated the anti-inflammatory potential of methyl 7-abieten-18-oate to suppress COX-2 enzyme activity. COX-2 protein and methyl 7-abieten-18-oate docked results were very good in terms of docked score (−7.8704 kcal/mol) and binding interactions with the functional residues; furthermore, when subjected to MD simulation (for 100 ns), they presented correlation with docking results with minor fluctuations. ADMET characteristics additionally supported that compound **47** could be recommended for further investigations in a clinical test; it could be a future drug for the management of inflammation, which is the cause of other metabolic disorders, so it is highly required to be controlled. Furthermore, more research is needed to identify the accountable constituents with the capacity to carry out the observed activities.

Author Contributions: M.S., W.M. and N.U.R. designed the project and wrote the original manuscript. S.B. performed the computational screening, and Z.K. carried out in vitro activities. J.N.A.-S. and T.A. extracted the essential oils, performed the GC-MS analysis, and interpreted the data. M.S., O.U., W.M. and N.U.R. conducted the in vivo activities. W.M., N.U.R. and A.A.-H. supervised the project and assisted in reviewing and editing the manuscript. All authors have read and agreed to the published version of the manuscript.

Funding: This research was funded by The Research Council (TRC) through the funded project (BFP/RGP/CBS/21/002). The APC was funded by the same project.

Institutional Review Board Statement: The study was conducted according to the guidelines of approved by the ethical committee (No: AWKUM/Bot/2019/1720, Date: 29 January 2019) of Abdul Wali Khan University Mardan, Mardan (AWKUM).

Informed Consent Statement: Not applicable.

Data Availability Statement: The data presented in this study are available in the article.

Acknowledgments: The authors would also like to give sincere appreciation to The Research Council (TRC) through the funded project (BFP/RGP/CBS/21/002) and the Natural and Medical Sciences Research Center, University of Nizwa, Nizwa, Oman for providing lab facilities.

Conflicts of Interest: The authors declare no conflict of interest.

References

1. Shah, M.; Murad, W.; Ur Rehman, N.; Halim, S.A.; Ahmed, M.; Rehman, H.; Zahoor, M.; Mubin, S.; Khan, A.; Nassan, M.A. Biomedical applications of *Scutellaria edelbergii* Rech. f.: In vitro and in vivo approach. *Molecules* **2021**, *26*, 3740. [CrossRef]
2. Shah, M.; Mubin, S.; Tagde, P.; Ullah, O.; Rahman, M.; Al-Harrasi, A.; Rehman, N.U.; Murad, W. Phytochemical Profiling and Bio-Potentiality of Genus *Scutellaria*: Biomedical Approach. *Biomolecules* **2022**, *12*, 936. [CrossRef]
3. Cazella, L.N.; Glamoclija, J.; Soković, M.; Gonçalves, J.E.; Linde, G.A.; Colauto, N.B.; Gazim, Z.C. Antimicrobial activity of essential oil of *Baccharis dracunculifolia* DC (Asteraceae) aerial parts at flowering period. *Front. Plant Sci.* **2019**, *10*, 27. [CrossRef]
4. Chebbac, K.; Moussaoui, A.E.; Bourhia, M.; Salamatullah, A.M.; Alzahrani, A.; Guemmouh, R. Chemical analysis and antioxidant and antimicrobial activity of essential oils from *Artemisia negrei* L. *Against drug-resistant microbes Evid. Based Complement. Alternat. Med.* **2021**, *2021*, 5902851.
5. Miguel, M.G. Antioxidant and anti-inflammatory activities of essential oils: A short review. *Molecules* **2010**, *15*, 9252–9287. [CrossRef]
6. De Cássia da Silveira e Sá, R.; Lima, T.C.; da Nobrega, F.R.; de Brito, A.E.M.; de Sousa, D.P. Analgesic-like activity of essential oil constituents: An update. *Int. J. Mol. Sci.* **2017**, *18*, 2392. [CrossRef]

7. Scuteri, D.; Hamamura, K.; Sakurada, T.; Watanabe, C.; Sakurada, S.; Morrone, L.A.; Rombolà, L.; Tonin, P.; Bagetta, G.; Corasaniti, M.T. Efficacy of essential oils in pain: A systematic review and meta-analysis of preclinical evidence. *Front. Pharmacol.* **2021**, *12*, 640128. [CrossRef]
8. Abdollahi, A.; Koohpayeh, S.; Najafipoor, S.; Mansoori, Y.; Abdollahi, S.; Jaafari, S. Evaluation of drug Resistance and *Staphylococcal cassette* chromosome (SCCmec) types among methicillin-Resistant *Staphylococcus aureus* (MRSA). *Alborz Univ. Med. J.* **2012**, *1*, 47–52. [CrossRef]
9. Shah, M.; Murad, W.; Mubin, S.; Ullah, O.; Rehman, N.U.; Rahman, M. Multiple health benefits of curcumin and its therapeutic potential. *Environ. Sci. Pollut. Res.* **2022**, *29*, 43732–43744. [CrossRef]
10. Jideani, A.I.; Silungwe, H.; Takalani, T.; Omolola, A.O.; Udeh, H.O.; Anyasi, T.A. Antioxidant-rich natural fruit and vegetable products and human health. *Int. J. Food Prop.* **2021**, *24*, 41–67. [CrossRef]
11. Ndoye Foe, F.M.-C.; Tchinang, T.F.K.; Nyegue, A.M.; Abdou, J.-P.; Yaya, A.J.G.; Tchinda, A.T.; Essame, J.-L.O.; Etoa, F.-X. Chemical composition, in vitro antioxidant and anti-inflammatory properties of essential oils of four dietary and medicinal plants from Cameroon. *BMC Complement. Altern. Med.* **2016**, *16*, 117. [CrossRef]
12. Kabir, M.T.; Rahman, M.H.; Shah, M.; Jamiruddin, M.R.; Basak, D.; Al-Harrasi, A.; Bhatia, S.; Ashraf, G.M.; Najda, A.; El-Kott, A.F. Therapeutic promise of carotenoids as antioxidants and anti-inflammatory agents in neurodegenerative disorders. *Biomed. Pharmacother.* **2022**, *146*, 112610. [CrossRef] [PubMed]
13. Ismael, J.; Dessalegn, E.; Fereja, W.M. In Vitro antioxidant and antibacterial activity of leaf extracts of *Measa lanceolata*. *Int. J. Food Prop.* **2021**, *24*, 702–712. [CrossRef]
14. Basit, A.; Shutian, T.; Khan, A.; Khan, S.M.; Shahzad, R.; Khan, A.; Khan, S.; Khan, M. Anti-inflammatory and analgesic potential of leaf extract of *Justicia adhatoda* L.(Acanthaceae) in Carrageenan and Formalin-induced models by targeting oxidative stress. *Biomed. Pharmacother.* **2022**, *153*, 113322. [CrossRef] [PubMed]
15. Sarmento-Neto, J.F.; Do Nascimento, L.G.; Felipe, C.F.B.; De Sousa, D.P. Analgesic potential of essential oils. *Molecules* **2015**, *21*, 20. [CrossRef] [PubMed]
16. De Sousa, D.P. Analgesic-like activity of essential oils constituents. *Molecules* **2011**, *16*, 2233–2252. [CrossRef]
17. Liaqat, I.; Riaz, N.; Saleem, Q.-u.-A.; Tahir, H.M.; Arshad, M.; Arshad, N. Toxicological evaluation of essential oils from some plants of Rutaceae family. *Evid. Based Complement. Alternat. Med.* **2018**, *2018*, 4394687. [CrossRef]
18. Kamte, S.L.N.; Ranjbarian, F.; Cianfaglione, K.; Sut, S.; Dall'Acqua, S.; Bruno, M.; Afshar, F.H.; Iannarelli, R.; Benelli, G.; Cappellacci, L. Identification of highly effective antitrypanosomal compounds in essential oils from the Apiaceae family. *Ecotoxicol. Environ. Saf.* **2018**, *156*, 154–165. [CrossRef]
19. Chopra, H.; Dey, P.S.; Das, D.; Bhattacharya, T.; Shah, M.; Mubin, S.; Maishu, S.P.; Akter, R.; Rahman, M.H.; Karthika, C. Curcumin nanoparticles as promising therapeutic agents for drug targets. *Molecules* **2021**, *26*, 4998. [CrossRef]
20. Zuzarte, M.; Sousa, C.; Cavaleiro, C.; Cruz, M.T.; Salgueiro, L. The Anti-Inflammatory Response of *Lavandula luisieri* and *Lavandula pedunculata* Essential Oils. *Plants* **2022**, *11*, 370. [CrossRef]
21. Cantino, P.; Harley, R.; Wagstaff, S. Genera of Labiatae: Status and classification. *Adv.in Labiate Sci.* **1992**, *11*, 511–522.
22. Ghavam, M.; Manca, M.L.; Manconi, M.; Bacchetta, G. Chemical composition and antimicrobial activity of essential oils obtained from leaves and flowers of *Salvia hydrangea* DC. ex Benth. *Sci. Rep.* **2020**, *10*, 15647. [CrossRef] [PubMed]
23. Zahra, G.; Khadijeh, B.; Hossein, D.; Ali, S. Essential oil composition of two *Scutellaria* species from Iran. *J. Trad. Chinese Med. Sci.* **2019**, *6*, 244–253. [CrossRef]
24. Mamadalieva, N.Z.; Sharopov, F.; Satyal, P.; Azimova, S.S.; Wink, M. Composition of the essential oils of three Uzbek *Scutellaria* species (Lamiaceae) and their antioxidant activities. *Nat. Prod. Res.* **2017**, *31*, 1172–1176. [CrossRef]
25. Georgieva, Y.; Katsarova, M.; Stoyanov, P.; Mladenov, R.; Denev, P.; Teneva, D.; Plotnikov, E.; Bozov, P.; Dimitrova, S. Metabolite profile and antioxidant activity of some species of Genus *Scutellaria* growing in Bulgaria. *Plants* **2020**, *10*, 45. [CrossRef]
26. Shah, M.; Rahman, H.; Khan, A.; Bibi, S.; Ullah, O.; Ullah, S.; Ur Rehman, N.; Murad, W.; Al-Harrasi, A. Identification of α-Glucosidase Inhibitors from *Scutellaria edelbergii*: ESI-LC-MS and Computational Approach. *Molecules* **2022**, *27*, 1322. [CrossRef]
27. Utegenova, G.A.; Pallister, K.B.; Kushnarenko, S.V.; Özek, G.; Özek, T.; Abidkulova, K.T.; Kirpotina, L.N.; Schepetkin, I.A.; Quinn, M.T.; Voyich, J.M. Chemical composition and antibacterial activity of essential oils from *Ferula* L. species against methicillin-resistant *Staphylococcus aureus*. *Molecules* **2018**, *23*, 1679. [CrossRef]
28. Surendran, S.; Qassadi, F.; Surendran, G.; Lilley, D.; Heinrich, M. Myrcene—what are the potential health benefits of this flavouring and aroma agent? *Front. Nutr.* **2021**, *400*, 1–14. [CrossRef]
29. Aati, H.; El-Gamal, A.; Kayser, O. Chemical composition and biological activity of the essential oil from the root of *Jatropha pelargoniifolia* Courb. native to Saudi Arabia. *Saudi Pharm. J.* **2019**, *27*, 88–95. [CrossRef]
30. Rehman, N.U.; Shah, M.; Ullah, S.; Khan, M.; Khan, A.; Ullah, O.; Hussain, J.; Al-Harrasi, A. Enzymes Inhibition and Antioxidant Potential of Medicinal Plants Growing in Oman. *BioMed Res. Int.* **2022**, *2022*, 7880387. [CrossRef]
31. Shah, M.; Murad, W.; Ur Rehman, N.; Mubin, S.; Al-Sabahi, J.N.; Ahmad, M.; Zahoor, M.; Ullah, O.; Waqas, M.; Ullah, S. GC-MS analysis and Biomedical therapy of oil from n-hexane fraction of *Scutellaria edelbergii* Rech. f.: In vitro, in vivo, and in silico approach. *Molecules* **2021**, *26*, 7676. [CrossRef] [PubMed]
32. Rehman, N.U.; Alsabahi, J.N.; Alam, T.; Khan, A.; Rafiq, K.; Khan, M.; Al-Harrasi, A. Chemical constituents and carbonic anhydrase II activity of essential oil of *Acridocarpus orientalis* A. Juss. in comparison with stem and leaves. *J. Essent. Oil Bear. Plants* **2021**, *24*, 68–74. [CrossRef]

33. Adams, R.P. *Identification of Essential Oil Components by Gas Chromatography/Mass Spectrometry*; Allured Publishing Corporation: Carol Stream, IL, USA, 2007.
34. Kedare, S.B.; Singh, R. Genesis and development of DPPH method of antioxidant assay. *J. Food Sci. Technol.* **2011**, *48*, 412–422. [CrossRef]
35. Mondal, M.; Hossain, M.S.; Das, N.; Khalipha, A.B.R.; Sarkar, A.P.; Islam, M.T.; Smrity, S.Z.; Biswas, S.; Kundu, S.K. Phytochemical screening and evaluation of pharmacological activity of leaf Methanolic extract of *Colocasia affinis* Schott. *Clin. Phytoscience* **2019**, *5*, 8. [CrossRef]
36. Vane, J.R.; Botting, R.M. Anti-inflammatory drugs and their mechanism of action. *Inflam. Res.* **1998**, *47*, 78–87. [CrossRef]
37. Bibi, S.; Sakata, K. Current status of computer-aided drug design for type 2 diabetes. *Curr. Comput. Aided Drug Des.* **2016**, *12*, 167–177. [CrossRef]
38. Bibi, S.; Hasan, M.M.; Wang, Y.-B.; Papadakos, S.P.; Yu, H. Cordycepin as a Promising Inhibitor of SARS-CoV-2 RNA dependent RNA polymerase (RdRp). *Curr. Med. Chem.* **2022**, *29*, 152–162. [CrossRef]
39. Milne, G.W. *Software Review of ChemBioDraw 12.0*; CambridgeSoft, 100 CambridgePark Drive; ACS Publications: Cambridge, MA, USA, 2010.
40. Bondhon, T.; Fatima, A.; Jannat, K.; Hasan, A.; Jahan, R.; Nissapatorn, V.; Wiart, C.; Pereira, M.; Rahmatullah, M. In silico screening of *Allium cepa* phytochemicals for their binding abilities to SARS and SARS-CoV-2 3C-like protease and COVID-19 human receptor ACE-2. *Trop. Biomed.* **2021**, *38*, 214–221.
41. Orlando, B.J.; Malkowski, M.G. Crystal structure of rofecoxib bound to human cyclooxygenase-2. *Acta Crystallogr. Sect. F Struct. Biol. Commun.* **2016**, *72*, 772–776. [CrossRef]
42. Claria, J. Cyclooxygenase-2 biology. *Curr. Pharm. Des.* **2003**, *9*, 2177–2190. [CrossRef]
43. Discovery Studio. *Accelrys*; BIOVIA: Cambridge, UK, 2008.
44. Katsila, T.; Spyroulias, G.A.; Patrinos, G.P.; Matsoukas, M.-T. Computational approaches in target identification and drug discovery. *Comput. Struct. Biotechnol. J.* **2016**, *14*, 177–184. [CrossRef]
45. Vilar, S.; Cozza, G.; Moro, S. Medicinal chemistry and the molecular operating environment (MOE): Application of QSAR and molecular docking to drug discovery. *Curr. Top. Med. Chem.* **2008**, *8*, 1555–1572. [CrossRef]
46. Aldeghi, M.; Heifetz, A.; Bodkin, M.J.; Knapp, S.; Biggin, P.C. Accurate calculation of the absolute free energy of binding for drug molecules. *Chem. Sci.* **2016**, *7*, 207–218. [CrossRef]
47. Ahmad, S.; Ranaghan, K.E.; Azam, S.S. Combating tigecycline resistant Acinetobacter baumannii: A leap forward towards multi-epitope based vaccine discovery. *Eur. J. Pharm. Sci.* **2019**, *132*, 1–17. [CrossRef]
48. Lee, T.-S.; Allen, B.K.; Giese, T.J.; Guo, Z.; Li, P.; Lin, C.; McGee, T.D., Jr.; Pearlman, D.A.; Radak, B.K.; Tao, Y. Alchemical binding free energy calculations in AMBER20: Advances and best practices for drug discovery. *J. Chem. Inf. Model.* **2020**, *60*, 5595–5623. [CrossRef]
49. Bergonzo, C.; Cheatham III, T.E. Improved force field parameters lead to a better description of RNA structure. *J. Chem. Theory Comput.* **2015**, *11*, 3969–3972. [CrossRef]
50. Kräutler, V.; Van Gunsteren, W.F.; Hünenberger, P.H. A fast Shake algorithm to solve distance constraint equations for small molecules in molecular dynamics simulations. *J. Comput. Chem.* **2001**, *22*, 501–508. [CrossRef]
51. Roe, D.R.; Cheatham, T.E., III. PTRAJ and CPPTRAJ: Software for processing and analysis of molecular dynamics trajectory data. *J. Chem. Theory Comput.* **2013**, *9*, 3084–3095. [CrossRef]
52. Humphrey, W.; Dalke, A.; Schulten, K. VMD: Visual molecular dynamics. *J. Mol. Graph.* **1996**, *14*, 33–38. [CrossRef]
53. Daina, A.; Michielin, O.; Zoete, V. iLOGP: A simple, robust, and efficient description of n-octanol/water partition coefficient for drug design using the GB/SA approach. *J. Chem. Inf. Model.* **2014**, *54*, 3284–3301. [CrossRef]
54. Sander, T.; Freyss, J.; von Korff, M.; Rufener, C. DataWarrior: An open-source program for chemistry aware data visualization and analysis. *J. Chem. Inf. Mod.* **2015**, *55*, 460–473. [CrossRef]
55. Khan, M.S.; Mehmood, B.; Yousafi, Q.; Bibi, S.; Fazal, S.; Saleem, S.; Sajid, M.W.; Ihsan, A.; Azhar, M.; Kamal, M.A. Molecular docking studies reveal rhein from rhubarb *(rheum rhabarbarum)* as a putative inhibitor of ATP-binding cassette super-family G member 2. *Med. Chem.* **2021**, *17*, 273–288. [CrossRef]
56. Bekana, D.; Kebede, T.; Assefa, M.; Kassa, H. Comparative phytochemical analyses of resins of *Boswellia* species (*B. papyrifera* (Del.) Hochst., *B. neglecta S. Moore*, and *B. rivae* Engl.) from northwestern, southern, and southeastern Ethiopia. *Int. Sch. Res. Not.* **2014**, *2014*, 374678. [CrossRef]
57. Kurkcuoglu, M.; Yildiz, G.; Kose, Y.B. Essential Oil Composition of Two *Scutellaria* species from Tokat, Turkey. *J. Turkish Chem. Soc. Sec. A Chem.* **2019**, *6*, 115–118. [CrossRef]
58. Cicek, M.; Demirci, B.; Yilmaz, G.; Baser, K.H.C. Essential oil composition of three species of *Scutellaria* from Turkey. *Nat. Prod. Res.* **2011**, *25*, 1720–1726. [CrossRef]
59. Formisano, C.; Rigano, D.; Senatore, F.; Piozzi, F.; Arnold, N.A. Analysis of essential oils from *Scutellaria orientalis* ssp. alpina and *S. utriculata* by GC and GC-MS. *Nat. Prod. Commun.* **2011**, *6*, 919–932. [CrossRef]
60. Hussain, H.; Al-Harrasi, A.; Al-Rawahi, A.; Hussain, J. Chemistry and biology of essential oils of genus *Boswellia*. *Evid. Based Complement. Alternat. Med.* **2013**, *2013*, 140509. [CrossRef]
61. Mahmood, A.; Ahmed, R.; Kosar, S. Phytochemical screening and biological activities of the oil components of *Prunus domestica* Linn. *J. Saudi Chem. Soc.* **2009**, *13*, 273–277. [CrossRef]

62. Kasaian, J.; Alesheikh, P.; Mohammadi, A. Chemical compositions and biological activities of *Scutellaria* genus essential oils (lamiaceae). *Jundishapur J. Nat. Pharm. Prod.* **2020**, *15*, 62279. [CrossRef]
63. Yilmaz, G.; Cice, M.; Demirci, B.; Baser, K.H.C. Composition of the Essential Oils of *Scutellaria galericulata* and *S. tortumensis* from Turkey. *Nat. Volat. Essent. Oils* **2019**, *6*, 1–7.
64. Lawson, S.K.; Satyal, P.; Setzer, W.N. Phytochemical analysis of the essential oils from aerial parts of four *Scutellaria* "Skullcap" species cultivated in south Alabama: *Scutellaria baicalensis* Georgi, *S. barbata* D. Don, *S. incana* Biehler, and *S. lateriflora* L. *Nat. Prod. Commun.* **2021**, *16*, 193–259. [CrossRef]
65. Seow, Y.X.; Yeo, C.R.; Chung, H.L.; Yuk, H.-G. Plant essential oils as active antimicrobial agents. *Crit. Rev. Food Sci. Nutr.* **2014**, *54*, 625–644. [CrossRef]
66. Legault, J.; Pichette, A. Potentiating effect of β-caryophyllene on anticancer activity of α-humulene, isocaryophyllene and paclitaxel. *J. Pharm. Pharmacol.* **2007**, *59*, 1643–1647. [CrossRef]
67. Malti, C.E.W.; Baccati, C.; Mariani, M.; Hassani, F.; Babali, B.; Atik-Bekkara, F.; Paoli, M.; Maury, J.; Tomi, F.; Bekhechi, C. Biological activities and chemical composition of *Santolina africana* Jord. et Fourr. aerial part essential oil from Algeria: Occurrence of polyacetylene derivatives. *Molecules* **2019**, *24*, 204. [CrossRef]
68. Adnan, M.; Nazim Uddin Chy, M.; Mostafa Kamal, A.; Azad, M.O.K.; Paul, A.; Uddin, S.B.; Barlow, J.W.; Faruque, M.O.; Park, C.H.; Cho, D.H. Investigation of the biological activities and characterization of bioactive constituents of *Ophiorrhiza rugosa* var. prostrata (D. Don) & Mondal leaves through in vivo, in vitro, and in silico approaches. *Molecules* **2019**, *24*, 1367.
69. Sonigra, P.; Meena, M. Metabolic profile, bioactivities, and variations in the chemical constituents of essential oils of the *Ferula* genus (Apiaceae). *Front. Pharmacol.* **2021**, *11*, 608649. [CrossRef]
70. Gonzalez, M.A. Synthetic derivatives of aromatic abietane diterpenoids and their biological activities. *Eur. J. Med. Chem.* **2014**, *87*, 834–842. [CrossRef]
71. Da Silva Rivas, A.C.; Lopes, P.M.; de Azevedo Barros, M.M.; Costa Machado, D.C.; Alviano, C.S.; Alviano, D.S. Biological activities of α-pinene and β-pinene enantiomers. *Molecules* **2012**, *17*, 6305–6316.
72. Darwish, R.S.; Shawky, E.; El Naggar, E.M.B.; Hammoda, H.M.; Harraz, F.M. Evaluation of the effect of seasonal variation and organ selection on the chemical composition and antimicrobial activity of the essential oil of oriental-cedar (*Platyclaudus orientalis* (L.) Franco). *J. Essent. Oil Res.* **2021**, *33*, 69–79. [CrossRef]
73. Ambrosio, C.M.; Diaz-Arenas, G.L.; Agudelo, L.P.; Stashenko, E.; Contreras-Castillo, C.J.; Da Gloria, E.M. Chemical composition and antibacterial and antioxidant activity of a citrus essential oil and its fractions. *Molecules* **2021**, *26*, 2888. [CrossRef]
74. Popovici, R.A.; Vaduva, D.; Pinzaru, I.; Dehelean, C.A.; Farcas, C.G.; Coricovac, D.; Danciu, C.; Popescu, I.; Alexa, E.; Lazureanu, V. A comparative study on the biological activity of essential oil and total hydro-alcoholic extract of *Satureja hortensis* L. *Exp. Ther. Med.* **2019**, *18*, 932–942. [CrossRef]
75. Skaltsa, H.D.; Lazari, D.M.; Kyriazopoulos, P.; Golegou, S.; Triantaphyllidis, S.; Sokovic, M.; Kypriotakis, Z. Composition and antimicrobial activity of the essential oils of *Scutellaria sieberia* Benth. and *Scutellaria rupestris* Boiss. et Heldr. ssp. adenotricha (Boiss. et Heldr.) Greuter et Burdet from Greece. *J. Essent. Oil Res.* **2005**, *17*, 232–235. [CrossRef]
76. Skaltsa, H.D.; Lazari, D.M.; Mavromati, A.S.; Tiligada, E.A.; Constantinidis, T.A. Composition and antimicrobial activity of the essential oil of *Scutellaria albida* ssp. albida from Greece. *Planta Med.* **2000**, *66*, 672–674. [CrossRef]
77. Bogdan, M.A.; Bungau, S.; Tit, D.M.; Zaha, D.C.; Nechifor, A.C.; Behl, T.; Chambre, D.; Lupitu, A.I.; Copolovici, L.; Copolovici, D.M. Chemical profile, antioxidant capacity, and antimicrobial activity of essential oils extracted from three different varieties (Moldoveanca 4, Vis Magic 10, and Alba 7) of Lavandula angustifolia. *Molecules* **2021**, *26*, 4381. [CrossRef]
78. Moisa, C.; Lupitu, A.; Pop, G.; Chambre, D.R.; Copolovici, L.; Cioca, G.; Bungau, S.; Copolovici, D.M. Variation of the chemical composition of Thymus vulgaris essential oils by phenological stages. *Rev. Chim.* **2019**, *70*, 633–637. [CrossRef]
79. Zhang, J.-H.; Sun, H.-L.; Chen, S.-Y.; Zeng, L.; Wang, T.-T. Anti-fungal activity, mechanism studies on α-Phellandrene and Nonanal against *Penicillium cyclopium*. *Bot. Stud.* **2017**, *58*, 13. [CrossRef]
80. Popova, V.; Ivanova, T.; Stoyanova, A.; Nikolova, V.; Hristeva, T.; Gochev, V.; Yonchev, Y.; Nikolov, N.; Zheljazkov, V.D. Terpenoids in the essential oil and concentrated aromatic products obtained from *Nicotiana glutinosa* L. leaves. *Molecules* **2019**, *25*, 30. [CrossRef]
81. Xu, C.; Zhao, S.; Li, M.; Dai, Y.; Tan, L.; Liu, Y. Chemical composition, antimicrobial and antioxidant activities of essential oil from flue-cured tobacco flower bud. *Biotech. Biotechnol. Equip.* **2016**, *30*, 1026–1030. [CrossRef]
82. Burcova, Z.; Kreps, F.; Greifová, M.; Jablonsky, M.; Haz, A.; Schmidt, S.; Surina, I. Antibacterial and antifungal activity of phytosterols and methyl dehydroabietate of Norway spruce bark extracts. *J. Biotechnol.* **2018**, *282*, 18–24. [CrossRef]
83. Singh, G.; Maurya, S.; Catalan, C.; De Lampasona, M. Studies on essential oils, Part 42: Chemical, antifungal, antioxidant and sprout suppressant studies on ginger essential oil and its oleoresin. *Flavour Fragr. J.* **2005**, *20*, 1–6. [CrossRef]
84. Yu, J.; Lei, J.; Yu, H.; Cai, X.; Zou, G. Chemical composition and antimicrobial activity of the essential oil of *Scutellaria barbata*. *Phytochemistry* **2004**, *65*, 881–884. [CrossRef] [PubMed]
85. Zhu, X.; Han, C.; Gao, T.; Shao, H. Chemical composition, phytotoxic and antimicrobial activities of the essential oil of *Scutellaria strigillosa* Hemsley. *J. Essent. Oil Bear. Plants* **2016**, *19*, 664–670. [CrossRef]
86. Ullah, O.; Shah, M.; Rehman, N.U.; Ullah, S.; Al-Sabahi, J.N.; Alam, T.; Khan, A.; Khan, N.A.; Rafiq, N.; Bilal, S.; et al. Aroma Profile and Biological Effects of *Ochradenus arabicus* Essential Oils: A Comparative Study of Stem, Flowers, and Leaves. *Molecules* **2022**, *27*, 5197. [CrossRef] [PubMed]

87. Mot, M.-D.; Gavrilaș, S.; Lupitu, A.I.; Moisa, C.; Chambre, D.; Tit, D.M.; Bogdan, M.A.; Bodescu, A.-M.; Copolovici, L.; Copolovici, D.M.; et al. *Salvia officinalis* L. Essential Oil: Characterization, Antioxidant Properties, and the Effects of Aromatherapy in Adult Patients. *Antioxidants* **2022**, *11*, 808. [CrossRef] [PubMed]
88. Glevitzky, I.; Dumitrel, G.A.; Glevitzky, M.; Pasca, B.; Otrisal, P.; Bungau, S.; Cioca, G.; Pantis, C.; Popa, M. Statistical analysis of the relationship between antioxidant activity and the structure of flavonoid compounds. *Rev. Chim.* **2019**, *70*, 3103–3107. [CrossRef]
89. Kamatou, G.P.; Viljoen, A.M. Linalool–A review of a biologically active compound of commercial importance. *Nat. Prod. Commun.* **2008**, *3*, 193–327. [CrossRef]
90. Ghasemi, G.; Fattahi, M.; Alirezalu, A.; Ghosta, Y. Antioxidant and antifungal activities of a new chemovar of cumin (*Cuminum cyminum* L.). *Food Sci. Biotechnol.* **2019**, *28*, 669–677. [CrossRef]
91. Khalilov, L.; Paramonov, E.; Khalilova, A.; Odinokov, V.; Muldashev, A.; Baltaev, U.; Dzhemilev, U. Identification and biological activity of volatile organic compounds emitted by plants and insects. IV. Composition of vapor isolated from certain species of *Artemisia* plants. *Chem. Nat. Compd.* **2001**, *37*, 339–342. [CrossRef]
92. Uritu, C.M.; Mihai, C.T.; Stanciu, G.-D.; Dodi, G.; Alexa-Stratulat, T.; Luca, A.; Leon-Constantin, M.-M.; Stefanescu, R.; Bild, V.; Melnic, S. Medicinal plants of the family Lamiaceae in pain therapy: A review. *Pain Res. Manag.* **2018**, *2018*, 7801543. [CrossRef]
93. Chen, Y.; Zhao, Y.-Y.; Wang, X.-Y.; Liu, J.-T.; Huang, L.-Q.; Peng, C.-S. GC-MS analysis and analgesic activity of essential oil from fresh rhizoma of *Cyperus rotundus*. *J. Chinese Med. Mat.* **2011**, *34*, 1225–1229.
94. Mishra, D.; Bisht, G.; Mazumdar, P.M.; Sah, S.P. Chemical composition and analgesic activity of *Senecio rufinervis* essential oil. *Pharm. Biol.* **2010**, *48*, 1297–1301. [CrossRef] [PubMed]
95. Hameed, I.H.; Altameme, H.J.; Mohammed, G.J. Evaluation of antifungal and antibacterial activity and analysis of bioactive phytochemical compounds of *Cinnamomum zeylanicum* (*Cinnamon* bark) using gas chromatography-mass spectrometry. *Orien. J. Chem.* **2016**, *32*, 1769. [CrossRef]
96. Liang, J.; Huang, B.; Wang, G. Chemical composition, antinociceptive and anti-inflammatory properties of essential oil from the roots of *Illicium lanceolatum*. *Nat. Prod. Res.* **2012**, *26*, 1712–1714. [CrossRef] [PubMed]
97. Hajhashemi, V.; Sajjadi, S.E.; Zomorodkia, M. Antinociceptive and anti-inflammatory activities of *Bunium persicum* essential oil, hydroalcoholic and polyphenolic extracts in animal models. *Pharm. Biol.* **2011**, *49*, 146–151. [CrossRef] [PubMed]
98. Santos, F.; Jeferson, F.; Santos, C.; Silveira, E.; Rao, V. Antinociceptive effect of leaf essential oil from *Croton sonderianus* in mice. *Life Sci.* **2005**, *77*, 2953–2963. [CrossRef] [PubMed]
99. Colares, A.V.; Almeida-Souza, F.; Taniwaki, N.N.; Souza, C.d.S.F.; da Costa, J.G.M.; Calabrese, K.d.S.; Abreu-Silva, A.L. In vitro antileishmanial activity of essential oil of *Vanillosmopsis arborea* (Asteraceae) baker. *Evid. Based Complement. Alternat. Med.* **2013**, *2013*, 727042. [CrossRef]
100. Mogosan, C.; Vostinaru, O.; Oprean, R.; Heghes, C.; Filip, L.; Balica, G.; Moldovan, R.I. A comparative analysis of the chemical composition, anti-inflammatory, and antinociceptive effects of the essential oils from three species of *Mentha* cultivated in Romania. *Molecules* **2017**, *22*, 263. [CrossRef]
101. Boukhatem, M.N.; Darwish, N.H.; Sudha, T.; Bahlouli, S.; Kellou, D.; Benelmouffok, A.B.; Chader, H.; Rajabi, M.; Benali, Y.; Mousa, S.A. In vitro antifungal and topical anti-inflammatory properties of essential oil from wild-growing *Thymus vulgaris* (Lamiaceae) used for medicinal purposes in Algeria: A new source of carvacrol. *Sci. Pharm.* **2020**, *88*, 33. [CrossRef]
102. Dragomanova, S.; Tancheva, L.; Georgieva, M. A review: Biological activity of myrtenal and some myrtenal-containing medicinal plant essential oils. *Scripta Sci. Pharm.* **2018**, *5*, 22–33. [CrossRef]
103. Adeosun, T.E.; Ogunwande, I.A.; Avoseh, O.N.; Raji, I.P.; Lawal, O.A. Composition and anti-inflammatory activity of essential oil of *Jatropha curcas*. *Nat. Prod. Commun.* **2017**, *12*, 439–440. [CrossRef]
104. Apel, M.A.; Lima, M.E.; Sobral, M.; Young, M.C.M.; Cordeiro, I.; Schapoval, E.E.; Henriques, A.T.; Moreno, P.R.H. Anti-inflammatory activity of essential oil from leaves of *Myrciaria tenella* and *Calycorectes sellowianus*. *Pharm. Biol.* **2010**, *48*, 433–438. [CrossRef] [PubMed]
105. Razzaghi-Asl, N.; Mirzayi, S.; Mahnam, K.; Sepehri, S. Identification of COX-2 inhibitors via structure-based virtual screening and molecular dynamics simulation. *J. Mol. Graph. Mod.* **2018**, *83*, 138–152. [CrossRef]
106. Lobanov, M.Y.; Bogatyreva, N.; Galzitskaya, O. Radius of gyration as an indicator of protein structure compactness. *Mol. Biol.* **2008**, *42*, 623–628. [CrossRef]
107. Raniolo, S.; Limongelli, V. Ligand binding free-energy calculations with funnel metadynamics. *Nat. Protoc.* **2020**, *15*, 2837–2866. [CrossRef] [PubMed]
108. Alkorta, I.; Elguero, J.; Mó, O.; Yáñez, M.; Del Bene, J.E. Are resonance-assisted hydrogen bonds 'resonance assisted'? A theoretical NMR study. *Chem. Phys. Lett.* **2005**, *411*, 411–415. [CrossRef]
109. Ismail, S.; Ahmad, S.; Azam, S.S. Immunoinformatics characterization of SARS-CoV-2 spike glycoprotein for prioritization of epitope based multivalent peptide vaccine. *J. Mol. Liq.* **2020**, *314*, 113612. [CrossRef]
110. Chandrasekaran, B.; Abed, S.N.; Al-Attraqchi, O.; Kuche, K.; Tekade, R.K. Computer-aided prediction of pharmacokinetic (ADMET) properties. In *Dosage Form Design Parameters*; Elsevier: Amsterdam, Netherlands, 2018; pp. 731–755.
111. Sudha, K.N.; Shakira, M.; Prasanthi, P.; Sarika, N.; Kumar, C.N.; Babu, P.A. Virtual screening for novel COX-2 inhibitors using the ZINC database. *Bioinformation* **2008**, *2*, 325. [CrossRef]
112. Lipinski, C.A. Lead-and drug-like compounds: The rule-of-five revolution. *Drug Dis. Today Technol.* **2004**, *1*, 337–341. [CrossRef]

113. Veber, D.F.; Johnson, S.R.; Cheng, H.-Y.; Smith, B.R.; Ward, K.W.; Kopple, K.D. Molecular properties that influence the oral bioavailability of drug candidates. *J. Med. Chem.* **2002**, *45*, 2615–2623. [CrossRef]
114. Kimura, T.; Higaki, K. Gastrointestinal transit and drug absorption. *Biol. Pharm. Bull.* **2002**, *25*, 149–164. [CrossRef]
115. Mayhan, W.G. Regulation of blood–brain barrier permeability. *Microcirculation* **2001**, *8*, 89–104. [PubMed]
116. Alonso, C.; Carrer, V.; Espinosa, S.; Zanuy, M.; Cordoba, M.; Vidal, B.; Domínguez, M.; Godessart, N.; Coderch, L.; Pont, M. Prediction of the skin permeability of topical drugs using in silico and in vitro models. *Eur. J. Pharm. Sci.* **2019**, *136*, 104945. [CrossRef] [PubMed]
117. Bibi, S.; Sakata, K. An integrated computational approach for plant-based protein tyrosine phosphatase non-receptor type 1 inhibitors. *Curr. Comput. Aided Drug Des.* **2017**, *13*, 319–335. [CrossRef] [PubMed]

antioxidants

MDPI

Review

Reviewing the Traditional/Modern Uses, Phytochemistry, Essential Oils/Extracts and Pharmacology of *Embelia ribes* Burm.

Vineet Sharma [1,†], Dev Nath Singh Gautam [1,*], Andrei-Flavius Radu [2,3,†], Tapan Behl [4], Simona Gabriela Bungau [2,5,*] and Cosmin Mihai Vesa [3]

1 Department of Rasa Shastra & Bhaishajya Kalpana, Faculty of Ayurveda, Institute of Medical Sciences, Banaras Hindu University, Varanasi 221005, India; vinitbhu93@gmail.com
2 Doctoral School of Biomedical Sciences, Faculty of Medicine and Pharmacy, University of Oradea, 410073 Oradea, Romania; andreiflavius.radu@gmail.com
3 Department of Preclinical Disciplines, Faculty of Medicine and Pharmacy, University of Oradea, 410073 Oradea, Romania; v_cosmin_15@yahoo.com
4 Department of Pharmacology, Chitkara College of Pharmacy, Chitkara University, Punjab 140401, India; tapanbehl31@gmail.com
5 Department of Pharmacy, Faculty of Medicine and Pharmacy, University of Oradea, 410028 Oradea, Romania
* Correspondence: drdnsgautam@gmail.com (D.N.S.G.); sbungau@uoradea.ro (S.G.B.)
† These authors contributed equally to this work.

Abstract: Objectives: *Embelia ribes* Burm. (*E. ribes*, Myrsinaceae), also known as Vidanga in Ayurveda, has been shown to have significant therapeutic benefits on several disorders, and its main chemical bioactive constituent, embelin, has the therapeutic potential to be converted into innovative drugs, which is why it has recently received considerable interest. In the present work, we provide a higher level of comprehension, awareness, and extensive knowledge of the traditional uses, phytochemistry, and pharmacological characteristics of *E. ribes* throughout the last several decades (February 1965 to June 2021), emphasizing the importance of the study of essential oils extracted from *E. ribes*, which show a major potential for exerting antioxidant and anti-inflammatory activity. Materials and Methods: Google Scholar, MEDLINE, EMBASE, Scifinder, Scopus, and ScienceDirect were used to conduct a thorough literature search. Results: *E. ribes* is high in essential oils, alkaloids, flavonoids, steroids, and phenolics, all of which have medicinal benefits. The essential oils/extracts and isolated chemical constituents exhibited antioxidant activity, wound healing, antidiabetic, central nervous system (CNS)-related disease, antiviral, antiobesity, cardioprotective, antifungal, antibacterial, and antifertility activity, among other promising pharmacological effects. Conclusion: The translation between traditional applications and modern medicine may make *E. ribes* a promising target for the implementation of innovative medication. To investigate the efficacy and safety profile of *E. ribes*, further high-quality preclinical studies using advanced methodologies are required.

Keywords: *Embelia ribes* Burm.; Vidanga; antioxidants; essential oils/extracts; embelin; vilangin; pharmacology; phytochemistry; plant-based compounds

Citation: Sharma, V.; Gautam, D.N.S.; Radu, A.-F.; Behl, T.; Bungau, S.G.; Vesa, C.M. Reviewing the Traditional/Modern Uses, Phytochemistry, Essential Oils/Extracts and Pharmacology of *Embelia ribes* Burm. *Antioxidants* **2022**, *11*, 1359. https://doi.org/antiox11071359

Academic Editors: Filomena Nazzaro and Monica Rosa Loizzo

Received: 1 June 2022
Accepted: 12 July 2022
Published: 13 July 2022

1. Introduction

Embelia ribes Burm. (*E. ribes*) is a woody shrub from the family Myrsinaceae, commonly known as Vidanga, and it has been used in traditional medicine due to several observed effects, including analgesic, anthelmintic, antioxidant, antibacterial, antidiabetic, anticancer, antihyperlipidemic, wound healing, and anti-spermatogenic activity, etc. [1,2]. It is a plant widely distributed in Cambodia, south China, India, Laos, Malaysia, Sri Lanka, Thailand, Vietnam, etc. In the mature fruits, the globular berries are dark red to almost black in color, with a tiny beak-like protrusion at the apex and five-lobed persisting calyxes. The epicarp's thin epidermis emerged warty, and the middle wide mesocarp is composed of large tabular parenchymatous tissues, widely dispersed groups of stony cells and fibrovascular bundles. Simple, druse crystals of calcium oxalate and starch grains (elliptical-shaped) were also identified. Layers of brachy sclereids with a pyramidal shape make up the endocarp. Endocarp containing a single seed is surrounded by stony endocarp. The seeds' bases are depressed inward incursions of the perisperm and ruminating endosperm, and they are mottled with yellowish brown dots [3].

Embelin (with IUPAC name 2,5 dihydroxy-3-undecyl-1,4 benzoquinone) is an important bioactive substance found in the *Embelia* genus. Hepatoprotective, anti-inflammatory, antioxidant, antimitotic, radioprotective, anticancer, contraceptive, anti-spermatogenic, anti-infective, antihyperlipidemic, antihyperglycemic, analgesic, antipyretic, and wound healing activity are included in embelin's therapeutic profile [4].

Traditional Chinese medicine includes ethnic medicine as an important component. Due to their consistent therapeutic results and low clinical toxicity, ethnic medicine and its formulations have gained wide acceptance in Asian and Western countries in recent years. Therefore, considerable research is focused on the evaluation of ethnic remedies. *E. ribes* exhibits a variety of biological actions, among the most promising of which are its anti-tumor, antioxidation, and anti-inflammatory properties, according to pharmacological examinations [5].

The literature on this topic can be described as insufficiently developed, with few published papers focusing exhaustively and extensively on *E. ribes* (as it can be also seen from the number of the final references mentioned in the PRISMA diagram), which led to the idea of conducting this research. The main purpose of our work is to provide up-to-date and in-depth information on the traditional uses, phytochemistry, and pharmacology of *E. ribes*, documented over the years, up to the present. In addition, research into the therapeutic potential of essential oils/extracts and isolated compounds obtained from *E. ribes*, correlated with their antioxidant potential, has highlighted the scientific connotation of traditional uses and described the value and importance of considering both contemporary therapeutic known actions and traditional uses in folk medicine.

2. Methodology

To perform a review of the pharmacological activity of *E. ribes*, emphasizing the antioxidant potential of the essential oils, its traditional usage, and phytochemistry, a thorough literature search was conducted. Between February 1965 to June 2021, we used scientific electronic databases, including SCiFinder, PubMed, MEDLINE, EMBASE, Scopus, ScienceDirect, and the Google Scholar library; medical subject heading (MeSH) terms were used for searching in PubMed. The keywords were searched alone and in conjunction, using the Boolean operator AND. The PRISMA flow chart describes in Figure 1 the methodology for the already published information selection, using the directions given by Page et al. [6,7].

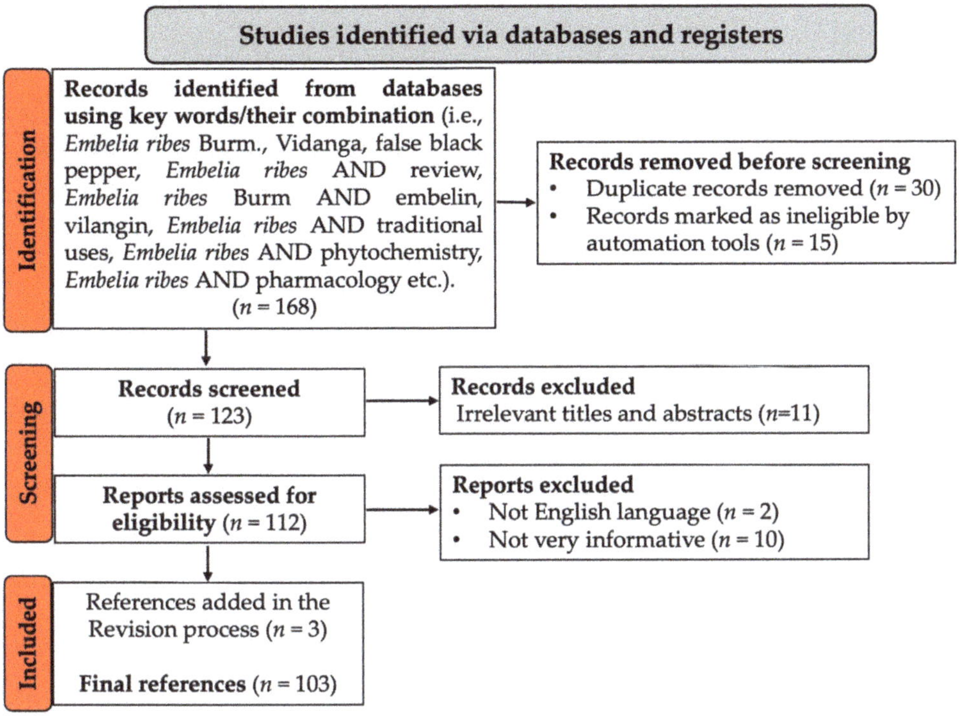

Figure 1. PRISMA flow chart describing the process of published data selection.

3. Traditional Uses

Traditional, ancient medicine was plant based, also using parts of animals and different substances from nature, being a mix between religion and science [8]. In this regard, different parts of *E. ribes* were used for their therapeutical actions [3,9,10] as is presented in Figure 2, and the pharmacological activity due to embelin is highlighted in Figure 3.

For more than 5000 years, *E. ribes* has been utilized traditionally for a variety of medical conditions. Traditional applications also varied according to the part of the plant that is used (e.g., fruits, seeds, root barks, leaves). Ancient uses suggested various forms of application (e.g., paste, powder, oil, and decoction). The paste was used for mouthwashes in preventing cavities, but also for skin-related disorders; the powder was used for various types of infections, indigestion, constipation, epilepsy, and as a blood purifier; the oil was used for dermatologic disorders and wound infections, and decoction of the roots was administered in cardiac diseases and insanity. Moreover, *E. ribes* has also been used for rejuvenation, bloating, vomiting, gastritis, and was frequently used in weight loss therapy and as a contraceptive. Among the most common effects obtained in traditional uses are carminative, anti-malignant, diuretic (e.g., fruit), anthelmintic (e.g., seeds), antibacterial, and pneumoprotective (e.g., leaves, root bark) [11,12].

Traditional uses have been translated to modern medicine and different pharmacological effects have been evaluated through experimental studies in order to elucidate mechanisms of action, dosages, precautions etc. Among the phytocompounds with the greatest potential are embelin, vilangin, embeliaflavosides and embelialkylresorcinols.

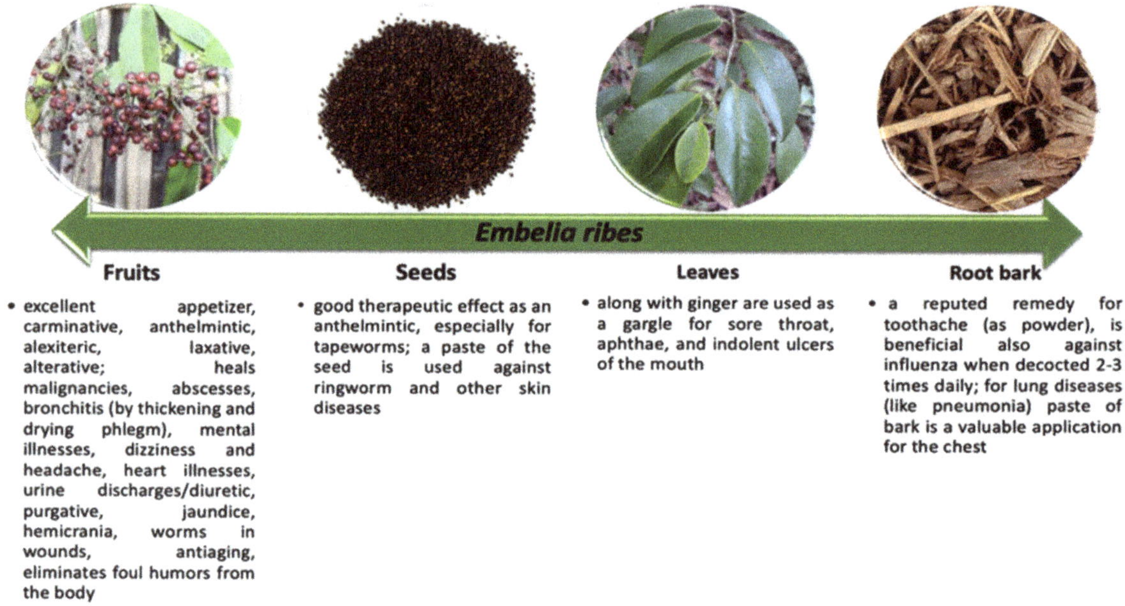

Figure 2. Summarizing the main effects of different parts of *E. ribes* plant.

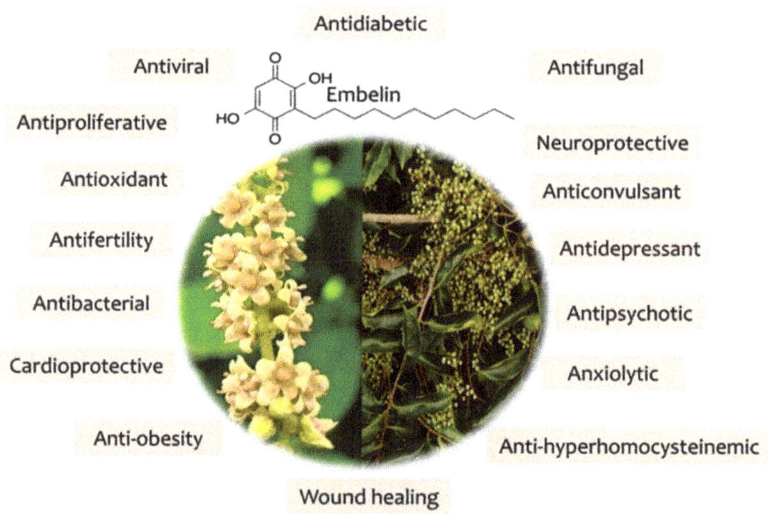

Figure 3. Pharmacological activities induced by embelin.

4. Essential Oils/Extracts and Nanoparticles of *E. ribes*

4.1. Aqueous Extract

Several experimental studies have been conducted in order to assess the correlation between the phytochemistry and pharmacology of *E. ribes*' phytocompounds, the importance of essential oils progressively increased with the development of analytical methods.

In 2006, scientists, mostly from the Department of Pharmaceutical Botany at Hamdard University, New Delhi, India, botanically authenticated dried fruits of *E. ribes* (voucher specimen no. UB 2) purchased at a local market. The output of the aqueous extract of *E. ribes* was about 5.261% on average. Since previous research indicated the *E. ribes* extracts efficiency in doses of 100 and 200 mg/kg b.w., the resulted aqueous extract of *E. ribes* was diluted in Tween 80 and provided orally to adult male Wistar albino rats [13]. Standardization of *E. ribes* aqueous extract was performed by using standard chemical tests, used to conduct preliminary phytochemical screening of *E. ribes* fruits' aqueous extract, for phytoconstituents detection. The extract contained essential oils, alkaloids, proteins, flavonoids, carbohydrates, phenolic components, and saponins, these compounds being considered responsible for its antioxidant action [13].

4.2. Embelin Isolation Method

The plant was procured from a local market located in Delhi, India. The berries of *E. ribes* were air dried, finely pulverized, and preserved in an airtight container. After the extraction with n-hexane using the Soxhlet equipment, the solvent was evaporated via distillation. The residue was further mixed with cold pet ether. Furthermore, the residue was then homogenized in a solution of methanol and dichloromethane (DCM), which was stored to crystallize. After crystallization, the particles were collected by filtration and washed firstly with n-hexane, then with DCM. The embelin was obtained in a yellowish colored crystalline structure described by morphology, coloration, and consistency using analytical techniques such as infrared (IR), high-performance thin-layer chromatography (HPTLC), and liquid chromatography–mass spectrometry (LC-MS) [14].

4.3. Ethanolic Extract

The leaf ethanol extract was used to isolate the phytocompound embelin. On a silica gel chromatographic column, the ethanolic extract was evaluated and eluted with methanol and chloroform. Thin layer chromatography (TLC) was utilized to analyze the eluted fractions. Furthermore, they were divided into five fractions. The higher-concentration determined in fraction two was recrystallized from chloroform to produce an orange red needle-like aromatic molecule. Quinones were qualitatively examined in the crystalline compound. IR and proton nuclear magnetic resonance (^{1}HNMR) were used to confirm the structure. The ethanolic extracts and the phytocompound embelin were used to generate two different types of drug formulations (e.g., 0.2% gel for topical application and oral suspension of 30 mg/mL of crude ethanolic extracts) [15].

4.4. Gold and Silver Nanoparticles

Green synthesis of gold (GNPs) and silver (SNPs) nanoparticles has stimulated the interest of researchers in the subject of nanomedicine in recent years. The seed extract of *E. ribes* (SEEr) was used as a capping and reducing agent, which resulted in an eco-friendly, cost-effective, fast, and simple technique for the synthesis of GNPs and SNPs. Ultraviolet-to-visible (UV-Vis) spectroscopy, dynamic light scattering (DLS), high-resolution transmission electron microscopy (HR-TEM), Fourier transform infrared (FT-IR), and X-ray powder diffraction (XRD) were used to characterize the synthesized GNPs and SNPs. The α,α-diphenyl-β-picrylhydrazyl (DPPH) and phosphomolybdenum assays were used to determine the ability of GNPs and SNPs to scavenge free radicals. Furthermore, the antibacterial activity of GNPs and SNPs against two microorganisms was measured using the disc diffusion method, and the cytotoxicity of GNPs and SNPs against MCF-7 cell lines at different doses was determined using the 3-[4,5-dimethylthiazol-2-yl]-2,5 diphenyl tetrazolium bromide (MTT) assay. Both GNPs and SNPs developed from *E. ribes* produced promising results, demonstrating their clinical significance [16].

UV-Vis spectrophotometer examination revealed the synthesis of GNPs and SNPs from a solution of silver nitrate and chloroauric acid. Due to the reduction of metal ions to neutral, the gold solution has changed from light yellow to wine red, with a specific peak in the region between 500 and 550 nm, whilst the silver solution has changed from colorless to yellowish-brown, with a peak between 400 and 440 nm.

Alkaloids, quinones, proteins, reducing sugars, and saponins, among other elements of *E. ribes* seed extract, may be responsible for metal ion reduction. SEEr-GNPs and -SNPs were 10–30 nm in size and 5–35 nm in size, respectively. GNPs and SNPs developed from SEEr are spherical and polydisperse in nature. SEEr-GNPs had a hydrodynamic diameter of 6 to 68 nm, while SEEr-SNPs had a diameter of 5 to 122 nm [16].

4.5. Vilangin

Even though certain well-known bioactive principles such as embelin have been comprehensively examined, the presence of other low polar and volatile components, as well as their bioactivities, has yet to be investigated. Furthermore, *E. ribes* fruits also contain an alkaloid (i.e., christembine) and a volatile oil in addition to embelin (i.e., vilangin) [17].

The antioxidant and anticancer properties of vilangin, a dimeric embelin derivative synthesized from embelin and isolated from *E. ribes* berries, were recently discovered in experimental studies [18].

Vilangin is synthesized by mixing embelin and formaldehyde in an acetic acid solution. By gradually warming, embelin was dissolved in glacial acetic acid. Moreover, formaldehyde was added, and the solution was heated on a water bath, then followed the cooling process to room temperature. The intense orange precipitate was filtered and crystallized from dioxane. The structure of vilangin is depicted in Figure 4. The IUPAC name is 2-[(2,5-dihydroxy-3,6-dioxo-4-undecylcyclohexa-1,4-dien-1-yl)methyl]-3,6-dihydroxy-5-undecylcyclohexa-2,5-diene-1,4-dione [17,18].

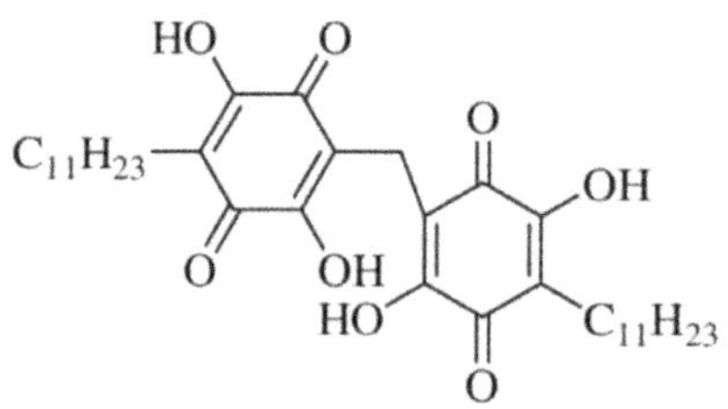

Figure 4. Structure of vilangin.

Several experimental studies were conducted in order to test the antioxidant potential (i.e., DPPH assay, cupric ion reducing antioxidant capacity assay, ferric reducing antioxidant power assay, and total antioxidant capacity) and anticancer activity (i.e., cytotoxicity assessment) of vilangin, which has been less studied [17,19].

Vilangin's antioxidant activity was evaluated and compared to a standard (butylated hydroxytoluene). At 1000 μg/mL, the radical scavenging activity was at its highest (72.35). Cupric ion decreasing antioxidant capacity of vilangin was based on concentration. At a concentration of 1 mg/mL, the reference had a higher cupric ion decreasing antioxidant ability than vilangin. In the presence of antioxidants, the ferric reducing antioxidant capacity assay measures the reduction of ferric iron Fe(3+) to ferrous iron Fe(2+). Moreover, the assay is extensively used to determine single antioxidant activity and total antioxidant activity in plant extracts [18].

At 40 μL, the activity was comparable to that of conventional butylated hydroxytoluene. When compared to butylated hydroxytoluene (2.206 $\pm$ 0.11 mM Fe(II)/g), vilangin (1.084 $\pm$ 0.02 mM Fe(II)/g) had an almost twofold superior ferric reducing capacity [17].

The phosphomolybdenum approach was used to assess vilangin's total antioxidant activity. This approach relies on the antioxidant substances reducing molybdenum Mo (VI)–Mo (V) and forming a green Mo (V)-antioxidant complex with a maximum absorption at 695 nm. Vilangin (0.842 $\pm$ 0.01 mg gallic acid equivalent/g) and butylated hydroxytoluene

1.137 ± 0.06 mg gallic acid equivalent/g) exhibited high absorbance values, indicating that the compounds showed an important antioxidant activity [18].

It has been reported that vilangin possessed cytotoxic effect against the A549 lung adenocarcinoma cancer cell line in vitro. It showed 61.95% activity at a dosage of 500 μg/mL, with an IC50 of 400 μg/mL (53.66%). In a concentration-dependent way, all of the concentrations employed in the experiment reduced cell viability significantly ($p < 0.05$).

Further studies are needed to establish the efficacy and safety profile of vilangin, an antioxidant phytocompound less investigated than embelin [17,18].

5. Phytochemistry of Bioactive Compounds

According to the scientific literature, the following substances were isolated from *E. ribes*:

- Three chemical compounds, identified as embelin (1), embeliaribyl ester (2), embeliol (3), and embelinol (4) were isolated from the seeds [20];
- 3-alkyl-1,4-benzoquinone (5), N-(3-carboxylpropyl)-5-amino-2-hydroxy-3-tridecyl-1,4 -benzoquinone (6), o-methyl rapanone (7) and rapanone (8) were isolated from the plant [21];
- Embelialkyl resorcinols A-I (9–17), virenol A (18), pentaketide (19), 1-(3,5-dihydroxyphenyl) heptan-1′-one (20), 1-(3,5-dihydroxyphenyl) nonan-1′-one (21), and 1-(3,5-dihydroxyphenyl) undecan-1′-one (22–26) were isolated from the ethanolic fruits extract [22];
- N-(3-carboxylpropyl)-5-amino-2-hydroxy-3-tridecyl-1,4-benzoquinone (27), 5,6-dihydroxy -7-tridecyl-3-[4-tridecyl-3-hydroxy-5-oxo-2(5H)-furylidene]-2-oxo-3(2H)-benzofuran (28), 2,5-dihydroxy-3-tridecyl-1,4-benzoquinone (29), 9 2,5-dihydroxyl-3-undecyl-1,4-benzoquinone (30), 10 2,5-dihydroxyl-3-pentadecyl-1,4-benzoquinone (31), 9 5-(8Z-pentadecenyl)-1,3-benzenediol (32), 11 5-(8Z-heptadecenyl)-1,3-benzenediol (33), 12 5-(8Z,11Z-heptadecadienyl)-1,3-benzenediol (34), 13 5-pentadecyl-1,3-benzenediol (35), 12 3-methoxy-5-pentylphenol, 14 3,5-dimethoxy-4-hydroxy phenyl-1-O-β-D-glucopyranoside (36), 15 2,6-dimethoxy-4-hydroxyphenyl-1-O-β-D-glucopyranoside,16 (37) (+)-catechin (38), 17 and (+)-lyoniresinol-3α-O-β-glucoside (39), sitosterol (40) and daucosterol (41) were isolated from the ethanolic extract of the roots [23,24];
- Five phenolic derivatives with various aliphatic chains and two phenolic glucosides, identified as 5-(8-pentadecenyl)-1, 3-benzenediol (42), 5-(8, 11-heptadecadienyl)-1, 3-benzenediol (43), 5-pentadecyl-1, 3-benzenediol (44), 5-(8-heptadecenyl)-1, 3-benzenediol (45), 3-methoxy-5-pentane-1-phenol (46), 3, 5-dimethoxy-4-hydroxyphenyl-1-O-β-D-glucopyranoside (47), and 2, 6-dimethoxy-4-hydroxyphenyl-1-O-β-D-glucopyranoside (48) respectively were isolated from the plant [24];
- Three flavonoid glycosides, embelia flavosides A-C (49–51) were isolated from the fruits [25];
- A new alkenyl resorcinol compound (namely embeliphenol A) was isolated from ethyl acetate stems extract [26].

The activity of the plant-based compounds identified above is shown in Figure 5, given that their chemical composition and structure imprint their interactions, as well as their multiple chemical, biochemical, physical, and pharmacological properties [27].

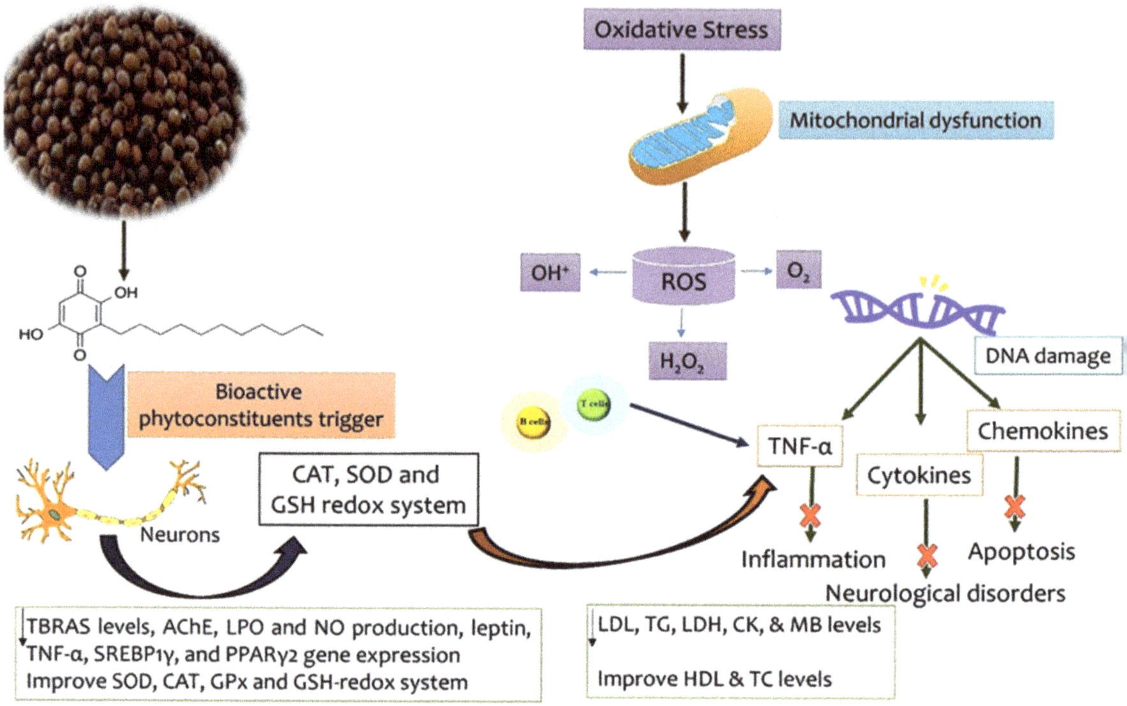

Figure 5. Mechanism of action of *E. ribes* and its bioactive compounds. ROS, reactive oxygen species; H_2O_2, hydrogen peroxide; DNA, deoxyribonucleic acid; TNF-α, tumor necrosis factor-alpha; CAT, catalase; SOD, superoxide dismutase; GSH, glutathione; LDL, low density lipoprotein; TG, triglyceride; LDH, lactate dehydrogenase; CK&MB, creatin kinase iso-enzyme; HDL, high density lipoprotein; TC, total cholesterol; TBRAS, thiobarbituric acid reactive substance; AchE, acetylcholine esterase; LPO, lactoperoxydase; NO, nitric oxide; SREBP1γ, sterol regulatory element-binding proteins 1 gamma; PPARγ2 peroxisome proliferator-activated receptor gamma 2; GPx, glutathione peroxidase.

6. Pharmacological Properties

6.1. Wound Healing Activity

Ethanol leaf extract (30 mg/mL) and embelin (4 mg/mL) isolated from *E. ribes* were evaluated for wound healing activity in albino rats. The wound healing activity of ethanol extract and embelin was found significant as compared with framycetin (standard skin ointment). The tested drug showed faster epithelialization of wounds with a high rate of wound contraction, higher tensile strength, and collagenation. The histopathological study exhibited improved cross-linking of collagen fibers and the absence of monocytes [15].

6.2. Antidiabetic Activity

Diabetes mellitus (DM) has been treated orally with natural remedies, based on folk medication, since ancient times. The Indian classical text of Ayurveda describes *E. ribes* as pungent, a therapeutic tool that produces an improvement in digestive fire, healing flatulence, and colic. Bhandari et al. presented the first pilot study concerning biochemical confirmation of the potential of *E. ribes* in diabetic dyslipidemia. The lipid-lowering and antioxidant activity of ethanol extract (200 mg/kg, p.o., for 20 days) of *E. ribes* were studied in streptozotocin (STZ)-induced (40 mg/kg) diabetes in rats. The ethanol extract treated diabetic rats and exhibited a significant ($p < 0.01$) reduction in blood glucose, serum total cholesterol, and triglycerides, with elevated HDL cholesterol levels as compared with STZ-induced diabetic rats. The ethanol extract reduced the thiobarbituric acid-reactive substances (TBARS) values of the liver and pancreas as compared to TBARS values of the liver and pancreas of STZ-induced diabetic rats. The results of the test drug were found to be similar to gliclazide (25 mg/kg, p.o.), a standard antihyperglycemic drug [28].

In another study, embelin (25 and 50 mg/kg, p.o., for 21 days) was administered to alloxan-induced diabetic rats, which showed a decrease in fasting serum blood glucose levels and an increase in the rat's body weight. The histological examination of the liver, kidney, and pancreas showed that normal architecture confirmed that embelin recovered the biological function of the liver, kidney, and pancreas [29]. In addition, embeliphenol A showed α-glucosidase inhibitory potency in a concentration-dependent manner [26].

Nephropathy linked with type 2 diabetes is the most prevalent cause of the end-stage renal disorder. The protective effect of ethanolic fruit extract of *E. ribes* (100 and 200 mg/kg, p.o., for 21 days) was evaluated against a high-fat diet and low dose STZ (35 mg/kg, i.p.)-induced diabetic nephrotoxicity in rats. Ethanolic fruit extracts of *E. ribes* displayed protective effects by a significant decrease ($p < 0.01$) of body weight, fasting blood glucose level, blood pressure (BP), alkaline phosphatase, serum lactate dehydrogenase, total cholesterol, creatinine, and triglycerides, whereas a rise in serum albumin and level of total protein was also detected. Furthermore, ethanolic fruit extracts of *E. ribes* treatment significantly ($p < 0.01$) reduced the kidney TBARS levels and elevated catalase (CAT), glutathione (GSH), and superoxide dismutase (SOD) levels in diabetic rats [30].

Aqueous extracts of *E. ribes* (100 and 200 mg/kg, p.o. for 40 days) alleviated renal damage in STZ-induced diabetic rats, by improvement in blood glucose, lipid metabolism, BP-lowering, inhibition of pancreatic lipid peroxidation process, and increased the levels of pancreatic SOD, CAT, and GSH [31]. Oral administration of ethanolic extracts of *E. ribes* (100 and 200 mg/kg, for 6 weeks) showed a significant reduction in the blood sugar levels, glycated hemoglobin, heart rate, and systolic BP in rats as compared to diseased rats [32].

The antidiabetic activity of ethanolic extract of *E. ribes* (100 and 200 mg kg/day, for 21 days) was performed in a high-fat diet and low dose STZ-induced diabetic rats. Antihyperglycemic potency of ethanolic extract of *E. ribes* was exhibited by significant reduction ($p < 0.01$) in blood sugar levels, hepatic glucose-6-phosphatase effect, and an elevated content of glycogen. Moreover, the ethanolic extract significantly ($p < 0.01$) reinstated the increased BP, and slowed down the degeneration of liver fat and oxidative alterations in diabetic rats [33]. A 50% ethanolic extract of berries of *E. ribes* reduced blood glucose levels in alloxan-induced diabetic rats [34].

The efficacy and safety profile of *E. ribes* and its active biomarker, embelin, in the treatment of DM were assessed in a systematic review and meta-analysis. The inverse-variance model was used to conduct a meta-analysis of *E. ribes*/embelin/derivatives of embelin versus diabetic control. The systematic review and meta-analysis comprised a total of 13 research studies, all of which were conducted on rat models. Blood glucose and glycosylated hemoglobin were considerably reduced by *E. ribes* extracts and embelin. Moreover, the results of the meta-analysis also revealed significant improvements in insulin and lipid profile, hemodynamic parameters, and oxidative stress markers. Two embelin derivatives that also relieved diabetes symptoms were 6-bromoembelin and vilangin. The therapeutic potential of *E. ribes* helped to reduce the effects of diabetes-related weight gain. The medical data supports *E. ribes*/embelin/embelin derivatives' anti-diabetic effectiveness. However, more clinical trial research is needed to confirm the current findings [35].

6.3. Anti-Obesity Activity

The anti-obesity potential of standardized ethanol extract of *E. ribes* (100 mg/kg, p.o., for 21 days) was evaluated in murine model of high-fat diet (HFD)-induced obesity. Standardized ethanolic extract displayed a preventive activity on the gain of body weight, accumulation of visceral fat, and higher BP. Moreover, standardized ethanolic extract of *E. ribes* treatment reduced the myocardial lipid peroxidation (LPO) and improved antioxidant levels in obese rats [36].

The preventive activity of embelin (50 mg/kg, p.o., for 21 days) was evaluated against hyperlipidemia and oxidative stress in HFD-induced obesity in male Wistar rats. To induce obesity, animals were fed with an HFD for a period of 28 days. Administration of embelin reduced body weight, BP, visceral fat pad weight, and lipid levels. In addition, embelin significantly ($p < 0.01$) reduced the level of hepatic TBARS, but elevated the level of CAT, SOD, and GSH levels in obese animals [37]. Aqueous extracts of *E. ribes* improved insulin resistance in a rat model with HFD-induced obesity, the possible mechanism of action being the downregulation of leptin, tumor necrosis factor-alpha (TNF-α), sterol regulatory element-binding proteins 1 gamma (SREBP1γ), and peroxisome proliferator-activated receptor gamma 2 (PPARγ2) gene expression. The leptin may contribute to hepatic steatosis by promoting insulin resistance and by altering insulin signaling in hepatocytes, which consequently promote increased intracellular fatty acid [38].

Embelin reduced adiposity in C57BL/6 mice by reducing the oxidative stress and inflammation caused by the HFD. Oral treatment of embelin resulted in a significant decrease in levels of nuclear factor erythroid 2-related factor and nuclear factor kappa-B (NF-κB) protein expression in liver tissue, as well as an improvement in obesity biomarkers. Thiobarbituric acid reactive substance (TBRAS), GSH, SOD, and CAT levels in liver tissue are also improved in embelin-treated HFD-fed mice along necrotic and inflammatory alterations that were significantly reduced in the liver tissue [14].

6.4. Cardioprotective Activity

The cardioprotective potential of *E. ribes* water fruits has been assessed in a rat model of acute myocardial infarction produced by isoproterenol (ISO) (5.25 and 8.5 mg/kg, s.c., for 2 days) and ISO-induced cardiomyopathy in STZ (40 mg/kg, i.v.)-induced diabetic rats. Water *E. ribes* extract (100 mg/kg, p.o., for 40 days) and alcoholic extract (200 mg/kg, p.o., for 40 days) lowered heart rate, systolic BP, elevated serum lactate dehydrogenase (LDH), serum creatine kinase (CK), and myocardial LPO, and markedly improved myocardial endogenous antioxidants (GSH, SOD, and CAT) levels [39–41].

6.5. Antioxidant Activity

A hydroxyl group is considered beneficial when examining radical scavenging properties since it releases a H atom, rendering the scavenged radical less reactive. It can also generate a quinone structure if an ortho or para-H-atom is available for abstraction, as in the case of the 3,4-dihydroxyphenyl moiety of quercetin. Embelin, on the other hand, is unable to react with superoxide in the classical manner, as recently demonstrated by structural and computational investigations [42].

Embelin scavenges the superoxide radical by extracting its electron and releasing molecular oxygen in a distinctive manner. The fact that both embelin hydroxyl groups form strong intramolecular H-bonds with the carbonyl groups in the surroundings, makes them less available for scavenging. Furthermore, embelin's ability to protect against paraquat-induced lung damage was investigated in correlation to its antioxidant and anti-inflammatory properties. Malondialdehyde (MDA), SOD, CAT, glutathione peroxidase (GPx), inflammatory cytokines (i.e., interleukin-1β, TNF-α, interleukin-6) and NF-κB/mitogen-activated protein kinase (NF-κB/MAPK) were evaluated in experimental assays on lung tissue. Embelin therapy reduced MDA while increasing SOD, CAT, and GPx levels. In paraquat-administered and paraquat-intoxicated rats, embelin significantly lowered inflammatory cytokine levels. The results showed the impact of embelin's activity on the oxidative effects of paraquat [43,44].

Embeliaflavosides A–C (flavonoid glycosides) showed significant 2,2′-azino-bis(3-ethylbenzothiazoline)-6-sulphonic acid (ABTS) radical scavenging potency (IC_{50} = 2.52–9.78 μM), and DPPH scavenging activity (IC_{50} = 7.56–26.47 μM) [25]. The antioxidant activity of ethanolic and aqueous extracts of *E. ribes* flower was investigated in vitro by determining nitric oxide (NO) scavenging activity, ferric thiocyanate, and total reduction efficiency, as well as by DPPH assays. Ethanolic and aqueous extracts of *E. ribes* flower showed a concentration-dependent escalation in NO, DPPH free radical, ferric thiocyanate inhibition/scavenging action, and equivalent overall reduction capacity. Compared to aqueous extracts, ethanolic extracts displayed greater scavenging action [45].

The scavenging capability of embelin has been demonstrated by utilizing hydrodynamic voltammetry, which produces the superoxide radical in situ. Embelin as a scavenger of superoxide outperforms the conventional food additive antioxidant 2,6-bis(1,1-dimethylethyl)-4-20methylphenol (butylated hydroxytoluene). Moreover, in the voltaic cell, embelin can even completely eliminate the superoxide radical. According to computational results, two types of embelin scavenging actions may be implicated, the first by π–π interaction mechanisms and the second through proton capture in the cell [2,46].

The biological activity of THP-1 human leukemic monocytes and BV-2 mouse microglia was examined in a study to assess its antioxidant properties. The antioxidant effects examined in MTT assay, proliferation curves, and antioxidant activity using a fluorescent probe showed promising results after 24 h. The long alkyl C10 tail of embelin may be essential for cell membrane insertion, which promotes the antioxidant defense mechanism and cytoprotection in microglia. Consequently, embelin may be a promising pharmaceutical tool for reducing the damage caused by metabolic and neurodegenerative illnesses [46].

E. ribes is a traditional Chinese herbal medicine that is used to treat a variety of ailments. However, there is no solid information on its chemical composition. The components of *E. ribes* were examined using ultra-high-performance liquid chromatography quadrupole time-of-flight tandem mass spectrometry (UHPLC-Q-TOF mass spectrometry). Numerous compounds were detected, including 16 phenolics, 16 flavonoids, 5 fatty acids, and 4 coumarins [2].

The total phenolic and total flavonoid content of the acetic ether extract of *E. ribes* was also analyzed. The results showed a major potential of *E. ribes* to be an important source of phenolics (308.16 mg gallic acid equivalents/g of extract) and flavonoids (62.00 mg rutin equivalents/g of extract). Furthermore, acetic ether extract had a significant antioxidant effect (ferric reducing activity power: 0.15 mg/mL; DPPH: 0.18 mg/mL; ABTS: 0.06 mg/mL). Acetic ether extracts also inhibited NO production in lipopolysaccharide (LPS)-stimulated macrophage RAW 264.7 cells and the release of pro-inflammatory cytokines. These results support the hypothesis that *E. ribes* can be an effective antioxidant and anti-inflammatory agent [5].

Pulse Radiolytic Assessment

The dynamics and molecular processes of embelin's reactivity with hydroxyl (•OH), one-electron oxidizing, peroxyl, and thiyl radicals were examined using nanosecond electron pulse radiolysis to better understand the mechanism for its antioxidant function. Because embelin is poorly soluble in neutral water solutions due to intramolecular hydrogen bonding, investigations in alkaline conditions in aqueous solutions have been conducted [47,48].

At pH 10, embelin interacts with •OH to produce a broad transitory absorption spectrum in the wavelength range of 380–480 nm that lasts up to 100 µs. The reactivity of embelin with one-electron oxidants, such as azidyl (N_3•) and trichloromethyl peroxyl radicals (•CCl_3O_2), has also been investigated in order to separate phenoxyl type radical absorption bands from OH-adduct and transitory due to H-abstraction. Adducts and phenoxyl type radicals are expected to be produced when embelin ($O_2E(OH)_2$) interacts with •OH and oxidants, respectively. The large absorption band in the 380–480 nm area can be attributed to transient conditions produced by OH-addition and/or H-abstraction, according to a comparison of transitory absorption spectra seen in its interaction with •OH to those recorded with one-electron oxidants [47].

Due to significant intramolecular hydrogen bond, radiolysis of embelin in non-polar solvents such as CH_2Cl_2 and 1-chlorobutane did not result in an oxidation reaction. A comparison of transient absorption data at 410 nm also reveals that only 35% of the •OH undergoes one-electron oxidation, with the rest forming •OH adducts on the ring or H-abstraction on the alkyl side chain. Furthermore, as related to the N_3• radical reaction, 85% of the •CCl3O2 radical induces one-electron oxidation [49].

The medical importance of embelin requires research into its interactions with other free radicals of biological significance, as well as the regeneration of embelin free radicals using a regularly used antioxidant, ascorbic acid. Thiyl radicals (RS•) are formed during cellular redox processes and as a result of GSH's antioxidant effect. These are reactive oxidants that target lipids, causing them to oxidize or isomerize. As a result, thiyl radical repair is required for the storage of thiols (RSH) for antioxidant action and to protect lipids [50,51].

The absorption spectrum of embelin with glutathiyl radical (GS•) at pH 3.9 is comparable to that of phenoxyl radical, with peaks at 430 and 320 nm. With GS•, however, there is no time resolution in the absorption spectrum. This indicates that GS• only produces its phenoxyl type radical when it combines with embelin [52].

Embelin has been discovered to scavenge biologically relevant oxidizing radicals significantly. Embelin scavenged •OH and GS• to form embelin phenoxyl radical (λ_{max} 410 nm), which is then scavenged by ascorbate anion. In vitro research shows that it can inhibit lipid peroxidation, reduce Fe(3+), and restore Mn-SOD, indicating that it could be an effective antioxidant in the biological system. Its free radical scavenging activity has also been reported to be superior to that of α-tocopherol. It can be assumed that, in addition to its metabolic functions, embelin's antioxidant activity is due to its therapeutic properties [47].

6.6. Neuroprotective Effect

Elevated plus maze (EPM) and Morris water maze model (MWM) were used to test the anti-Alzheimer's activity of embelin versus diazepam (1 mg/kg, b.w., i.p.)-induced amnesia. In the EPM model, embelin reduced the transfer latency time in a dose-dependent way, while in the MWM model, it reduced the escape latency time. The diazepam-treated group exhibited substantial increases in escape and transfer delay, indicating impairment in learning and memory. On the other hand, embelin significantly reversed the diazepam-induced amnesia and enhanced learning and memory in mice models in a dosage and time-dependent way [53].

Clinically recognized targets of Alzheimer's disease (AD) include beta-secretase (BACE-1) and cholinesterase, both of which have benefited greatly from natural products [54]. Considering AD is a significant public health concern, it needs the use of multi-targeted medications to treat it [55,56]. The enzymes namely acetylcholinesterase (AChE), butyrylcholinesterase (BChE), and β-site of amyloid precursor protein cleaving enzyme (BACE-1) were inhibited by embelin, with IC50 values of 2.5, 5.4, and 2.1 μM, respectively. Embelin also increased the activity of P-gp, an efflux pump implicated in the clearance of amyloid-beta from the AD brain, in LS-180 cells. In addition, a study of cell viability in a neural cell line revealed that embelin had a wide therapeutic profile. These results suggested that embelin is a multi-targeted drug that reduces amyloid oligomer formation through BACE-1 inhibition [57], enhances cholinergic transmission through AChE/BChE inhibition, and promotes amyloid clearance (through P-gp induction) [58].

Antioxidants have been studied in the addition to developing neuroprotective medicines that could be used in stroke treatment. Stroke is an acute and progressive neurological condition that is the second largest cause of death. Ethanolic and aqueous fruits extracts of *E. ribes* (100 and 200 mg/kg, p.o., for 30 days) increased antioxidant activity and demonstrated neuroprotective potential against middle cerebral artery occlusion (MCAO)-induced localized cerebral ischemia in male Wistar rats. When compared to MCAO+ vehicle group rats, ethanolic and aqueous fruits extract of *E. ribes* gradually improved grip strength behavior, GSH, GPx, glutathione reductase (GR), and glutathione-S-transferase (GST) levels in the brain, as well as a significant decline in LDH levels in serum and TBARS levels in the brain [13,59]. As compared to ischemia control, pretreatment with embelin (25 and 50 mg/kg, p.o.) enhanced locomotor activity, holding latency time, and falling beam walking latency. Embelin decreased LPO and enhanced total thiol content and GST activity [60].

6.7. Anxiolytic Activity

The anxiolytic effect of embelin was evaluated in EPM and open field test (OFT) models. In the EPM model, embelin significantly improved the percentage of time spent and the number of entries. Embelin increased the time spent, crossing number, and reduced the time of immobility in the lightbox. Embelin showed significant growth in the rearing number, assisted rearing, and squares crossed in OFT [61].

The anxiolytic effect of the methanolic extract of *E. ribes* (100 and 300 mg/kg p.o.) was examined using the hole board test, EPM, mirrored chamber equipment, and gamma-aminobutyric acid (GABA) estimation. The number of entries and time spent in the open arm of EPM was increased by methanolic extract of *E. ribes*. In the hole board test, methanolic extract of *E. ribes* showed a substantial increase in the number of head dips. Furthermore, methanolic extract of *E. ribes* exhibited an increase in the number of entries and time spent in the mirrored chamber, as well as a reduction in the time it took to enter the mirrored room. GABA concentration in the brain was significantly increased by methanolic extract of *E. ribes* at the oral dose level of 300 mg/kg [62].

According to Afzal et al., embelin possesses strong anxiolytic action that is dose dependent. It has been hypothesized that the observed activity was due to an antagonistic effect on the GABA receptor complex, since most anxiolytic and antidepressant drugs selectively bind to the GABA receptor's high-affinity benzodiazepine binding region. A small dark safe compartment and a large lighted preference chamber compose the test apparatus. In the EPM device, embelin at a dose of 5 mg/kg and embedelin (2.5 mg and 5 mg/kg) significantly enhanced the percentage of time spent and the number of entries in the open arm. In the EPM-test on mice, the results showed that embelin had dose-dependent anxiolytic action. In an open field experimental test, embelin showed a considerable rise in the number of crossing, rearing, and assisted rearing [61].

6.8. Antidepressant Activity

Under the tail suspension test (TST) and forced swimming test (FST) experimental models, embelin (2.5 and 5 mg/kg, i.p.) showed therapeutic potential for controlling depression by reducing immobility in a dose-dependent manner. In both animal models, the efficacy of embelin (5 mg/kg) was found equivalent to the standard drug (i.e., imipramine, 15 mg/kg) [63].

The anti-depressant effect of embelin has been demonstrated in animal model experiments using two widely known assessment methods: TST and FST. In academic research, the TST is used to detect stress in rodents as an experimental approach. In mouse models, the FST is used to assess the anti-depressant efficacy of novel chemicals, anti-depressant agents, and experimental development targeted at translating or preventing depressive-like conditions. It is founded on the observation that an animal will become immobile if it is subjected to short-term continuous stress. It has been described as generating a circumstance in which the animal experiences "behavioral despair", whereby the animal loses hope of surviving the stressful environment [52].

6.9. Antipsychotic Activity

Antipsychotic action of embelin (5 and 10 mg/kg, p.o., for 15 days) was investigated against stereotyped activity and apomorphine-induced climbing activity in rats and mice, respectively. Embelin inhibited apomorphine-induced climbing and stereotyped activity in rodents. In the brain of experimental animals, embelin also reversed increased levels of dopamine, noradrenaline, and serotonin neurotransmitters. In both animal models, embelin exhibited more substantial results at a dosage of 10 mg/kg [64].

6.10. Anticonvulsant Activity

Embelin (2.5, 5, and 10 mg/kg, i.p.) showed anticonvulsant activity by seizures inhibition, decrease in locomotion induced by electroshock, and pentylenetetrazole (PTZ) in a dose-dependent manner [65].

6.11. Antifertility Activity

Embelin, isolated from berries of *E. ribes*, altered the testicular histology and glycogen, gametogenic counts, and accessory sex gland fructose at the dose levels 0.3, 0.4 and 0.5 mg/kg body weight administered subcutaneously for 35 days [66].

6.12. Antibacterial Activity

Both Gram-positive and Gram-negative bacteria were used to investigate the antibacterial efficacy of embelin, an isolated component from *E. ribes* berries. The microdilution method and the agar plate method were used to investigate the minimal inhibitory and bactericidal concentrations of embelin. Embelin was shown to be bactericidal against Gram-positive bacteria and bacteriostatic against Gram-negative bacteria [67]. The antimicrobial activity of SEEr was evaluated against *Escherichia coli, Pseudomonas aeruginosa, Staphylococcus aureus*, and *Bacilus subtilis. Escherichia coli* exhibited the highest inhibition zone followed by *P. aeruginosa*, whereas *S. aureus* and *B. subtilis* both showed lower inhibitory zones [68].

6.13. Antifungal Activity

The SEEr showed inhibitory effects of the fungal growth, with the highest activity seen at 2.0 mg seed extract concentration. *Colletotrichum crassipes* displayed maximum inhibition zones, while *Cladosporium, Armillaria mellea, Colletotricum capsici, Aspergillus niger*, and *Rhizopus oryzae* were the next species tested for inhibition of the fungal growth. In comparison to other species, *Aspergillus terreus* and *Candida albicans* displayed smaller inhibitory zones after interaction with SEEr [69]. Furthermore, the standard in vitro antifungal susceptibility test techniques such as the national committee for clinical laboratory standard M27-A2 protocol (NCCLS) was used to assess the antifungal activity of various *E. ribes* extracts and its bioactive component, embelin. In 96 well plates, four different types of extracts (i.e., petroleum ether, solvent ether extract, methanol extract, and water extract) were tested, and detection was performed with a colorimetric plate reader at 530 nm. The NCCLS approach demonstrated that methanol extract and embelin showed the minimum inhibitory concentration required to inhibit the growth of 50% of organisms (MIC50) range of 120 mg/L against *Candida albicans* (MTCCno.183), and embelin had reported MIC50 values < 700 mg/L among the four *Candida* species studied. Except for the water extract, the percentage growth increased when the concentration of the plant extracts dropped [70].

6.14. Antiproliferative Activity

Embelin exhibited chemo-preventive activity against phenobarbital-induced hepatocarcinogenesis in Wistar rats. The crude hexane fruits extract of *E. ribes* demonstrated cytotoxicity toward human leukemic cells (K562) and Dalton's lymphoma ascites (DLA) cells [71]. Embelialkylresorcinols C, E, F, and H showed moderate cytotoxicity (half maximal inhibitory concentration, IC50 = 23.06 to 41.49 μM) against three human cancer cell lines (i.e., Hep3B, A549, and HCC1806) [22]. The anticancer efficacy of embelin was determined using MCF-7 breast cancer cells in a cytotoxicity and apoptosis study. The IC50 value of embelin was found to be 80μg/mL in MCF-7 breast cancer cells. It was also shown that embelin was able to alter cell viability and trigger apoptosis in MCF-7 breast cancer cells in a dose-dependent way [72].

Rai et al. analyzed the survival time of cancer patients in the presence of various causes of mortality. The data of surveillance, epidemiology, and end results (SEER) were used to perform the study. According to the SEER program of the US National Cancer Institute (2000–2014), a total of 2875 people were diagnosed with breast cancer, with 577 of them dying as a result of the breast cancer disease [73]. In mice xenograft breast cancer models of

MDA-MB-231 cells, embelin (10 mg/kg b.w., i.p. twice weekly for 28 days) reduced tumor volume and caspase-3 activation. Embelin triggers apoptosis via interacting with a variety of signaling pathways, which vary depending on the cancer's origin. Embelin exhibited a variety of biological traits that are important to human cancer chemoprevention, and growing data suggests that embelin may regulate many tumor cell characteristics. NF-κB, tumor suppressor gene (p53), phosphoinositide 3-kinase/protein kinase B (PI3K/Akt), and signal transducer and activator of transcription 3 (STAT3) signaling pathways are all modulated by embelin, which causes cell apoptosis that can be intrinsic or extrinsic. Embelin stimulates autophagy in cancer cells, although these autophagic cell-death processes have received less attention than apoptotic cell-death processes [74].

The embelin effect on the development of human prostate cancer cells was studied. In comparison with the case of breast cancer (MCF-7, MDA-MB-231, and T47D), hepatoma (Hep3B, HepG2, and HuH-7), and choriocarcinoma (JEG-3), embelin significantly suppressed cell growth in human prostate cancer cell lines, including PC3, LNCaP-LN3, DU145 and normal prostate epithelial cell, RWPE-1. It has been discovered that embelin caused apoptosis in PC3 cells in a time-dependent approach, which was associated with decreased expression of B-cell lymphoma 2 (Bcl-2), B-cell lymphoma-extra-large (Bcl-xL), and myeloid cell leukemia 1 (Mcl-1), enhanced Bcl-2 associated x, apoptosis regulator (Bax) translocation into mitochondria, and a decrease in mitochondrial membrane potential [2].

Embelin increased the expression of the voltage-dependent anion channel (VDAC) 1 and oligomerization, which may stimulate cytochrome c and apoptosis-inducing factor (AIF) release. The effects on wingless-type (Wnt)/β-catenin signaling were investigated since embelin was able to decrease Akt activation and cyclooxygenase-2 (COX-2) production. Embelin inhibited β-catenin expression while activating glycogen synthase kinase (GSK)-3. In embelin-treated cells, there was a decrease in β-catenin-mediated T-cell factor (TCF) transcriptional activity and gene transcription, including cyclin D1, c-myelocytomatosis (c-myc), and matrix metalloproteinase (MMP)-7. Lithium chloride, a GSK-3 inhibitor, inhibited changes in β-catenin levels in response to embelin, suggesting that embelin may reduce β-catenin expression through GSK-3 activation. Moreover, when PC3 cells were exposed to embelin, cell migration and invasion were significantly reduced. In conclusion, these results imply that the pro-apoptotic impact of embelin in prostate cancer cells is partially mediated by suppression of Akt signaling and activation of GSK-3 [75].

The potential of embelin to provide a therapeutic effect on glioma was examined. Human glioma cells were discovered to be inhibited by embelin, although normal immortalized human astrocytes were not. In addition, embelin promoted apoptosis in human glioma cells by suppressing NF-κB, a critical transcription factor linked to a variety of human disorders, including cancer, that regulates a number of genes involved in tumor progression, including cell proliferation and survival. Even though embelin was found as an X-linked inhibitor of apoptosis protein (XIAP) inhibitor, it had no effect on XIAP in glioma cells, but instead reduced NF-κB activity by hindering the nuclear translocation of p65, as a result of decreasing phosphorylation and proteasomal degradation of nuclear factor of kappa light polypeptide gene enhancer in B-cells inhibitor, and alpha (IκBα) in glioma cells. Additionally, p65 overexpression in glioma cells reduced embelin-induced apoptosis. These results showed that embelin could be a promising new treatment for glioma by reducing NF-κB activity and thereby limiting cancer cell proliferation and inducing apoptosis [76].

Investigations seeking to define embelin's precise molecular target resulted in the discovery of embelin as an inhibitor of the XIAP's baculovirus inhibitory repeat domain (BIR3). In addition, embelin has been shown to inhibit 5-lipoxigenase (5-LOX) and microsomal prostaglandin E2 synthase-1 (mPGES-1), as well as plasminogen activator inhibitor-1 (PAI-1) and P300/CBP associated factor (PCAF). Furthermore, embelin has been observed to interfere with mitochondrial oxidative phosphorylation through both redox and non-redox processes [77].

Even though embelin has been shown to possess a wide range of therapeutic effects, the mechanisms by which it exerts anticancer benefits are as yet unclear. The critical role of oxidative stress-induced MAPK signaling as a primary mechanism for its anticancer effects has been discovered by tracking the molecular alterations associated with early apoptotic phase. Embelin treatment of A549 lung cancer cells resulted in an increase in phospho-p38 and phospho-c-jun n-terminal kinase (JNK) levels as early as 4 h. The activation of caspase-3 by embelin was blocked by pretreatment of cells with particular inhibitors of p38 (PD169316) and JNK (SP600125). The observed changes in phosphorylation levels of p38, JNK, and ERK 1/2 in the presence or absence of specific MAP kinase inhibitors are mainly attributable to embelin and not due to cross talk between MAP kinases, according to experimental studies. Pretreatment of cells with 5,10,15,20-tetrakis(N-methyl-4′-pyridyl) porphyrinato iron III (FeTMPyP) reduced embelin-induced changes in MAPK phosphorylation and apoptosis, indicating that reactive oxygen species (ROS) play a key role. The observations are not related to embelin′s inhibitory action on XIAP, as second mitochondria-derived activator of caspases N7 peptide (SMAC-N7-Ant peptide), a selective inhibitor of XIAP′s BIR3 domain, did not mimic embelin-induced apoptotic effects. The results of this study showed the significance of p38 and JNK pathways in embelin-induced apoptosis and provided scientific information for improving the therapeutic profile of embelin [78].

6.15. Antiviral Activity

Ethyl acetate fruits extract of *E. ribes* and the isolated compound embelin showed maximum antiviral potency with virus-inhibiting activity IC50 of 0.2 μg/mL, selectivity index (SI) = 32 and IC50 of 0.3 μM and SI = 10. Furthermore, embelin was tested against influenza viruses (i.e., influenza A and B viruses). The results of the experimental study showed that H5N2 virus was the most susceptible to embelin (SI = 31), whereas H3N2 virus had the greatest resistance (SI = 5) [79].

The antiviral activity of embelin (5, 10, and 20 mg/kg/day, i.p., for 4 days) was evaluated in LPS-induced acute respiratory syndrome in murine models. Pretreatment with embelin showed its potential as a therapeutic drug for acute respiratory distress syndrome by decreasing inflammation of the lung, mononucleated cellular infiltration, nitrate/nitrite, total protein, albumin concentrations, TNF-α in the bronchoalveolar lavage fluid, and myeloperoxidase activity in lung homogenate. Embelin protected the partial pressure of oxygen (pO_2) down-regulation and the partial pressure of carbon dioxide (pCO_2) augmentation [80].

In the human population, the herpes simplex virus-1 (HSV-1) causes a wide spectrum of infections, from minor to life-threatening. HSV-1 infections have effective therapies, although they are generally limited to HSV-1 latency and the development of resistance to current treatments. An experimental study investigated the impact of embelin on HSV-1 in cultivated Vero cells in terms of antioxidant and antiviral effects. A large amount of ROS (i.e., hydrogen peroxide, H_2O_2) was produced, indicating oxidative stress processes. Antiviral assays, antioxidant assays, penetration, and post penetration assays, confocal microscopy, quantitative polymerase chain reaction (qPCR), and time course attachment were performed after Vero cells were infected with a recombinant strain of HSV-1. The results of the conducted study showed that embelin is noncytotoxic at doses ranging from 20 to 70 μM. HSV-1 virions were treated with embelin, which inhibited the adhesion and penetration of HSV-1 virions to host cells, resulting in a 98.7–100% inhibition and affecting the early stages of HSV-1 infection in Vero cells. The production of H_2O_2 was substantially reduced when virions were treated with embelin concentrations ranging from 35 to 60 μM. Furthermore, embelin lowers oxidative damage induced by HSV-1 infection and is an efficient antiviral for reducing HSV-1 infection in Vero cells in culture. More research is needed to conclude if embelin may be used as a therapeutic agent [81].

6.16. Anti-Hyper Homocysteinemic Activity

The anti-hyper homocysteinemic and lipid-lowering potential of ethanol fruits extract (100 and 200 mg/kg, p.o., for 1 month) of *E. ribes* was studied in methionine (1 g/kg, p.o., for 1 month)-induced hyper homocysteinemia in male albino rats. Ethanol fruits extract significantly decreased levels of homocysteine, LDH, total cholesterol, triglycerides, LDL levels in serum, and LPO levels in homogenates of the heart, with an increase in serum HDL and myocardial GSH levels. All the findings of ethanol extract were compared with the standard drug (folic acid (100 mg/kg, p.o.)) [82]. Another study showed that aqueous fruits extract of *E. ribes* (100 and 200 mg/kg, p.o., for 1 month) significantly ($p < 0.01$) reduced the homocysteine levels, LDH, total cholesterol, triglycerides, LDL-C, and VLDL-C and elevated the HDL-C levels in serum. Furthermore, a significant ($p < 0.01$) reduction in LPO levels with improvement in GSH level was also found in hyper homocysteinemic rats [83].

6.17. Protection against Liver Injury

Embelin's protective actions against acute liver injury have also been explored in the literature. Adult mice were administered a single injection of thioacetamide (TAA) (300 g/g body weight) to generate an animal model of acute liver damage. Embelin was given as a 50 g/g body weight intragastric gavage starting two days before TAA injection and continued throughout the research. The log-rank test was used to examine the mice's survival using the Kaplan–Meier method. The acute liver injury procedure was repeated, and the mice were evaluated at the appropriate times. Necrosis/inflammation and liver regeneration were examined using hematoxylin and eosin staining and picrosirius red staining, respectively. The activity of serum alanine aminotransferase/alkaline phosphatase was used to measure liver function. Immunohistochemistry was used to assess the amounts of cleaved caspase-3 and F4/80 expression in the liver. GraphPad Software was used to conduct the statistical analysis to validate the experiment. The mice's survival and liver function were significantly better in the group given embelin before TAA exposure than in the TAA exposure-only group. TAA-induced hepatic necrosis/apoptosis was diminished significantly by embelin. In the embelin-treated recovery group, massive inflammatory cell infiltration, which is associated with hepatic fibrogenesis, occurred earlier than in the spontaneous recovery group. Furthermore, embelin therapy boosted macrophage activity quickly and efficiently. Consequently, embelin may effectively prevent acute liver injury. Its therapeutic potential requires further investigation [84].

6.18. Anti-Aging

Embelin, a benzoquinone derivative with anticancer, anti-inflammatory, and antioxidative properties, has been investigated for potential anti-aging properties. However, little research has been conducted on the use of cosmetic raw components for embelin. Using a H_2O_2-induced cellular aging model of human dermal fibroblasts (HDFs), the antioxidant and anti-senescence properties of embelin were investigated. The water-soluble tetrazolium salt (WST-1) assay was used to determine cell viability. By using quantitative reverse transcription polymerase chain reaction (qRT-PCR), the gene expression model in HDFs by embelin has been quantified.

Changes in intracellular ROS concentration were assessed using dichlorofluorescein diacetate (DCF-DA). The senescence-associated β-galactosidase (SA-β-galactosidase) protocol, a method of staining β-galactosidase, was used to determine cell senescence. Cell mobility was measured using a wound healing technique. The WST-1 experiment revealed that embelin restored cell viability in a dose-dependent way after H_2O_2 reduced cell viability.

Furthermore, embelin elevated the expression of SOD1, GPx1, and CAT genes, implying that embelin-induced antioxidant potential may be strengthened by overexpression of intracellular antioxidant-related genes. The SA-β-galactosidase assay was used to examine whether embelin decreases cell senescence in an H_2O_2-induced senescence model of HDFs. Embelin reduced SA-β-galactosidase activity in H_2O_2-treated HDFs in a dose-dependent way. Moreover, embelin reduced the expression of the p21 and MMP1 genes in H_2O_2-treated HDFs in a dose-dependent approach. In H_2O_2-treated HDFs, however, embelin elevated COL1A1 genes in a dose-dependent manner. The study showed that embelin could be used as an anti-aging cosmetic component with anti-senescence and antioxidant characteristics [85].

6.19. Toxicological Profile

Experimental studies have demonstrated that not all herbal products are safe for direct human administration, especially in children and pregnant women, despite the fact that medicinal plants have less adverse effects than synthetic drugs. According to the results of the toxicity tests, administering rats and mice oral doses of embelin ranging from 10 mg to 3 g/kg was safe [86].

The spleen, kidney, and liver masses were unaltered by embelin administering in female cyclic rats at a dose of 120 mg/kg body weight, while the initial mass of the adrenals significantly increased. By administering embelin for 6 weeks, certain pathophysiological alterations can be observed. Whereas the spleen's histological properties remained relatively constant, the adrenals exhibited enlargement. It has been documented that chicken model organisms given an oral dose of 1.25 g/kg of *E. ribes* experienced ganglion cell degeneration in their retinal cells. However, retinotoxicity and visual impairment have not been reported at a dose of 0.25 g/kg [2].

Two groups of pregnant rats were administered 2.5 times and 5 times the recommended dose of an Ayurvedic contraceptive known as pippalyadi vati, which contained equal portions of powdered seeds or fruit berries of *E. ribes*, fruit of *Piper longum*, and borax powder. The pregnant rats gave birth to shorter and lower birth weighted infants. Moreover, soft tissue abnormalities and intestinal herniation into the umbilical cord have been reported in the developing fetuses of pippalyadi-treated mothers. Similar herniation was not observed in the control groups [87].

The use of *E. ribes* should be monitored, especially when administered in higher doses for children and pregnant women, as experimental tests using these herbs in animal models have demonstrated potential toxicological effects.

7. Experimental and Clinical Studies

Experimental studies describing the effects of *E. ribes* Burm. on different types of cancer are summarized in Table 1.

Table 1. In vivo studies depicting the main pharmacological proprieties of *E. ribes*, administered as aqueous/ethanolic extract mainly from berries or other parts of the plant and as pure form.

Animal/Dose	Observation/Outcomes	Ref.
	Anticancer	
AOM/DSS-induced colon cancer in C57BL/6 male mice/50 mg/d/kg b.w. mixed in diet for 10 days before the CAC challenge, then for 19 or 85 days	↓ Tumor incidence and volume, ↓ IL-6; ↓ STAT3	[88]
DENA/PB-induced liver carcinogenesis in Wistar male rats/50 mg/kg b.w. (p.o.), 14 weeks	↓ Neoplastic nodules	[89]
DMH models in C57 mice, both sexes/100 mg/d/kg b.w. mixed in diet for 30 weeks	↓ Tumor incidence and multiplicity, ↓ PCNA; ↓ Cox-2; ↓ c-Myc; ↓ Survivin	[90]
Female C57BL/6 with H7 or Panc 02 cells - ectopic mouse model/50 mg/kg b.w. i.p. daily, for 2 weeks - orthotopic mouse model/50 mg/kg b.w. i.p. every other day for 1 week	↓ Tumor volume and metastasis	[91]

Table 1. *Cont.*

Animal/Dose	Observation/Outcomes	Ref.
Male Swiss albino mice solid tumor model with EAC cells/Photodynamic therapy with embelin, 12.5 mg/kg b.w. i.p.	↓ Tumor incidence and volume, ↓ myeloperoxidase, ↓ β-d-glucuronidase and Bcl-2; ↑ Rhodanese and Bax	[92]
	In disorders of the CNS	
Swiss albino rats and mice/2.5, 5, and 10 mg/kg	Anticonvulsant: ↓ in the duration of HLTE in MES (2.5 and 5 mg/kg, i.p.) Electroshock,100% protection against mortality ↑ Clonic + tonic onsets at all doses	[65]
Swiss albino mice/2.5 and 5 mg/kg	Antidepressant-like effect in TST ↓ Immobility in the FST Exhibited significant activity in mice TST + FST experimental models	[63]
Male Wistar rats/50, 75, 100 mg/kg	Focal cerebral ischemia brain: ↓ infarction and edema (100 mg/kg) Decreased MDA level (75 and 100 mg/kg) ↑ SOD and CAT (100 mg/kg)	[93]
Swiss albino mice/2.5 and 5 mg/kg	Anxiolytic: ↑ Time spent and number of entries in open arm (elevated plus maze) ↓ Duration of immobility in light box (light/dark model) ↑ Rearing assisted rearing and number of squares crossed (open field test) Embelin: anxiolytic effect in dose-dependent manner	[61]
Male Swiss albino mice/10 and 20 mg/kg	Sickness: Embelin prevented anorexia, anhedonia, Ameliorated brain oxidative stress markers Protective effect in LPS-induced sickness behavior	[94]
Adult Wistar rats/10 and 20 mg/kg/day	Huntington's disease: Loss of b.w. ↓ Oxidative stress ↓ 69–76% brain lesion Protect the neurons from 3-NP toxicity	[95]
Female C57BL/6 mice/25 and 50 mg/kg	Multiple sclerosis (autoimmune encephalomyelitis, CNS inflammation): ↓ Human CD14+ monocyte-derived dendritic cell differentiation ↓ Duction in the EAE clinical score. ↓ Inflammatory Th1 and Th17 cells in EAE.	[96]
Female Sprague–Dawley rats, male C57BL/6 mice/200 nM	Traumatic brain injury: Inhibition of NF-κB expression of XIAP increases in PFT-treated animals. p53 and NF-κB dependent mechanisms delayed neurodegeneration	[97]
Wistar rats' pups/20 mg/kg	HI-induced neurological injury: Confirm sex differences in behavioral and anatomical outcome XIAP protect the female brain from the early HI injury	[98]
C57BL/6 male, GI female, and Ovx female mice/20 mg/kg	Cerebral ischemia: Inhibitor of XIAP exacerbated stroke-induced injury in females, no effect in males	[99]
Human glioma cell lines T98G, U87MG, and H4. IM-PHFA/0–50 μM	Apoptosis in human glioma cells via NF-κB inhibition: Embelin suppressed proliferation of human glioma cells Apoptosis in human glioma cells by inhibiting NF-κB. ↓ NF-κB activity by reducing nuclear translocation of p65	[76]
Human brain glioma U87 cells/0, 50, and 100 μg/mL	Apoptosis in human glioma cells via the mitochondrial pathway: Time- + dose-dependent apoptosis of brain glioma cells Arrest the cell cycle in the G0/G1 phase Changes in brain glioma cell mitochondrial membrane potential Shifting of Bax and Bcl-2 to cause apoptosis	[100]

Table 1. *Cont.*

Animal/Dose	Observation/Outcomes	Ref.
Male Wistar rats/25 and 50 mg/kg	Global ischemia/reperfusion-induced brain injury: ↑ Locomotor activity and hanging latency time ↓ Beam walking latency ↓ Lipid peroxidation ↑ Total thiol content and glutathione-S-transferase neuroprotective agent and useful in the treatment of stroke	[60]
	Antidiabetics	
Albino rats of either sex/100 and 200 mg/kg Wistar rats of either sex/200 mg/kg	Hemodynamic measurement (heart rate, systolic BP), blood glucose, HbA1c, blood GSH, serum marker enzymes (LDH and CK), oxidative stress markers in pancreatic tissue (SOD, CAT, GSH, and LPO), histopathology of pancreatic tissue	[31,41]
Wistar rats of either sex/200 mg/kg	Blood glucose, serum lipid profile (TC, TG and HDL), LPO and protein contents in liver and pancreas	[28]
Wistar rats of either sex/100 mg/kg	HbA1c, blood glucose and GSH, serum marker enzymes (CK, LDH), oxidative stress markers in pancreatic tissue (CAT, SOD, GSH, TBARS), histopathology exam of pancreatic tissue	[101]
Wistar albino rats of either sex/100 mg/kg	Hemodynamic measurement (systolic BP, heart rate), HbA1c, blood glucose	[32]
Male Wistar rats/100 mg/kg	Liver weight, b.w., fasting blood glucose, OGTT, hemodynamic measurement (systolic/diastolic BP, heart rate,); serum adiponectin, insulin, leptin, lipase levels; HOMA-IR values; hepatic glucose-6-phosphatase activity/glycogen content; serum lipid profile (AI, CRI, HDL, LDL, TC, TG, VLDL); oxidative stress markers in liver tissue (CAT, GSH, SOD, TBARS), histopathology exam of liver	[33]

3-NP, 3-nitropropionic acid; AI, atherogenic index; AOM, azoxymethane; ApoB, apolipoprotein B; Bax, Bcl-2 associated x, apoptosis regulator; Bcl-2, B-cell lymphoma 2; BMI, body mass index; BP, blood pressure; b.w., body weight; C57B6, C57 black 6; CAC, colitis-associated cancer; CAT, catalase; CK, creatine kinase; c-Myc, c-myelocytomatosis; CNS, central, nervous system; COX-2, cyclooxygenase-2; CRI, coronary risk index; DENA, Diethyl nitrosamine; DMH, 1,2-dimethylhydrazine; DSS, dextran sodium sulfate; EAC, Ehrlich's ascites carcinoma; EAE, experimental autoimmune encephalomyelitis; FST, forced Swimming Test; GPx, glutathione peroxidase; GSH, glutathione; HbA1c, glycosylated hemoglobin; HDL, high-density lipoprotein; HFD, high-fat diet; H7, human embryonic stem cell (hESC) line 7; HI, hypoxia-ischemia; HLTE, hind limb tonic extension; HOMA-IR, homeostasis model assessment of insulin resistance; HPLC, high-performance liquid chromatography; IM-PHFA, immortalized primary human fetal astrocytes; IL-6, interleukin-6; i.p., intra peritoneal; ISO, isoproterenol; LDH, lactate dehydrogenase; LDL, low-density lipoprotein; LPO, lipid peroxidation; LPS, lipopolysaccharide; MDA, malondialdehyde; MES, maximal electroshock-induced seizure; NF-κB, nuclear factor kappa-light-chain-enhancer of activated B cells; NMR, nuclear magnetic resonance; NPD, normal pellet diet; OGTT, oral glucose tolerance test; PANC02, pancreatic adenocarcinoma epithelial cell line; p-Akt, phosphorylated protein kinase B; PB, phenobarbital; PCNA, proliferating cell nuclear antigen; PFT, pifithrin-alpha; p.o., per os; SOD, superoxide dismutase; STAT3, signal transducer and activator of transcription 3; STZ, streptozotocin; TBARS, thiobarbituric acid-reactive substances; TC, total cholesterol; TG, total triglyceride; TLC, thin layer chromatography; TNF-α, tumor necrosis factor-alpha; TST, tail suspension test; UV, ultraviolet; VLDL, very low-density lipoprotein; XIAP, X-linked inhibitor of apoptosis; ↓, decreasing/reducing; ↑, increasing.

Hypothyroidism is a hypometabolic clinical state resulting from inadequate production of thyroid hormones. Vidanga Vati was given thrice a day after meals with lukewarm water for the duration of 8 weeks. *E. ribes* showed statistically significant results on a 21.93% reduction in the level of thyroid-stimulating hormone-sensitive (s-TSH) and on almost all the signs and symptoms of hypothyroidism [102].

In children afflicted with ascarids, clinical trials using alcoholic and aqueous extracts of *E. ribes* fruits were conducted. Alcoholic extract was shown to be successful in 80% of instances, while aqueous extract was shown to be successful in 55% of cases, resulting in ova-free stools. No adverse effects were identified neither during nor after treatment. When the fruits of *E. ribes* were given to worm-infested individuals at dosages of 200 mg/kg, they had positive outcomes. In cases of infections with tapeworm, giardia, and nana, there was a significant improvement. It was enough to take a single dose of up to 8 g. Within 6 to 24 h after taking the medicine, the worms were ejected. The medicine was well tolerated and was considered to have a good safety profile [103].

8. Future Perspectives and Conclusions

As of now, the available medical literature indicates that after providing all the previous data, it can be concluded that *E. ribes* has been used to treat digestive, carminative, laxative, anti-helminthic, and other disorders as a traditional folk medicine in several Asiatic countries and cultures (India, Sri Lanka, Malaya, Singapore, and China). This review highlights the traditional usage, bioactive constituents, and pharmacological characteristics of *E. ribes* in order to explain the scientific connotation and boost the application of the medicinal value of *E. ribes*. Chemical components such as essential oils, alkaloids, phenols, and flavonoids abound in the plant, resulting as significant components with antioxidant, antidiabetic, anticancer, and other relevant therapeutical properties [2].

Embelin and vilangin are two natural compounds with major antioxidant potential, established by experimental studies, while further studies are needed to detail the molecular mechanisms and efficacy and safety profiles [1,18]. Consequently, *E. ribes* may become a useful medical tool in the future therapeutic management of numerous pathologies. Furthermore, the importance of chemical ingredients, current pharmacological effects, and classic usage in folk remedies are all detailed in the present paper. In-depth research on bioactive constituents, mechanisms, toxicity, pharmacokinetics, and clinical trials will be conducted in the future to give a more scientific understanding of *E. ribes*. These review's findings will aid in the judicious use of *E. ribes* medication and its future development.

Funding: Publication of this paper was supported by the University of Oradea, Oradea, Romania, through an internal project. The research was funded by the Romanian Ministry of Research, Innovation and Digitization through Programme 1—Development of the National Research and Development System, Subprogramme 1.2—Institutional Performance—Projects for funding the excellence in RDI, Contract No. 29 PFE/30.12.2021 with University of Oradea, Oradea, Romania.

Acknowledgments: The authors wish to thank to the University of Oradea for the financial support in publishing this paper. Also, the authors express their sincere gratitude to Bharat Ratna Mahamana Pandit Madan Mohan Malviya, the founder of the Banaras Hindu University, Varanasi, for his services to humanity, great vision, and blessings.

Conflicts of Interest: The authors declare no conflict of interest.

References

1. Vasu, S. Botanical Pharmacognosy of the Fruit of *Embelia ribes* Burm. F. *Pharmacogn. Nat. Prod.* **2015**, *1*, 2015. [CrossRef]
2. Mishra, N. Importance of *Embelia ribes*: An update. *IJPSR* **2013**, *4*, 3823–3838.
3. Harish, G.U.; Danapur, V.; Jain, R.; Patell, V.M. Endangered Medicinal Plant *Embelia ribes* Burm.f.-A Review. *Pharmacogn. J.* **2012**, *4*, 6–19. [CrossRef]
4. Xu, Y.; Liu, D.; Hu, J.; Ding, P.; Chen, M. Hyaluronic acid-coated pH sensitive poly (β-amino ester) nanoparticles for co-delivery of embelin and TRAIL plasmid for triple negative breast cancer treatment. *Int. J. Pharm.* **2020**, *573*, 118637. [CrossRef] [PubMed]
5. Guo, S.; He, M.; Liu, M.; Huang, W.; Ouyang, H.; Feng, Y.; Zhong, G.; Yang, S. Chemical Profiling of *Embelia ribes* by Ultra-High-Performance Liquid Chromatography Quadrupole Time-of-Flight Tandem Mass Spectrometry and Its Antioxidant and Anti-inflammatory Activities In Vitro. *J. Chromatogr. Sci.* **2020**, *58*, 241–250. [CrossRef] [PubMed]
6. Page, M.J.; McKenzie, J.E.; Bossuyt, P.M.; Boutron, I.; Hoffmann, T.C.; Mulrow, C.D.; Shamseer, L.; Tetzlaff, J.M.; Akl, E.A.; Brennan, S.E.; et al. The PRISMA 2020 statement: An updated guideline for reporting systematic reviews. *J. Clin. Epidemiol.* **2021**, *134*, 178–189. [CrossRef]
7. Page, M.J.; McKenzie, J.E.; Bossuyt, P.M.; Boutron, I.; Hoffmann, T.C.; Mulrow, C.D.; Shamseer, L.; Tetzlaff, J.M.; Moher, D. Updating guidance for reporting systematic reviews: Development of the PRISMA 2020 statement. *J. Clin. Epidemiol.* **2021**, *134*, 103–112. [CrossRef]
8. Bungau, S.G.; Popa, V.-C. Between Religion and Science Some Aspects Concerning Illness and Healing in Antiquity. *Transylv. Rev.* **2015**, *24*, 3–18.
9. Kirtikar, K.; Basu, B. *Indian Medicinal Plants*; International Book Distribution: Dehradun, India, 2006; Volume 2, pp. 1496–1497.
10. Nadkarni, A. *Nadkarni's (2007) Indian Materia Medica*; Popular Prakashan Pvt Ltd.: Mumbai, India, 2007; Volume 1, p. 478.
11. Rajasekharan, P.E.; Karpakal, S. Ethnopharmacological Uses of *Embelia ribes* Burm.f.-A Review. *J. Pharm. Biol. Sci.* **2014**, *9*, 23–30. [CrossRef]
12. Bist, M.; Prasad, D.S. *Embelia ribes*: A valuable medicinal plant. *J. Chem. Pharm. Res.* **2016**, *2016*, 1229–1233.
13. Bhandari, U.; Ansari, M.N. Protective effect of aqueous extract of *Embelia ribes* Burm fruits in middle cerebral artery occlusion-induced focal cerebral ischemia in rats. *Indian J. Pharmacol.* **2008**, *40*, 215. [CrossRef] [PubMed]

14. Bansal, P.; Bhandari, U.; Ahmad, S. Embelin from *Embelia ribes* ameliorates oxidative stress and inflammation in high-fat diet-fed obese C57BL/6 mice. *Pharmacogn. Mag.* **2020**, *16*, 443.
15. Swamy, H.K.; Krishna, V.; Shankarmurthy, K.; Rahiman, B.A.; Mankani, K.; Mahadevan, K.; Harish, B.; Naika, H.R. Wound healing activity of embelin isolated from the ethanol extract of leaves of *Embelia ribes* Burm. *J. Ethnopharmacol.* **2007**, *109*, 529–534. [CrossRef] [PubMed]
16. Dhayalan, M.; Denison, M.I.J.; Krishnan, K. In vitro antioxidant, antimicrobial, cytotoxic potential of gold and silver nanoparticles prepared using *Embelia ribes*. *Nat. Prod. Res.* **2017**, *31*, 465–468. [CrossRef] [PubMed]
17. Vadivel, V.; Abirami, K.; Natesan, R.; Brindha, P. Antioxidant, anti-cancer and phytochemical constituents of hexane extract fractions of an indian medicinal plant *Embelia ribes* burm. *Int. J. Pharm. Phytochem. Res.* **2016**, *8*, 228–234.
18. Balachandran, C.; Duraipandiyan, V.; Balakrishna, K.; Sundaram, R.; Vijayakumar, A.; Ignacimuthu, S.; Al-Harbi, N. Synthesis and medicinal properties of plant-derived vilangin. *Environ. Chem. Lett.* **2013**, *11*, 303–308. [CrossRef]
19. Stalin, A.; Irudayaraj, S.S.; Gandhi, G.R.; Balakrishna, K.; Ignacimuthu, S.; Al-Dhabi, N.A. Hypoglycemic activity of 6-bromoembelin and vilangin in high-fat diet fed-streptozotocin-induced type 2 diabetic rats and molecular docking studies. *Life Sci.* **2016**, *153*, 100–117. [CrossRef]
20. Hao, K.; Ali, M.; Siddiqui, A.W. New compounds from the seeds of *Embelia ribes* Burm. *Die Pharm.* **2005**, *60*, 69–71.
21. McErlean, C.S.; Moody, C.J. First synthesis of N-(3-carboxylpropyl)-5-amino-2-hydroxy-3-tridecyl-1, 4-benzoquinone, an unusual quinone isolated from *Embelia ribes*. *J. Org. Chem.* **2007**, *72*, 10298–10301. [CrossRef]
22. Chen, J.-P.; Zhu, L.-J.; Su, X.-X.; Zhang, K.-X.; Zhang, X.; Wang, J.-H.; Yao, X.-S. New alkylresorcinols from the fruits of *Embelia ribes*. *Fitoterapia* **2018**, *128*, 66–72. [CrossRef]
23. Lin, P.; Li, S.; Wang, S.; Yang, Y.; Shi, J. A nitrogen-containing 3-alkyl-1, 4-benzoquinone and a gomphilactone derivative from *Embelia ribes*. *J. Nat. Prod.* **2006**, *69*, 1629–1632. [CrossRef] [PubMed]
24. Lin, P.-C.; Li, S.; Wang, S.-J.; YANG, Y.-C.; SHI, J.-G. Phenolic constituents in *Embelia ribes*. *Chin. Tradit. Herb. Drugs* **2006**, *37*, 818.
25. Qin, Y.; Chen, J.-P.; Li, C.-Y.; Zhu, L.-J.; Zhang, X.; Wang, J.-H.; Yao, X.-S. Flavonoid glycosides from the fruits of *Embelia ribes* and their anti-oxidant and α-glucosidase inhibitory activities. *J. Asian Nat. Prod. Res.* **2021**, *23*, 724–730. [CrossRef] [PubMed]
26. Dang, P.H.; Nguyen, H.X.; Nguyen, N.T.; Le, H.N.T.; Nguyen, M.T.T. α-Glucosidase Inhibitors from the Stems of *Embelia ribes*. *Phytother. Res.* **2014**, *28*, 1632–1636. [CrossRef]
27. Glevitzky, I.; Dumitrel, G.A.; Glevitzky, M.; Pasca, B.; Otrisal, P.; Bungau, S.; Cioca, G.; Pantis, C.; Popa, M. Statistical Analysis of the Relationship Between Antioxidant Activity and the Structure of Flavonoid Compounds. *Rev. Chim.* **2019**, *70*, 3103–3107. [CrossRef]
28. Bhandari, U.; Kanojia, R.; Pillai, K. Effect of ethanolic extract of *Embelia ribes* on dyslipidemia in diabetic rats. *Int. J. Exp. Diabetes Res.* **2002**, *3*, 159–162. [CrossRef]
29. Mahendran, S.; Badami, S.; Maithili, V. Evaluation of antidiabetic effect of embelin from *Embelia ribes* in alloxan induced diabetes in rats. *Biomed. Prev. Nutr.* **2011**, *1*, 25–31. [CrossRef]
30. Chaudhari, H.S.; Bhandari, U.; Khanna, G. *Embelia ribes* extract reduces high fat diet and low dose streptozotocin-induced diabetic nephrotoxicity in rats. *EXCLI J.* **2013**, *12*, 858.
31. Bhandari, U.; Ansari, M.N. Antihyperglycaemic activity of aqueous extract of *Embelia ribes* Burm in streptozotocin-induced diabetic rats. *Indian J. Exp. Biol.* **2008**, *46*, 607–613.
32. Bhandari, U.; Jain, N.; Ansari, M.; Pillai, K. Beneficial effect of *Embelia ribes* ethanolic extract on blood pressure and glycosylated hemoglobin in streptozotocin-induced diabetes in rats. *Fitoterapia* **2008**, *79*, 351–355. [CrossRef]
33. Bhandari, U.; Chaudhari, H.S.; Khanna, G.; Najmi, A.K. Antidiabetic effects of *Embelia ribes* extract in high fat diet and low dose streptozotocin-induced type 2 diabetic rats. *Front. Life Sci.* **2013**, *7*, 186–196. [CrossRef]
34. Purohit, A.; Vyas, K.B.; Vyas, S.K. Hypoglycaemic activity of *Embelia ribes* berries (50% etoh) extract in alloxan induced diabetic rats. *Anc. Sci. Life* **2008**, *27*, 41.
35. Durg, S.; Veerapur, V.P.; Neelima, S.; Dhadde, S.B. Antidiabetic activity of *Embelia ribes*, embelin and its derivatives: A systematic review and meta-analysis. *Biomed. Pharmacother.* **2017**, *86*, 195–204. [CrossRef] [PubMed]
36. Bhandari, U.; Chaudhari, H.; Bisnoi, A.; Kumar, V.; Khanna, G.; Javed, K. Anti-obesity effect of standardized ethanol extract of *Embelia ribes* in murine model of high fat diet-induced obesity. *PharmaNutrition* **2013**, *1*, 50–57. [CrossRef]
37. Chaudhari, H.S.; Bhandari, U.; Khanna, G. Preventive effect of embelin from *Embelia ribes* on lipid metabolism and oxidative stress in high-fat diet-induced obesity in rats. *Planta Med.* **2012**, *78*, 651–657. [CrossRef] [PubMed]
38. Nazish, I.; Ansari, S.H.; Arora, P. Antiobesity actions of *Embelia ribes*. *Pharmacogn. J.* **2012**, *4*, 73–80. [CrossRef]
39. Ansari, M.N.; Bhandari, U. Effect of an ethanol extract of *Embelia ribes* fruits on isoproterenol-induced myocardial infarction in albino rats. *Pharm. Biol.* **2008**, *46*, 928–932. [CrossRef]
40. Bhandari, U.; Ansari, M.N.; Islam, F. Cardioprotective effect of aqueous extract of *Embelia ribes* Burm fruits against isoproterenol-induced myocardial infarction in albino rats. *Indian J. Exp. Biol.* **2008**, *46*, 35–40.
41. Bhandari, U.; Ansari, M.N. Ameliorative effect of an ethanol extract of *Embelia ribes* fruits on isoproterenol-induced cardiotoxicity in diabetic rats. *Pharm. Biol.* **2009**, *47*, 669–674. [CrossRef]
42. Caruso, F.; Paumier, S.; Rossi, M. X-Ray Crystal Structure of Embelin and Its DFT Scavenging of Superoxide Radical. *J. Comput. Chem.* **2018**, *39*, 1143–1148. [CrossRef]

43. Chen, X.; Gao, M.; Jian, R.; Hong, W.D.; Tang, X.; Li, Y.; Zhao, D.; Zhang, K.; Chen, W.; Zheng, X. Design, synthesis and α-glucosidase inhibition study of novel embelin derivatives. *J. Enzym. Inhib. Med. Chem.* **2020**, *35*, 565–573. [CrossRef]
44. SreeHarsha, N. Embelin impact on paraquat-induced lung injury through suppressing oxidative stress, inflammatory cascade, and MAPK/NF-κB signaling pathway. *J. Biochem. Mol. Toxicol.* **2020**, *34*, e22456. [CrossRef] [PubMed]
45. Basavaraj, H.; Ashok, P. In vitro antioxidant activity of aqueous and ethanolic extract of *Coscinium fenestratum* root and *Embelia ribes* flower. *Res. J. Pharm. Technol.* **2012**, *5*, 513–517.
46. Caruso, F.; Rossi, M.; Kaur, S.; Garcia-Villar, E.; Molasky, N.; Belli, S.; Sitek, J.D.; Gionfra, F.; Pedersen, J.Z.; Incerpi, S. Antioxidant Properties of Embelin in Cell Culture. Electrochemistry and Theoretical Mechanism of Scavenging. Potential Scavenging of Superoxide Radical through the Cell Membrane. *Antioxidants* **2020**, *9*, 382. [CrossRef] [PubMed]
47. Joshi, R.; Kamat, J.P.; Mukherjee, T. Free radical scavenging reactions and antioxidant activity of embelin: Biochemical and pulse radiolytic studies. *Chem.-Biol. Interact.* **2007**, *167*, 125–134. [CrossRef]
48. Ferreira, G.; Laddha, K.S. Stress degradation studies on embelin. *Indian J. Pharm. Sci.* **2013**, *75*, 246–250.
49. Joshi, R.; Adhikari, S.; Patro, B.S.; Chattopadhyay, S.; Mukherjee, T. Free radical scavenging behavior of folic acid: Evidence for possible antioxidant activity. *Free Radic. Biol. Med.* **2001**, *30*, 1390–1399. [CrossRef]
50. Schöneich, C.; Asmus, K.D. Reaction of thiyl radicals with alcohols, ethers and polyunsaturated fatty acids: A possible role of thiyl free radicals in thiol mutagenesis? *Radiat. Env. Biophys.* **1990**, *29*, 263–271. [CrossRef]
51. Schöneich, C.; Dillinger, U.; von Bruchhausen, F.; Asmus, K.-D. Oxidation of polyunsaturated fatty acids and lipids through thiyl and sulfonyl radicals: Reaction kinetics, and influence of oxygen and structure of thiyl radicals. *Arch. Biochem. Biophys.* **1992**, *292*, 456–467. [CrossRef]
52. Packer, J.E.; Slater, T.F.; Willson, R.L. Direct observation of a free radical interaction between vitamin E and vitamin C. *Nature* **1979**, *278*, 737–738. [CrossRef]
53. Saini, P.; Lakshmayya, L.; Bisht, V.S. Anti-Alzheimer activity of isolated karanjin from *Pongamia pinnata* (L.) pierre and embelin from *Embelia ribes* Burm.f. *Ayu* **2017**, *38*, 76. [PubMed]
54. Karthika, C.; Appu, A.P.; Akter, R.; Rahman, M.H.; Tagde, P.; Ashraf, G.M.; Abdel-Daim, M.M.; ul Hassan, S.S.; Abid, A.; Bungau, S. Potential innovation against Alzheimer's disorder: A tricomponent combination of natural antioxidants (vitamin E, quercetin, and basil oil) and the development of its intranasal delivery. *Environ. Sci. Pollut. Res.* **2022**, *29*, 10950–10965. [CrossRef] [PubMed]
55. Sharma, V.; Firdaus, Z.; Rai, H.; Nayak, P.K.; Singh, T.D.; Gautam, D.N.S. Consumption of Ashtanga Ghrita (clarified cow butter added with herb extracts) improves cognitive dysfunction induced by scopolamine in rats via regulation of acetylcholinesterase activity and oxidative stress. *Drug Metab. Pers. Ther.* **2021**, *36*, 337–350. [CrossRef] [PubMed]
56. Sharma, V.; Sharma, R.; Gautam, D.S.; Kuca, K.; Nepovimova, E.; Martins, N. Role of Vacha (Acorus calamus Linn.) in neurological and metabolic disorders: Evidence from ethnopharmacology, phytochemistry, pharmacology and clinical study. *J. Clin. Med.* **2020**, *9*, 1176. [CrossRef] [PubMed]
57. Behl, T.; Kaur, I.; Fratila, O.; Brata, R.; Bungau, S. Exploring the Potential of Therapeutic Agents Targeted towards Mitigating the Events Associated with Amyloid-β Cascade in Alzheimer's Disease. *Int. J. Mol. Sci.* **2020**, *21*, 7443. [CrossRef]
58. Nuthakki, V.K.; Sharma, A.; Kumar, A.; Bharate, S.B. Identification of embelin, a 3-undecyl-1, 4-benzoquinone from *Embelia ribes* as a multitargeted anti-Alzheimer agent. *Drug Dev. Res.* **2019**, *80*, 655–665. [CrossRef] [PubMed]
59. Nazam Ansari, M.; Bhandari, U.; Islam, F.; Tripathi, C. Evaluation of antioxidant and neuroprotective effect of ethanolic extract of *Embelia ribes* Burm in focal cerebral ischemia/reperfusion-induced oxidative stress in rats. *Fundam. Clin. Pharmacol.* **2008**, *22*, 305–314. [CrossRef] [PubMed]
60. Thippeswamy, B.; Nagakannan, P.; Shivasharan, B.; Mahendran, S.; Veerapur, V.; Badami, S. Protective effect of embelin from *Embelia ribes* Burm. against transient global ischemia-induced brain damage in rats. *Neurotox. Res.* **2011**, *20*, 379–386. [CrossRef]
61. Afzal, M.; Gupta, G.; Kazmi, I.; Rahman, M.; Upadhyay, G.; Ahmad, K.; Imam, F.; Pravez, M.; Anwar, F. Evaluation of anxiolytic activity of embelin isolated from *Embelia ribes*. *Biomed. Aging Pathol.* **2012**, *2*, 45–47. [CrossRef]
62. Ghaisas, M.; Wadikar, A.; Gulati, T.; Limaye, R. Anxiolytic Effect of a Methanolic Extract of the *Embelia ribes* Burm F. in Mice. *Res. J. Pharm. Technol.* **2010**, *3*, 1136–1139.
63. Gupta, G.; Kazmi, I.; Afzal, M.; Upadhyay, G.; Singh, R.; Habtemariam, S. Antidepressant-like activity of Embelin isolated from *Embelia ribes*. *Phytopharmacology* **2013**, *4*, 87–95.
64. Durg, S.; Kumar, N.; Vandal, R.; Dhadde, S.B.; Thippeswamy, B.S.; Veerapur, V.P.; Badami, S. Antipsychotic activity of embelin isolated from *Embelia ribes*: A preliminary study. *Biomed. Pharm.* **2017**, *90*, 328–331. [CrossRef]
65. Mahendran, S.; Thippeswamy, B.S.; Veerapur, V.P.; Badami, S. Anticonvulsant activity of embelin isolated from *Embelia ribes*. *Phytomedicine* **2011**, *18*, 186–188. [CrossRef]
66. Agrawal, S.; Chauhan, S.; Mathur, R. Antifertility effects of embelin in male rats. *Andrologia* **1986**, *18*, 125–131. [CrossRef] [PubMed]
67. Narayanaswamy, R.; Gnanamani, A.; Mandal, A. A potential antibacterial agent Embelin, a natural benzoquinone extracted from *Embelia ribes*. *Biol. Med.* **2011**, *3*, 1–7.
68. Rani, A.S.; Sulakshana, G.; Nagamani, V. Evaluation of antibacterial potential of *Embelia ribes*. *Med. Plants Int. J. Phytomed. Relat. Ind.* **2011**, *3*, 71. [CrossRef]
69. Rani, A.S.; Saritha, K.; Nagamani, V.; Sulakshana, G. In vitro Evaluation of Antifungal Activity of the Seed Extract of *Embelia Ribes*. *Indian J. Pharm. Sci.* **2011**, *73*, 247–249. [CrossRef]

70. Rathi, S.G.; Bhaskar, V.H.; Patel, P.G. Antifungal activity of *Embelia ribes* plant extracts. *Int. J. Pharm. Biol. Res.* **2010**, *6*, 10.
71. Joy, B.; Lakshmi, S. Antiproliferative properties of *Embelia ribes*. *Open Proc. Chem. J.* **2010**, *3*, 17–22.
72. Kaur, V.; Hallan, S.S.; Nidhi, A.; Mishra, N. Isolation of embelin from and evaluation of its anti-cancer potential in *Embelia ribes* breast cancer. *Asian J. Pharm. Pharmacol.* **2015**, *1*, 33–39.
73. Rai, H.; Panwar, M.S.; Tomer, S.K. Analysis of masked data with Lindley failure model. *Commun. Stat. Simul. Comput.* **2021**, 1–25. [CrossRef]
74. Ko, J.-H.; Lee, S.-G.; Yang, W.M.; Um, J.-Y.; Sethi, G.; Mishra, S.; Shanmugam, M.K.; Ahn, K.S. The Application of Embelin for Cancer Prevention and Therapy. *Molecules* **2018**, *23*, 621. [CrossRef] [PubMed]
75. Park, N.; Baek, H.S.; Chun, Y.J. Embelin-Induced Apoptosis of Human Prostate Cancer Cells Is Mediated through Modulation of Akt and β-Catenin Signaling. *PLoS ONE* **2015**, *10*, e0134760. [CrossRef] [PubMed]
76. Park, S.-Y.; Lim, S.-L.; Jang, H.-J.; Lee, J.-H.; Um, J.-Y.; Kim, S.-H.; Ahn, K.S.; Lee, S.-G. Embelin induces apoptosis in human glioma cells through inactivating NF-κB. *J. Pharmacol. Sci.* **2013**, *121*, 192–199. [CrossRef]
77. Joshi, R.; Ghanty, T.K.; Mukherjee, T. Formation of semiquinone radical in the reaction of embelin (2,5-dihydroxy-3-undecyl-1,4-benzoquinone) with reductants as well as oxidants. Characterization by pulse radiolysis and structure investigation by quantum chemical study. *J. Mol. Struct.* **2009**, *928*, 46–53. [CrossRef]
78. Avisetti, D.R.; Babu, K.S.; Kalivendi, S.V. Activation of p38/JNK pathway is responsible for embelin induced apoptosis in lung cancer cells: Transitional role of reactive oxygen species. *PLoS ONE* **2014**, *9*, e87050. [CrossRef]
79. Hossan, M.S.; Fatima, A.; Rahmatullah, M.; Khoo, T.J.; Nissapatorn, V.; Galochkina, A.V.; Slita, A.V.; Shtro, A.A.; Nikolaeva, Y.; Zarubaev, V.V.; et al. Antiviral activity of *Embelia ribes* Burm. f. against influenza virus in vitro. *Arch. Virol.* **2018**, *163*, 2121–2131. [CrossRef]
80. Shirole, R.L.; Shirole, N.L.; Saraf, M.N. *Embelia ribes* ameliorates lipopolysaccharide-induced acute respiratory distress syndrome. *J. Ethnopharmacol.* **2015**, *168*, 356–363. [CrossRef]
81. Elias, T.; Lee, L.; Rossi, M.; Caruso, F.; Adams, S. In Vitro Analysis of the Antioxidant and Antiviral Activity of Embelin Against Herpes Simplex Virus-1. *Microorganisms* **2021**, *9*, 434. [CrossRef]
82. Ansari, M.N.; Bhandari, U. Antihyperhomocysteinemic activity of an Ethanol Extract from *Embelia ribes*. in albino Rats. *Pharm. Biol.* **2008**, *46*, 283–287. [CrossRef]
83. Bhandari, U.; Ansari, M.N.; Islam, F.; Tripathi, C.D. The effect of aqueous extract of *Embelia ribes* Burm on serum homocysteine, lipids and oxidative enzymes in methionine induced hyperhomocysteinemia. *Indian J. Pharmacol.* **2008**, *40*, 152–157. [CrossRef] [PubMed]
84. Wang, H.; Zhang, H.; Wang, Y.; Yang, L.; Wang, D. Embelin can protect mice from thioacetamide-induced acute liver injury. *Biomed. Pharmacother.* **2019**, *118*, 109360. [CrossRef] [PubMed]
85. Ahn, J.; Yong, H. Antioxidant and inhibition of senescence effects of embelin. *Biomed. Dermatol.* **2018**, *2*, 11. [CrossRef]
86. Li, Z.; Chen, S.-J.; Yu, X.-A.; Li, J.; Gao, X.-M.; He, J.; Chang, Y.-X. Pharmacokinetic and Bioavailability Studies of Embelin after Intravenous and Oral Administration to Rats. *Evid.-Based Complement. Altern. Med.* **2019**, *2019*, 9682495. [CrossRef] [PubMed]
87. Sharma, A.; Kumar, R.; Mishra, A.; Gupta, R. Problems associated with clinical trials of Ayurvedic medicines. *Rev. Bras. Farmacogn.* **2010**, *20*, 276–281. [CrossRef]
88. Dai, Y.; Jiao, H.; Teng, G.; Wang, W.; Zhang, R.; Wang, Y.; Hebbard, L.; George, J.; Qiao, L. Embelin reduces colitis-associated tumorigenesis through limiting IL-6/STAT3 signaling. *Mol. Cancer Ther.* **2014**, *13*, 1206–1216. [CrossRef]
89. Sreepriya, M.; Bali, G. Chemopreventive effects of embelin and curcumin against N-nitrosodiethylamine/phenobarbital-induced hepatocarcinogenesis in Wistar rats. *Fitoterapia* **2005**, *76*, 549–555. [CrossRef]
90. Dai, Y.; Qiao, L.; Chan, K.W.; Yang, M.; Ye, J.; Ma, J.; Zou, B.; Gu, Q.; Wang, J.; Pang, R.; et al. Peroxisome proliferator-activated receptor-gamma contributes to the inhibitory effects of Embelin on colon carcinogenesis. *Cancer Res.* **2009**, *69*, 4776–4783. [CrossRef]
91. Peng, M.; Huang, B.; Zhang, Q.; Fu, S.; Wang, D.; Cheng, X.; Wu, X.; Xue, Z.; Zhang, L.; Zhang, D.; et al. Embelin inhibits pancreatic cancer progression by directly inducing cancer cell apoptosis and indirectly restricting IL-6 associated inflammatory and immune suppressive cells. *Cancer Lett.* **2014**, *354*, 407–416. [CrossRef]
92. Joy, B.; Nishanth Kumar, S.; Soumya, M.S.; Radhika, A.R.; Vibin, M.; Abraham, A. Embelin (2,5-dihydroxy-3-undecyl-p-benzoquinone): A bioactive molecule isolated from *Embelia ribes* as an effective photodynamic therapeutic candidate against tumor in vivo. *Phytomedicine* **2014**, *21*, 1292–1297. [CrossRef]
93. Patel, R.S.; Gohil, P. Effect of embelin in middle cerebral artery occlusioninduced focal cerebral ischemia in rats. *Oxid. Antioxid. Med. Sci.* **2014**, *3*, 135–139. [CrossRef]
94. Shaikh, A.; Dhadde, S.B.; Durg, S.; Veerapur, V.; Badami, S.; Thippeswamy, B.; Patil, J.S. Effect of Embelin Against Lipopolysaccharide-induced Sickness Behaviour in Mice. *Phytother. Res.* **2016**, *30*, 815–822. [CrossRef]
95. Dhadde, S.B.; Nagakannan, P.; Roopesh, M.; Kumar, S.A.; Thippeswamy, B.; Veerapur, V.P.; Badami, S. Effect of embelin against 3-nitropropionic acid-induced Huntington's disease in rats. *Biomed. Pharmacother.* **2016**, *77*, 52–58. [CrossRef] [PubMed]
96. Xue, Z.; Ge, Z.; Zhang, K.; Sun, R.; Yang, J.; Han, R.; Peng, M.; Li, Y.; Li, W.; Zhang, D.; et al. Embelin suppresses dendritic cell functions and limits autoimmune encephalomyelitis through the TGF-β/β-catenin and STAT3 signaling pathways. *Mol. Neurobiol.* **2014**, *49*, 1087–1101. [CrossRef] [PubMed]

97. Plesnila, N.; Von Baumgarten, L.; Retiounskaia, M.; Engel, D.; Ardeshiri, A.; Zimmermann, R.; Hoffmann, F.; Landshamer, S.; Wagner, E.; Culmsee, C. Delayed neuronal death after brain trauma involves p53-dependent inhibition of NF-κB transcriptional activity. *Cell Death Differ.* **2007**, *14*, 1529–1541. [CrossRef]
98. Hill, C.; Alexander, M.; McCullough, L.; Fitch, R. Inhibition of X-linked inhibitor of apoptosis with embelin differentially affects male versus female behavioral outcome following neonatal hypoxia-ischemia in rats. *Dev. Neurosci.* **2011**, *33*, 494–504. [CrossRef]
99. Siegel, C.; Li, J.; Liu, F.; Benashski, S.E.; McCullough, L.D. miR-23a regulation of X-linked inhibitor of apoptosis (XIAP) contributes to sex differences in the response to cerebral ischemia. *Proc. Natl. Acad. Sci. USA* **2011**, *108*, 11662–11667. [CrossRef]
100. Wang, A.; Zhang, B.; Zhang, J.; Wu, W. Embelin-induced brain glioma cell apoptosis and cell cycle arrest via the mitochondrial pathway. *Oncol. Rep.* **2013**, *29*, 2473–2478. [CrossRef]
101. Bhandari, U.; Jain, N.; Pillai, K. Further studies on antioxidant potential and protection of pancreatic β-cells by *Embelia ribes* in Experimental diabetes. *Exp. Diabetes Res.* **2007**, *2007*, 015803. [CrossRef]
102. Kanzaria, H.; Dave, A.; Manani, Y.; Agravat, P. A clinical study on hypothyroidism and its management with Vidanga. *Eur. J. Biomed. Pharmaceut. Sci.* **2017**, *4*, 241–244.
103. Guru, L.; Mishra, D. Effect of *Embelia ribes* (Burm) in patients infested by ascarides. *Antiseptic* **1965**, *62*, 273–278.

antioxidants

MDPI

Article

Valeriana pilosa Roots Essential Oil: Chemical Composition, Antioxidant Activities, and Molecular Docking Studies on Enzymes Involved in Redox Biological Processes

Patricia Minchán-Herrera [1], Roberto O. Ybañez-Julca [1,*], Ivan M. Quispe-Díaz [1], Edmundo A. Venegas-Casanova [1], Rafael Jara-Aguilar [1], Felipe Salas [2], Liz Zevallos-Escobar [3], Osvaldo Yáñez [4], Ricardo Pino-Rios [5], Pedro Buc Calderon [2,6] and Julio Benites [1,2,*]

1 Facultad de Farmacia y Bioquímica, Universidad Nacional de Trujillo, Trujillo 13011, Peru; pminchan@unitru.edu.pe (P.M.-H.); iquispe@unitru.edu.pe (I.M.Q.-D.); evenegas@unitru.edu.pe (E.A.V.-C.); djara@unitru.edu.pe (R.J.-A.)

2 Química y Farmacia, Facultad de Ciencias de la Salud, Universidad Arturo Prat, Casilla 121, Iquique 1100000, Chile; felisala@unap.cl (F.S.); pedro.buccalderon@uclouvain.be (P.B.C.)

3 Escuela de Farmacia y Bioquímica, Universidad Católica Los Ángeles de Chimbote, Chimbote 02801, Peru; lzevallose@uladech.edu.pe

4 Facultad de Ingeniería y Negocios, Universidad de las Américas, Santiago 7500000, Chile; oyanez@udla.cl

5 Instituto de Ciencias Químicas Aplicadas, Facultad de Ingeniería, Universidad Autónoma de Chile, Santiago 7500912, Chile; ricardopinor@gmail.com

6 Research Group in Metabolism and Nutrition, Louvain Drug Research Institute, Université Catholique de Louvain, 73 Avenue E. Mounier, GTOX 7309, 1200 Brussels, Belgium

* Correspondence: rybanez@unitru.edu.pe (R.O.Y.-J.); juliob@unap.cl (J.B.); Tel.: +51-0449-7634-5993 (R.O.Y.-J.); +56-57-2252-6275 (J.B.)

Citation: Minchán-Herrera, P.; Ybañez-Julca, R.O.; Quispe-Díaz, I.M.; Venegas-Casanova, E.A.; Jara-Aguilar, R.; Salas, F.; Zevallos-Escobar, L.; Yáñez, O.; Pino-Rios, R.; Calderon, P.B.; et al. *Valeriana pilosa* Roots Essential Oil: Chemical Composition, Antioxidant Activities, and Molecular Docking Studies on Enzymes Involved in Redox Biological Processes. *Antioxidants* **2022**, *11*, 1337. https://doi.org/10.3390/antiox11071337

Academic Editors: Delia Mirela Tit and Simona Bungau

Received: 6 June 2022
Accepted: 4 July 2022
Published: 7 July 2022

Abstract: *Valeriana pilosa* is usually employed in Peruvian folk medicine in the form of infusion to treat stomach pain, and has antispasmodic, relaxing, sleep-promoting, and sedative properties, as well as is an anti-inflammatory. In this study, *Valeriana pilosa* essential oil (VPEO) was obtained by hydrodistillation, analyzed by GC and GC/MS, and 47 compounds were identified. Major oil components were α-patchoulene (5.8%), α-humulene (6.1%), seychellene (7.6%), and patchoulol (20.8%). Furthermore, we assessed the in vitro antioxidant activities, molecular docking, and Ligand Efficiency studies on enzymes involved in cellular redox pathways such as CYP2C9, catalase, superoxide dismutase, and xanthine oxidase. Essential oil antioxidant activities were assessed by FRAP, $ABTS^{\bullet+}$, and $DPPH^{\bullet}$ radical scavenging activity. VPEO displays high antioxidant activity as compared to essential oils of *Valeriana jatamansi* and *Valeriana officinalis* oil roots. In addition, molecular docking and ADMET prediction was employed to compare the absorption, metabolism, and toxicity properties of *Valeriana pilosa* compounds. In the molecular docking studies, limonene, *p*-cimene, carvone, α-cubebene, cyclosativene, α-guaiene, allo-aromadendrene, valencene, and eremophyllene were the compounds with the best docking score on CYP2C9 and xanthine oxidase. Thus, volatile components of *Valeriana pilosa* could be associated with the detected antioxidant activity, acting as putative inhibitors of CYP2C9 and xanthine oxidase.

Keywords: *Valeriana pilosa*; antioxidant activities; molecular docking; antioxidant enzyme; oxidative stress

1. Introduction

Sociocultural and health care necessities of rural people of emerging and developing countries are mainly assured by the use of curative and aromatic plants. Despite advances in modern medicine, it is estimated that over 80% of the developing world's population still relies on traditional medicines (mainly herbs) to provide their health care needs, a tendency essentially imputed to strong cultural beliefs, accessibility, and low costs [1].

Natural compounds such as essential oils are complex mixtures, including typically volatile compounds, which have been employed in traditional and modern medicines as well as in perfumes and cosmetics manufacturing, aside from pharmaceutical therapies and herbal beverages [2,3].

The sub-family Caprifoliaceae of the genus *Valeriana* contains more than 350 species distributed throughout the temperate Northern Hemisphere, Africa, and South America. In the Andean region, it represents an important center of secondary diversification. Numerous species are currently employed worldwide as medicines [4,5]. In Peru, about 73 species have been reported, and among them 45 species are endemic [6]. *Valeriana pilosa* has been widely used by local inhabitants for stomach distress and epilepsy, for its antispasmodic, relaxing, sleep-promoting, and sedative properties, and even as an anti-inflammatory [7–9]. Peruvian folk medicine often refers to *Valeriana pilosa* as "Valeriana", "Coche coche", "Valeriana de paramo", "Ornamo", or "Babilla".

The volatile components of plant species from Chilean and Peruvian Andean highlands communities have been one of the focuses of our chemical and biological research projects [10–15]. Since the chemical composition of VPEO has not yet been investigated, the aim of this study was to identify and quantify the components of VPEO from roots of *Valeriana pilosa*. In addition to the chemical composition, we investigated its antioxidant activities, and performed molecular docking studies on some redox enzymes and ADMET profiles.

Briefly, the present investigation aims to (i) determine the chemical composition of VPEO; (ii) evaluate the in vitro antioxidant activities; (iii) carry out in silico studies about the inhibitory effect of VPEO volatile phytochemicals on the crystal structure of some critical proteins; and (iv) perform ADMET prediction of VPEO compounds.

The results obtained in this study may supply further guidance for the correct use of *Valerian pilosa*. Moreover, we advocate for better protection of this herb within the context of a growing demand in the market.

2. Materials and Methods

2.1. Plant Material

Valeriana pilosa R & P plants were collected in January 2021 in the Community of San Juan de Corralpampa at 3500 m above sea level, in the Hualgayoc Province, Department of Cajamarca, Peru. The specimen was identified and deposited in the "Herbarium Truxillense de la Facultad de Ciencias Biologicas de la Universidad Nacional de Trujillo" (voucher specimen HUT 61241–61242).

2.2. Essential Oil Isolation

Essential oil of the roots (50 g) was extracted through hydrodistillation for 3 h using a Clevenger-type apparatus. The yield was determined based on a moisture-free basis as 0.20% (*w*/*w*). The oil obtained was dried over anhydrous Na_2SO_4. VPEO was filtered, and the sample container was tightly sealed and stored at +4 °C until analysis.

2.3. Gas Chromatography Analysis (GC)

All chemicals were of analytical reagent grade, and they were obtained from Sigma-Aldrich-Fluka (St. Louis, MO, USA), Merck (Darmstadt, Germany) and were employed as supplied. The VPEO was studied on a Perkin Elmer Clarus 600 gas chromatograph according to procedures reported by Benites et al. [15].

2.4. Gas Chromatography-Mass Spectrometry (GC-MS)

The GC MS analysis of the VPEO was carried out as reported previously [15]. Briefly, analyses were performed on a Perkin Elmer Clarus 600 gas chromatograph, consisting of a DB-1 fused-silica column (30 m × 0.25 mm i.d., film thickness 0.25 μM; J & W Scientific, Folsom, CA, USA), and interfaced with a Perkin Elmer Clarus 600T mass spectrometer (software version 4.1, Perkin Elmer, Shelton, CT, USA). Both injector and oven temperatures were as above; transfer line temperature, 280 °C; ion source temperature, 220 °C; carrier gas, helium, adapted to a linear rate of 30 cm/s; split ratio, 1:40; ionization energy, 70 eV; scan range, 40–300 m/z; scan time, 1 s. The identification of components was achieved by comparing their retention indices, relative to C_9–C_{21} n-alkane indices and GC-MS spectra from a homemade library, made by analyses of reference oils, laboratory-synthesized components, and commercial sample standards.

2.5. Antioxidant Capacity Assays

2.5.1. Ferric-Reducing Antioxidant Power (FRAP) Assay

The FRAP assay was performed as previously reported [16] with the following adjustments. Briefly, the FRAP stock solutions included 300 mM acetate buffer pH 3.6, 10 mM TPTZ (2,4,6-tripyridyl-s-triazine) solution in 40 mM HCl, and 20 mM $FeCl_3 \times 6H_2O$. The working solution was made by mixing 25 mL of acetate buffer, 2.5 mL of TPTZ, and 2.5 mL of $FeCl_3 \times 6H_2O$. Prior to use the solution was heated at 3 °C. A stock solution of 0.5 mM of the Trolox® (Sigma-Aldrich-Fluka, St. Louis, MO, USA) was further prepared by serial dilutions (0.05, 0.1, 0.2, 0.3, 0.4, and 0.5 mM).

Aliquots of VPEO (8 μL) were let to react with 200 μL of the fresh FRAP solution for 30 min in the dark. Afterwards, the absorbance of colored product ferrous tripyridyltriazine complex was read at 593 nm (n = 3). The standard curve was made with the standard antioxidant Trolox®. Results are expressed as mM of TEAC (Trolox® equivalents)/mL of VPEO.

2.5.2. $ABTS^{\bullet+}$ Free Radical Scavenging Activity

The $ABTS^{\bullet+}$ radical assay was performed according to procedures reported by Re et al. [17]. First, Trolox® solution (1 mg/mL) was prepared by dissolving it in ethanol (EtOH) and further stored in the dark. Stock solutions were successively diluted in 96-well microplates to final concentrations of 50, 100, 200, 300, 400, 600, 700, and 800 μM. The radical discoloration was initiated by adding 10 μL of each dilution into 300 μL ABTS radical cation solution, and the resulting absorbance was measured at 750 nm. For essential oil analysis, 10 μL of VPEO was used instead of 10 μL of Trolox®. Afterwards, a curve of % $ABTS^{\bullet+}$ radical versus concentration was plotted and IC_{50} values were extrapolated. IC_{50} implies the concentration of sample required to scavenge 50% of ABTS radical cation.

2.5.3. DPPH Free Radical Scavenging Activity

Free radical scavenging activity was assessed by using a stable free radical, namely DPPH (2,2-diphenyl-1-picrylhydrazyl), following the modified method reported by Baran et al. [18]. Trolox® solution (1 mg/mL) was made by dissolution in EtOH and further kept in the dark. Final concentrations of 1, 0.8, 0.6, 0.4, 0.2, and 0.1 mM were made in 96-well microplates by consecutive dilution of stock solutions. A mix of 300 μL DPPH radical solution and 20 μL of each dilution was incubated for 30 min at room temperature, and the absorbance was measured at 517 nm. For the essential oil analysis, 20 μL of VPEO was used instead of 20 μL of Trolox®. Results are expressed as IC_{50}. All experiments were conducted in triplicate, and data are expressed as mean values ± SD.

2.6. In Silico Studies

2.6.1. Molecular Docking and Ligand Efficiency

To explore the ability of VPEO to act as potential protein inhibitor, compounds **1** to **47** were subjected to a molecular docking analysis looking for their binding modes on the following proteins: CYP2C9 [19], catalase [20], superoxide dismutase [21], and xanthine oxidase [22]. For all docking studies conducted in this study, AutoDock (v 4.2.1, Scripps Research Institute, San Diego, CA, USA) and AutoDock Vina (v 1.0.2, Scripps Research Institute, San Diego, CA, USA) were employed [23]. The three-dimensional coordinates of all structures were optimized using MOPAC2016 [24] software by PM6-D3H4 semi-empirical method [25,26] (see Tables S1, S2, and S6 for smiles and mol2 files in the Supplementary Material). AutoDockTools package was used to prepare the ligand files [27]. The crystal structure of CYP2C9 (PDB Code: 1OG5), catalase (PDB Code: 1TGU), superoxide dismutase (PDB Code: 2SOD), and xanthine oxidase (PDB Code: 3NRZ) were downloaded from the Protein Data Bank [28]. These four proteins were treated with the Schrödinger's Protein Preparation Wizard [29]; polar hydrogen atoms were included, non-polar hydrogen atoms were merged, and charges were assigned. Docking was treated as rigid and performed using the empirical free energy function and the Lamarckian Genetic Algorithm provided by AutoDock Vina [30]. The grid map dimensions were $20 \times 20 \times 20$ Å^3. The creation of the binding pocket of superoxide dismutase was based on the center coordinates −13.910, 34.868, and 14.639, while the binding pocket of xanthine oxidase was based on the coordinates 19.480, 19.305, and 18.151. These binding sites were established in previous literature [31–34]. All other parameters were set as default defined by AutoDock Vina. Dockings were repeated 20 times with space search exhaustiveness set to 50. The best binding energy (kcal·mol^{-1}) was selected for evaluation. For docking result 3D representations, the Discovery Studio [35] 3.1 (Accelrys, San Diego, CA, USA) molecular graphics system was used.

Ligand Efficiency (*LE*) was calculated by using K_d, a dissociation constant indicating the bond strength between the ligand/protein [36–38]. K_d was calculated by applying the following equations:

$$\Delta G^0 = -2.303\mathrm{RT}\log(K_d) \tag{1}$$

$$K_d = 10^{\frac{\Delta G^0}{2.303\mathrm{RT}}} \tag{2}$$

where ΔG^0 is the binding energy (BE, in kcal·mol^{-1}) found from docking experiments, R is the gas constant, and T is the temperature in Kelvin, in standard conditions of aqueous solution at 298.15 K, neutral pH, and remaining concentrations of 1 M. As indicated in Equation (3), *LE* allows the comparison of molecules according to their average binding energy [38,39]. Thus, it is determined as the ratio of binding energy per non-hydrogen atom, as follows [36–38,40]:

$$LE = -\frac{2.303\mathrm{RT}}{\mathrm{HAC}}\log(K_d) \tag{3}$$

where K_d is obtained from Equation (2) and HAC denotes the heavy atom count (i.e., number of non-hydrogen atoms) in a ligand.

On the other hand, Binding Efficiency Index (*BEI*) and Lipophilic Ligand Efficiency (*LLE*) are calculated using the K_d obtained from molecular docking. *BEI* allows to calculate the binding capacity weighted by molar mass (Equation (4)), whereas *LLE* (Equation (5)) determines the binding capacity with respect to its lipophilicity (clogP obtained from SwissADME webserver) [41,42].

$$BEI = \frac{-\log(K_d)}{MW} \tag{4}$$

$$LLE = -\log(K_d) - \mathrm{clogP} \tag{5}$$

To complement this Ligand Efficiency study, an additional analysis of the size of the molecules in relation to the binding energy was implemented, the score normalization based on the number of non-hydrogen atoms. This score-based approach ($IE_{norm,\,\mathrm{binding}}$) is biased toward the selection of high molecular weight compounds because of the contribution of the compound size to the energy score [43]. Such biasing behavior was observed to depend on the shape and chemical properties of the binding pocket. The procedure starts with the normalization of the binding energy (IE_{binding}) by the number of heavy atoms (HAC) or by a selected power of HAC in each respective compound. This normalization approach shifts the MW distribution of selected compounds into better agreement with that of the VPEO database. In the present study, the following equation was used to calculate the normalized binding energy value.

$$IE_{norm,\ \text{binding}} = \frac{IE_{\text{binding}}}{\text{HAC}^{\frac{1}{2}}} \quad (6)$$

2.6.2. Non-Covalent Interactions

To qualitatively identify regions where intermolecular interactions such as steric repulsion, hydrogen bonds, and Van der Waals interactions predominate in the structural protein–ligand, the non-covalent interaction index (NCI) was employed [44,45]. For biological systems studies, the NCI is based on the promolecular electron density, its derivatives, and the reduced density gradient, as reported elsewhere [46,47]. Molecular visualization of the systems was conducted by using the VMD software package [48].

2.7. *ADMET Prediction*

The pkCSM online tool (http://biosig.unimelb.edu.au/pkcsm/prediction, accessed on 7 February 2022) [49], was utilized to predict absorption, distribution, metabolism, excretion, and toxicity (ADMET) of VPEO.

2.8. *Statistical Analysis*

GraphPad Prism 8.0.2 software (San Diego, CA, USA) was used for statistical analysis. The IC_{50} value was established by a nonlinear regression analysis.

3. Results and Discussion

3.1. *Chemical Composition of VPEO Roots*

Data from gas chromatography (GC) and GC-MS analysis of plant root VPEO are shown in Table 1. The VPEO components were identified by comparing the GC retention indices (RI) on polar and non-polar columns. Such constituents were determined according to the retention time of a series of *n*-alkanes with linear interpolation with those standards and our essential oils database. Forty-seven compounds were revealed by the GC analysis of the essential oil, accounting for 87.5% of the total composition. The major constituents were sesquiterpene hydrocarbons (37.7%), while the monoterpene hydrocarbons were present in concentrations of 9.5%. The oxygen-containing sesquiterpenes were prevalent (26.6%) as compared to oxygen-containing monoterpenes (8.3%). In addition, other compounds were present in low concentrations in oil (5.7%). Notable differences in valerian root oil composition have been reported, a fact likely due to a different geographical environment, crop type, season, plant physiological age, and the method of oil isolation [50,51].

Figure 1 shows the major constituents of the VPEO, which included natural sesquiterpenes such as α-patchoulene (5.8%), α-humulene (6.1%), seychellene (7.6%), and patchoulol (20.8%). However, different VPEO constituents have been shown by chemical analysis; for instance, the essential oil of *Valeriana jatamansi* roots from India contained only seven major sesquiterpene components, which were identified as β-vatirenene (28.07%), β-patchoulene (20.18%), dehydroaromadendrene (15.92%), β-gurjunene (13.0%), patchouli alcohol (11.72%), β-guaiene (5.88%), and α-muurolene (5.20%) [52]. In Vietnam, root essential oils of *Valeriana hardwickii* reported sixty-two components representing 81.6% of total oil, and the major compounds in the root oil were camphene (12.9%), bornyl acetate (17.6%), and maaliol (10.6%) [53].

Table 1. Percentage composition of the essential oil isolated from *Valeriana pilosa* R and P roots collected in Cajamarca, Peru.

N°	Components	RI [a]	Relative Content (%)	Identification Method	RI Data [b]
1	Isovaleric acid	886	2.6	RI,MS	827–888
2	Tricyclene	921	t	RI,MS	914–930
3	α-Thujene	924	0.1	RI,MS	905–948
4	α-Pinene	930	3.7	RI,MS	909–956
5	Camphene	938	1.4	RI,MS	929–978
6	3-Methyl valeric acid	947	3.1	RI,MS	941–968
7	Sabinene	958	0.4	RI,MS	944–980
8	1-Octen-3-ol	961	t	RI,MS	958–986
9	β-Pinene	963	0.6	RI,MS	952–986
10	Myrcene	975	0.1	RI,MS	962–993
11	Limonene	1009	3.2	RI,MS	995–1044
12	p-Cymene	1013	t	RI,MS	992–1072
13	1,8-Cineole	1015	4.3	RI,MS	1007–1046
14	Linalool	1074	0.1	RI,MS	1078–1107
15	Isopentyl isovalerate	1094	t	RI,MS	1094–1105
16	Camphor	1102	0.2	RI,MS	1105–1150
17	Menthone	1120	0.8	RI,MS	1124–1142
18	Isomenthone	1126	0.2	RI,MS	1132–1159
19	Borneol	1134	t	RI,MS	1140–1188
20	Neomenthol	1139	t	RI,MS	1153–1176
21	Menthol	1148	1.2	RI,MS	1141–1185
22	Carvone	1210	0.1	RI,MS	1210–1246
23	Menthyl acetate	1278	1.4	RI,MS	1276–1294
24	α-Cubebene	1345	0.2	RI,MS	1340–1360
25	Cyclosativene	1363	0.1	RI,MS	1363–1368
26	α-Copaene	1375	1.0	RI,MS	1351–1407
27	β-Patchoulene	1378	0.4	RI,MS	1375–1380
28	β-Bourbonene	1379	0.4	RI,MS	1346–1396
29	β-Elemene	1388	0.8	RI,MS	1362–1410
30	β-Caryophyllene	1414	3.5	RI,MS	1411–1421
31	Seychellene	1431	7.6	RI,MS	1457–1461
32	α-Guaiene	1437	4.1	RI,MS	1409–1490
33	α-Humulene	1447	6.1	RI,MS	1428–1489
34	allo-Aromadendrene	1456	2.2	RI,MS	1442–1474
35	α-Patchoulene	1457	5.8	RI,MS	1457–1486
36	γ-Muurolene	1469	1.0	RI,MS	1449–1502
37	Germacrene-D	1474	0.4	RI,MS	1451–1519
38	Valencene	1484	0.3	RI,MS	1458–1495
39	Eremophyllene	1490	0.3	RI,MS	1490–1492
40	γ-Cadinene	1500	0.2	RI,MS	1480–1531
41	7-epi-α-Selinene	1503	2.5	RI,MS	1503–1540
42	δ-Cadinene	1505	0.8	RI,MS	1486–1563
43	Spathulenol	1552	1.6	RI,MS	1552–1622
44	β-Caryophyllene oxide	1561	2.9	RI,MS	1549–1617
45	T-Cadinol	1616	0.5	RI,MS	1611–1644
46	δ-Cadinol	1618	0.5	RI,MS	1618–1652
47	Patchoulol	1625	20.8	RI,MS	1625–1666

[a] RI—retention index as determined on the DB-1 column using the homologous series of *n*-alkanes (C_9–C_{21}); t—trace (<0.05). [b] RI data—retention index data reported in plant essential oils on non-polar column (www.webbook.nist.gov, accessed on 21 March 2022).

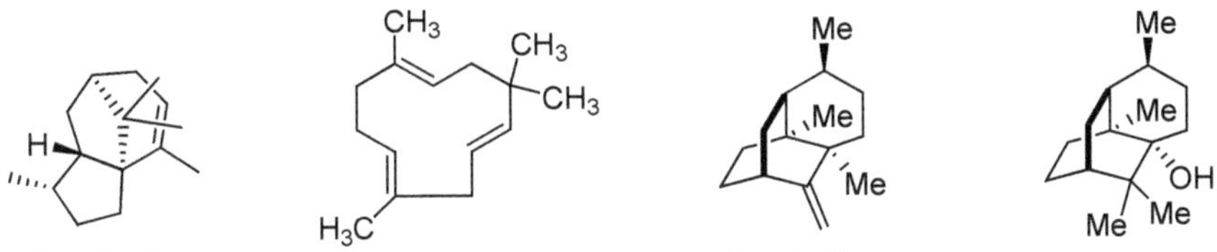

Figure 1. Chemical structures of abundant compounds identified in the essential oil of *Valeriana pilosa* roots.

Several molecules have been identified as basic oil constituents from about 15 studied *Valerian officinalis* root samples from different European countries (Belgium, Czech, Estonia, France, Germany, Greece, Hungary, Latvia, Lithuania, Moldova, Russia, Scotland, Ukraine). They contain 86 identified compounds (>90% of the total oil) such as bornyl acetate (2.9–33.7%), α-fenchene (0–28.3%), valerianol (0.2–18.2%), valerenal (tr-15.6%), isovaleric acid (0–13.1%), camphene (0–11.1%), valeranone (0.5–10.9%), valerenic acid (0–9.8%), sesquiterpene alcohol C (tr-8.0%), spathulenol (0.3–7.3%), and allo-aromadendrene (0.3–6.9%) [54]. Kessyl acetate, kessanyl acetate, and patchouli alcohol are the main constituents in many Valeriana species [55], but only the latter molecule is present in the Valeriana species under study in this work.

3.2. *Antioxidant Capacity of VPEO*

Table 2 includes VPEO antioxidant activities, which were determined by using various chemical-based methodologies. These assays have been developed on different approaches providing evidence about free radicals and essential oil interactions. Herein, the antioxidant activity of essential oils was assessed using three different assays, namely FRAP, $ABTS^{\bullet +}$, and DPPH.

Table 2. Antioxidant activities of essential oil of *Valeriana pilosa*.

Samples	FRAP (mM TEAC)	$ABTS^{\bullet +}$ IC_{50}	DPPH IC_{50}
VPEO	0.0421 ± 0.02	0.30 ± 0.05	0.38 ± 0.07
Quercetin	143.00 ± 0.04	0.07 ± 0.03	0.06 ± 0.02
Trolox®	-	0.012 ± 0.07	0.011 ± 0.04

FRAP = ferric-reducing antioxidant power; $ABTS^{\bullet +}$ = 2,2′-azinobis (3-ethylbenzothiazoline-6-sulfonic acid); DPPH = 2,2-diphenyl-1-picrylhydrazyl radical; GAE = gallic acid equivalent; TEAC = Trolox® equivalent antioxidant capacity. Results are expressed as mean values ± SEM (n = 3).

In the FRAP assay, when the colorless Fe^{3+}-TPTZ complex interacts with a potential antioxidant, it is reduced to an intense blue Fe^{2+}-TPTZ. This assay has been shown to be suitable for screening antioxidant capacities and to compare the efficacy of different compounds [56]. Results from the FRAP assay show a low reducing activity of the essential oil (TEAC = 0.0421 mM) as compared to quercetin (TEAC = 143 mM), used as an antioxidant standard molecule.

The $ABTS^{\bullet +}$ coloring assay is currently employed to determine the antioxidant activity of a wide variety of compounds, such as hydrogen-donating antioxidants or scavengers of aqueous phase radicals and chain-breaking antioxidants or scavengers of lipid peroxyl radicals [57]. In this radical scavenging assay, the VPEO displayed a good activity (IC_{50} of 0.30 μg/mL) when compared to IC_{50} values of standards, quercetin (IC_{50} of 0.07 μg/mL) and Trolox® (IC_{50} of 0.012 μg/mL).

In the DPPH assay, the reduction of the stable radical DPPH to the yellow-colored DPPH-H is employed to measure the capability of an antioxidant molecule to act as a donor of hydrogen atoms or electrons. Table 2 shows that VPEO reduced DPPH with a IC_{50} of 0.38 μg/mL; a high value as compared with essential oils of *Valeriana jatamansi* and *Va-*

leriana officinalis oil roots, displaying a weak radical scavenging activity with IC_{50} values 876 µg/mL [52] and 493.40 µg/mL, respectively [58]. Such high antioxidant activity of VPEO is likely due to the presence of functionalized sesquiterpenes such as patchoulol (one of the major constituents), as well as spathulenol, T-cadinol, and γ-cadinol (minor constituents).

Since essential oils are complicated mixtures composed of huge amounts of molecules, their whole biological activity is hard to be explained. Thus, numerous reports about antioxidant activity of essential oils usually refer to concepts such as synergism, antagonism, and additivity [59]. In addition, discerning the real mechanism of antioxidant activity is not an easy task. To this end, several mechanism-based explanations have been provided: free radicals scavenging; hydrogen donation; and metallic ion chelation by antioxidants [60]. Due to the high reduction ability displayed by VPEO, it can be inferred that their components might be potent natural antioxidants.

3.3. Molecular Docking and Ligand Efficiency Analysis of VPEO

Molecular docking is a key tool that can show insights for understanding plausible mechanisms of action displayed by in vitro biological active molecules. In this context, molecular docking was used to find a protein target as a possible mechanism that could be correlated with (and likely explain) the observed in vitro antioxidant activity of VPEO. An in-silico-based approach was used to explore whether some VPEO constituents may inhibit some proteins involved in redox biological processes. The targeted proteins were CYP2C9 (a phase I enzyme involved in oxidation of xenobiotics), catalase (enzyme regulating hydrogen peroxide intracellular levels), superoxide dismutase (enzyme catalyzing the dismutation of superoxide anion into hydrogen peroxide and molecular oxygen), and xanthine oxidase (a key enzyme regulating the formation of uric acid and superoxide anion).

Figure 2 shows the heat map of the intermolecular docking energy values of 47 VPEO components. The values are listed as a three-colored scheme (red-yellow-green) showing a clear trend of a set of compounds acting as putative inhibitors for a given protein. For each protein target, the range was set from red color (as the energy value corresponding to the native ligand) to green, spanning a 5 kcal·mol^{-1} interval. This approach is appropriate especially for sets of compounds sharing high structural resemblance.

Molecular dockings, K_d values, Ligand Efficiency (*LE*), Binding Efficiency Index (*BEI*), and Lipophilic Ligand Efficiency (*LLE*) analyses are summarized in Table 3, as well as in Tables S2–S5. Results show that all VPEO constituents act as potential inhibitors of CYP2C9 while about 65% may be considered inhibitors of xanthine oxidase. Less than 50% of VPEO constituents appeared as rather weak inhibitors of catalase, but the vast majority of them have no effect on superoxide dismutase.

CYP2C9 appeared as the best protein target for all VPEO constituents, as shown by their intermolecular docking energy and Ligand Efficiency values. Indeed, by using the values obtained by all compounds, for CYP2C9 analysis the average values of both Binding Efficiency (*BE*) and Ligand Efficiency were −6.56 and 0.52 kcal·mol^{-1}, respectively, while compared to the xanthine oxidase protein target, the average values of *BE* and Ligand Efficiency were −6.33 and 0.50 kcal·mol^{-1}, respectively. Therefore, it may be assumed that CYP2C9 and xanthine oxidase are targeted proteins likely involved in VPEO effects. These results are supplemented by the calculations obtained from score normalization based on the number of non-hydrogen atoms (Figure 3). The proteins CYP2C9 and xanthine oxidase appear with a similar score of −1.8 kcal·mol^{-1}, which coincide as the best protein targets for all VPEO compounds, meaning that they might be involved in the antioxidant effects of VPEO compounds.

An important aspect of normalizing binding energy is the ability to bias selection towards lower molar weight (MW) compounds, thereby identifying compounds more appropriate for lead optimization. Ligand-based postdocking structural clustering leads to the selection of diverse compounds, and many of them would have been lost through selection based on binding energy alone. Then, it is important to establish a relationship between binding energy and MW of VPEO components. Comparing the unnormalized energetic values in Figure 2 and the normalized energetic values, Figure 3 shows that there are three compounds that stand out in Figure 3, namely **11** (Limonene), **12** (*p*-Cymene), and **22** (Carvone). Figure 4 shows compounds that have low normalized energetic values of interaction with CYP2C9 (−2.0, −1.9, and −1.9 kcal·mol^{-1}) and xanthine oxidase (−2.1, −2.2, and −2.2 kcal·mol^{-1}). In addition, they can be considered as suitable lead molecules for a drug candidate, as they have low MW. This feature makes them advantageous because they generally exhibit better properties of being good lead candidates due to their simpler intrinsic chemical structures, rendering them suitable for further drug optimization.

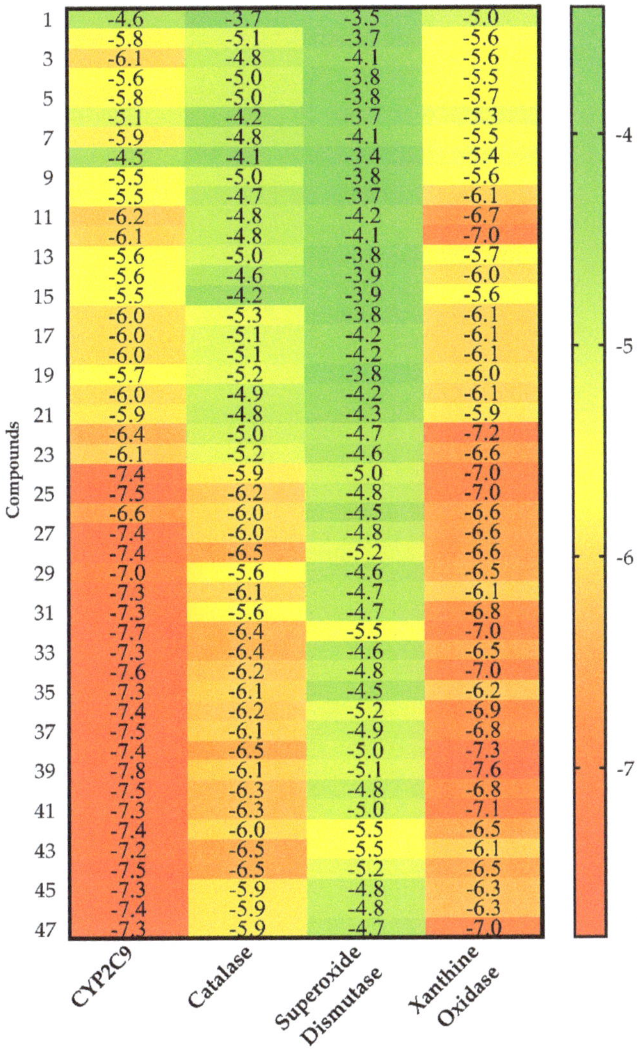

Figure 2. Heat map of the intermolecular docking energy values (kcal·mol^{-1}) of VPEO components on CYP2C9, catalase, superoxide dismutase, and xanthine oxidase proteins. Values are listed as a three-colored scheme from red (high energy) to green (low energy).

Table 3. Molecular docking results for the best six compounds of VPEO regarding CYP2C9 and xanthine oxidase. Intermolecular docking energy values ($\Delta E_{binding}$), K_d values, Ligand Efficiency (*LE*), Binding Efficiency Index (*BEI*), and Lipophilic Ligand Efficiency (*LLE*) for the CYP2C9 and xanthine oxidase complexes.

Docking and Ligand Efficiency Analysis					
Compounds	**$\Delta E_{binding}$ (kcal·mol^{-1})**	**K_d**	**LE (kcal·mol^{-1})**	**BEI (kDa)**	**LLE**
			CYP2C9		
24	−7.40	3.77×10^{-6}	0.49	26.54	1.15
25	−7.50	3.19×10^{-6}	0.50	26.90	1.54
32	−7.70	2.27×10^{-6}	0.51	27.61	0.92
34	−7.60	2.69×10^{-6}	0.51	27.26	1.30
38	−7.40	3.77×10^{-6}	0.49	26.54	0.70
39	−7.80	1.92×10^{-6}	0.52	27.97	0.99
			Xanthine Oxidase		
24	−7.00	7.41×10^{-6}	0.47	25.10	0.86
25	−7.00	7.41×10^{-6}	0.47	25.10	1.17
32	−7.00	7.41×10^{-6}	0.47	25.10	0.40
34	−7.00	7.41×10^{-6}	0.47	25.10	0.86
38	−7.30	4.47×10^{-6}	0.49	26.18	0.62
39	−7.60	2.69×10^{-6}	0.51	27.26	0.84

In the case of non-covalent interactions between compounds **11**, **12**, and **22** with CYP2C9 and xanthine oxidase proteins (see Figure 4), weak interactions such as Van der Waals type interactions and aromatic (π—π stacking) and hydrophobic interactions are included, except for compound **22**, which forms a hydrogen bond at the binding site of xanthine oxidase.

In order to check the binding modes of the VPEO components, molecular docking simulations were performed with the co-crystallized ligand pose of warfarin bound to CYP2C9 (PDBID: 1OG5) and the co-crystallized ligand pose of quercetin bound to xanthine oxidase (PDBID: 3NVY). Such co-crystallized ligands were re-docked into the binding site with specific docking parameters and scoring functions, to check whether the docking software is reliable for the systems (Figures S1–S3). The conformation with the lowest binding energy of warfarin and quercetin was compared to the co-crystallized ligand pose. The binding energy value for warfarin was −9.8 kcal·mol^{-1} and for quercetin was −8.1 kcal·mol^{-1}. The root mean square deviation (RMSD) value of the docked conformation with respect to the experimental conformation was 1.06 Å for warfarin and 1.51 Å for quercetin (Figures S1 and S2), indicating the reliability of the docking protocol, as the threshold of reliability is 2.0 Å for a good docking protocol.

Specifically, Table 3 shows the six molecules of the VPEO constituents having the best affinity for both proteins (CYP2C9 and xanthine oxidase): they include **24** (α-cubebene), **25** (cyclosativene), **32** (α-guaiene), **34** (allo-aromadendrene), **38** (valencene), and **39** (eremophyllene). Such compounds have low $\Delta E_{binding}$ values in the range of −7.0 and −7.80 kcal·mol^{-1} but they are not abundant compounds of VPEO. In this context, we would like to stress that compounds **31** (seychellene), **33** (α-humulene), **35** (α-patchoulene), and **47** (patchoulol), which have a high abundance in VPEO, also have stable $\Delta E_{binding}$ values: −7.3 kcal·mol^{-1} for the CYP2C9 protein and $\Delta E_{binding}$ values ranging from −6.2 kcal·mol^{-1} to −7.0 kcal·mol^{-1} for the xanthine oxidase protein. These results show their binding tendency with regard to the CYP2C9 protein.

In addition to displaying a best affinity for both proteins, the six molecules of the VPEO constituents have strong ligand binding to the protein, as shown by their low K_d values. Regarding *LE* values, they are in the range of 0.47 and 0.51 kcal·mol^{-1}, compared to *LE* values greater than 0.3 kcal·mol^{-1} required to be considered as a reference [61].

According to such descriptors, compounds **24**, **25**, **32**, **34**, **38**, and **39** may be considered suitable lead molecules to a drug candidate due to their *LE*, binding energies, and their affinity for the cellular targets CYP2C9 and xanthine oxidase. Note that orally administered drugs have *LE* values between 0.50 and 0.52 kcal·mol^{-1} [62].

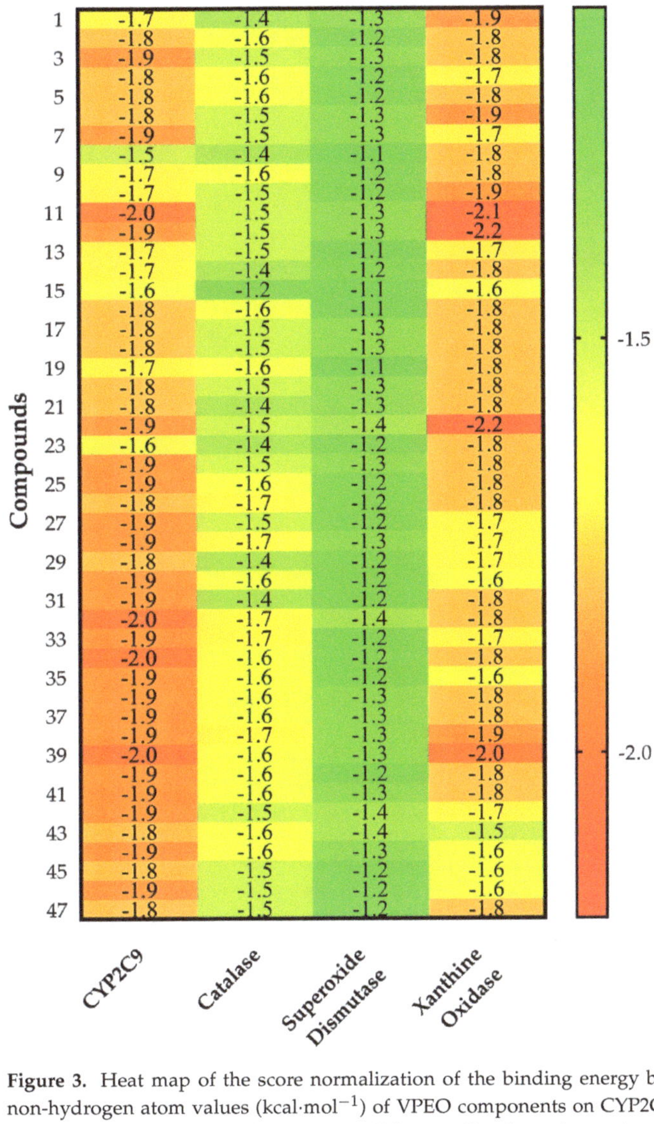

Figure 3. Heat map of the score normalization of the binding energy based on the number of non-hydrogen atom values (kcal·mol^{-1}) of VPEO components on CYP2C9, catalase, superoxide dismutase, and xanthine oxidase proteins. Values are listed as a three-colored scheme from red (high energy) to green (low energy).

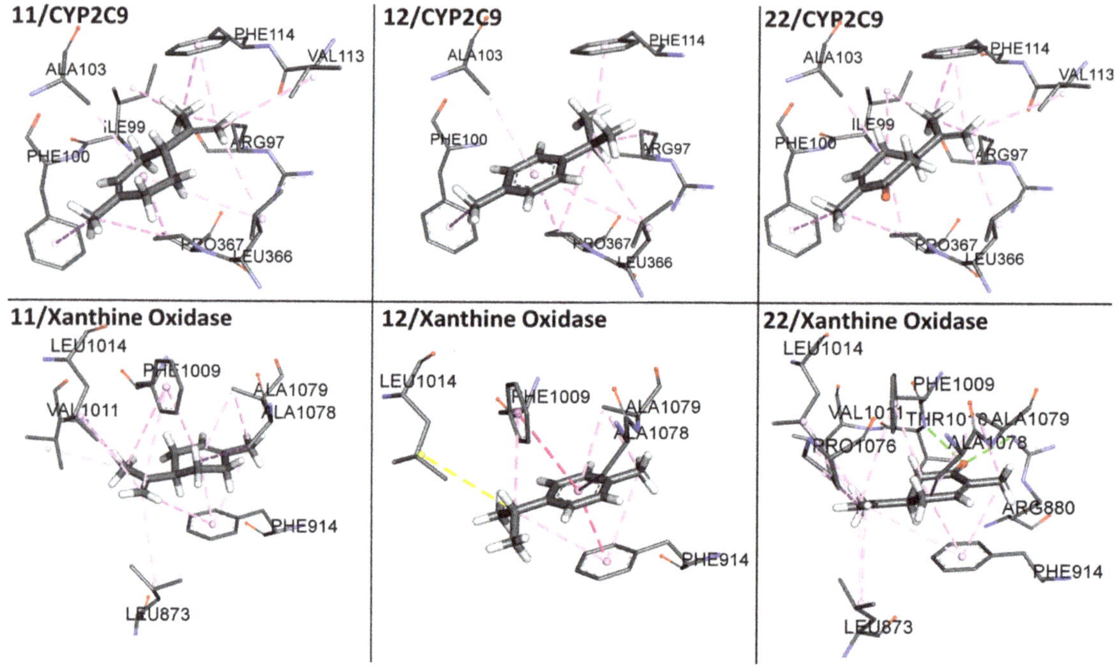

Figure 4. Schematic representation for the best three compounds (**11**, **12**, and **22**) of the score normalization of the binding energy based on the number of non-hydrogen atoms of VPEO bound to CYP2C9 and xanthine oxidase. The surrounding amino acid residues in the binding pocket of CYP2C9 and xanthine oxidase within 3 Å are shown.

Regarding the Binding Efficiency Index (*BEI*), the reference values should be in the range of 20 and 27 kDa. Since *BEI* values of compounds **24**, **25**, **31**, **32**, **33**, **34**, **35**, **38**, **39**, and **47** are within such reference range, it appears that the ligands reveal a high structure–activity relationship with CYP2C9 and xanthine oxidase (see Tables S2 and S5).

Another essential parameter to be considered is the Lipophilic Ligand Efficiency (*LLE*) index, which determines ligand-binding capacity to the protein and its lipophilic power [62]. Based on the properties of a standard oral drug, with a calculated LogP (cLogP) of ~2.5–3.0, ideal *LLE* values for an optimized drug candidate are in the range of $5 < LLE < 7$, and were calculated based on oral administration of known drugs [63]. The *LLE* values for selected compounds, namely **24**, **25**, **31**, **32**, **33**, **34**, **35**, **38**, **39**, and **47**, are out of such range, having values lower than 5. Since cLogP had relatively high values, the ligands therefore display lipophilic properties.

Figure 5 shows interactions of compounds **24**, **25**, **32**, **34**, **38**, and **39** with the surrounding amino acid residues in the binding pocket of CYP2C9 and xanthine oxidase within 3 Å. They included non-covalent interactions, which are associated with weak Van der Waals type interactions and aromatic (π—π stacking) and hydrophobic interactions. It should be stressed that the main non-covalent interactions of compounds **24**, **25**, **32**, **34**, **38**, and **39** with CYP2C9 and xanthine oxidase binding sites are based on weak Van der Waals and hydrogen bond interactions. In the case of non-covalent interactions between compounds **31**, **33**, **35** and **47** with CYP2C9 and xanthine oxidase proteins (see Figure S3), weak interactions such as Van der Waals and hydrophobic interactions are included. Although compound **47** has a hydroxyl group, it does not form a hydrogen bond to stabilize this interaction, with weak Van der Waals type interactions prevailing.

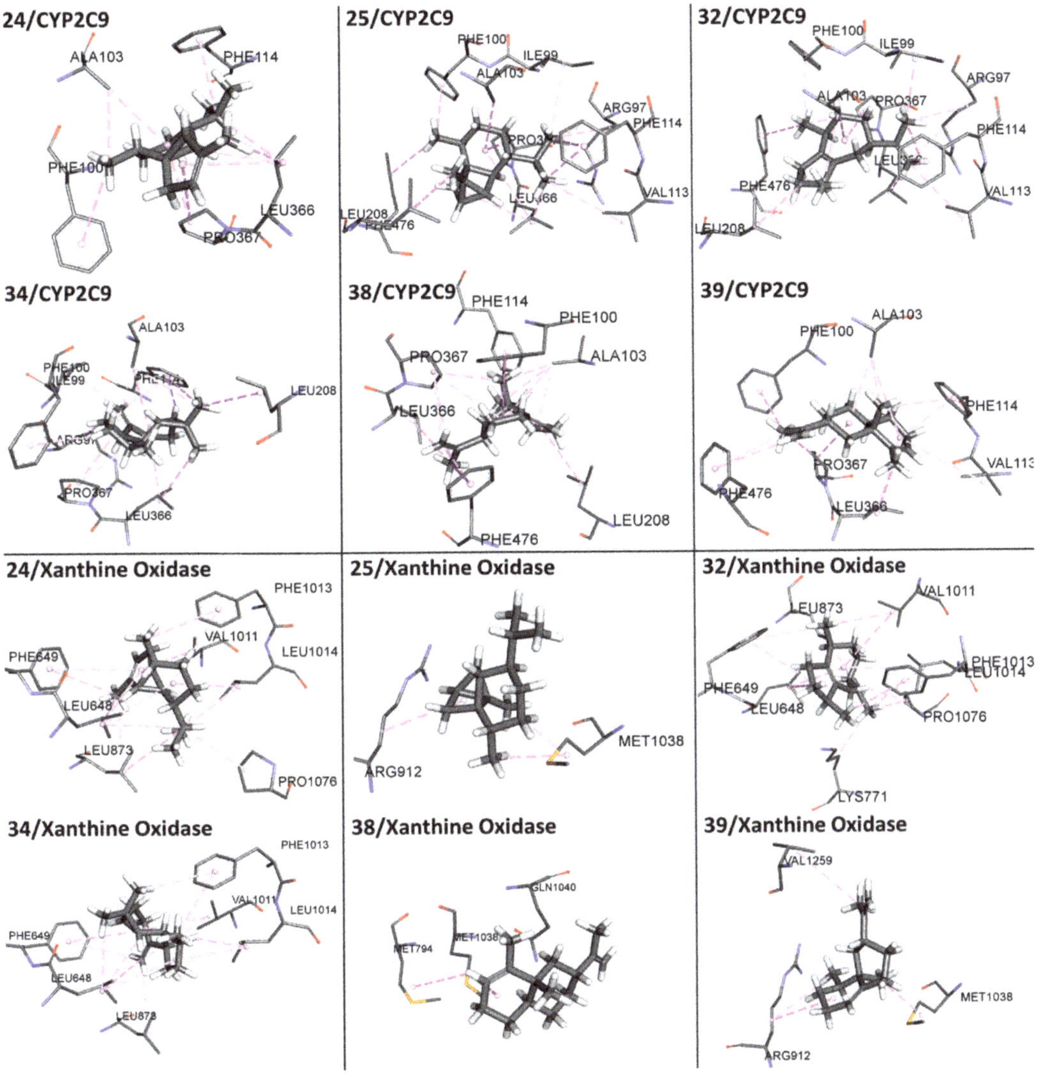

Figure 5. Molecular docking visualization for the best six compounds (**24**, **25**, **32**, **34**, **38**, and **39**) of VPEO bound to CYP2C9 and xanthine oxidase. The surrounding amino acid residues in the binding pocket of CYP2C9 and xanthine oxidase within 3 Å are shown.

Regarding aromatic and hydrophobic interactions, the most notable interaction of VPEO selected constituents with CYP2C9 residues included the following amino acids: Arg97, Ile99, Phe100, Leu102, Ala103, Val113, Phe114, Leu208, Ile213, Leu366, Pro367, and Phe476. In particular, residues Leu208, Leu366, and Phe476 form a hydrophobic patch in the active site [19].

In the case of xanthine oxidase, such interactions between VPEO selected constituents and enzyme residues were established with the following amino acids: Leu648, Phe649, Lys771, Met794, Leu873, Arg912, Val1011, Phe1013, Leu1014, Met1038, Gln1040, Pro1076, Ala1078, and Val1259.

In addition to the identification of amino acid residues involved in non-covalent interactions between protein targets and selected VPEO constituents, Figure 6 shows the strengths of main non-covalent interactions for compounds **38** and **39** on the CYP2C9 and xanthine oxidase binding sites according to the NCI analysis. These interactions are mainly based on strong attraction, weak attraction, and strong repulsion (blue, green, and red colors, see bottom scale of Figure 6).

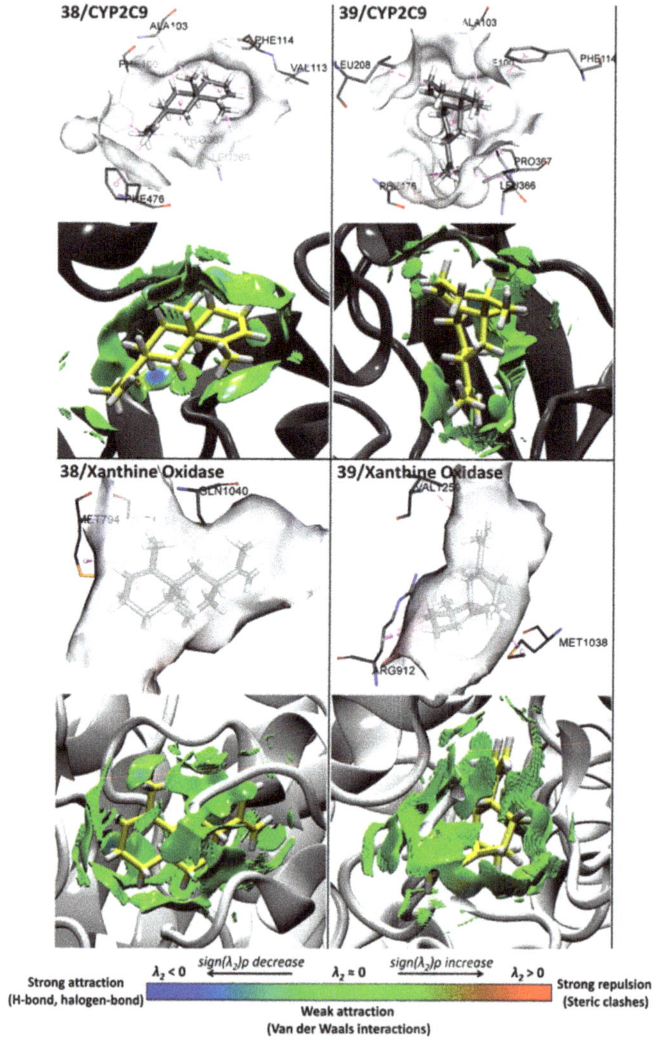

Figure 6. Docking and NCI analysis for the best two compounds, **38** and **39**, of VPEO essential oils bound to CYP2C9 and xanthine oxidase.

The activity of compound **38** at the CYP2C9 binding site involves Van der Waals type interactions with residues Phe100, Ala103, Phe114, Leu208, Leu366, Pro367, and Phe476. Compound **39** shows Van der Waals type interactions with residues Phe100, Ala103, Val113, Phe114, Leu366, Pro367, and Phe476. For compounds interacting at the xanthine oxidase binding site, compounds **38** and **39** display Van der Waals type interactions with residues Met794, Met1038, Gln1040, Arg912, and Val1259. Altogether, it may be concluded that hydrophobic forces emerge as the main interactions playing a key role in the mechanism of action of VPEO constituents.

Regarding the interaction between CYP2C9 (as protein target) and compound **38** (as the most active VPEO constituent), it is interesting to note that such a non-covalent interaction involves six amino acids, namely Phe100, Ala103, Phe114, Leu 366, Pro367, and Phe 476. Most of such amino acids are involved in CYP2C9 activity. Indeed, Phe114 points into the active site, playing an important role in forming interactions with substrates. Moreover, residues Phe100, Leu366, and Phe476 have been reported to form a hydrophobic patch in the active site. On the other hand, Pro367 is somehow involved in CYP2C9 hem stabilization [18].

3.4. ADMET Profiles of VPEO

During the processes of discovery and development of drugs, pesticides, food additives, and consumer and industrial chemicals, both pharmacokinetic and toxicity properties have a significant influence. This information is especially useful during environmental and human hazard assessment. Pharmacokinetic parameters and toxicity data were obtained by using the pkCSM Online Tool, and they are reported in Table 4.

Table 4. ADMET properties of chemical constituents of the *Valeriana pilosa* essential oil.

		Property									
		Absorption			Distribution			Metabolism	Excretion	Toxicity	
		Model Name									
N°	Components	Caco−2	IA	SP	VD ss	BBB	CNS	CYP2D6/ CYP3A4 Inhibitor	TC	Oral Rat Acute Tox.(LD_{50})	Oral Rat Chronic Tox.- LOAEL
1	Isovaleric acid	1.578	88.820	−2.730	−0.937	−0.227	−2.229	No/No	0.391	1.644	2.691
2	Tricyclene	1.353	93.922	−1.912	0.781	0.849	−1.924	No/No	−0.073	1.608	2.103
3	α-Thujene	1.386	95.256	−1.371	0.575	0.810	−1.793	No/No	0.077	1.589	2.243
4	α-Pinene	1.38	96.041	−1.827	0.667	0.791	−2.201	No/No	0.043	1.770	2.262
5	Camphene	1.387	94.148	−1.435	0.547	0.787	−1.710	No/No	0.049	1.554	2.247
6	3-Methyl valeric acid	1.574	95.413	−2.732	−0.752	−0.198	−2.512	No/No	0.441	1.656	2.632
7	Sabinene	1.404	95.356	−1.342	0.566	0.836	−1.463	No/No	0.071	1.549	2.309
8	1-Octen-3-ol	1.481	93.214	−1.760	0.134	0.514	−2.291	No/No	0.461	1.722	1.915
9	β-Pinene	1.385	95.525	−1.653	0.685	0.818	−1.857	No/No	0.030	1.673	2.28
10	Myrcene	1.400	94.696	−1.043	0.363	0.781	−1.902	No/No	0.438	1.643	2.406
11	Limonene	1.401	95.898	−1.721	0.396	0.732	−2.370	No/No	0.213	1.880	2.336
12	p-Cymene	1.527	93.544	−1.192	0.697	0.478	−1.397	No/No	0.239	1.827	2.328
13	1,8-Cineole	1.485	96.505	−2.437	0.491	0.368	−2.972	No/No	1.009	2.010	2.029
14	Linalool	1.493	93.163	−1.737	0.152	0.598	−2.339	No/No	0.446	1.704	2.024
15	Isopentyl isovalerate	1.182	95.333	−1.745	−0.036	0.602	−1.818	No/No	0.481	1.582	2.271
16	Camphor	1.499	95.965	−2.002	0.331	0.612	−2.158	No/No	0.109	1.653	1.981
17	Menthone	1.520	96.739	−1.909	0.201	0.593	−2.117	No/No	0.244	1.691	2.095
18	Isomenthone	1.229	97.324	−1.872	0.174	0.607	−2.155	No/No	0.244	1.796	2.028
19	Borneol	1.484	93.439	−2.174	0.337	0.646	−2.331	No/No	1.035	1.707	1.877
20	Neomenthol	1.505	94.213	−2.087	0.207	0.573	−2.290	No/No	1.182	1.733	1.991
21	Menthol	1.376	95.257	−1.919	0.137	0.584	−2.119	No/No	1.182	1.946	2.017
22	Carvone	1.413	97.702	−2.145	0.179	0.588	−2.478	No/No	0.225	1.860	1.972
23	Menthyl acetate	1.698	96.497	−2.208	0.125	0.539	−2.390	No/No	1.207	1.823	2.040
24	α-Cubebene	1.389	95.964	−1.997	0.717	0.860	−1.552	No/No	0.980	1.568	1.364
25	Cyclosativene	1.360	95.698	−2.526	0.747	0.946	−1.422	No/No	0.771	1.689	1.366

Table 4. *Cont.*

		Property									
		Absorption			Distribution			Metabolism	Excretion	Toxicity	
		Model Name									
N°	**Components**	**Caco−2**	**IA**	**SP**	**VD ss**	**BBB**	**CNS**	**CYP2D6/ CYP3A4 Inhibitor**	**TC**	**Oral Rat Acute Tox.(LD$_{50}$)**	**Oral Rat Chronic Tox.- LOAEL**
26	α-Copaene	1.374	96.221	−2.225	0.806	0.887	−1.659	No/No	0.950	1.644	1.356
27	β-Patchoulene	1.400	95.658	−1.730	0.786	0.791	−1.959	No/No	0.941	1.569	1.387
28	β-Bourbonene	1.395	95.668	−2.205	0.624	0.879	−1.218	No/No	0.967	1.601	1.431
29	β-Elemene	1.410	94.359	−1.279	0.601	0.809	−1.714	No/No	0.251	1.535	1.309
30	β-Caryophyllene	1.423	94.845	−1.580	0.652	0.733	−2.172	No/No	1.088	1.617	1.416
31	Seychellene	1.386	96.161	−2.249	0.787	0.866	−1.606	No/No	0.983	1.675	1.409
32	α-Guaiene	1.420	95.512	−1.538	0.682	0.763	−2.235	No/No	1.219	1.679	1.365
33	α-Humulene	1.421	94.682	−1.739	0.505	0.663	−2.555	No/No	1.282	1.766	1.336
34	Allo-Aromadendrene	1.395	95.302	−1.828	0.753	0.822	−1.769	No/No	0.926	1.526	1.332
35	α-Patchoulene	1.394	94.515	−1.833	0.751	0.818	−1.759	No/No	0.973	1.552	1.334
36	γ-Muurolene	1.427	96.475	−1.561	0.67	0.809	−1.631	No/No	1.188	1.540	1.473
37	Germacrene-D	1.436	95.59	−1.429	0.544	0.723	−2.138	No/No	1.420	1.634	1.413
38	Valencene	1.434	96.587	−1.473	0.692	0.779	−1.955	No/No	1.205	1.604	1.480
39	Eremophyllene	1.401	94.127	−1.461	0.686	0.776	−1.865	No/No	1.211	1.543	1.351
40	γ-Cadinene	1.427	96.475	−1.561	0.67	0.809	−1.631	No/No	1.188	1.540	1.473
41	7-epi-α-Selinene	1.373	94.846	−1.989	0.674	0.804	−3.226	No/No	1.183	1.912	1.129
42	δ-Cadinene	1.422	96.128	−1.462	0.689	0.773	−1.945	No/No	1.182	1.552	1.448
43	Spathulenol	1.388	93.235	−2.141	0.522	0.600	−2.447	No/No	0.895	1.687	1.390
44	β-Caryophyllene oxide	1.414	95.669	−3.061	0.564	0.647	−2.521	No/No	0.905	1.548	1.224
45	T-Cadinol	1.352	96.460	−2.285	0.543	0.565	−3.299	No/No	1.147	2.065	0.895
46	δ-Cadinol	1.479	94.296	−1.923	0.420	0.596	−2.151	No/No	1.085	1.918	1.475
47	Patchoulol	1.475	92.467	−2.397	0.668	0.649	−2.303	No/No	0.871	1.707	1.238

Caco−2: Caucasian colon adenocarcinoma permeability (Log Papp in 10^{-6}cm/s). **IA**: intestinal absorption (% Absorbed). **SP**: skin permeability (logKp). **VDss**: steady state Volume of Distribution (Log L/kg). **BBB**: blood–brain barrier permeability (log BB). CNS: central nervous system (Log PS). **CYP2D6**: Cytochrome P450 2D6 inhibitor; **CYP3A4**: Cytochrome P450 3A4 inhibitor. **TC**: total clearance (Log mL/min/kg). **LD$_{50}$**: lethal dose, 50% (mol/Kg). **LOAEL**: Lowest Observed Adverse Effect Level (Log mg/kg bw/day).

From the ADMET results, it is found that all the structures had a molecular weight ranging between 102 and 222 g/mol, a good indicator for penetrability, because the upper limit to obtain this ability is molecular weight values of less than 500 g/mol [64]. All the molecules of VPEO show Caco−2 permeability values above 1.18 and high intestinal absorption (88.8–97.7%) as well, predicting that they would be absorbed in the small intestine [65]. VPEO compounds have skin permeability values ranging from −1.043 to −3.061 cm/h, but only the compounds **1**, **6**, **25**, and **44** have values >−2.5, suggesting that most volatile phytochemicals easily penetrate the skin adequately. Note that molecules will penetrate the skin with difficulty if the logKp value is greater than −2.5 cm/hour [66].

All the compounds have acceptable volume distribution (VDss), with values above −0.15. Since compounds **1** and **6** have log BB < 0.3 they are likely unable to penetrate the blood–brain barrier (BBB). All other compounds display mean values greater than 0.3 and are therefore able to access the brain. The data available so far indicate that permeability into the central nervous system (CNS) occurs with values ranging from −2.555 to −1.218; therefore, 46.8% of the VPEO compounds would be able to permeate the central nervous system [67].

Regarding metabolism, none of the VPEO components appeared to be CYP2D6 and CYP3A4 inhibitors and they will not interfere with CYP450 biotransformation reactions.

Excretion parameters are illustrated as total clearance. They showed that only compound **2** (tricyclene) reached a negative value (−0.073 log mL/min/kg), while the rest of the compounds have positive values, indicating a rapid excretion. In addition, the adverse interactions of all VPEO constituents with the organic cation transport 2 (OCT2) showed no potential contraindication (data not shown).

Finally, the acute oral toxicity in rats (LD_{50}) ranged from 1.526 to 2.065 mol/kg, corresponding to a low toxicity. The hepatotoxicity descriptor showed that all compounds are devoid of liver toxicity.

4. Conclusions

A double GC and GC-MS approach was used to identify the chemical components of the essential oil from *Valeriana pilosa*. The VPEO showed to contain sesquiterpene hydrocarbons, monoterpene hydrocarbons, oxygenated monoterpenes, and oxygenated sesquiterpenes. After isolation of the in vitro antioxidant activities, molecular docking studies on enzymes involved in redox balance proteins such as CYP2C9, catalase, superoxide dismutase, and xanthine oxidase, as well as ADMET properties, were investigated. A high antioxidant activity of the oil was found as compared to values obtained with the essential oils of *Valeriana jatamansi* and *Valeriana officinalis* oil roots. In the molecular docking studies, α-cubebene, cyclosativene, α-guaiene, allo-aromadendrene, valencene, and eremophyllene were the compounds with the best docking score on CYP2C9 and xanthine oxidase. Nevertheless, when the molar weight was taken into account and energy values were normalized, three compounds, namely limonene, *p*-cymene, and carvone, were highlighted. Eremophyllene may be also included due to its improved binding to these proteins, making it suitable for further drug optimization.

Additionally, according to the ADMET prediction using the pkCSM online tool, it appeared that most of compounds display suitable pharmacokinetic properties, as shown by absorption, distribution, metabolism, excretion parameters, and low toxicities.

Supplementary Materials: The following supporting information can be downloaded at: https://www.mdpi.com/article/10.3390/antiox11071337/s1. Figure S1: Superposition of the co-crystal warfarin with its docked pose bound to CYP2C9. Figure S2: Superposition of the co-crystal quercetin with its docked pose bound to xanthine oxidase. Figure S3: Molecular docking visualization for the abundant compounds identified in the VPEO bound to CYP2C9 and xanthine oxidase. Table S1: Canonical SMILES of 47 *Valeriana pilosa* essential oils used for Ligand Efficiency studies. Table S2: Complete results for essential oils from *Valeriana pilosa* with CYP2C9 target: intermolecular docking energy values ($\Delta E_{binding}$), K_d values, Ligand Efficiency (*LE*), Binding Efficiency Index (*BEI*), and Lipophilic Ligand Efficiency (*LLE*). Table S3: Complete results for essential oils from *Valeriana pilosa* with catalase target: intermolecular docking energy values ($\Delta E_{binding}$), K_d values, Ligand Efficiency (*LE*), Binding Efficiency Index (*BEI*), and Lipophilic Ligand Efficiency (*LLE*). Table S4: Complete results for essential oils from *Valeriana pilosa* with superoxide dismutase target: intermolecular docking energy values ($\Delta E_{binding}$), K_d values, Ligand Efficiency (*LEB*), Binding Efficiency Index (*BEI*), and Lipophilic Ligand Efficiency (*LLE*). Table S5: Complete results for essential oils from *Valeriana pilosa* with xanthine oxidase target: intermolecular docking energy values ($\Delta E_{binding}$), K_d values, Ligand Efficiency (*LE*), Binding Efficiency Index (*BEI*), and Lipophilic Ligand Efficiency (*LLE*) and Table S6: Mol2 files for all compounds studied in this work.

Author Contributions: Conceptualization, R.P.-R. and J.B.; methodology, J.B.; formal analysis, P.M.-H., R.O.Y.-J. and I.M.Q.-D.; investigation, P.M.-H., F.S., E.A.V.-C., R.J.-A., L.Z.-E., O.Y., R.P.-R. and J.B.; writing—original draft preparation, O.Y., P.B.C. and J.B.; writing—review and editing, P.B.C. and J.B.; funding acquisition, P.M.-H. and J.B. All authors have read and agreed to the published version of the manuscript.

Funding: This work was supported by the Concytec project and the World Bank "Improvement and Expansion of Services of the National System of Science, Technology and Technological Innovation" through the executing unit ProCiencia 8682-PE, contract 07-2018 FONDECYT-BM-PDAEG-UNT.

Institutional Review Board Statement: Not applicable.

Informed Consent Statement: Not applicable.

Data Availability Statement: The data presented in this study are available in the article.

Acknowledgments: The authors thank A. Mantilla and O. Caballero, coordinators of the doctoral program of the Concytec project, and the World Bank for their management of the necessary resources for its development and execution.

Conflicts of Interest: The authors declare no conflict of interest.

References

1. Bodeker, G.; Kronenberg, F. A public health agenda for traditional, complementary, and alternative medicine. *Am. J. Public Health* **2002**, *92*, 1582–1591. [CrossRef] [PubMed]
2. Tripathi, P.; Dubey, N.K. Exploitation of natural products as an alternative strategy to control postharvest fungal rotting of fruit and vegetables. *Postharvest Biol. Technol.* **2004**, *32*, 235–245. [CrossRef]
3. Figueiredo, A.C.; Barroso, J.; Pedro, L.; Scheffer, J. Factors affecting secondary metabolite production in plants: Volatile components and essential oils. *Flavour Fragr. J.* **2008**, *23*, 213–226. [CrossRef]
4. Jugran, A.K.; Rawat, S.; Bhatt, I.D.; Rawal, R.S. Essential oil composition, phenolics and antioxidant activities of Valeriana jatamansi at different phenological stages. *Plant Biosyst.* **2021**, *155*, 891–898. [CrossRef]
5. Bell, C.D. Preliminary phylogeny of Valerianaceae (Dipsacales) inferred from nuclear and chloroplast DNA sequence data. *Mol. Phylogenet. Evol.* **2004**, *31*, 340–350. [CrossRef]
6. Kutschker, A. Revision of genus *Valeriana* (Valerianaceae) in Southern South America. *Gayana Bot.* **2011**, *68*, 244–296. [CrossRef]
7. Galan, A.; Sanchez, I.; Montoya, J.; Linares, E.; Campos, J.; Vicente, J. La vegetacion del norte del Peru: De los bosques a la jalca en Cajamarca. *Acta Bot. Malacit.* **2015**, *40*, 157–190. [CrossRef]
8. Seminario-Cunya, J.; Rumay-Sanchez, L.; Seminario-Cunya, A. Biologia de *Valeriana pilosa* R. &. P. (Valerianaceae): Una especie en peligro de extinción de las altas montañas de Peru. *Bol. Latinoam. Caribe Plants Med. Aromat.* **2016**, *15*, 337–351.
9. Sánchez, I. *Especies Medicinales de Cajamarca I: Contribución Etnobotánica, Morfológica y Taxonómica*; UPAGU: Cajamarca, Peru; Lumina Copper Fondo Editorial: Vancouver, BC, Canada, 2011; pp. 166–169.
10. Rojo, L.; Benites, J.; Rodriguez, A.; Venancio, F.; Ramalho, L.; Teixeira, A.; Feio, S.; do Ceu Costa, M. Composition and antimicrobial screening of the essential oil of *Acantholippia deserticola* (Phil.ex F. Phil.) Moldenke. *J. Essent. Oil Res.* **2006**, *18*, 695–697. [CrossRef]
11. Rojo, L.E.; Benites, J.; López, J.; Rojas, M.; Díaz, P.; Pastene, E.; Ordoñez, J. Comparative study on the antioxidant effects and phenolic content of twelve highly consumed medicinal plants from South American Andes. *Bol. Latinoam. Caribe Plants Med. Aromat.* **2009**, *8*, 498–508.
12. Benites, J.; Moiteiro, C.; Miguel, G.; Rojo, L.; López, J.; Venâncio, F.; Ramalho, L.; Feio, S.; Dandlen, S.; Casanova, H.; et al. Composition and biological activity of the essential oil of Peruvian Lantana camara. *J. Chil. Chem. Soc.* **2009**, *54*, 379–384. [CrossRef]
13. Benites, J.; López, J.; Rojo, L.; Díaz, P.; Rojas, M.; Venâncio, F.; Moiteiro, M. Chemical composition of the essential oil of the leaves and stems of *Xenophillum poposum*. *Chem. Nat. Compound.* **2011**, *46*, 988–989. [CrossRef]
14. Benites, J.; Moiteiro, C.; Figueiredo, A.C.; Rijo, P.; Buc-Calderon, P.; Bravo, F.; Gajardo, S.; Sánchez, I.; Torres, I.; Ganoza, M. Chemical composition and antimicrobial activity of essential oil of Peruvian *Dalea strobilacea* Barneby. *Bol. Latinoam. Caribe Plants Med. Aromat.* **2016**, *15*, 429–435.
15. Benites, J.; Ríos, D.; Guerrero-Castilla, A.; Enríquez, C.; Zavala, E.; Ybañez-Julca, R.O.; Quispe-Díaz, I.; Jara-Aguilar, R.; Buc Calderon, P. Chemical Composition and Assessment of Antimicrobial, Antioxidant and Antiproliferative Activities of Essential oil from *Clinopodium sericeum*, a Peruvian Medicinal Plant. *Rec. Nat. Prod.* **2021**, *15*, 175–186. [CrossRef]
16. Baki, S.; Tufan, A.N.; Altun, M.; Özgökçe, F.; Güçlü, K.; Özyürek, M. Microwave-Assisted Extraction of Polyphenolics from Some Selected Medicinal Herbs Grown in Turkey. *Rec. Nat. Prod.* **2018**, *12*, 29–39. [CrossRef]
17. Re, R.; Pellegrini, N.; Proteggente, A.; Pannala, A.; Yang, M.; Rice-Evans, C. Antioxidant activity applying an improved ABTS radical cation decolorization assay. *Free Radic. Biol. Med.* **1999**, *26*, 1231–1237. [CrossRef]
18. Baran, A.; Karakılıç, E.; Faiz, Ö.; Özer, F. Synthesis of chalcone-containing zinc and cobalt metallophthalocyanines; investigation of their photochemical, DPPH radical scavenging and metal chelating characters. *Org. Commun.* **2020**, *13*, 65–78. [CrossRef]
19. Williams, P.A.; Cosme, J.; Ward, A.; Angove, H.C.; Vinković, D.M.; Jhoti, H. Crystal structure of human cytochrome P450 2C9 with bound warfarin. *Nature* **2003**, *424*, 464–468. [CrossRef]
20. Longley, W. The crystal structure of bovine liver catalase: A combined study by x-ray diffraction and electron microscopy. *J. Mol. Biol.* **1967**, *30*, 323–327. [CrossRef]

21. Tainer, J.A.; Getzoff, E.D.; Beem, K.M.; Richardson, J.S.; Richardson, D.C. Determination and analysis of the 2 Å structure of copper, zinc superoxide dismutase. *J. Mol. Biol.* **1982**, *160*, 181–217. [CrossRef]
22. Cao, H.; Pauff, J.M.; Hille, R. Substrate orientation and catalytic specificity in the action of xanthine oxidase: The sequential hydroxylation of hypoxanthine to uric acid. *J. Biol. Chem.* **2010**, *285*, 28044–28053. [CrossRef] [PubMed]
23. Trott, O.; Olson, A. AutoDock Vina: Improving the speed and accuracy of docking with a new scoring function, efficient optimization and multithreading. *J. Comput. Chem.* **2010**, *31*, 455–461. [CrossRef] [PubMed]
24. Stewart, J.J.P. *Stewart Computational Chemistry*; MOPAC: Colorado Springs, CO, USA, 2016.
25. Stewart, J.J.P. Optimization of parameters for semiempirical methods V: Modification of NDDO approximations and application to 70 elements. *J. Mol. Model.* **2007**, *13*, 1173–1213. [CrossRef] [PubMed]
26. Řezáč, J.; Hobza, P. Advanced corrections of hydrogen bonding and dispersion for semiempirical quantum mechanical methods. *J. Chem. Theory Comput.* **2012**, *8*, 141–151. [CrossRef] [PubMed]
27. Sanner, M.F. Python: A programming language for software integration and development. *J. Mol. Graph. Model.* **1999**, *17*, 55–84. [CrossRef]
28. Berman, H.M.; Westbrook, J.; Feng, Z.; Gilliland, G.; Bhat, T.N.; Weissig, H.; Shindyalov, I.N.; Bourne, P.E. The Protein Data Bank. *Nucleic Acids Res.* **2000**, *28*, 235–242. [CrossRef]
29. Madhavi Sastry, G.; Adzhigirey, M.; Day, T.; Annabhimoju, R.; Sherman, W. Protein and ligand preparation: Parameters, protocols, and influence on virtual screening enrichments. *J. Comput. Aided Mol. Des.* **2013**, *27*, 221–234. [CrossRef]
30. Morris, G.M.; Goodsell, D.S.; Halliday, R.S.; Huey, R.; Hart, W.E.; Belew, R.K.; Olson, A.J. Automated docking using a Lamarckian genetic algorithm and an empirical binding free energy function. *J. Comput. Chem.* **1998**, *19*, 1639–1662. [CrossRef]
31. Vadabingi, N.; Avula, V.K.R.; Zyryanov, G.V.; Vallela, S.; Anireddy, J.S.; Pasupuleti, V.R.; Mallepogu, V.; Chamarthi, N.R.; Ponne, V.C. Multiple molecular targets mediated antioxidant activity, molecular docking, ADMET, QSAR and bioactivity studies of halo substituted urea derivatives of α-Methyl-l-DOPA. *Bioorg. Chem.* **2020**, *97*, 103708–103728. [CrossRef]
32. Yang, B.; Hao, F.; Li, J.; Chen, D.; Liu, R. Binding of chrysoidine to catalase: Spectroscopy, isothermal titration calorimetry and molecular docking studies. *J. Photochem. Photobiol. B Biol.* **2013**, *128*, 35–42. [CrossRef]
33. Singaram, K.; Marimuthu, D.; Baskaran, S.; Chinaga, S. A Facile Synthesis of New 4,6-Dichloropyridine Derivatives, Their Biological Evaluation for Antimicrobial and Antioxidant Activity, and Docking Studies. *J. Chin. Chem. Soc.* **2016**, *63*, 758–769. [CrossRef]
34. Alsaraf, S.; Hadi, Z.; Al-Lawati, W.M.; Al Lawati, A.A.; Khan, S.A. Chemical composition, in vitro antibacterial and antioxidant potential of Omani Thyme essential oil along with in silico studies of its major constituent. *J. King Saud Univ. Sci.* **2020**, *32*, 1021–1028. [CrossRef]
35. Dassault Systèmes BIOVIA. *Discovery Studio Modeling Environment*; Dassault Systèmes BIOVIA: San Diego, CA, USA, 2017.
36. Abad-Zapatero, C. Ligand efficiency indices for effective drug discovery. *Expert Opin. Drug Discov.* **2007**, *2*, 469–488. [CrossRef]
37. Abad-Zapatero, C.; Perišić, O.; Wass, J.; Bento, A.P.; Overington, J.; Al-Lazikani, B.; Johnson, M.E. Ligand efficiency indices for an effective mapping of chemico-biological space: The concept of an atlas-like representation. *Drug Discov. Today* **2010**, *15*, 804–811. [CrossRef] [PubMed]
38. Abad-Zapatero, C. (Ed.) *Ligand Efficiency Indices for Drug Discovery*; Academic Press: San Diego, CA, USA, 2013; ISBN 978-0-12-404635-1.
39. Reynolds, C.H.; Tounge, B.A.; Bembenek, S.D. Ligand Binding Efficiency: Trends, Physical Basis, and Implications. *J. Med. Chem.* **2008**, *51*, 2432–2438. [CrossRef] [PubMed]
40. Cavalluzzi, M.M.; Mangiatordi, G.F.; Nicolotti, O.; Lentini, G. Ligand efficiency metrics in drug discovery: The pros and cons from a practical perspective. *Expert Opin. Drug Discov.* **2017**, *12*, 1087–1104. [CrossRef]
41. Veber, D.F.; Johnson, S.R.; Cheng, H.-Y.; Smith, B.R.; Ward, K.W.; Kopple, K.D. Molecular properties that influence the oral bioavailability of drug candidates. *J. Med. Chem.* **2002**, *45*, 2615–2623. [CrossRef]
42. Daina, A.; Michielin, O.; Zoete, V. SwissADME: A free web tool to evaluate pharmacokinetics, drug-likeness and medicinal chemistry friendliness of small molecules. *Sci. Rep.* **2017**, *7*, 42717. [CrossRef]
43. Pan, Y.; Huang, N.; Cho, S.; MacKerell, A.D. Consideration of Molecular Weight during Compound Selection in Virtual Target-Based Database Screening. *J. Chem. Inf. Comput. Sci.* **2003**, *43*, 267–272. [CrossRef]
44. Johnson, E.R.; Keinan, S.; Mori-Sánchez, P.; Contreras-García, J.; Cohen, A.J.; Yang, W. Revealing Noncovalent Interactions. *J. Am. Chem. Soc.* **2010**, *132*, 6498–6506. [CrossRef]
45. Contreras-García, J.; Johnson, E.R.; Keinan, S.; Chaudret, R.; Piquemal, J.-P.; Beratan, D.N.; Yang, W. NCIPLOT: A Program for Plotting Noncovalent Interaction Regions. *J. Chem. Theory Comput.* **2011**, *7*, 625–632. [CrossRef] [PubMed]
46. Cantero-López, P.; Robledo Restrepo, S.M.; Yañez, O.; Zúñiga, C.; Santafé-Patiño, G.G. Theoretical study of new LmDHODH and LmTXNPx complexes: Structure-based relationships. *Struct. Chem.* **2021**, *32*, 167–177. [CrossRef]
47. Sharma, V.; Cantero-López, P.; Yañez-Osses, O.; Kumar, A. Effect of Cosolvents DMSO and Glycerol on the Self-Assembly Behavior of SDBS and CPC: An Experimental and Theoretical Approach. *J. Chem. Eng. Data* **2018**, *63*, 3083–3096. [CrossRef]
48. Humphrey, W.; Dalke, A.; Schulten, K. VMD: Visual molecular dynamics. *J. Mol. Graph.* **1996**, *14*, 33–38. [CrossRef]
49. Pires, D.E.V.; Blundell, T.L.; Ascher, D.B. PkCSM: Predicting Small-Molecule Pharmacokinetic and Toxicity Properties Using Graph-Based Signatures. *J. Med. Chem.* **2015**, *58*, 4066–4072. [CrossRef]

50. Letchamo, W.; Ward, W.; Heard, B.; Heard, D. Essential oil of *Valeriana officinalis* L. cultivars and their antimicrobial activity as influenced by harvesting time under commercial organic cultivation. *J. Agric. Food Chem.* **2004**, *52*, 3915–3919. [CrossRef]
51. Lunz, K.; Stappen, I. Back to the Roots—An Overview of the Chemical Composition and Bioactivity of Selected Root-Essential Oils. *Molecules* **2021**, *26*, 3155. [CrossRef]
52. Thusoo, S.; Gupta, S.; Sudan, R.; Kour, J.; Bhagat, S.; Hussain, R.; Bhagat, M. Antioxidant Activity of Essential Oil and Extracts of *Valeriana jatamansi* Roots. *BioMed Res. Int.* **2014**, *2014*, 614187. [CrossRef]
53. Huynha, L.; Pacherb, T.; Trana, H.; Novak, J. Comparative analysis of the essential oils of *Valeriana hardwickii* Wall. from Vietnam and *Valeriana officinalis* L. from Austria. *J. Essent. Oil Res.* **2013**, *25*, 408–413. [CrossRef]
54. Raal, A.; Arak, E.; Orav, A.; Kailas, T.; Müürisepp, M. Variation in the composition of the essential oil of commercial *Valeriana officinalis* L. roots from different countries. *J. Essent. Oil Res.* **2008**, *20*, 524–529. [CrossRef]
55. Houghton, P.J. The biological activity of valerian and related plants. *J. Ethnopharmacol.* **1988**, *22*, 121–142. [CrossRef]
56. Benzie, I.F.F.; Strain, J.J. The Ferric Reducing Ability of Plasma (FRAP) as a Measure of "Antioxidant Power": The FRAP Assay. *Anal. Biochem.* **1996**, *239*, 70–76. [CrossRef] [PubMed]
57. Shahidi, F.; Zhong, Y. Measurement of antioxidant activity. *J. Funct. Foods* **2015**, *18*, 757–781. [CrossRef]
58. Wang, J.; Zhao, J.; Liu, H.; Zhou, L.; Liu, Z.; Wang, J.; Han, J.; Yu, Z.; Yang, F. Chemical Analysis and Biological Activity of the Essential Oils of Two Valerianaceous Species from China: *Nardostachys chinensis* and *Valeriana officinalis*. *Molecules* **2010**, *15*, 6411–6422. [CrossRef]
59. Ben, H.A.; Ben, H.N. Essential oil from *Artemisia phaeolepis*: Chemical composition and antimicrobial activities. *J. Oleo Sci.* **2013**, *62*, 973–980. [CrossRef]
60. Rawat, S.; Jugran, A.K.; Bhatt, I.D.; Rawal, R.S.; Andola, H.C.; Dhar, U. Essential oil composition and antioxidant activity in *Valeriana jatamansi* Jones: Influence of seasons and growing sources. *J. Essent. Oil Res.* **2016**, *29*, 101–107. [CrossRef]
61. Lamazares, E.; Macleod-Carey, D.; Miranda, F.P.; Mena-Ulecia, K. Theoretical evaluation of novel thermolysin inhibitors from bacillus thermoproteolyticus. Possible antibacterial agents. *Molecules* **2021**, *26*, 386. [CrossRef]
62. Murray, C.W.; Erlanson, D.A.; Hopkins, A.L.; Keserü, G.M.; Leeson, P.D.; Rees, D.C.; Reynolds, C.H.; Richmond, N.J. Validity of Ligand Efficiency Metrics. *ACS Med. Chem. Lett.* **2014**, *5*, 616–618. [CrossRef]
63. Hopkins, A.L.; Keserü, G.M.; Leeson, P.D.; Rees, D.C.; Reynolds, C.H. The role of ligand efficiency metrics in drug discovery. *Nat. Rev. Drug Discov.* **2014**, *13*, 105–121. [CrossRef]
64. Ekowati, J.; Diyah, N.W.; Nofianti, K.A.; Hamid, I.S. Molecular Docking of Ferulic Acid Derivatives on P2Y12 Receptor and their ADMET Prediction. *J. Math. Fundam. Sci.* **2018**, *50*, 203–219. [CrossRef]
65. Angelis, I.D.; Turco, L. Caco-2 Cells as a Model for Intestinal Absorption. *Curr. Protoc. Toxicol.* **2011**, *47*, 20–26. [CrossRef] [PubMed]
66. Cheng, F.; Li, W.; Zhou, Y.; Shen, J.; Wu, Z.; Liu, G.; Lee, P.W.; Tang, Y. admetSAR: A Comprehensive Source and Free Tool for Assessment of Chemical ADMET Properties. *J. Chem. Inf. Model.* **2012**, *52*, 3099–3105. [CrossRef] [PubMed]
67. Nau, R.; Sorgel, F.; Eiffert, H. Penetration of Drugs through the Blood-Cerebrospinal Fluid/Blood-Brain Barrier for Treatment of Central Nervous System Infections. *Clin. Microbiol. Rev.* **2010**, *23*, 858–883. [CrossRef] [PubMed]

antioxidants

Article

Heracleum persicum Essential Oil Nanoemulsion: A Nanocarrier System for the Delivery of Promising Anticancer and Antioxidant Bioactive Agents

Shima Ghareh Bashlouei [1], Ehsan Karimi [1,*], Mohsen Zareian [2], Ehsan Oskoueian [3,*] and Majid Shakeri [4,*]

1 Department of Biology, Mashhad Branch, Islamic Azad University, Mashhad 917568, Iran; shimagharabashlou@gmail.com
2 Department of Biology and Biological Engineering, Chalmers University of Technology, 41296 Göteborg, Sweden; mohsen.zareian@chalmers.se
3 Department of Research and Development, Arka Industrial Cluster, Mashhad 9188944586, Iran
4 Department of Medicine, University of Washington, Seattle, WA 98109, USA
* Correspondence: ehsankarimi@mshdiau.ac.ir (E.K.); e.oskoueian@gmail.com (E.O.); mshakeri@uw.edu (M.S.)

Abstract: Essential oils are important compounds for the prevention and/or treatment of various diseases in which solubility and bio-accessibility can be improved by nanoemulsion systems. *Heracleum persicum* oil nanoemulsion (HAE-NE) was prepared and biological properties were investigated against human breast cancer cells and normal human fibroblasts foreskin. Particle size, zeta potential and poly dispersity index were 153 nm, −47.9 mV and 0.35, respectively. (E)anethole (57.9%), terpinolene (13.8%), γ-terpinene (8.1%), myrcene (6.8%), hexyl butyrate (5.2%), octyl butanoate (4.5%) and octyl acetate (3.7%) was detected in nanoemulsion. Proliferation of cancer cells at IC_{50} = 2.32 μg/mL was significantly ($p < 0.05$) inhibited, and cell migration occurred at 1.5 μL/mL. The HAE-NE at 1.5, 2.5 and 3.5 μg/concentration up-regulated caspase 3 and enhanced sub-G1 peak of cell cycle with nil cytotoxic effects in the liver, kidney and jejunum of mice. Villus height, villus width, crypt depth and goblet cells in mice group fed with 10 and 20 mg/kg body weight of HAE-NE improved. Cellular redox state in the liver indicated 10 and 20 mg/kg body weight of nanoemulsion significantly up-regulated the expression of SOD, CAT and GPx genes. *Heracleum persicum* oil nanoemulsion could be an eco-friendly nanotherapeutic option for pharmaceutical, cosmetological and food applications.

Keywords: phytochemicals; medicinal plants; plant-based nanostructures; eco-friendly nanotherapeutics; volatile compounds; therapeutics; phytobiotics; apiaceae family

Citation: Bashlouei, S.G.; Karimi, E.; Zareian, M.; Oskoueian, E.; Shakeri, M. *Heracleum persicum* Essential Oil Nanoemulsion: A Nanocarrier System for the Delivery of Promising Anticancer and Antioxidant Bioactive Agents. *Antioxidants* **2022**, *11*, 831. https://doi.org/10.3390/antiox11050831

Academic Editors: Delia Mirela Tit and Simona Bungau

Received: 1 March 2022
Accepted: 21 April 2022
Published: 25 April 2022

1. Introduction

Heracleum persicum, a flowering plant of the family Apiaceae, naturally grows under humid conditions and has traditionally been used as a medicinal herb and flavoring agent. The potential of *H. persicum* for use in biomedicine, food and pharmaceutical industries has recently attracted much attention [1,2]. *Heracleum persicum* has been extensively used as food additives, food preservatives, flavoring agents and spices (i.e., to flavor pickles) and for the treatment purposes of gastrointestinal, neurological, respiratory, urinary and rheumatologically dysfunctions [1,3]. Physiological functions of *H. persicum* can be antioxidant, antihyperlipidemic, antidiabetic, anti-inflammatory, anticonvulsant, antimicrobial, analgesic and cardio- and gastro-protective properties [4–7]. The major bioactive compounds identified to induce such physiological effects are volatiles, ethyl esters, n-alkenes, phenolics, flavonoids, alkaloids, terpenoids and triterpenes [1]. Hexyl butyrate, anethole, octyl acetate, hexyl-2-methylbutanoate and hexyl isobutyrate were found to be the main natural phytochemicals of *H. persicum* essential oil [1,7].

In traditional medicine, essential oils are known to act as pharmaceutical agents, and thereby, can be used for the prevention and/or treatment of various diseases [8]. A wide spectrum of volatile bioactive phytochemicals, e.g., phenolics, flavonoids and terpenes, can be found in the essential oils of plant origins [9,10]. The solubility and bio-accessibility of essential oils can be improved by nanoemulsion systems, which ultimately can be developed as drug delivery systems [11,12]. Nanoemulsions are nanoscaled emulsion-based systems containing therapeutic compounds intended to keep pharmaceutical ingredients active and deliver components to the target cells in a more effective fashion [13]. Nanoemulsions are typically made of a stable isotropic system containing two immiscible liquid types, which are basically mixed together to form stable nanodroplets with a 20–200 nm range [13]. Characteristics and details of various nanoemulsion systems were broadly described earlier [13,14] and will not be discussed here. Recently, nanoscale emulsion systems have gained much interest and attention—thanks to special characteristics, i.e., high performance and stability [15]. As such, nanoemulsions continue to be developed as one of the main formulation systems in the pharmaceutical and cosmeceutical industries [15]. The present study, therefore aimed to synthesize a nanoscale emulsion system from the essential oils of *Heracleum persicum*. Cellular and molecular characteristics of the nanoemulsion against cancer cell line, cytotoxicity and antioxidant properties in vitro and in vivo were also investigated. To the best of our knowledge, this is the first report of a nanoscale emulsion system made from the Persian herbal medicine *Heracleum persicum* for future development in the pharmaceutical and cosmeceutical industries.

2. Materials and Methods

2.1. Chemicals, Reagents and Cell Lines

Polysorbate 20 and 80, Tween 80 (polyethylene glycol sorbitan monooleate), polyethylene glycol (PEG), DMEM culture medium trypsin, MTT [3-(4, 5-dimethylthiazol-2-yl)-2, 5-diphenyltetrazolium bromide] and Fetal Bovine Serum (FBS), were purchased from (Merck, Darmstadt, Germany). Cancer cell lines (MDA-MB-231, ATCC HTB 26) and human fibroblasts foreskin (ATCC SCRC-1041) as normal cell line were obtained from the Pasteur Institute of Iran. The PCR Master Mix and SYBR Green Master mix were purchased from Vazyme Biotech (Nanjing, China). The RNA extraction kit and cDNA Synthesis Kit were purchased from CinnaGene (Tehran, Iran).

2.2. Essential Oil Extraction Procedure

The essential oils in the aerial parts of *H. persicum* were extracted by a Soxhlet apparatus. In brief, 200 g dried *H. persicum* was mix with 1000 mL ethanol and boiled for 2 h, after which the extracted solution was separated by a vacuum controller as previously described [16].

2.3. Nanoemulsion Preparation, Identification and Characterization

Nanoemulsion was prepared as described previously [17] with slight modifications. Briefly, 3 g of *H. persicum* oil extract were mixed with 97 mL deionized water and sonicated for 30 min at 20 kHz ultrasonic frequency. The ratio of essential oil to non-ionic surfactant was 1:3 *v/v*. The HAE-NE characteristics, i.e., size, polydispersity index (PDI) and dynamic light scattering (DLS), was also measured. The Z-average values and morphology of nanodroplets were analyzed according to the distribution intensity and field emission scanning electron microscopy (FESEM), respectively. Volatile bioactive compounds present in HAE-NE were detected by Gas Chromatography-Mass Spectrometry as previously described [18]. One microgram of the oil sample was injected into GC (Agilent 6890) coupled with an Agilent 5973N Mass Spectrometer (Santa Clara, CA, USA). Essential oils were analyzed on a BPX5 fused silica column (30 m $\times$ 0.25 mm i.d. $\times$ 0.25 μm) under electronic impact mode (70 eV), split injection ratio (1:35) and He as carrier gas (initial pressure: 110 kPa), with 0.5 mL/min flow rate under a scan range of 40–500 amu. The temperature program of the oven was set as follows. Initially, 50 °C for 5 min, then a gradient of 3 °C/min up to 240 °C, after which it was raised at 15 °C/min up to 300 °C and finally held at 300 °C for 3 min. Compounds were identified by comparing retention time

(RT) with the literature and mass spectra library, i.e., the National Institute of Standards and Technology (NIST).

2.4. Anticancer Assay

The anticancer effects of HAE-NE were assessed against human breast cancer cell line (MDA-MB-231), whereas human fibroblasts foreskin (HFF) was used as normal cell line [19]. Briefly, 5×10^3 cells were seeded for 24 h after which a varying concentration of HAE-NE (0.7, 1.5, 3.1 and 6.2 μg/mL) were treated against cell lines. After 48 h incubation, the medium was refreshed with the media containing MTT (0.5 mg/mL) incubated for 3 h at 37 °C, after which the medium was substituted with the media containing dimethylsulfoxide (DMSO, 100 μL). Optical density (OD) of the samples were recorded at 570 nm and cell viability was calculated as follows:

$$\text{Cell viability (\%)} = (\text{OD Sample}/\text{OD Control}) \times 100$$

2.5. Cell Migration Assay

The impact of the HAE-NE on the migration ability of human breast cancer cell line was conducted in vitro as previously described [19]. In brief, cells were first cultured to reach 80–100% confluency and a series of lines were made using a sterile 10-μL micropipette tip. Medium was replaced and washed completely with new medium to remove detached cells. The wounds were photographed (0 h) and the medium containing 0 and 1.5 μg/mL HAE-NE were added, after which cells were incubated at 37 °C under 5% CO_2 for 10 h and photos of the wounds were taken. The migration distance was calculated by deducting the distance of wound edges at 0 and 10 h.

2.6. Gene Expression of Caspase 3

The profiling of the caspase 3 was evaluated using the Real-Time PCR as previously described [20]. The RNA was extracted at the end of the cell incubation using RNeasy Mini Kit (Qiagen, Hilden, Germany) and cDNA was synthesized using cDNA synthesis kit (Qiagen, Hilden, Germany) as per the manufacturer's protocols. The PCR reaction conditions were optimized and set as follows: 94 °C for 5 min (1×), 94 °C for 25 s, 58 °C for 30 s and 72 °C for 25 s (40×). All RT-PCR amplifications were performed in triplicate and primers are listed in Table 1.

Table 1. Details of the primer sets used in the study.

	Gene	Forward (5′→3′)	Reverse (5′→3′)	Accession Number
MDA-MB231 cell line	Caspase-3	CTGGACTGTGGCATTGAGAC	ACAAAGCGACTGGATGAACC	NM_001284409
	GAPDH [1]	GAAGGTGAAGGTCGGAGTC	GAAGATGGTGATGGGATTTC	NM002046
Mice tissue	SOD	GAGACCTGGGCAATGTGACT	GTTTACTGCGCAATCCCAAT	NM_011434
	CAT	ACATGGTCTGGGACTTCTGG	CAAGTTTTTGATGCCCTGGT	NM_009804
	GPx	CACAGTCCACCGTGTATGCC	GTGTCCGAACTGATTGCACG	NM_008160
	GAPDH	GACTTCAACAGCAACTCCCAC	TCCACCACCCTGTTGCTGTA	NM_001289726

[1] GAPDH: glyceraldehyde 3-phosphate dehydrogenase; SOD: superoxide dismutase; CAT: catalase; GPx: glutathione peroxidase.

2.7. Flow Cytometry-Based Assay

Various concentrations of HAE-NE (0.7, 1.5, 3.1, and 6.2 μg/mL) were treated against the cancer cell line for 48 h, after which cells were washed and mixed with Propidium Iodide (PI), 0.2% Triton X-100 and 0.1% sodium citrate and maintained for 10 min at 37 °C under dark conditions as previously described [20]. The AO/PI dyes were mixed with the harvested and washed cells for staining purposes and were observed under a fluorescence microscope (Olympus BX41, Tokyo, Japan). A laser flow cytometer (FACSCalibur, Becton Dickinson, TX, USA) was also performed to determine the cell cycle alterations.

2.8. In Vivo Assay

The toxicity of HAE-NE was determined using 15 female mice (22–25 g) kept in individual cages (24 °C with 50% humidity) and maintained for 1 week to adapt to the laboratory condition. Animals were randomly divided into three groups ($n = 5$) and treated as follows: a control group were orally gavaged with distilled water, whereas the other two groups orally received HAE-NE at 10 and 20 mg/kg body weight, respectively. The experiment lasted for 30 days. All animal experiments were conducted according to the ethical principles approved by Azad University ethics codes (IR.IAU.MSHD.REC.1399.141).

2.9. Histopathological Analysis and Tissue Staining

For histopathological analyses and tissue staining, animals, at the end of animal trials, were first euthanized and the liver, kidney and jejunum were obtained and washed with 0.9% NaCl serum, maintained in formalin 10%, after which paraffinized and cut into thin pieces (5 µm) [21]. Tissue samples were stained with hematoxylin and eosin as previously described [17,22]. Cell morphologies were investigated using an inverted microscope and results were interpreted by an experienced pathologist for further tissue characterization.

2.10. Antioxidant Gene Expression in Mice Liver

The expressions of SOD, CAT and GPx, which are basically recognized to be important antioxidant biomarkers, were investigated in mice liver. The effect of different concentration of nanoemulsion (0, 10 and 20 mg/kg body weight) were investigated as previously described [23]. In brief, the liver tissues of the mice were crushed and prepared for RNA extraction using the RNeasy Mini kit (Qiagen, Hilden, Germany), after which a Quantitect Reverse Transcription Kit (Qiagen, Hilden, Germany) was employed for the synthesis of cDNA libraries and primer sequences targeted for CAT, SOD and GPx genes and a house keeping (GAPDH) gene were also designed (Table 1). To perform a comparative Real-Time PCR, a SYBR Green PCR Master Mix (Qiagen, Hilden, Germany) was employed and designated genes were amplified under the following program: 95 °C for 5 min (1×), 95 °C for 20 sec, 55 °C for 20 sec and 72 °C for 25 sec (35×). The gene expressions were normalized to GAPDH as a reference gene, after which they were normalized to the expression of the respective genes in the control group.

2.11. Statistical Analyses

Data were reported as mean ± standard deviation. All experiments were performed in triplicate. The statistical analyses were carried out using one-way ANOVA by the statistical package SPSS (version 21, SPSS Inc., Chicago, IL, USA) and α at 95% was defined as a statistically significant difference. Symbols on figures indicated differences between groups—ns: $p \geq 0.05$, *: $p < 0.05$, **: $p < 0.01$ and ***: $p < 0.001$.

3. Results and Discussion

3.1. HAE-NE Characterization and Identification

The particle size, poly dispersity index and zeta potential of HAE-NE were 153.64 nm, 0.35 ± 0.07 and −47.9 mV, respectively. The poly dispersity index in our study (0.35 ± 0.07) is, in principle, less than 0.3, which indicates homogenous dispersion [24]. The zeta potential is also an index which demonstrates the stability of colloidal dispersions. A zeta potential value outside −30 to +30 mV is typically recognized as satisfactory repulsive force to achieve and maintain physical colloidal stability [24]. The dynamic light scattering results revealed that the nanoemulsion size dispersion was 153.64 nm (Figure 1, left). The Field Emission Scanning Electron Microscopy indicated the spherical shape of HAE-NE, which was in accordance with particle size data (Figure 1, right).

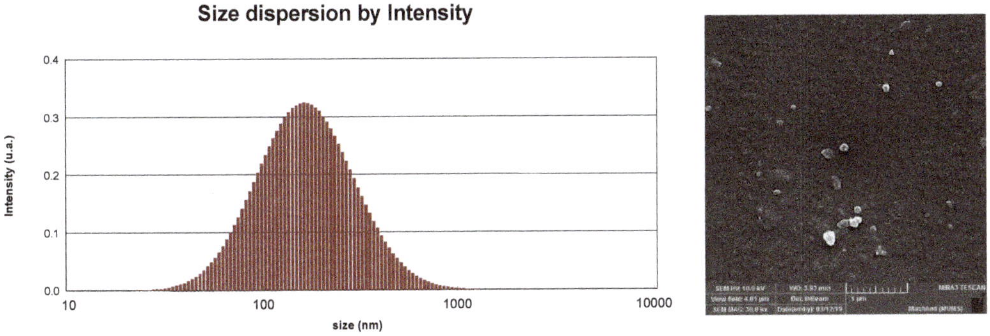

Figure 1. The characteristics of *Heracleum persicum* essential oil nanoemulsion and anticancer effect of HAE-NE. The dynamic light scattering results revealed that the nanoemulsion size dispersion was 153.64 nm (**left**). The Field Emission Scanning Electron Microscopy indicated the spherical shape of HAE-NE, which was in accordance with particle size data (**right**).

Bioactive compounds were also detected by GC-MS in HAE-NE, in which the main volatile compounds were anethole (57.9%) and terpinolene (13.8%) (Table 2). Anethole was also identified as major compound in the oil composition of leaf (47.5%) and flower (38.6%) of *H. persicum* [25]. Similar volatiles identified in our study (Table 2) were previously reported to be present in various parts of *Heracleum persicum* [25–28]. Variation in the percentage of volatiles in the essential oil content within various *Heracleum persicum* species is mostly influenced by environmental factors [27]. The HAE-NE in our study (Table 2) also contained hexyl butyrate (5.2%), octyl butanoate (4.5%) and octyl acetate (3.7%). Essential oils rich in octyl acetate and hexyl butyrate are valuable for medicinal and commercial purposes [27,28]. Other bioactive compounds identified in HAE-NE (Table 2) were reported to exhibit a variety of biological properties, e.g., bactericidal, antioxidant and anti-inflammatory activities [7,28–31].

Table 2. Major compounds present in the nanoemulsion synthesized using essential oil obtained from aerial parts of *H. persicum*.

Peak	Compound	Percentage
1	(E)-anethole	57.9
2	terpinolene	13.8
3	ɣ-terpinene	8.1
4	myrcene	6.8
5	hexyl butyrate	5.2
6	octyl butanoate	4.5
7	octyl acetate	3.7

The cytotoxicity potential of HAE-NE against human cancer cell line (MDA-MB-231) and normal cell line (HDF) indicated that HAE-NE inhibited the cancer cell proliferation significantly ($p < 0.05$) with IC_{50} = 2.32 µg/mL (Figure 2). Furthermore, various concentrations of HAE-NE did not show toxicity affects against the normal cell (HFF). Similar findings on the effect of *H. persicum* essential oils against human cancer lines of HeLa, LS180, were also reported [28]; however, the IC_{50} in the present work was found to be lower (Figure 2) than previous reports.

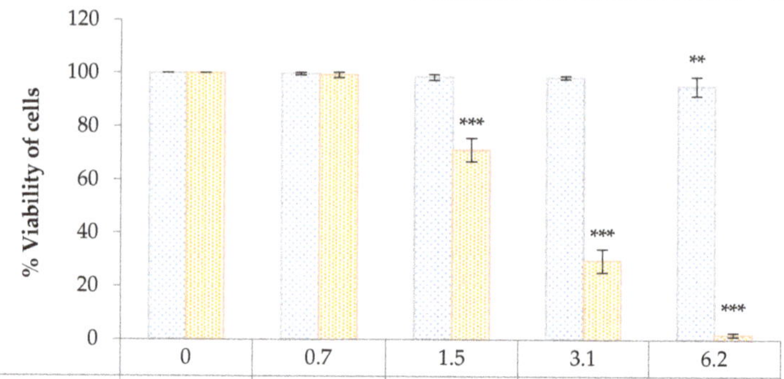

	0	0.7	1.5	3.1	6.2
HFF	100	99.7	98.6	98.5333	95.2667
MDA-MB231	100	99.3333	71.4667	29.7333	1.8667

Figure 2. Effect of *Heracleum persicum* oil nanoemulsion (HAE-NE) on human fibroblasts foreskin (HFF) and human breast cancer cell (MDA-MB-231) viability. Values represent the mean ± standard deviation from three independent experiments. Symbols provided on bars indicated differences between groups—no symbol: $p \geq 0.05$, **: $p < 0.01$ and ***: $p < 0.001$.

3.2. Migration Analysis

The potential of HAE-NE on MDA-MB231 cell migration at the concentration of 1.5 µL/mL incubation indicated the untreated MDA-MB231 cells demonstrated complete wound closure after 20 h (Figure 3). The HAE-NE at 1.5 µL/mL concentration exhibited strong inhibition on the migration of breast cancer cells. Cell migration is deemed to be important in cancer progression to incurable metastatic disease [32]. Therapeutics which potentially block the migration of cancer cells may inhibit/reduce metastasis, and thereby, can profoundly improve cancer therapies [33]. Such functionality is of particular interest and prime importance in cancer treatments such as triple negative breast cancer, in which targeted drugs are currently not reasonably developed [33].

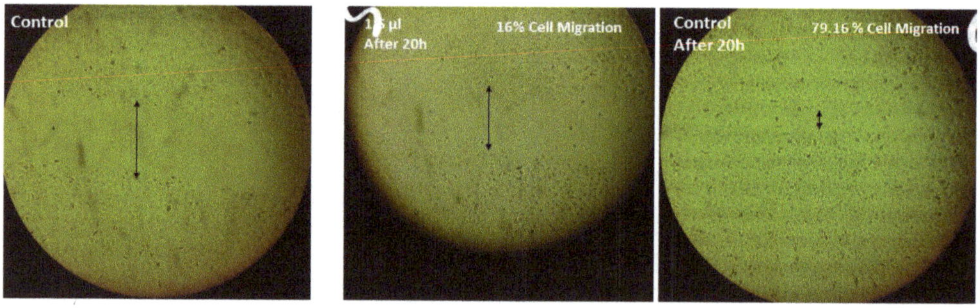

Figure 3. Migration inhibition percentage of HAE-NE in MDA-MB231 cells after 20 h of incubation. Error bar indicates the standard error of the mean of three independent experiments. Cells were visualized at 200× magnification.

3.3. In Vitro Gene Expression Profiling of Caspase 3

Different concentration of HAE-NE (1.5, 2.5 and 3.5 µg/mL) significantly ($p < 0.05$) up-regulated Caspase 3 gene (Figure 4), indicating the HAE-NE can induce apoptosis death in breast cancer cell line. Caspase-3—a cysteine-aspartic protease—is known to play a key role in apoptotic pathways, which basically interferes with the apoptosis response. The expression of Caspase-3 is typically influenced by the status of oxidative stress and antioxidant defense systems. The overexpression of Caspase-3 may ultimately induce apoptosis responses. Essential oils present in plant extracts can remarkably induce Caspase-3-dependent apoptosis properties and examples of such plant extracts include, but are not limited to, *Afrostyrax lepidophyllus* Mildbr, *Monodora myristica*, *Arachis hypogaea*, *Ricinus communis* L. *Garcinia epunctata* and *Ptycholobium contortum* [34–37].

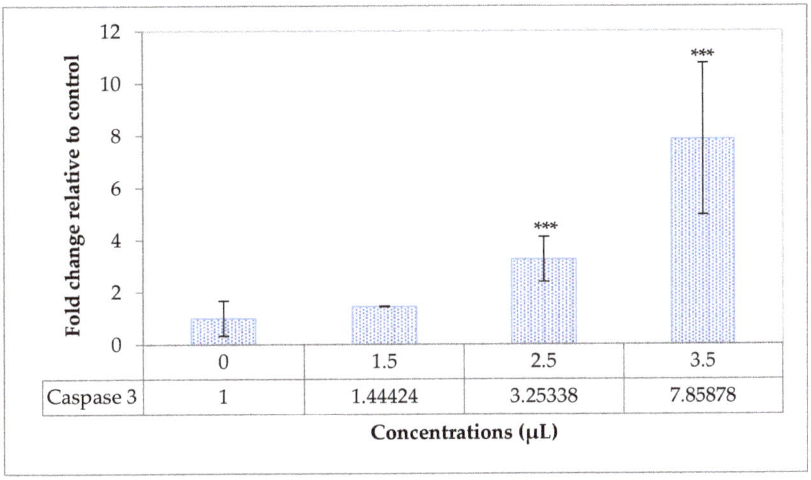

Figure 4. The expression of caspase-3 gene in human breast cancer cell line (MDA-MB231) with different concentrations of *Heracleum persicum* oil nanoemulsion (HAE-NE), indicating the HAE-NE can induce apoptosis death in breast cancer cell line. Symbols provided on bars indicate differences between groups—***: $p < 0.001$.

3.4. Flow Cytometry Analysis

The flow cytometry data showed the increase in HAE-NE concentration and enhanced sub-G1 peak of the cell cycle in MDA-MB-231 cells, and confirmed cell apoptotic death, which could be recognized as a potential apoptosis inducer (Figure 5). Similar to our findings, essential oils present in lemongrass (*Cymbopogon citratus* Stapf) were also reported [38] to induce cell cycle arrest in cancer cells. Similarly phytochemicals of *Bulbine frutescens* demonstrated cell cycle arrest at G1 phase and induced apoptosis in breast cancer cells [39].

3.5. Histopathological Alterations and Morphometric Analysis

The histopathological features of the liver, kidney and jejunum tissue upon treatment with HAE-NE show changes in the histomorphology and that cellular death were not detected in tissues (Figure 6), and the morphometric analyses such as villus height, villus width, crypt depth and goblet cells (Table 3) were in accordance with the findings of the jejunum histopathology. The mice group fed 10 and 20 mg/kg body weight of HAE-NE also showed an enhancement in the morphometric parameters. Goblet cells secrete mucin to protect the internal wall of the intestine and HAE-NE in our study increased the number of goblet cells (Table 3).

3.6. Antioxidant Gene Profiling In Vitro

The profiling of antioxidant genes, i.e., SOD, CAT and GPx, as the cellular redox state in liver indicates 10 and 20 mg/kg body weight of nanoemulsion significantly ($p < 0.05$) up-regulated the expression of antioxidant-related genes (Table 4). Similar antioxidant capacities of *H. persicum* aqueous extract in diabetes-induced oxidative stress rats were also reported [6], in which antioxidant defense systems in the brain, kidney and liver were enhanced upon supplementation with plant extract or antioxidants [40]. Another observation made in a human trial with minimal coronary artery disease confirmed antioxidant effects of *H. persicum* fruit extract in modulating oxidative stress biomarkers, such as SOD, CAT and GPx activities [41]. Such antioxidant properties were mainly attributed to (E)-anethole present in *H. persicum* extracts [28]. Cancer cells tend to counteract the effects of free radicals via antioxidant enzymes, such as SOD, CAT and GPx as a mechanism to increase cell viability under stressful conditions caused by augmented metabolism.

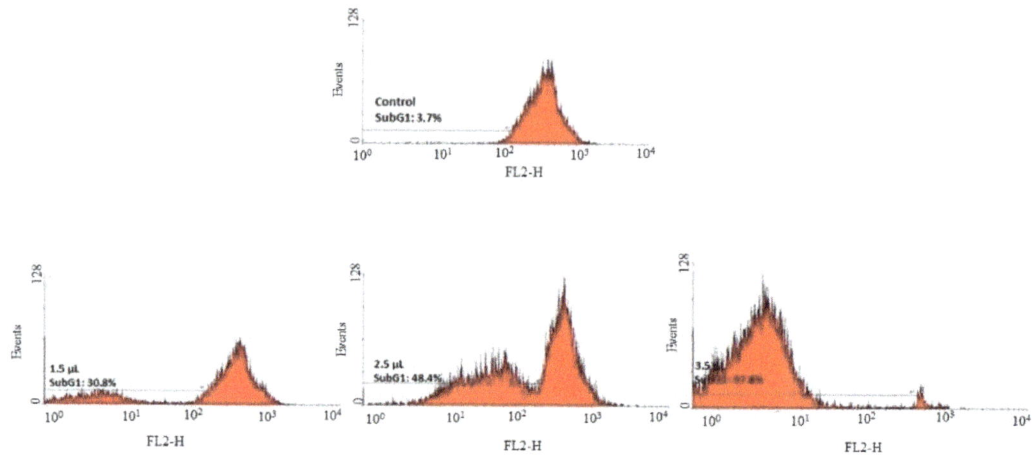

Figure 5. Flow cytometric analysis of cell cycle in human breast cancer cells (MDA-MB-231) upon treatment with different concentrations of *Heracleum persicum* oil nanoemulsion (HAE-NE) during 48 h. Increase in HAE-NE concentration enhanced the sub G1 peak of the cell cycle in MDA-MB-231 cells, thereby confirming cell apoptotic death, which could be recognized as a potential apoptosis inducer.

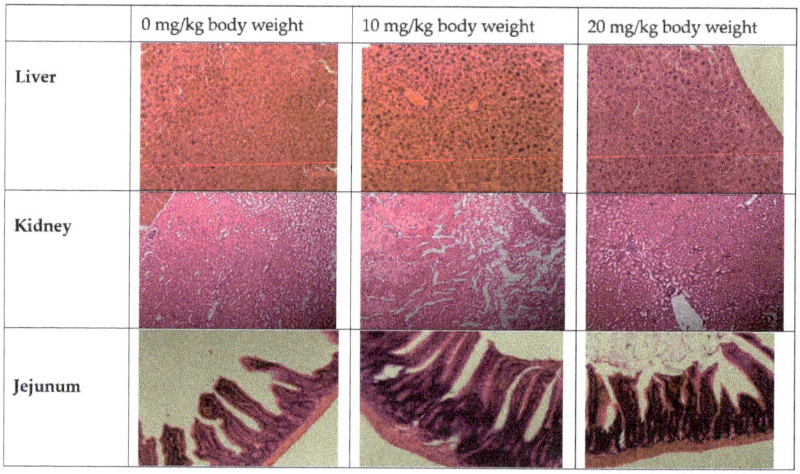

Figure 6. Histopathological analysis of live, kidney and jejunum of the mice treated with different concentration of *Heracleum persicum* oil nanoemulsion. Treatment with *Heracleum persicum* oil nanoemulsion (HAE-NE) showed that cellular death was not detected in tissues.

Table 3. Morpho-characteristics of mice jejunum treated with different concentration of *Heracleum persicum*.

	0 mg/kg Body Weight	10 mg/kg Body Weight	20 mg/kg Body Weight	SEM
Villus Height (μm)	322	341	362	3.89
Villus Width (μm)	68	76	94	4.38
Crypt Depth (μm)	71	76	75	5.94
Goblet Cells	2.3	3.1	3.8	0.29

SEM: Standard Error of the Mean. Means with different superscript letters indicate significant difference at $p < 0.05$.

Table 4. The expression analyses of antioxidant-related genes in the liver of mice.

	0 mg/kg Body Weight	10 mg/kg Body Weight	20 mg/kg Body Weight	SEM [1]
SOD [2]	1^{c}	1.9^{b}	2.8^{a}	0.06
CAT	1^{c}	1.6^{b}	2.3^{a}	0.08
GPx	1^{c}	1.5^{b}	1.9^{a}	0.05

[1] SEM: Standard Error of the Mean. [2] SOD: Superoxide dismutase; CAT: Catalase; GPx: Glutathione peroxidase. Means with different superscript letters indicate significant difference at $p < 0.05$.

In the light of these research, it may be concluded that *H. Persicum* Essential Oil Nanoemulsion could considered as a promising antioxidant and anticancer natural drug in breast cancer therapy. Further research on the various phytochemicals of individual groups of natural components may demonstrated the exact capacity of *H. Persicum* Essential Oil to inhibit different cancer cells and encourage the improvement as novel broad-spectrum herbal antioxidant and anticancer formulation in future.

4. Conclusions

The present study demonstrated preparation and characterization of *H. persicum* oil nanoemulsion with biological properties in vitro and in vivo. Bioactive compounds present in the nanoemulsion were also detected. In addition to nanoemultion properties against normal and cancer cell lines, the cytotoxic effects in the liver, kidney and jejunum of mice and the enhancement of the morphometric parameters were also demonstrated and discussed. Given the several side effects of synthetic drugs for the treatment purposes of intestinal diseases which could ultimately deteriorate patients' conditions, *Heracleum persicum* oil nanoemulsion, in our study, exhibited the potentials to be developed as therapeutic drugs. Our finding suggested *H. persicum* oil nanoemulsion could be an eco-friendly nanotherapeutic option for pharmaceutical, cosmetological and food application purposes.

Author Contributions: Conceptualization, E.K. and E.O.; methodology, E.O. and M.S.; formal analysis, E.K. and M.Z.; investigation, S.G.B.; resources, E.K.; data curation, E.K.; writing—original draft preparation, S.G.B. and M.Z.; writing—review and editing, E.K., M.S. and M.Z.; supervision, E.K. and E.O.; project administration, E.K.; funding acquisition, E.K. and E.O. All authors have read and agreed to the published version of the manuscript.

Funding: This research received no external funding.

Institutional Review Board Statement: The animal study protocol was approved by the Institutional Review Board (or Ethics Committee) of Azad University Iran ethics codes (IR.IAU.MSHD.REC.1399.141, 2020).

Informed Consent Statement: Not applicable.

Data Availability Statement: Data is contained within the article.

Acknowledgments: We thank Islamic Azad University for providing animal facility for the study.

Conflicts of Interest: The authors declare no conflict of interest.

References

1. Majidi, Z.; Lamardi, S.S. Phytochemistry and biological activities of *Heracleum persicum*: A review. *J. Integr. Med.* **2018**, *16*, 223–235. [CrossRef]
2. Mehmandost, N.; García-Valverde, M.T.; Soriano, M.L.; Goudarzi, N.; Lucena, R.; Chamjangali, M.A.; Cardenas, S. *Heracleum persicum* based biosorbent for the removal of paraquat and diquat from waters. *J. Environ. Chem. Eng.* **2020**, *8*, 104481. [CrossRef]
3. Bahadori, M.B.; Dinparast, L.; Zengin, G. The genus *Heracleum*: A comprehensive review on its phytochemistry, pharmacology, and ethnobotanical values as a useful herb. *Compr. Rev. Food Sci. Food Saf.* **2016**, *15*, 1018–1039. [PubMed]
4. Majidi, Z. Effects of *Heracleum persicum* hydroalcoholic extract on insulin, serum anti-oxidant enzymes, glucose, and lipid profiles in alloxan-induced diabetic rats. *Iran. J. Med. Sci.* **2020**, *45*, 199–206. [PubMed]
5. Akbaribazm, M.; Goodarzi, N.; Rahimi, M.; Naseri, L.; Khazaei, M. Anti-inflammatory, anti-oxidative and anti-apoptotic effects of *Heracleum persicum* L. extract on rats with gentamicin-induced nephrotoxicity. *Asian Pac. J. Trop. Biomed.* **2021**, *11*, 47–58.
6. Alkan, E.E.; Celik, I. The therapeutics effects and toxic risk of *Heracleum persicum* Desf. extract on streptozotocin-induced diabetic rats. *Toxicol. Rep.* **2018**, *5*, 919–926. [CrossRef]
7. Hajhashemi, V.; Sajjadi, S.E.; Heshmati, M. Anti-inflammatory and analgesic properties of *Heracleum persicum* essential oil and hydroalcoholic extract in animal models. *J. Ethnopharmacol.* **2009**, *124*, 475–480. [CrossRef] [PubMed]
8. Panda, S.; Sahoo, S.; Tripathy, K.; Singh, Y.D.; Sarma, M.K.; Babu, P.J.; Singh, M.C. Essential oils and their pharmacotherapeutics applications in human diseases. *Adv. Tradit. Med.* **2020**, *22*, 1–15.
9. Ahmad, A.; Elisha, I.L.; van Vuuren, S.; Viljoen, A. Volatile phenolics: A comprehensive review of the anti-infective properties of an important class of essential oil constituents. *Phytochemistry* **2021**, *190*, 112864.
10. Wani, A.R.; Yadav, K.; Khursheed, A.; Rather, M.A. An updated and comprehensive review of the antiviral potential of essential oils and their chemical constituents with special focus on their mechanism of action against various influenza and coronaviruses. *Microb. Pathog.* **2021**, *152*, 104620. [CrossRef] [PubMed]
11. Gonçalves, A.C.S.; Ferreira, A.S.M.; Dias, A.; Sárria, M.P.; Gomes, A.C. Nanotechnology-inspired bionanosystems for valorization of natural origin extracts. *Sustain. Agric. Rev.* **2020**, *44*, 47–71.
12. Sharma, M.; Grewal, K.; Jandrotia, R.; Batish, D.R.; Singh, H.P.; Kohli, R.K. Essential oils as anticancer agents: Potential role in malignancies, drug delivery mechanisms, and immune system enhancement. *Biomed. Pharmacother.* **2022**, *146*, 112514.
13. Sheth, T.; Seshadri, S.; Prileszky, T.; Helgeson, M.E. Multiple nanoemulsions. *Nat. Rev. Mater.* **2020**, *5*, 214–228.
14. Das, S.S.; Bharadwaj, P.; Bilal, M.; Barani, M.; Rahdar, A.; Taboada, P.; Bungau, S.; Kyzas, G.Z.J.P. Stimuli-responsive polymeric nanocarriers for drug delivery, imaging, and theragnosis. *Polymers* **2020**, *12*, 1397. [CrossRef]
15. Marzuki, N.H.C.; Wahab, R.A.; Hamid, M.A. An overview of nanoemulsion: Concepts of development and cosmeceutical applications. *Biotechnol. Biotechnol. Equip.* **2019**, *33*, 779–797. [CrossRef]
16. López-Bascón, M.; De Castro, M.L. Soxhlet extraction. In *Liquid-Phase Extraction*; Elsevier: Amsterdam, The Netherlands, 2020; pp. 327–354.
17. Shoorvarzi, S.N.; Shahraki, F.; Shafaei, N.; Karimi, E.; Oskoueian, E. Citrus aurantium L. bloom essential oil nanoemulsion: Synthesis, characterization, cytotoxicity, and its potential health impacts on mice. *J. Food Biochem.* **2020**, *44*, e13181. [CrossRef] [PubMed]
18. Mehrabanjoubani, P.; Nohooji, M.G.; Karimi, E.; Abdolzadeh, A. The differences between *Froriepia subpinnata* (Ledeb.) Baill. and *Pimpinella anisum* L. commonly named as anarijeh based on major components of the essential oil; a marker for resolve ambiguities. *J. Med. Plants* **2021**, *20*, 59–71. [CrossRef]
19. Charmforoshan, E.; Karimi, E.; Oskoueian, E.; Es-Haghi, A.; Iranshahi, M. Inhibition of human breast cancer cells (MCF-7 cell line) growth via cell proliferation, migration, and angiogenesis by auraptene of Ferula szowitsiana root extract. *J. Food Meas. Charact.* **2019**, *13*, 2644–2653. [CrossRef]
20. Sajjadi, M.; Karimi, E.; Oskoueian, E.; Iranshahi, M.; Neamati, A. Galbanic acid: Induced antiproliferation in estrogen receptor-negative breast cancer cells and enhanced cellular redox state in the human dermal fibroblasts. *J. Biochem. Mol. Toxicol.* **2019**, *33*, e22402. [CrossRef] [PubMed]
21. Shakeri, M.; Cottrell, J.J.; Wilkinson, S.; Zhao, W.; Le, H.H.; McQuade, R.; Furness, J.B.; Dunshea, F.R. Dietary betaine improves intestinal barrier function and ameliorates the impact of heat stress in multiple vital organs as measured by evans blue dye in broiler chickens. *Animals* **2020**, *10*, 38. [CrossRef]
22. Cardiff, R.D.; Miller, C.H.; Munn, R.J. Manual hematoxylin and eosin staining of mouse tissue sections. *Cold Spring Harb. Protoc.* **2014**, *2014*, 655–658. [CrossRef] [PubMed]
23. Beyrami, M.; Karimi, E.; Oskoueian, E. Synthesized chrysin-loaded nanoliposomes improves cadmium-induced toxicity in mice. *Environ. Sci. Pollut. Res.* **2020**, *27*, 40643–40651. [CrossRef] [PubMed]
24. Danaei, M.; Dehghankhold, M.; Ataei, S.; Hasanzadeh Davarani, F.; Javanmard, R.; Dokhani, A.; Khorasani, S.; Mozafari, M. Impact of particle size and polydispersity index on the clinical applications of lipidic nanocarrier systems. *Pharmaceutics* **2018**, *10*, 57. [CrossRef]
25. Sefidkon, F.; Dabiri, M.; Mohammad, N. Analysis of the oil of *Heracleum persicum* L. (leaves and flowers). *J. Essent. Oil Res.* **2002**, *14*, 295–297. [CrossRef]
26. Asgarpanah, J.; Mehrabani, G.D.; Ahmadi, M.; Ranjbar, R.; Ardebily, M.S.-A. Chemistry, pharmacology and medicinal properties of *Heracleum persicum* Desf. Ex Fischer: A review. *J. Med. Plants Res.* **2012**, *6*, 1813–1820.

27. Radjabian, T.; Salimi, A.; Rahmani, N.; Shockravi, A.; Mozaffarian, V. Essential oil composition of some wild populations of *Heracleum persicum* Desf. Ex Fischer growing in Iran. *J. Essent. Oil Bear. Plants* **2013**, *16*, 841–849. [CrossRef]
28. Firuzi, O.; Asadollahi, M.; Gholami, M.; Javidnia, K. Composition and biological activities of essential oils from four *Heracleum* species. *Food Chem.* **2010**, *122*, 117–122. [CrossRef]
29. Bakkali, F.; Averbeck, S.; Averbeck, D.; Idaomar, M. Biological effects of essential oils—A review. *Food Chem. Toxicol.* **2008**, *46*, 446–475.
30. Chouhan, S.; Sharma, K.; Guleria, S. Antimicrobial activity of some essential oils—Present status and future perspectives. *Medicines* **2017**, *4*, 58. [CrossRef]
31. Rios, J.-L. Essential oils: What they are and how the terms are used and defined. In *Essential Oils in Food Preservation, Flavor and Safety*; Elsevier: Amsterdam, The Netherlands, 2016; pp. 3–10.
32. Koedoot, E.; Fokkelman, M.; Rogkoti, V.-M.; Smid, M.; van de Sandt, I.; de Bont, H.; Pont, C.; Klip, J.E.; Wink, S.; Timmermans, M.A. Uncovering the signaling landscape controlling breast cancer cell migration identifies novel metastasis driver genes. *Nat. Commun.* **2019**, *10*, 1–16.
33. Thakuri, P.S.; Gupta, M.; Singh, S.; Joshi, R.; Glasgow, E.; Lekan, A.; Agarwal, S.; Luker, G.D.; Tavana, H. Phytochemicals inhibit migration of triple negative breast cancer cells by targeting kinase signaling. *BMC Cancer* **2020**, *20*, 4. [CrossRef] [PubMed]
34. Moukette, B.M.; Castelão-Baptista, J.P.; Ferreira, L.; Silva, A.M.; Simões, R.F.; Cabral, C.; Pieme, C.A.; Ngogang, J.Y.; Sardão, V.A.; Oliveira, P.J. *Afrostyrax lepidophyllus* Mildbr. and *Monodora myristica* (Gaertn.) dunal extracts decrease doxorubicin cytotoxicity on H9c2 cardiomyoblasts. *Evid.-Based Complement. Altern. Med.* **2021**, *2021*, 8858165. [CrossRef] [PubMed]
35. Fazelifar, P.; Tabrizi, M.H.; Rafiee, A. The *Arachis hypogaea* essential oil nanoemulsion as an efficient safe apoptosis inducer in human lung cancer cells (A549). *Nutr. Cancer* **2021**, *73*, 1059–1067. [CrossRef] [PubMed]
36. Javanshir, A.; Karimi, E.; Maragheh, A.D.; Tabrizi, M.H. The antioxidant and anticancer potential of *Ricinus communis* L. essential oil nanoemulsions. *J. Food Meas. Charact.* **2020**, *14*, 1356–1365. [CrossRef]
37. Mbaveng, A.T.; Fotso, G.W.; Ngnintedo, D.; Kuete, V.; Ngadjui, B.T.; Keumedjio, F.; Andrae-Marobela, K.; Efferth, T. Cytotoxicity of epunctanone and four other phytochemicals isolated from the medicinal plants *Garcinia epunctata* and *Ptycholobium contortum* towards multi-factorial drug resistant cancer cells. *Phytomedicine* **2018**, *48*, 112–119. [CrossRef] [PubMed]
38. Trang, D.T.; Hoang, T.K.V.; Nguyen, T.T.M.; Van Cuong, P.; Dang, N.H.; Dang, H.D.; Quang, T.N.; Dat, N.T. Essential oils of lemongrass (*Cymbopogon citratus* Stapf) induces apoptosis and cell cycle arrest in A549 lung cancer cells. *BioMed Res. Int.* **2020**, *2020*, 5924856. [CrossRef] [PubMed]
39. Kushwaha, P.P.; Vardhan, P.S.; Kapewangolo, P.; Shuaib, M.; Prajapati, S.K.; Singh, A.K.; Kumar, S. *Bulbine frutescens* phytochemical inhibits notch signaling pathway and induces apoptosis in triple negative and luminal breast cancer cells. *Life Sci.* **2019**, *234*, 116783. [CrossRef]
40. Shakeri, M.; Cottrell, J.J.; Wilkinson, S.; Ringuet, M.; Furness, J.B.; Dunshea, F.R. Betaine and antioxidants improve growth performance, breast muscle development and ameliorate thermoregulatory responses to cyclic heat exposure in broiler chickens. *Animals* **2018**, *8*, 162. [CrossRef]
41. Panahi, Y.; Dadjou, Y.; Pishgoo, B.; Akbari, A.; Sahebkar, A. Antioxidant activity of *Heracleum persicum* fruit extract: Evidence from a randomized controlled trial. *J. Diet. Suppl.* **2016**, *13*, 530–537. [CrossRef]

antioxidants

MDPI

Article

Effect of Diet and Essential Oils on the Fatty Acid Composition, Oxidative Stability and Microbiological Profile of Marchigiana Burgers

Isa Fusaro [1], Damiano Cavallini [2], Melania Giammarco [1,*], Annalisa Serio [3], Ludovica M. E. Mammi [2,*], Julio De Matos Vettori [1], Lydia Lanzoni [1], Andrea Formigoni [2] and Giorgio Vignola [1]

1 Faculty of Veterinary Medicine, University of Teramo, Località Piano D'Accio, 64100 Teramo, Italy; ifusaro@unite.it (I.F.); jdematosvettori@unite.it (J.D.M.V.); llanzoni@unite.it (L.L.); gvignola@unite.it (G.V.)
2 Department of Veterinary Medical Sciences, Alma Mater Studiorum University of Bologna, Via Tolara di Sopra 50, Ozzano dell'Emilia, 40064 Bologna, Italy; damiano.cavallini@unibo.it (D.C.); andrea.formigoni@unibo.it (A.F.)
3 Faculty of Bioscience and Technology for Food, Agriculture and Environment, University of Teramo, Campus Coste Sant'Agostino, Via R. Balzarini 1, 64100 Teramo, Italy; aserio@unite.it
* Correspondence: mgiammarco@unite.it (M.G.); ludovica.mammi@unibo.it (L.M.E.M.); Tel.: +39-086-1266-816 (M.G.); +39-051-2097-015 (L.M.E.M.)

Abstract: The objective of this study is to evaluate the effects of including linseed (L) or linseed plus vitamin E (LE) in the diet of Marchigiana young bulls on the oxidative stability, color measurements, microbiological profile and fatty acid composition (FA) of burgers treated with and without a blend of essential oils (*Rosmarinus officinalis and Origanum vulgare var. hirtum*) (EOs). For this aim, the burgers were analysed for pH, thiobarbituric-acid-reactive substance (TBARS) content, Ferric Reducing/Antioxidant Power Assay (FRAP), vitamin E and colour measurements (L, a*, b) at 3, 6, 9, 12 days of storage: the TBARs were the highest in group L compared to C and LE after 12 days of storage (0.98, 0.73, and 0.63 mg MDA/kg, respectively). The TBARS content was also influenced by the use of EO compared to burgers not treated with EO ($p < 0.05$). The vitamin E content was influenced by the diet ($p < 0.01$), but not by the EO. The meat of the L group showed the lowest value of redness (a*) compared to C and LE ($p < 0.01$), while the use of EO did not affect colour parameters. The microbiological profile of the burgers showed a lower Pseudomonas count for L and LE at T0 (2.82 ± 0.30 and 2.30 ± 0.52 Log CFU/g, respectively) compared to C (3.90 ± 0.38 Log CFU/g), while the EO did not influence the microbiological profile. The FA composition was analysed at 0 and 12 days. The burgers from the LE group showed the highest value of polyunsaturated FA compared to the L and C groups ($p < 0.05$). Our findings suggest that the inclusion of vitamin E in a concentrate rich in polyunsaturated fatty acids is useful to limit intramuscular fat oxidation and to preserve the colour stability of burgers from young Marchigiana bulls enriched with healthy fatty acids. Moreover, linseed and vitamin E had a positive effect on microbial loads and growth dynamics, containing microbial development through time.

Keywords: TBARS; FRAP; colour parameter; linseed and vitamin E; oregano and rosemary; microbiological profile; beef meat

Citation: Fusaro, I.; Cavallini, D.; Giammarco, M.; Serio, A.; Mammi, L.M.E.; De Matos Vettori, J.; Lanzoni, L.; Formigoni, A.; Vignola, G. Effect of Diet and Essential Oils on the Fatty Acid Composition, Oxidative Stability and Microbiological Profile of Marchigiana Burgers. *Antioxidants* **2022**, *11*, 827. https://doi.org/10.3390/antiox11050827

Academic Editors: Delia Mirela Tit and Simona Bungau

Received: 24 March 2022
Accepted: 22 April 2022
Published: 24 April 2022

1. Introduction

In Italy, as in other European countries, the purchase of processed foods, such as meat burgers and patties, has tripled in the last 50 years probably due to changes in the Italian lifestyle, preferring food that is faster to cook [1]. This is also probably due to their price and preparation versatility as well as changes in eating habits, the availability of a different kind of meat product and sociodemographic changes [2]. Nonetheless, the meat from ruminants is considered a major source of saturated fatty acids (SFAs) because red meat consists of approximately 40% of SFAs, 50% monounsaturated fatty acids (MUFAs), 5% trans fatty acids and 4% polyunsaturated fatty acids (PUFAs) [3]. The main daily meals in Western countries include a meat-containing dish that contributes to approximately half of the maximal recommended intake of SFAs [4,5].

Several large observational studies have reported strict associations between SFA concentrations in the diet and several health problems, ranging from cardiovascular disease (CVD) to cancer risk [6,7]. Conversely, there is emerging evidence that a diet rich in PUFAs, in particular with three double bonds like n-3 fatty acids (n-3FAs) acids and conjugated linoleic acids (CLAs), has beneficial effects on human health, such as decreasing low-density lipoprotein and cholesterol, and conferring anti-inflammatory, anti-atherogenic and anti-carcinogenic effects [8]. The feeding strategies were demonstrated to be the best approach to reduce SFAs in meat and, in general, in ruminants' products [9]. It is well known that the use of pasture or the introduction of vegetable oils in the diet of ruminants can induce the switch in meat and milk fatty acids (FAs) from saturated to unsaturated, as demonstrated in lamb [10], ewe [11], dairy cattle [12] and beef [13]. The use of feed rich in polyunsaturated fatty acids represents an innovation and enhancement in the breeding methods of typical local breeds such as Marchigiana, Chianina and Romagnola, in Italy. The husbandry techniques for the Marchigiana breed plan a diet rich in cereal-based feed during the finishing period [14]. For this reason, it could be interesting to add feed that can enrich beef meat with PUFAs. However, adding PUFAs to the meat makes lipid oxidation easier, with abrupt consequences on meat sensorial and chemical characteristics [15]. In particular, the meat's color is a fundamental sensory attribute that influences the consumers' choice of purchasing it, and, for this reason, it is necessary to maintain its attractive nature for the entire shelf-life period [16]. In this context, the use of natural antioxidant spices may be useful to contrast the increased oxidation due to the high concentration of unsaturated fatty acids, especially in the ground meat products. For example, the application of essential oils (EOs) has been demonstrated as helpful to protect meat products against oxidation [16]. Essential oils are extracts obtained by the distillation of a wide variety of plant materials, and they exert both antioxidant and antimicrobial effects, depending on their chemical composition. The interest of researchers and industries in their exploitation has recently increased, as EOs are considered natural compounds, capable of substituting synthetic preservatives, and are thus particularly appreciated by the consumers [17].

In the light of these considerations, the aim of this research is to evaluate the effects of a concentrate enriched with linseed and vitamin E on oxidative stability colour measurements, microbiological profile and fatty acids composition of meat burgers from young Marchigiana bulls. Moreover, aiming to improve the quality attributes of the burgers, *Rosmarinus officinalis* and *Origanum vulgare* var. *hirtum* essential oils are included in the burger formulations and their effects are evaluated during the meat storage.

2. Materials and Methods

2.1. Animals and Diets

The Marchigiana young bulls were managed during the experiment according to the Council Directive 98/58/EC of 20 July 1998 concerning the protection of animals kept for farming purposes, and were slaughtered according to the Council Regulation (EC) No 1099/2009 of 24 September 2009 on the protection of animals at the time of slaughtering. The experiment was conducted, as previously described by Fusaro et al., 2021 [18]. Briefly, a total of 36 Marchigiana young bulls bred in a commercial farm approximately 15 km northeast of Teramo, Abruzzo, Southern Italy were allotted in three groups receiving three different experimental diets (Table 1): Control (C), Linseed (L) and Linseed plus Vitamin E (LE). The groups were homogeneous for body weights (441.9; 438.8 and 440.1, for C, L and LE, respectively). Samples of Total Mixed Ration (TMR) were collected every week and analysed according to the standard methods of AOAC (2002) [19] for dry matter (DM), crude protein (CP), ether extract (EE) and ash. Neutral detergent fibre (NDF), acid detergent fibre (ADF) and acid detergent lignin (ADL) were determined, as previously described [19,20]. All feed samples were also analysed for FA composition, as described in the research [21]. The ingredients and chemical compositions of the diets are shown in Table 1.

Table 1. Ingredients of the three experimental diets: C (control), L (linseed) and LE (linseed + vitamin E).

Dietary Ingredients (Kg)	C	L	LE
Dehydrated Alfa Hay	2.86	2.85	2.85
Straw	1.14	1.14	1.14
Corn meal	4.57	4.56	4.56
Extruded linseed	0.00	0.97	0.97
Beans	1.14	0.46	0.46
Cereal bran	2.29	2.28	2.28
Hydrogenate fat	0.29	0.00	0.00
Sodium chloridae	0.06	0.06	0.06
Sodium bicarbonate	0.11	0.11	0.11
Vitamin E	0.00	0.00	0.02
Chemical composition (% DM)			
Dry matter	87.52	87.79	87.79
Crude fiber	6.82	6.86	7.91
Crude protein	13.87	13.91	13.91
Ether Extract	6.84	6.84	7.06
Ash	2.18	2.15	2.25
Fatty Acids Composition (% total fatty acids)			
SFA	51.16	14.84	16.72
MUFA	15.75	19.18	18.76
PUFA	33.09	65.98	64.52

2.2. Preparation of the Burgers

The meat from the left of the *Longissimus dorsi* muscle between the last rib and the 6th lumbar vertebra of each carcass was ground in a meat grinder with a 3 mm disk. From each carcass, 8 burgers 25 mm thick were obtained. Four burgers were kneaded with a blend of *Rosmarinus officinalis* and *Origanum vulgare* var. *hirtum* (1:1) EO (0.05 mL rosemary + 0.05 mL oregano per kg of meat) for 60 s, and the remaining burgers were kneaded with an equal amount of PBS (Phosphate Buffer Saline) as a control. PBS is isotonic and non-toxic. All samples (n = 288) underwent modified atmosphere packaging (MAP) (66% O_2; 25% CO_2; 9% N_2) and were then stored at +4 °C for 12 days and sampled for the subsequent analysis. Whole trays were placed in a dark chamber at 4 °C and then removed from the chamber for the analysis.

2.3. Meat Quality: Chemical Analyses, pH and Colour Measurements

At time 0, all the burgers were analysed for moisture, fat, protein and ash [19]. The pH, TBARs, FRAP, Vitamin E and colour parameter were analysed at 0, 3, 6, 9 and 12 days of storage [22]. The pH of the meat samples was measured with a penetrating electrode adapted to a portable pH meter (Crison pHmeter 507 and a 52–32 spear electrode, Crison Instruments, Spain). Burger samples from each treatment were subjected to pH recording during storage (0, 3, 6, 9 and 12 days of storage). At the same time points, meat colour parameters of lightness (L) redness (a*) and yellowness (b) were measured for the burgers, according to the CIELab system, with a Minolta Chroma Meter CR-300 (Minolta Camera Co., Osaka, Japan) with a D65 illuminant and an 8 mm aperture. The burgers were allowed to bloom in direct contact with air for 1 h before colorimetric measurements were performed on the burger surfaces and reported as the mean of three measurements.

2.4. Burgers Oxidative Stability: Thiobarbituric-Acid-Reactive Substances (TBARS), Ferric Reducing Antioxidant Power (FRAP) and Vitamin E

The lipid oxidation of meat samples was evaluated by TBARS measurement with the method of Inserra et al., (2014) [23] at 0, 3, 6, 9 and 12 days of storage. Three replicates were run for each sample. At the same time points, the spectrophotometric FRAP method was used for the antioxidant capacity determination according to previous studies [24]. Muscle vitamin E (α-tocopherol) concentration was measured at the same time points according to Koprivnjak et al. (1996) [25].

2.5. Fatty Acid Analysis

The fatty acid analysis was conducted on burgers from the three experimental groups at time point 0 and 12. Briefly, intramuscular lipids were extracted according to the protocol of Folch et al. (1957) [26]. After cold methylation with the technique of Frega and Lerker (1984) [27], the FAs were determined by gas chromatography with a Chrompack CP-SIL 88 capillary column. Before statistical analysis, the data on FA composition were processed to calculate the following FA classes: MUFA, PUFA and SFA; n-3 (Σ n-3: sum of C18:3n-3, C20:5n-3, C22:5n-3 and C22:6n-3); and n-6 (Σn-6: sum of C18:2n-6t9,t12, C18:2n-6, CLAt10,c12, C18:3n-6, C20:3n-6, C20:4n-6 and C22:4n-6) [28]. The I-Harris index [29] was calculated as the sum of EPA and DHA. FA quantities were expressed as mg FA/100 g sample.

2.6. Microbiological Analysis

To evaluate the microbial population during the burgers' shelf-life, 10 g of meat from each burger were homogenised with a sterile saline solution (NaCl 9 g/L in deionised water) in a ratio of 1:10 (*w*/*v*) for 300 s in stomacher. Afterwards, ten-fold serial dilutions were prepared, and the following microbial groups were determined as follows: Mesophilic and Psychrotrophic Aerobic Counts were determined on Plate Count Agar at 30 °C for 48 h and at 8 °C for 7 d, respectively; presumptive lactobacilli in MRS at 30 °C for 48 h; presumptive lactococci on M17 agar at 30 °C for 48 h; coliforms in violet red bile agar (VRBA) at 37 °C for 48 h; staphylococci on Mannitol Salt Agar at 37 °C for 48 h; *Pseudomonas* spp. on Pseudomonas Isolation Agar (Acumedia, Dot Scientific, Burton, MI, USA) at 22 °C for 48 h and *Brochothrix thermosphacta* on STAA agar base added with STAA selective supplement at 25 °C for 48 h. Where not differently specified, all the culture media were from Oxoid Thermo Fisher (Rodano, Italy). The samples were analysed in triplicate and the mean of the results was calculated.

2.7. Statistical Analysis

All data were analysed with a GLM procedure using JMPpro v16.0 (SAS Institute Inc., Cary, NC, USA). Meat quality composition at time point zero was analysed using diet as the only fixed effect. Data relative to oxidative stability, colour measurements, microbiological profile and fatty acid composition were analysed, including the fixed effects of diet, EO and time of storage and their interactions.

Results for interactions between the main effects are shown in Tables and Figures. Tukey's test was performed to assess significant differences between means. p-value ≤ 0.05 was considered the threshold for significant differences. Results are presented as treatment mean and standard error of the mean.

3. Results

3.1. Beef Burger Characteristics: Chemical Analyses, pH, Color Measurements and Oxidative Stability

The chemical composition of the burgers is presented in Table 2. The moisture contents ranged from 74.79 to 75.45% showing no differences among the meat samples from three groups. The different diets did not influence the lipid, protein and ash content of the meat ($p > 0.05$).

Table 2. Effect of three experimental diets on meat quality chemical composition at time point 0.

	Diet				
	C	L	LE	*p*-Value	SEM
Moisture (%)	74.79	75.43	75.45	0.72	9.33
Protein (%)	21.70	21.05	21.02	0.66	3.32
Fat (%)	2.29	2.36	2.35	0.98	0.24
Ash (%)	1.22	1.16	1.18	0.68	0.15

C = Control; L = Linseed; LE = Linseed plus vitamin E; SEM = Standard error of the mean.

Table 3 presents the pH, oxidative stability (TBARS, FRAP and vitamin E) and colour parameters (L, a* and b) of the burgers from the three experimental groups, treated with or without EO and their interactions. The experimental diets significantly affected the burgers' pH ($p < 0.05$), showing a lower value in the meat of group L (5.60) compared to group LE (5.67), while a higher value was registered for group C (5.70). The burges' pH was also affected by the use of EOs ($p < 0.03$), showing a higher value for the samples treated with EOs compared to the meat without EOs (5.47 vs. 5.65, respectively). The TBARS content of the burgers showed a higher value in group L (0.60 MDA/kg of meat) compared to the C and LE groups (0.47 and 0.39 MDA/kg of meat) ($p < 0.05$). Moreover, the use of EOs influenced the TBARS value ($p < 0.05$), but not the vitamin E content ($p > 0.05$). Vitamin E content was higher in the LE group compared to C, while the L group showed an intermediate value ($p < 0.05$). The FRAP showed differences among the dietary treatment and between the burgers treated and not treated with EOs. The burgers of the LE group had a higher FRAP (0.74 µmolFe/g) compared to those of the C and L groups (0.54 and 0.55 µmol Fe/g, respectively, in C and L). The FRAP was also influenced by the use of EOs ($p < 0.05$), showing the highest value in the samples treated with EOs (0.67 µmolFe/g) compared to the burgers without EOs (0.52 µmolFe/g).

Table 3. Effects of diets (C, L and LE) and essential oils (O and WO) on Marchigiana beef burgers' colour parameters (L, a*, b), pH, FRAP and vitamin E content.

	Diet			EO		*p*-Value						
	C	L	LE	O	WO	Diet	EO	Time	D*EO	D*T	EO*T	SEM
pH	5.70 ^{A}	5.60 ^{C}	5.67 ^{B}	5.47 ^{b}	5.65 ^{a}	<0.01	0.03	<0.01	0.67	0.54	0.55	0.09
TBARS	0.47 ^{ab}	0.60 ^{a}	0.39 ^{b}	0.43 ^{B}	0.55 ^{A}	0.05	<0.01	<0.01	0.41	<0.01	<0.01	0.01
Vitamin E mg/kg	0.64 ^{c}	0.81 ^{b}	1.38 ^{a}	1.12	1.11	0.01	0.56	0.74	0.13	0.78	0.54	0.05
FRAP µmolFe/g	0.56 ^{b}	0.55 ^{b}	0.74 ^{a}	0.67 ^{b}	0.52 ^{a}	0.03	0.05	0.51	0.24	0.56	0.38	1.02
L	43.85 ^{B}	45.29 ^{A}	43.40 ^{B}	44.68 ^{A}	43.68 ^{B}	<0.01	0.77	<0.01	0.18	<0.01	0.12	1.05
a	19.39 ^{B}	17.83 ^{C}	20.68 ^{A}	19.27	19.34	<0.01	0.92	<0.01	0.38	<0.01	0.57	1.18
b	12.65 ^{A}	11.63 ^{B}	12.90 ^{A}	12.36	12.43	<0.01	0.70	<0.01	0.64	<0.01	0.88	1.21

C = Control; L = Linseed; LE = Linseed plus vitamin E; O = with essential oil; WO = without essential oil. D*EO = Diet*Essential Oils; D*T = Diet*Time; EO*T = Essential Oils*Time. Different letters on the same row indicate significant differences within diet or EO ($^{a–c}$: $p < 0.05$; $^{A–C}$: $p < 0.01$). SEM = Standard error of the mean.

The lightness was higher ($p < 0.01$) in the L group compared to the LE and C groups. Conversely, the redness was lower in the L group and higher in the LE and C groups ($p < 0.01$). The yellowness was lower in the L group compared to the C and LE groups ($p < 0.01$). The use of essential oils did not influence the colour parameter ($p > 0.05$)

Figure 1a–c shows the differences between the colour parameters during the storage time. Figure 1a shows that the lightness value was higher from day 3 to day 12 of storage in group L compared to that of group LE, while the C group had an intermediate value on days 0 and 9 of storage ($p < 0.01$). Moreover, the interaction between time of storage and diet was significantly correlated ($p < 0.01$). The yellowness of the burgers (Figure 1b) was affected by the diet ($p < 0.01$), showing a higher value in the LE group compared to the C and L groups on day 0; on day 3 ($p < 0.01$), the b value was similar between the C and LE groups and lower for the L group, while on day 6, the yellowness of the meat of the L was similar with that of the LE and C groups, even if C and LE showed different yellowness value ($p < 0.05$). The redness shown in Figure 1c has a higher value for group LE (storage time point "0") compared to groups C and L ($p < 0.01$). On day 3 of storage, the C and L groups showed a similar value compared to the LE group ($p < 0.01$), while on days 9 and 12 of storage, the redness was lower in the L group compared to the C and LE groups ($p < 0.01$). The interaction between the storage time and experimental diets significantly affected the redness of the meat ($p < 0.01$).

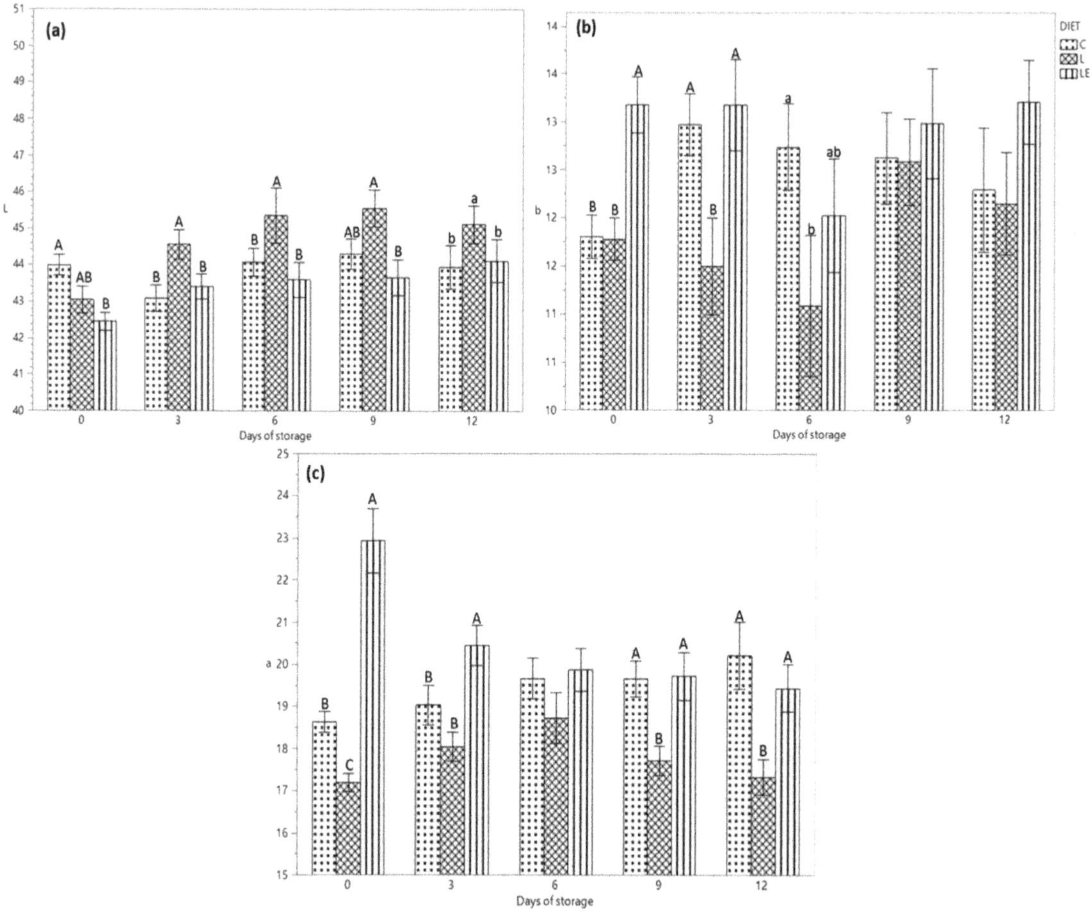

Figure 1. (**a**) Lightness, (**b**) Yellowness and (**c**) Redness of burgers from Marchigiana beef from three different dietary treatments, according to the storage time. Different letters indicate significant differences (a, b: $p < 0.05$; A, B, C: $p < 0.01$) within each time point. C = Control; L = Linseed; LE = Linseed plus vitamin E.

Figure 2a shows the TBARS content during the storage time according to the three different dietary treatments. On days 0 and 3 of storage, a higher concentration of TBARS content was observed in group C compared to group LE ($p < 0.05$). Conversely, on day 6 ($p < 0.05$) and 9 ($p < 0.01$) of storage, the TBARS of group L was higher compared to that of groups C and LE; on day 12, the C group showed the lowest value compared to group LE, while group L had the highest value (0.73, 0.98 and 0.63 mg MDA/kg in group C, L and LE, respectively).

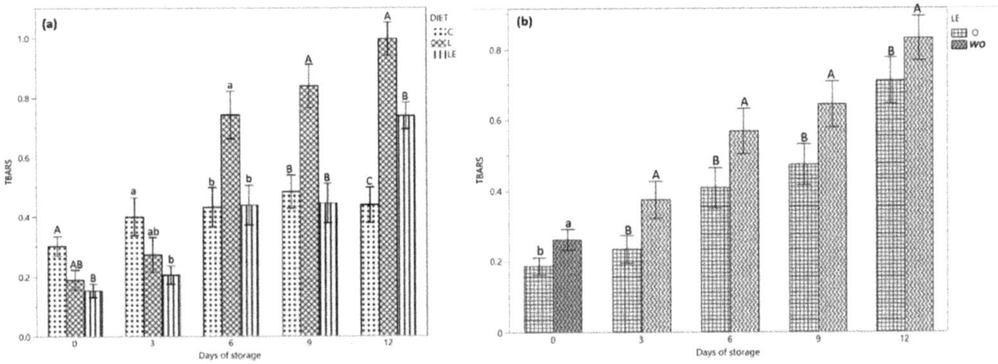

Figure 2. (**a**) TBARS values of the burgers from Marchigiana beef from three different dietary treatments according to the storage time. (**b**) TBARS values of the burgers treated with and without EOs according to the storage time. Different letters indicate significant differences (a, b: $p < 0.05$; A, B, C: $p < 0.01$) within each time point. C = Control; L = Linseed; LE = Linseed plus vitamin E. O = with essential oil; WO = without essential oil.

Figure 2b and Table 3 show the TBARS content according to the storage time of burgers treated with and without essential oils. At all the time points, the use of EOs was correlated with a lower concentration of TBARS compared to the samples not treated with EOs.

3.2. Microbiological Profile

The microbiological profile of the samples is depicted in Figure 3. Regarding the microbiological profile of the products, a difference in mesophilic (Figure 3a) and psichrotrophic (Figure 3c) counts was revealed for all the samples at all the time points of the analysis. In particular, LE samples showed lower microbial counts in comparison with the other groups. The difference was already clear on day 0, until 12 days of refrigerated storage, when the mesophilic count only slightly increased for LE samples; it rose sharply for the L and C samples without and with essential oils. Smaller differences among the samples were observed for the psychrotrophic count; nevertheless, additionally in this case, LE samples showed the smallest counts, particularly in the first six days of storage. As regards the specific microbial groups analysed, the count trend during time reflected the previous behaviour, with the LE samples showing the lowest counts, particularly for presumptive lactobacilli (Figure 3b) and lactococci and then also for *Bhrochotrix thermosphacta* and *Pseudomonas* spp. (data not shown). In particular, the lower *Pseudomonas* counts observed already at T0 for the L and LE groups (2.82 ± 0.30 and 2.30 ± 0.52 Log CFU/g) with respect to group C (3.90 ± 0.38 Log CFU/g) allowed the control of this microbial group below the 5.00 Log CFU/g acceptable counts until the end of storage. On the contrary, microbial loads greater than 6.00 Log CFU/g were observed for the control samples starting from 9 days of storage.

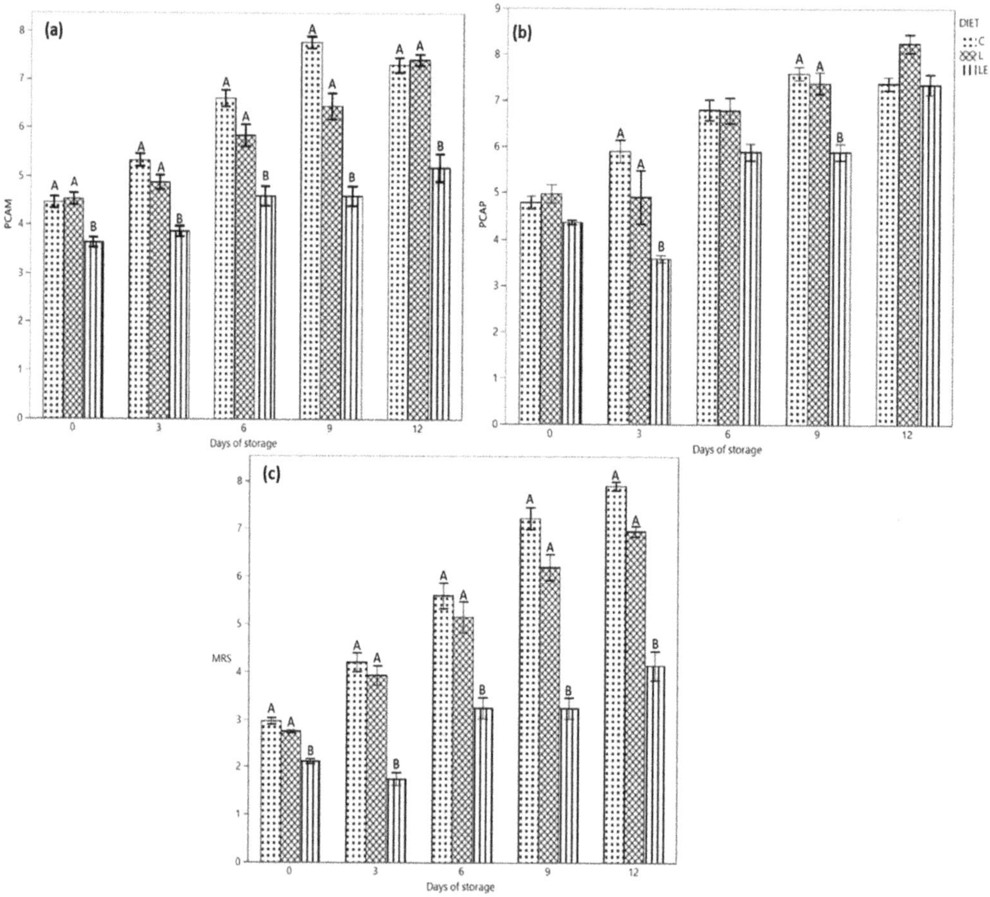

Figure 3. (**a**) Evolution of the mesophilic aerobic population, (**b**) psychrotrophic aerobic population and (**c**) lactic acid bacteria during the refrigerated storage of Marchigiana beef burgers in a modified atmosphere packaging. C = Control; L = Linseed; LE = Linseed plus vitamin E. Different letters indicate significant differences (A, B: $p < 0.01$) within each time point.

For each sampling time, coliforms and staphylococci were below the detection limit for samples L and LE.

Regarding EO application, not significant ($p > 0.05$) differences were observed for EO-treated and -untreated samples within the same kind of samples in the first 3–6 days of storage, while differences increased during longer storage times, with different behaviour depending on the animal diet and the target microbial groups, which could present counts even higher in the EO-treated samples (data not shown). In any case, the differences were not statistically significant.

3.3. Fatty Acid Profile

The FA composition of burgers is shown in Table 4. Dietary linseed supplementation significantly increased the content of PUFA to 253.63 and 226.48 mg/100 g in the LE and L samples, while the lowest value was recorded for the C group (203.22 mg/100 g of sample). Moreover, the lowest value for MUFA was 595.53 mg/100 g in the sample in the C group, while the LE group sample had the highest value (699.22 mg/100 g of sample) and the intermediate vale was observed for the L group (595.53 mg/100 g of sample). SFA had an opposite trend, showing the highest value for group C (956.99 mg/100 g of sample) compared to groups L (721.53 mg/100 g of sample) and LE (679.51 mg/100 g of sample). Groups L and LE also showed a higher content of n-3 PUFA than group C, although these FA were higher in group LE (68.40 mg/100 g of sample) than in group L (49.43 mg/100 g of sample; $p < 0.05$). The total n-6 PUFA value was not affected by the diet ($p = 0.68$). The burgers from groups L and LE, compared with group C, had higher percentages of EPA and DHA (I-Harris index; 8.21 and 8.85 mg/100 g sample in L and LE, respectively, and 7.25 mg/100 g of sample in C; $p < 0.05$). In contrast, the CLA value was higher in the LE group (15.60 mg/100 g of meat) and lower in the C group (5.79 mg/100 g of sample; $p < 0.05$), whereas the L group had an intermediate value (11.22 mg/100 g sample). The saturation index was different among the dietary treatments, showing a higher value for the C group (1.02 mg/100 g sample) than L and LE groups (0.7 mg/100 g sample). The use of essential oils had no difference in the treatments.

Table 4. Effects of experimental diets (C, L and LE) and treatment (EO) with (O) and without essential oils (WO) on health indices and pro-oxidant fatty acids on Marchigiana beef meat on days 0 and 12 of storage.

	Diet			EO		*p*-Value							
	C	L	LE	O	WO	Diet	EO	Time	D*EO	D*T	EO*T	SEM
SFA	956.99 ^{a}	721.53 ^{ab}	679.51 ^{b}	766.78	705.24	0.05	0.06	0.27	0.15	0.58	0.60	1.06
MUFA	595.53 ^{b}	642.10 ^{ab}	699.22 ^{a}	681.75	609.48	0.05	0.17	0.14	0.16	0.45	0.98	0.90
PUFA	203.22 ^{b}	226.48 ^{ab}	253.63 ^{a}	221.31	234.25	0.05	0.40	0.42	0.90	0.33	0.45	1.26
n-3 PUFA	31.70 ^{C}	49.43 ^{B}	68.40 ^{A}	52.16	47.52	0.01	0.15	0.14	0.56	0.20	0.68	0.14
n-6 PUFA	121.88	125.00	121.54	122.89	122.72	0.68	0.96	0.68	0.57	0.06	0.68	1.13
CLA	5.79 ^{C}	11.22 ^{B}	15.60 ^{A}	11.90	10.84	0.01	0.06	0.58	0.55	0.06	0.27	0.94
I-HARRIS	7.25 ^{b}	8.85 ^{a}	8.21 ^{a}	7.80	8.40	0.05	0.42	0.10	0.64	0.42	0.27	0.03

Fatty acids are expressed as mg/100 gr of samples. SFA = Saturated fatty acids. MUFA = Monounsaturated fatty acids. PUFA = Polyunsaturated fatty acids. N3-PUFA = Σ C18:2 t11, c15 + C18:2 c9, c15 + C18:3 c9, c12, c15 + C:22:5 c7, c10, c13, c16, c19 + EPA + DHA. N6-PUFA = Σ C18:2t9, t12 + C18:2 c9, t12 + C18:2 t9, c12 + C18:2 c9, c12 + CLA t10, c12 + C20:2 c11, c14 + C20:3 c8, c11, c14 + C20:4 c5, c8, c11, c14. CLA = conjugated linoleic acid. I-Harris = (EPA + DHA). Different letters in the same row indicate significant differences (^{a,b}: $p < 0.05$; $^{A-C}$: $p < 0.01$). SEM = Standard error of the mean. C = Control; L = Linseed; LE = Linseed plus vitamin E; O = with essential oil; WO = without essential oil. D*EO = Diet*Essential Oils; D*T = Diet*Time; EO*T = Essential Oils*Time.

4. Discussion

The experimental diets did not affect the meat's chemical composition, showing similar results in each experimental group regarding protein, fat and ash content (Table 2). Generally, the meat composition is strictly related to the diet composition and, as described by other authors [30,31], the similar energetic and protein content of the experimental diets used in this study did not reveal differences in the protein, fat and ash content of meat.

Table 3 shows the effects of diets (C, L and LE) and the use of essential oils (O and WO) on Marchigiana beef burgers' oxidative stability, pH, colour measurements and vitamin E content. In accordance with Juárez et al. (2012) [32], our results show that vitamin E had a role on the oxidative status of meat in the LE group, which showed a decrease in TBARS and an increase in FRAP values compared to the burgers from beef on diets without vitamin E. Additionally, the use of EOs influenced the oxidative parameters, demonstrating the ability of EOs to prevent oxidation [33,34].

In the present study, the oxidative stability and colour measurements of the burgers were significantly influenced by the three different experimental diets, showing that the burgers from the LE group were characterised by a lower TBARS content and a higher redness during storage. This effect could be due to the use of vitamin E during the finishing period, which seems to be considered an appropriate pre-slaughter feed strategy. In contrast to our study, a recent work on Normand cull cows showed that a diet supplemented with vitamin E (155 IU/kg of diet DM) and plant extracts rich in polyphenols did not prevent the lipid oxidation of the meat, with an increase in MDA concentration [35]. This discrepancy could be probably due to the greater amount of vitamin E in the diet of the LE group in the present experimental trial than in previous studies [35,36]. Indeed, as we expected, the concentration of vitamin E was higher in group LE compared to the other two groups, showing that the use of vitamin E in the diet influences their concentration in the meat. A large number of studies have observed the antioxidant effects of vitamin E to prevent the oxidation [36,37].

Figure 1a shows that the lightness was influenced ($p < 0.01$) by the diet and time of storage. In particular, the higher value of lightness was recorded in the burgers of the L group from day 3 to day 12 of storage compared to the LE and C groups. This result agrees with that reported by Juárez et al. (2012) [32], which found the lightness in patties increased when flaxseed without vitamin E was included in the diet. On the contrary, a greater pigment content in the diet implies a stronger light absorption and consequently lower reflectance or transmittance in the meat, making the product opaquer. Moreover, the colour parameters could be also influenced by the increase in oxidation level (TBARS), which in the burgers of group L was higher, due to an abrupt reduction in mitochondrial respiratory activity, which determines a great production of metmyoglobin [38]. A recent review on the effect of bioactive dietary nutrients on meat oxidation and colour stability has shown that in most ruminant studies the greatest colour stability and the least peroxidation were obtained in the meat of animals fed a super-nutritional dose of vitamin E. [39]. Our results also demonstrated for the first time that the oxidative stability and colour measurements of burgers from young Marchigiana bulls could be preserved when a high dosage of vitamin E is added during the finishing period. Some authors argued that the lightness could be partially influenced by differences in the intramuscular fat content [40]. In our study, we supposed that the difference in lightness could be due to the FA composition of the burgers because the content of intramuscular fat was similar between the diets (Table 3), as also demonstrated in dairy goat kids [41]. Moreover, in the current study, the higher concentration of vitamin E in the meat of the LE group increases stability against oxidation and makes the meat more desirable [32]. In contrast, Fusaro et al. (2021) [18] have reported that a dietary regimen with linseed and vitamin E was strictly related to a higher L value in steak from Marchigiana beef. Probably, in the ground beef as the one that was used in this study, the cellular integrity is likely to be disrupted, owing to the greater exposure of tissues to oxygen and the simultaneous dilution of antioxidant concentration.

Both storage time and experimental diets also affected ($p < 0.05$) the yellowness. In the control and LE burgers, the b value was higher on days 3 and 6 of storage. Conversely, in group L, the b fluctuated during the storage time with a lower value compared to that of the C and LE groups, probably due to auto-oxidation to brown metamyoglobin (Figure 1b), induced by higher lipid oxidation. This result was not observed in the LE samples, which is consistent with the antioxidant role of vitamin E.

The redness decreased from day 0 to day 12 in the burgers from all the three experimental groups. A decrease was registered in the L group at the end of the storage time (Figure 1c), which is probably due to the easier oxidation of red oxymyoglobin to brownish metmyoglobin determined by the presence of reducing systems and on lipid oxidation, as also demonstrated in burgers from lamb meat [42]. Moreover, the a* value of the burgers at time point 0 was significantly higher in group LE compared to the other two groups, probably due the antioxidant effect of vitamin E that could have protected the heme pigments of meat from oxidation [43].

The TBARS test was used to predict the oxidative stability of lipids in burger samples. It was demonstrated that the rancid smell and taste became detectable by the consumers when a value of 2 mg/kg of TBARS is found in the meat or meat products [32]. The TBARS values were significantly influenced by the diet, storage time and the use of EOs (Table 2 and Figure 2a,b). Our results show a higher degree of oxidation in burgers from the meat of animals fed with linseed, a lower degree in the control group and animals that received linseed and vitamin E. The time of storage also influenced ($p < 0.01$) the concentration of TBARS in all the burgers from the three dietary treatments, even if on days 6 and 9 of storage, the concentration of TBARS was constant in the LE and C groups, while higher in the samples from the L group. Conversely, during the first three days of storage, the group LE showed a lower level of TBARS compared to the L and C groups, while after 12 days, the TBARS had an intermediate value in groups L and C. Our results show a significant increase in TBARS in the burgers of animals that received a diet rich in linseed without the addition of vitamin E as also demonstrate by Juarez et al. (2012) [32], who found that beef from animals fed only with flax seeds showed a higher oxidation level compared to the animals that received a supplementation of vitamin E. The antioxidant capacity of vitamin E was evident in the burgers of the LE group compared to group L [44]; the results from the current research suggest that the oxidative stability of burgers from Marchigiana beef may be directly influenced by the use of antioxidants compounds, such as vitamin E and EOs, more than the fatty acid profile of intramuscular fat. To the best of our knowledge, this is the first study in which minced meat was used without adding fat in burger preparation. Indeed, as reported by Wang et al. [45], the beef patties experienced a greater rate of lipid oxidation and discoloration when a greater amount of fat is used for their preparation.

Our results also suggest that the use of EOs may be a favourable strategy to limit the oxidation process in meat burgers. This result could be attributed to the high concentration of phenolic compounds in EOs as well as other substances, such as flavonoids, which are responsible for antioxidant activity [46]. The food industry widely utilises a rosemary extract for its antioxidant properties [47]. The efficacy of rosemary in decreasing lipid oxidation has been reported by several authors in meat from poultry, pork [48] and beef [49]. Oregano extracts have also been demonstrated to be effective in inhibiting peroxidation because it blocks free radical formation [50].

In our experiments, burgers from control animals and those fed with linseed showed similar microbial loads and growth dynamics. On the contrary, LE samples generally had a better microbiological profile with lower counts. Therefore, our data demonstrate that the animal's diet could significantly affect both the microbiological profile of the meat and its evolution during storage, having a direct impact on the improvement of the whole quality profile.

As regards the details, the most relevant data are associated with the mesophilic population as well as lactic acid bacteria (LAB). According to the results, the MAP packaging selected LAB and *Brochothrix thermosphacta*, which are usually the dominant spoiling microorganisms in the applied conditions [16]. As a consequence, the significantly lower counts ($p < 0.05$) observed in the LE samples are an important result to delay the spoilage. As the gas mixture applied contained a high percentage of oxygen, *Pseudomonas* spp. were also able to grow during burger storage, again with lower counts in the LE and L samples. *Brochothrix thermosphacta* and *Pseudomonas* spp. greatly contribute to the development of off-odours and off-flavours during meat storage, as the first microorganism produces cheesy and acid odours, while the latter is responsible for sulphuric, putrid, sweet and fruity odours [51]. The different diet supplementation of the animals seems to have a significant effect on the microbial population of the derived burgers, in particular in terms of microbial load. In fact, the supplementation with linseed and vitamin E improves the microbiological quality of the products, with lower counts not only at time 0, but also during the whole storage at 4 °C. Depending on the bacterial group considered, the meat from animals fed with linseed without vitamin E also shows lower counts with respect to the controls, particularly for the mesophilic aerobic population, lactic acid bacteria and *Pseudomonas* spp. While a diet supplemented with linseed oil has been demonstrated to affect the qualitative composition of the rumen microbiota, with a selective effect on specific population [52], no literature data are available on the effect of linseed on meat microbiota. Some authors proved the antimicrobial effect of selected fatty acids in vitro [53], while no results on the correlation between fatty acid composition and meat microbiota are actually available. Nevertheless, analysing the whole dataset, it appears that the compositional and chemical-physical changes determined on the meat characteristics by the linseed alone, and most of all by the linseed associated with vitamin E supplementation, improve the microbiological profile of the final products, extending their shelf-life. This is a crucial point, as minced-meat-derived products are highly exposed to microbial spoilage and can be stored for a very short time. On the contrary, including the essential oils within the burger's formulation did not help to contain the microbial growth during time. In fact, in the tested conditions, the biopreservatives were ineffective in reducing the microbial load, and in several cases, they even stimulate the microbial growth. In accordance with Serio et al. (2010) [54], after a lag phase extension, sub-lethal concentrations of essential oils stimulate the growth of several bacteria. Moreover, some Specific Spoilage Organisms (SSOs), such as *Pseudomonas* spp., are particularly resistant to phenolic compounds, such as thymol [55], contained in *O. hirtum*, the EO applied in our study. Therefore, while the applied EO concentration is useful to reduce lipid oxidation, higher EO amounts could be necessary to contain the microbial growth.

Table 4 summarises the results of meat fatty acid composition. According to Albertì et al. (2013) [56], the levels of SFA, MUFA and PUFA in the meat and meat products could be ascribable to multifactorial effects. Moreover, other authors found different levels of FA in the meat of young bulls fed with flaxseed [57] compared to our results, or linseed [58] showing as breeds, production systems or slaughter weight can influence these parameters.

The levels of MUFA were affected by diet, but not by the essential oil treatment. This result agrees with previous studies [57,59] in which an increase in MUFA in meat of animals supplemented with unsaturated fatty acids were recorded. Even if the quantity of linseed was the same between the L and LE groups, the significant increase in MUFA and PUFA in the burgers of the LE group recorded here was mainly ascribable to the use of vitamin E. We observed the same trend for PUFA ($p < 0.05$), n-3 ($p < 0.01$) and CLA ($p < 0.01$) in the LE group compared to groups L and C. Other authors [58,59] confirmed that the diet enriched with linseed can affect the concentration of n-3 FA in meat as demonstrated also in our research between the L and LE groups compared to group C. The higher concentration of n-3 FA also contributed to increase the availability of linolenic acid in the muscle, resulting in an enhanced synthesis of its elongation and desaturation products, such as EPA (Eicosapentaenoic acid) and DHA (Docosahexaenoic acid). The higher value of I-Harris index is due to the higher EPA and DHA level in the muscle of animals that received a diet enriched with linseed. This is a very important aspect on the human health view because the higher level of healthy FA (I-Harris index and n-3) and lower SFA suggests that the hamburger from the L and LE groups are healthier compared to the C group [29]. We also registered a different concentration of these long-chain FA between groups L and LE, probably due to the use of vitamin E added to extruded linseed in the diet of the LE group compared to group L. The action of vitamin E in the rumen is not well known, but seems to influence the activity of Butyrivibrio Fibrisolvens, modifying the rumen PUFA biohydrogenation with a consequent increase in fatty acids, such as CLA, and the same PUFA [60,61].

The burgers from groups L and LE showed a higher CLA concentration compared to C group. These results are coherent with those reported in other studies [40,62,63], where diets rich in linoleic acids were demonstrated to increase the levels of C18:1 t11 acid in meat.

5. Conclusions

This study demonstrated that the use of linseed and vitamin E in the diet of Marchigiana young bulls had an important effect on colour and lipid stability of the burgers. Furthermore, EOs seem to protect the burgers from lipid oxidation through the antioxidant effects during storage. On the other hand, the applied EO concentration was not effective in reducing microbial spoilage over time. Nevertheless, linseed and vitamin E supplementation had a positive effect on microbial loads and growth dynamics, containing microbial development over time. These results seem to suggest that, to optimise the minced meat for burger preparation from Marchigiana bulls, the use of vitamin E is recommended. Indeed, the use of vitamin E in high doses (2,1 gr/head/d) with about 1 kg of extruded linseed in the finishing diet of young Marchigiana bulls improves the profile of fatty acids with beneficial effects also on the consumer's health.

Author Contributions: Conceptualization, I.F., M.G. and A.S.; methodology, M.G., A.S. and I.F.; software, D.C.; validation, I.F. and M.G., A.S.; formal analysis, D.C.; investigation, M.G. and A.S.; data curation, I.F. and D.C.; writing—original draft preparation, I.F., A.S., M.G. and L.M.E.M.; writing—review and editing, I.F., M.G., L.M.E.M., L.L. and J.D.M.V.; visualization, I.F. and M.G.; supervision, A.F. and G.V.; project administration, I.F., M.G. and A.S.; funding acquisition, M.G., I.F. and G.V. All authors have read and agreed to the published version of the manuscript.

Funding: This study has been funded by the Rural Development Programme 2007–2013-Abruzzo Region-Measure 1.2.4. "Cooperation for the development of new products, technologies in the agriculture and food sector and in the forestry sector" and by the Project Demetra (Dipartimenti di Eccellenza 2018–2022, CUP_C46C18000530001) funded by the Italian Ministry for Education, University and Research.

Institutional Review Board Statement: Ethical approval was not required because this study was conducted within the normal breeding activities of the farm and the animals were slaughtered at the end of their productive cycle.

Informed Consent Statement: Not applicable.

Data Availability Statement: The data supporting the findings of this study are available within the article.

Acknowledgments: The authors thank the "Azienda Agricola Sanlorenzo" Sardinara (Teramo) for farm support.

Conflicts of Interest: The authors declare no conflict of interest.

References

1. Food and Agriculture Organization. World Livestock 2011—Livestock in Food Security. Rome, Italy. 2011. FAO. Available online: http://www.fao.org/docrep/014/i2373e/i2373e00.htm (accessed on 20 January 2022).
2. Leclercq, C.; Arcella, D.; Piccinelli, R.; Sette, S.; Le Donne, C.; Turrini, A. The Italian National Food Consumption Survey INRAN-SCAI 2005–06: Main Results in Terms of Food Consumption. *Pub. Health Nutr.* **2009**, *12*, 2504–2532. [CrossRef] [PubMed]
3. Bellini, G.; Lipizzi, F.; Cosentino, F.; Giordano, P. *Atlas of Italian Agriculture: 6th Census of Agriculture*; National Institute of Statistics: Rome, Italy, 2013.
4. Nordic Council. *Nordic Nutrition Recommendations 2012: Integrating Nutrition and Physical Activity*, 5th ed.; Nordic Co-Operation: Copenhagen, Denmark, 2014; p. 629.
5. Flowers, S.; McFadden, B.R.; Carr, C.C.; Mateescu, R.G. Consumer preferences for beef with improved nutrient profile. *J. Anim. Sci.* **2019**, *97*, 4699–4709. [CrossRef] [PubMed]
6. Astrup, A.; Magkos, F.; Bier, D.M.; Brenna, T.; de Oliveira Otto, M.C.; Hill, J.O.; King, J.C.; Mente, A.; Ordovas, J.M.; Volek, J.S.; et al. Saturated Fats and Health: A Reassessment and Proposal for Food-Based Recommandations. *J. Am. Coll. Cardiol.* **2020**, *76*, 844–857. [CrossRef]
7. Ashraful, I.; Nurul, M.; Shafayet, A.; Siddiqui, A.; Hossain, P.; Farhana Sultana, F.; Kabir, R. Trans fatty acids and lipid profile: A Serious Risk Factor to Cardiovascular Disease, Cancer and Diabetes Diabetes & Metabolic Syndrome. *Clinic. Res. Rev.* **2019**, *13*, 1643–1647. [CrossRef]
8. Chen, J.; Liu, H. Nutritional Indices for Assessing Fatty Acids: A Mini-Review. *Int. J. Mol. Sci.* **2020**, *21*, 5695. [CrossRef] [PubMed]
9. Giaretta, E.; Mordenti, A.; Palmonari, A.; Brogna, N.; Canestrari, G.; Belloni, P.; Cavallini, D.; Mammi, L.; Cabbri, R.; Formigoni, A. NIRs calibration models for chemical composition and fatty acid families of raw and freeze-dried beef: A comparison. *J. Food Comp. Anal.* **2019**, *83*, 103257. [CrossRef]
10. Fusaro, I.; Giammarco, M.; Chincarini, M.; Vaintrub, M.O.; Palmonari, A.; Mammi, L.M.E.; Formigoni, A.; Giuseppe, L.; Vignola, G. Effect of ewe diet on milk and muscle fatty acid composition of suckling lambs of the protected geographical origin abbacchio romano. *Animals* **2020**, *10*, 25. [CrossRef]
11. Gómez-Cortés, P.; Bach, A.; Luna, P.; Juárez, M.; Fuente, M.A. Effects of extruded linseed supplementation on n-3 fatty acids and conjugated linoleic acid in milk and cheese from ewes. *J. Dairy Sci.* **2009**, *92*, 4122–4134. [CrossRef]
12. Oliveira, X.S.M.; Palma, A.S.V.; Reis, B.R.; Franco, C.S.R.; Marconi, A.P.S.; Shiozaki, F.A.; Reis, L.G.; Salles, M.S.V.; Netto, A.S. Inclusion of soybean and linseed oils in the diet of lactating dairy cows makes the milk fatty acid profile nutritionally healthier for the human diet. *PLoS ONE* **2021**, *16*, e0246357. [CrossRef]

13. Marino, R.; Della Malva, A.; Caroprese, M.; De Palo, P.; Santillo, A.; Sevi, A.; Albenzio, M. Effects of whole linseed supplementation and treatment duration on fatty acid profile and endogenous bioactive compounds of beef muscle. *Animal* **2019**, *13*, 444–452. [CrossRef]
14. Lasagna, E.; Ceccobelli, S.; Di Lorenzo, P.; Albera, A.; Filippini, F.; Sarti, F.M.; Panella, F.; Di Stasio, L. Comparison of four italian beef cattle breeds by means of functional genes. *Ital. J. Anim. Sci.* **2015**, *14*, 86–89. [CrossRef]
15. Domínguez, R.; Pateiro, M.; Gagaoua, M.; Barba, F.J.; Wangang Zhang, W.; Lorenzo, J.M. A Comprehensive Review on Lipid Oxidation in Meat and Meat Products. *Antioxidants* **2019**, *8*, 429. [CrossRef] [PubMed]
16. Paparella, A.; Mazzarrino, G.; Chaves-López, C.; Rossi, C.; Sacchetti, G.; Guerrieri, O.; Serio, A. Chitosan boosts the antimicrobial activity of Origanum vulgare essential oil in modified atmosphere packaged pork. *Food Microbiol.* **2016**, *59*, 23–31. [CrossRef] [PubMed]
17. Pelaes Vital, A.C.; Guerrero, A.; Guarnido, P.; Severino, I.C.; Olleta, J.L.; Blasco, M.; Prado, I.N.; Maggi, F.; Campo, M.M. Effect of active-edible coating and essential oils on lamb patties oxidation during display. *Foods* **2021**, *10*, 263. [CrossRef]
18. Fusaro, I.; Cavallini, D.; Giammarco, M.; Manetta, A.C.; Martuscelli, M.; Mammi, L.M.E.; Lanzoni, L.; Formigoni, A.; Vignola, G. Oxidative status of Marchigiana beef enriched in n-3 fatty acids and vitamin E, treated with a blend of oregano and rosemary essential oils. *Front. Vet. Sci.* **2021**, *8*, 662079. [CrossRef]
19. AOAC. *Official Methods of Analysis*, 16th ed.; Association of Official Analytical Chemists: Arlington, VA, USA, 2002.
20. Van Soest, P.J.; Robertson, J.B.; Lewis, B.A. Methods for dietary fiber, neutral detergent fiber, and nonstarch polysaccharides in relation to animal nutrition. *J. Dairy Sci.* **1991**, *74*, 3583–3597. [CrossRef]
21. Lock, A.L.; Teles, B.M.; Perfield, J.W.; Bauman, D.E.; Sinclair, L.A. A conjugated linoleic acid supplement containing trans-10, cis-12 reduces milk fat synthesis in lactating sheep. *J. Dairy Sci.* **2006**, *89*, 1525–1532. [CrossRef]
22. Zequan, X.; Zirong, W.; Jiankun, L.; Xin, M.; Hopkins, D.L.; Holman, B.W.B.; Bekhit, A.E.-D.A. The effect of freezing time on the quality of normal and pale, soft and exudative (PSE)-like pork. *Meat Sci.* **2019**, *152*, 1–7. [CrossRef]
23. Inserra, L.; Priolo, A.; Biondi, L.; Lanza, M.; Bognanno, M.; Gravador, R. Dietary citrus pulp reduces lipid oxidation in lamb meat. *Meat Sci.* **2014**, *96*, 1489–1493. [CrossRef]
24. Szydłowska-Czerniaka, A.; Bartkowiak-BrodabIgor, I.; GyörgyKarlovitsd, K.; Szłyka, E. Antioxidant capacity, total phenolics, glucosinolates and colour parameters of rapeseed cultivars. *Food Chem.* **2011**, *127*, 556–563. [CrossRef]
25. Koprivnjak, J.F.; Lum, K.R.; Sisak, M.M.; Saborowski, R. Determination of α-, $\gamma(+\beta)$-, and δ-tocopherols in a variety of liver tissues by reverse-phase high pressure liquid chromatography. *Comp. Bioch. Phys.* **1996**, *113*, 143–148. [CrossRef]
26. Folch, J.; Lees, M.; Sloane Stanley, G.H. A simple method for the isolation and purification of total lipids from animal tissues. *J. Biol. Chem.* **1957**, *226*, 497–509. [CrossRef]
27. Frega, N.; Lerker, G. Rapid methods for the quality control of food oils. *Riv. Ital. Sost. Gr.* **1984**, *61*, 385–391.
28. Vannice, G.; Rasmussen, H. Position of the academy of nutrition and dietetics: Dietary Fatty Acids for Healthy Adults. *J. Acad. Nutr. Diet.* **2014**, *114*, 136–153. [CrossRef]
29. Harris, W.S.; Assaad, B.; Poston, W.C. Tissue omega-6/omega-3 fatty acid ratio and risk for coronary artery disease. *Am. J. Cardiol.* **2006**, *98*, 19–26. [CrossRef]
30. Mordenti, A.L.; Brogna, N.; Canestrari, G.; Bonfante, E.; Eusebi, S.; Mammi, L.M.E.; Giaretta, E.; Formigoni, A. Effects of breed and different lipid dietary supplements on beef quality. *Anim. Sci. J.* **2019**, *90*, 619–627. [CrossRef]
31. Cavallini, D.; Mammi, L.M.E.; Fustini, M.; Palmonari, A.; Heinrichs, A.J.; Formigoni, A. Effects of ad libitum or restricted access to total mixed ration with supplemental long hay on production, intake, and rumination. *J. Dairy Sci.* **2018**, *101*, 10922–10928. [CrossRef]
32. Juárez, M.; Dugan, M.E.R.; Aldai, N.; Basarab, J.A.; Baron, V.S.; McAllister, T.A.; Aalhus, J.L. Beef quality attributes as affected by increasing the intramuscular levels of vitamin E and omega-3 fatty acids. *Meat Sci.* **2012**, *90*, 764–769. [CrossRef]
33. Berdahl, D.R.; McKeague, J. Rosemary sage extracts as antioxidants for food preservation. In *Handbook of Antioxidants for Food Preservation*; Shahidi, F., Ed.; Woodhead Publishing: Cambridge, UK, 2015; pp. 177–217. [CrossRef]
34. Ruiz-Hernández, K.; Sosa-Morales, M.R.; Cerón-García, A.; Gómez-Salazar, J.A. Physical, Chemical and Sensory Changes in Meat and Meat Products Induced by the Addition of Essential Oils: A Concise Review. *Food Rev. Int.* **2021**, *8*, 1–30. [CrossRef]
35. Delosièrea, M.; Duranda, D.; Bourguetb, C.; Terlouw, E.M.C. Lipid oxidation, pre-slaughter animal stress and meat packaging: Can dietary supplementation of vitamin E and plant extracts come to the rescue? *Food Chem.* **2020**, *309*, 125668. [CrossRef]
36. Albertí, P.; Campo, M.M.; Beriain, M.J.; Ripoll, G.; Sañudo, C. Effect of including whole linseed and vitamin E in the diet of young bulls slaughtered at two fat covers on the sensory quality of beef packaged in two different packaging systems. *J. Sci. Food Agric.* **2017**, *97*, 753–760. [CrossRef] [PubMed]
37. Bellésa, M.; Lealb, L.N.; Díaza, V.; Alonsoa, V.; Roncalésa, P.; Beltrána, J.A. Effect of dietary vitamin E on physicochemical and fatty acid stability of fresh and thawed lamb. *Food Chem.* **2018**, *239*, 1–8. [CrossRef] [PubMed]
38. Ramanathan, R.; Mancini, R.A. Role of Mitochondria in Beef Color: A Review. *Meat Musc. Biol.* **2018**, *2*, 309–320. [CrossRef]
39. Álvarez-Rodríguez, J.; Urrutia, O.; Lobón, S.; Guillermo Ripoll, G.; Bertolín, J.R.; Joy, M. Insights into the role of major bioactive dietary nutrients in lamb meat quality: A Review. *J. Anim. Sci. Biotechnol.* **2020**, *13*, 20. [CrossRef]
40. Bernardi, D.M.; Bertol, T.M.; Pflanzer, S.B.; Sgarbieri, V.C.; Pollonio, M.A.R. ω-3 in meat products: Benefits and Effects on Lipid Oxidative Stability. *J. Sci. Food Agric.* **2016**, *96*, 2620–2634. [CrossRef]

41. De Palo, P.; Maggiolino, A.; Centoducati, N.; Tateo, A. Effects of different milk replacers on carcass traits, meat quality, meat color and fatty acids profile of dairy goat kids. *Small Rumin. Res.* **2015**, *131*, 6–11. [CrossRef]
42. Fernandes, R.P.P.; Trindade, M.A.; Tonin, F.G.; Pugine, S.M.P.; Lima, C.G.; Lorenzo, J.M.; de Melo, M.P. Evaluation of oxidative stability of lamb burger with Origanum vulgare extract. *Food Chem.* **2017**, *233*, 101–109. [CrossRef]
43. Faustman, C.; Sun, Q.; Mancini, R.; Suman, S.P. Myoglobin and lipid oxidation interactions: Mechanistic Bases and Control. *Meat Sci.* **2010**, *86*, 86–94. [CrossRef]
44. Idamokoro, E.M.; Falowo, A.B.; Oyeagu, C.E.; Afolayan, A.J. Multifunctional activity of vitamin E in animal and animal products: A review. *Anim. Sci. J.* **2020**, *91*, e13352. [CrossRef]
45. Wang, Y.; Domínguez, R.; Lorenzo, J.M.; Bohrer, B.M. The Relationship between Lipid Content in Ground Beef Patties with Rate of Discoloration and Lipid Oxidation during Simulated Retail Display. *Foods* **2021**, *10*, 1982. [CrossRef]
46. Kaur, R.; Gupta, T.B.; Bronlund, J.; Kaur, L. The potential of rosemary as a functional ingredient for meat products—A review. *Food Rev.* **2021**, *9*, 4517–4528. [CrossRef]
47. Ribeiro, J.S.; Santos, M.J.M.C.; Silva, L.K.R.; Pereira, L.C.L.; Santos, I.A.; Lannes, S.C.S.; Silva, M.V. Natural antioxidants used in meat products: A brief review. *Meat Sci.* **2019**, *148*, 181–188. [CrossRef] [PubMed]
48. Zhai, H.; Liu, H.; Wang, S.; Wu, J.; Kluenter, M.A. Potential of essential oils for poultry and pigs. *Anim. Nutr.* **2018**, *4*, 179–186. [CrossRef]
49. Pelaes Vital, A.C.; Guerrero, A.; Barbosa Carvalho Kempinski, E.M.; de Oliveira Monteschio, J.; Sary, C.; Ramos, T.R.; MarCampo, M.; do Prado, I.N. Consumer profile and acceptability of cooked beef steaks with edible and active coating containing oregano and rosemary essential oils. *Meat Sci.* **2018**, *143*, 153–158. [CrossRef]
50. Pateiro, M.; Munekata, P.E.S.; Sant'Ana, A.S.; Domínguez, R.; Rodríguez-Lázaro, D.; Lorenzo, J.M. Application of essential oils as antimicrobial agents against spoilage and pathogenic microorganisms in meat products. *Int. J. Food Microbiol.* **2021**, *16*, 337. [CrossRef]
51. Iulietto, M.F.; Sechi, P.; Borgogni, E.; Cenci-Goga, B.T. Meat spoilage: A Critical Review of a Neglected Alteration due to Ropy slime Producing Bacteria. *Ital. J. Anim. Sci.* **2015**, *14*, 315–326. [CrossRef]
52. Yang, S.L.; Bu, D.P.; Wang, J.Q.; Hu, Z.Y.; Li, D.; Wei, H.Y.; Zhou, L.Y.; Loor, J.J. Soybean oil and linseed oil supplementation affect profiles of ruminal microorganisms in dairy cows. *Animal* **2009**, *3*, 1562–1569. [CrossRef]
53. Ouattara, B.; Simard, R.E.; Holley, R.A.; Piette, G.J.P.; Bégin, A. Antibacterial activity of selected fatty acids and essential oils against six meat spoilage organisms. *Int. J. Food Microb.* **1997**, *37*, 155–162. [CrossRef]
54. Serio, A.; Chiarini, M.; Tettamanti, E.; Paparella, A. Electronic paramagnetic resonance investigation of the activity of Origanum vulgare L. essential oil on the Listeria monocytogenes membrane. *Lett. Appl. Microbiol.* **2010**, *51*, 149–157. [CrossRef]
55. Skandamis, P.N.; Nychas, G.J.E. Preservation of fresh meat with active and modified atmosphere packaging conditions. *Int. J. Food Microbiol.* **2010**, *79*, 35–45. [CrossRef]
56. Albertí, P.; Gómez, I.; Mendizabal, J.A.; Ripoll, G.; Barahona, M.; Sarriés, V.; Insausti, K.; Beriain, M.J.; Purroy, A.; Realini, C. Effect of whole linseed and rumen-protected conjugated linoleic acid enriched diets on feedlot performance, carcass characteristics, and adipose tissue development in young Holstein bulls. *Meat Sci.* **2013**, *94*, 208–214. [CrossRef] [PubMed]
57. Renna, M.; Brugiapaglia, A.; Zanardi, E.; Prandini, A.; Moschini, M.; Sigolo, S. Fatty acid profile, meat quality and flavour acceptability of beef from double-muscled Piemontese young bulls fed ground flaxseed. *Ital. J. Anim. Sci.* **2019**, *18*, 355–365. [CrossRef]
58. Morittu, V.M.; Spina, A.A.; Iommelli, P.; Poerio, A.; Oliverio, F.V.; Britti, D.; Tudisco, R. Effect of Integration of Linseed and Vitamin E in Charolaise × Podolica Bulls' Diet on Fatty Acids Profile, Beef Color and Lipid Stability. *Agriculture* **2021**, *11*, 1032. [CrossRef]
59. Mordenti, A.L.; Brogna, N.; Merendi, F.; Formigoni, A.; Sardi, L.; Cardenia, V.; Rodriguez-Estrada, M.T. Effect of dietary inclusion of different lipid supplements on quality and oxidative susceptibility of beef meat. *Ital. J. Anim. Sci.* **2018**, *18*, 105–110. [CrossRef]
60. Pottier, J.; Focant, M.; Debier, C.; De Buysser, G.; Goffe, C.; Mignolet, E.; Froidmont, E.; Larondelle, Y. Effect of Dietary Vitamin E on Rumen Biohydrogenation Pathways and Milk Fat Depression in Dairy Cows Fed High-Fat Diets. *J. Dairy Sci.* **2006**, *89*, 685–692. [CrossRef]
61. Yoshimura, E.H.; Santos, N.W.; Machado, E.; Agustinho, B.C.; Pereira, L.M.; De Aguiar, S.C.; Franzolin, R.; Gasparino, E.; Santos, G.T.; Zeoula, L.M. Effects of dairy cow diets supplied with flaxseed oil and propolis extract, with or without vitamin E, on the ruminal microbiota, biohydrogenation, and digestion. *Anim. Feed Sci. Technol.* **2018**, *241*, 163–172. [CrossRef]
62. Klieve, A.V.; Hennessy, D.; Ouwerkerk, D.; Forster, R.J.; Mackie, R.I.; Attwood, G.T. Establishing populations of Megasphaera elsdenii YE 34 and Butyrivibrio fibrisolvens YE 44 in the rumen of cattle fed high grain diets. *J. Appl. Microbiol.* **2003**, *95*, 621–630. [CrossRef]
63. Griinari, J.M.; Corl, B.A.; Lacy, S.H.; Chouinard, P.Y.; Nurmela, K.V.V.; Bauman, D.E. Conjugated linoleic acid is synthesized endogenously in lactating dairy cows by 19-desaturase. *J. Nutr.* **2000**, *130*, 2285–2291. [CrossRef]

antioxidants

MDPI

Article

Salvia officinalis L. Essential Oil: Characterization, Antioxidant Properties, and the Effects of Aromatherapy in Adult Patients

Maria-Daniela Mot [1,†], Simona Gavrilaș [2,†], Andreea I. Lupitu [2,†], Cristian Moisa [2], Dorina Chambre [2], Delia Mirela Tit [1,3], Mihaela Alexandra Bogdan [1], Adina-Maria Bodescu [4], Lucian Copolovici [2], Dana Maria Copolovici [2,*] and Simona Gabriela Bungau [1,3]

1 Doctoral School of Biomedical Sciences, University of Oradea, 410087 Oradea, Romania; mariadanielamot@gmail.com (M.-D.M.); dtit@uoradea.ro (D.M.T.); mihaela.alexandra.bogdan@gmail.com (M.A.B.); sbungau@uoradea.ro (S.G.B.)
2 Faculty of Food Engineering, Tourism and Environmental Protection, Institute for Research, Development and Innovation in Technical and Natural Sciences, "Aurel Vlaicu" University of Arad, 310330 Arad, Romania; simona.gavrilas@uav.ro (S.G.); pag.andreea@yahoo.com (A.I.L.); moisa.cristian@yahoo.com (C.M.); dorinachambree@yahoo.com (D.C.); lucian.copolovici@uav.ro (L.C.)
3 Department of Pharmacy, Faculty of Medicine and Pharmacy, University of Oradea, 410028 Oradea, Romania
4 "Adam Müller Guttenbrunn" High School, 310245 Arad, Romania; adina.bodescu@ltamg.ro
* Correspondence: dana.copolovici@uav.ro
† These authors contributed equally to this work.

Citation: Mot, M.-D.; Gavrilaș, S.; Lupitu, A.I.; Moisa, C.; Chambre, D.; Tit, D.M.; Bogdan, M.A.; Bodescu, A.-M.; Copolovici, L.; Copolovici, D.M.; et al. *Salvia officinalis* L. Essential Oil: Characterization, Antioxidant Properties, and the Effects of Aromatherapy in Adult Patients. *Antioxidants* **2022**, *11*, 808. https://doi.org/10.3390/antiox11050808

Academic Editors: Filomena Nazzaro and Dimitrios Stagos

Received: 20 March 2022
Accepted: 18 April 2022
Published: 21 April 2022

Publisher's Note: MDPI stays neutral with regard to jurisdictional claims in published maps and institutional affiliations.

Abstract: The purpose of this study is to reveal the chemical and biochemical characteristics and the potential aromatherapy applications of the essential oil (EO) of *Salvia officinalis* (common sage) within a hospital environment. The chemical composition was determined by gas chromatography with mass spectrometry and ATR-FTIR spectroscopy. Three types of sage EOs were included in this study: two commercial oils and one oil obtained by in-house hydrodistillation. Based on the findings, these EOs were included in different chemotypes. The first two samples were similar to the most common chemotype (α-thujone > camphor > 1,8-cineole > β-thujone), while the in-house sage EO revealed a high content of 1,8-cineole, borneol, α-thujone, similar to the Dalmatian type. The latter sample was selected to be evaluated for its antioxidant and medical effects, as borneol, a bicyclic monoterpene, is known as a substance with anesthetic and analgesic effects in traditional Asian medicine. The study suggests that the antioxidant capacity of the sage EO is modest (33.61% and 84.50% inhibition was determined by DPPH and ABTS assays, respectively), but also that the inhalation of sage EO with high borneol content by hospitalized patients could improve these patients' satisfaction.

Keywords: *Salvia officinalis*; essential oil; chemical analysis; antioxidant activity; aromatherapy; inhalation

1. Introduction

Essential oils (EOs) have proven their specific applications in different areas, including in human health, as a treatment for various disorders (migraine [1], skin disorders [2], fatigue [3,4], stroke [5], sleep disorders [6,7], endocrine disorders [8], depressive disorders [9], etc.). In medicine, EOs are considered good candidates as complementary and alternative treatment components due to their antimicrobial properties [10], possible anesthetic [11], or immunostimulatory effects [12]. An EO can be used alone and/or in synergetic associations with other EOs [13,14].

The qualitative composition of essential oils is determined by the genome of the particular plant and the variation caused by several environmental conditions such as daylight, temperature, and light. The conditions mentioned above select a pattern of seasonal fluctuations in plant metabolites. These conditions vary predictably and significantly throughout the vegetation interval, usually recurrent each year [15].

Salvia officinalis L. is one of the most widely used sources for EOs in traditional medicine. Even though the chemical composition of *Salvia officinalis* EOs has already been determined, the composition of EOs is very complex, depending on the plant part, time of harvesting, season, genetic diversity, climate, and meteorological conditions. Different studies have used gas-chromatography techniques [16–19] or FTIR [20] to determine *Salvia officinalis* L. EOs' chemical composition. It is claimed that the antimicrobial properties of EOs are related to an extended alcoholic profile. Also, it was assessed that skin injury restoration is enhanced in the case of using plant extracts containing increased amounts of ketones [2,21]. EOs are rapidly absorbed via the skin into the bloodstream and thus travel to the brain and other organs [1,4,5,9,22–28]. A reduction of migraine discomfort is the result of the combination of the following factors: neurogenic inflammation, as well as a decrease in distress perception, along with blood vessel dilation [1]. The same processes may be responsible for the positive response to aromatherapy and synergic musical therapy obtained for anxiousness and/or suffering states recorded in different post-procedural phases of adult patients [29,30]. For insomnia improvement, the olfactory route is considered based on the comparative results regarding inhalation versus the transdermal pathway [31]. The inhalation of EOs may stimulate the immune and limbic systems responsible for the body's state of well-being and emotional integration. The quality of life in burnout cases, determined by different diseases, such as chronic hemodialysis [32] or long periods of hard work, could be improved through aromatherapy [33,34]. Several reports are focused on evaluating essential oils' toxicity profile [35–37]. For example, sage essential oil toxicity has been assessed on rat hepatocytes [36], and it is not toxic at concentrations below 200 nL mL^{-1}. A report presents an accidental exposure to a newborn and a toddler to sage EO that lead to generalized tonic-clonic seizures [38]. Not all medical aspects are fully elucidated so far, as relevant elements such as the toxicological profile and different interactions still need to be clarified and studied.

It is used to treat colds, tuberculosis, bronchitis, gastrointestinal diseases, inflammation, and presents antibacterial, antifungal [39], antitumor [37,40–42], and antioxidant properties [43]. With respect to *Salvia officinalis,* different mechanisms of action have been proposed. Most of the research has been focused on the influence of *Salvia officinalis* on mental functions. Memory enhancement is thought to be a consequence of *Salvia officinalis'* stimulating effect on the nicotinic and muscarinic receptors. Studies have also been conducted on the inhibitory effect on acetylcholinesterase, a compound considered to have a major role in Alzheimer-specific symptom progress [44]. At 100 mg/mL concentration, *Salvia officinalis* extract revealed high activity towards *Escherichia coli*, which is ordinary with respect to *Staphylococcus aureus* and *Pseudomonas aeruginosa*, and low in the case of *Bacillus subtilis* [45]. The use of the extracts in the presence of Hp-2-Minh ligand improved their biocidal characteristic even at lower concentrations. Research has proven the antimicrobial potential of EOs. This finding has medical significance, as it justifies reducing the use of synthetic medicinal products and utilizing EOs as a viable alternative in cases of drug-resistant pathogens.

Aromatherapy refers to the use of volatile compounds or fragrances from EOs obtained from plants, typically by inhalation, to prevent or cure diseases, infections, and indispositions. For alleviating pain, nausea, and anxiety, an ancient practice—clinical aromatherapy—is gaining attention in contemporary health services, where the intention of both the consumers and clinicians is to minimize the usage of medications. Thus, using integrative and complementary therapies is increasing, and scientific proof for the integrative therapies continues to flourish [46].

Different EOs were included in complementary and alternative treatment (such as aromatherapy) and were evaluated in a hospital setting in the following instances: peppermint EO for nausea and vomiting in women [25], lavender EO [47], *Salvia officinalis* EO for reducing nausea and vomiting in patients with cancer undergoing chemotherapy [48], *Rosa damascene* EO to decrease anxiety and increase sleep quality in cardiac patients [49], and *Citrus aurantium* to reduce anxiety in patients with acute coronary syndrome [50]. In the case of hospitalized patients, one of the main issues that arise is the decrease of peace of mind and the increase in anxiety during the hospitalization period [51].

Therefore, the present investigation aimed to assess the chemical composition and antioxidant activity of the essential oil of *S. officinalis* obtained from aerial parts of the plant and evaluate aromatherapy's influence on patients in a hospital setup. The results of this study can be a starting point for the development of Aromatherapy Programs in Patient Care Settings in Romania.

2. Materials and Methods

All the reagents and solvents used in the experimental part were of adequate analytical or chromatographical grade (Sigma Aldrich, Fluka, Switzerland, and Merck, Darmstadt, Germany).

2.1. Sample Collection and Preparation

The EO named L-SEO was obtained by hydro-distillation from the dried aerial parts of *Salvia officinalis* plants grown in Arad County, Romania (coordinates 46°10′30″ N 21°18′45″ E, sample harvested in June 2019). The EO was stored in glass vials at +4 °C until further analysis. The samples B-SEO and EG-SEO are sage EOs commercially available on the Romanian market.

2.2. GC-MS Determination of the Chemical Composition of the Essential Oil from Salvia officinalis

The constituents of the EOs were determined by the gas chromatography method, using a gas chromatograph (GC) (Shimadzu2010, Kyoto, Japan) coupled with a triple quadruple mass spectrometer (MS) (TQ 8040, Shimadzu, Kyoto, Japan). The column used was an Optima 1 MS (30 m × 0.25 mm i.d., with a film thickness of 0.25 mm). At a flow rate of 1 mL/min, Helium was used as a carrier gas. The procedure for separating and quantifying EO components is described above [52]. The compounds from the analyzed samples were identified based on their mass spectra using the NIST 14 and Wiley 09 mass spectrum libraries (Scientific Instrument Services, Palmer, MA, USA). The Kovats retention indices for the identified compounds were calculated using the C8-C40 alkane standard. All analyses were performed in triplicate.

2.3. ATR-FTIR Spectroscopy

The ATR-FTIR spectra of *Salvia officinalis* L. EOs (L-SEO, B-SEO, and EG-SEO) were obtained on the wavelength range between 600 and 4000 cm^{-1} with a Bruker Vertex 70 (Bremen, Germany) spectrophotometer equipped with a Pike Miracle ATR cell. A sample volume of ~10 µL from each EO was placed directly on the surface of the ZnSe ATR crystal in the Teflon depression and covered with a metal cover. To avoid evaporation of the sample, a black outer ring was screwed on to press the Teflon depression tightly on the crystal. At the same time, the upper handle of the ATR cell was rotated so that its slightly concave stainless steel tip applied pressure against the metal cover.

The experimental spectra of the samples were recorded with a resolution of 4 cm^{-1}, and 32 scans were accumulated per spectrum. The spectra were obtained in duplicate for each sample, and the average spectrum of the two measurements was calculated. Before each ATR measurement, the ZnSe crystal was carefully cleaned with isopropyl alcohol, and an air background spectrum was performed.

The OPUS software, version 6.5 from Bruker (Germany), was employed for spectra acquisition, minimum–maximum normalization, baseline correction, and identifying the wavelength value corresponding to the maximum absorbance of the recorded FT-IR bands.

2.4. Antioxidant Activity (DPPH and ABTS Assays)

The L-SEO sample's antioxidant capacity was evaluated using two spectrophotometric assays (DPPH assay and ABTS assay), as reported earlier [22]. A UV-VIS spectrophotometer Specord 200 (Analytik Jena, Jena, Germany) and a 10 mm quartz cuvette were used in the procedure described above. Briefly, for the DPPH assay, we mixed a 0.1 mL control sample with 3 mL of 0.2 mM ethanolic DPPH• solution. The absorbance was recorded after

60 min of reaction in the dark at λ = 517 nm. Positive controls containing 0.02–4 mM Trolox solutions were prepared. The data are expressed in mg Trolox/L and inhibition (%).

For the ABTS assay, we mixed 0.5 mL sample or control with 1 mL ABTS* solution, prepared 16 h before from ABTS reagent and 2.45 mM aqueous solution of sodium persulfate. The absorbance was recorded at λ = 734 nm, after 10 min of reaction time in the dark. Positive controls containing 0.02–1.0 mM Trolox solutions were prepared. The data are expressed in mmol TEAC/L (TEAC: Trolox Equivalent Antioxidant Capacities) and inhibition (%). All the experiments were performed in triplicate.

2.5. Aromatherapy Effects: Clinical Application

This study was performed between August 2019 and August 2020, as a randomized, single-blind study involving adult patients at the Lipova City Hospital, Arad County, Romania.

In total, 174 hospitalized patients aged 23–85 years were enrolled, meeting the following inclusion criteria: adult patients with chronic conditions who have had at least one hospitalization in the same section in the past, an unimpaired sense of smell, no psychiatric pathologies, ability to communicate, and minimum 4 days and maximum 5 days hospital stay. Exclusion criteria included surgery intervention and unwillingness to participate.

Patients were randomized into two groups: 50 in the control group and 124 in the aromatherapy group; no patients were excluded, and no patients quit the study. Random allocation of the patients was made by hospital personnel who were not involved in the data collection or data analysis, depending on the number of patients hospitalized in each room. During the hospitalization period, patients admitted to the same ward were included in either the aromatherapy group or the control group (there were no patients from both groups at the same time in a ward). For the patients enrolled in the aromatherapy group, on a daily basis, a hospital staff member prepared the cotton disc with two drops of sage EO (L-SEO sample) that was kept on the patient's pillow for a minimum of 30 min as the patient inhaled the volatile compounds. The patients from the control group received only routine care.

Only one EO (L-SEO) was used in aromatherapy, which was selected after determining the chemical composition.

Descriptive information form: A questionnaire was prepared by the authors of this study, including the patients' related characteristics such as gender, age, weight, educational level, habits, personal evaluation, and health status. The questionnaire was completed by respondents on the last day of the hospitalization and was used to determine the following: (a) the demographic profile of the participants in terms of age, gender, weight, educational level, social status, health status in the last year, marital status, the use of aroma indoor and of perfumes, the use of sedatives/anxiolytics and the presence of some allergies; (b) the evaluation on the quality of in-hospital services in the present hospitalization stay, with seven possible answers: very weak, weak, acceptable, improved, good, very good, and excellent.

The study was approved by the Ethics Committee of Lipova City Hospital (no. 62/1 July 2019) and was conducted following the ethical principles of the Declaration of Helsinki [53]. The patients provided their written informed consent after a verbal and written explanation of the study protocol.

2.6. Statistical Analysis

Data were processed via GraphPad Prism (version 5.0 for Windows, GraphPad Software, San Diego, CA, USA), and F values (at $p < 0.05$) were considered statistically significant. The t-test, chi-squared test, Kruskal-Wallis test, Fisher's exact test, Friedman test, and the Mann-Whitney test were used to compare qualitative variables in two groups.

3. Results

3.1. Chemical Analyses

3.1.1. GC-MS Analyses

Figure 1 depicts the chromatograms obtained for the three *Salvia officinalis* essential oils samples investigated in this study, and Table 1 shows their composition as determined by GC-MS method. The EOs' constitutive elements are listed according to the elution time. In the liquid phase, 47 compounds were separated.

Table 1. *Salvia officinalis* L. essential oil chemical composition (% TIC) determined by GC-MS analyses.

No.	KI	Compound/Class	Present Study			Turkey [17]	Poland [18]	Morocco [37]	Sudan [19]	Brazil [16]
			B-SEO	EG-SEO	L-SEO					
1	925	Unidentified	0.21 ± 0.01	0.24 ± 0.01						
2	926	Tricyclene/MH			0.42 ± 0.01					
3	930	α-Thujene/MH	0.17 ± 0.01	0.15 ± 0.01	0.45 ± 0.01			0.31		
4	939	α-Pinene/MH	5.96 ± 0.03	4.23 ± 0.02	7.8 ± 0.05	7.17	0.02	3.18	8.96	3.07
5	954	Camphene/MH	5.59 ± 0.07	6.92 ± 0.10	8.73 ± 0.12	8.40		3.67	5.09	4.40
6	975	Sabinene/MH	1.70 ± 0.01	1.55 ± 0.03	0.14 ± 0.01			0.41	0.15	
7	979	β-Pinene/MH	0.78 ± 0.03	0.55 ± 0.02	10.52 ± 0.15	2.92	0.40	2.57		
8	990	β-Myrcene/MH			0.72 ± 0.06	1.16		1.94	3.65	
9	1012	4-Carene/MH	0.42 ± 0.02	0.07 ± 0.01						
10	1024	p-Cymene/MH	0.85 ± 0.02	1.01 ± 0.03		1.33				
11		α-Terpinene/MH			0.34 ± 0.02	0.18		0.2	0.22	
12	1029	1,8-Cineole/MH	13.39 ± 0.21	14.22 ± 0.19	17.98 ± 0.23	18.54	23.72	17.52		14.8
13	1050	trans-β-Ocimene/MH			0.52 ± 0.06					
14	1051	D-Limonene/MH				2.46		1.7	0.37	
15	1056	Linalool/MH							0.79	
16	1059	γ-Terpinene/MH	0.60 ± 0.05	0.15 ± 0.06	0.62 ± 0.08	0.18		0.42		
17	1088	α-Terpinolene/MH	0.60 ± 0.10	0.20 ± 0.05		0.12				
18	1089	2-Carene/MH			0.19 ± 0.01					
19	1102	α-Thujone/MO	26.03 ± 0.25	26.73 ± 0.26	8.74 ± 0.12	22.30		21.22	0.91	24.8
20	1109	β-Thujone/MO	4.65 ± 0.11	4.19 ± 0.12	1.34 ± 0.11	14.28		13.45		3.97
21	1146	Camphor/MO	20.09 ± 0.24	22.64 ± 0.27	2.56 ± 0.12	14.40	18.22	21.23	11.57	10.9
22	1169	Borneol/MO	3.1 ± 0.20	3.31 ± 0.19	15.86 ± 0.23	0.37	2.42	1.67	0.81	11.1
23	1188	α-Terpineol/MO	0.50 ± 0.07	0.46 ± 0.08						
24	1287	Bornyl acetate/MO	1.93 ± 0.11	2.27 ± 0.13	0.88 ± 0.06	0.32	0.22			
25	1290	Thymol/MO			0.33 ± 0.01					
26	1351	α-Cubebene/SH			0.2 ± 0.06					
27	1380	α-Copaene/SH			0.08 ± 0.01				0.13	
28	1381	α-Ylangene/SH			0.26 ± 0.11					
29	1389	β-Bourbonene/SH			0.15 ± 0.03					
30	1393	Unidentified		0.21 ± 0.03						
31	1402	Unidentified	0.23 ± 0.07							
32	1419	β-Caryophyllene/SH	4.54 ± 0.23	3.28 ± 0.21	5.66 ± 0.22	0.58			3.76	2.89
33	1429	Unidentified	0.39 ± 0.01	0.29 ± 0.3						
34	1432	γ-Cadinen/SH			0.24 ± 0.09					
35	1439	α-Guaiene/SH			0.14 ± 0.01					
36	1456	α-Humulene/SH	4.89 ± 021	4.29 ± 0.21	8.64 ± 0.25	0.94		1.45		1.47
37	1479	γ-Muurolene/SH	0.40 ± 0.08	0.26 ± 0.02	0.63 ± 0.06					
38	1500	α-Muurolene/SH			0.43 ± 0.07					
39	1512	Unidentified			0.06 ± 0.01					
40	1523	δ-Cadinene/SH	0.10 ± 0.01	0.03 ± 0.01	0.17 ± 0.01					
41	1576	Isoledene/SH			0.61 ± 0.05					
42	1583	Caryophyllene oxide/SO	0.19 ± 0.01	0.29 ± 0.03	0.22 ± 0.07					
43	1592	Viridiflorol/SO			3.09 ± 0.19					0.6
44	1593	Unidentified	2.34 ± 0.19	1.88 ± 0.18						
45	1594	Unidentified	0.35 ± 0.03	0.61 ± 0.08	0.6 ± 0.09					
46	1603	Unidentified			0.08 ± 0.01					
47	1607	Unidentified			0.6 ± 0.08					

MH—monoterpene hydrocarbons; MO—oxygenate monoterpene; SH—sesquiterpene hydrocarbons; SO—oxygenated sesquiterpene, KI—Kovats retention index.

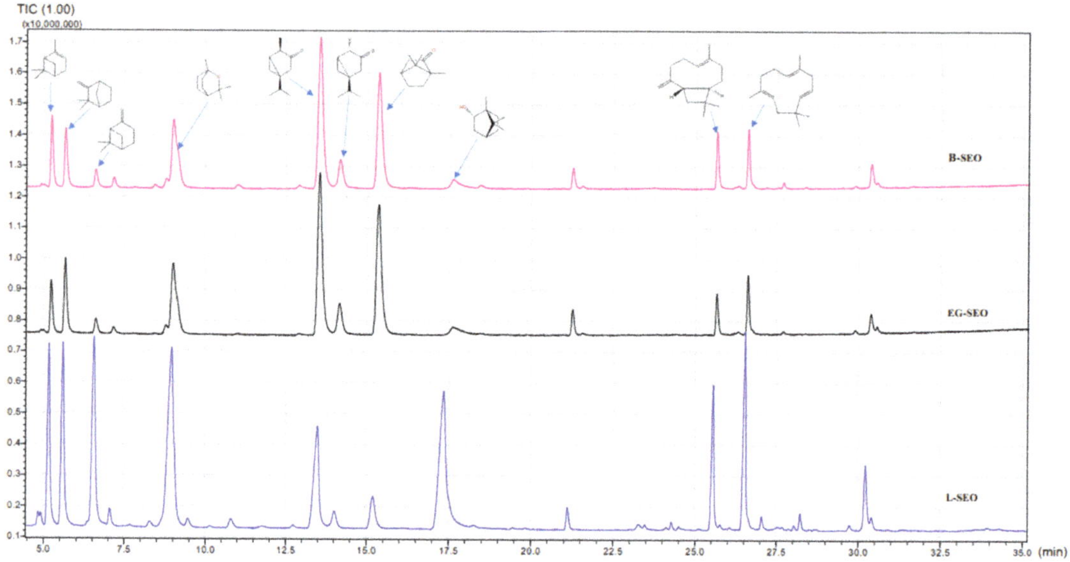

Figure 1. GC-MS chromatograms obtained for the investigated *Salvia officinalis* essential oil samples: B-SEO, pink; EG-SEO, black; L-SEO, blue.

The results obtained follow the data presented in the literature [43]. The monoterpenes are the most represented, whilst the differences between the hydrocarbons (1,8-cineole, 17.98%; β-pinene, 10.52%; camphene, 8.73%) and the oxygenated (borneol, 15.86%; β-thujone, 1.34%) ones are modest. The sesquiterpenes have a prevalence lower than 9% (α-humulene, 8.64%; β-caryophyllene, 5.66%; caryophyllene oxide, 0.22%). The analyzed samples have a different composition in comparison to major compounds. B-SEO and EG-SEO have two major compounds (namely α-thujone and camphor), whilst L-SEO has four major compounds (namely 1,8-cineol, borneol, β-pinene, and α-thujone). L-SEO has a closer composition to the marocain one, possibly due to the plant growing conditions. B-SEO and EG-SEO have a composition similar to the EO obtained in Turkey, as shown in Table 1.

The differences in the chemical composition of the analyzed EOs could be attributed to several factors such as environmental factors, growth conditions, and time of harvesting [19,43,54]. The results obtained in this present study agree with previous research that concluded that 1,8-cineole, camphor, and α- and β-thujone are the main components of *Salvia officinalis* EOs [17–19,37]. As shown earlier, (+)-borneol is the only enantiomer found in the EO of Salvia species [55]; this compound is associated with analgesic and sedative effects [56].

3.1.2. ATR-FTIR Spectroscopy

The obtained Attenuated Total Reflectance-Fourier Transform Infrared (ATR-FTIR) spectra for the investigated *Salvia officinalis* L. EOs are depicted in Figure 2, and the wavelengths values (in the 600–4000 cm^{-1} range) for all recorded peaks are presented in Table 2. The different chemical compositions of the investigated EOs samples led to obtaining major differences in the intensity of the peaks located at the following wavelengths (L-SEO/B-SEO/ED-SEO): 3459/3474/3474 cm^{-1}, 1741/1741/1741 cm^{-1}, 1642/1637/1638 cm^{-1}, 1054/-/- cm^{-1}, -/1045/1045 cm^{-1}, 982/983/983 cm^{-1}, and 877/880/879 cm^{-1}.

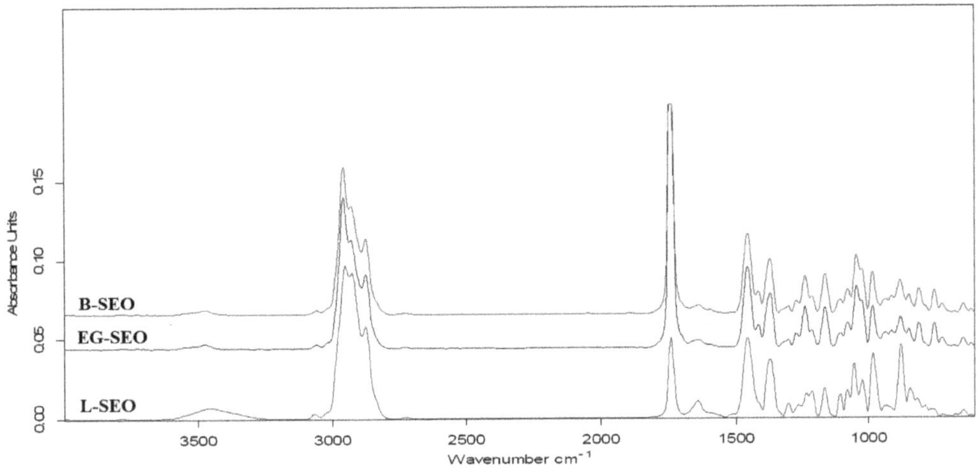

Figure 2. The ATR-FTIR spectra of the *Salvia officinalis* L. essential oil samples.

Table 2. The ATR-FTIR absorption band for SEO samples and vibrational assignments.

Wavenumber (cm^{-1}) of ATR-FTIR Recorded Bands			Vibrational Assignment
L-SEO	**B-SEO**	**EG-SEO**	
3459	3474	3474	(O-H) stretching vibration [57]
3066 2951 2926 2875	3058 2958 2929 2875	3058 2958 2930 2875	(CH) stretching vibration of CH_3 and CH_2 (Csp^3 and Csp^2) [57]
1741 1642	1741 1637	1741 1638	(C=O) stretching vibration in carbonyl group; (C=C) stretching vibration in (>C=CH_2), (-CH=CH-), and (-CH=C<) alkyl groups [20,56–59]
1456 - 1371 1303 1266 1236 1214 1166 1106 1080 1054 - 1022	1455 1415 1371 1303 1274 1241 1215 1165 1104 1077 - 1045 1022	1455 1415 1371 1303 1274 1241 1215 1165 1104 1078 - 1045 1022	(C-H) symmetric and asymmetric bending vibration of (CH_3) and (CH_2) groups, (C-H) in-plane bending, (C-O) symmetric and asymmetric stretching vibration, (O-H) in-plane bending, (CH_3(CO)) symmetric bending, (C-O-C) symmetric and asymmetric stretching [20,57,59,60]
982 877 844 817 759 642	983 880 847 810 750 642	983 879 848 810 750 642	(CH_2) and (CH) out-of-plane wagging, (O-H) out-of-plane banding vibration [57,59–61]

3.1.3. Antioxidant Capacity

The antioxidant capacity of the sage EO (sample L-SEO) was evaluated by using two methods: DPPH assay (inhibition 33.61 ± 2.12%, antioxidant activity 0.81 ± 0.11 mg Trolox/L) and ABTS assay (inhibition 84.50 ± 2.23%, antioxidant activity 0.81 ± 0.03 mmol TEAC/L). All analyses were performed in triplicate. The data are expressed as mean ± STDEV.

3.2. Patients' Characteristics

The mean age of the patients from the control group was 58.74 years whilst that of the salvia EO group was 52.34 years, with a p-value of 0.3530 (t-test). There are no significant differences between the two groups with respect to the following characteristics: gender, weight, educational level, social status, health status in the last year, marital status, the use of aroma indoor and perfumes, the use of sedatives/anxiolytics, and the presence of some allergies (Table 3). Between the groups, there is a significant difference in the variables of the consumption of alcohol, smoking, and the presence of chronic disease.

Table 3. Patients' demographic and clinical characteristics.

Variable	Control Group	Salvia EO Group	Total	p-Value
	n (%)			
Sex (Female/Male)	14/36	46/78	174 (100)	
Weight (Kg)	76.80 ± 13.74	74.98 ± 14.74		0.2173 *
Educational level				0.5877 **
Middle school	7 (14)	11 (9)	18 (10)	
High school	25 (50)	68 (55)	93 (53)	
University	18 (36)	45 (36)	63 (36)	
Social status				0.3032 **
Social aid	3 (6)	7 (6)	10 (17)	
Active	24 (48)	75 (60)	99 (57)	
Retired	23 (46)	42 (34)	65 (37)	
Health status in the last year				0.2839 **
Excellent	13 (26)	34 (27)	47 (27)	
Very good	5 (10)	24 (19)	29 (17)	
Good	17 (34)	36 (29)	53 (30)	
Bad	15 (30)	26 (21)	41 (24)	
Very bad	0 (0)	4 (3)	4 (2)	
Do you use fragrances/aroma in rooms/cars?				0.1135 **
Daily	22 (44)	63 (51)	85 (49)	
1–3 times/week	5 (10)	23 (19)	28 (16)	
No	23 (46)	38 (31)	61 (35)	
Do you use perfumes?				0.4354 **
Daily	18 (36)	57 (46)	75 (43)	
1–3 times/week	22 (44)	49 (40)	71 (41)	
No	10 (20)	18 (15)	28 (16)	
Do you smoke?				0.0046 **
Daily	7 (14)	46 (37)	53 (30)	
Occasionally	18 (36)	42 (34)	60 (34)	
No	25 (50)	36 (29)	61 (35)	
Do you drink alcoholic drinks?				0.1561 ***
Daily	0 (0)	1 (1)	1 (1)	
Occasionally	10 (20)	51 (41)	61 (35)	
No	40 (80)	72 (58)	112 (64)	
Do you use sedatives/anxiolytics?				0.0951 ***
Daily	0 (0)	0 (0)	0 (0)	
Occasionally	15 (30)	31 (25)	46 (26)	
No	35 (70)	93 (75)	128 (74)	
Do you suffer from a chronic disease?				0.6036 ****
Yes	19 (38)	42 (34)	61 (35)	
No	31 (62)	82 (66)	113 (65)	
Do you suffer from any drug/food allergies?				0.0186 ****
Yes	9 (18)	46 (37)	55 (32)	
No	41 (82)	78 (63)	119 (68)	

n—number of patients; * t-test; ** chi-squared test; *** Kruskal-Walis test; **** Fisher's exact test.

The patients' perception of the quality of in-hospital services and, thus, their well-being after receiving treatment while hospitalized were assessed by answering questions designed with seven possible answers: very weak, weak, acceptable, improved, good, very good, and excellent. Group comparison was made using the Mann-Whitney U test. No significant differences were observed in the between-group comparison ($p = 0.8969$, U = 23). Most patients in both groups rated in-hospital services as excellent and very good (64% in the control group and 76% in the aromatherapy group) (Figure 3). No adverse effects were reported by patients or observed by medical staff in this study.

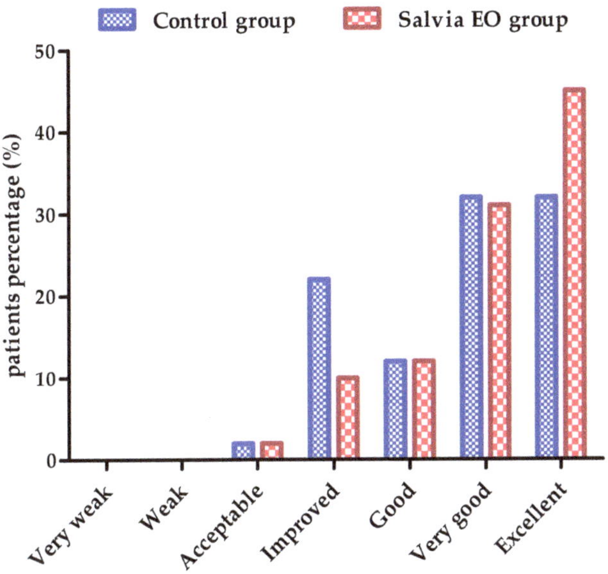

Figure 3. Patients' satisfaction with the treatment they received during their current hospitalization (Mann-Whitney U test).

4. Discussion

Much research has focused on investigating sage EO extracted from sage leaves harvested from diverse parts of the world. The percentage of the constituents varies widely due to geographical region, season, environmental conditions, genetic differences, phenological stages, sampling, and extraction methods. The number of detected constituents varies, being around 14–67 (14 with concentrations > 1% [41], 15 [62], 22 [17], 23 [43], 42 [19], 44 [63], 67 [64]). Also, sage EO's composition is dependent on the growing ecosystem conditions, as determined in an exhaustive study reported by Russo et al., where 18 samples were collected in 2008–2009 in the south-central part of Italy and investigated [41]. The EOs obtained from sage leaves contained α-thujone (7.8–20.1%), camphor (8.4–20.8%), borneol (2.5–16.9%), γ-murolene (2.9–13.8%), and sclareol (5.9–23.1%) as the major compounds. Recently, another study identified oxygenated monoterpenes (67.7%) and monoterpenes hydrocarbons (19.1%) in the EO from sage leaves harvested in June 2020 in Tuscany, Italy. The main components identified in this EO batch were 1,8-cineole (30.3%), camphor (17.1%), α-thujone (9.7%), camphene (7.9%), and chrysanthenone (6.8%) [43].

Due to their chemical composition (shown in Table 1), two of the investigated samples (B-SEO and EG-SEO) were included in the more common C1c chemotype, as classified by Craft et al. [65]: α-thujone > camphor > 1,8-cineole > β-thujone, which was reported in samples from Turkey [17], Brazil [16], Mexico [65], and Croatia [66]. Meanwhile, the L-SEO sample is a chemotype found in the sage EO from the Dalmatian region, where 1,8-cineole is the main component and borneol is in high quantity [67].

The reported FTIR results of the EOs show that the components found in higher concentrations dominate the resulting vibrational spectra, while the components found in low concentrations do not have significant influence [54,68,69]. Ciko et al. [20] showed that FTIR analysis of the EO obtained from *Salvia officinalis* L. leaves indicated the presence of monoterpenes such as thujones, camphor, 1,8-cineole, and pinene.

The GC-MS results (Table 1) obtained for L-SEO sample indicate 1,8-cineol (17.98%) and borneol (15.86%) as their main components, which can influence the features of the recorded ATR-FTIR spectrum and the intensity of the obtained bands (Figure 2). The 1,8-cineole (eucalyptol) can be distinguished by the following key bands: 1371 cm^{-1} (δ_{sym}(CH_3-CO) symmetric in plan bending/scissoring), 1266 cm^{-1} (ν(C-O) stretching from alkyl ether, 1214 cm^{-1} (ν_{asym}(C-O-C) asymmetric stretching), 1080 cm^{-1} (ν_{sym}(C-O-C) symmetric stretching), and 982 cm^{-1} (ω_{asym}(CH_2) asymmetric out-of-plane bending/wagging vibration). Due to the higher content of 1,8-cineole in L-SEO, the 982 cm^{-1} band is more intense than in B-SEO and EG-SEO, with 844 cm^{-1} corresponding to ω_{asym}(C-H) asymmetric out-of-plane bending/wagging vibration [20,70,71].

For borneol, the characteristics bands are 3459 cm^{-1} (ν_{sym}(O-H) stretching from H-bonded alcohols), 1303 cm^{-1} (δ_{sym}(O-H) scissoring vibration), 1166 cm^{-1} (δ_{sym}(C-OH) in-plane bending of alcohol groups), 1054 cm^{-1} (ν_{asym}(C-O) asymmetric stretching from primary alcohols), 1106 cm^{-1} (ν_{sym}(C-O) symmetric stretching deformation), and 642 cm^{-1} (ω(C-O) out-of-plane bending) [20,30].

Other classes of compounds that occur in high proportions in the composition of L-SEO are bicyclic monoterpenes with =CH_2 alkyl groups and sesquiterpenes with C=C bonds (sabinene + camphene, 8.87%; β-caryophyllene + α-humulene, 14.3%), with the key bands located at 3066 cm^{-1} (ν_{sym}(=C-H), Csp^2 from the terminal vinyl group) and 1642 cm^{-1} (ν(C=CH_2) stretching vibration). Schultz et. al. reported a band due to the stretching vibration of (C=C) located at 1653 cm^{-1} for sabinene and 1635 cm^{-1} for β-caryophyllene [70], while Agatonovik-Kustin et al. mentioned a vinyl group vibration at 1635–1650 cm^{-1}, although the intensity of this peak was very low [70].

Compared to B-SEO and EG-SEO samples, in the case of L-SEO, the 1642 cm^{-1} band is more evident due to the higher content of compounds with =CH_2 and C=C groups, with 1456 cm^{-1} corresponding to δ_{sym}(C-H) and δ_{asym}(C-H) in-plane bending of CH_3 and -CH_2 groups. The C=CH_2 in-plane deformation vibration was not recorded as a separate band near ~1410–1420 cm^{-1} [20], probably being hidden under the -CH_3 and -CH_2 absorption bands, with 877 cm^{-1} corresponding to ν(=CH_2) methylene out-of-plane deformation due to the strained ring structure with an exocyclic =CH_2 group [70]. The higher intensity of the 877 cm^{-1} band, for the L-SEO compared with B-SEO and EG-ESO, is due to the higher content of compounds with the exocyclic =CH_2 group.

In addition, it is mentioned that in the 2850–3000 cm^{-1} wavelength range, a group of characteristic bands with strong intensity were recorded (2951 cm^{-1}, 2926 cm^{-1}, 2875 cm^{-1}) due to the symmetric and asymmetric stretching vibration of (C-H) bond from CH_3, CH_2, and CH. The intensities of these bands are higher for L-SEO than for B-SEO and EG-SEO.

For the B-SEO and EG-SEO samples, the obtained GC-MS results (Table 2) reveal that thujone and (+) camphor are the major components (50.77%) that influence the absorption band intensity. These compounds are monoterpenes with a carbonyl group (>C=O), belonging to the ketones class. In the obtained B-SEO and EG-SEO ATR-FTIR spectra (Figure 3), the band positioned at 1741 cm^{-1} is attributed to thujones and camphor being specific for (ν(C=O)) stretching vibration of 5-membered cyclic ketones [18,23]. Compared with the L-SEO sample, in the case of B-SEO and EG-SEO, the intensity of the 1741 cm^{-1} band is higher due to the different chemical compositions of the EOs.

Other absorption bands that showed changes in intensity compared to L-SEO sample, due to the difference in the chemical composition of B-SEO and EG-SEO, were observed at 3474 cm^{-1} due to (ν(O-H)) stretching vibration, weak band; 1415 cm^{-1} due to (ν(C=CH_2) in-plane deformation vibration; 1241 cm^{-1} due to (ν(C-O)) stretching vibration; and 1045 cm^{-1} due to (ν(C-O)) stretching vibration from primary alcohols [69].

The main efficient constituents of *Salvia essential* oils are terpenes and polyphenols, with one major bioactivity being represented by their antioxidant action. The antioxidant capacity of the investigated sage EO (L-SEO) reached 33.61% and 84.50% by DPPH and ABTS analyses, respectively, which are values specific to samples rich in 1,8-cineole (60–70% inhibition in [69–72]).

The use of plant-derived essential oils for wellness and to sustain human health is called aromatherapy. In British herbal encyclopedias, the species of the Salvia family have long been renowned for being characterized as compounds that improve memory. Salvia essential oils can be helpful for the enhancement of cognition and mood, as demonstrated by various studies [73].

In this randomized, controlled study, the participants were adult men and women undergoing hospitalization who received or did not receive EOs for daily inhalation. The demographic and clinical data (shown in Table 3) revealed no statistical differences between the groups.

Our results revealed that the level of patient perception of the quality of in-hospital services, when sage EO is inhaled daily by patients, during hospitalization, is not statistically significant, as compared to that of the patients from the control group. The aromatherapy confirmed its beneficial action as a complementary treatment to increase the well-being of patients during hospitalization, as it showed a significant effect on hospitalized patients who evaluated their level of satisfaction as excellent (Figure 3).

The main limitations of this study consist of the somewhat limited number of patients in each group and a lack of medium/long-term follow-up. Further studies with a more extended follow-up period and more in-depth research on the use of sage EO in massage are needed to confirm our findings. Additionally, it is worth noting that the study did not include post-surgery patients. Therefore, our future goal is to develop randomized clinical trials to evaluate the large-scale efficacy of aromatherapy as an alternative therapy in patients with diverse pathologies. On the other hand, some strengths of this study must be highlighted, including the use of a chemically characterized EO in aromatherapy and the increase in some of the patients' satisfaction regarding their hospitalization experience with this approach.

5. Conclusions

The GC-MS analysis of three samples of *Salvia officinalis* L. EOs indicated the presence of terpenes such as 1,8-cineole, thujones, borneol, camphor, sabinene, camphene, and caryophyllenes as the main components. The FTIR results agree with the GS-MS data of investigated samples, which leads to the conclusion that two commercial samples are part of the α-thujone > camphor > 1,8-cineole > β-thujone chemotype. In contrast, the other, in-house sample is part of the 1,8-cineole, borneol chemotype (Dalmatian type). The chemical analysis of EOs used in aromatherapy mechanism is very important, as different biological outcomes could arise based on the major compounds identified. The present results did not reveal a statistically significant difference between groups that inhaled sage EO during hospitalization versus the group that did not. However, more research is needed to investigate the mechanisms responsible for specific effects attributed to EOs to justify the therapeutic effect of sage EO and identify the side effects related to this product.

Author Contributions: Conceptualization, D.M.T. and D.M.C.; methodology, M.-D.M., A.I.L., C.M., D.C. and D.M.C.; software, A.I.L., C.M., D.C. and L.C.; validation, A.I.L., C.M., D.C., D.M.C. and L.C.; formal analysis, A.I.L., C.M., A.-M.B., D.C. and L.C.; investigation, M.-D.M., A.I.L., C.M., D.C. and D.M.C.; data curation, A.I.L., D.C., A.-M.B. and D.M.C.; writing—original draft preparation, M.-D.M., S.G., M.A.B. and D.M.C.; writing—review and editing, D.M.T., S.G.B. and D.M.C.; supervision, S.G.B. and D.M.C.; project administration, D.M.C.; funding acquisition, A.I.L. and L.C. All authors have read and agreed to the published version of the manuscript.

Funding: This work was supported by a grant of the Romanian Ministry of Education and Research, CNCS—UEFISCDI, project number PN-III-P1-1.1-PD-2019-0349, within PNCDI III.

Institutional Review Board Statement: The study was conducted according to the guidelines of the Declaration of Helsinki and was approved by the Ethics Committee of Lipova City Hospital (Spitalul Orasenesc Lipova), Romania (no. 62/1 July 2019).

Informed Consent Statement: Informed consent was obtained from all subjects involved.

Data Availability Statement: Data is contained within the article.

Acknowledgments: The authors wish to thank Lisa Craiut for the professional editing of this paper.

Conflicts of Interest: The authors declare no conflict of interest.

References

1. Yuan, R.; Zhang, D.; Yang, J.; Wu, Z.; Luo, C.; Han, L.; Yang, F.; Lin, J.; Yang, M. Review of Aromatherapy Essential Oils and Their Mechanism of Action Against Migraines. *J. Ethnopharmacol.* **2021**, *265*, 113326. [CrossRef] [PubMed]
2. Winkelman, W.J. Aromatherapy, Botanicals, and Essential Oils in Acne. *Clin. Dermatol.* **2018**, *36*, 299–305. [CrossRef] [PubMed]
3. Lizarraga-Valderrama, L.R. Effects of Essential Oils on Central Nervous System: Focus on Mental Health. *Phytother. Res.* **2021**, *35*, 657–679. [CrossRef]
4. Zeinalian, M.; Eshaghi, M.; Hadian, M.; Naji, H.; Marandi, S.M.M.; Asgary, S. Eight Essential Foods in Iranian Traditional Medicine and their Role in Health Promotion and Well-being. *Int. J. Prev. Med.* **2017**, *8*, 2. [CrossRef] [PubMed]
5. Contrada, M.; Cerasa, A.; Tonin, P.; Bagetta, G.; Scuteri, D. Aromatherapy in Stroke Patients: Is It Time to Begin? *Front. Behav. Neurosci.* **2021**, *15*, 749353. [CrossRef] [PubMed]
6. Fismer, K.L.; Pilkington, K. Lavender and Sleep: A systematic Review of the Evidence. *Eur. J. Integr. Med.* **2012**, *4*, E436–E447. [CrossRef]
7. Lillehei, A.S.; Halcon, L.L. A Systematic Review of the Effect of Inhaled Essential Oils on Sleep. *J. Altern. Complement. Med.* **2014**, *20*, 441–451. [CrossRef]
8. Hawkins, J.; Hires, C.; Dunne, E.; Baker, C. The Relationship between Lavender and Tea Tree Essential Oils and Pediatric Endocrine Disorders: A Systematic Review of the Literature. *Complement. Ther. Med.* **2020**, *49*, 102288. [CrossRef]
9. Zhang, Y.; Long, Y.; Yu, S.; Li, D.; Yang, M.; Guan, Y.; Zhang, D.; Wan, J.; Liu, S.; Shi, A.; et al. Natural Volatile Oils Derived from Herbal Medicines: A Promising Therapy Way for Treating Depressive Disorder. *Pharmacol. Res.* **2021**, *164*, 105376. [CrossRef]
10. Almanaa, T.N.; Alharbi, N.S.; Ramachandran, G.; Chelliah, C.K.; Rajivgandhi, G.; Manoharan, N.; Kadaikunnan, S.; Khaled, J.M.; Alanzi, K.F. Anti-biofilm Effect of *Nerium oleander* Essential Oils Against Biofilm Forming *Pseudomonas aeruginosa* on Urinary Tract Infections. *J. King Saud Univ. Sci.* **2021**, *33*, 101340. [CrossRef]
11. Brandao, F.R.; Farias, C.F.S.; Souza, D.C.D.; de Oliveira, M.I.B.; de Matos, L.V.; Majolo, C.; de Oliveira, M.R.; Chaves, F.C.M.; O'Sullivan, F.L.D.; Chagas, E.C. Anesthetic Potential of the Essential Oils of *Aloysia triphylla, Lippia sidoides* and *Mentha piperita* for *Colossoma macropomum*. *Aquaculture* **2021**, *534*, 736275. [CrossRef]
12. Valdivieso-Ugarte, M.; Plaza-Diaz, J.; Gomez-Llorente, C.; Gomez, E.L.; Sabes-Alsina, M.; Gil, A. In Vitro Examination of Antibacterial and Immunomodulatory Activities of Cinnamon, White Thyme, and Clove Essential Oils. *J. Funct. Foods* **2021**, *81*, 104436. [CrossRef]
13. Chraibi, M.; Fadil, M.; Farah, A.; Lebrazi, S.; Fikri-Benbrahim, K. Antimicrobial Combined Action of *Mentha pulegium, Ormenis mixta* and *Mentha piperita* Essential Oils Against *S. aureus, E. coli* and *C. tropicalis*: Application of Mixture Design Methodology. *LWT-Food Sci. Technol.* **2021**, *145*, 111352. [CrossRef]
14. Rocha, R.R.; Matos, M.N.C.; Guerrero, J.A.P.; Cavalcante, R.M.B.; Melo, R.S.; Azevedo, A.M.A.; Pereira, A.M.G.; Lopes, P.H.R.; Rodrigues, T.H.S.; Bandeira, P.N.; et al. Comparative Study of the Chemical Composition, Antibacterial Activity and Synergic Effects of the Essential Oils of *Croton tetradenius* Baill. and *C. pulegiodorus* Baill. Against *Staphylococcus aureus* Isolates. *Microb. Pathog.* **2021**, *156*, 104934. [CrossRef]
15. Bogdan, M.A.; Bungau, S.; Tit, D.M.; Zaha, D.C.; Nechifor, A.C.; Behl, T.; Chambre, D.; Lupitu, A.I.; Copolovici, L.; Copolovici, D.M. Chemical Profile, Antioxidant Capacity, and Antimicrobial Activity of Essential Oils Extracted from Three Different Varieties (Moldoveanca 4, Vis Magic 10, and Alba 7) of *Lavandula angustifolia*. *Molecules* **2021**, *26*, 4381. [CrossRef] [PubMed]
16. Delamare, A.P.L.; Moschen-Pistorello, I.T.; Artico, L.; Atti-Serafini, L.; Echeverrigaray, S. Antibacterial Activity of the Essential Oils of *Salvia officinalis* L. and *Salvia triloba* L. Cultivated in South Brazil. *Food Chem.* **2007**, *100*, 603–608. [CrossRef]
17. Kulak, M.; Gul, F.; Sekeroglu, N. Changes in Growth Parameter and Essential Oil Composition of Sage (*Salvia officinalis* L.) Leaves in Response to Various Salt Stresses. *Ind. Crops Prod.* **2020**, *145*, 112078. [CrossRef]
18. Rzepa, J.; Wojtal, L.; Staszek, D.; Grygierczyk, G.; Labe, K.; Hajnos, M.; Kowalska, T.; Waksmundzka-Hajnos, M. Fingerprint of Selected Salvia Species by HS-GC-MS Analysis of Their Volatile Fraction. *J. Chromatogr. Sci.* **2009**, *47*, 575–580. [CrossRef]
19. Mohamed, A.; Mustafa, A. Gas Chromatography-Mass Spectrometry (GC-MS) Analysis of Essential Oil *Salvia officinalis* in Sudan. *J. Multidis. Res. Rev.* **2019**, *1*, 43–45.
20. Ciko, L.; Andoni, A.; Ylli, F.; Plaku, E.; Taraj, K.; Armand, C. Extraction of Essential Oil from Albanian *Salvia officinalis* L. and its Characterization by FTIR Spectroscopy. *Asian J. Chem.* **2016**, *28*, 1401–1402. [CrossRef]
21. Glevitzky, I.; Dumitrel, G.A.; Glevitzky, M.; Pasca, B.; Otrisal, P.; Bungau, S.; Cioca, G.; Pantis, C.; Popa, M. Statistical Analysis of the Relationship between Antioxidant Activity and the Structure of Flavonoid Compounds. *Rev. Chim.* **2019**, *70*, 3103–3107. [CrossRef]
22. Rafi, N.; Khodadadizadeh, A.; Nematabad, M.S.; Sayadi, A.R. The Evaluation of the Effect of Aromatherapy with Lavender Essential Oil on the Quality of Sleep of Cardiac Patients Candidate for Angiography. *Pak. J. Med. Health Sci.* **2020**, *14*, 1143–1147.

23. Russo, P.; Frustaci, A.; Del Bufalo, A.; Fini, M.; Cesario, A. From Traditional European Medicine to Discovery of New Drug Candidates for the Treatment of Dementia and Alzheimer's Disease: Acetylcholinesterase Inhibitors. *Curr. Med. Chem.* **2013**, *20*, 976–983. [CrossRef] [PubMed]
24. Bagheri, H.; Salmani, T.; Nourian, J.; Mirrezaie, S.M.; Abbasi, A.; Mardani, A.; Vlaisavljevic, Z. The Effects of Inhalation Aromatherapy Using Lavender Essential Oil on Postoperative Pain of Inguinal Hernia: A Randomized Controlled Trial. *J. Perianesthesia Nurs.* **2020**, *35*, 642–648. [CrossRef]
25. Mohr, C.; Jensen, C.; Padden, N.; Besel, J.M.; Brant, J.M. Peppermint Essential Oil for Nausea and Vomiting in Hospitalized Patients: Incorporating Holistic Patient Decision Making Into the Research Design. *J. Holist. Nurs.* **2021**, *39*, 126–134. [CrossRef]
26. Bertone, A.C.; Dekker, R.L. Aromatherapy in Obstetrics: A Critical Review of the Literature. *Clin. Obstet. Gynecol.* **2021**, *64*, 572–588. [CrossRef]
27. Karan, N.B. Influence of lavender oil inhalation on vital signs and anxiety: A randomized clinical trial. *Physiol. Behav.* **2019**, *211*, 112676. [CrossRef]
28. Lordani, T.V.A.; de Lara, C.E.; Ferreira, F.B.P.; Monich, M.D.T.; da Silva, C.M.; Lordani, C.R.F.; Bueno, F.G.; Teixeira, J.J.V.; Lonardoni, M.V.C. Therapeutic Effects of Medicinal Plants on Cutaneous Wound Healing in Humans: A Systematic Review. *Mediat. Inflamm.* **2018**, *2018*, 7354250. [CrossRef]
29. Deng, C.; Xie, Y.; Liu, Y.; Li, Y.; Xiao, Y. Aromatherapy Plus Music Therapy Improve Pain Intensity and Anxiety Scores in Patients with Breast Cancer during Perioperative Periods: A Randomized Controlled Trial. *Clin. Breast Cancer* **2021**, *22*, 115–120. [CrossRef]
30. Huang, H.; Wang, Q.; Guan, X.; Zhang, X.; Kang, J.; Zhang, Y.; Zhang, Y.; Zhang, Q.; Li, X. Effect of Aromatherapy on Preoperative Anxiety in Adult Patients: A Meta-Analysis of Randomized Controlled Trials. *Complement. Ther. Clin. Pract.* **2021**, *42*, 101302. [CrossRef]
31. Tang, Y.; Gong, M.; Qin, X.; Su, H.; Wang, Z.; Dong, H. The Therapeutic Effect of Aromatherapy on Insomnia: A Meta-Analysis. *J. Affect. Disord.* **2021**, *288*, 1–9. [CrossRef] [PubMed]
32. Mohammadpourhodki, R.; Sadeghnezhad, H.; Ebrahimi, H.; Basirinezhad, M.H.; Maleki, M.; Bossola, M. The Effect of Aromatherapy Massage with Lavender and Citrus Aurantium Essential Oil on Quality of Life of Patients on Chronic Hemodialysis: A Parallel Randomized Clinical Trial Study. *J. Pain Symptom Manag.* **2021**, *61*, 456–463.e1. [CrossRef] [PubMed]
33. Varney, E.; Buckle, J. Effect of Inhaled Essential Oils on Mental Exhaustion and Moderate Burnout: A Small Pilot Study. *J. Altern. Complement. Med.* **2013**, *19*, 69–71. [CrossRef] [PubMed]
34. Wu, C.-Y.; Lee, H.-F.; Chang, C.W.; Chiang, H.-C.; Tsai, Y.-H.; Liu, H.-E. The Immediate Effects of Lavender Aromatherapy Massage versus Massage in Work Stress, Burnout, and HRV Parameters: A Randomized Controlled Trial. *Evid.-Based Complement. Altern. Med.* **2020**, *2020*, 8830083. [CrossRef] [PubMed]
35. Mesquita, K.D.; Feitosa, B.D.; Cruz, J.N.; Ferreira, O.O.; Franco, C.D.; Cascaes, M.M.; Oliveira, M.S.; Andrade, E.H. Chemical composition and preliminary toxicity evaluation of the essential oil from *Peperomia circinnata* Link var. *circinnata*. (*Piperaceae*) in *Artemia salina* Leach. *Molecules* **2021**, *26*, 7359. [CrossRef] [PubMed]
36. Lima, C.F.; Carvalho, F.; Fernandes, E.; Bastos, M.L.; Santos-Gomes, P.C.; Fernandes-Ferreira, M.; Pereira-Wilson, C. Evaluation of toxic/protective effects of the essential oil of *Salvia officinalis* on freshly isolated rat hepatocytes. *Toxicol. In Vitro Int. J. Publ. Assoc. BIBRA* **2004**, *18*, 457–465. [CrossRef]
37. El Hadri, A.; del Rio, M.G.; Sanz, J.; Coloma, A.G.; Idaomar, M.; Ozonas, B.R.; González, J.B.; Reus, M.I.S. Cytotoxic Activity of alpha-Humulene and Transcaryophyllene from *Salvia officinalis* in Animal and Human Tumor Cells. *An. R. Acad. Nac. Farm.* **2010**, *76*, 343–356.
38. Halicioglu, O.; Astarcioglu, G.; Yaprak, I.; Aydinlioglu, H. Toxicity of *Salvia officinalis* in a newborn and a child: An alarming report. *Pediatr. Neurol.* **2011**, *45*, 259–260. [CrossRef]
39. Ahangari, F.; Farshbaf-Khalili, A.; Javadzadeh, Y.; Adibpour, M.; Oskouei, B.S. Comparing the Effectiveness of *Salvia officinalis*, Clotrimazole and their Combination on Vulvovaginal Candidiasis: A Randomized, Controlled Clinical Trial. *J. Obstet. Gynaecol. Res.* **2019**, *45*, 897–907. [CrossRef]
40. Luca, T.; Napoli, E.; Privitera, G.; Musso, N.; Ruberto, G.; Castorina, S. Antiproliferative Effect and Cell Cycle Alterations Induced by *Salvia officinalis* Essential Oil and Its Three Main Components in Human Colon Cancer Cell Lines. *Chem. Biodivers.* **2020**, *17*, e2000309. [CrossRef]
41. Russo, A.; Formisano, C.; Rigano, D.; Senatore, F.; Delfine, S.; Cardile, V.; Rosselli, S.; Bruno, M. Chemical Composition and Anticancer Activity of Essential Oils of Mediterranean sage (*Salvia officinalis* L.) Grown in Different Environmental Conditions. *Food Chem. Toxicol.* **2013**, *55*, 42–47. [CrossRef] [PubMed]
42. Sertel, S.; Eichhorn, T.; Plinkert, P.K.; Efferth, T. Anticancer Activity of *Salvia officinalis* Essential Oil against HNSCC Cell Line (UMSCC1). *Hno* **2011**, *59*, 1203–1208. [CrossRef] [PubMed]
43. Ovidi, E.; Masci, V.L.; Zambelli, M.; Tiezzi, A.; Vitalini, S.; Garzoli, S. *Laurus nobilis*, *Salvia sclarea* and *Salvia officinalis* Essential Oils and Hydrolates: Evaluation of Liquid and Vapor Phase Chemical Composition and Biological Activities. *Plants* **2021**, *10*, 707. [CrossRef]
44. Ghorbani, A.; Esmaeilizadeh, M. Pharmacological Properties of *Salvia officinalis* and its Components. *J. Tradit. Complement. Med.* **2017**, *7*, 433–440. [CrossRef] [PubMed]
45. Ghaedi, M.; Naghiha, R.; Jannesar, R.; Dehghanian, N.; Mirtamizdoust, B.; Pezeshkpour, V. Antibacterial and Antifungal Activity of Flower Extracts of *Urtica dioica*, *Chamaemelum nobile* and *Salvia officinalis*: Effects of $Zn[OH]_2$ Nanoparticles and Hp-2-minh on their Property. *J. Ind. Eng. Chem.* **2015**, *32*, 353–359. [CrossRef]

46. Pearson, A.C.S.; Cutshall, S.M.; Hooten, W.M.; Rodgers, N.J.; Bauer, B.A.; Bhagra, A. Perspectives on the Use of Aromatherapy from Clinicians Attending an Integrative Medicine Continuing Education Event. *BMC Complement. Altern. Med.* **2019**, *19*, 174 [CrossRef] [PubMed]
47. Cho, E.H.; Lee, M.-Y.; Hur, M.-H. The Effects of Aromatherapy on Intensive Care Unit Patients' Stress and Sleep Quality: A Nonrandomised Controlled Trial. *Evid.-Based Complement. Altern. Med.* **2017**, *2017*, 2856592. [CrossRef] [PubMed]
48. Nasab, F.R.S.; Shahrbabaki, P.M.; Dehghan, M.; Tajadini, H.; Baniasadi, H.; Sabzevari, S. Effect of Abdominal Massage with and without *Salvia officinalis* on Nausea and Vomiting in Patients with Cancer Undergoing Chemotherapy: A Randomized Clinical Trial. *J. Oncol.* **2021**, *2021*, 9989228. [CrossRef]
49. Jodaki, K.; Abdi, K.; Mousavi, M.S.; Mokhtari, R.; Asayesh, H.; Vandali, V.; Golitaleb, M. Effect of *Rosa damascene* Aromatherapy on Anxiety and Sleep Quality in Cardiac Patients: A Randomized Controlled Trial. *Complement. Ther. Clin. Pract.* **2021**, *42*, 101299. [CrossRef]
50. Moslemi, F.; Alijaniha, F.; Naseri, M.; Kazemnejad, A.; Charkhkar, M.; Heidari, M.R. *Citrus aurantium* Aroma for Anxiety in Patients with Acute Coronary Syndrome: A Double-Blind Placebo-Controlled Trial. *J. Altern. Complement. Med.* **2019**, *25*, 833–839. [CrossRef]
51. Gudmundsson, G.; Gislason, T.; Janson, C.; Lindberg, E.; Ulrik, C.S.; Brondum, E.; Nieminen, M.M.; Aine, T.; Hallin, R.; Bakke, P. Depression, Anxiety and Health Htatus after Hospitalisation for COPD: A Multicentre Study in the Nordic Countries. *Respir. Med.* **2006**, *100*, 87–93. [CrossRef] [PubMed]
52. Beicu, R.; Alexa, E.; Obistioiu, D.; Cocan, I.; Imbrea, F.; Pop, G.; Circioban, D.; Moisa, C.; Lupitu, A.; Copolovici, L.; et al. Antimicrobial Potential and Phytochemical Profile of Wild and Cultivated Populations of Thyme (*Thymus* sp.) Growing in Western Romania. *Plants* **2021**, *10*, 1833. [CrossRef] [PubMed]
53. WMA Declaration of Helsinki-Ethical Principles for Medical Research Involving Human Subjects. Available online: https://www.wma.net/policies-post/wma-declaration-of-helsinki-ethical-principles-for-medical-research-involving-human-subjects/ (accessed on 23 June 2019).
54. Moisa, C.; Lupitu, A.; Pop, G.; Chambre, D.R.; Copolovici, L.; Cioca, G.; Bungau, S.; Copolovici, D.M. Variation of the Chemical Composition of Thymus Vulgaris Essential Oils by Phenological Stages. *Rev. De Chim.* **2019**, *70*, 633–637. [CrossRef]
55. Demirci, B.; Tabanca, N.; Baser, K.H.C. Enantiomeric Distribution of Some Monoterpenes in the Essential Oils of some Salvia Species. *Flavour Fragr. J.* **2002**, *17*, 54–58. [CrossRef]
56. Granger, R.E.; Campbell, E.L.; Johnston, G.A.R. (+)- And (−)-Borneol: Efficacious Positive Modulators of GABA Action at Human Recombinant $\alpha1\beta2\gamma2L$ GABAA Receptors. *Biochem. Pharmacol.* **2005**, *69*, 1101–1111. [CrossRef] [PubMed]
57. Coates, J. Interpretation of Infrared Spectra, A Practical Approach. In *Encyclopedia of Analytical Chemistry*; Meyers, R.A., McKelvy, M.L., Eds.; John Wiley & Sons, Ltd.: New York, NY, USA, 2006. [CrossRef]
58. Schulz, H.; Baranska, M. Identification and quantification of valuable plant substances by IR and Raman spectroscopy. *Vib. Spectrosc.* **2007**, *43*, 13–25. [CrossRef]
59. Schulz, H.; Özkan, G.; Baranska, M.; Krüger, H.; Özcan, M. Characterisation of essential oil plants from Turkey by IR and Raman spectroscopy. *Vib. Spectrosc.* **2005**, *39*, 249–256. [CrossRef]
60. Chambre, D.R.; Moisa, C.; Lupitu, A.; Copolovici, L.; Pop, G.; Copolovici, D.M. Chemical composition, antioxidant capacity, and thermal behavior of *Satureja hortensis* essential oil. *Sci. Rep.* **2020**, *10*, 21322. [CrossRef]
61. Agatonovic-Kustrin, S.; Ristivojevic, P.; Gegechkori, V.; Litvinova, T.M.; Morton, D.W. Essential Oil Quality and Purity Evaluation via FT-IR Spectroscopy and Pattern Recognition Techniques. *Appl. Sci.* **2020**, *10*, 7294. [CrossRef]
62. Samani, M.R.; Pirbalouti, A.G.; Moattar, F.; Golparvar, A.R. L-Phenylalanine and Bio-Fertilizers Interaction Effects on Growth, Yield and Chemical Compositions and Content of Essential Oil From the Sage (*Salvia officinalis* L.) Leaves. *Ind. Crops Prod.* **2019**, *137*, 1–8. [CrossRef]
63. Ed-Dra, A.; Filali, F.R.; Lo Presti, V.; Zekkori, B.; Nalbone, L.; Bouymajane, A.; Trabelsi, N.; Lamberta, F.; Bentayeb, A.; Giuffrida, A.; et al. Chemical Composition, Antioxidant Capacity and Antibacterial Action of Five Moroccan Essential Oils Against Listeria Monocytogenes and Different Serotypes of *Salmonella enterica*. *Microb. Pathog.* **2020**, *149*, 104510. [CrossRef] [PubMed]
64. Hayouni, E.A.; Chraief, I.; Abedrabba, M.; Bouix, M.; Leveau, J.Y.; Mohammed, H.; Hamdi, M. Tunisian *Salvia officinalis* L. and *Schinus molle* L. Essential Oils: Their Chemical Compositions an Their Preservative Effects Against *Salmonella* Inoculated in Minced Beef Meat. *Int. J. Food Microbiol.* **2008**, *125*, 242–251. [CrossRef]
65. Craft, J.D.; Satyal, P.; Setzer, W.N. The Chemotaxonomy of Common Sage (*Salvia officinalis*) Based on the Volatile Constituents. *Medicines* **2017**, *4*, 47. [CrossRef] [PubMed]
66. Jug-Dujaković, M.; Ristić, M.; Pljevljakušić, D.; Dajić-Stevanović, Z.; Liber, Z.; Hančević, K.; Radić, T.; Šatović, Z. High Diversity of Indigenous Populations of Dalmatian Sage (*Salvia officinalis* L.) in Essential-Oil Composition. *Chem. Biodivers.* **2012**, *9*, 2309–2323. [CrossRef] [PubMed]
67. Kuštrak, D.; Kuftinec, J.; Blažević, N. Yields and Composition of Sage Oils from Different Regions of the Yugoslavian Adriatic Coast. *J. Nat. Prod.* **1984**, *47*, 520–524. [CrossRef]
68. Valderrama, A.C.S.; Rojas De, G.C. Traceability of Active Compounds of Essential Oils in Antimicrobial Food Packaging Using a Chemometric Method by ATR-FTIR. *Am. J. Anal. Chem.* **2017**, *8*, 726–741. [CrossRef]
69. Miguel, G.; Cruz, C.; Faleiro, M.L.; Simoes, M.T.F.; Figueiredo, A.C.; Barroso, J.G.; Pedro, L.G. *Salvia officinalis* L. Essential Oils: Effect of Hydrodistillation Time on the Chemical Composition, Antioxidant and Antimicrobial Activities. *Nat. Prod. Res.* **2011**, *25*, 526–541. [CrossRef]

70. Abou Baker, D.H.; Amarowicz, R.; Kandeil, A.; Ali, M.A.; Ibrahim, E.A. Antiviral Activity of *Lavandula angustifolia* L. and *Salvia officinalis* L. Essential Oils Against Avian Influenza H5N1 Virus. *J. Agric. Food Res.* **2021**, *4*, 100135. [CrossRef]
71. Ben Farhat, M.; Jordan, M.J.; Chaouech-Hamada, R.; Landoulsi, A.; Sotomayor, J.A. Variations in Essential Oil, Phenolic Compounds, and Antioxidant Activity of Tunisian Cultivated *Salvia officinalis* L. *J. Agric. Food Chem.* **2009**, *57*, 10349–10356. [CrossRef]
72. Bozin, B.; Mimica-Dukic, N.; Samojlik, I.; Jovin, E. Antimicrobial and antioxidant properties of rosemary and sage (*Rosmarinus officinalis* L. and *Salvia officinalis* L., Lamiaceae) essential oils. *J. Agric. Food Chem.* **2007**, *55*, 7879–7885. [CrossRef]
73. Fu, Z.; Wang, H.; Hu, X.; Sun, Z.; Han, C. The Pharmacological Properties of Salvia Essential Oils. *J. App. Pharm. Sci.* **2013**, *3*, 122–127. [CrossRef]

antioxidants

MDPI

Article

Modulation of Antioxidant Defense in Farmed Rainbow Trout (*Oncorhynchus mykiss*) Fed with a Diet Supplemented by the Waste Derived from the Supercritical Fluid Extraction of Basil (*Ocimum basilicum*)

Gabriele Magara [1], Marino Prearo [2], Cristina Vercelli [3], Raffaella Barbero [4], Marco Micera [5,6], Alfonso Botto [5], Christian Caimi [7], Barbara Caldaroni [1], Cinzia Margherita Bertea [6], Giuseppe Mannino [6,*], Damià Barceló [8,9], Monia Renzi [10], Laura Gasco [7], Giovanni Re [3], Alessandro Dondo [2], Antonia Concetta Elia [1] and Paolo Pastorino [2,*]

1. Department of Chemistry Biology and Biotechnology, University of Perugia, 06123 Perugia, Italy; magara.gabriele@gmail.com (G.M.); bcaldaroni@gmail.com (B.C.); antonia.elia@unipg.it (A.C.E.)
2. The Veterinary Medical Research Institute for Piemonte, Liguria and Valle d'Aosta, 10154 Turin, Italy; marino.prearo@izsto.it (M.P.); alessandro.dondo@izsto.it (A.D.)
3. Department of Veterinary Sciences, University of Torino, Grugliasco, 10095 Turin, Italy; cristina.vercelli@unito.it (C.V.); giovanni.re@unito.it (G.R.)
4. ASL TO4, Servizio Veterinario-Igiene Degli Allevamenti e Delle Produzioni Zootecniche, Settimo Torinese, 10036 Turin, Italy; rbarbero@aslto4.piemonte.it
5. Exenia Group S.R.L, Pinerolo, 10064 Turin, Italy; marco.micera@unito.it (M.M.); info@exeniagroup.it (A.B.)
6. Department of Life Sciences and Systems Biology, University of Turin, 10123 Turin, Italy; cinzia.bertea@unito.it
7. Department of Agricultural, Forest and Food Sciences, University of Torino, Grugliasco, 10095 Turin, Italy; christian.caimi@unito.it (C.C.); laura.gasco@unito.it (L.G.)
8. Catalan Institute for Water Research (ICRA), 17003 Girona, Spain; dbcqam@cid.csic.es
9. Institute of Environmental Assessment and Water Research (IDAEA-CSIC), 08034 Barcelona, Spain
10. Department of Life Sciences, University of Trieste, 34127 Trieste, Italy; mrenzi@units.it

* Correspondence: giuseppe.mannino@unito.it (G.M.); paolo.pastorino@izsto.it (P.P.); Tel.: +39-0116706362 (G.M.); +39-0112686251 (P.P.)

Citation: Magara, G.; Prearo, M.; Vercelli, C.; Barbero, R.; Micera, M.; Botto, A.; Caimi, C.; Caldaroni, B.; Bertea, C.M.; Mannino, G.; et al. Modulation of Antioxidant Defense in Farmed Rainbow Trout (*Oncorhynchus mykiss*) Fed with a Diet Supplemented by the Waste Derived from the Supercritical Fluid Extraction of Basil (*Ocimum basilicum*). *Antioxidants* **2022**, *11*, 415. https://doi.org/10.3390/antiox11020415

Academic Editors: Delia Mirela Tit, Simona Bungau and Evangelos Zoidis

Received: 16 January 2022
Accepted: 16 February 2022
Published: 18 February 2022

Publisher's Note: MDPI stays neutral with regard to jurisdictional claims in published maps and institutional affiliations.

Abstract: Phytotherapy is based on the use of plants to prevent or treat human and animal diseases. Recently, the use of essential oils and polyphenol-enriched extracts is also rapidly increasing in the aquaculture sector as a means of greater industrial and environmental sustainability. Previous studies assessed the antibacterial and antiparasitic effects of these bioactive compounds on fish. However, studies on the modulation of oxidative stress biomarkers are still scant to date. Thus, in this study, the modulation of antioxidant defense against oxidative stress exerted by fish diets supplemented with a basil supercritical extract (F1-BEO) was assessed in rainbow trout *Oncorhynchus mykiss*. The F1-BEO extracted with supercritical fluid extraction was added to the commercial feed flour (0.5, 1, 2, 3% w/w) and mixed with fish oil to obtain a suitable compound for pellet preparation. Fish were fed for 30 days. The levels of stress biomarkers such as superoxide dismutase, catalase, glutathione peroxidase, glutathione S-transferase, glutathione reductase, glyoxalase I, glyoxalase II, lactate dehydrogenase, glutathione and malondialdehyde showed a boost in the antioxidant pathway in fish fed with a 0.5% F1-BEO-supplemented diet. Higher F1-BEO supplementation led to a failure of activity of several enzymes and the depletion of glutathione levels. Malondialdehyde concentration suggests a sufficient oxidative stress defense against lipid peroxidation in all experimental groups, except for a 3% F1-BEO-supplemented diet (liver 168.87 ± 38.79 nmol/mg prot; kidney 146.86 ± 23.28 nmol/mg prot), compared to control (liver 127.76 ± 18.15 nmol/mg prot; kidney 98.68 ± 15.65 nmol/mg prot). Our results suggest supplementing F1-BEO in fish diets up to 0.5% to avoid potential oxidative pressure in farmed trout.

Keywords: circular economy; volatile compounds; antioxidants; pro-oxidants; enzymatic activity; gas chromatography; proanthocyanidins; polyphenols

1. Introduction

The expansion of aquaculture provides an alternative way to meet rising global demands for fish and currently contributes to 52% of the fish volume consumed [1]. For three decades, this food sector has recorded the highest annual growth rate; approximately 10% in the 1990s and 5.8% between 2000 and 2016 [2]. The wide use of antimicrobial agents in modern food animal production has led to the emergence of antimicrobial resistance worldwide [3,4]. In aquaculture, this has resulted in the emergence of antibiotic-resistant bacteria in aquatic environments, increase in antibiotic resistance in fish pathogens, transfer of resistance determinants to bacteria of land animals and to human pathogens and alterations in the bacterial flora both in sediments and in the water column [4,5].

Phytotherapy is based on the use of plants to prevent or treat human or animal diseases [6]. At the beginning of the 20th century, phytotherapy was in competition with modern medicine and, in particular, with antibiotic molecules [7]. Recently, the use of essential oils (EOs) and polyphenol-enriched extracts (PEEs) is also rapidly increasing in the aquaculture sector as a means of greater industrial and environmental sustainability [8]. Previous studies showed beneficial effects of these bioactive compounds on growth, immunity, antibacterial and antiparasitic activities in fish crops [9–18]. In particular, *Ocimum basilicum* was effective against *Streptococcus agalactiae* and *Pseudomonas fluorescens*, pathogens of particular concern for farmed fish [19].

Bioactive compounds included in both EOs and PEEs have shown a positive effect of increasing the growth of fish species and modulating the antioxidant and immune defense pathways [10,17,20–23]. For example, the administration of oregano *Origanum heracleoticum* essential oil for eight weeks via a diet with a substitution of 0.05% led to an increase in the body indices (hepatosomatic, viscerosomatic and condition factor indices) in the channel catfish *Ictalurus punctatus* and an enhancement in the activity of the antioxidant enzymes superoxide dismutase (SOD) and catalase (CAT) associated with EO [23]. A similar result was observed by Abdel-Latif et al. [20] and Zhang et al. [22] in specimens of common carp *Cyprinus carpio* fed with a diet based on *Origanum vulgare* essential oil for eight weeks. The essential oil also increased the fish's immune properties and resistance against *Aeromonas hydrophila*, with increased transcription levels of interleukin (IL)-1β and IL-10 and down-regulated tumor necrosis factor (TNF)-α and transforming growth factor (TGF)-down. On the other hand, in the rainbow trout *Oncorhynchus mykiss*, a similar positive immune and antioxidant enhancement was induced by *Origanum onites* EO [21].

The extraction of volatile compounds (VOCs) included in OEs can be performed using various methodologies, such as solvent extraction, Clevenger apparatus, or supercritical fluid extraction (CO_2-SFE) [24]. CO_2-SFE is one of the green technologies that has emerged as an environmentally friendly, efficient and scalable process for the production of both oils and VOC-enriched extracts. Compared to other conventional methods, SFE has a lower solvent recovery, thermal degradation of molecules and extraction time [25]. However, during the extraction of VOCs from aromatic plant matrices, a series of fractions characterized by VOCs mixed with polyphenolic compounds and lipophilic compounds are produced [26]. These side fractions, not reaching the quality standards of traditional OEs, are considered waste and represent an economic problem for the company that produces them.

The studies conducted reported encouraging results on the biocidal properties of VOCs and other functional molecules against fish pathogens and the ability to enhance the defenses of organisms in response to diseases. Basil contains biologically active compounds, including VOCs and polyphenols, that have displayed antimicrobial [27–29], insecticidal [30], nematocidal [31], fungistatic and antioxidant [29,32]. However, the biochemical mechanisms of basil bioactive compounds for increasing fish immunity and especially for the modulation of antioxidant and detoxifying pathways related to oxidative stress are still lacking.

In biological systems, free oxygen radicals or Reactive Oxygen Species (ROS) are generated and eliminated continuously [33]. Although in normal conditions, their generation is counterbalanced by antioxidant molecules, the level of ROS increases when organisms are exposed to xenobiotics, following a physiological condition called oxidative stress [34]. An oxidative burst may be important for the destruction of viral and pathogens components but may also exacerbate mortality in aquatic organisms [35]. The levels of antioxidant and detoxifying molecules, such as superoxide dismutase (SOD), catalase (CAT), glutathione peroxidase (GPx), glutathione S-transferase (GST), glutathione reductase (GR), glyoxalase I and II (GI and GII), lactate dehydrogenase (LDH) and total glutathione (GSH + 2GSSG), can be modulated as defense against oxidative pressure [36–44]. While SOD, CAT and GPx can dismutate the harmful ROS superoxide anion (O_2^-) and hydrogen peroxide (H_2O_2) in water, GST and glyoxalases metabolize xenobiotics and potentially toxic compounds. Glutathione reductase ensures the regeneration of reduced glutathione, the most powerful biological antioxidant molecule, from the oxidized form. The LDH enzyme is involved in the anaerobic generation pathway of ATP. If the antioxidant pathway works correctly, oxidative damage resulting from ROS overproduction, such as lipid peroxidation, is avoided. In this context, high malondialdehyde levels are considered a reliable marker of lipid peroxidation.

Investigating the effects of potential bioactive compounds on farmed fish species currently represents a thriving and extremely necessary research topic to fill the scientific gaps that will allow optimizing the guidelines, also at the European level, to maximize fish productivity in a sustainable manner. The aim of this study was to assess the time-course modulation of antioxidant defense against oxidative stress exerted by fish diets supplemented with a waste derived from the supercritical fluid extraction of basil (F1-BEO) enriched in polyphenols and VOCs in rainbow trout *Oncorhynchus mykiss*.

2. Materials and Methods

2.1. Preparation of the Basil Supercritical Fluid Extract (SFE)

The basil supercritical fluid extracts were obtained from dried, clean basil leaves (size between 0.3 and 0.5 cm; residual humidity: <10%) by the Exenia group s.r.l. (Pinerolo, Italy) using a supercritical fluid extractor (SCF-100; Separeco s.r.l., Pinerolo, Italy). Briefly, CO_2 supplied from a gas cylinder was cooled by ethanol to −5 °C. Then, liquid CO_2 and 200 g of ethanol, used as a co-solvent, were pressurized at 140 bar, mixed and heated to 50 °C. The flow rate was set at 18 kg/h for a total extraction time equal to 3 h. For the experiment, approximately 700 g of basil leaves were filled into a steel cylinder equipped with a mesh filter (80 μm) and then introduced into the extraction vessel, which was placed in a heating autoclave. At the end of the extraction process, two different fractions were obtained. The first fraction (F1-BEO; yield: 3.7% *w*/*w*), considered a waste of production, contained mainly lipophilic compounds mixed with VOCs, while the second one (F2-BEO; yield: 4.8% *w*/*w*) was almost exclusively composed of VOCs. In order to valorize the waste derived from the extraction process via CO_2-SFE, the first fraction was sampled and stored at −20 °C until subsequent experiments.

2.2. Characterization of VOCs in F1-BEO

In order to identify and quantify VOCs in the production process waste, F1-BEO was extracted using Hexane as a solvent in a 1:10 (*w*/*v*) ratio. The extraction process consisted of 2 min of vortexing, 30 min of ultrasound at room temperature, and then centrifugation (5000× *g* for 10 min, at 4 °C). The extract obtained as previously described was diluted, and 1 μL was injected in GC-MS (5975T, Agilent Technologies, Santa Clara, CA, USA) for the identification of VOCs and in GC-FID (GC-2010 Plus, SHIMADZU, Kyoto, Japan) for quantification. The chromatographic separation was obtained in ZB5-MS (30 m length, 250 μm diameter and stationary phase thickness of 0.25 μm, 5% phenyl-arylene and 95% poly-dimethyl siloxane) column (Phenomenex, Torrance, CA, USA). Chromatographic temperature conditions used for the separation were: 60 °C held for 1 min and then raised 3 °C per minute until 250 °C. The temperature was then brought to 300 °C in 20 min. This temperature was held for an additional 5 min. VOCs were identified by both comparing the mass fragmentation spectra with reference NIST 98 software and the

Retention indices. Quantification of VOCs was performed, building a β-caryophyllene (Sigma-Aldrich, Waltham, MA, USA) calibration curve by injecting different concentrations of it in the range of 0.01 and 0.1 mg/mL).

2.3. Content and Composition of Fatty Acids in F1-BEO

Total fatty acid content was estimated after cold extraction using a mixture composed of chloroform and methanol in a 1:2 (*v*/*v*) ratio [45]. Briefly, the extraction mixture was added to 5 g of F1-BEO using a 1:10 (*w*/*v*) ratio. After vortexing the sample for 5 min, one-third of the volume of water was added in order to create a two-phase system. The tubes were then centrifuged (5000× *g*, 10 min, 25 °C), and the aqueous phase containing methanol was discarded. In order to carry out an exhaustive extraction of the fat portion, the extraction process was repeated two more times, and the different chloroform phases were combined. Finally, the organic phases were dried by using a rotavapor, and the total fat content was calculated as a weight ratio to the originally weighed grams.

Concerning the fatty acid profile, 10 mg of the fat dry portion were trans-esterified by incubating 1 mL of 10% (*w*/*v*) Boron tri-fluoride solubilized in methanol for 60 min at 80 °C. Before incubation, 50 μg of heptadecanoic acid (C17:0) was added to each sample as the internal standard. The obtained fatty acid methyl esters (FAMEs) were then purified by the serial addition of 1 mL water and 1 mL hexane following a short centrifugation step (5000× *g*, 10 min, 25 °C). After each centrifugation, the organic phases were combined, dehydrated by using anhydrous $MgSO_4$ and employed for the Gas Chromatographic (GC) analysis (GCMS-TQ8040, Shimadzu, Kyoto, Japan). The chromatographic separation was obtained in the ZB5-MS (30 m length, 250 μm diameter and stationary phase thickness of 0.25 μm, 5% phenyl-arylene and 95% poly-dimethyl siloxane) column (Phenomenex, Torrance, CA, USA). Chromatographic temperature conditions employed for the separation consisted of 60 °C held for 1 min and then raised 10 °C per minute until 180 °C. The temperature was then brought to 230 °C after 40 min and 320 °C after 20 min. This temperature was held for an additional 5 min. Fatty acids were identified by comparing the mass fragmentation spectra to the reference NIST 98 database and the linear retention indices calculated versus the C9–C25 hydrocarbon mixtures and by injecting pure standards (Sigma-Aldrich, Waltham, MA, USA). Quantification of fatty acids was performed, building a calibration curve of pure fatty acids already trans-methylated (range of 0.01 and 0.1 mg/mL).

2.4. Determination of Bioactive Compounds in F1-BEO

2.4.1. Extract Preparation

Bioactive compounds were extracted from F1-BEO by adding 90% (*v*/*v*) methanol directly to F1-BEO using a 1:10 (*v*/*v*) ratio. Samples were vortexed for 5 min and sonicated at room temperature for 30 min. After sonication, they were centrifuged (20 min at 6000× *g*, 4 °C), and the hydrophilic layer was separated. In order to carry out an exhaustive extraction of bioactive compounds, the extraction process was repeated two more times, and the different hydroalcoholic phases were combined. Methanolic extracts were stored at −20 °C until further chemical analysis.

2.4.2. Total Polyphenol Content

The Total Polyphenol Content (TPC) was estimated via Folin–Ciocalteu assay, as previously reported [46]. Briefly, 5 μL of a suitable dilution of F1-BEO was incubated in a 96-well plate with 3 μL of Folin–Ciocalteu Reagent, 6 μL of 20% (*w*/*v*) sodium carbonate and 86 μL of water. After incubating the mixture for 1 min at 80 °C, samples were kept at room temperature for 20 min in the dark. Consequently, the absorbances of each well were read at 720 nm against a blank containing all the reagents except the sample. Quantification was performed using an external calibration curve of gallic acid (GA). Analyses were performed in triplicate, and data were expressed as mmol of GA equivalents (GAE) per 100 g of extract.

2.4.3. Total Flavan-3-ols Content

The Total Flavan-3-ols Content (TFC) was estimated via BL-DMAC assay, as previously reported [46]. Briefly, 28 μL of a proper dilution of F1-BEO was incubated in a 96-well plate with 84 μL of 1% (*w*/*v*) 4-(Dimethylamino)cinnamaldehyde dissolved in 75% (*v*/*v*) ethanol and acidified with 12.5% (*v*/*v*) hydrochloric acid. After 20 min of incubation at room temperature in the dark, the absorbances were measured at 640 nm against a blank containing the same reagents except for the sample. Quantification was performed using an external calibration curve of A2-type Proanthocyanidin (A2-PAC). Analyses were performed in triplicate, and data were expressed as mmol of A2-PAC equivalents (A2-PACE) per 100 g of extract.

2.5. Spectrophotometric Evaluation of Antioxidant Power

The same methanolic extracts used for the quantification of TPC and TFC were also employed for the evaluation of the antioxidant properties of F1-BEO. This analysis included both the evaluation of radical scavenging activity and the evaluation of the reducing power. For each assay, the inhibition percentage was calculated using Equation (1):

$$AA\% = [(A_{blank} - As_{ample})/A_{blank}] \times 100 \quad (1)$$

where AA% is the percentage of color reduction in the reagent; A_{blank} is the absorbance of blank containing all reagents required to perform the assays except the sample; A_{sample} is the absorbance of the sample read at the specific wavelength for each assay. 6-Hydroxy-2,5,7,8-tetramethylchroman-2-carboxylic acid (Trolox) was employed as a reference standard for all the assays, and the antioxidant activity (AOA) was expressed as mmol of the Trolox equivalent (TE) per 100 g of Fresh Weight (FW).

2.5.1. Radical Scavenging Activity

The radical scavenging activity of F1-BEO was measured both via 2,2′-azino-bis(3-ethylbenzothiazoline-6-sulphonic acid (ABTS) and 2,2-diphenyl-1-picrylhydrazyl (DPPH) assay [47]. For ABTS, 7 mM ABTS was incubated for 16 h with 2.45 mM $K_2S_2O_8$ at room temperature using a 1:10 (*v*/*v*) ratio. After the formation of the radical $ABTS^+$, the reaction solution was diluted in methanol until reaching a final absorbance between 0.70 and 0.80 at 734 nm. For the assay, 90 μL of the diluted ABTS mixture was incubated for 5 min with 10 μL of different methanolic dilutions of F1-BEO. The decay of the radical $ABTS^+$ was then monitored by reading the color decrease at 734 nm.

Regarding the DPPH assay, 10 μL of different methanolic dilutions of F1-BEO were added to 90 μL of 0.1 DPPH. After vigorous shaking, the mixture was incubated for 30 min in the dark and at 25 °C. The decay of the radical DPPH˙ was then monitored by reading the color decrease at 517 nm.

2.5.2. Reducing Activity

The reducing activity was measured via ferric reducing antioxidant power (FRAP) assay [47]. To perform the assay, 300 mM CH_3COONa (pH = 3.6), 10 mM 2,4,6-Tris(2-pyridyl)-s-triazine (TPTZ), and 20 mM $FeCl_3$ were mixed in 8:1:1 (*v*/*v*/*v*) ratio. Consequently, 90 μL of the reagent mixture was incubated for 20 min at 37 °C with different methanolic dilutions of F1-BEO. The color development indicating the reduction of Fe(III) in Fe(II) was then monitored at 595 nm.

2.6. Identification of Polyphenolic Compounds

The identification of polyphenolic compounds was performed analyzing the same extract used for spectrophotometric determination via High-Pressure Liquid Chromatography (HPLC) coupled with Electron Spray Ionization (ESI) and Ion Trap Mass-Spectrometer (MS). The binary solvent system employed for the analysis was MilliQ H_2O acidified with 0.1% (*v*/*v*) formic acid (Solvent A) and acetonitrile acidified with 0.1% (*v*/*v*) formic acid

(Solvent B) in a C18 Luna column (5 µm, 150 × 4.6 mm i.d., Phenomenex, USA) maintained at 27 °C. The elution was performed according to the following method: 0–10 min 5% B; 10–13 min linear increase to 95% B; 13–20 min hold 95% B; 21 min linear decrease 5% B and 22 min back to the initial conditions, as previously described [45]. Polyphenol identification was performed by analyzing each compound retention time and fragmentation pattern and by comparing them to compounds already identified in basil extracts and reported in the literature [48].

2.7. Diet Formulation and Rainbow Trout Nutrition

The diet used for the trial was prepared at the experimental facility of the Department of Agricultural, Forest and Food Sciences (Carmagnola, Turin Province, Italy). The supplementation with basil extract was made with commercial feed flour (Alterna Eel, Skretting; ingredients: fish meal, fish oil, wheat red dog, wheat gluten, blood meal from poultry, a soya bean protein concentrate, swine hemoglobin and whey powder; proximate composition: protein 48%, lipid 11%, ash 8%, fiber 1%). The waste fraction derived from the supercritical fluid extraction of basil was added to the commercial feed flour in the proportions of 0.5% (*w*/*w*), 1% (*w*/*w*), 2% (*w*/*w*) and 3% (*w*/*w*). A control diet without basil (only feed flour Alterna Eel, Skretting) was also made. Then, the mixture was subsequently mixed with fish oil in order to obtain a suitable material for pellet preparation. The pellets were obtained using a 4.0 mm die meat grinder and dried at 30 °C for 48 h. The five diets (A: control; B: 0.5%; C: 1%; D: 2%; E: 3%) were finally stored in dark bags at 4 °C until their utilization.

Fish feeding experimentation was carried out on 430 sex-reversed females of rainbow trout exhibiting a sterile filiform gonad purchased from a private fish farm in northwest Italy. Thirty individuals were randomly selected for anatomopathological, parasitological, bacteriological and virologic examination following methods previously reported [49] to ensure that the fish were in optimal health condition. Fish were conditioned for 20 days before the beginning of the experiment. A 30-day trial was carried out using 20 square fiberglass tanks of 400 L supplied by artesian well water (constant temperature of 13 ± 1 °C) in an open system (flow-through), with each tank having a water inflow of 8 L min^{-1}. Dissolved oxygen was measured every day and ranged between 8.5 and 9.3 mg L^{-1}, while water pH was equal to 7.6 ± 0.5. The fish were exposed to natural photoperiod (12 h light/12 h dark). After the acclimatization period, 400 fishes were lightly anesthetized (tricaine methanesulfonate (MS-222) 70 mg L^{-1}; Sigma-Aldrich, Milano, Italy), individually weighed (mean body weight: 250 ± 50 g) and randomly equally distributed to each tank (20 fish per tank). The experimental diets (A, B, C, D, E) were randomly assigned to the tanks (four replicate tanks per diet). The fish were fed by hand to apparent visual satiation six days per week. The daily feed quantity was set at 1% of tank biomass. The tank biomass was kept constant at 20 kg m^{-3} by lowering the water level in each tank, based on the fish biomass removed for analysis. Mortality was checked every day. For the purpose of this study, 32 fish from each experimental group (eight fish per tank; four replicates per diet) were sampled at the middle (15 days; T1) and at the end (30 days; T2) of the experiment. At each sampling campaign, fish were captured using a landing, immediately suppressed using an overdose (170 mg kg^{-1}) of MS-222 and necropsied. Livers and kidneys were immediately sampled from each specimen and stored at −80 °C for biochemical analysis.

2.8. Oxidative Stress Biomarkers

2.8.1. Preparation of Fish Tissue Extracts

Biochemical analyses were conducted according to previous studies on rainbow trout [50]. Liver and kidney from 32 fish per treatment (8 fish from each replicate) and endpoint were individually diluted (1:10 and 1:5, respectively) in 100 mM potassium phosphate (KP) buffer pH 7.5 added with 0.008 tiu/mL aprotinin, 0.1 mg/mL bacitracin and 2.5% sodium chloride (NaCl). Tissues were homogenized and centrifuged for 45 min at 50,000× *g* at 4 °C. The supernatant was divided into aliquots and used for the enzymatic assay. The total protein concentration was assessed according to Lowry et al. [51] and used to normalize the enzyme activities. The absorbance of each oxidative stress biomarker was measured in triplicate by spectrophotometry (Varian Cary 100) at 25 °C.

2.8.2. Evaluation of SOD Activity

For the determination of SOD levels, 50 mM of Na_2CO_3 buffer (pH 10) containing 0.1 mM of EDTA, 500 mM of cytochrome C, 1 mM of hypoxanthine and xanthine oxidase were used (λ = 550 nm).

2.8.3. Evaluation of CAT Activity

The activity of CAT was measured using sodium phosphate buffer (100 mM, pH 7) and 24 mM of H_2O_2 (λ = 240 nm).

2.8.4. Evaluation of GPx Activity

GPx was determined in 100 mM sodium phosphate buffer (pH 7.5) added to 1 mM of EDTA, 0.12 mM of NADPH, 2 mM of GSH, GR 1U and 0.6 mM of H_2O_2 (λ = 340 nm).

2.8.5. Evaluation of GST Activity

To assess GST activity, 100 mM of sodium phosphate buffer (pH 6.5) with 2 mM of GSH and CDNB was used (λ = 340 nm).

2.8.6. Evaluation of GR Activity

GR assay was performed at 340 nm in 100 mM of sodium phosphate buffer at pH 7 using 1 mM of GSSG and 0.06 mM of NADPH (λ = 340 nm).

2.8.7. Evaluation of GI Activity

GI was measured at 240 nM in 100 mM of sodium phosphate buffer at pH 6.8 added to a solution of 2 mM of GSH + methylglyoxal (λ = 240 nm).

2.8.8. Evaluation of GII Activity

GII activity was determined at 340 nm in 100 mM of MOPS buffer at pH 7, 0.2 mM of DTNB and 0.4 mM of LSG (λ = 412 nm).

2.8.9. Evaluation of LDH Activity

LDH levels were assessed at 340 nm in 50 mM of imidazole buffer at pH 7.2, 1 mM of pyruvate and 0.15 mM of NADH (λ = 340 nm).

2.8.10. Evaluation of GSH + 2GSSG Concentration

GSH + 2GSSG concentration was measured according to the method of Akerboom and Sies [52].

2.8.11. Determination of MDA Levels

MDA levels were assessed according to the method described in Pacini et al. [53].

2.9. Statistical Analysis

Normality and homoscedasticity of data were assessed through the Shapiro–Wilk and Levene tests, respectively. As the data were normally distributed, two-way ANOVA following Tukey's post hoc test was used to check statistically significant differences in oxidative stress biomarkers between the experimental groups at the same time point. Statistical significance was set at $p < 0.05$. R software (RStudio, Inc., version 3.5.2., Boston, MA, USA) was used to perform the statistical analyses.

3. Results

3.1. Chemical Profiling of F1-BEO

The spectrophotometric evaluation of F1-BEO included the quantification of the total amount of polyphenols (TPC) and flavan-3-ols (TFC), along with the determination of their antioxidant properties. Data related to UV/Vis estimation are reported in Table 1. F1-BEO showed a very high content of bioactive compounds, as demonstrated by the TPC value. Moreover, a high amount of TFC was detected. Concerning the antioxidant properties, the extract displayed strong antioxidant activity both in terms of radical scavenging and reducing activity, as measured by ABTS, DPPH and FRAP assays. In particular, F1-BEO showed the highest ABTS and DPPH values with respect to FRAP value, suggesting the prevalence of polyphenolic compounds operating through a radical exchange of one or two unpaired electrons.

Table 1. Total Polyphenol Content, Total Flavan-3-ol Content, Radical Scavenging Activity (ABTS and DPPH) and Reducing Activity (FRAP) of F1-BEO. DPPH: 2,2-diphenyl-1-picrylhydrazyl; ABTS: 2,2′-azino-bis(3-ethylbenzothiazoline-6-sulphonic acid; FRAP: Ferric Reducing Antioxidant Power; GAE: Gallic Acid Equivalent; TE: Trolox Equivalent.

Total Polyphenol Content	32.97 ± 1.63	*mmol GAE per 100 g of FW*
Total Flavan-3-ol Content	21.21 ± 1.04	*mmol A2-PACE per 100 g of FW*
Radical Scavenging Activity		
DPPH	70.32 ± 1.39	*mmol TE per 100 g*
ABTS	29.92 ± 0.99	*mmol TE per 100 g*
Reducing Activity		
FRAP	28.62 ± 2.05	*mmol TE per 100 g*

FW: fresh weight; A2-PACE: A2-type Proanthocyanidin content equivalent; TE: Trolox Equivalent.

In order to identify the most representative polyphenol compounds present in F1-BEO, HPLC-ESI-MS/MS analysis was carried out. Based on both the fragmentation pattern of each compound and their retention time, HPLC-ESI-MS/MS analysis allowed the putative identification of 23 different polyphenols (Figure 1). Among the identified compounds, six are flavones (Scolymoside, Isomyricetin, Myricetin diglucoside, Cynaroside, Myricetin and Luteolin); six are flavonols (Nicotiflorin, Isoquercitrin, Astragalin, Kaempferol, Quercetin and Rutin); three are flavanols (Aromadendrin, Arthromerin B and Taxifolin); and eight are polyphenolic acids. Among the polyphenolic acids, one is a derivative of hydroxycinnamic acid (Chicoric Acid), and seven belong to the salvianolic acid family.

In order to identify VOCs in F1-BEO, GC-MS analysis was carried out, whereas GC-FID was used to quantify the constituents in the same fraction (Table 2). In F1-BEO, the GC analysis identified several volatile compounds. Linalool, Estragol and α-Bergamotene accounted for about 60% of the total volatile content. Other identified VOCs included 1,8-Cineole, Methylcinnamate, and β-Caryophyllene.

Finally, our analyses showed that the F1-BEO fraction was also composed of about 10% of fats. The GC-MS and GC-FID analyses made it possible to identify the different fatty acids that compose this portion (Table 3). Among them, Palmitic Acid, Linoleic Acid and Oleic Acid are the most abundant, accounting for 77% of the total content of fatty acids.

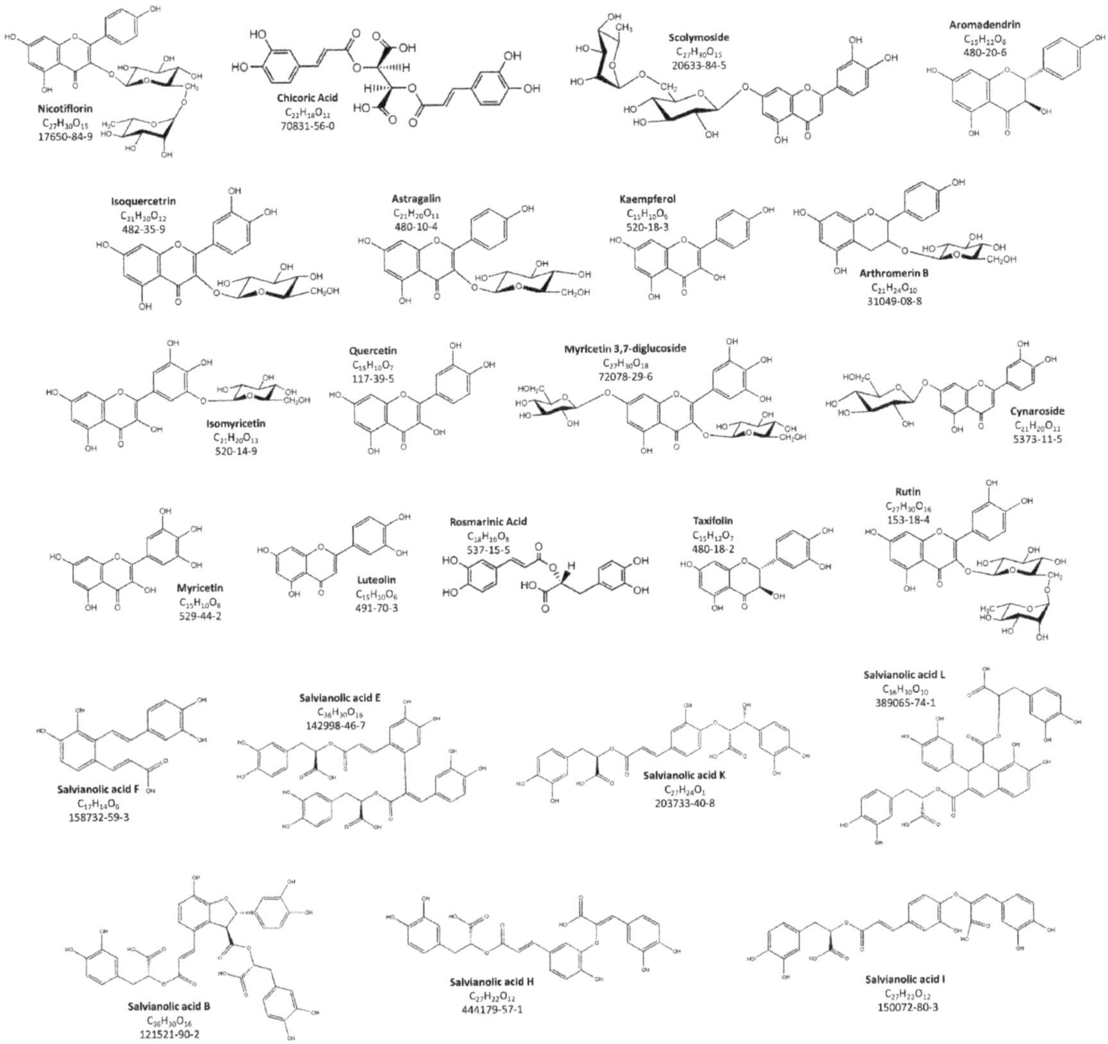

Figure 1. Polyphenolic compounds identified in F1-BEO via HPLC-ESI-MS/MS.

Table 2. Identification and Quantification of Volatile Organic Compounds (VOCs) in F1-BEO. CAS ID: Chemical Abstracts Service Identification number.

	Total Volatile Content		73.18 ± 2.79	*mg per 100 g of FW*
Formula	**Compound**	**CAS ID**		
$C_{10}H_{18}O$	*1,8-Cineole*	470-82-6	9.33 ± 0.45	%
$C_{10}H_{18}O$	*Linalool*	78-70-6	25.29 ± 0.81	%
$C_{10}H_{12}O$	*Estragol*	140-67-0	18.79 ± 0.78	%
$C_{10}H_{12}O_2$	*Eugenol*	97-53-0	4.49 ± 0.12	%
$C_{10}H_{10}O_2$	*Methylcinnamylate*	103-26-4	8.71 ± 0.15	%
$C_{11}H_{14}O_2$	*Methyleugenol*	93-15-2	6.58 ± 0.08	%
$C_{15}H_{24}$	*b-Caryophyllene*	87-44-5	7.47 ± 0.29	%
$C_{15}H_{24}$	*α-Bergamotene*	17699-05-7	19.34 ± 1.09	%

Table 3. Identification and Quantification of Fatty Acids in F1-BEO. CAS ID: Chemical Abstracts Service Identification number.

	Total Fat Content		9.97 ± 0.25	g per 100g of FW
Formula	**Compound**	**CAS ID**		
C14:0	*Myristic acid*	544-63-8	*3.05 ± 0.12*	%
C16:1 trans	*Palmitoleic acid*	373-49-9	*1.89 ± 0.05*	%
C16:1 cis	*Palmitovaccenic acid*	373-49-9	*2.46 ± 0.08*	%
C16:0	*Palmitic acid*	57-10-3	*37.28 ± 1.52*	%
C18:2	*Linoleic acid*	60-33-3	*10.82 ± 0.41*	%
C18:1 cis	*Oleic acid*	112-80-1	*28.95 ± 1.32*	%
C18:1 trans	*Elaidic acid*	112-79-8	*1.06 ± 0.04*	%
C18:0	*Stearic acid*	57-11-4	*6.53 ± 0.12*	%
C20:0	*Arachidic acid*	506-30-9	*5.16 ± 0.15*	%
C22:0	*Behenic acid*	112-85-6	*2.75 ± 0.06*	%

3.2. Changes in Oxidative Stress Biomarkers in Rainbow Trout after Feeding with F1-BEO

During the experimental trial, no mortality was observed.

The results of the two-way ANOVA of the treatment, time and interaction on the livers and kidneys of rainbow trout are reported in Table 4.

Table 4. Results of two-way ANOVA of time (15 and 30 days), treatment (F1-BEO substitution in fish diets) and interaction (time x treatment) on livers and kidneys of *O. mykiss*.

	F Time (dfn, dfd)		**Time (F-Value)**		**F Treatment-Interaction (dfn, dfd)**		**Treatment (F-Value)**		**Interaction (F-Value)**	
	Liver	*Kidney*	*Liver*	*Kidney*	*Liver*	*Kidney*	*Liver*	*Kidney*	*Liver*	*Kidney*
SOD	(1; 70)	(1; 70)	16.34 ***	6.02 *	(4; 70)	(4; 70)	1.00	0.55	0.98	3.72 **
CAT	(1; 70)	(1; 70)	0.002	41.36 ***	(4; 70)	(4; 70)	2.12	3.36 *	1.54	9.62 ***
GPx	(1; 70)	(1; 70)	12.92 ***	2.94	(4; 70)	(4; 70)	6.58 ***	0.67	1.21	8.80 ***
GST	(1; 70)	(1; 70)	65.07 ***	1.91	(4; 70)	(4; 70)	23.81 ***	34.10 ***	4.88 **	3.92 **
GR	(1; 70)	(1; 70)	156.9 ***	0.41	(4; 70)	(4; 70)	21.34 ***	4.05 **	14.69 ***	1.55
GI	(1; 70)	(1; 70)	19.99 ***	1.63 *	(4; 70)	(4; 70)	10.82 ***	113.3 ***	4.98 **	16.08 ***
GII	(1; 70)	(1; 70)	8.01 **	29.48 ***	(4; 70)	(4; 70)	2.52 *	4.46 **	1.31	6.82 ***
LDH	(1; 70)	(1; 70)	27.49 ***	8.98 **	(4; 70)	(4; 70)	24.93 ***	7.27 ***	25.52 ***	1.13
Glut	(1; 70)	(1; 70)	0.04 **	0.16	(4; 70)	(4; 70)	18.71 ***	34.78 ***	13.24 ***	10.48 ***
MDA	(1; 70)	(1; 70)	0.05	6.83 *	(4; 70)	(4; 70)	3.88 **	16.11 ***	2.69 *	7.69 ***

Degrees of freedom (dfn = numerator, dfd = denominator) and F-statistics (F) are provided. Significant code *** $p < 0.001$; ** $p < 0.01$; * $p < 0.05$ (Two-way ANOVA with Tukey's multiple comparisons test, $p < 0.05$).

SOD activity in treated fish was comparable with those of the control group in both tissues (Figure 2, Panel A and B), whereas CAT levels increased in T2 in the liver (50%) and in T1 in the kidney in the groups fed with F1-BEO (Figure 2, Panel C and D). On the contrary, GPx activity was lower than the control in the livers of trout treated with 3% (*w*/*w*) F1-BEO (T1), following decreased activity (40%) in fish treated with 1–2% (*w*/*w*) in T2 (Figure 2, Panel E). On the same endpoint, a similar enzymatic trend was observed in kidneys (Figure 2, Panel F).

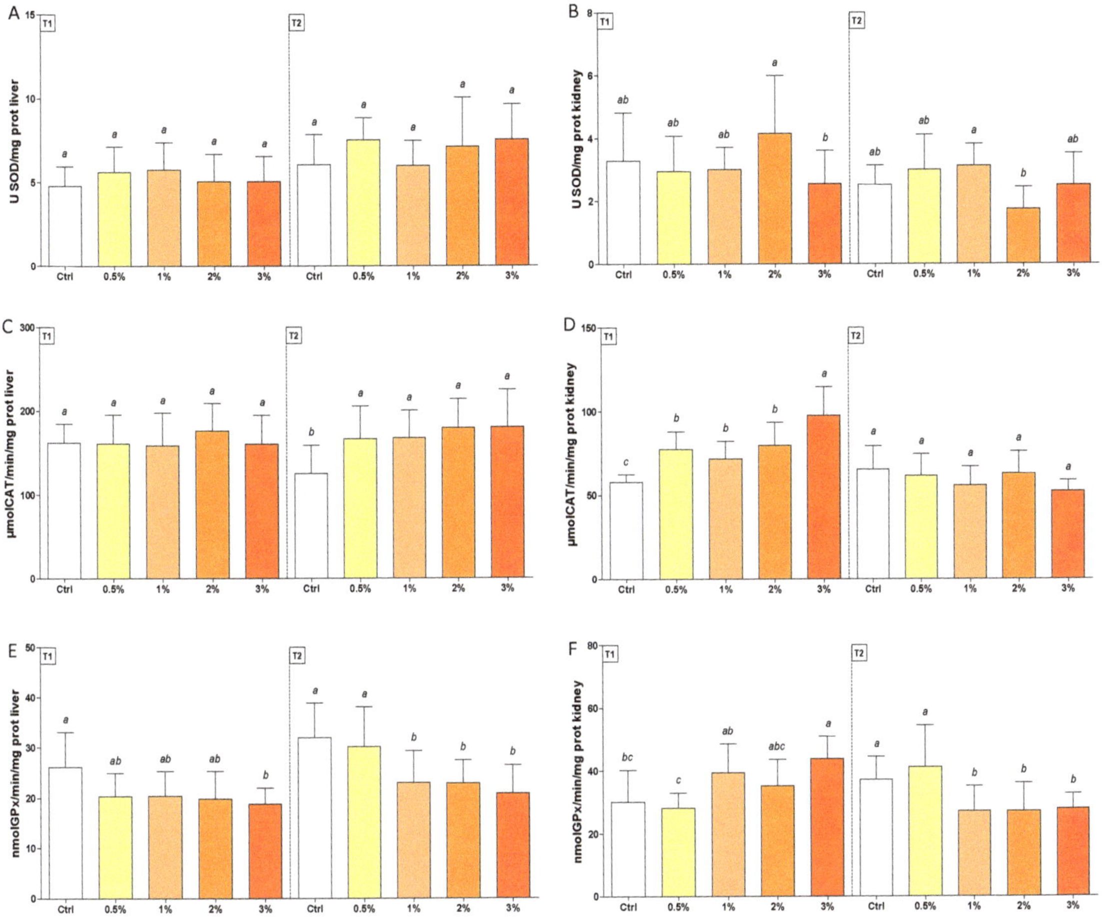

Figure 2. Enzymatic activity of SOD (Panel (**A**,**B**)), CAT (Panel (**C**,**D**)) and GPx (Panel (**E**,**F**)) in livers (on left, Panel (**A**,**C**,**E**)) and kidneys (on the right, Panel (**B**,**D**,**F**)) of rainbow trout *Oncorhynchus mykiss* fed with F1-BEO-supplemented diets (0.5%, 1%, 2%, 3% (*w*/*w*) supplementation) after 15 (T1) and 30 (T2) days. Different lowercase letters indicate statistically significant differences among experimental groups ($p < 0.05$) for each experimental endpoint.

GST levels were consistently lower than control (30%) in 1–2–3% (*w*/*w*)-fed trout in the liver, whereas enzyme activity was enhanced (50%) in fish exposed to 0.5% (*w*/*w*) basil essential oil diet in T2 (Figure 3, Panel A). In kidneys, enzyme activity was restored in T2, except in trout fed with the 3% (*w*/*w*) F1-BEO diet (Figure 3, Panel B). GR activity increased (2-fold) in T2 in livers of trout treated with 1–2–3% (*w*/*w*) F1-BEO diets, whereas only an early and transient depletion (30%) was observed in kidneys in fish fed with 3% (*w*/*w*) F1-BEO diet (Figure 3, Panel C and D).

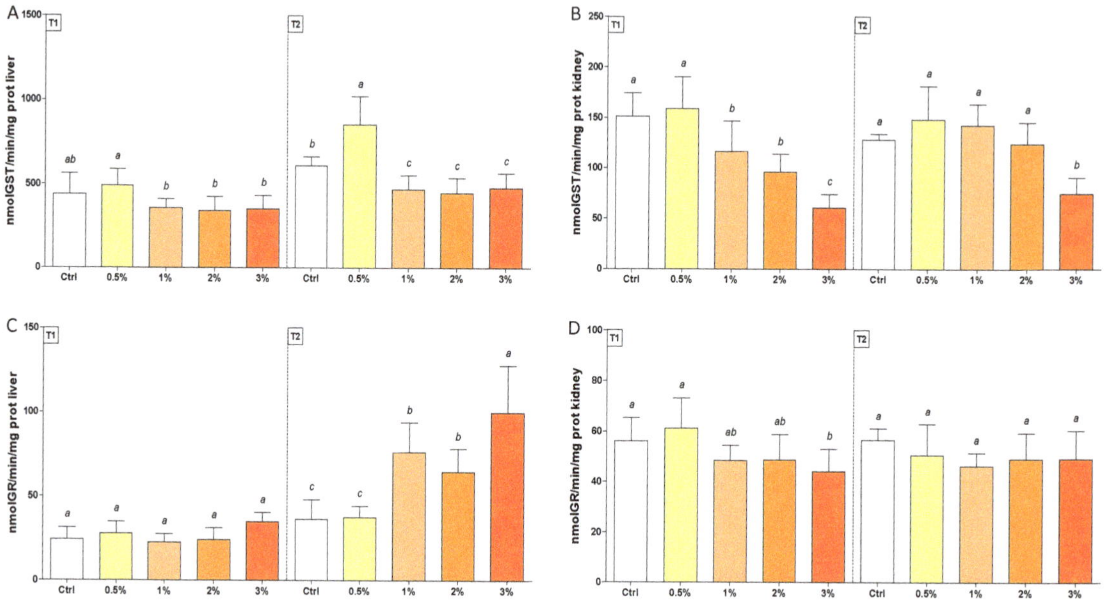

Figure 3. Enzymatic activity of GST (Panel (**A**,**B**)) and GR (Panel (**C**,**D**)) in livers (on the left, Panel (**A**,**C**)) and kidneys (on the right, Panel (**B**,**D**)) of rainbow trout *Oncorhynchus mykiss* fed with F1-BEO-supplemented diets (0.5%, 1%, 2%, 3% (*w*/*w*) supplementation) after 15 (T1) and 30 (T2) days. Different lowercase letters indicate statistically significant differences among experimental groups ($p < 0.05$) for each experimental endpoint.

In the same tissue, GI levels were consistently lower than control (80%) in trout exposed to 1–2–3% (*w*/*w*) F1-BEO diets. On the contrary, enhanced enzyme activity was observed in 0.5% (*w*/*w*) F1-BEO-treated trout in T2. In livers, a consistent depletion in GI activity was measured in T2 in fish fed with 1–2–3% (*w*/*w*) F1-BEO diets (Figure 4, Panel A and B). A transient decrease (30%) in GII activity was observed in kidneys of trout treated with 1–2–3% (*w*/*w*) F1-BEO diets, whereas a late depletion (30%) occurred in fish exposed to 0.5% (*w*/*w*) of the same extract (Figure 4, Panel D). LDH levels were enhanced (1.5-fold) in trout fed with 0.5% (*w*/*w*) and lowered (80%) in the other treated groups in the liver. In kidneys, enzyme activity was consistently depleted (40%) in trout fed with 3% (*w*/*w*) F1-BEO (Figure 4, Panel E and F).

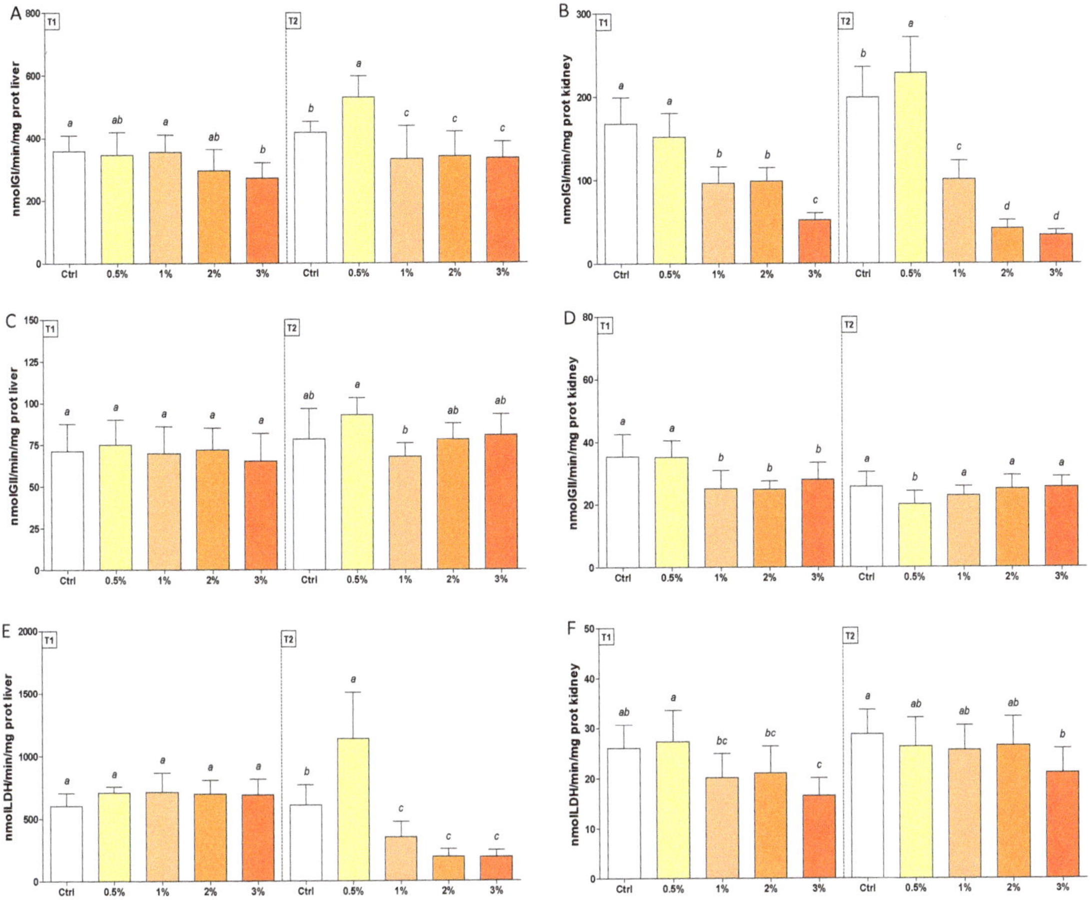

Figure 4. Enzymatic activity of GI (Panel (**A**,**B**)), GII (Panel (**C**,**D**)), and LDH (Panel (**E**,**F**)) in livers (on the left, Panel (**A**,**C**,**E**)) and kidneys (on the right, Panel (**B**,**D**,**F**)) of rainbow trout *Oncorhynchus mykiss* fed with F1-BEO-supplemented diets (0.5%, 1%, 2%, 3% (*w*/*w*) supplementation) after 15 (T1) and 30 (T2) days. Different lowercase letters indicate statistically significant differences among experimental groups ($p < 0.05$) for each experimental endpoint.

A late decrease (50%) in the total glutathione concentration was observed in livers of trout treated with 1–2–3% (*w*/*w*) F1-BEO, whereas in kidneys, an early depletion followed by a thiol restoration was measured, except in diets supplemented with 3% (*w*/*w*) F1-BEO. In both tissues, GSH + 2GSSG levels were enhanced (30%) in trout exposed to 0.5% (*w*/*w*) F1-BEO (Figure 5, Panel A and B). MDA concentration increased in both tissues only in T2 in fish fed with 3% (*w*/*w*) F1-BEO (Figure 5, Panel C and D).

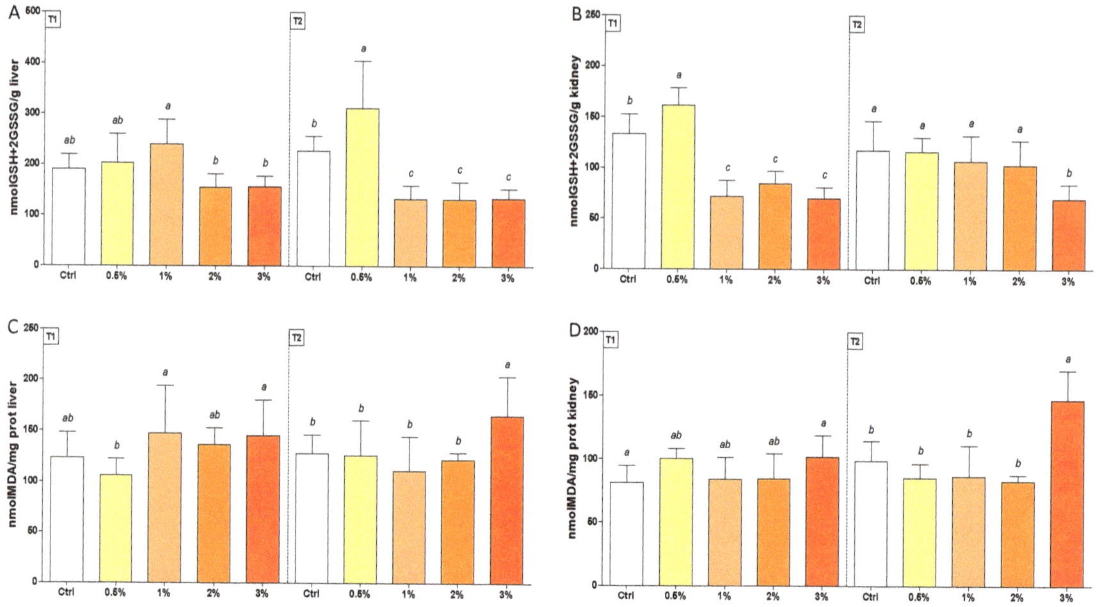

Figure 5. GSH + 2GSSG (Panel (**A**,**B**)) and MDA (Panel (**C**,**D**)) levels in livers (on the left, Panel (**A**,**C**)) and kidneys (on the right, Panel (**B**,**D**)) of rainbow trout *Oncorhynchus mykiss* fed with F1-BEO-supplemented diets (0.5%, 1%, 2%, 3% (w/w) supplementation) after 15 (T1) and 30 (T2) days. Different lowercase letters indicate statistically significant differences among experimental groups ($p < 0.05$) for each experimental endpoint.

4. Discussion

In this study, the addition of F1 BEO (0.5–3% w/w) in fish meal led to the modulation of oxidative stress biomarkers. Although SOD activity remained unchanged in both tissues during the whole experimental period, the increased levels of CAT suggested a SOD-independent overproduction of H_2O_2 in livers and kidneys of rainbow trout. Specific investigations on fish are lacking, but it is known that *Achillea millefolium* essential oil can stimulate the production of H_2O_2 in a murine model [54]. This evidence, along with the results herein reported, allow hypothesizing a similar mechanism of production of hydrogen peroxide in fish in order to boost the non-specific immune system against pathogens. To the best of our knowledge, very scant studies have assessed the effects of a few specific polyphenolic compounds on the activity of antioxidant enzymes in fish [55–61]. CAT levels were enhanced in the livers of zebrafish *Danio rerio* following exposure to phenolic compounds, including quercetin and isoquercitin, from mango peel extract [55]. On the other hand, a significant decrease in enzyme activity was observed in the spotted snakehead *Channa punctata* treated with 0.14 g/L quercetin for 21 days [56] and olive flounder *Paralichthys olivaceus* fed with quercetin-supplemented diets [57,58]. Salvianolic acid can regulate the Keap1/Nrf2 pathway and enzyme activity, such as CAT in *D. rerio*, protecting fish against oxidative stress [59]. Increased levels of CAT were reported in livers of rainbow trout fed with diets supplemented with origanum essential oil [21] and in silver catfish exposed to *Aloysia triphylla* [62,63] or tambaqui *Colossoma macropomum* treated with *Myosotis sylvatica* and *Curcuma longa* essential oil [64]. A similar enzymatic trend was observed in channel catfish *Ictalurus punctatus* exposed to *Origanum vulgare* essential oil [23]. The improvement in non-specific immune parameters and decreased mortality in Mozambique tilapia *Oreochromis mossambicus* fed with diets supplemented with *Citrus limon* peels essential oil was shown in a previous study [65]. Moreover, CAT activity was increased in silver catfish *Rhamdia quelen* treated with linalool chemotype of *Lippia alba* essential oil [66] and in rainbow trout *O. mykiss* fed with cineole-supplemented diets [67]. The expression of CAT was also up-regulated by dietary limonene in Nile tilapia *Oreochromis niloticus* [68]. Elevated levels of catalase were also observed in *O. mykiss* following dietary exposure to polyunsaturated fatty acids (PUFA), suggesting that variation of lipid profile and especially unsaturated fatty acids in diets can modulate the shield against oxidative stress in freshwater fish [69].

Despite the enhancement of catalase activity herein reported, the levels of total glutathione and related enzymes suggested a weakening antioxidant defense. This outcome is in contrast with previous studies on *Rhamdia quelen* exposed to quercetin and rutinin [60,61] and on silver catfish fed with diets supplemented with *L. alba* essential oil, in which enhanced glutathione levels were observed following increased GPx and GST activities [70]. The published and present results suggest glutathione's key role in modulating the antioxidant response. Likely, the effects of different essential oil on thiol can forge an effective defense against oxidative pressure. Previous studies showed that essential oils generally improved the redox state in fish species [63,69,71,72]. The activity of GPx was increased in livers of rainbow trout fed with 0.5–1.0 g/kg diet of *Salvia officinalis*, *Mentha spicata* and *Thymus vulgaris* [73], whereas *A. triphylla* and *Melaleuca alternifolia* essential oils enhanced the levels of GST in silver catfish [62,74]. Enhanced enzyme levels were also observed in common carp *Cyprinus carpio* exposed to cineole [67], whereas no altered activity was reported in Nile tilapia following treatment with limonene [68]. In this study, although the high Radical Scavenging Activity and Reducing Activity measured in F1 BEO suggest basil extract's high antioxidant properties likely due to the presence of polyphenols recognized as antioxidant molecules [75], the low levels of glutathione and related enzymes in fish fed diets with higher percentages of F1 BEO suggests the antioxidant threshold was exceeded for trout. Previous studies showed that pro-oxidant effects could be observed when organisms are exposed to higher essential oil doses [71,72,76,77]. A high concentration (>30 μL/L) of *L. alba* essential oil caused a decrease in non-protein thiol groups, GPx and GST levels [76], while the same essential oil led to the enhancement of the same biomarkers when silver catfish were exposed to lower concentrations [70]. The variability of GPx and GST following the exposure to different essential oils and fish species can be related to the change in lipid profile of diets based on the percentage of the essential oil supplement. Indeed, the modulation of enzymes in the GST family, including GPx, has previously been associated with changes in the lipid panel and PUFA in diets [69,78]. A previous study showed that GPx increased in trout fed with PUFA-supplemented diets [69]. The expression of GST was down-regulated in salmon fed vegetable oil compared to those treated with fish oil, which is consistent with the higher auto-oxidative potential of LC-PUFA [78]. Moreover, the decreased activity of GPx has been previously associated with linalool exposure in *C. carpio* [79]. Therefore, the decreased GPx and GST activity herein observed in fish fed with the higher supplements (1–3%) of BEO advises for oxidative challenge likely due to the changed lipid profile and VOCs in basil-extract-supplemented feedings. On the contrary, the increased GST activity observed in rainbow trout fed with diets supplemented with 0.5% BEO suggests a boosted detoxifying defense. Similar results were observed in GI levels. To the best of our knowledge, this is the first scientific contribution to the biochemical response of glyoxalase systems in fish exposed to F1-BEO. The increase in GI levels in fish fed with 0.5% BEO-supplemented diets and the depletion in all other experimental treated groups, along with the decrease in GII activity in kidneys, suggest an impediment of the metabolization of alpha-ketoaldehydes, leading to potential oxidative damage in rainbow trout. On the other hand, the increased GR activity can be related to the attempt to regenerate GSH to mitigate the oxidative damage promoted by the high concentrations of BEO in diets. A previous study showed that 30 mg/L clove oil reduced GR levels in the brain and gills of common carp [80], whereas the supplement of 0.2 g/kg and 0.4 g/kg of *Cymbopogon citratus* and *Pelargonium graveolens* essential oil, respectively, increased GR activity in Nile tilapia [81]. Despite the increase in GR levels, the activity of LDH enzyme tested for the first time in fish exposed to F1-BEO suggest an enhancement of the anaerobic energy-producing pathway only in fish fed a 0.5% BEO-supplemented diet, while a failure of this emergency pathway was observed in diets with higher substitutions (1–3%). Overall, the concentration of MDA in livers and kidneys indicates lipid peroxidation only for rainbow trout fed with 3% BEO-supplemented diets. Contrasting results have been reported in the literature about the role of polyphenols and essential oils in lipid peroxidation. Quercitin, rutinin and resveratrol lowered the levels of lipid peroxidation in *R. quelen* [60,61] and *Oreochromis niloticus* [82]. Exposure to a 0.5–1.0 g/kg diet of *Salvia officinalis*, *Mentha spicata* and *Thymus vulgaris* essential oils decreased lipid peroxidation in rainbow trout [73]. *Myosotis sylvatica* and *C. longa* essential oils protected tambaqui from lipid peroxidation, boosting the enzymatic antioxidant pathway [64]. Clove oil did not cause lipid peroxidation, although the activity of several main oxidative stress biomarkers was impaired [80]. On the other hand, increased lipid peroxidation was reported by Salbego et al. [76] in fish treated with high concentrations of *L. alba* essential oil.

5. Conclusions

In conclusion, the higher supplements of F1-BEO in fish diets (1–3% *w*/*w*) promoted the failure of several key antioxidant enzymes' activity, such as GPx, GST, GI and GII, and the lowering of glutathione levels. However, the levels of MDA suggest a sufficient oxidative stress defense promoted by antioxidant pathways during the experimental period in fish fed with 0.5–2% (*w*/*w*) F1-BEO-supplemented diets. The decrease in pivotal stress-shielding molecules alerts for potential oxidative damage when trout are fed highly substituted diets with F1-BEO for longer periods of time. Further studies including challenges for fish, such as exposure to air, crowding and bacterial challenges, are needed to understand the pro-oxidant role of F1-BEO as a substitute in fish diets for sustainable aquaculture.

Author Contributions: Conceptualization, G.M. (Giuseppe Mannino) and P.P.; Data curation, G.M. (Gabriele Magara), C.M.B., G.M. (Giuseppe Mannino), A.C.E. and P.P.; Investigation, G.M. (Gabriele Magara), M.P., C.V., R.B., M.M., A.B., C.C., B.C., G.M. (Giuseppe Mannino), D.B., M.R., L.G., G.R., A.D., A.C.E. and P.P.; Methodology, G.M. (Gabriele Magara), M.P., C.V., R.B., M.M., A.B., B.C., C.M.B., G.M. (Giuseppe Mannino), G.R., A.D., A.C.E. and P.P.; Project administration, P.P.; Supervision, P.P.; Writing—original draft, G.M. (Gabriele Magara), G.M. (Giuseppe Mannino) and P.P.; Writing—review and editing, M.P., C.V., R.B., M.M., C.C., B.C., C.M.B., D.B., M.R., L.G., G.R., A.D., A.C.E. and P.P. All authors have read and agreed to the published version of the manuscript.

Funding: This research was funded by Italian Ministry of Health, Ricerca Finalizzata, grant number GR-2013-02355796.

Institutional Review Board Statement: The experimental protocol was designed according to the guidelines of the European Union Council 2010/63/EU for the use and care of experimental animals. The animal study protocol was approved by the Institutional Review Board of the Italian Ministry of Health (authorization n. 196/2020-PR).

Informed Consent Statement: Not applicable.

Data Availability Statement: Data is contained within the article.

Conflicts of Interest: MC was a Ph.D. student of the Ph.D. Program in Pharmaceutical and Biomolecular Sciences of the University of Turin and in an apprenticeship in the Exenia Group s.r.l. The remaining authors declare that the research was conducted in the absence of any commercial or financial relationships that could be construed as a potential conflict of interest. The funders had no role in the design of the study; in the collection, analyses, or interpretation of data; in the writing of the manuscript; or in the decision to publish the results.

References

1. Food and Agriculture Organization Sustainability in Action. *State of World Fisheries and Aquaculture*; Food and Agriculture Organization of the United Nations: Rome, Italy, 2020; Volume 200.
2. Ahmed, N.; Thompson, S.; Glaser, M. Global Aquaculture Productivity, Environmental Sustainability, and Climate Change Adaptability. *Environ. Manag.* **2019**, *63*, 159–172. [CrossRef] [PubMed]

3. Park, Y.H.; Hwang, S.Y.; Hong, M.K.; Kwon, K.H. Use of Antimicrobial Agents in Aquaculture. *Rev. Sci. Tech. (Int. Off. Epizoot.)* **2012**, *31*, 189–197. [CrossRef] [PubMed]
4. Sicuro, B.; Pastorino, P.; Barbero, R.; Barisone, S.; Dellerba, D.; Menconi, V.; Righetti, M.; De Vita, V.; Prearo, M. Prevalence and Antibiotic Sensitivity of Bacteria Isolated from Imported Ornamental Fish in Italy: A Translocation of Resistant Strains? *Prev. Vet. Med.* **2020**, *175*, 104880. [CrossRef] [PubMed]
5. Schar, D.; Klein, E.Y.; Laxminarayan, R.; Gilbert, M.; Van Boeckel, T.P. Global Trends in Antimicrobial Use in Aquaculture. *Sci. Rep.* **2020**, *10*, 21878. [CrossRef] [PubMed]
6. Enioutina, E.Y.; Teng, L.; Fateeva, T.V.; Brown, J.C.; Job, K.M.; Bortnikova, V.V.; Krepkova, L.V.; Gubarev, M.I.; Sherwin, C.M. Phytotherapy as an Alternative to Conventional Antimicrobials: Combating Microbial Resistance. *Expert Rev. Clin. Pharmacol.* **2017**, *10*, 1203–1214. [CrossRef]
7. Chevalley, A. Utilisation de la Phytothérapie et de L'aromathérapie Dans le Cadre du Conseil Vétérinaire Chez le Chat, le Chien et le Cheval. Ph.D. Thesis, Université de Lorraine, Nancy, France, 2016.
8. Dawood, M.A.; El Basuini, M.F.; Zaineldin, A.I.; Yilmaz, S.; Hasan, M.; Ahmadifar, E.; El Asely, A.M.; Abdel-Latif, H.M.; Alagawany, M.; Abu-Elala, N.M. Antiparasitic and Antibacterial Functionality of Essential Oils: An Alternative Approach for Sustainable Aquaculture. *Pathogens* **2021**, *10*, 185. [CrossRef]
9. Gholipourkanani, H.; Buller, N.; Lymbery, A. In Vitro Antibacterial Activity of Four Nano-encapsulated Herbal Essential Oils against Three Bacterial Fish Pathogens. *Aquac. Res.* **2019**, *50*, 871–875. [CrossRef]
10. Jahazi, M.A.; Hoseinifar, S.H.; Jafari, V.; Hajimoradloo, A.; Van Doan, H.; Paolucci, M. Dietary Supplementation of Polyphenols Positively Affects the Innate Immune Response, Oxidative Status, and Growth Performance of Common Carp, *Cyprinus carpio* L. *Aquaculture* **2020**, *517*, 734709. [CrossRef]
11. Lee, S.; Najiah, M.; Wendy, W.; Nadirah, M. Chemical Composition and Antimicrobial Activity of the Essential Oil of *Syzygium aromaticum* Flower Bud (Clove) against Fish Systemic Bacteria Isolated from Aquaculture Sites. *Front. Agric. China* **2009**, *3*, 332–336. [CrossRef]
12. Majolo, C.; Pilarski, F.; Chaves, F.C.M.; Bizzo, H.R.; Chagas, E.C. Antimicrobial Activity of Some Essential Oils against *Streptococcus agalactiae*, an Important Pathogen for Fish Farming in Brazil. *J. Essent. Oil Res.* **2018**, *30*, 388–397. [CrossRef]
13. Meguro, S.; Hasumura, T.; Hase, T. Coffee Polyphenols Exert Hypocholesterolemic Effects in Zebrafish Fed a High-Cholesterol Diet. *Nutr. Metab.* **2013**, *10*, 61. [CrossRef] [PubMed]
14. Navarrete, P.; Toledo, I.; Mardones, P.; Opazo, R.; Espejo, R.; Romero, J. Effect of *Thymus vulgaris* Essential Oil on Intestinal Bacterial Microbiota of Rainbow Trout, *Oncorhynchus mykiss* (Walbaum) and Bacterial Isolates. *Aquac. Res.* **2010**, *41*, e667–e678. [CrossRef]
15. Schiewe, M.H.; Crosa, J.H. *Vibrio ordalii* sp. Nov.: A Causative Agent of Vibriosis in Fish. *Curr. Microbiol.* **1981**, *6*, 343–348. [CrossRef]
16. Wei, L.S.; Wee, W. Chemical Composition and Antimicrobial Activity of *Cymbopogon nardus* Citronella Essential Oil against Systemic Bacteria of Aquatic Animals. *Iran. J. Microbiol.* **2013**, *5*, 147. [PubMed]
17. Van Doan, H.; Hoseinifar, S.H.; Hung, T.Q.; Lumsangkul, C.; Jaturasitha, S.; El-Haroun, E.; Paolucci, M. Dietary Inclusion of Chestnut (*Castanea sativa*) Polyphenols to Nile Tilapia Reared in Biofloc Technology: Impacts on Growth, Immunity, and Disease Resistance against *Streptococcus agalactiae*. *Fish Shellfish Immunol.* **2020**, *105*, 319–326. [CrossRef]
18. Öntaş, C.; Baba, E.; Kaplaner, E.; Küçükaydin, S.; Öztürk, M.; Ercan, M.D. Antibacterial Activity of *Citrus limon* Peel Essential Oil and *Argania spinosa* Oil against Fish Pathogenic Bacteria. *Kafkas Üniv. Vet. Fak. Derg.* **2016**, *22*, 741–749.
19. El-Ekiaby, W.T. Basil Oil Nanoemulsion Formulation and Its Antimicrobial Activity against Fish Pathogen and Enhance Disease Resistance against *Aeromonas hydrophila* in Cultured Nile Tilapia. *Egypt. J. Aquac.* **2019**, *9*, 13–33. [CrossRef]
20. Abdel-Latif, H.M.; Abdel-Tawwab, M.; Khafaga, A.F.; Dawood, M.A. Dietary Origanum Essential Oil Improved Antioxidative Status, Immune-Related Genes, and Resistance of Common Carp (*Cyprinus carpio* L.) to *Aeromonas hydrophila* Infection. *Fish Shellfish Immunol.* **2020**, *104*, 1–7. [CrossRef]
21. Diler, O.; Gormez, O.; Diler, I.; Metin, S. Effect of Oregano (*Origanum onites* L.) Essential Oil on Growth, Lysozyme and Antioxidant Activity and Resistance against *Lactococcus garvieae* in Rainbow Trout, *Oncorhynchus mykiss* (Walbaum). *Aquac. Nutr.* **2017**, *23*, 844–851. [CrossRef]
22. Zhang, R.; Wang, X.W.; Liu, L.L.; Cao, Y.C.; Zhu, H. Dietary Oregano Essential Oil Improved the Immune Response, Activity of Digestive Enzymes, and Intestinal Microbiota of the Koi Carp, *Cyprinus carpio*. *Aquaculture* **2020**, *518*, 734781. [CrossRef]
23. Zheng, Z.L.; Tan, J.Y.; Liu, H.Y.; Zhou, X.H.; Xiang, X.; Wang, K.Y. Evaluation of Oregano Essential Oil (*Origanum heracleoticum* L.) on Growth, Antioxidant Effect and Resistance against *Aeromonas hydrophila* in Channel Catfish (*Ictalurus punctatus*). *Aquaculture* **2009**, *292*, 214–218. [CrossRef]
24. Charles, D.J.; Simon, J.E. Comparison of Extraction Methods for the Rapid Determination of Essential Oil Content and Composition of Basil. *J. Am. Soc. Hortic. Sci.* **1990**, *115*, 458–462. [CrossRef]
25. Orio, L.; Alexandru, L.; Cravotto, G.; Mantegna, S.; Barge, A. UAE, MAE, SFE-CO_2 and Classical Methods for the Extraction of *Mitragyna speciosa* Leaves. *Ultrason. Sonochem.* **2012**, *19*, 591–595. [CrossRef] [PubMed]
26. Devani, B.M.; Jani, B.L.; Balani, P.C.; Akbari, S.H. Optimization of Supercritical CO_2 Extraction Process for Oleoresin from Rotten Onion Waste. *Food Bioprod. Process.* **2020**, *119*, 287–295. [CrossRef]

27. Carović-Stanko, K.; Orlić, S.; Politeo, O.; Strikić, F.; Kolak, I.; Milos, M.; Satovic, Z. Composition and Antibacterial Activities of Essential Oils of Seven *Ocimum* Taxa. *Food Chem.* **2010**, *119*, 196–201. [CrossRef]
28. Lachowicz, K.J.; Jones, G.P.; Briggs, D.R.; Bienvenu, F.E.; Wan, J.; Wilcock, A.; Coventry, M.J. The Synergistic Preservative Effects of the Essential Oils of Sweet Basil (*Ocimum basilicum* L.) against Acid-Tolerant Food Microflora. *Lett. Appl. Microbiol.* **1998**, *26*, 209–214. [CrossRef]
29. Tenore, G.C.; Campiglia, P.; Ciampaglia, R.; Izzo, L.; Novellino, E. Antioxidant and Antimicrobial Properties of Traditional Green and Purple "Napoletano" Basil Cultivars (*Ocimum basilicum* L.) from Campania Region (Italy). *Nat. Prod. Res.* **2017**, *31*, 2067–2071. [CrossRef]
30. Umerie, S.C.; Anaso, H.U.; Anyasoro, L.J.C. Insecticidal Potentials of *Ocimum basilicum* Leaf-Extract. *Bioresour. Technol.* **1998**, *64*, 237–239. [CrossRef]
31. Vieira, R.F.; Simon, J.E. Chemical Characterization of Basil (*Ocimum* Spp.) Found in the Markets and Used in Traditional Medicine in Brazil. *Econ. Bot.* **2000**, *54*, 207–216. [CrossRef]
32. Politeo, O.; Jukic, M.; Milos, M. Chemical Composition and Antioxidant Capacity of Free Volatile Aglycones from Basil (*Ocimum basilicum* L.) Compared with Its Essential Oil. *Food Chem.* **2007**, *101*, 379–385. [CrossRef]
33. Di Giulio, R.T.; Washburn, P.C.; Wenning, R.J.; Winston, G.W.; Jewell, C.S. Biochemical Responses in Aquatic Animals: A Review of Determinants of Oxidative Stress. *Environ. Toxicol. Chem. Int. J.* **1989**, *8*, 1103–1123. [CrossRef]
34. Ji, L.L. Exercise and Oxidative Stress: Role of the Cellular Antioxidant Systems. *Exerc. Sport Sci. Rev.* **1995**, *23*, 135–166. [CrossRef] [PubMed]
35. Guo, X.; He, Y.; Zhang, L.; Lelong, C.; Jouaux, A. Immune and Stress Responses in Oysters with Insights on Adaptation. *Fish Shellfish Immunol.* **2015**, *46*, 107–119. [CrossRef] [PubMed]
36. Elia, A.C.; Pacini, N.; Fioravanti, M.L.; Dörr, A.J.M.; Zaccaroni, A.; Parmeggiani, A.M.; Gustinelli, A.; Mordenti, O.; Abete, M.C.; Prearo, M. Assessment of Detoxifying Markers for Florfenicol in Rainbow Trout Liver. *J. Aquat. Anim. Health* **2016**, *28*, 258–265. [CrossRef]
37. Elia, A.C.; Ciccotelli, V.; Pacini, N.; Dörr, A.J.M.; Gili, M.; Natali, M.; Gasco, L.; Prearo, M.; Abete, M.C. Transferability of Oxytetracycline (OTC) from Feed to Carp Muscle and Evaluation of the Antibiotic Effects on Antioxidant Systems in Liver and Kidney. *Fish Physiol. Biochem.* **2014**, *40*, 1055–1068. [CrossRef]
38. Elia, A.C.; Fanetti, A.; Dörr, A.J.M.; Taticchi, M.I. Effects of Concentrated Drinking Water Injection on Glutathione and Glutathione-Dependent Enzymes in Liver of *Cyprinus carpio* L. *Chemosphere* **2008**, *72*, 791–796. [CrossRef]
39. Khessiba, A.; Hoarau, P.; Gnassia-Barelli, M.; Aissa, P.; Roméo, M. Biochemical Response of the Mussel *Mytilus galloprovincialis* from Bizerta (Tunisia) to Chemical Pollutant Exposure. *Arch. Environ. Contam. Toxicol.* **2001**, *40*, 222–229. [CrossRef]
40. Magara, G.; Elia, A.C.; Dörr, A.J.M.; Abete, M.C.; Brizio, P.; Caldaroni, B.; Righetti, M.; Pastorino, P.; Scoparo, M.; Prearo, M. Metal Load and Oxidative Stress Driven by Organotin Compounds on Rainbow Trout. *Environ. Sci. Pollut. Res.* **2021**, *28*, 35012–35022. [CrossRef]
41. Magara, G.; Sangsawang, A.; Pastorino, P.; Oddon, S.B.; Caldaroni, B.; Menconi, V.; Kovitvadhi, U.; Gasco, L.; Meloni, D.; Dörr, A.J.M. First Insights into Oxidative Stress and Theoretical Environmental Risk of Bronopol and Detarox® AP, Two Biocides Claimed to Be Ecofriendly for a Sustainable Aquaculture. *Sci. Total Environ.* **2021**, *778*, 146375. [CrossRef]
42. Torricelli, P.; Elia, A.C.; Magara, G.; Feriotto, G.; Forni, C.; Borromeo, I.; De Martino, A.; Tabolacci, C.; Mischiati, C.; Beninati, S. Reduction of Oxidative Stress and Ornithine Decarboxylase Expression in a Human Prostate Cancer Cell Line PC-3 by a Combined Treatment with α-Tocopherol and Naringenin. *Amino Acids* **2021**, *53*, 63–72. [CrossRef]
43. Elia, A.C.; Prearo, M.; Dörr, A.J.M.; Pacini, N.; Magara, G.; Brizio, P.; Gasco, L.; Abete, M.C. Effects of Astaxanthin and Canthaxanthin on Oxidative Stress Biomarkers in Rainbow Trout. *J. Toxicol. Environ. Health Part A* **2019**, *82*, 760–768. [CrossRef] [PubMed]
44. Elia, A.C.; Burioli, E.; Magara, G.; Pastorino, P.; Caldaroni, B.; Menconi, V.; Dörr, A.J.M.; Colombero, G.; Abete, M.C.; Prearo, M. Oxidative Stress Ecology on Pacific Oyster *Crassostrea gigas* from Lagoon and Offshore Italian Sites. *Sci. Total Environ.* **2020**, *739*, 139886. [CrossRef] [PubMed]
45. Mannino, G.; Campobenedetto, C.; Vigliante, I.; Contartese, V.; Gentile, C.; Bertea, C.M. The Application of a Plant Biostimulant Based on Seaweed and Yeast Extract Improved Tomato Fruit Development and Quality. *Biomolecules* **2020**, *10*, 1662. [CrossRef] [PubMed]
46. Campobenedetto, C.; Agliassa, C.; Mannino, G.; Vigliante, I.; Contartese, V.; Secchi, F.; Bertea, C.M. A Biostimulant Based on Seaweed (*Ascophyllum nodosum* and *Laminaria digitata*) and Yeast Extracts Mitigates Water Stress Effects on Tomato (*Solanum lycopersicum* L.). *Agriculture* **2021**, *11*, 557. [CrossRef]
47. Gentile, C.; Mannino, G.; Palazzolo, E.; Gianguzzi, G.; Perrone, A.; Serio, G.; Farina, V. Pomological, Sensorial, Nutritional and Nutraceutical Profile of Seven Cultivars of Cherimoya (*Annona cherimola* Mill). *Foods* **2021**, *10*, 35. [CrossRef]
48. Prinsi, B.; Morgutti, S.; Negrini, N.; Faoro, F.; Espen, L. Insight into Composition of Bioactive Phenolic Compounds in Leaves and Flowers of Green and Purple Basil. *Plants* **2020**, *9*, 22. [CrossRef]
49. Pastorino, P.; Vela Alonso, A.I.; Colussi, S.; Cavazza, G.; Menconi, V.; Mugetti, D.; Righetti, M.; Barbero, R.; Zuccaro, G.; Fernández-Garayzábal, J.F. A Summer Mortality Outbreak of Lactococcosis by *Lactococcus garvieae* in a Raceway System Affecting Farmed Rainbow Trout (*Oncorhynchus mykiss*) and Brook Trout (*Salvelinus fontinalis*). *Animals* **2019**, *9*, 1043. [CrossRef]

50. Elia, A.C.; Magara, G.; Righetti, M.; Dörr, A.J.M.; Scanzio, T.; Pacini, N.; Abete, M.C.; Prearo, M. Oxidative Stress and Related Biomarkers in Cupric and Cuprous Chloride-Treated Rainbow Trout. *Environ. Sci. Pollut. Res.* **2017**, *24*, 10205–10219. [CrossRef]
51. Lowry, O.H.; Rosebrough, N.J.; Farr, A.L.; Randall, R.J. Protein Measurement with the Folin Phenol Reagent. *J. Biol. Chem.* **1951**, *193*, 265–275. [CrossRef]
52. Akerboom, T.P.; Sies, H. Assay of Glutathione, Glutathione Disulfide, and Glutathione Mixed Disulfides in Biological Samples. In *Methods in Enzymology*; Elsevier: Amsterdam, The Netherlands, 1981; Volume 77, pp. 373–382. ISSN 0076-6879.
53. Pacini, N.; Prearo, M.; Abete, M.C.; Brizio, P.; Dörr, A.J.M.; Reimschuessel, R.; Andersen, W.; Gasco, L.; Righetti, M.; Elia, A.C. Antioxidant Responses and Renal Crystal Formation in Rainbow Trout Treated with Melamine Administered Individually or in Combination with Cyanuric Acid. *J. Toxicol. Environ. Health Part A* **2013**, *76*, 491–508. [CrossRef]
54. Lopes, F.C.M.; Benzatti, F.P.; Jordão Junior, C.M.; Moreira, R.R.D.; Carlos, I.Z. Effect of the Essential Oil of *Achillea millefolium* L. in the Production of Hydrogen Peroxide and Tumor Necrosis Factor-Alpha in Murine Macrophages. *Rev. Bras. Ciênc. Farm.* **2005**, *41*, 401–405. [CrossRef]
55. Lizárraga-Velázquez, C.E.; Hernández, C.; González-Aguilar, G.A.; Heredia, J.B. Effect of Dietary Intake of Phenolic Compounds from Mango Peel Extract on Growth, Lipid Peroxidation and Antioxidant Enzyme Activities in Zebrafish (*Danio rerio*). *Lat. Am. J. Aquat. Res.* **2019**, *47*, 602–611. [CrossRef]
56. Bhattacharjee, P.; Das, S. Deltamethrin Induced Alteration of Biochemical Parameters in *Channa punctata*, Bloch and Its Amelioration by Quercetin. *Bull. Environ. Contam. Toxicol.* **2017**, *98*, 763–769. [CrossRef] [PubMed]
57. Shin, H.S.; Yoo, J.H.; Min, T.S.; Lee, K.Y.; Choi, C.Y. The Effects of Quercetin on Physiological Characteristics and Oxidative Stress Resistance in Olive Flounder, *Paralichthys olivaceus*. *Asian-Australas. J. Anim. Sci.* **2010**, *23*, 588–597. [CrossRef]
58. Shin, H.S.; Yoo, J.H.; Min, T.S.; Lee, J.; Choi, C.Y. Effect of Quercetin on the Activity and MRNA Expression of Antioxidant Enzymes and Physiological Responses in Olive Flounder (*Paralichthys olivaceus*) Exposed to Cadmium. *Asian-Australas. J. Anim. Sci.* **2010**, *23*, 742–749. [CrossRef]
59. Li, E.; Wang, Y.; Li, Q.; Li, L.; Wei, L. Protective Effects of Sal B on Oxidative Stress-Induced Aging by Regulating the Keap1/Nrf2 Signaling Pathway in Zebrafish. *Molecules* **2021**, *26*, 5239. [CrossRef]
60. Pês, T.S.; Saccol, E.M.; Ourique, G.M.; Londero, É.P.; Gressler, L.T.; Golombieski, J.I.; Glanzner, W.G.; Llesuy, S.F.; Gonçalves, P.B.; Neto, J.R. Quercetin in the Diet of Silver Catfish: Effects on Antioxidant Status, Blood Parameters and Pituitary Hormone Expression. *Aquaculture* **2016**, *458*, 100–106. [CrossRef]
61. Pês, T.S.; Saccol, E.M.; Ourique, G.M.; Londero, É.P.; Gressler, L.T.; Finamor, I.A.; Rotili, D.A.; Golombieski, J.I.; Glanzner, W.G.; Llesuy, S.F. Effect of Diets Enriched with Rutin on Blood Parameters, Oxidative Biomarkers and Pituitary Hormone Expression in Silver Catfish (*Rhamdia quelen*). *Fish Physiol. Biochem.* **2016**, *42*, 321–333. [CrossRef]
62. Gressler, L.T.; Riffel, A.P.K.; Parodi, T.V.; Saccol, E.M.H.; Koakoski, G.; da Costa, S.T.; Pavanato, M.A.; Heinzmann, B.M.; Caron, B.; Schmidt, D. Silver Catfish *Rhamdia quelen* Immersion Anaesthesia with Essential Oil of *Aloysia triphylla* (L'Hérit) Britton or Tricaine Methanesulfonate: Effect on Stress Response and Antioxidant Status. *Aquac. Res.* **2014**, *45*, 1061–1072. [CrossRef]
63. Zeppenfeld, C.C.; Saccol, E.M.H.; Pês, T.S.; Salbego, J.; Koakoski, G.; dos Santos, A.C.; Heinzmann, B.M.; da Cunha, M.A.; Barcellos, L.J.G.; Pavanato, M.A. *Aloysia triphylla* Essential Oil as Food Additive for *Rhamdia quelen*–Stress and Antioxidant Parameters. *Aquac. Nutr.* **2017**, *23*, 1362–1367. [CrossRef]
64. Saccol, E.M.; Toni, C.; Pês, T.S.; Ourique, G.M.; Gressler, L.T.; Silva, L.V.; Mourão, R.H.; Oliveira, R.B.; Baldisserotto, B.; Pavanato, M.A. Anaesthetic and Antioxidant Effects of *Myrcia sylvatica* (G. Mey.) DC. and *Curcuma longa* L. Essential Oils on Tambaqui (*Colossoma macropomum*). *Aquac. Res.* **2017**, *48*, 2012–2031. [CrossRef]
65. Baba, E.; Acar, Ü.; Öntaş, C.; Kesbiç, O.S.; Yılmaz, S. Evaluation of *Citrus limon* Peels Essential Oil on Growth Performance, Immune Response of Mozambique Tilapia *Oreochromis mossambicus* Challenged with *Edwardsiella tarda*. *Aquaculture* **2016**, *465*, 13–18. [CrossRef]
66. De Freitas Souza, C.; Baldissera, M.D.; Bianchini, A.E.; da Silva, E.G.; Mourão, R.H.V.; da Silva, L.V.F.; Schmidt, D.; Heinzmann, B.M.; Baldisserotto, B. Citral and Linalool Chemotypes of *Lippia alba* Essential Oil as Anesthetics for Fish: A Detailed Physiological Analysis of Side Effects during Anesthetic Recovery in Silver Catfish (*Rhamdia quelen*). *Fish Physiol. Biochem.* **2018**, *44*, 21–34. [CrossRef] [PubMed]
67. Mirghaed, A.T.; Hoseini, S.M.; Ghelichpour, M. Effects of Dietary 1, 8-Cineole Supplementation on Physiological, Immunological and Antioxidant Responses to Crowding Stress in Rainbow Trout (*Oncorhynchus mykiss*). *Fish Shellfish Immunol.* **2018**, *81*, 182–188. [CrossRef]
68. Aanyu, M.; Betancor, M.B.; Monroig, O. Effects of Dietary Limonene and Thymol on the Growth and Nutritional Physiology of Nile Tilapia (*Oreochromis niloticus*). *Aquaculture* **2018**, *488*, 217–226. [CrossRef]
69. Köprücü, K.; Yonar, M.E.; Özcan, S. Effect of Dietary N-3 Polyunsaturated Fatty Acids on Antioxidant Defense and Sperm Quality in Rainbow Trout (*Oncorhynchus mykiss*) under Regular Stripping Conditions. *Anim. Reprod. Sci.* **2015**, *163*, 135–143. [CrossRef]
70. Saccol, E.M.; Uczay, J.; Pês, T.S.; Finamor, I.A.; Ourique, G.M.; Riffel, A.P.; Schmidt, D.; Caron, B.O.; Heinzmann, B.M.; Llesuy, S.F. Addition of *Lippia alba* (Mill) NE Brown Essential Oil to the Diet of the Silver Catfish: An Analysis of Growth, Metabolic and Blood Parameters and the Antioxidant Response. *Aquaculture* **2013**, *416*, 244–254. [CrossRef]
71. Azambuja, C.R.; Mattiazzi, J.; Riffel, A.P.K.; Finamor, I.A.; de Oliveira Garcia, L.; Heldwein, C.G.; Heinzmann, B.M.; Baldisserotto, B.; Pavanato, M.A.; Llesuy, S.F. Effect of the Essential Oil of *Lippia alba* on Oxidative Stress Parameters in Silver Catfish (*Rhamdia quelen*) Subjected to Transport. *Aquaculture* **2011**, *319*, 156–161. [CrossRef]

72. Salbego, J.; Toni, C.; Becker, A.G.; Zeppenfeld, C.C.; Menezes, C.C.; Loro, V.L.; Heinzmann, B.M.; Baldisserotto, B. Biochemical Parameters of Silver Catfish (*Rhamdia quelen*) after Transport with Eugenol or Essential Oil of *Lippia alba* Added to the Water. *Braz. J. Biol.* **2017**, *77*, 696–702. [CrossRef]
73. Sönmez, A.Y.; Bilen, S.; Alak, G.; Hisar, O.; Yanık, T.; Biswas, G. Growth Performance and Antioxidant Enzyme Activities in Rainbow Trout (*Oncorhynchus mykiss*) Juveniles Fed Diets Supplemented with Sage, Mint and Thyme Oils. *Fish Physiol. Biochem.* **2015**, *41*, 165–175. [CrossRef]
74. Souza, C.F.; Baldissera, M.D.; Silva, L.d.L.; Geihs, M.A.; Baldisserotto, B. Is Monoterpene Terpinen-4-Ol the Compound Responsible for the Anesthetic and Antioxidant Activity of *Melaleuca alternifolia* Essential Oil (Tea Tree Oil) in Silver Catfish? *Aquaculture* **2018**, *486*, 217–223. [CrossRef]
75. Perron, N.R.; Brumaghim, J.L. A Review of the Antioxidant Mechanisms of Polyphenol Compounds Related to Iron Binding. *Cell Biochem. Biophys.* **2009**, *53*, 75–100. [CrossRef]
76. Salbego, J.; Becker, A.G.; Gonçalves, J.F.; Menezes, C.C.; Heldwein, C.G.; Spanevello, R.M.; Loro, V.L.; Schetinger, M.R.C.; Morsch, V.M.; Heinzmann, B.M. The Essential Oil from *Lippia alba* Induces Biochemical Stress in the Silver Catfish (*Rhamdia quelen*) after Transportation. *Neotrop. Ichthyol.* **2014**, *12*, 811–818. [CrossRef]
77. Zeppenfeld, C.C.; Toni, C.; Becker, A.G.; dos Santos Miron, D.; Parodi, T.V.; Heinzmann, B.M.; Barcellos, L.J.G.; Koakoski, G.; da Rosa, J.G.S.; Loro, V.L. Physiological and Biochemical Responses of Silver Catfish, *Rhamdia quelen*, after Transport in Water with Essential Oil of *Aloysia triphylla* (L'Herit) Britton. *Aquaculture* **2014**, *418*, 101–107. [CrossRef]
78. Morais, S.; Pratoomyot, J.; Taggart, J.B.; Bron, J.E.; Guy, D.R.; Bell, J.G.; Tocher, D.R. Genotype-Specific Responses in Atlantic Salmon (*Salmo salar*) Subject to Dietary Fish Oil Replacement by Vegetable Oil: A Liver Transcriptomic Analysis. *BMC Genom.* **2011**, *12*, 255. [CrossRef] [PubMed]
79. Yousefi, M.; Vatnikov, Y.A.; Kulikov, E.V.; Ghelichpour, M. Change in Blood Stress and Antioxidant Markers and Hydromineral Balance of Common Carp (*Cyprinus carpio*) Anaesthetized with Citronellal and Linalool: Comparison with Eugenol. *Aquac. Res.* **2019**, *50*, 1313–1320. [CrossRef]
80. Velíšek, J.; Stejskal, V.; Kouřil, J.; Svobodová, Z. Comparison of the Effects of Four Anaesthetics on Biochemical Blood Profiles of Perch. *Aquac. Res.* **2009**, *40*, 354–361. [CrossRef]
81. Al-Sagheer, A.A.; Mahmoud, H.K.; Reda, F.M.; Mahgoub, S.A.; Ayyat, M.S. Supplementation of Diets for *Oreochromis niloticus* with Essential Oil Extracts from Lemongrass (*Cymbopogon citratus*) and Geranium (*Pelargonium graveolens*) and Effects on Growth, Intestinal Microbiota, Antioxidant and Immune Activities. *Aquac. Nutr.* **2018**, *24*, 1006–1014. [CrossRef]
82. Jia, R.; Li, Y.; Cao, L.; Du, J.; Zheng, T.; Qian, H.; Gu, Z.; Jeney, G.; Xu, P.; Yin, G. Antioxidative, Anti-Inflammatory and Hepatoprotective Effects of Resveratrol on Oxidative Stress-Induced Liver Damage in Tilapia (*Oreochromis niloticus*). *Comp. Biochem. Physiol. Part C Toxicol. Pharmacol.* **2019**, *215*, 56–66. [CrossRef]

antioxidants

MDPI

Article

Chemical Composition and Antioxidant Activity of *Ammi visnaga* L. Essential Oil

Fatima Zahra Kamal [1,2], Gabriela Dumitrita Stanciu [3,*], Radu Lefter [4], Valeriu V. Cotea [5], Marius Niculaua [6], Daniela Carmen Ababei [7], Alin Ciobica [8,*] and Abdellah Ech-Chahad [1]

1 Laboratory of Physical Chemistry of Processes and Materials, Faculty of Sciences and Techniques, Hassan First University, B.P. 539, Settat 26000, Morocco; fatimzahra.kamal@gmail.com (F.Z.K.); echchahad@gmail.com (A.E.-C.)
2 Laboratory of Agri-Food and Health, Faculty of Sciences and Techniques, Hassan First University, B.P. 539, Settat 26000, Morocco
3 Advanced Research and Development Center for Experimental Medicine (CEMEX), Grigore T. Popa University of Medicine and Pharmacy, 16 Universitatii Street, 700115 Iași, Romania
4 Biomedical Research Center, Romanian Academy, Iași Branch, 8th Carol I Avenue, 700506 Iași, Romania; radu_lefter@yahoo.com
5 Department of Oenology, "Ion Ionescu de la Brad" University of Life Sciences, 3rd M. Sadoveanu Alley, 700490 Iași, Romania; vcotea@uaiasi.ro
6 Research Centre for Oenology Iași, Romanian Academy, Iași Branch, 9th M. Sadoveanu Alley, 700505 Iași, Romania; niculaua@gmail.com
7 Pharmacodynamics and Clinical Pharmacy Department, Grigore T. Popa University of Medicine and Pharmacy, 16 Universitatii Street, 700115 Iași, Romania; dana.ababei@gmail.com
8 Department of Biology, Faculty of Biology, Alexandru Ioan Cuza University, 11th Carol I Avenue, 700506 Iași, Romania
* Correspondence: gabriela-dumitrita.s@umfiasi.ro (G.D.S.); alin.ciobica@uaic.ro (A.C.)

Citation: Kamal, F.Z.; Stanciu, G.D.; Lefter, R.; Cotea, V.V.; Niculaua, M.; Ababei, D.C.; Ciobica, A.; Ech-Chahad, A. Chemical Composition and Antioxidant Activity of *Ammi visnaga* L. Essential Oil. *Antioxidants* **2022**, *11*, 347. https://doi.org/10.3390/antiox11020347

Academic Editors: Delia Mirela Tit, Simona Bungau and Filomena Nazzaro

Received: 22 December 2021
Accepted: 8 February 2022
Published: 10 February 2022

Abstract: The present study evaluated the chemical composition and the in vitro and in vivo antioxidant potential of *Ammi visnaga* L. essential oil to provide a scientific basis for the use of this plant in the traditional pharmacopoeia. Gas chromatography-mass spectrometry was used to identify the volatile constituents present of the oil. The in vitro antioxidant capacity was evaluated by the DPPH and the reducing power assays. For the in vivo tests, oral administration of *Ammi visnaga* L. oil (600 and 1200 mg/kg body weight) was performed in Swiss albino mice treated with acetaminophen (400 mg/kg). The toxic effect of acetaminophen and the action of the essential oil were measured by determining the levels of lipid peroxidation and antioxidant enzymes in liver and kidneys homogenates. The major components identified were butanoic acid, 2-methyl-, pentyl ester, (Z)-β-ocimene, D-limonene, linalool, pulegone and lavandulyl-butyrate. The in vitro DPPH and reducing power assays showed moderate to low free radical scavenging activity and the antioxidant power was positively correlated with the polyphenols' concentration. In vivo, the *Ammi visnaga* L. essential oil showed a high antioxidant capacity at both concentrations (600 and 1200 mg/kg), effectively increasing the levels of reduced glutathione, superoxide dismutase, and catalase and significantly reducing the lipid peroxidation. The results obtained from this study suggest that *Ammi visnaga* L. could represent a source of molecules with antioxidant potential in the prevention of free radical-related diseases.

Keywords: *Ammi visnaga* L.; chemical composition; antioxidant activity; essential oil; gas chromatography-mass spectrometry; oxidative stress

1. Introduction

The reactive oxygen species (ROS) have been shown to act as a backbone of the immune system and to form the key component of the phagocytic cells and apoptotic processes, when produced at moderate concentrations through endogenous processes by mitochondria and various intracellular enzymes (glucose oxidase, myeloperoxidase, phospholipase A, etc.) and exogenous sources, such as environmental agents (pollution, radiation, and UV), cigarette smoking, drugs, and certain food [1]. However, when the ROS production rate exceeds the disposal rate, this leads to oxidative stress, a harmful molecular state, responsible for the oxidation of proteins, lipids, and DNA [1]. The damage to the structure and functioning of these macro-molecules, and the misinterpretation of protein translation mechanisms and genetic information arising thereof, can end up in several human diseases, including coronary heart disease, Alzheimer's disease, and aging [2–4].

To thwart the disease-mediator oxidative damage, the tissues possess endogenous antioxidant defense systems of non-enzymatic type (glutathione, bilirubin, and coenzyme Q10) and enzymatic type, such as superoxide dismutase (SOD), catalase (CAT), and glutathione peroxidase (GSH-Px) [5]. These molecules ensure the scission of the auto-oxidative chain reaction, the conversion of radical species to non-radical or less toxic species and diminish the localized oxygen concentrations [5]. However, under the pressure of a wide variety of cellular homeostasis disrupting factors, ranging from metabolic [6] and mental stress conditions [7], to lifestyle [8], dietary [9], and aging conditions [10,11], endogenous antioxidants may prove insufficient to neutralize ROS generation and, therefore, call for exogenous antioxidant intake (dietary) to maintain optimal cellular functions [9,12]. Among the natural antioxidants, secondary and primary metabolites present in aromatic and medicinal plants have been shown to have a powerful antioxidant potential [13].

The antioxidant capacity of the Apiaceae family (Umbelliferae) has been extensively studied, showing very promising potential [14–16]. *Ammi visnaga* L. (*Daucus visnaga* L., *Visnaga daucoides* Gaertn.), a member of the Apiaceae, commonly known as Toothpick weed in England, or Khella in Arab countries, has only recently been reviewed for its numerous curative properties, out of which the antioxidant ones have been approached by very few studies [17]. A Northern hemisphere widespread biennial or annual herbaceous plant [17,18], *A. visnaga* L. presents an erect, cylindrical, and highly branched stem reaching 130 cm in height, covered with greyish green foliage. Its white flowers are grouped into umbels of 6–10 cm in diameter and the fruits are tiny, ovoid, and smooth formed of two partial mericarps, each 2 mm long greyish brown in color when the plant is dry [17]. In the popular pharmacopoeia, the plant is used as dried powder or boiled in water to treat renal colic, mild angina symptoms, asthma, spastic bronchitis, abdominal cramps, urinary calculi, vitiligo and psoriasis, vertigo, diabetes, and kidney stones [18]. In the pharmaceutical industry, various prescription drugs, such as amiodarone (ventricular arrhythmias), nifedipine (treatment of stable, variant, and unstable angina, mild to severe hypertension, and Raynaud's phenomenon), and cromolyn (mastocytosis) are derived from *A. visnaga* L. [18].

In the frame of the abovementioned literature data, the present study aimed to investigate/establish the chemical composition of the essential oil of *A. visnaga* L. (from northwest Morocco), testing (by the way for the first time to our knowledge) both its in vitro and in vivo antioxidant activity.

2. Materials and Methods

2.1. Chemical Characterization of *A. visnaga* Essential Oil

2.1.1. Plant Material

Healthy samples of *A. visnaga* L., without signs of contamination (fungal, bacterial, or viral), were collected at dawn during the June–July months, 2021, at flowering stage in their natural habitat, from Ouazzane (North Morocco) (34°48′ North, 5°36′ West, Altitude 614 m). After taxonomic identification, a specimen was deposited in the herbarium for future reference (Voucher n° 0356/M) at the Faculty of Sciences and Technologies, Hassan 1st University, Settat, Morocco. The plant was carefully washed with sterile distilled water, to remove dust and foreign matter, then shade dried and ground to fine powder using an electric grinder.

2.1.2. Isolation of Essential Oil

A quantity of 500 g of *A. visnaga* powder were placed in a flask with 4 L of distilled water, and then subjected to hydro distillation for 6 h using a Clevenger type apparatus (JP Selecta, Barcelona, Spain). The golden-yellow essential oils obtained were dried on sodium anhydrous sulphate, and then stored at +4 °C in amber glass bottles with screw caps (to avoid the negative effect of light until the test).

2.1.3. Chemicals and Reagents

Ascorbic acid (AA), aluminum chloride ($AlCl_3$), iron chloride ($FeCl_3$), Folin-Ciocalteu reagent, 2,2′-Diphenyl-1-picrylhydrazyl (DPPH), nitro blue tetrazolium (NBT), reduced glutathione (GSH), superoxide dismutase (SOD), catalase (CAT) from bovine liver, 1,2-dithiobis nitro benzoic acid (DTNB), thiobarbituric acid (TBA) and trichloroacetic acid (TCA), acetaminophen (APAP), catechin, quercetin, ethylenediaminetetraacetic acid (EDTA), gallic acid, vanillin, nicotinamide-adenine dinucleotide phosphate (NADPH), butylated hydroxytoluene, phenazine methosulfate, sodium carboxymethyl cellulose (CMC), and n-alkanes (C6-C30) were purchased from Sigma Co. (St. Louis, MO, USA). Potassium phosphate monobasic, sodium pyrophosphate dibasic, sodium carbonate (Na_2CO_3), sodium hydroxide (NaOH), sodium nitrite ($NaNO_2$), disodium hydrogen phosphate (Na_2HPO_4), hydrogen peroxide (H_2O_2), potassium ferricyanide [$K_3Fe(CN)_6$], sodium sulphate anhydrous (Na_2SO_4), hydrochloric acid, acetic acid (ACA), and n-butanol (99.8%) were of analytical grade and purchased from Merck (Nottingham, UK).

2.1.4. Gas Chromatography–Mass Spectrometry of Essential Oil

The gas chromatography–mass spectrometry (GC-MS) analysis was performed without derivatization using gas chromatograph (Agilent 7890A Series) coupled to a mass spectrometer (MS) (5975C) (Agilent Technologies, Santa Clara, CA, USA) equipped with a multimode injector and a 123-BD11 column (ASTM D6584) (Agilent Technologies, Santa Clara, CA, USA) with a dimension of 15 m × 320 μm × 0.1 μm at the Moroccan Foundation for Advanced Science, Innovation and Research (MAScIR) Institute. In total, 10 μL of the liquid samples were dissolved in an appropriate volume of chloroform. Then, 4 μL of the soluble extract was injected into the column by 1:5 split mode using helium as carrier gas at a flow rate of 2 mL min^{-1}. The composition of the essential oil determined from the peak areas was calculated as a percentage of the total compounds existing in the sample detection using full scan mode between 30–1000 m/z, with a gain factor of 5 and electron impact ionization. The temperatures of the ion source and the quadrupoles were 230 and 150 °C, respectively. The oven temperature was programmed at 30 °C for 3 min and then increased by 10 °C min^{-1} to 250 °C. The compounds identification was carried out using the NIST 2017 MS Library (https://chemdata.nist.gov/dokuwiki/doku.php?id=chemdata:start (accessed on 25 January 2022)). A series of n-alkanes (C6–C30) were used in this experiment to calculate the Retention Index.

2.2. The In Vitro Antioxidant Assay

2.2.1. Total Polyphenol Determination

The total phenolic content was determined by the colorimetric method of Folin–Ciocalteu described by Singleton and Rossi [19]. A measured 2 mL of sodium carbonate Na_2CO_3 (75 g/L) was added to the mixture of 0.5 mL of sample and 2.5 mL of 10% Folin-Ciocalteu reagent. The absorbance was measured at 760 nm, after 30 min of incubation at room temperature, and in the dark against the blank. The results were expressed as gallic acid equivalent per gram of extract (mg GAE/g). Three tests were performed for each sample.

2.2.2. Total Flavonoid Determination

Total flavonoid content was determined by the method described by Zhishen et al. and Kim et al. [20,21]. A volume of 500 μL of oil (prepared in methanol) was mixed with 120 μL of 5% $NaNO_2$. After 5 min, 120 μL of 10% $AlCl_3$ (freshly prepared) was added; the medium was shaken to homogenize the contents. After 6 min of incubation at room temperature and in the dark, 800 μL of NaOH (1M) was added. The solution was thoroughly homogenized, and the absorbance was read immediately at 430 nm against a blank. The standard curve for total flavonoids was made using quercetin standard solution. The results were expressed as milligrams of quercetin equivalent per gram of extracts (mg Quercetin Equivalent (QE)/g).

2.2.3. Condensed Tannins Determination

The content determination of condensed tannins (known as proanthocyanidins) was done by using the colorimetric method; this method is based on the depolymerization of these molecules in the presence of sulfuric acid, followed by the formation of anthocyanidols of a specific red color, in the presence of vanillin, analyzable at 500 nm. In a test tube, 1 mL aliquot of the extract is added to 3 mL of 4% vanillin and 1.5 mL of concentrated HCl. After 15 min of incubation at room temperature, the absorbance was measured at 500 nm. The tannin content was determined from the calibration curve, performed in parallel under the same operating conditions using catechin as positive control [22].

2.2.4. 2,2′-Diphenyl-1-picrylhydrazyl (DPPH) Assay

The free radical scavenging activity of *A. visnaga* L. was measured using the DPPH assay. Quantitative estimation of radical scavenging activity was measured according to the protocol described by Bougandoura et al. [23]. In total, 2 mL of freshly prepared 0.1 mM methanolic DPPH was added to essential oil (0.5 mL; 0–1 mg/mL), and the mixture was shaken vigorously for 5 min using a vortex, and incubated 30 min in the dark, at room temperature. Absorbance was measured at 517 nm with a VR-2000 spectrophotometer (JP Selecta, Barcelona, Spain) against a blank, and compared with standard (butylated hydroxytoluene (BHT)). Scavenging activity was expressed as the percentage inhibition (I%) calculated by the following equation [24]:

$$I\% = (A_{blank} - A_{sample}/A_{blank}) \times 100$$

The IC_{50} was calculated graphically by linear regression of I% versus concentrations (C).

2.2.5. Reducing Power Assay

The reducing power of *A. visnaga* essential oil was determined by the method of Oyaizu [25]. A volume of 1 mL of samples or standard antioxidants was mixed with phosphate buffer (2.5 mL, 0.2 M, pH 6.6) and 2.5 mL of 1% *w/v* potassium ferricyanide $K_3[Fe(CN)_6]$. After 20 min of incubation at 50 °C in the dark, 2.5 mL of 10% TCA was added to stop the reaction; the tubes were centrifuged at 3000 rpm for 10 min. Then, 2.5 mL of supernatant was mixed with 2.5 mL of distilled water and 0.5 mL of 0.1% aqueous $FeCl_3$ solution. The absorbance was read at 700 nm against a blank and compared with standard (AA 0–1 mg/mL). The results were expressed as mg AA per g sample. All tests were performed in triplicates.

2.3. *The In Vivo Antioxidant Assay*

2.3.1. Animal Models and Induction of Oxidative Stress

Thirty adult male Swiss albino mice, with an initial body weight of 25–35 g, were housed in polypropylene cages (5 mice/cage) containing wood shavings bedding, provided with a label holder mentioning the name of the batch, and placed in a controlled environment (T = 22 ± 3 °C, relative humidity 40–70%, 12 h light/dark cycles "lights on at 7:00 a.m.", with standard mice chow and water ad libitum). For the induction of oxidative stress, mice received APAP (400 mg/kg) by intraperitoneal injection (ip) once, as an acute toxic dose.

2.3.2. Study Design

The test consisted of measuring the effects of different doses of the essential oil of *A. visnaga* L. on the biochemical parameters of the tissue homogenates of mice GSH, CAT, SOD, and malondialdehyde (MDA). For this purpose, mice were randomly divided into six groups ($n = 5$):

- Group I was designated as vehicle and was treated with 0.1% CMC;
- Group II (negative control) received no treatment but had free access to water and food;
- Groups III (toxic control), IV, V, and VI received a single intraperitoneal injection of APAP (400 mg/kg, ip) before the start of the experiment to induce hepato-renal oxidative injury. Group IV served as the standard and received AA, 200 mg/kg body weight. Groups V and VI received *A. visnaga* L. essential oil at doses of 600 and 1200 mg/kg body weight.

These doses were selected following a screening procedure in which we tested three doses (600, 1200, and 1800 mg/kg) of *A. visnaga* L. essential oil by oral administration in rats for 2 weeks (unpublished results). At the highest concentration (1800 mg/kg), we observed installation of the LD50, whereas the other two doses did not provoke signs or symptoms of toxicity and were selected for further antioxidant studies. The animals were treated orally, once a day (at 9 a.m.) for two weeks (14 days). A 10-day quarantine was observed before treatment [26].

2.3.3. Body and Organ Weight

The mice were weighed twice (at baseline to endpoint), and the changes in body weight were recorded. After the mice were sacrificed, the liver and kidneys were weighed accurately, and the relative weight was calculated as follows:

Relative organ weight = (absolute organ weight (g) × 100)/body weight of mice on sacrifice day (g).

2.3.4. Preparation of Tissue Homogenates

After euthanasia of the mice, the liver and kidney samples were collected and rinsed immediately in ice-cold saline, dried, and weighed. Then, the samples were ground and homogenized in cold potassium phosphate (50 mM, pH 7.0, containing 1 mM EDTA) per gram of tissue. The homogenate was centrifuged at 10,000× *g* for 15 min at 4 °C, to purge cellular debris and the supernatants were collected [27]. The levels of MDA, GSH, CAT, and SOD were measured.

2.3.5. Assessment of Oxidative Stress Biomarkers in Tissue Homogenate

Evaluation of the Enzymatic Activity of Catalase

CAT activity was determined spectrophotometrically according to the method described by Aebi [28]. The decomposition of hydrogen peroxide into water and oxygen by the CAT enzyme was observed using a spectrophotometer (240 nm, 1 min, 25 °C). Enzyme activity was determined as the unit of activity corresponding to nmol H_2O_2 destroyed/min/mL [29]. Measured 100 µL of tissue homogenate was added to 1.9 mL of phosphate buffer (50 mmol/L, pH 7.0) and 1 mL of freshly prepared H_2O_2 (2 mmol/L) and the mixture was transferred to the cuvette. The blank and standard were performed in the same manner, without tissue homogenate and with CAT instead of tissue homogenate, respectively.

The reading was taken spectrophotometrically at wavelength $\lambda = 240$ nm against the blank for 1 min. The results were expressed as units/mg of protein [28].

Evaluation of the Enzymatic Activity of Superoxide Dismutase

The evaluation of SOD is based on the reduction of nitro blue tetrazolium (NBT) to water-insoluble formazan blue; SOD inhibits the reduction of NBT, which can be measured at 560 nm by spectrophotometry [30]. A reaction mixture consisting of 200 μL of tissue homogenate, 1.2 mL of sodium pyrophosphate buffer (0.025 mol/L, pH 8.3), 100 μL of phenazine methosulfate (186 μmol/L), and 300 μL of NBT (300 μmol/L) was introduced into a test tube. The reaction was triggered by the addition of 200 μL of NADPH (780 μmol/L). After incubation at 30 °C for 90 s, a volume of 1 mL of glacial ACA was added to stop this reaction. Then, the reaction mixture was vigorously stirred with 4 mL n-butanol, allowed to stand for 10 min, and centrifuged. The butanol layer was separated and the intensity of the chromogen in the butanol layer was read at 560 nm against a blank. The calibration curve was performed with concentrations ranging from 10 to 100 μL of SOD as a standard. A linear regression equation was used to measure the activity of SOD [31].

Evaluation of the Enzymatic Activity of Reduced Glutathione

For the determination of GSH, the method described by Ellman was used [32]. In total, 50 μL of tissue homogenates were diluted in a phosphate buffer (10 mL, 0.1 M, pH 8). Then, 3 mL of the dilution mixture was transferred to cleaned test tubes, and 20 μL of DTNB (0.01 M) was added. The mixture was allowed to stand for the next 5 min, and then the absorbance was measured at a wavelength $\lambda = 412$ nm against a blank prepared under the same conditions without tissue homogenate [27,32].

Estimation of Lipid Peroxidation

The lipid peroxidation reflected by MDA levels was assessed using the method of Sastre et al. [33]. The MDA, the product of polyunsaturated fatty acid peroxidation in cells, is a good marker for the free radical induced damage and oxidative stress. The principle of this assay is based on the condensation of one molecule of MDA in a warm acidic medium (pH 2 to 3, 90–100 °C) with two molecules of TBA for a pink-colored chromogen that can be measured at 532 nm. A total of 500 μL of liver homogenate samples were added to 500 μL of 10% TCA acid and 1 mL of 0.67% TBA and centrifuged at 3000 rpm for 15 min. The collected supernatant was heated to 100 °C, in a water bath, for 15 min. After cooling, n-butanol was added to neutralize the mixture and the optical density of the samples was measured at 535 nm. The amount of MDA in the sample was expressed as mmol/g of tissue extract (liver and kidneys).

2.4. Statistical Analysis

All in vitro experiments were performed at least in triplicate. Analyses of all samples were run in triplicate and averaged. The data were statistically analyzed using one-way ANOVA followed by Dunnett's test using GraphPad Prism 8.4.3 (GraphPad Software Inc., San Diego, CA, USA), and the values were expressed as mean ± SD. Pearson correlation analysis was used to evaluate the relationship between body and organ weight. A probability of $p < 0.05$ was considered as significant.

3. Results

3.1. Chemical Characterization of *A. visnaga* L. Essential Oil

The hydro distillation of *A. visnaga* L. plant gave a golden yellow essential oil, and the percentage yield was calculated to be 1.4% (*v*/*w* fresh material). The GC-MS analysis of the essential oil of *A. visnaga* L. resulted in a total of 33 compounds identified and eluted from between 1.195 and 23.902 min; β-ocimene, D-limonene, propanoic acid, 2-methyl-2-methylbutyl ester, linalool, pulegone, lavandulyl isobutyrate, and β-myrcene were the main compounds identified in the essential oil of *A. visnaga* L. with 86% comparison with the Wiley and NIST libraries, as it is presented in Table 1.

Table 1. Chemical composition of *A. visnaga* L. essential oil.

Compounds	[a] IR	[b] IR	Area %
Propanoic acid, 2-methyl-, butyl ester	1011	944	3.40
Bicyclo[3.1.0]hexane, 4-methylene-1-(1-methylethyl)-	1018	987	1.23
β-myrcene	1022	1003	1.36
Butanoic acid, 2-methyl-, 2-methyl propyl ester	1025	1015	4.39
Butane, 1-(ethenyloxy)-3-methyl-	1030	-	12.28
4-Carene	1034	1030	1.61
(Z)-β-ocimene	1037	1037	4.03
Butanoic acid, 2-methyl-, pentyl ester	1059	1121	16.13
Linalool	1064	1097	22.94
D-Limonene	1086	1030	0.69
Pulegone	1209	1209	5.45
Lavandulyl isobutyrate	1275	1404	0.48

[a] RI (retention index) measured relative to n-alkanes (C6–C30) on the non-polar 123 DB11 column. [b] Linear retention index taken from the NIST 05 library and the literature.

3.2. Total Phenol, Total Flavonoid, and Total Condensed Tannins Contents of A. visnaga L. Essential Oil

The total phenol contents (TPC) values in *A. visnaga* L. essential oil were determined based on the absorbance values, being expressed in terms of gallic acid equivalents (GAE) in reference to the standard curve equation:

$$y = 0.0133x - 0.01, R^2 = 0.9902$$

The total flavonoid contents (TFC) were expressed in terms of quercetin equivalent using the standard curve equation:

$$y = 0.0175x - 0.0136, R^2 = 0.9911$$

and the values ranged between 5.82 ± 0.79 g QE/g extract.

The total condensed tannin (TCT) content was 15.08 ± 1.76 catechin/g extract:

$$y = 0.0156x - 0.0184, R^2 = 0.9902$$

TPC were in the range of 7.26 ± 1.68 extract (mg GAE/g extract), TFC ranged between 5.82 ± 0.79 (mg QE/g extract), and TCT ranged between 15.08 ± 1.76 (mg CE/g extract) (data are expressed as means ± standard deviation of triplicate samples).

3.3. Determination of the In Vitro Antioxidant Activities

The antioxidant potential of *A. visnaga* L. essential oil was evaluated using two different test assays: the free radical scavenging activity (DPPH) and the reducing power. Known antioxidant compounds, BHT and AA, were used as reference.

The free radical scavenging capacities of the corresponding oil were measured by the DPPH test. According to the results obtained, the studied essential oil showed a low ability to reduce the DPPH free radical concentration with an IC_{50} value of 4.13 ± 0.22 mg/mL, in comparison with the reference antioxidant BHT (IC_{50} = 0.17 ± 0.01 mg/mL). The reduction in DPPH absorbance was dose-dependent. The inhibition rate of the DPPH-produced free radicals reached 12.53% at 1 mg/mL (Figure 1); this ability decreased with decreasing oil concentration, at 800 (9.68%), 600 (6.83%), 400 (4.55%), and 200 µL/mL (3.13%), respectively.

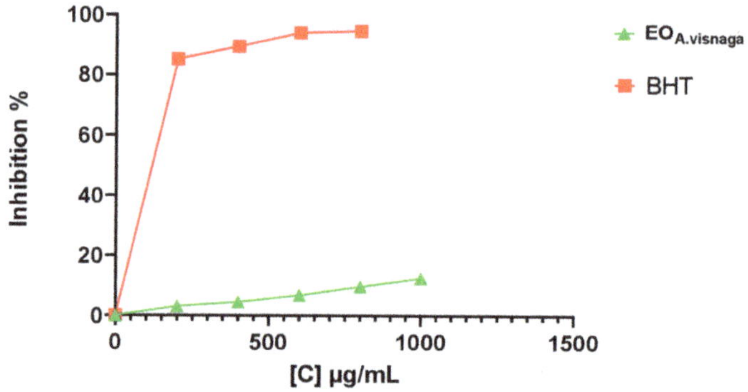

Figure 1. Percentage inhibition of the DPPH by different concentrations of essential oil of *A. visnaga* L. and standard (results expressed as means ± SD, of three parallel measurements ($p < 0.05$)). EO: essential oil; BHT: butylated hydroxytoluene.

In the reducing power assay, the *A. visnaga* L. essential oil showed increased absorbance with the increase in the concentration, a dose-dependent effect similar to the AA standard (Table 2). In terms of the EC_{50} value, *A. visnaga* L. essential oil exhibited lower reducing power (0.730 ± 0.017 mg/mL) compared to that of the AA standard (EC_{50} = 0.034 ± 0.28 mg/mL) (values expressed are means ± SD, $n = 3$).

Table 2. Reducing power of *A. visnaga* L. essential oil at different concentrations.

Sample	Sample Concentration (µg/mL)				
	200	400	600	800	1000
Essential oil	0.13 ± 0.01 *	0.29 ± 0.02 *	0.38 ± 0.04 *	0.57 ± 0.005 *	0.64 ± 0.005 *
Ascorbic acid	0.43 ± 0.01	0.71 ± 0.01	0.95 ± 0.03	1.21 ± 0.01	1.52 ± 0.005

Values expressed are means ± SD, $n = 3$, * $p < 0.05$.

Pearson correlation coefficients were calculated to identify the relationship between the antioxidant activity and the total phenolic, condensed tannins, and total flavonoid contents (Figure 2). The analysis revealed a positive correlation between DPPH/IC_{50} value and the total flavonoids, condensed tannins, and total phenolic content, with $r = 0.958$; $R^2 = 0.917$, $r = 0.851$; $R^2 = 0.724$, and $r = 0.971$; $R^2 = 0.942$, respectively. A very strong positive correlation was observed for the phenols and flavonoids, suggesting that 94.2% and 91.7%, respectively, of the variability in antioxidant activity of *A. visnaga* essential oil is explained by the variability in TPC and TFC. The other 5.8 and 8.3% of variance is explained by unknown factors that were not measured in the experiment.

3.4. Determination of In Vivo Antioxidant Potential

3.4.1. The Effects of the Essential Oil of *A. visnaga* L. on the Body Weight of the Treated Mice

The body weight of the mice was measured at the beginning (day 0) and at the end of the test (day 14); the results are presented in Table 3. For the control group, the AA group, and the two essential oil-treated groups (600 mg/kg and 1200 mg/kg, respectively), the body weight registered slight and normal increases throughout the treatment period without any statistical differences between day 0 and day 14 ($p > 0.05$). However, for the APAP group, there was a significant reduction in body weight at the end of the treatment ($p < 0.05$) compared to day 0.

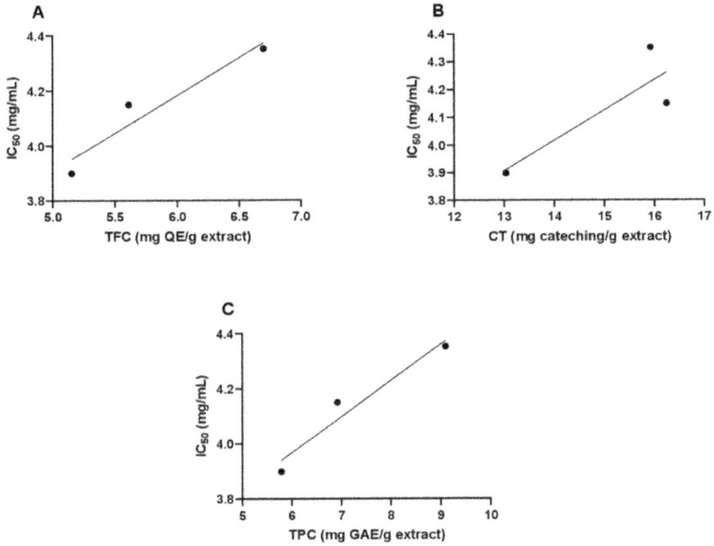

Figure 2. Pearson correlation graphics: (**A**) Pearson correlation between DPPH IC_{50} and flavonoids, (**B**) Pearson correlation between DPPH IC_{50} and Condensed Tannins, and (**C**) Pearson correlation between DPPH IC_{50} and phenols of *A. visnaga* L. TFC: total flavonoid content; CT: condensed tannins; TPC: total phenolic content; QE: quercetin equivalent.

Table 3. Body weight of the mice during the treatment period.

Groups	Mean Body Weight in Gram ± SD	
	Day 0	Day 14
C	29.39 ± 0.29	29.58 ± 0.24
CMC	30.48 ± 0.31	30.71 ± 0.30
APAP	32.54 ± 0.43	29.78 ± 0.65 *
AA	27.47 ± 0.28	27.92 ± 0.72
AV 1	34.53 ± 0.29	35.09 ± 0.25
AV 2	25.71 ± 0.39	26.42 ± 0.88

C: Normal control; CMC: vehicle group–carboxymethyl cellulose 0.1%; APAP: Toxic control treated with acetaminophen 400 mg/kg body weight (ip); AV1: essential oil of *A. visnaga* 600 mg/kg body weight; AV2: essential oil of *A. visnaga* 1200 mg/kg body weight; AA: Ascorbic acid 200 mg/kg body weight. All data are mean ± S.D (n = 5/group), * $p < 0.05$ APAP at Day 0 vs. APAP at Day 14.

The presence of body weight differences between treatment and control groups reported in Table 3 make the organ weight interpretation more complicated. To detect target organ damage, the relative organ weight to body weight was used (Table 4).

Table 4. Organ weights relative to body weight of Swiss Albinos mice.

Organs	Groups					
	C	CMC	AA	APAP	AV1	AV2
Kidneys	1.37 ± 0.11	1.36 ± 0.17	1.24 ± 0.08 *;#	1.04 ± 0.14 **	1.15 ± 0.10 *	1.19 ± 0.07 *;#
Liver	5.26 ± 0.26	5.05 ± 0.11	4.66 ± 0.16 **;###	3.88 ± 0.13 ***	4.04 ± 0.18 ***	4.83 ± 0.35 *;##

All values are expressed as mean ± SD. C: Normal control; CMC: vehicle group–carboxymethyl cellulose 0.1%; APAP: Toxic control treated with acetaminophen 400 mg/kg body weight (ip); AV1: essential oil of *A. visnaga* 600 mg/kg body weight; AV2: essential oil of *A. visnaga* 1200 mg/kg body weight; AA: Ascorbic acid 200 mg/kg body weight (significant differences as compared with the normal control group * $p < 0.05$; ** $p < 0.01$; *** $p < 0.001$; significant differences as compared with the toxic control group # $p < 0.05$; ## $p < 0.01$; ### $p < 0.001$).

Based on the results in Table 4, a statistically significant decrease in the organ weights relative to body weight was observed in the kidneys and liver of mice exposed to APAP treatment (AA, APAP alone, AV1, and AV2) compared with the control group. In addition, a significant increase in the organ weights relative to body weight was recorded in the kidneys and liver of mice exposed to APAP (AA, AV1 and AV2) compared with the APAP toxic group ($^{\#}$ $p < 0.05$). These changes in the organ weights relative to body weight indicate a toxic effect of APAP on the animals' organs (liver and kidneys) ($p < 0.05$).

Supplementation with AA and *A. visnaga* L. essential oil, except in mice models receiving an oral dose of AVI (i.e., 600 mg/kg PC of *A. visnaga* essential oil), showed a significant ability to thwart the toxic effect of APAP ($^{\#}$ $p < 0.05$).

Interestingly, a higher dose essential oil of *A. visnaga* (i.e., 1200 mg/kg PC) showed a comparable protective effect to the positive control AA in mice models ($^{\#}$ $p < 0.05$). The above observations indicate the hepato/renal protective activity of *A. visnaga* L. essential oil in a dose-dependent manner.

3.4.2. Antioxidant Activity Assay

The GSH, SOD, MDA, and CAT levels were measured in the liver and kidney tissues of the mice and are presented in Figure 3.

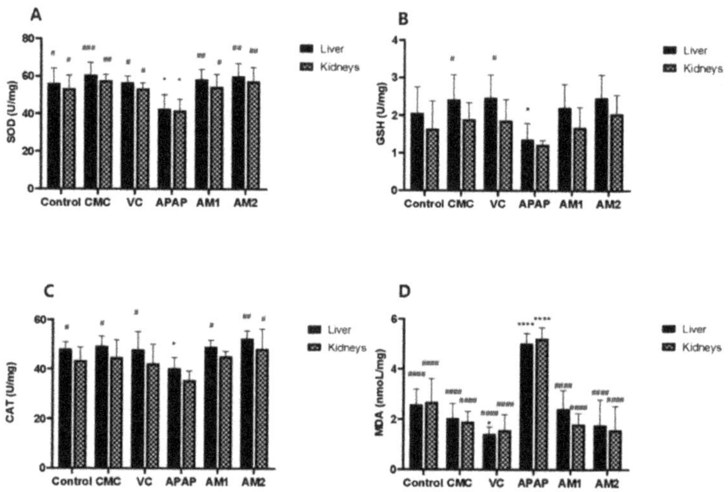

Figure 3. Effect of *A. visnaga* L. essential oil on antioxidant enzymes and MDA levels against acetaminophen-induced liver and kidneys injury in mice. Values are expressed as the mean ± SD ($n = 5$), * $p < 0.05$; **** $p < 0.0001$ vs. the normal control group and $^{\#}$ $p < 0.05$; $^{\#\#}$ $p < 0.01$; $^{\#\#\#}$ $p < 0.001$; $^{\#\#\#\#}$ $p < 0.0001$ vs. the toxic control group. (**A**) Effect of *A. visnaga* L. on the SOD level in APAP-treated mice liver and kidneys; (**B**) Effect of *A. visnaga* L. on the GSH level in APAP-treated mice liver and kidneys; (**C**) Effect of *A. visnaga* L. on the CAT level in APAP-treated mice liver and kidneys; (**D**) Effect of *A. visnaga* L. on the MDA level in APAP-treated mice liver and kidneys. SOD: superoxide dismutase; GSH: reduced glutathione; CAT: catalase; MDA: malondialdehyde; AA: ascorbic acid; CMC: sodium carboxymethyl cellulose; APAP: acetaminophen; AV1: essential oil of *A. visnaga* 600 mg/kg body weight; AV2: essential oil of *A. visnaga* 1200 mg/kg body weight.

In the case of the liver homogenates, the toxic control group (III) showed significantly decreased levels of SOD, CAT, and GSH compared to the control, the standard AA, and the essential oil (600 and 1200 mg/kg body weight) -treated groups ($p < 0.05$). However, in the kidney homogenates, only SOD was significantly lower in the toxic group compared to the normal control ($p < 0.05$). The MDA content as a marker of lipid peroxidation, reflecting that the degree of oxidative stress was significantly higher in the groups receiving APAP in the homogenates of both tissues studied compared to the negative control and all other treated groups, including the essential oil-treated groups ($p < 0.05$).

Treatment with essential oil preserved the levels of SOD, CAT, and GSH at values comparable to the normal control ($p > 0.05$) and showed a significant dose-dependent decrease in the level of MDA when compared to the APAP toxic control in liver and kidney tissue ($p < 0.05$). The *A. visnaga* L. essential oil at 1200 mg/kg had the lowest MDA level compared to the normal and CMC control, but it was higher than the AA positive control. CAT activity did not show any significant differences between the groups.

4. Discussion

To date, very few studies exist on the *A. visnaga* L. essential oils and their antioxidant activity [34]. This study species showed a yield of essential oil of 1.4% *v/w* during the flowering period, higher than those mentioned by Khadhri et al. [35] and Khalfallah et al. [36] for Tunisian and Algerian species, respectively (yields ranging from 0.2–1.3%), or than those obtained during the fruiting period by Satrani et al. for the same Moroccan species [37] and by Günaydin et al. for a Turkish species (yields between 0.1–0.27%) [38]. The difference in yield of extracts can be influenced by factors such as origin, species, extraction methods, time of harvest, developmental stages, and geo climatic factors [39–41].

Chromatographic analysis showed that the main compound of the obtained essential oils were the terpenes. (Z)-β-ocimene was the main constituent, followed by the oxygenated monoterpenes D-limonene, linalool, pulegone and lavandulyl-butyrate. This composition differs from the one Günaydin et al. previously reported, characterized by nerol (29.98%), a-bisabolol (20.86%), and butylated hydroxytoluene (18.55%) [38]. While the Iranian *A. visnaga* L. essential oil was reported to be dominated by cis-pinene hydrate (42.83%), methyl octadecanoate (14.73%), and a-terpinene (9.20%), the composition of the essential oil resulted at the final of the present research was characterized by different monoterpenes as previously described by Sellami et al., with linalool (22.7–32%) as the main compound [35,42]. The presence of terpene alcohols such as the α-terpineol and of fatty acids, such as valeric acids, was mentioned by Feirouz and Salima [39]. The essential oils extracted from the North African *A. visnaga* L. plants (Algeria, Tunisia, and Morocco) show an analogy in their composition, but are dissimilar from those collected of the Iranian or Turkish species. These variations in chemical composition may be linked to ecological and genetic factors, such as soil type and proprieties, soil management, or environmental stress, which can lead to the modulation of certain enzymatic groups, responsible for the regulation of biosynthetic pathways [43–45].

The in vitro tests have shown that the essential oil of *A. visnaga* L. presents less antioxidant and phenolic activity than the reference agents. Our results revealed a lower or a relatively low content of total phenols, flavonoids, and condensed tannins (TPC = 7.26 ± 1.68 mg GAE/g extract; TFC = 5.82 ± 0.79 mg QE/g extract; CT = 15.08 ± 1.76 mg CE/g extract). El Karkouri et al. recently analyzed the same Moroccan species, and reported similar values to ours (of 7.16 mg GAE/g) for the crude extract obtained by Soxhlet extraction in methanol 70%, and lower values for the crude extract obtained by hydro-acetone, hydro-methanol, and Soxhlet extraction in ethanol [46]. Conversely, other studies analyzing the composition of the ethanolic and methanolic extracts of *A. visnaga* L. have reported higher TP, TF, and CT contents comparatively with our results, although this may partially be caused by the differences between extracts and essential oil [47,48].

The different concentrations of essential oil determined in this research showed low in vitro DPPH radical scavenging activity and low reducing power efficiency. These results are consistent with those of Aourabi et al. [47] and Keddad et al. [34], who reported an IC_{50} of 3.05 mg/mL at 1 mg/mL for the Moroccan *A. visnaga* aqueous extract and a 9.16% DPPH inhibition at 1 mg/mL for the Algerian *A. visnaga* L. In contrast to these results, Darkaoui et al. reported for the Moroccan species an IC_{50} of 56.053 ± 0.856 μg/mL [49], lower than that obtained in our study, while the antioxidant activity of the methanolic extract of the Palestinian *A. visnaga* L. showed better results (IC_{50} = 6.07 ± 2.14 µg/mL) with an inhibition percentage of 90.63% at 100 µg/mL [50].

Concerning the in vitro reducing power, essential oil showed good activity at an EC_{50} = 730.53 ± 0.017 µg/mL, although this is relatively lower when compared to other species within the *Apiaceae* family, such as the hydroalcoholic extract of the aerial parts of *Ammoides atlantica* with an IC_{50}: 92.70 ± 1.00 µg/mL, the Tunisian *Pimpinella anisum* during maturation (EC_{50} = 0.523 mg/mL), or the ethanolic seed extract of *Trachyspermum ammi*, with a reducing power ability at an optic density of 0.91 at 25 µg/mL [51–53].

The slightly lower antiradical activity of our sample can be explained by the deficiency in phenolic content, which is suggested by their non-dominance in the chemical composition. In this regard, strong correlations were observed between the antioxidant activity, TPC, TFC, and CT at a 95% confidence level. These results agree with the reports of the literature that show that the bleaching of the DPPH solution is strongly correlated with the amount of total phenolic content [54–56].

To our knowledge, this is the first study to evaluate the antioxidant potential of *A. visnaga* L. essential oil in vivo. The results of this study revealed that APAP treatment induced oxidative stress in the liver and kidney of mice, which was characterized by increased levels of MDA, a marker of lipid peroxidation, as well as reductions in the antioxidant activities SOD, CAT, and GSH. An upregulation of lipid peroxidation is reported to induce subcellular and tissue disruption in liver and kidney [57,58]. Other studies have reported APAP-induced oxidative stress, with massive lipid peroxides production and decreased levels of SOD, CAT, and GPx [59–61]. The decrease in the antioxidant defense status of SOD, CAT, and GSH accompanied by the increase in LPO in liver and kidney tissues after APAP exposure could be attributed to ROS excessive generation and subsequent cellular damage of the healthy cells capable of halting oxidative attack [62,63]. The endogenous detoxification provided by the lowered SOD, CAT, and GSH activities would, thus, be insufficient in scavenging the superoxide anions and hydroxyl ions [1,64].

The administration of *A. visnaga* L. essential oil against APAP resulted in an increase in SOD, CAT, and GSH activities and a statistical decrease in MDA levels in mouse liver and kidney tissues. In this context, we have found an inconsistency between the in vivo very good antioxidant efficiency and the in vitro low-moderate radical scavenging activity. This interesting in vivo/in vitro difference has been reported elsewhere for other antioxidants, such as inulin [65], berry juices [66], or milk-derived antioxidative peptides [67], where the structure of the compound imprints on each of the substances a unique behavior [68]. The potential synergy with other compounds along the metabolic pathways could be one cause to trigger the significant increases in the antioxidant activity, less predictable by the indirect in vitro assays executed under specific conditions [67]. However, as Slatnar et al. suggested, these differences in antioxidant capacity could rest in the molecular nature of the bioactive compounds [64]. Pietta et al. reported, similarly to us, ineffective in vitro antioxidant activity but important antioxidant in vivo effects in the case of a ginkgo extract with standardized content of polyphenols [69]. However, not the polyphenols, but the terpenic ginkgolides present in the ginkgo extracts and insensitive to the used in vitro assay were later found to play a major antioxidant role in vivo [69,70]. In our case, the antioxidant activity may also derive, in large part, from the specific terpene composition that was identified by GC-MS analysis: the monoterpenes (Z)-β-ocimene, linalool, pulegone, and lavandulyl butyrate.

Significant antioxidant properties of monoterpenic compounds were determined previously in the essential oil extracted from leaves and fruit of *Schinus molle* L. [64]. Similar antioxidant activity was reported by Ruberto et al., for the monoterpenes α-pinene, β-pinene, limonene, β-myrcene, sabinene, and terpinolene [71]. Both α- and β-terpinolene exhibit antioxidant activity comparable to that of α-tocopherol [70,72]. Moreover, Edziri et al. reported that the high percentage of monoterpenes identified in the essential oils of *Retama raetam* (Forssk.) Webb (Fabaceae) shows good antioxidant activity [73]. In line with our results, ocimene and linalool measured by a Total TRAP assay and piperitenone from *Mentha longifolia* L. essential oil exhibit a high antioxidant potential [74,75]. Administration of linalool significantly reduces liver MDA excess levels in mice treated with lipopolysaccharide (LPS)/D-galactosamine by downregulating myeloperoxidase activity [76]. The pulegone is also reported for its significant role in lipid peroxidation decrease (MDA marker) coupled with increasing antioxidant levels [76].

The present study clearly demonstrates that intraperitoneal injection of APAP at a dose of 400 mg/kg body weight significantly decreased the body weight of mice and the relative organ weight at the end of the experiment. In this context, Abdul Hamid et al. also observed that treatment of male Sprague-Dawley rats with APAP decreased body weight and kidney weight [77]. In the same way, Jaeschke et al. have indicated that APAP overdose is strongly associated with hepatotoxicity and nephrotoxicity in humans and animal models [78]. In humans, when the glucuronidation and sulfation pathways are saturated, the 10% of APAP that is not metabolized and excreted in the urine is oxidized by cytochrome P450 2E1 to form the toxic intermediate N-acetyl-p-benzoquinone imine (NAPQI) [79]. At therapeutic doses, this unstable metabolite is detoxified by conjugation to glutathione (GSH) and eliminated via the urine or bile [79]. Upon APAP overdose, overproduced NAPQI results in GSH depletion and binds to proteins to form an APAP–protein adduct, thereby inducing cellular oxidative stress at the hepatic level [79]. A similar mechanism was observed in renal tissue, in male Wistar albino rats: an overdose of APAP estimated by 900 mg/kg body weight induced nephrotoxicity, which was reflected by a significant increase in renal MDA, in addition to significant reductions in renal reduced glutathione (GSH) [80]. Supplementation with *A. visnaga* L. essential oil provided protection against APAP-induced hepatotoxicity and nephrotoxicity and restored relative organ weights in mice in a dose-dependent manner, through downregulation of MDA levels and upregulation of GSH and other antioxidant enzymes.

With regard to the risk of toxicity by administration of the essential oil, it is reported in previous toxicity studies on *A. visnaga* L. seeds that neither ethanolic extract nor aqueous extract reveal any signs or symptoms of toxicity in rats at a limit dose of 5000 mg/kg [81,82]. To obtain the human equivalent dose (HED) according to the FDA guidelines [83], it was calculated as follows:

$$\text{HED} = \text{animal dose} \times (\text{Animal Km}/\text{Human Km})$$

where Km is the factor for converting mg/kg dose to mg/m^2 dose, considering an average human body weight at 60 kg (Km = 37) and an average mouse weight at 20 g (Km = 3). This resulted in a HED value of 0.097 g/kg for our highest dose of 1200 mg/kg, which is too low to provoke toxicity leading to mortality, according to the LD50 values reported in the literature (toxic doses at 60 g for an adult weighing 60 kg) [84].

Regarding the absorption and metabolism of *A. visnaga* essential oil, its major compounds, including limonene and linalool, were reported to have good dermal, pulmonary, and oral absorption [85–87]. Due to its lipophilic nature, the limonene is quickly and easily diffused and absorbed by the skin, with a rapid increase in plasma concentration at 10 min after dermal application in both humans and mice models [86,87]. In humans, repeated exposure to limonene in inhalation chambers, at 10, 225, and 450 mg/m^3, resulted in a 70% absorption rate into the blood from the pulmonary alveoli. A blood clearance of 1.1 l/kg/h was observed after 4 h, indicating easy metabolism of limonene by the body [88,89]. Linalool is also reported to have an up to 90% absorption rate at the gut level in only 7 h after oral administration in rats, at a dose of 500 mg/kg [89]. In addition, the quantification of human whole blood showed the appearance of linalool 25 min after inhalation and decreasing around 40–45 min. In total, 58% of the ingested dose of linalool was eliminated and excreted in urine, 25% as $^{14}CO_2$ in exhaled air, and 16% in faces after 72 h, via the induction of cytochrome P450-dependent monooxygenase and liver microsomal uridine diphosphate glucuronyl transferase [90]. This suggests that linalool metabolites were able to enter intermediate metabolic pathways without risk of long-term exposure due to tissue accumulation [90].

At present, very few studies have provided evidence regarding the safety or efficacy of this essential oil through clinical trials. Translational studies are necessary to predict the metabolism and potential toxicity, as well as systematic clinical trials to test the action of *A. visnaga* L. essential oil in targeted or generalized therapies.

5. Conclusions

The present study examined the chemical composition and antioxidant activities in vitro and in vivo of *A. visnaga* L. plant. GC-MS analysis of the essential oil of *A. visnaga* L. showed a strong presence of terpenoids, with Z-(β)-ocimene and pulegone as two major components. In vitro, the natural essential oil expressed low DPPH radical scavenging activity and reduced antioxidant power; this is due to the low phenolic composition. The in vivo study on Swiss albino mice showed that dietary supplementation of *A. visnaga* L. essential oil significantly improved the antioxidant status marked by an increase in the antioxidant enzymatic activities of SOD, CAT, and GSH-Px, and the decrease in the concentration of MDA. Based on the above results, the high antioxidant activity observed in vivo may be due to the richness of *A. visnaga* L. essential oil in terpene compounds, which may be useful in the treatment of oxidative stress related diseases.

Author Contributions: Conceptualization, F.Z.K., G.D.S., V.V.C., and M.N.; Data curation, R.L., V.V.C., M.N., D.C.A., A.C., and A.E.-C.; Formal analysis, F.Z.K., G.D.S., R.L., D.C.A., and A.C.; Investigation, F.Z.K., R.L., and M.N.; Methodology, F.Z.K., G.D.S., and V.V.C.; Project administration, F.Z.K.; Supervision, R.L. and A.E.-C.; Writing—original draft, F.Z.K., G.D.S., R.L., and A.C.; Writing—review and editing, F.Z.K. and R.L. All authors have read and agreed to the published version of the manuscript.

Funding: This research received no external funding.

Institutional Review Board Statement: The experiment was carried out according to the Guide for the Care and Use of Laboratory Animals and approved by the Bioethics consultative Commission Faculty of sciences Agadir (CCBE-FSA Ref. No: FCR-AE-10/2021-0001).

Informed Consent Statement: Not applicable.

Data Availability Statement: Data is contained within this manuscript.

Acknowledgments: The authors would like to thank Lim Vuanghao of the Integrative Medicine Cluster, Advanced Medical and Dental Institute, Sains Malaysia University, Penang, Malaysia and Latifa Ez-zaher of the Laboratory of Cell Biology and Molecular Genetics, Ibn Zohr University, Faculty of Sciences, Agadir, Morocco for their invaluable assistance in conducting this study.

Conflicts of Interest: The authors declare no conflict of interest.

References

1. Zahra, K.F.; Lefter, R.; Ali, A.; Abdellah, E.-C.; Trus, C.; Ciobica, A.; Timofte, D. The Involvement of the Oxidative Stress Status in Cancer Pathology: A Double View on the Role of the Antioxidants. *Oxid. Med. Cell. Longev.* **2021**, *2021*, 1–25. [CrossRef] [PubMed]
2. Lobo, V.; Patil, A.; Phatak, A.; Chandra, N. Free radicals, antioxidants and functional foods: Impact on human health. *Pharmacogn. Rev.* **2010**, *4*, 118–126. [CrossRef] [PubMed]
3. Stanciu, G.D.; Luca, A.; Rusu, R.N.; Bild, V.; Chiriac, S.I.B.; Solcan, C.; Bild, W.; Ababei, D.C. Alzheimer's Disease Pharmacotherapy in Relation to Cholinergic System Involvement. *Biomolecules* **2020**, *10*, 40. [CrossRef] [PubMed]
4. Stanciu, G.; Rusu, R.; Bild, V.; Filipiuc, L.; Tamba, B.-I.; Ababei, D. Systemic Actions of SGLT2 Inhibition on Chronic mTOR Activation as a Shared Pathogenic Mechanism between Alzheimer's Disease and Diabetes. *Biomedicines* **2021**, *9*, 576. [CrossRef] [PubMed]
5. Halliwell, B. Free radicals and antioxidants–quo vadis? *Trends Pharmacol. Sci.* **2011**, *32*, 125–130. [CrossRef]
6. Devaraj, S.; Leonard, S.; Traber, M.G.; Jialal, I. Gamma-tocopherol supplementation alone and in combination with al-pha-tocopherol alters biomarkers of oxidative stress and inflammation in subjects with metabolic syndrome. *Free Radic. Biol. Med.* **2008**, *44*, 1203–1208. [CrossRef] [PubMed]
7. Bouayed, J.; Rammal, H.; Soulimani, R. Oxidative Stress and Anxiety: Relationship and Cellular Pathways. *Oxidative Med. Cell. Longev.* **2009**, *2*, 63–67. [CrossRef] [PubMed]

8. Teixeira, K.R.C.; dos Santos, C.P.; de Medeiros, L.A.; Mendes, J.A.; Cunha, T.M.; De Angelis, K.; Penha-Silva, N.; de Oliveira, E.P.; Crispim, C.A. Night workers have lower levels of antioxidant defenses and higher levels of oxidative stress damage when compared to day workers. *Sci. Rep.* **2019**, *9*, 4455. [CrossRef] [PubMed]
9. Ávila-Escalante, M.L.; Coop-Gamas, F.; Cervantes-Rodríguez, M.; Méndez-Iturbide, D.; Aranda-González, I.I. The effect of diet on oxidative stress and metabolic diseases—Clinically controlled trials. *J. Food Biochem.* **2020**, *44*, e13191. [CrossRef] [PubMed]
10. Guemouri, L.; Artur, Y.; Herbeth, B.; Jeandel, C.; Cuny, G.; Siest, G. Biological variability of superoxide dismutase, glutathione peroxidase, and catalase in blood. *Clin. Chem.* **1991**, *37*, 1932–1937. [CrossRef]
11. Gorni, D.; Finco, A. Oxidative stress in elderly population: A prevention screening study. *Aging Med.* **2020**, *3*, 205–213. [CrossRef] [PubMed]
12. Hassan, W.; Noreen, H.; Rehman, S.; Gul, S.; Amjad Kamal, M.; Paul Kamdem, J.; Zaman, B.; da Rocha, J.B.T. Oxidative Stress and Antioxidant Potential of One Hundred Medicinal Plants. *Curr. Top. Med. Chem.* **2017**, *17*, 1336–1370. [CrossRef] [PubMed]
13. Xu, D.-P.; Li, Y.; Meng, X.; Zhou, T.; Zhou, Y.; Zheng, J.; Zhang, J.-J.; Li, H.-B. Natural Antioxidants in Foods and Medicinal Plants: Extraction, Assessment and Resources. *Int. J. Mol. Sci.* **2017**, *18*, 96. [CrossRef]
14. Jeyabalan, J.; Aqil, F.; Soper, L.; Schultz, D.J.; Gupta, R.C. Potent Chemopreventive/Antioxidant Activity Detected in Common Spices of the Apiaceae Family. *Nutr. Cancer* **2015**, *67*, 1201–1207. [CrossRef] [PubMed]
15. Önder, A.; Çinar, A.S.; Sarialtin, S.Y.; Izgi, M.N.; Çoban, T. Evaluation of the Antioxidant Potency of Seseli L. Species (Apiaceae). *Turk. J. Pharm. Sci.* **2020**, *17*, 197–202. [CrossRef]
16. Thiviya, P.; Gamage, A.; Piumali, D.; Merah, O.; Madhujith, T. Apiaceae as an Important Source of Antioxidants and Their Applications. *Cosmetics* **2021**, *8*, 111. [CrossRef]
17. Khalil, N.; Bishr, M.; Desouky, S.; Salama, O. Ammi Visnaga L., a Potential Medicinal Plant: A Review. *Molecules* **2020**, *25*, 301. [CrossRef]
18. Bhagavathula, A.S.; Al-Khatib, A.J.M.; Elnour, A.A.; Al Kalbani, N.M.; Shehab, A. Ammi Visnaga in treatment of urolithiasis and hypertriglyceridemia. *Pharm. Res.* **2014**, *7*, 397–400.
19. Singleton, V.L.; Rossi, J.A. Colorimetry of total phenolics with phosphomolybdic-phosphotungstic acid reagents. *Am. J. Enol. Vitic.* **1965**, *16*, 144–158.
20. Zhishen, J.; Mengcheng, T.; Jianming, W. The determination of flavonoid contents in mulberry and their scavenging effects on superoxide radicals. *Food Chem.* **1999**, *64*, 555–559. [CrossRef]
21. Kim, D.-O.; Chun, O.K.; Kim, Y.J.; Moon, H.-Y.; Lee, C.Y. Quantification of Polyphenolics and Their Antioxidant Capacity in Fresh Plums. *J. Agric. Food Chem.* **2003**, *51*, 6509–6515. [CrossRef]
22. Price, M.L.; Van Scoyoc, S.; Butler, L.G. A critical evaluation of the vanillin reaction as an assay for tannin in sorghum grain. *J. Agric. Food Chem.* **1978**, *26*, 1214–1218. [CrossRef]
23. Bougandoura, N.; Bendimerad, N. Evaluation de l'activité antioxydante des extraits aqueux et méthanolique de Satureja calamintha ssp.Nepeta (L.) Briq. *Rev. Nat. Technol.* **2013**, *5*, 14–19.
24. Samarth, R.M.; Panwar, M.; Kumar, M.; Soni, A.; Kumar, M.; Kumar, A. Evaluation of antioxidant and radical-scavenging activities of certain radioprotective plant extracts. *Food Chem.* **2008**, *106*, 868–873. [CrossRef]
25. Oyaizu, M. Studies on products of browning reaction. Antioxidative activities of products of browning reaction prepared from glucosamine. *Jpn. J. Nutr. Diet.* **1986**, *44*, 307–315. [CrossRef]
26. Chen, Y.; Huang, B.; He, J.; Han, L.; Zhan, Y.; Wang, Y. In vitro and in vivo antioxidant effects of the ethanolic extract of Swertia chirayita. *J. Ethnopharmacol.* **2011**, *136*, 309–315. [CrossRef]
27. Merghem, M.; Dahamna, S. In-Vitro Antioxidant Activity and Total Phenolic Content of Ruta montana L. Extracts. *J. Drug Deliv. Ther.* **2020**, *10*, 69–75. [CrossRef]
28. Aebi, H. Catalase. Methods of enzymatic analysis. In *Methods of Enzymatic Analysis*; Bergmeyer, H.U., Ed.; Verlag Chemie, Academic Press Inc.: Cambridge, MA, USA, 1974; pp. 673–684.
29. Aebi, H. Catalase in vitro. In *Methods in Enzymology*; Elsevier: Amsterdam, The Netherlands, 1984; pp. 121–126. [CrossRef]
30. Beauchamp, C.; Fridovich, I. Superoxide dismutase: Improved assays and an assay applicable to acrylamide gels. *Anal. Biochem.* **1971**, *44*, 276–287. [CrossRef]
31. Kakkar, P.; Das, B.; Viswanathan, P.N. A modified spectrophotometric assay of superoxide dismutase. *Indian J. Biochem. Biophys.* **1984**, *21*, 130–132.
32. Ellman, G.L. Tissue sulfhydryl groups. *Arch. Biochem. Biophys.* **1959**, *82*, 70–77. [CrossRef]
33. Sastre, J.; Pallardó, F.V.; de la Asunción, J.G.; Viña, J. Mitochondria, oxidative stress and aging. *Free Radic. Res.* **2000**, *32*, 189–198. [CrossRef] [PubMed]
34. Keddad, A.; Baaliouamer, A.; Hazzit, M. Chemical Composition and Antioxidant Activity of Essential Oils from Umbels of Algerian Ammi visnaga (L.). *J. Essent. Oil Bear. Plants* **2016**, *19*, 1243–1250. [CrossRef]
35. Khadhri, A.; El Mokni, R.; Mguis, K.; Araujo, M.E.M. Variability of two essential oils of Ammi visnaga (L.) Lam. a traditional Tunisian medicinal plant. *J. Med. Plant Res.* **2011**, *5*, 5079–5082.
36. Khalfallah, A.; Labed, A.; Semra, Z.; Kaki, B.; Kabouche, A.; Touzani, R.; Kabouche, Z. Antibacterial activity and chemical composition of the essential oil of Ammi visnaga L. (Apiaceae) from Constantine, Algeria. *Int. J. Med. Aromat. Plants* **2011**, *1*, 302–305.

37. Satrani, B.; Farah, A.; Talbi, M. Composition chimique et activité antibactérienne et antifongique de l'huile essentielle extraite du bois de Tetraclinis articulata du Maroc. *Ann. Falsif. Expert. Chim. Toxicol.* **2004**, *97*, 75–84.
38. Günaydin, K.; Beyazit, N. The chemical investigations on the ripe fruits of Ammi visnaga (Lam.) Lamarck growing in Turkey. *Nat. Prod. Res.* **2004**, *18*, 169–175. [CrossRef]
39. Feirouz, B.; Salima, K.-G. Antibacterial Activity and Chemical Composition ofAmmi visnaga, L. Essential Oil Collected from Boumerdes (Algeria) During Three Periods of the Plant Growth. *J. Essent. Oil Bear. Plants* **2014**, *17*, 1317–1328. [CrossRef]
40. Bungau, S.; Behl, T.; Aleya, L.; Bourgeade, P.; Aloui-Sossé, B.; Purza, A.L.; Abid, A.; Samuel, A.D. Expatiating the impact of anthropogenic aspects and climatic factors on long-term soil monitoring and management. *Environ. Sci. Pollut. Res.* **2021**, *28*, 30528–30550. [CrossRef]
41. Gitea, M.A.; Gitea, D.; Tit, D.M.; Purza, L.; Samuel, A.D.; Bungau, S.; Badea, G.E.; Aleya, L. Orchard management under the effects of climate change: Implications for apple, plum, and almond growing. *Environ. Sci. Pollut. Res.* **2019**, *26*, 9908–9915. [CrossRef]
42. Sellami, H.K.; Flamini, G.; Cioni, P.L.; Smiti, S. Composition of the Essential Oils in Various Organs at Different Developmental Stages of Ammi visnaga (L.) Lam. from Tunisia. *Chem. Biodivers.* **2011**, *8*, 1990–2004. [CrossRef]
43. Hashim, S.; Jan, A.; Marwat, K.B.; Khan, M.A. Phytochemistry and medicinal properties of Ammi visnaga (Apiacae). *Pak. J. Bot.* **2014**, *46*, 861–867.
44. Samuel, A.D.; Bungau, S.; Tit, D.M.; Melinte, C.E.; Purza, L.; Badea, G.E. Effects of Long Term Application of Organic and Mineral Fertilizers on Soil Enzymes. *Rev. Chim.* **2018**, *69*, 2608–2612. [CrossRef]
45. Samuel, A.D.; Tit, D.M.; Melinte, C.E.; Iovan, C.; Purza, L.; Gitea, M.; Bungau, S. Enzymological and physico-chemical evaluation of the effects of soil management practices. *Rev. Chim.* **2017**, *68*, 2243–2247. [CrossRef]
46. El Karkouri, J.; Drioiche, A.; Soro, A.; Ailli, A.; Benhlima, N.; Bouzoubaa, A.; El Makhoukhi, F.; Oulhaj, H.; Elombo, F.K.; Zair, T. Identification and antioxidant activity of Ammi visnaga L. polyphenols from the Middle Atlas in Morocco. *Mediterr. J. Chem.* **2020**, *10*, 649–658. [CrossRef]
47. Aourabi, S.; Driouch, M.; Ammor, K.; Sfaira, M.; Touhami, M.E.; Mahjoubi, F. Evaluation of anticorrosion and antioxidant activities of ethanolic extract of Ammi visnaga. *Anal. Bioanal. Electrochem.* **2018**, *10*, 912–929.
48. Imane, B.; Ouafa, R.; Rachid, D. Antimicrobial and antioxidant activity of Ammi visnaga (L) phenolic extracts and their effects on planktonic and biofilm growth of food spoilage Bacillus cereus. *Int. J. Biosci.* **2016**, *9*, 32–47.
49. Darkaoui, N.; Benbrahim, K.F.; Chaibi, M.; Boukhira, S.; Ez-Zriouli, R.; Ouaritini, B. Phytochemical and Antioxidant Activity of the Essential Oil of Ammi visnaga L. from Morocco. *Pharma Chem.* **2017**, *9*, 68–72.
50. Amin, J.N.; Murad, A.; Motasem, A.M.; Ibrahem, S.R.; Ass'ad, J.M.; Ayed, A.M. Phytochemical screening and in-vitro evaluation of antioxidant and antimicrobial activities of the entire Khella plant (Ammi visnaga L.) a member of palestinian flora. *Int. J. Pharmacogn. Phytochem. Res.* **2015**, *7*, 137–143.
51. Benteldjoune, M.; Boudiar, T.; Bakhouche, A.; del Mar Contreras, M.; Lozano-Sánchez, J.; Bensouici, C.; Kabouche, Z.; Segura-Carretero, A. Antioxidant activity and characterization of flavonoids and phenolic acids of Ammoides atlantica by RP–UHPLC–ESI–QTOF–MSn. *Nat. Prod. Res.* **2021**, *35*, 1639–1643. [CrossRef] [PubMed]
52. Bettaieb Rebey, I.; Aidi Wannes, W.; Kaab, S.B.; Bourgou, S.; Tounsi, M.S.; Ksouri, R.; Fauconnier, M.L. Bioactive compounds and antioxidant activity of Pimpinella anisum L. accessions at different ripening stages. *Sci. Hortic.* **2019**, *246*, 453–461. [CrossRef]
53. Bajpai, V.K.; Agrawal, P. Studies on Phytochemicals, Antioxidant, Free Radical Scavenging and Lipid Peroxidation Inhibitory effects of Trachyspermum ammi seeds. *Indian J. Pharm. Educ. Res.* **2015**, *49*, 58–65. [CrossRef]
54. Lim, Y.Y.; Lim, T.T.; Tee, J.J. Antioxidant properties of several tropical fruits: A comparative study. *Food Chem.* **2007**, *103*, 1003–1008. [CrossRef]
55. Rufino, M.D.S.M.; Alves, R.E.; de Brito, E.S.; Pérez-Jiménez, J.; Saura-Calixto, F.; Mancini-Filho, J. Bioactive compounds and antioxidant capacities of 18 non-traditional tropical fruits from Brazil. *Food Chem.* **2010**, *121*, 996–1002. [CrossRef]
56. Zargoosh, Z.; Ghavam, M.; Bacchetta, G.; Tavili, A. Effects of ecological factors on the antioxidant potential and total phenol content of Scrophularia striata Boiss. *Sci. Rep.* **2019**, *9*, 1–15. [CrossRef] [PubMed]
57. Knight, T.R.; Fariss, M.W.; Farhood, A.; Jaeschke, H. Role of lipid peroxidation as a mechanism of liver injury after aceta-minophen overdose in mice. *Toxicol. Sci.* **2003**, *76*, 229–236. [CrossRef] [PubMed]
58. Song, S.J.; Kim, S.-M.; Lee, S.-H.; Moon, J.-Y.; Hwang, H.S.; Kim, J.S.; Park, S.-H.; Jeong, K.H.; Kim, Y.G. Rhabdomyolysis-Induced AKI Was Ameliorated in NLRP3 KO Mice via Alleviation of Mitochondrial Lipid Peroxidation in Renal Tubular Cells. *Int. J. Mol. Sci.* **2020**, *21*, 8564. [CrossRef]
59. Jodynis-Liebert, J.; Matławska, I.; Bylka, W.; Murias, M. Protective effect of Aquilegia vulgaris (L.) on APAP-induced oxidative stress in rats. *J. Ethnopharmacol.* **2005**, *97*, 351–358. [CrossRef]
60. McGill, M.R.; Williams, C.D.; Xie, Y.; Ramachandran, A.; Jaeschke, H. Acetaminophen-induced liver injury in rats and mice: Comparison of protein adducts, mitochondrial dysfunction, and oxidative stress in the mechanism of toxicity. *Toxicol. Appl. Pharmacol.* **2012**, *264*, 387–394. [CrossRef]
61. Wang, Z.; Hao, W.; Hu, J.; Mi, X.; Han, Y.; Ren, S.; Jiang, S.; Wang, Y.; Li, X.; Li, W. Maltol improves APAP-induced hepatotoxicity by inhibiting oxidative stress and inflammation response via NF-κB and PI3K/Akt signal pathways. *Antioxidants* **2019**, *8*, 395. [CrossRef]

62. Takemoto, K.; Hatano, E.; Iwaisako, K.; Takeiri, M.; Noma, N.; Ohmae, S.; Toriguchi, K.; Tanabe, K.; Tanaka, H.; Seo, S.; et al. Necrostatin-1 protects against reactive oxygen species (ROS)-induced hepatotoxicity in acetaminophen-induced acute liver failure. *FEBS Open Bio* **2014**, *4*, 777–787. [CrossRef]
63. Mongi, S.; Riadh, B.; Noura, B.; Fatma, R.; Kamel, J.; Abdelfattah, E.F. Antioxidant and protective effects of Artemisia campestris essential oil against chlorpyrifos-induced kidney and liver injuries in rats. *Front. Physiol.* **2021**, *12*, 194.
64. Martins, M.D.R.; Arantes, S.; Candeias, F.; Tinoco, M.T.; Cruz-Morais, J. Antioxidant, antimicrobial and toxicological properties of Schinus molle L. essential oils. *J. Ethnopharmacol.* **2014**, *151*, 485–492. [CrossRef] [PubMed]
65. Shang, H.; Zhou, H.-Z.; Yang, J.-Y.; Li, R.; Song, H.; Wu, H.-X. In vitro and in vivo antioxidant activities of inulin. *PLoS ONE* **2018**, *13*, e0192273. [CrossRef] [PubMed]
66. Slatnar, A.; Jakopic, J.; Stampar, F.; Veberic, R.; Jamnik, P. The Effect of Bioactive Compounds on In Vitro and In Vivo Antioxidant Activity of Different Berry Juices. *PLoS ONE* **2012**, *7*, e47880. [CrossRef]
67. Power, O.; Jakeman, P.; Fitzgerald, R.J. Antioxidative peptides: Enzymatic production, in vitro and in vivo antioxidant activity and potential applications of milk-derived antioxidative peptides. *Amino Acids* **2013**, *44*, 797–820. [CrossRef] [PubMed]
68. Glevitzky, I.; Dumitrel, G.A.; Glevitzky, M.; Pasca, B.; Otřísal, P.; Bungau, S.; Cioca, G.; Pantis, C.; Popa, M. Statistical Analysis of the Relationship Between Antioxidant Activity and the Structure of Flavonoid Compounds. *Rev. Chim.* **2019**, *70*, 3103–3107. [CrossRef]
69. Pietta, P.; Simonetti, P.; Mauri, P. Antioxidant Activity of Selected Medicinal Plants. *J. Agric. Food Chem.* **1998**, *46*, 4487–4490. [CrossRef]
70. Pietta, P.G.; Rava, A.; Mauri, P.L. Rapid HPLC method for the analysis of terpenes from Ginkgo biloba in extracts, capsules and syrups by reversed phase. *J. Pharm. Biomed. Anal.* **1992**, *10*, 1077–1080. [CrossRef]
71. Ruberto, G.; Baratta, M.T. Antioxidant activity of selected essential oil components in two lipid model systems. *Food Chem.* **2000**, *69*, 167–174. [CrossRef]
72. Kamal-Eldin, A.; Appelqvist, L.-Å. The chemistry and antioxidant properties of tocopherols and tocotrienols. *Lipids* **1996**, *31*, 671–701. [CrossRef]
73. Edziri, H.; Mastouri, M.; Chéraif, I.; Aouni, M. Chemical composition and antibacterial, antifungal and antioxidant activities of the flower oil ofRetama raetam(Forssk.) Webb from Tunisia. *Nat. Prod. Res.* **2010**, *24*, 789–796. [CrossRef] [PubMed]
74. Dorman, H.J.D.; Surai, P.; Deans, S.G. In Vitro Antioxidant Activity of a Number of Plant Essential Oils and Phytoconstituents. *J. Essent. Oil Res.* **2000**, *12*, 241–248. [CrossRef]
75. Iqbal, T.; Hussain, A.I.; Chatha, S.A.S.; Naqvi, S.A.R.; Bokhari, T.H. Antioxidant Activity and Volatile and Phenolic Profiles of Essential Oil and Different Extracts of Wild Mint (Mentha longifolia) from the Pakistani Flora. *J. Anal. Methods Chem.* **2013**, *2013*, 1–6. [CrossRef] [PubMed]
76. Li, J.; Zhang, X.; Huang, H. Protective effect of linalool against lipopolysaccharide/d-galactosamine-induced liver injury in mice. *Int. Immunopharmacol.* **2014**, *23*, 523–529. [CrossRef]
77. Hamid, Z.A.; Budin, S.B.; Jie, N.W.; Hamid, A.; Husain, K.; Mohamed, J. Nephroprotective effects of Zingiber zerumbet Smith ethyl acetate extract against paracetamol-induced nephrotoxicity and oxidative stress in rats. *J. Zhejiang Univ. Sci. B* **2012**, *13*, 176–185. [CrossRef]
78. Jaeschke, H.; Xie, Y.; McGill, M. Acetaminophen-induced Liver Injury: From Animal Models to Humans. *J. Clin. Transl. Hepatol.* **2014**, *2*, 153–161. [CrossRef]
79. Dargan, P.; Kalsi, S.S.; Wood, D.; Waring, W.S. Does cytochrome P450 liver isoenzyme induction increase the risk of liver toxicity after paracetamol overdose? *Open Access Emerg. Med.* **2011**, *3*, 69–76. [CrossRef]
80. Emam, H.T.; Madboly, A.G. Ameliorative effects of hesperidin and melatonin against acetaminophen-induced nephrotoxicity in adult albino rats. *Egypt. J. Forensic Sci. Appl. Toxicol.* **2021**, *21*, 31–46. [CrossRef]
81. Ez-zahir, A.; Naya, A.; Seddik, N.; Marnissi, F.; Belghmi, K.; Oudghiri, M. In vivo anti-psoriasis, anti-inflammatory activities and oral toxicity studies of aqueous extract from seeds of Ammi visnaga L. *Int. J. Res. Pharm. Sci.* **2020**, *11*, 1025–1037. [CrossRef]
82. Koriem, K.M.M.; Arbid, M.S.; El-Attar, M.A. Acute and subacute toxicity of Ammi visnaga on rats. *Interdiscip. Toxicol.* **2019**, *12*, 26–35. [CrossRef]
83. FDA. Food and Drug Administration Guidance for Industry: Estimating the Maximum Safe Starting dose in Initial Clinical Trials for Therapeutics in Adult Healthy Volunteers. 2005. Available online: https://www.fda.gov/regulatory-information/search-fda-guidance-documents/estimating-maximum-safe-starting-dose-initial-clinical-trials-therapeutics-adult-healthy-volunteers (accessed on 20 December 2021).
84. Guba, R. Toxicity myths: The actual risks of essential oil use. *Perfum. Flavorist* **2000**, *25*, 10–28.
85. Kohlert, C.; Van Rensen, I.; März, R.; Schindler, G.; Graefe, E.U.; Veit, M. Bioavailability and Pharmacokinetics of Natural Volatile Terpenes in Animals and Humans. *Planta Med.* **2000**, *66*, 495–505. [CrossRef] [PubMed]
86. Schäfer, R.; Schäfer, W. Die perkutane Resorption verschiedener Terpene-Menthol, Campher, Limo-nen, Isobornylacetat, α-Pinen-aus Badezusätzen. *Drug Res.* **1982**, *32*, 56–58.
87. Schuster, O.; Haag, F.; Priester, H. Transdermale Absorption von Terpenen aus den etherischen Ölen der Pinimenthol-S-Salbe. *Die Med. Welt* **1986**, *37*, 100–102.

88. Bogdan, M.A.; Bungau, S.; Tit, D.M.; Zaha, D.C.; Nechifor, A.C.; Behl, T.; Chambre, D.; Lupitu, A.I.; Copolovici, L.; Copolovici, D.M. Chemical Profile, Antioxidant Capacity, and Antimicrobial Activity of Essential Oils Extracted from Three Different Varieties (Moldoveanca 4, Vis Magic 10, and Alba 7) of *Lavandula angustifolia*. *Molecules* **2021**, *26*, 4381. [CrossRef]
89. Falk-Filipsson, A.; Löf, A.; Hagberg, M.; Hjelm, E.W.; Wang, Z. d-Limonene exposure to humans by inhalation: Uptake, distribution, elimination, and effects on the pulmonary function. *J. Toxicol. Environ. Health* **1993**, *38*, 77–88. [CrossRef]
90. Parke, D.V.; Rahman, K.M.Q.; Walker, R. The absorption, distribution and excretion of linalool in the rat. *Biochem. Soc. Trans.* **1974**, *2*, 612–615. [CrossRef]

antioxidants

Article

Bio-Evaluation of the Wound Healing Activity of *Artemisia judaica* L. as Part of the Plant's Use in Traditional Medicine; Phytochemical, Antioxidant, Anti-Inflammatory, and Antibiofilm Properties of the Plant's Essential Oils

Hamdoon A. Mohammed [1,2,*], Kamal A. Qureshi [3], Hussein M. Ali [4,5], Mohsen S. Al-Omar [1,6], Omar Khan [7] and Salman A. A. Mohammed [4,*]

1 Department of Medicinal Chemistry and Pharmacognosy, College of Pharmacy, Qassim University, Buraydah 51452, Saudi Arabia; m.omar@qu.edu.sa
2 Department of Pharmacognosy, Faculty of Pharmacy, Al-Azhar University, Cairo 11371, Egypt
3 Department of Pharmaceutics, Unaizah College of Pharmacy, Qassim University, Unaizah 51911, Saudi Arabia; ka.qurishe@qu.edu.sa
4 Department of Pharmacology and Toxicology, College of Pharmacy, Qassim University, Buraydah 51452, Saudi Arabia; hu.ali@qu.edu.sa
5 Department of Biochemistry, Faculty of Medicine, Al-Azhar University, Assiut 71524, Egypt
6 Department of Medicinal Chemistry and Pharmacognosy, Faculty of Pharmacy, Jordan University of Science and Technology (JUST), Irbid 22110, Jordan
7 Department of Pharmaceutics, College of Pharmacy, Qassim University, Buraydah 51452, Saudi Arabia; ok.mohamad@qu.edu.sa
* Correspondence: ham.mohammed@qu.edu.sa (H.A.M.); m.azmi@qu.edu.sa (S.A.A.M.)

Citation: Mohammed, H.A.; Qureshi, K.A.; Ali, H.M.; Al-Omar, M.S.; Khan, O.; Mohammed, S.A.A. Bio-Evaluation of the Wound Healing Activity of *Artemisia judaica* L. as Part of the Plant's Use in Traditional Medicine; Phytochemical, Antioxidant, Anti-Inflammatory, and Antibiofilm Properties of the Plant's Essential Oils. *Antioxidants* **2022**, *11*, 332. https://doi.org/10.3390/antiox11020332

Academic Editors: Delia Mirela Tit and Simona Bungau

Received: 9 January 2022
Accepted: 4 February 2022
Published: 8 February 2022

Abstract: *Artemisia judaica* (ArJ) is a Mediterranean aromatic plant used traditionally to treat gastrointestinal ailments, skin diseases, atherosclerosis, and as an immuno-stimulant. This study describes ArJ essential oil constituents and investigates their wound healing activity. The in vitro antioxidant and antibiofilm activities of ArJ essential oil were investigated. The in vivo pro/anti-inflammatory and oxidative/antioxidant markers were compared with standard silver sulfadiazine (SS) in a second-degree skin burn experimental rat model. The gas chromatography-equipped flame ionization detector (GC-FID) analysis of ArJ essential oil revealed the major classes of compounds as oxygenated monoterpenes (>57%) and cinnamic acid derivatives (18.03%). The antimicrobial tests of ArJ essential oil revealed that *Bacillus cereus*, *Candida albicans*, and *Aspergillus niger* were the most susceptible test organisms. Two second-degree burns (each 1 inch square in diameter) were created on the dorsum of rats using an aluminum cylinder heated to 120 °C for 10 s. The wounds were treated either with ArJ or SS ointments for 21 days, while the negative control remained untreated, and biopsies were obtained for histological and biochemical analysis. The ArJ group demonstrated a significant increase in antioxidant superoxide dismutase (SOD) and catalase (CAT) enzymatic activities, while lipid peroxide (LP) levels remained insignificant compared to the negative control group. Additionally, ArJ and SS groups demonstrated a significant decrease in inflammatory levels of tumor necrosis factor α (TNF-α) compared to the negative group, while interleukin 1 beta (IL-1b) and IL-6 were comparable to the negative group. At the same time, anti-inflammatory IL-10 and transforming growth factor beta 1 (TGF-b1) markers increased significantly in the ArJ group compared to the negative control. The ArJ results demonstrated potent wound healing effects, comparable to SS, attributable to antioxidant and anti-inflammatory effects as well as a high proportion of oxygenated monoterpenes and cinnamate derivatives.

Keywords: *Artemisia judaica*; essential oil; wound healing; inflammatory markers; antibiofilm activity; antioxidant activity

1. Introduction

Artemisia is a genus of annual, perennial, and biennial herbs in the Asteraceae (Compositae) family [1]. The plants of the genus *Artemisia* are frequently used in traditional medicine as remedies for human and animal ailments. For instance, *Artemisia* species have been used in traditional medicine for respiratory disorders, including coughs and phlegm, as a pain killer, worm expelling agent, diaphoretic and diuretic agent, and for the treatment of wounds, hypertension, and allergies [2]. In addition, some of the *Artemisia* plants are traditionally used to treat seizures, and the activity is confirmed through in vivo animal experiments [3–5]. *Artemisia* species have been reported in in vitro and in vivo experiments and in clinical trials evaluating their anticancer, antimalarial, antimicrobial, and antiviral activities [6,7].

Furthermore, several side effects and misuses have also been reported for some of the genus' plants. For instance, *A. monosperma* leaves are not recommended in pregnancy and are used to induce abortion in Jordan [8]. However, this plant, in addition to other plants of *Artemisia*, e.g., *A. vulgaris*, has been used in folklore medicine for labor induction [8–10]. Besides abortion, vomiting, diarrhea, headache, pruritus, and rashes have been reported among young children and pregnant women who used *A. annua* to treat malaria [11]. The *Artemisia* plants' biological activities were attributed to the presence of essential oils, sesquiterpene lactones, flavonoids, bitter principles, coumarins, and phenolic acids [1,2,12,13]. Several *Artemisia* species grow wildly or as cultivated plants for their use as medication and as a herbal tea preparation in the Mediterranean region [9,14,15].

Artemisia judaica L. (ArJ) is widely grown in the Mediterranean region, including Algeria, Libya, Egypt, Jordan, and Saudi Arabia [16–20]. In Saudi Arabia, ArJ grows in the kingdom's northern region, including the border area of the Hail-Qassim regions [21]. ArJ has been reported for several traditional uses, e.g., healing external wounds and repairing snake and scorpion bites [22]. In addition, ArJ is traditionally used to treat gastrointestinal disorders, sexual inability, hyperglycemia, heart diseases, inflammatory disorders, arthritis, cancers [1,20], skin diseases, atherosclerosis, and enhance vision and immunity [23,24]. The Bedouins in Egypt (Sinai) and Saudi Arabia also use the plant as a herbal tea in treating GIT disorders [16]. Biologically, ArJ demonstrated antidiabetic, antioxidant, hepatoprotective, and anti-inflammatory activities in experimental animals [22,25,26] due to the properties inherent in the chemical structure of the compounds it contains [27]. The plant also exhibited weak antimicrobial activity against Gram-positive and Gram-negative bacteria [28,29]. In vitro studies reported the plant extract's potential antioxidant and anticancer activities [28,29].

ArJ chemical analysis revealed the presence of flavonoids, e.g., glycosides and aglycones of apigenin, luteolin, and quercetin [22]. Other natural classes, such as phenolics, triterpenes, bitter principles, and sesquiterpene lactones, i.e., judaicin, have also been reported from the plant [22,30]. Additionally, ArJ is an aromatic plant. Its essential constituents have been identified from the plant species growing in different areas and climatic regions [18,20,23,24,31]; as well known as the anthropogenic factors, environmental conditions primarily affect the composition of the plant [32]. The overall analysis of the essential oil constituents of ArJ indicated that the monoterpene, i.e., piperitone, is the major chemotypic constituent in the plant from different genotypes [1,24]. In addition, other essential constituents of the plants, such as camphor, ethyl cinnamate, and spathulenol, have also been identified in relatively high concentrations in individual plant genotypes [24]. In addition, environmental conditions and the geographical locations of the plant growing areas have been reported to affect the major chemotypic constituents of ArJ essential oils. Table 1 demonstrates the major constituents of the plant essential oils from different locations.

Table 1. Major constituents of the ArJ essential oils from plant species growing in different areas.

Locations	Major Constituents	Y%	Ref.
Saudi Arabia	cis-Thujone (2.5%), thymol (3.5%), trans-sabinyl acetate (3.3%), carvacrol (3.5%), b-eudesmol (13.1%), eudesma-4 (15), 7-dien-1-b-ol (3.5%), and hexadecanoic acid (5.7%)	0.18% (*v*/*w*)	[24]
Algeria	Piperitone (66.17%), ethyl cinnamate isomer (6.11%), spathulenol (2.34%), E-longipinane (2.55%)	1.7% (*w*/*w*)	[33]
Egypt	Piperitone (49.1%) and camphor (34.5%), borneol (3.90%)		[34]
Sinai, Egypt	Camphor (31.4%), endo-borneol (5.72%), piperitone (29.9%)	0.28%	[18]
Jordan	Artemisia ketone (9–24%), chrysanthenone (4–31%), piperitone (3–15%), camphor (0.3–16%), cinnamate (11.0%)	0.4–0.9% (*w*/*w*)	[20,35]
Libya	cis-Chrysanthenol (9.1%), piperitone (30.2%), ethyl cinnamate (3.8%).	0.62% (*w*/*w*)	[23]

Y% refers to the yield of the essential oil.

The methods used for essential oil production from aromatic plants vary and mostly depend on the nature of the volatile constituents, the amount of the essential oils, and the nature of the plant samples [36]. Thereby, distillation procedures are primarily used for the plants containing a considerable amount of the thermostable volatile constituents; however, volatile (e.g., diethyl ether) and non-volatile (e.g., lard) solvent extraction processes are used for the extraction of the highly delicate aromatic plants which contain heat-sensitive and small quantities of the essential oils [37]. In addition, modern extraction techniques, such as CO_2, supercritical CO_2 extraction and microwave-assisted extraction techniques, are used for the industrial-scale production of the essential oils with specific advantages, e.g., time- and quantity-based efficiency and environmentally friendly properties [38–40].

Burn injury traumas occur by friction, cold, heat, radiation, chemical, or electric sources, but hot liquids, solids, and fire contribute significantly towards burn injuries [41]. In Saudi Arabia, 52% of all burns occur in young children, and males are more prone to burns than females (1.42:1). Burn wounds require immediate attention to avoid hypovolemic shock and sepsis [42]. New approaches and drugs are being researched to facilitate faster burn wound healing [43], thereby minimizing adverse reactions, like allergy or irritation, due to topical agents that increase the rehabilitation period [44]. In addition to their general availability, herbal medicines have demonstrated a promising role in wound healing compared to silver sulfadiazine (SS) [45–47]. Nevertheless, modern approaches and methodologies are required to validate claims for herbal compounds [48].

The current study is designed to demonstrate the wound healing properties of ArJ essential oils as part of the plant's use in traditional medicine. Therefore, a phytochemical analysis of the ArJ essential oil for the species growing in the Northern Qassim region of Saudi Arabia was conducted. The study also investigated the antioxidant, antimicrobial, and antibiofilm activities of the plant essential oil as associated analyses related to the wound healing potential of the plant.

2. Materials and Methods

2.1. Plant Materials and Distillation Procedure

The aerial plant parts were collected during March 2020, in the morning, from the Northern Qassim region of Saudi Arabia and identified as *Artemisia judaica* L. by the taxonomists in the Department of Plant Production and Protection, College of Agriculture, Qassim University (Buraydah, Saudi Arabia). A sample of the plant with the registered number #090 was kept at the herbarium of the College of Pharmacy, Qassim University (Buraydah, Saudi Arabia). The plant was dried in the shade at room temperature for ten days before the distillation process. The plant materials, 200 g, were reduced to coarse powder form, backed to a 2 L conical flask with a stopper, and thoroughly mixed with 700 mL of distilled water. The flask was connected to the Clevenger apparatus and fixed over the heating mantel. The flask contents were allowed to boil for a continuous 5 h. The distillate essential oil was collected over anhydrous sodium sulfate and stored in an opaque glass vial in a −20 °C freezer.

2.2. GC-FID Analysis of the Essential Oil

A gas chromatography (Perkin Elmer Auto System XL, Waltham, MA, USA) equipped flame ionization detector (GC-FID) was used to analyze the essential oil of ArJ. The chromatographic separation of the oil samples was achieved on a fused silica capillary column ZB5 (60 m × 0.32 mm i.d. × 0.25 μm film thickness). The oven temperature was maintained initially at 50 °C and programmed from 50 to 240 °C at a rate of 3 °C/min. ArJ essential oil sample was dissolved in analytical grade diethyl ether (2.9 mg of the oil in 100 μL of the solvent). Then, 1 μL of the mixture was injected with a 1/20 split ratio. The helium was used as the carrier gas at a 1.1 mL/min flow rate. The injector and detector temperatures were 220 and 250 °C, respectively.

2.3. Gas Chromatography–Mass Spectroscopy Analysis of the Volatile Oil

The GC–MS analysis of ArJ essential oil was conducted using an Agilent 8890 GC system attached to a PAL RTC 120 auto-sampler and equipped with a mass detector, Agilent 9977B GC/MSD mass spectrometer (Agilent technology, Santa Clara, CA, USA). An HP-5 capillary column (30 m, 250 μm i.d., 0.25 μm film thickness) was used to separate target molecules. The initial column temperature (50 °C for 2 min, isothermal) was programmed up to 220 °C at a rate of 5 °C/min, and then 10 °C/min up to 280 °C and kept constant at 280 °C for 10 min (isothermal). The injector temperature was 230 °C. Helium was used as a carrier gas at 1 mL/min flow rate. All the mass spectra were recorded using the following conditions. The run time was about 65 min. The transfer line was set at 280 °C, and the ionization source and the mass analyzer temperatures were set at 230 and 150 °C, respectively. Diluted samples (1% v/v) were injected with split mode (split ratio 1:15).

2.4. Identification of the Essential Oil Constituents

The constituents of the oil were identified based on the experimental retention index (RI) calculated with references to a series of standard n-alkenes series (C8–C40) and the retention indexes reported for the ArJ essential constituents besides the reported retention indexes obtained for the analysis of different essential oils under similar GC experimental conditions. In addition, the National Institute of Standards and Technology (NIST-11) and mass fragmentation patterns of the peaks were also used to identify the compounds. The relative percentages of the constituents were calculated from the area under the peak obtained from the GC-FID chromatogram.

2.5. Antioxidant Activity of ArJ Essential Oil

2.5.1. Total Antioxidant Capacity (TAC)

The method described by Aroua et al. [49] was followed to conduct this experiment. In brief, sulfuric acid (0.6 M) and ammonium molybdate (4 mM) in sodium phosphate buffer (28 mM) were mixed to prepare the molybdate reagent. Then, 3.6 mL of the molybdate reagent was mixed with 0.4 mL of ArJ essential oil (containing 200 μg of the oil) in a stoppered glass test tube. The tube was vortexed and warmed for 30 min at 90 °C in a water bath. After cooling, the absorbance of the developed blue color was recorded at 695 nm using a spectrophotometer against a blank prepared essential oil. The TAC of ArJ essential oil was calculated equivalent to the Trolox using the standard calibration curve.

2.5.2. DPPH (2,2-Diphenyl-1-Picrylhydrazyl) Scavenging Activity (DPPH-SA)

The method was conducted according to Shimada et al. [50]: 1 mL of the diluted ArJ essential oil (containing 200 μg of the oil in methanol) was mixed with 1 mL of DPPH (prepared by dissolving 6 mg of the DPPH in 50 mL of methanol). The mixture absorbance was measured at 517 nm after 30 min of standing at room temperature in a dark place. The DPPH-SA was calculated equivalent to Trolox from three independent measurements.

2.5.3. Ferric Reducing Antioxidant Power (FRAP) Assay

Minor modifications to the method of Benzie and Strain [51] were carried out to measure the FRAP of ArJ essential oil. FRAP working reagent was freshly prepared by adding TPTZ (2,4,6-Tris(2-pyridyl)-s-triazine, 10 mM prepared in 40 mM HCl) to $FeCl_3 \cdot 6H_2O$ (20 mM) and acetate buffer (300 mM, pH 3.6) in a ratio 1:1:10. Then, 2 mL of the FRAP reagent was added to 0.1 mL of the ArJ essential oil (containing 200 μg the oil), the mixtures were incubated for 30 min at room temperature, and the absorbance was recorded at 593 nm. The procedure was conducted in triplicate, and the prepared FRAP–Trolox calibration curve was used to calculate the extract activity as mg Trolox equivalent per gram of the used plant's dried extract.

2.5.4. Metal Chelating Activity Assay (MCA)

The ArJ essential oil ability to chelate iron compared to the EDTA was evaluated using Zengin et al.'s method [52]. Briefly, a mixture of the ArJ essential oil (2 mL of ethanol containing 200 μg of the oil) and ferrous chloride (25 μL, 2 mM) was added to 100 μL of ferrozine to inchoate the color. The mixture's absorbance was recorded at 562 nm against a blank (2 mL of the ArJ essential oil plus 200 μL of the ferrous chloride without ferrozine). The standard calibration curve of EDTA was prepared, and the chelating activity of the ArJ essential oil was calculated in equivalents of the EDTA.

2.6. Antimicrobial Activity of ArJ Essential Oil

2.6.1. Preliminary Antimicrobial Activity

The preliminary antimicrobial activity of ArJ essential oil was determined by the disc diffusion method [53]. Modified Mueller–Hinton agar (MMHA) and potato dextrose agar (PDA) were used as test media. MMHA plates were prepared according to the protocol mentioned in the literature [54]. The sterile paper discs (6 mm in diameter) were impregnated with 20 μL of pure ArJ essential oil and then used to evaluate the antimicrobial potential of ArJ essential oil against the selected human pathogens. Levofloxacin (5 μg/disc) and clotrimazole (50 μg/disc) were used as antibacterial and antifungal control (C) drugs. Each test organism's inoculum was prepared in sterile tryptic soy broth (TSB), and the turbidity of each suspension was adjusted equal to 0.5 MacFarland standard, which is equal to 1.5×10^8 colony forming units (CFU/mL) for bacteria, 1×10^6–5×10^6 CFU/mL for yeast and 4×10^5 to 5×10^6 CFU/mL for mold. Following that, 100 μL suspensions of each adjusted inoculum were poured individually over the surface of the test agar plates and then uniformly spread using sterile swabs. The prepared discs of ArJ and control drugs were then put on the inoculated plates. The plates were incubated at 35 °C for 24 h for bacteria and 48 h for fungi. After incubation, the diameters of inhibitory zones were calculated on a millimeter (mm) scale. Each test was performed in triplicate. The results are expressed in mm ± standard deviation (SD).

2.6.2. Minimum Inhibitory Concentration (MIC) and Minimum Biocidal Concentration (MBC)

MIC was determined by the resazurin-based micro-broth dilution method, while MBC was performed following the standard spot inoculation method [53,55]. The inocula of each test bacteria were prepared in TSB, following the Clinical and Laboratory Standards Institute (CLSI) guidelines (https://clsi.org/, accessed on 1st December 2021), where the OD_{600} value (0.08–0.12) was adjusted, resulting in ~1×10^8 CFU/mL. Then, adjusted inocula were further diluted by 1:100 in TSB, resulting in ~1×10^6 CFU/mL. In contrast, the inocula of test fungi were prepared in potato dextrose broth (PDB) following the CLSI guidelines, where the OD_{600} value (0.08–0.12) was adjusted, the resulting stock suspension contained 1×10^6 to 5×10^6 CFU/mL for yeast and 4×10^5 to 5×10^6 CFU/mL for mold. A working yeast suspension was prepared by a 1:100 dilution followed by a 1:20 dilution of the stock suspension with PDB, resulting in 5.0×10^2 to 2.5×10^3 cells/mL, while a working mold suspension was prepared by a 1:50 dilution of the stock suspension with

PDB, resulting in 0.8×10^4 to 1×10^5 cells/mL. The initial stock solution of ArJ essential oil was prepared in DMSO (dimethyl sulfoxide) at a 200 μL/mL concentration. Each well in column 1 was dispensed with 200 μL of stock solution of ArJ essential oil. At the same time, each well of columns 2 to 10 contained 100 μL of tryptic soy broth (TSB) for antibacterial evaluation, while for antifungal assessment, 100 μL of potato dextrose broth (PDB) was used. A two-fold serial dilution of ArJ essential oil was made from columns 1 to 10 using a multichannel micropipette, resulting in concentrations of ArJ essential oil ranging from 200–0.39 μL/mL in columns 1 to 10. Column 11 had 200 μL of standardized inoculum suspensions, which served as negative control (NC), and column 12 had 200 μL of sterile broth, which served as sterility control (SC). Each organism's adjusted inoculum was dispensed, 100 μL into each test well in columns 1–10, respectively. The 100 μL of adjusted microbial inocula were dispensed in all the wells of columns 1 to 10, resulting in ~5×10^5 CFU/mL for bacteria and ~2.5×10^2 to 1.25×10^3 CFU/mL for *C. albicans*, and 0.4×10^4 to 5×10^4 CFU/mL for *A. niger*. At this stage, the final concentrations of ArJ essential oil were 100 to 0.195 μL/mL in columns 1 to 10. The time taken to prepare and dispense the OD-adjusted microbial inocula did not exceed 15 min. The inoculated plates were incubated at 35 °C for 24 h for bacteria and 48 h for fungi. Following incubation, 30 μL of sterile resazurin dye (0.015% w/v) was dispensed into each well of columns 1 to 12, and then plates were re-incubated for 1–2 h to observe color change. After incubation, columns with the lowest concentrations showing no color change (blue resazurin color stayed intact) were scored as MIC.

MBC was determined by directly plating the contents of wells with concentrations above the MIC on sterile tryptic soy agar (TSA) plates for bacteria, while potato dextrose agar (PDA) plates were used for fungi. The contents from the wells, which did not change from blue to pink, were inoculated on sterile tryptic soy agar (TSA) plates and incubated at 35 °C for 24 h for bacteria and 48 h for fungi. The lowest concentration of ArJ did not produce isolated colonies of the test organisms on inoculated agar plates considered as the MBC. The results are recorded in μL/mL.

2.6.3. Minimum Biofilm Inhibitory Concentration (MBIC) and Minimum Biofilm Eradication Concentration (MBEC)

MBIC Assay

MBIC is defined as the lowest concentration of the antimicrobial agent (ArJ), preventing the biofilm formation of the tested organism. MBIC was conducted against the bacteria only. The 96-well microtiter plate was used to evaluate the anti-biofilm activity of ArJ [54]. The inocula of the test organisms were prepared in TSB equal to 0.5 MacFarland standard (1–2×10^8 CFU/mL). An aliquot of 100 μL from the adjusted inocula was dispensed into each test well of a 96-well plate. Then 100 μL of different concentrations of ArJ were dispensed into test wells. Thus, the final concentrations for MBIC assessment were MIC, $2 \times$ MIC, and $4 \times$ MIC. The wells containing only 200 μL of TSB served as a blank control (BC), whereas those containing bacterial cultures without ArJ served as negative control (NC). The plates were incubated in a shaking water bath at 35 °C for 24 h at 100 rpm shaking speed. After incubation, the supernatants from each well were decanted gently by reversing the plates on a tissue paper bed/or removed by a pipette without disturbing the biofilms. The plates were dried in air for 30 min, stained with 0.1% (w/v) crystal violet at room temperature for 30 min, and then washed three times with distilled water. Subsequently, the crystal violet was solubilized by adding 200 μL of 95% ethanol to each test well. The absorbance was recorded in a microplate reader (xMark™ Microplate Absorbance Spectrophotometer-Bio-Rad, Hercules, CA, USA) at 650 nm. The lowest concentration of ArJ at which the absorbance equals or falls below the negative control is considered MBIC. Each test was performed in triplicate. The mean of three independent tests was taken. The results are expressed in μL/mL.

MBEC Assay

MBEC is defined as the minimum concentration of an antimicrobial agent (ArJ) that eradicates the biofilm of the test organism [54]. A 200 μL (1–2 × 10^8 CFU/mL) inoculum of each test organism was inoculated into each test well of a flat-bottom 96-well microtiter plate. The plates were incubated at 35 °C for 48 h in a shaking water bath at 100 rpm shaking speed for biofilm formation. After the biofilms had formed, the contents of the test wells were decanted gently by reversing the plates on a tissue paper bed/or removed by a pipette without disturbing the biofilms. The various concentrations, i.e., MIC, 2 × MIC, and 4 × MIC of ArJ, were added to different test wells (200 μL/well). The inoculated plates were re-incubated at 35 °C for 24 h. After incubation, the contents of each test well were discarded by inverting the plates on a tissue bed. The plates were dried in air for 30 min, and then 200 μL of sterile TSB was dispensed in each test well. Then, 30 μL of 0.015% *w/v* resazurin dye was added into each test well. The plates were re-incubated for 1–2 h. After re-incubation, the MBEC was recorded by observing the color change from blue to pink. The column with no color change (blue resazurin color stayed intact) was scored MBEC. Biofilm without ArJ served as a negative control (NC). Each test was performed in triplicate. The mean of three independent tests was taken. The results are expressed in μL/mL.

2.7. Preparation of Ointment Formulation Loaded ArJ Essential Oil

Ointment formulation of 5% *w/w* strength of ArJ essential oil was prepared. The simple ointment base was prepared by the fusion method according to the *British Pharmacopoeia 1988* [56]. Briefly, 100 g of simple ointment base was prepared by melting hard paraffin (5 g) in a beaker at 61 °C. The other ingredients, i.e., cetostearyl alcohol (5 g), wool fat (5 g), and soft white paraffin (85 g), were added in descending order of melting point. The homogenous mixture was removed from the heat and stirred until cold. Then, 5% *w/w* strength ArJ essential oil ointment was prepared by incorporating 5 g of the essential oil into 95 g of a simple ointment base in small portions by mixing with trituration using an ointment mortar and pestle. Finally, the ArJ ointment was transferred to a clean container. The control ointment, 50 g of the entire base ingredients, was taken and treated in the same way to formulate without the essential oil. The prepared ArJ ointment was physically examined and was consistent, homogenous, and stable for the measured one month.

2.8. In Vivo Wound Healing Animal Experiment

Twenty healthy 3-month-old Sprague Dawley female rats weighing about 150 ± 50 g were individually maintained in the cage under 25 ± 2°, 65% humidity, 12:12 light/dark cycle. Animals were fed with a standard chow diet with water ad libitum, and the wound healing study was conducted following the guidelines of the Institutional Animal Ethics Committee (Registration # 21-04-06). The animal groups involved intact, negative control, positive control (1% silver sulfadiazine, SS), and ArJ 5% ointment.

2.9. Skin Burn Induction Model

Briefly, the animals were anesthetized using xylazine 5 mg/kg and ketamine 50 mg/kg, and the rat's dorsum was shaved with a hair trimmer (GEEPAS®, Guangzhou, China) at a 45° angle to minimize the angle skin injury during shaving and disinfected using 70% ethanol. An aluminum cylinder (1-inch square diameter, 86 g weight) was heated using a hot water bath at 120 °C for at least 60 min to ensure thermal equilibrium with the water. The exact temperature of the cylinder and the water was measured before inducing the burns using a dual probe thermometer (UT320D Mini Contact Type Thermometer Dual Channel K/J Thermocouple, UNI-T, Dongguan, China). Second-degree burns of 1-inch square diameter were induced on the rat's shaved dorsum by patching the aluminum cylinder on the rat's dorsum for 10 s, allowing it to stand on its own weight to ensure symmetrical burns across all rats [57–59]. The animals were administered with 0.9% normal saline i.p injection 10 mL/kg. The treated groups were applied topically twice daily for three weeks with ArJ 5% ointment or 1% SS cream topically on the wound area.

2.10. Biopsy

At the end of the experiment on day 21, the animals were euthanized and a biopsy measuring 1 × 1 cm diameter was collected using scissors and tweezers from the underlying tissue. One part of the biopsy was fixed in 3.7% formalin for paraffin embedding, while another part was homogenized and the supernatant isolated and stored at −20 °C for biochemistry analysis.

2.11. Histological Staining

Tissue was processed within 48 h of collection by dehydration with increasing ethanol percentages before being cleared with xylene and embedded in paraffin wax. Tissue sections of 5 microns were cut using a microtome (MEDIMEAS, Haryana, India), allowing simultaneous sectioning of the epidermis and dermis. Sections were stained using hematoxylin and eosin (H&E) and visualized under a light microscope at 40× magnification. Five fields/sections were counted for the amount of fibroblast, collagen, inflammation, and neovascularization, and the data was scored from 0–4, where 0, 1, 2, 3, and 4 represented normal, low, moderate, high, and very high, respectively, as described previously [60].

2.12. Determination of Oxidants and Antioxidants

The catalase (CAT, Serial No. 24IF07D5A0) and superoxide dismutase (SOD, Serial No. 745402C55B) activity, and lipid peroxide (LP, malondialdehyde, Serial No. 1F4346D808) levels were determined in skin wound tissue homogenate by enzyme-linked immunosorbent assay (ELISA) kits (Cloud Clone Corp Company, Houston, TX, USA), according to the manufacturer's instructions. The absorbance was measured at 450 nm by a microplate ELISA reader and the concentration was calculated using a standard curve.

2.13. Determination of Pro-Inflammatory and Anti-Inflammatory Cytokine Levels: Interleukins, TGF-b, and TNF-α Levels

The pro-inflammatory cytokines: interleukin 1 beta (IL-1b, Serial No. 282D397BBC), IL-6 (Serial No. 51D9580378), and TNF-α (Serial No. 3898289A45) and the anti-inflammatory cytokines (IL-10 (Serial No. 6AB644B25F), transforming growth factor beta 1 (TGF-b1, Serial No. 34997E20C3) were assayed in tissue homogenate by ELISA kits (Cloud Clone Corp Company, Houston, TX, USA) according to the manufacturer's instructions. The microplates were measured with a 450 nm filter by a microplate reader.

2.14. Wound Area Measurement

Using a standard camera, images of skin burn for all the animals were captured on the day of burn induction and at different time points (week 1, 2, and 3), while the wound measurement was performed before the treatment and 2 weeks after the treatment using freely available Image J software (version 1.8.1, Public Domain, Madison, WI, USA). Due to hair regrowth on the wound area, wound size could not be measured accurately after 2 weeks.

2.15. Statistical Analysis

Data were expressed as the mean ± standard error of the mean (SEM) (n = 5). Differences between groups were analyzed using one-way ANOVA, except for wound area measurement at different time-points, which was analyzed using two-way ANOVA followed by a post hoc test using Tukey's multi-group comparison on GraphPad Prism 8.0.2 (GraphPad Software, San Diego, CA, USA). The data were considered significant if $p < 0.05$ [61]. The superscripts (A–C) describing significance among the groups in the tables were obtained using Minitab 19.1 (Minitab LLC, State College, PA, USA).

3. Results and Discussion

3.1. Essential Oil Constituents of *A. judaica*

Several parameters have been reported as influencing factors affecting essential oil production, constituents, and quality; the parameters include the maturity stage of the plant, the oil extraction processes, and the drying methods applied to the aromatic plant samples, as well as the environmental conditions where the aromatic plant grows [62–65]. The essential oil of wild ArJ growing in the Northern Qassim region of Saudi Arabia has been isolated by the hydro-distillation technique using a Clevenger apparatus from the shade-dried aerial parts of the plant. Three different distillation experiments have been used to calculate the essential oil production percentage of 1.71 ± 0.3% *w/w* of the dried plant aerial parts. The percentage yield was higher than the reported yields for the cultivated species of the plant growing in Saudi Arabia (0.18% *v/w*) [24], indicating the higher capacity of the wild species of ArJ to biosynthesize essential oils. In addition, the nature of the plant sample, i.e., fresh or dried, and the conditions of the drying process could be factors affecting oil production percentage. The reported oil production percentage (0.18% *w/w*) has been calculated for the fresh plant samples [24]. However, the current percentage (1.71 ± 0.3% *w/w*) of essential oil production resulted from the distillation of the ten-day dried plant sample, which is consistent with the reported percentages of the essential oil production from dried samples of the aromatic plants, i.e., rosemary and sage [66,67]. Moreover, the current essential oil recovery percentage (1.71 ± 0.3% *w/w*) was nearly similar to the recorded data reported for the wild species of ArJ growing in the Southern region of Jordan (1.62%) [20].

The produced oil samples obtained from each distillation experiment were independently subjected to GC-FID analysis (Supplementary file, Figure S1). The results expressed in Table 1 show the mean relative percentage of the individual compounds plus standard deviations obtained from the three GC-FID spectroscopic runs. Kovats retention index was calculated with the C_8–C_{40} series of *n*-alkenes analyzed under identical extermination conditions. The reported retention indexes were also used to identify the ArJ essential constituents. The results shown in Table 2 indicated that oxygenated monoterpenes represented ≈ 57% of the plants' essential constituents among all essential oil classes. The higher percentage of oxygenated monoterpenes was attributed to the presence of piperitone in a high concentration (31.99% of the total essential oils in the plant). In addition, other oxygenated monoterpenes, e.g., terpinene-4-ol, α-thujone, β-thujone, 1,8-cineole, camphor, and linalool, were represented at relatively high concentrations of 6.42, 5.94, 3.61, 2.56, 1.92, and 1.21%, respectively, with a total percentage of 21.66%. The concentration of piperitone (31.99%) and the total oxygenated monoterpene concentrations (57%) among the total essential oil constituents (Figure 1) were consistent with the reported chemotypic properties of the plant [24] that have been found, 30–70% of piperitone in the essential oil of ArJ growing wildly in different regions of the Mediterranean countries, such as Egypt, Algeria, and Jordan [20,31,68,69].

Table 2. Essential oil constituents of *A. judaica* growing in the Northern Qassim region of Saudi Arabia.

RT	Chemical Compounds	Area Mean	RI^{cal}	RI^{rep}	m/z	Weight g/100 g of the Plant
12.096	(Z)-3-Hexenol	0.50 ± 0.08	845	845		0.0085
12.209	2-Methyl-ethylbutanoate	0.4 ± 0.06	850	853		0.0068
16.403	Sabinene	0.13 ± 0.11	953	954	59.04 (100%), 81.05 (96.37%), 96.07 (83.12%)	0.0022
18.449	α-Phellandrene	1.30 ± 0.16	1000	999	68.04 (100%), 79.03 (42.69%), 93.04 (93.23%)	0.0222
19.795	Limonene	0.72 ± 0.01	1029	1028	85.04 (100%), 55.03 (12.29%), 70.07 (7.28%)	0.0123

Table 2. *Cont.*

RT	Chemical Compounds	Area Mean	RI^{cal}	RI^{rep}	m/z	Weight g/100 g of the Plant
20.239	1,8-Cineole	2.56 ± 0.05	1038	1040	69.04 (100%), 110.08 (70.37%), 95.06 (46.98%)	0.0437
21.359	γ-Terpinene	3.58 ± 0.18	1062	1063	135.05 (100%), 91.03 (20.36%), 107.03 (11.26%)	0.0612
23.548	Linalool	1.21 ± 0.04	1108	1104	91.04 (100%), 92.04 (98.97%), 55.04 (47.04%)	0.0207
23.739	α-Thujone	5.94 ± 0.09	1112	1112	95.06 (100%), 81.04 (68.49%), 109.04 (35.48%)	0.1016
24.24	β-Thujone	3.61 ± 0.03	1123	1124	84.0 (100%), 55.02 (80.71%), 126.05 (47.17%)	0.0617
24.49	α-Campholenal	0.91 ± 0.02	1128		82.04 (100%), 110.06 (91.66%), 95.04 (43.11%)	0.0156
24.639	Terpinene-4-ol	6.42 ± 0.17	1132	1140	70.04 (100%), 83.03 (71.89%), 71.03 (29.17)	0.1098
25.443	Isothujol	0.27 ± 0.47	1149	1145		0.0046
25.651	Camphor	1.92 ± 0.09	1154	1155	68.01 (100%), 81.02 (26.97%), 55.04 (18.33%)	0.0328
26.454	Borneol	0.47 ± 0.01	1170	1170	95.04 (100%), 110.04 (48.36%), 54.06 (25.12%)	0.0080
27.156	p-Cymene-8-ol	0.08 ± 0.14	1186	1185	135.06 (100%), 150.08 (40.42), 91.03 (32.05)	0.0014
29.403	Neral	0.36 ± 0.00	1235	1236	69.02 (100%), 68.01 (17.21%), 83.01 (11.68%)	0.0062
30.034	Linalyl acetate	0.10 ± 0.17	1249	1250	107.06 (100%), 95.04 (34.18), 55.03 (14.79%)	0.0017
30.845	Piperitone	31.99 ± 0.50	1268	1260	82.04 (100%), 110.02 (37.44%), 95.06 (19.20%)	0.5470
31.485	Phellandral	0.38 ± 0.02	1281			0.0065
31.965	p-Cymen-7-ol	0.08 ± 0.13	1292	1290		0.0014
32.146	Thymol	1.81 ± 0.03	1296	1297	135.06 (100%), 107.03 (11.26), 77.01 (10.75%)	0.0309
33.406	Carvacrol	0.10 ± 0.12	1325	1324		0.0017
33.643	Citronellyl acetate	0.89 ± 0.02	1330	1334	107.06 (100%), 91.04 (39.11), 122.08 (15.51)	0.0152
34.649	(*E*)-Methyl cinnamate	0.35 ± 0.02	1354	1355	131.02 (100%), 103.03 (61.07), 162.04 (49.31%)	0.0060
35.894	*Cis*-Ethyl cinnamate	4.02 ± 0.06	1383	1376	131.04 (100%), 103.04 (48.29%), 77.03 (30.40%)	0.0687

Table 2. *Cont.*

RT	Chemical Compounds	Area Mean	RI^{cal}	RI^{rep}	m/z	Weight g/100 g of the Plant
36.303	Jasmone	0.71 ± 0.02	1392	1396	91.03 (100%), 95.03 (66.88%), 79.03 (60.65%)	0.0121
36.658	β-Bourbounene	4.06 ± 0.17	1401	1401	111.02 (100%), 137.07 (41.99%), 180.09 (21.39%)	0.0694
37.754	β-Caryophyllene	0.43 ± 0.08	1427	1429	161.12 (100%), 105.04 (57.31%), 93.05 (27.09%)	0.0075
39.75	*Trans*-Ethyl cinnamate	13.67 ± 0.55	1477	1455	131.04 (100%), 103.04 (48.29%), 77.03 (30.40%)	0.2337
40.542	Valencene	3.24 ± 0.09	1497	1497	161.12 (100%), 105.04 (57.31%), 91.04 (53.35%)	0.0554
41.138	γ-Cadinene	0.79 ± 0.14	1511	1513	161.11 (100%), 133.07 (30.58%), 120.07 (27.73%)	0.0135
41.968	σ-Cadinene	2.37 ± 0.09	1532	1526	91.04 (100%), 205.11 (86.11%), 77.02 (46.25%)	0.04052476
44.321	Spathulenol	3.33 ± 0.07	1593	1575	91.04 (100%), 93.05 (73.39), 77.02 (46.25%)	0.0569
47.198	β-Eudesmol	0.22 ± 0.19	1671	1672	59.04 (100%), 149.11 (67.04%), 146.14 (33.10%)	0.0038
48.344	a-Caryophylene acetate	1.11 ± 0.31	1702	1696	67.04 (100%), 95.06 (62.38%), 96.07 (41.92%)	0.0190
	Total			100		1.71
	Monoterpene hydrocarbons				5.74	
	Oxygenated monoterpenes				57.20	
	Sesquiterpene hydrocarbons				10.88	
	Oxygenated sesquiterpenes				4.66	
	Phenolics				1.87	
	Cinnamic acid derivatives				18.03	

RT, Retention time; RI^{cal}, Calculated retention index; RI^{rep}, Reported retention index; *m/z*, mass to charge ratio.

Besides the monoterpenes, GC-FID analysis of ArJ also showed a comparatively high percentage of cinnamic acid derivatives (18.03%), represented by the presence of three essential constituents, i.e., (*E*)-methyl cinnamate (0.35%), *cis*-ethyl cinnamate (4.02%), and *trans*-ethyl cinnamate (13.67%). Notably, ethyl cinnamate has been reported as one of the major chemotypes of the plant [24]. Monoterpene hydrocarbons, sesquiterpene hydrocarbons, oxygenated sesquiterpenes, and phenolic essential oils were also represented in the essential oil to a lesser extent, with 5.74, 10.88, 4.66, and 1.87%, respectively (Table 2).

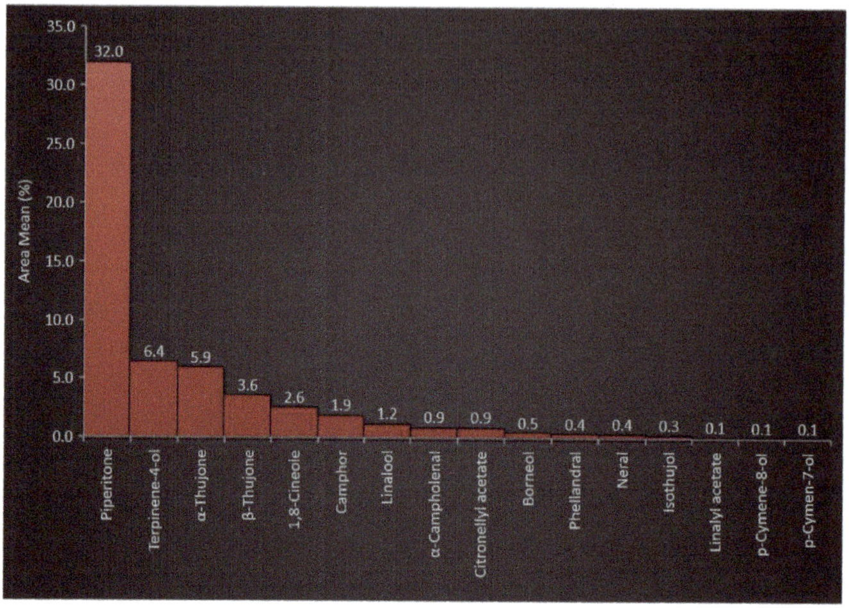

Figure 1. Representation of the oxygenated monoterpenes in *A. judaica* essential oil.

3.2. *In Vitro Antioxidant Activity of the ArJ Essential Oil*

Antioxidants are promising therapeutic agents in wound healing [70]. Most of the reported plants with wound healing activity possess noticeable antioxidant potency, which has been examined by different in vitro and in vivo assays [71]. The essential oils obtained from several plants of the genus *Artemisia*, e.g., *A. diffusa* and *A. herba-alba*, have exhibited potential free radical scavenging, reducing, and metal-chelating properties [72,73]. The measurements, i.e., TAC, DPPH-SA, FRAP, and MCA, were conducted for the essential oil of ArJ quantitatively. The plant's essential oil reduced the molybdate ions (VI) to molybdenum (V) in the TAC assay at a level of 59.32 mg of Trolox equivalents per gram of the plant essential oil.

Moreover, the ArJ essential oil exhibited notable reducing characteristics towards the ferric ion measured by the FRAP assay (22.34 mg of Trolox equivalent per gram of the essential oil). This ArJ essential oil-reducing characteristic in the TAC and FRAP contributes to the overall activity of this oil as an antioxidant agent [74]. The noticeable reducing characteristic of the ArJ essential oil could be attributed to the presence of camphor (1.92%), ethyl cinnamate (4.02%), and piperitone (31.99%) in relatively high concentrations [75–77]. The results also revealed that the essential oil of ArJ can chelate iron by 26.99 mg of EDTA equivalents per gram of essential oil, which was consistent with the reported ferrous ion-chelating activity of ArJ [35]. As iron has a primary role in the Fenton reaction involving the conversion of the oxidizing agent hydrogen peroxide (H2O2) into the more reactive hydroxyl radical (HO·), the iron-chelating agents, such as the ArJ essential oil, interfere with the progression and exaggeration of the oxidative stress [78].

Furthermore, scavenging activity of ArJ essential oil has been reported [35,69]. In the current study, the essential oil of the ArJ also exhibited scavenging activity, measured as 10.70 mg of Trolox equivalent per gram of the essential oil against the stable free radical DPPH. The results of the antioxidant activity of the plant essential oil seems also to be attributable to the presence of considerable percentages of oxygenated monoterpenes, cinnamate derivatives, and phenolics in the essential oils of the plants, all of which are known for their antioxidant activity [79–81]. The overall results obtained from quantitative in vitro antioxidant assays confirmed the antioxidant activity of the ArJ essential oil and supported the association between the wound healing potential of the plant and its antioxidant activity.

3.3. Antimicrobial Profile of ArJ Essential Oil

3.3.1. Preliminary Antimicrobial Activity

The results of preliminary antibacterial activity demonstrated that all the tested organisms, including Gram-positive and Gram-negative bacteria, are susceptible to ArJ essential oil, except *Pseudomonas aerugenosa* ATCC 9027, which showed resistance at the given concentration of ArJ essential oil, i.e., 20 μL/disc (Figure 2 and Table 3). The results further demonstrated that *Bacillus cereus* is a highly susceptible test organism, with an inhibition zone of 12.9 ± 0.10 mm at the given concentration of ArJ essential oil. In contrast, the lowest antibacterial activity was observed against *Klebsiella pneumoniae* and *Shigella flexneri*, with inhibition zones of 6.2 ± 0.10 mm and 6.2 ± 0.10 mm in diameter, respectively (Figure 2 and Table 3). Additionally, the findings indicated that the range for the mean zone of inhibition for Gram-positive bacteria is 7.2–12.9 mm, while for Gram-negative bacteria it is 6.2–10.0 mm, indicating that Gram-positive bacteria are more susceptible than Gram-negative bacteria to a given dose of ArJ essential oil.

Figure 2. Preliminary antimicrobial activity of ArJ essential oil. (**a**) *Staphylococcus aureus* (*S. aureus*) ATCC 29213; (**b**) *Staphylococcus saptophyticus* (*S. saptophyticus*) ATCC 43867; (**c**) *Streptococcus pyogenes* (*S. pyogenes*)-A ATCC 27736; (**d**) *Streptococcus pneumoniae* (*S. pneumoniae*) ATCC 49619; (**e**) *Enterococcus faecalis* (*E. faecalis*) ATCC 29212; (**f**) *Bacillus cereus* (*B. cereus*) ATCC 10876; (**g**) *Escherichia coli* (*E. coli*)

ATCC 25922; (**h**) *Klebsiella pneumonie* (*K. pneumoniae*) ATCC 27736; (**i**) *Salmonella typhimurium* (*S. typhimurium*) ATCC 13311; (**j**) *Shigella flexneri* (*S. flexneri*) ATCC 12022; (**k**) *Proteus vulgaris* (*P. vulgaris*) ATCC 6380; (**l**) *Proteus mirabilis* (*P. mirabilis*) ATCC 29906; (**m**) *Candida albicans* (*C. albicans*) ATCC 10231; (**n**) *Aspergillus niger* (*A. niger*) ATCC 6275. AJ referred to Artemisia judaica essential oil, while C referred to the drug control.

Table 3. Preliminary antimicrobial activity of ArJ essential oil.

Microorganisms	Zone of Inhibition (mm)	
	ArJ Essential Oil	Control Drugs
S. aureus ATCC 29213	7.7 ± 0.20	14.2 ± 0.20
S. saprophyticus ATCC 43867	8.8 ± 0.20	12.8 ± 0.20
S. pyogenes (A) ATCC 27736	7.4 ± 0.30	11.7 ± 0.10
S. pneumoniae ATCC 49619	7.2 ± 0.17	11.7 ± 0.20
E. faecalis ATCC 29212	8.7 ± 0.17	11.9 ± 0.10
B. cereus ATCC 10876	12.9 ± 0.10	19.6 ± 0.35
E. coli ATCC 25922	6.4 ± 0.10	23.1 ± 0.20
K. pneumoniae ATCC 27736	6.2 ± 0.10	21.1 ± 0.10
S. typhimurium ATCC 13311	10.0 ± 0.20	16.3 ± 0.30
S. flexneri ATCC 12022	6.2 ± 0.10	17.9 ± 0.17
P. vulgaris ATCC 6380	8.1 ± 0.17	16.2 ± 0.35
P. mirabilis ATCC 29906	7.7 ± 0.20	18.7 ± 0.20
C. albicans ATCC 10231	25.2 ± 0.20	25.0 ± 0.20
A. niger ATCC 6275	15.0 ± 0.20	13.1 ± 0.35

Note: All results are in mean ± SD. Each test was performed in triplicate. Control drugs = levofloxacin (antibacterial), 5 µg/disc; and clotrimazole (antifungal), 50 µg/disc.

The results for preliminary antifungal activity indicate that both the tested fungal strains are highly susceptible to ArJ essential oil. The results also indicated that the highest antifungal activity was observed against *Candida albicans* with an inhibition zone of 25.2 ± 0.20 mm, while *Aspergillus niger* had an inhibition zone of 15.0 ± 0.20 mm at the given concentration of ArJ essential oil. The control antibiotics inhibited the growth of all the tested organisms at the given concentrations, i.e., 5 µg/disc for levofloxacin and 50 µg/disc for clotrimazole, respectively (Figure 2 and Table 3).

3.3.2. Minimum Inhibitory Concentration (MIC), Minimum Biocidal Concentration (MBC), Minimum Biofilm Inhibitory Concentration (MBIC), and Minimum Biofilm Eradication Concentration (MBEC)

The MIC and MBC results for the tested bacteria revealed that the MIC values ranged from 6.25 to 100 µL/mL, while MBC values ranged from 12.5 to >100 µL/mL (Table 4). The MIC and MBC results for the tested fungi demonstrated that *Candida albicans* had MIC and MBC values of 3.125 µL/mL and 6.25 µL/mL, respectively, whereas *Aspergillus niger* had values of 6.25 µL/mL and 12.5 µL/mL, respectively. The MBIC and MBEC results revealed that the MBIC values for the tested bacteria ranged from 6.25 to 100 µL/mL, whereas the MBEC values ranged from 12.5 to 200 µL/mL (Table 4).

Table 4. Results of MIC, MBC, MBIC, and MBEC of ArJ essential oil.

Microorganisms	MIC	MBC	MBIC	MBEC
S. aureus ATCC 29213	50	100	50	100
S. saprophyticus ATCC 43867	50	100	50	100
S. pyogenes (A) ATCC 27736	100	>100	100	200
S. pneumoniae ATCC 49619	100	>100	100	200
E. faecalis ATCC 29212	100	>100	100	200
B. cereus ATCC 10876	6.25	12.5	6.25	12.5
E. coli ATCC 25922	50	100	50	100
K. pneumoniae ATCC 27736	25	50	25	50
S. typhimurium ATCC 13311	12.5	25	12.5	25
S. flexneri ATCC 12022	12.5	25	12.5	25
P. vulgaris ATCC 6380	25	50	25	50
P. mirabilis ATCC 29906	100	>100	100	200
C. albicans ATCC 10231	6.25	12.5	NT	NT
A. niger ATCC 6275	3.125	6.25	NT	NT

Note: All the results are in μL/mL. NT = Not tested. MIC = Minimum Inhibitory Concentration, MBC = Minimum Biocidal Concentration, MBIC = Minimum Biofilm Inhibitory Concentration, MBEC = Minimum Biofilm Eradication Concentration. All the results are in μL/mL.

Our findings for ArJ essential oil antimicrobial activity are consistent with previously published data [17,24,82–85]. Benmansour et al. demonstrated that ArJ essential oil had an excellent inhibitory effect against tested MRSA (methicillin-resistant *Staphylococcus aureus*), *S. aureus*, and *B. subtilis* [17], which is consistent with our results. Benderradji et al. showed that petroleum ether and ethyl acetate extracts of *A. sahariensis* leaves had the highest inhibitory activity against most tested strains. The most reported significant inhibition zone was obtained with chloroform extract of the plant against *Pseudomonas* [82]. These findings partially corroborate our results, since ArJ essential oil could not kill *Pseudomonas*, which might be a consequence of the essential oil and extract's differing phytochemical contents, as well as species variations. Elazzouzia et al. demonstrated that the essential oil of *A. ifranensis* had highly potent antibacterial activity against the tested *S. aureus* [55], which is, again, consistent with our results. Kazemi et al. demonstrated that the essential oil of the aerial parts of *A. kermanensis* had highly potent antibacterial activity against *B. subtilis*, *P. aeruginosa*, and *S. aureus*, which is partially consistent with our results [85]. Al-Wahaibi et al. demonstrated that essential oils derived from *A. judaica* and *A. herba-alba* had potent antimicrobial potential against the tested organisms, including *Aspergillus fumigatus*, *Syncephalastrum racemosum*, *Geotricum candidum Candida albicans*, *Streptococcus pneumoniae*, *Bacillus subtilis*, and *Escherichia coli*, except *Pseudomonas aeruginosa*; these results are consistent with our results [24]. The results of our study indicated that ArJ essential oil has highly potent antimicrobial activity, demonstrating that ArJ essential oil could be a promising antimicrobial drug candidate and can cure various human infections, e.g., wound infections, boils, acne, etc., caused by various life-threatening pathogens, including bacteria and fungi. These results encouraged us to conduct wound healing testing on an animal model to verify the antimicrobial properties of ArJ essential oil in-vivo.

3.4. In Vivo Skin Burn Wound Healing

In the current study, the second-degree burn was induced on female rats based on the recent publication that observed significant wound healing in second and third-degree wounds [57]. The choice of three-month-old female rats was due to their quicker wound healing and greater wound contraction ability as compared to males [86].

3.4.1. Morphological Appearance and Histological Analysis of the Wounds

The observations of wounds over three weeks of treatment revealed the significant progression in the healing process among the treated groups. At the time of burn induction, the skin burns produced were whitish in color and round in shape. After one week of the treatment, a crust developed on the wound along with the disappearance of edema in the treated groups (ArJ and SS groups). The wound area started decreasing by the second week of the treatment; however, edema formation was still prominent in the untreated burn area. At the end of 3 weeks, the treatment wound area for both the ArJ and SS groups demonstrated recovery, while the untreated zone did not recover completely (Figure 3).

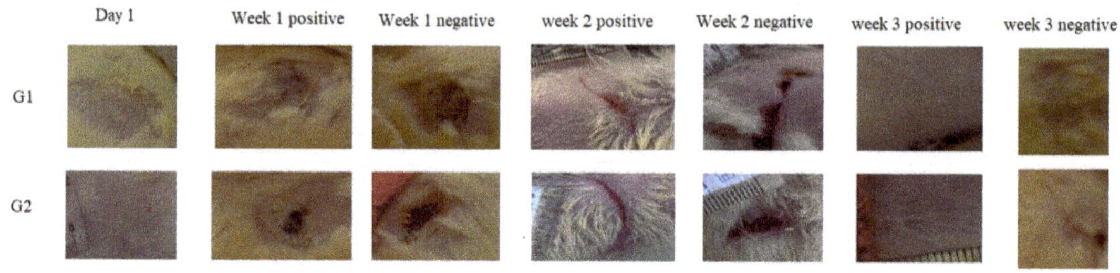

Figure 3. Morphological appearances of *Artemisia judaica* (G1)- and silver sulfadiazine (SS) (G2)-treated wounds at various time points.

Wound areas were measured using freely available Image J software. No significant differences were observed among the groups after the induction of skin burn. Two weeks after the treatment, the ArJ group's wound area decreased significantly ($p = 0.04$), while the SS group's wound size remained insignificant compared to the negative group (Supplementary file, Figure S2).

Wound healing is a complex restorative process of injured tissue to its original state [87], and it involves hemostasis, inflammation, proliferation, and remodeling [88] to prevent complex metabolic alteration affecting body organ systems. During the process of hemostasis, blood coagulation occurs, while the inflammation process ensures safety from invasive pathogens [88], thereby facilitating the proliferation step [89] towards remodeling the tissue maturation process [89,90]. During skin burn, cells and tissues are damaged substantially, thereby involving a complicated healing network compared to wound incision [87]. Based on the deepness of the burn wounds, they are categorized as first-, second-, and third-degree burns. The first-degree burn is generally red or gray without any blisters and normal capillary network, while in a second-degree burn, blisters and partial-thickness damage to the dermis are observed. Second- and third-degree burns are treated similarly [91]. In the current study, second-degree skin burn wounds were induced, which healed over 3 weeks for the ArJ and SS groups.

H&E staining was performed for all the animal groups (Figure 4). The H&E staining demonstrated epidermis integrity and the degree of neutrophilic infiltration in the dermis and capillaries of the ArJ and SS groups compared to the untreated wound zone. Wound healing visual appearance for ArJ was not enlarged, most probably due to the balm effect of paraffin (Figure 4B). Tissue sections were analyzed qualitatively for the amount of fibroblast, collagen, inflammation, and neovascularization in SS and ArJ groups. The data demonstrated increased collagen, fibroblast, and neovascularization, with decreased inflammation in the SS and ArJ groups compared with the negative control group (Supplementary file, Figure S3).

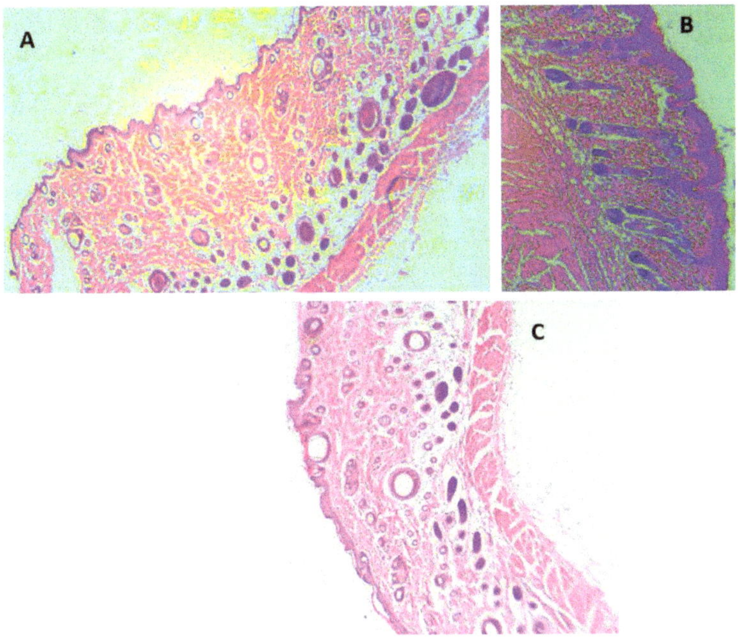

Figure 4. Histological analysis using hematoxylin and eosin staining of control (**A**), *Artemisia judaica* (**B**), and SS (**C**). Arrows (black) indicate neutrophil infiltration; magnification 40×.

3.4.2. Role of Antioxidants and Oxidative Stress Markers in Wound Healing

The ArJ ointment group demonstrated significantly increased antioxidant SOD ($p = 0.03$) and CAT ($p < 0.01$) enzymatic activities compared to the negative group. The SOD and CAT activities were comparable in the intact and negative control. The SS treated group demonstrated a significant difference in CAT activity compared to the negative group ($p = 0.01$) and was comparable to the ArJ group, while the differences were insignificant for SOD activity. The antioxidant activity contributing towards wound healing is in accordance with previous studies reporting enhanced wound healing due to potent antioxidant activities [92]. LP significantly increased in the negative control group ($p < 0.0001$) compared to the intact group, which accords with previously recorded data [93]. No significant differences in LP were observed in treatment groups with either ArJ or SS groups compared to the negative control, while the ArJ and SS treated groups exhibited significant increases ($p < 0.0001$) compared to the intact group (Table 5).

Table 5. Effect of *Artemisia judaica* ointment on antioxidant and oxidant levels in skin burn rat model.

Groups	CAT	SOD	LP
		ng/g	
I. Intact control	1.35 ± 0.05 ^{A,B}	0.04 ± 0.00 ^{A}	836.9 ± 37.75 ^{A}
II. Negative control (skin burn without treatment)	1.11 ± 0.06 ^{A}	0.04 ± 0.01 ^{A}	1214 ± 51.46 ^{B}
III. Silver sulfadiazine	1.79 ± 0.204 ^{B}	0.19 ± 0.06 ^{A,B}	1197 ± 30.30 ^{B}
IV. *Artemisia judaica*	1.82 ± 0.17 ^{B}	0.37 ± 0.13 ^{B}	1291 ± 18.85 ^{B}

Values are denoted as means ± SEM (Supplementary file, Tables S1–S3). Statistical significance was performed using one-way ANOVA, followed by a post hoc test on GraphPad Prism 8.0.2. CAT = catalase, LP = lipid peroxide, SOD = Superoxide dismutase. The mean values that do not share a superscript letter (A,B) in the respective columns of superoxide dismutase (SOD), catalase (CAT), and lipid peroxide (LP) are significantly different ($p < 0.05$) using Tukey's multi-group comparisons.

3.4.3. Role of Pro- and Anti-Inflammatory Markers in Wound Healing

The pro-inflammatory markers IL-1b and IL-6 values were comparable among all the studied groups, similar to the previously published articles [94,95]. At the same time, tumor necrosis factor α (TNF-α) significantly increased in the negative control group compared to the intact group ($p < 0.0001$). TNF-α values decreased significantly after treatment with SS ($p < 0.02$) and ArJ ($p < 0.002$) compared to the negative control group. TNF-α, the inflammatory cytokine activated during acute inflammation by macrophages/monocytes, plays a vital role in cell signaling, leading to necrosis or apoptosis. TNF-α participates in vasodilatation and edema formation and leukocyte adhesion to the epithelium through the expression of adhesion molecules. Furthermore, TNF-α regulates blood coagulation, contributes to oxidative stress at sites of inflammation, and indirectly induces fever. The data conforms with Gushiken and Periera, where TNF-α decreased after two weeks of treatment in skin wound tissue compared to the negative control [94,95].

The anti-inflammatory or the pro-angiogenic markers IL-10 and transforming growth factor beta 1 (TGF-b1) increased significantly in both ArJ ($p < 0.0001$ and $p < 0.0001$) and SS ($p < 0.0001$ and $p < 0.0001$) groups compared to the negative and intact control groups. However, differences in IL-10 and TGF-b1 levels in the negative control group were insignificant compared to the intact group (Table 6). The data confirm previous studies where IL-10 increased significantly in skin wound healing tissue after two weeks of treatment compared to the negative control group [94,95].

Table 6. Effect of *Artemisia judaica* ointment on inflammatory and pro-angiogenic markers in skin burn rat model.

Groups	IL-1b	IL-6	TNF-α	TGF-b1	IL-10
			ng/g		
I. Intact control	20.77 ±1.95 ^{A}	806.1 ±10.20 ^{A}	9.57 ±0.55 ^{A}	4.19 ±0.24 ^{A}	3.54 ±0.19 ^{A}
II. Negative control (skin burn without treatment)	19.37 ± 2.33 ^{A}	776.2 ±32.77 ^{A}	15.54 ± 0.92 ^{B}	3.87 ± 0.09 ^{A}	2.99 ± 0.25 ^{A}
III. Sulfadiazine	23.55± 0.88 ^{A}	789.4 ±18.02 ^{A}	12.85± 0.26 ^{C}	19.28± 0.30 ^{B}	12.68± 0.15 ^{B}
IV. *Artemisia judaica*	19.55 ± 1.34 ^{A}	869.2 ±51.91 ^{A}	11.96 ± 0.34 ^{C}	19.18 ± 0.33 ^{B}	13.39 ± 0.35 ^{B}

Values are denoted as means ± SEM (Supplementary file, Tables S4–S8). Statistical significance was performed using one-way ANOVA, followed by a post hoc test on GraphPad Prism 8.0.2. The mean values that do not share a superscript letter (A–C) in the respective columns of interleukin-1 (IL-1b), IL-6, IL-10, transforming growth factor beta 1 (TGF-b1), and tumor necrosis factor α (TNF-α) are significantly different ($p < 0.05$) using Tukey's multi-group comparisons.

Various plant and herbal products are economically cheap to procure and demonstrate modest therapeutic potency with minimum toxicity relative to synthetic drugs [45–47,96]. Eupolin ointment, derived from an aqueous extract of *C. odorata* leaves, is the first Vietnam FDA-approved product [59]. The wound healing mechanism, even though it remains unclear, nevertheless enhanced blood flow, decreased inflammatory response, and reduced infection rates, all of which are contributing factors to angiogenesis. Rats are loose-skinned, in contrast to the tight human skin, with quicker constriction of the wound than the epithelization process; as such, rat wound healing, even though it is resemblant, is not entirely similar to wound healing in human skin [97,98]. However, rats are widely used animal models due to their genetic and behavioral similarity, ease of handling, and being economically viable. Thus, the rat skin burn model serves as a vital knowledge resource.

Mortality has been one of the major concerns in patients with deep burns due to infections, and researchers have attempted to minimize wound infection risk and accelerate the healing process [58]. Topical antimicrobial ointments are commonly employed for such purposes as SS 1% with low toxicity and have a potent antibacterial effect in burn wound therapy management [99–101].

Increased antioxidant levels have demonstrated wound healing potency [102] by protecting tissue from oxidative stress [103,104]. In the current study, ArJ demonstrated antioxidant (augmented SOD and CAT) enzymatic activities, validating the role of antioxidant enzymes, in addition to potentiating wound healing through increased anti-inflammatory levels. TGF-b1 is involved in wound healing, angiogenesis, immune regulation, and cancer. On the contrary, TGF-b1, along with inflammatory marker IL-6, helps in T helper 17 differentiation (Th17), aggravating inflammation [105]. In our study, the levels of TGF-b1 increased significantly while no statistical difference was observed for IL-1b and IL-6 in the ArJ group compared to the control group.

Moreover, IL-10 is a key regulator of the immune system by limiting the inflammatory response, which could otherwise cause tissue damage. In one study, IL-10 knockout mice were found prone to colitis [106,107], and blocking of IL-10 resulted in severe pathology. On the contrary, increased IL-10 levels cause chronic infection, and IL-10 blocking paved the way for pathogen clearance [108]. Mucosal secretion of IL-10 and TNF-α were augmented during wound healing, demonstrating the protective effect of IL-10 against inflammation. On the contrary, treatment with *P. pinnata* increased the serum IL-10 concentration while downregulating TNF-α and IL-6 [45]. Our study showed a significant increase in IL-10 concomitant with a significant decrease in TNF-α in ArJ compared to the negative control group, thus confirming the potential role of IL-10 in promoting wound healing.

The current findings for the in vivo and in vitro antioxidant activity and the anti-inflammatory effect of ArJ could be attributed to the presence of higher percentages of oxygenated monoterpenes (57.2%) in the plant essential oils [109]. Oxygenated monoterpenes have been found as major constituents in the plants used traditionally to accelerate wound healing, e.g., the plant essential oil of *Helichrysum italicum* [110,111] has exhibited a primary role in potential antimicrobial activity and anti-inflammatory effects [111]. Furthermore, some of the major ArJ essential oils reported as antioxidant and anti-inflammatory agents, e.g., thujone (both α and β-thujone, 9.55%), 1,8-cineole (2.56%), camphor (1.92%), and borneol (0.47%), have chiefly contributed to the rosemary anti-inflammatory effect [112]. Furthermore, 1,8-cineole antioxidant and anti-inflammatory effects have been reported, and the compound effect as an inhibitor for the inflammatory markers, TNF-α, IL-6, IL-8, LTB 4, PGE 2, and IL-1β, as well as down-regulation of 5-lipoxygenase (LOX) and cyclooxygenase (COX) pathways, are well documented [113,114]. Moreover, methyl cinnamate, a major constituent in ArJ essential oil (4.02%), has demonstrated potent anti-inflammatory activity [115]. All these compounds participated in the antioxidant and anti-inflammatory effects of the ArJ essential oil as well as in the wound healing activity. However, other identified constituents in this article could also be playing a role in the demonstrated plant activities.

4. Conclusions

The phytochemical analysis of the ArJ essential oils was conducted and revealed the dominance of the highly active antioxidant volatile compounds, oxygenated monoterpenes, and cinnamic acid derivatives in the essential oil constituents of the plant. Such classes of compounds were reflected in the in vitro and in vivo potential antioxidant activity of the ArJ essential oil. In the current study, wound treatment with ArJ also demonstrated significantly increased SOD and CAT enzymatic activities, with insignificant LP levels compared to the negative control group. In addition, ArJ reduced the pro-inflammatory marker TNF-α and augmented pro-angiogenic/anti-inflammatory TGF-b1 and IL-10 levels. The antimicrobial and antibiofilm potential of ArJ essential oil against *Bacillus cereus, Candida albicans,* and *Aspergillus niger* confirmed in the study supports the effectiveness of ArJ essential oil as a wound healing candidate. These results validate the curative role of ArJ in the treatment of skin wounds, which is attributed to its antioxidant and anti-inflammatory effects, as well as its high proportion of oxygenated monoterpenes and cinnamate derivatives.

Supplementary Materials: The following supporting information can be downloaded at: https://www.mdpi.com/article/10.3390/antiox11020332/s1, Figure S1: Gas Chromatography (GC), chromatograms of the ArJ essential oil samples; Figure S2: Wound area measurement; Figure S3: Microscopic analysis of skin wound after treatment with SS and ArJ and compared with negative control; Tables S1–S8: The original data of SOD, CAT, LP, IL-1b, IL-6, IL-10, TNF-α, and TGF-b1, respectively.

Author Contributions: Conceptualization, H.A.M. and S.A.A.M.; methodology, H.A.M., K.A.Q., H.M.A., O.K. and S.A.A.M.; software, H.A.M., H.M.A., M.S.A.-O. and S.A.A.M.; validation, H.A.M., K.A.Q., H.M.A. and S.A.A.M.; formal analysis, H.A.M.; investigation, H.A.M., K.A.Q. and H.M.A.; resources, H.A.M.; data curation, K.A.Q. and H.M.A.; writing—original draft preparation, H.A.M., K.A.Q., H.M.A., O.K. and S.A.A.M.; writing—review and editing, H.A.M., K.A.Q., H.M.A., M.S.A.-O. and S.A.A.M.; visualization, H.A.M. and M.S.A.-O.; supervision, H.A.M.; project administration, H.A.M.; funding acquisition, H.A.M. All authors have read and agreed to the published version of the manuscript.

Funding: This work was supported by the Qassim University, represented by the Deanship of Scientific Research, under the grant number (10106-pharmacy-2020-1-3-I), during the academic year 1441 AH/2020 AD.

Institutional Review Board Statement: The study was conducted in accordance with the Institutional Animal Ethics Committee (Registration # 21-04-06).

Informed Consent Statement: Not applicable.

Data Availability Statement: Data are available in the manuscript and supplementary file.

Acknowledgments: The authors gratefully acknowledge Qassim University, represented by the Deanship of Scientific Research, on the financial support for this research under the number (10106-pharmacy-2020-1-3-I) during the academic year 1441 AH/2020 AD.

Conflicts of Interest: The authors declare no conflict of interest.

References

1. Bora, K.S.; Sharma, A. The genus *Artemisia*: A comprehensive review. *Pharm. Biol.* **2011**, *49*, 101–109. [CrossRef] [PubMed]
2. Tan, R.X.; Zheng, W.F.; Tang, H.Q. Biologically active substances from the genus *Artemisia*. *Planta Med.* **1998**, *64*, 295–302. [CrossRef]
3. Miño, J.H.; Moscatelli, V.; Acevedo, C.; Ferraro, G. Psychopharmacological effects of *Artemisia copa* aqueous extract in mice. *Pharm. Biol.* **2010**, *48*, 1392–1396. [CrossRef] [PubMed]
4. Vijayalakshmi, A.; Ravichandiran, V.; Prasad, K.V.D.; Kumar, A.K.; Rakesh, K.; Naidu, K.R.; Vardhan, K.S. Anticonvulsant activity of *Artemisia nilagirica* leaves. *J. Pharm. Res.* **2011**, *4*, 2881–2883.
5. Sayyah, M.; Nadjafnia, L.; Kamalinejad, M. Anticonvulsant activity and chemical composition of *Artemisia dracunculus* L. essential oil. *J. Ethnopharmacol.* **2004**, *94*, 283–287. [CrossRef] [PubMed]
6. Mahboubi, M. *Artemisia sieberi* Besser essential oil and treatment of fungal infections. *Biomed. Pharmacother.* **2017**, *89*, 1422–1430. [CrossRef] [PubMed]
7. Das, S. *Artemisia annua* (Qinghao): A pharmacological review. *Int. J. Pharm. Sci. Res.* **2012**, *3*, 4573–4577.
8. Hijazi, A.M.; Salhab, A.S. Effects of *Artemisia monosperma* ethanolic leaves extract on implantation, mid-term abortion and parturition of pregnant rats. *J. Ethnopharmacol.* **2010**, *128*, 446–451. [CrossRef] [PubMed]
9. Abad, M.J.; Bedoya, L.M.; Apaza, L.; Bermejo, P. The *Artemisia* L. genus: A review of bioactive essential oils. *Molecules* **2012**, *17*, 2542–2566. [CrossRef] [PubMed]
10. Nigam, M.; Atanassova, M.; Mishra, A.P.; Pezzani, R.; Devkota, H.P.; Plygun, S.; Salehi, B.; Setzer, W.N.; Sharifi-Rad, J. Bioactive compounds and health benefits of *Artemisia* species. *Nat. Prod. Commun.* **2019**, *14*, 1934578–19850354.
11. Willcox, M.L.; Burton, S.; Oyweka, R.; Namyalo, R.; Challand, S.; Lindsey, K. Evaluation and pharmacovigilance of projects promoting cultivation and local use of *Artemisia annua* for malaria. *Malar. J.* **2011**, *10*, 1–7. [CrossRef] [PubMed]
12. Hayat, M.Q.; Khan, M.A.; Ashraf, M.; Jabeen, S. Ethnobotany of the genus *Artemisia* L. (Asteraceae) in Pakistan. *Ethnobot. Res. Appl.* **2009**, *7*, 147–162. [CrossRef]
13. Kelsey, R.G.; Shafizadeh, F. Sesquiterpene lactones and systematics of the genus *Artemisia*. *Phytochemistry* **1979**, *18*, 1591–1611. [CrossRef]
14. Martínez, M.J.A.; Del Olmo, L.M.B.; Ticona, L.A.; Benito, P.B. The *Artemisia* L. genus: A review of bioactive sesquiterpene lactones. *Stud. Nat. Prod. Chem.* **2012**, *37*, 43–65.
15. Vouillamoz, J.F.; Carlen, C.; Taglialatela-Scafati, O.; Pollastro, F.; Appendino, G. The génépi *Artemisia* species. Ethnopharmacology, cultivation, phytochemistry, and bioactivity. *Fitoterapia* **2015**, *106*, 231–241. [CrossRef] [PubMed]
16. El-Amier, Y.A.; Al Borki, A.E. N.S.; Elagami, S.A. Potential of wild plant *Artemisia judaica* L. as sustainable source of antioxidant and antimicrobial compounds. *J. Exp. Sci.* **2019**, *10*, 4–8. [CrossRef]
17. Benmansour, N.; Benmansour, A.; El Hanbali, F.; González-Mas, M.C.; Blázquez, M.A.; El Hakmaoui, A.; Akssira, M. Antimicrobial activity of essential oil of *Artemisia judaica* L. from Algeria against multi-drug resistant bacteria from clinical origin. *Flavour Fragr. J.* **2016**, *31*, 137–142. [CrossRef]

18. Amin, S.M.; Hassan, H.M.; El Gendy, A.E.G.; El-Beih, A.A.; Mohamed, T.A.; Elshamy, A.I.; Bader, A.; Shams, K.A.; Mohammed, R.; Hegazy, M.F. Comparative chemical study and antimicrobial activity of essential oils of three *Artemisia* species from Egypt and Saudi Arabia. *Flavour Fragr. J.* **2019**, *34*, 450–459. [CrossRef]
19. Guetat, A.; Al-Ghamdi, F.A.; Osman, A.K. The genus *Artemisia* L. in the northern region of Saudi Arabia: Essential oil variability and antibacterial activities. *Nat. Prod. Res.* **2017**, *31*, 598–603. [CrossRef]
20. Abu-Darwish, M.S.; Cabral, C.; Gonçalves, M.J.; Cavaleiro, C.; Cruz, M.T.; Zulfiqar, A.; Khan, I.A.; Efferth, T.; Salgueiro, L. Chemical composition and biological activities of *Artemisia judaica* essential oil from southern desert of Jordan. *J. Ethnopharmacol.* **2016**, *191*, 161–168. [CrossRef]
21. Badr, A.; El-Shazly, H.H.; Helail, N.S.; El Ghanim, W. Genetic diversity of *Artemisia* populations in central and north Saudi Arabia based on morphological variation and RAPD polymorphism. *Plant Syst. Evol.* **2012**, *298*, 871–886. [CrossRef]
22. Moharram, F.A.; Nagy, M.M.; El Dib, R.A.; El-Tantawy, M.M.; El Hossary, G.G.; El-Hosari, D.G. Pharmacological activity and flavonoids constituents of *Artemisia judaica* L. aerial parts. *J. Ethnopharmacol.* **2021**, *270*, 113777. [CrossRef]
23. Janacković, P.; Novaković, J.; Soković, M.; Vujisić, L.; Giweli, A.A.; Dajić-Stevanović, Z.; Marin, P.D. Composition and antimicrobial activity of essential oils of *Artemisia judaica*, A. herba-alba and A. arborescens from Libya. *Arch. Biol. Sci.* **2015**, *67*, 455–466. [CrossRef]
24. Al-Wahaibi, L.H.N.; Mahmood, A.; Khan, M.; Alkhathlan, H.Z. Comparative study on the essential oils of *Artemisia judaica* and A. herba-alba from Saudi Arabia. *Arab. J. Chem.* **2020**, *13*, 2053–2065. [CrossRef]
25. Hussain, A.; Hayat, M.Q.; Sahreen, S.; Ain, Q.U.; Bokhari, S.A.I. Pharmacological promises of genus *Artemisia (Asteraceae)*: A review. *Proc. Pak. Acad. Sci. B Life Environ. Sci.* **2017**, *54*, 265–287.
26. Nofal, S.M.; Mahmoud, S.S.; Ramadan, A.; Soliman, G.A.; Fawzy, R. Anti-diabetic effect of *Artemisia judaica* extracts. *Res. J. Med. Med. Sci.* **2009**, *4*, 42–48.
27. Glevitzky, I.; Dumitrel, G.A.; Glevitzky, M.; Pasca, B.; Otrisal, P.; Bungau, S.; Cioca, G.; Pantis, C.; Popa, M. Statistical analysis of the relationship between antioxidant activity and the structure of flavonoid compounds. *Rev. Chim.* **2019**, *70*, 3103–3107. [CrossRef]
28. Nasr, F.A.; Noman, O.M.; Mothana, R.A.; Alqahtani, A.S.; Al-Mishari, A.A. Cytotoxic, antimicrobial and antioxidant activities and phytochemical analysis of *Artemisia judaica* and *A. sieberi* in Saudi Arabia. *Afr. J. Pharm. Pharmacol.* **2020**, *14*, 278–284.
29. El-Sayed, M.A.; BaAbbad, R.; Balash, A.; Al-Hemdan, N.A.; Softah, A. The potential anti *Helicobacter pylori* and antioxidant effects of *Artemisia judaica*. *Funct. Foods Health Dis.* **2013**, *3*, 332–340. [CrossRef]
30. Abd-Alla, H.I.; Aly, H.F.; Shalaby, N.M.M.; Albalawy, M.A.; Aboutabl, E.A. Hunting for renal protective phytoconstituents in *Artemisia judaica* L. and *Chrysanthemum coronarium* L. (*Asteraceae*). *Egypt. Pharm. J.* **2014**, *13*, 46. [CrossRef]
31. Charchari, S. The essential oil of *Artemisia judaica* L. from Algeria. *J. Essent. Oil Res.* **2002**, *14*, 16–17. [CrossRef]
32. Bungau, S.; Behl, T.; Aleya, L.; Bourgeade, P.; Aloui-Sossé, B.; Purza, A.L.; Abid, A.; Samuel, A.D. Expatiating the impact of anthropogenic aspects and climatic factors on long-term soil monitoring and management. *Environ. Sci. Pollut. Res.* **2021**, *28*, 3528–3550. [CrossRef] [PubMed]
33. Farah, R.; El Ouassis Dahmane, H.M.; Rym, E.; Amira, S.; el Houda, H.N.; Selma, B.A.; Nadia, F. Chemical composition and bilogical effects of essential oil of *Artemisia judaica* an endemic plant from central Sahara of Algeria Hoggar. *Int. J. Biosci.* **2017**, *10*, 16–23.
34. Abdallah Sallam, S.M.; Mohamed Abdelgaleil, S.A.; da Silva Bueno, I.C.; Abdelwahab Nasser, M.E.; Araujo, R.C.; Abdalla, A.L. Effect of some essential oils on in vitro methane emission. *Arch. Anim. Nutr.* **2011**, *65*, 203–214. [CrossRef] [PubMed]
35. Al-Qudah, M.A.; Onizat, M.A.; Alshamari, A.K.; Al-Jaber, H.I.; Bdair, O.M.; Muhaidat, R.; Al Zoubi, M.; Al-Bataineh, N. Chemical composition and antioxidant activity of Jordanian *Artemisia judaica* L. as affected by different drying methods. *Int. J. Food Prop.* **2021**, *24*, 482–492. [CrossRef]
36. Mengesha, B.; Mohammed, O.; Tessema, T.; Abate, S. *Production, Processing and Utilization of Aromatic Plants*; Ethiopian Institute of Agricultural Research: Addis Ababa, Ethiopia, 2010.
37. Piesse, G.W.S. *The Art of Perfumery and the Methods of Obtaining the Odors of Plants: With Instructions for the Manufacture of Perfumes for the Handkerchief, Scented Powders, Odorous Vinegars, Dentifrices, Pomatums, Cosmetics, Perfumed Soap, etc., to Which is Added an a*; Lindsay & Blakiston: Philadelphia, PA, USA, 1867.
38. Hanif, M.A.; Nisar, S.; Khan, G.S.; Mushtaq, Z.; Zubair, M. Essential Oils. In *Essential Oil Research*; Springer: Cham, Switzerland, 2019; pp. 3–17.
39. Zhi-ling, C.; Jian-ping, C.; Hui-lin, C.; Wei-tao, B.; Hai-yan, C.; Mo-lin, L. Research on the extraction of plant volatile oils. *Procedia Environ. Sci.* **2011**, *8*, 426–432. [CrossRef]
40. Sievers, A.F. *Methods of Extracting Volatile Oils from Plant Material and the Production of Such Oils in the United States*; US Department of Agriculture: Washington, DC, USA, 1952.
41. Kruger, E.; Kowal, S.; Bilir, S.P.; Han, E.; Foster, K. Relationship between patient characteristics and number of procedures as well as length of stay for patients surviving severe burn injuries: Analysis of the American Burn Association National Burn Repository. *J. Burn Care Res.* **2020**, *41*, 1037–1044. [CrossRef]
42. Rowan, M.P.; Cancio, L.C.; Elster, E.A.; Burmeister, D.M.; Rose, L.F.; Natesan, S.; Chan, R.K.; Christy, R.J.; Chung, K.K. Burn wound healing and treatment: Review and advancements. *Crit. Care* **2015**, *19*, 1–12. [CrossRef]

43. Ullah, A.; Munir, S.; Badshah, S.L.; Khan, N.; Ghani, L.; Poulson, B.G.; Emwas, A.-H.; Jaremko, M. Important flavonoids and their role as a therapeutic agent. *Molecules* **2020**, *25*, 5243. [CrossRef]
44. Guaouguaou, F.-E.; Taghzouti, K.; Oukabli, M.; Es-Safi, N.E. The Effect of *Salvia verbenaca* Extracts for Healing of Second-Degree Burn Wounds in Rats. *Curr. Bioact. Compd.* **2018**, *14*, 419–427. [CrossRef]
45. Dwivedi, D.; Dwivedi, M.; Malviya, S.; Singh, V. Evaluation of wound healing, anti-microbial and antioxidant potential of *Pongamia pinnata* in wistar rats. *J. Tradit. Complement. Med.* **2017**, *7*, 79–85. [CrossRef]
46. Fahimi, S.; Abdollahi, M.; Mortazavi, S.A.; Hajimehdipoor, H.; Abdolghaffari, A.H.; Rezvanfar, M.A. Wound healing activity of a traditionally used poly herbal product in a burn wound model in rats. *Iran. Red Crescent Med. J.* **2015**, *17*. [CrossRef]
47. Akhoondinasab, M.R.; Khodarahmi, A.; Akhoondinasab, M.; Saberi, M.; Iranpour, M. Assessing effect of three herbal medicines in second and third degree burns in rats and comparison with silver sulfadiazine ointment. *Burns* **2015**, *41*, 125–131. [CrossRef]
48. Herman, A.; Herman, A.P. Herbal products for treatment of burn wounds. *J. Burn Care Res.* **2020**, *41*, 457–465. [CrossRef] [PubMed]
49. Aroua, L.M.; Almuhaylan, H.R.; Alminderej, F.M.; Messaoudi, S.; Chigurupati, S.; Al-Mahmoud, S.; Mohammed, H.A. A facile approach synthesis of benzoylaryl benzimidazole as potential α-amylase and α-glucosidase inhibitor with antioxidant activity. *Bioorg. Chem.* **2021**, *114*, 105073. [CrossRef] [PubMed]
50. Shimada, K.; Fujikawa, K.; Yahara, K.; Nakamura, T. Antioxidative properties of xanthan on the autoxidation of soybean oil in cyclodextrin emulsion. *J. Agric. Food Chem.* **1992**, *40*, 945–948. [CrossRef]
51. Benzie, I.F.F.; Strain, J.J. The ferric reducing ability of plasma (FRAP) as a measure of "antioxidant power": The FRAP assay. *Anal. Biochem.* **1996**, *239*, 70–76. [CrossRef]
52. Zengin, G.; Nithiyanantham, S.; Locatelli, M.; Ceylan, R.; Uysal, S.; Aktumsek, A.; Selvi, P.K.; Maskovic, P. Screening of in vitro antioxidant and enzyme inhibitory activities of different extracts from two uninvestigated wild plants: *Centranthus longiflorus* subsp. *longiflorus* and *Cerinthe minor* subsp. *auriculata*. *Eur. J. Integr. Med.* **2016**, *8*, 286–292. [CrossRef]
53. Schwalbe, R.; Steele-Moore, L.; Goodwin, A.C. *Antimicrobial Susceptibility Testing Protocols;* CRC Press: Boca Raton, FL, USA, 2007; ISBN 1420014498.
54. Qureshi, K.A.; Imtiaz, M.; Parvez, A.; Rai, P.K.; Jaremko, M.; Emwas, A.-H.; Bholay, A.D.; Fatmi, M.Q. In vitro and in silico approaches for the evaluation of antimicrobial activity, time-kill kinetics, and anti-biofilm potential of thymoquinone (2-Methyl-5-propan-2-ylcyclohexa-2,5-diene-1,4-dione) against selected human pathogens. *Antibiotics* **2022**, *11*, 79. [CrossRef]
55. Qureshi, K.A.; Bholay, A.D.; Rai, P.K.; Mohammed, H.A.; Khan, R.A.; Azam, F.; Jaremko, M.; Emwas, A.-H.; Stefanowicz, P.; Waliczek, M.; et al. Isolation, characterization, anti-MRSA evaluation, and in-silico multi-target anti-microbial validations of actinomycin X_2 and actinomycin D produced by novel *Streptomyces smyrnaeus* UKAQ_23. *Sci. Rep.* **2021**, *11*, 1–21. [CrossRef]
56. The British Pharmacopoeia Commission. *British Pharmacopoeia;* Her Majesty's Stationary Office: London, UK, 1998.
57. Akhoondinasab, M.R.; Akhoondinasab, M.; Saberi, M. Comparison of healing effect of aloe vera extract and silver sulfadiazine in burn injuries in experimental rat model. *World J. Plast. Surg.* **2014**, *3*, 29–34.
58. Manafi, A.; Kohanteb, J.; Mehrabani, D.; Japoni, A.; Amini, M.; Naghmachi, M.; Zaghi, A.H.; Khalili, N. Active immunization using exotoxin A confers protection against *Pseudomonas aeruginosa* infection in a mouse burn model. *BMC Microbiol.* **2009**, *9*, 1–5. [CrossRef]
59. Phan, T.T.; Hughes, M.A.; Cherry, G.W. Enhanced proliferation of fibroblasts and endothelial cells treated with an extract of the leaves of *Chromolaena odorata* (Eupolin), an herbal remedy for treating wounds. *Plast. Reconstr. Surg.* **1998**, *101*, 756–765. [CrossRef] [PubMed]
60. Al-Omar, M.S.; Naz, M.; Mohammed, S.A.A.; Mansha, M.; Ansari, M.N.; Rehman, N.U.; Kamal, M.; Mohammed, H.A.; Yusuf, M.; Hamad, A.M.; et al. Pyrethroid-Induced Organ Toxicity and Anti-Oxidant-Supplemented Amelioration of Toxicity and Organ Damage: The Protective Roles of Ascorbic Acid and α-Tocopherol. *Int. J. Environ. Res. Public Health* **2020**, *17*, 6177. [CrossRef] [PubMed]
61. Kirkwood, B.R.; Sterne, J.A.C. *Essential Medical Statistics;* Essentials; Wiley: Hoboken, NJ, USA, 2010; ISBN 9781444392845.
62. Moisa, C.; Lupitu, A.; Pop, G.; Chambre, D.R.; Copolovici, L.; Cioca, G.; Bungau, S.; Copolovici, D.M. Variation of the chemical composition of *Thymus vulgaris* essential oils by phenological stages. *Rev. Chim.* **2019**, *70*, 633–637. [CrossRef]
63. Fathi, E.; Sefidkon, F. Influence of drying and extraction methods on yield and chemical composition of the essential oil of *Eucalyptus sargentii*. *J. Agric. Sci. Technol.* **2012**, *14*, 1035–1042.
64. Bogdan, M.; Bungau, S.; Tit, D.M.; Copolovici, L.; Behl, T.; Otrisal, P.; Aleya, L.; Cioca, G.; Berescu, D.; Uivarosan, D. Variations in the chemical composition of the essential oil of *Lavandula angustifolia* Mill., Moldoveanca 4 Romanian variety. *Rev. Chim.* **2020**, *71*, 307–315. [CrossRef]
65. Gil, A.; De La Fuente, E.B.; Lenardis, A.E.; López Pereira, M.; Suárez, S.A.; Bandoni, A.; Van Baren, C.; Di Leo Lira, P.; Ghersa, C.M. Coriander essential oil composition from two genotypes grown in different environmental conditions. *J. Agric. Food Chem.* **2002**, *50*, 2870–2877. [CrossRef]
66. Mohammed, H.A.; Al-Omar, M.S.; Mohammed, S.A.A.; Aly, M.S.A.; Alsuqub, A.N.A.; Khan, R.A. Drying Induced Impact on Composition and Oil Quality of Rosemary Herb, *Rosmarinus Officinalis* Linn. *Molecules* **2020**, *25*, 2830. [CrossRef]

67. Mohammed, H.A.; Eldeeb, H.M.; Khan, R.A.; Al-Omar, M.S.; Mohammed, S.A.A.; Sajid, M.S.M.; Aly, M.S.A.; Ahmad, A.M.; Abdellatif, A.A.H.; Eid, S.Y. Sage, *Salvia officinalis* L., Constituents, Hepatoprotective Activity, and Cytotoxicity Evaluations of the Essential Oils Obtained from Fresh and Differently Timed Dried Herbs: A Comparative Analysis. *Molecules* **2021**, *26*, 5757. [CrossRef]
68. Dob, T.; Chelghoum, C. Chemical composition of the essential oil of *Artemisia judaica* L. from Algeria. *Flavour Fragr. J.* **2006**, *21*, 343–347. [CrossRef]
69. El-Massry, K.F.; El-Ghorab, A.H.; Farouk, A. Antioxidant activity and volatile components of Egyptian *Artemisia judaica* L. *Food Chem.* **2002**, *79*, 331–336. [CrossRef]
70. Fitzmaurice, S.D.; Sivamani, R.K.; Isseroff, R.R. Antioxidant therapies for wound healing: A clinical guide to currently commercially available products. *Skin Pharmacol. Physiol.* **2011**, *24*, 113–126. [CrossRef] [PubMed]
71. Süntar, I.; Akkol, E.K.; Nahar, L.; Sarker, S.D. Wound healing and antioxidant properties: Do they coexist in plants? *Free Radic. Antioxid.* **2012**, *2*, 1–7. [CrossRef]
72. Younsi, F.; Trimech, R.; Boulila, A.; Ezzine, O.; Dhahri, S.; Boussaid, M.; Messaoud, C. Essential oil and phenolic compounds of *Artemisia herba-alba* (Asso.): Composition, antioxidant, antiacetylcholinesterase, and antibacterial activities. *Int. J. Food Prop.* **2016**, *19*, 1425–1438. [CrossRef]
73. Taherkhani, M. Chemical composition, antimicrobial, antioxidant activity, tyrosinase inhibition and chelating ability of the leaf essential oil of *Artemisia diffusa*. *J. Essent. Oil Bear. Plants* **2016**, *19*, 1600–1613. [CrossRef]
74. Yen, G.-C.; Duh, P.-D.; Chuang, D.-Y. Antioxidant activity of anthraquinones and anthrone. *Food Chem.* **2000**, *70*, 437–441. [CrossRef]
75. Cheraif, K.; Bakchiche, B.; Gherib, A.; Bardaweel, S.K.; Çol Ayvaz, M.; Flamini, G.; Ascrizzi, R.; Ghareeb, M.A. Chemical composition, antioxidant, anti-tyrosinase, anti-cholinesterase and cytotoxic activities of essential oils of six Algerian plants. *Molecules* **2020**, *25*, 1710. [CrossRef]
76. De Araújo Couto, H.G.S.; Blank, A.F.; de Oliveira e Silva, A.M.; de Lima Nogueira, P.C.; de Fátima Arrigoni-Blank, M.; de Castro Nizio, D.A.; de Oliveira Pinto, J.A. Essential oils of basil chemotypes: Major compounds, binary mixtures, and antioxidant activity. *Food Chem.* **2019**, *293*, 446–454. [CrossRef]
77. Hellali, N.; Mahammed, M.H.; Ramdane, F.; Talli, A. Antimicrobial and antioxidant activities of *Cymbopogon schoenanthus* (L.) spreng. essential oil, growing in Illizi-Algeria. *J. Med. Plants Res.* **2016**, *10*, 188–194.
78. Gulcin, İ. Antioxidants and antioxidant methods: An updated overview. *Arch. Toxicol.* **2020**, *94*, 651–715. [CrossRef] [PubMed]
79. Ćavar, S.; Maksimović, M. Antioxidant activity of essential oil and aqueous extract of *Pelargonium graveolens* L'Her. *Food Control* **2012**, *23*, 263–267. [CrossRef]
80. Ruberto, G.; Baratta, M.T. Antioxidant activity of selected essential oil components in two lipid model systems. *Food Chem.* **2000**, *69*, 167–174. [CrossRef]
81. Behl, T.; Upadhyay, T.; Singh, S.; Chigurupati, S.; Alsubayiel, A.M.; Mani, V.; Vargas-De-La-Cruz, C.; Uivarosan, D.; Bustea, C.; Sava, C. Polyphenols Targeting MAPK Mediated Oxidative Stress and Inflammation in Rheumatoid Arthritis. *Molecules* **2021**, *26*, 6570. [CrossRef]
82. Benderradji, L.; Ghadbane, M.; Messaoudi, N.; El Okki, L.E. *In Vitro Multiple Solution Extracts from Leaves of Artemisia judaica L. Var. Sahariensis (L. Chevall.) Collected from the Algerian Sahara and Its Antimicrobial Activities Against Pathogenic Microorganisms*; Springer International Publishing: Cham, Switzerland, 2021; ISBN 9783030512095.
83. Elazzouzi, H.; Khabbal, Y.; Bouachrine, M.; Zair, T.; Alaoui El Belghiti, M. Chemical composition and in vitro antibacterial activity of *Artemisia ifranensis* J. Didier essential oil Growing Wild in Middle Moroccan Atlas. *J. Essent. Oil Res.* **2018**, *30*, 142–151. [CrossRef]
84. Zeng, H.Y.; Alan, A.R.; Saxena, P.K. Evaluation of in vitro shoots of *Artemisia judaica* for allelopathic potential. *Acta Physiol. Plant.* **2009**, *31*, 1237–1248. [CrossRef]
85. Kazemi, M.; Dakhili, M.; Dadkhah, A.; Fadaeian, M.; Shafizadeh, S. Composition, antimicrobial and antioxidant activities of artemisia deserti kracsh essential oil from Iran. *Asian J. Chem.* **2013**, *25*, 47–51. [CrossRef]
86. Stipcevic, T.; Piljac, A.; Piljac, G. Enhanced healing of full-thickness burn wounds using di-rhamnolipid. *Burns* **2006**, *32*, 24–34. [CrossRef]
87. Aliabadi, A.; Valadaan, V.; Branch, K.; Kazerun, I. Comparison between the effect of silymarin and silver sulfadiazine on burned wound healing in rats. *Bulg. J. Vet. Med.* **2016**, *19*, 224–232. [CrossRef]
88. Nasiri, E.; Hosseinimehr, S.J.; Akbari, J.; Azadbakht, M.; Azizi, S. The effects of Punica granatum flower extract on skin injuries induced by burn in rats. *Adv. Pharmacol. Sci.* **2017**, *2017*, 3059745.
89. Ashrafi, M.; Baguneid, M.; Bayat, A. The role of neuromediators and innervation in cutaneous wound healing. *Acta Derm. Venereol.* **2016**, *96*, 587–597. [CrossRef] [PubMed]
90. Olczyk, P.; Mencner, Ł.; Komosinska-Vassev, K. The role of the extracellular matrix components in cutaneous wound healing. *Biomed. Res. Int.* **2014**, *2014*, 747584. [CrossRef] [PubMed]
91. Papadakis, M.A.; McPhee, S.J.; Rabow, M.C. *Medical Diagnosis & Treatment*; Mc Graw Hill: San Francisco, CA, USA, 2019.
92. Phan, T.-T.; Wang, L.; See, P.; Grayer, R.J.; Chan, S.-Y.; Lee, S.T. Phenolic compounds of Chromolaena odorata protect cultured skin cells from oxidative damage: Implication for cutaneous wound healing. *Biol. Pharm. Bull.* **2001**, *24*, 1373–1379. [CrossRef] [PubMed]

93. Ahmad, S.U.; Binti Aladdin, N.-A.; Jamal, J.A.; Shuid, A.N.; Mohamed, I.N. Evaluation of Wound-Healing and Antioxidant Effects of Marantodes pumilum (Blume) Kuntze in an Excision Wound Model. *Molecules* **2021**, *26*, 228. [CrossRef]
94. Pereira Beserra, F.; Sérgio Gushiken, L.F.; Vieira, A.J.; Augusto Bérgamo, D.; Luísa Bérgamo, P.; Oliveira de Souza, M.; Alberto Hussni, C.; Kiomi Takahira, R.; Henrique Nóbrega, R.; Monteiro Martinez, E.R. From Inflammation to Cutaneous Repair: Topical Application of Lupeol Improves Skin Wound Healing in Rats by Modulating the Cytokine Levels, NF-κB, Ki-67, Growth Factor Expression, and Distribution of Collagen Fibers. *Int. J. Mol. Sci.* **2020**, *21*, 4952. [CrossRef]
95. Gushiken, L.F.S.; Hussni, C.A.; Bastos, J.K.; Rozza, A.L.; Beserra, F.P.; Vieira, A.J.; Padovani, C.R.; Lemos, M.; Polizello Junior, M.; da Silva, J.J.M. Skin wound healing potential and mechanisms of the hydroalcoholic extract of leaves and oleoresin of Copaifera langsdorffii Desf. Kuntze in rats. *Evid.-Based Complement. Altern. Med.* **2017**, *2017*, 6589270. [CrossRef]
96. Bungau, S.; Vesa, C.M.; Abid, A.; Behl, T.; Tit, D.M.; Purza, A.L.; Pasca, B.; Todan, L.M.; Endres, L. Withaferin A—A Promising Phytochemical Compound with Multiple Results in Dermatological Diseases. *Molecules* **2021**, *26*, 2407. [CrossRef]
97. Kesarwani, A.; Nagpal, P.S.; Chhabra, H.S. Experimental animal modelling for pressure injury: A systematic review. *J. Clin. Orthop. Trauma* **2021**, *17*, 273–279. [CrossRef]
98. Sanapalli, B.K.R.; Yele, V.; Singh, M.K.; Krishnamurthy, P.T.; Karri, V.V.S.R. Preclinical models of diabetic wound healing: A critical review. *Biomed. Pharmacother.* **2021**, *142*, 111946. [CrossRef]
99. Hosseini, S.V.; Tanideh, N.; Kohanteb, J.; Ghodrati, Z.; Mehrabani, D.; Yarmohammadi, H. Comparison between Alpha and silver sulfadiazine ointments in treatment of Pseudomonas infections in 3rd degree burns. *Int. J. Surg.* **2007**, *5*, 23–26. [CrossRef]
100. Hosseini, S.V.; Niknahad, H.; Fakhar, N.; Rezaianzadeh, A.; Mehrabani, D. The healing effect of mixture of honey, putty, vitriol and olive oil in *Pseudomonas aeroginosa* infected burns in experimental rat model. *Asian J. Anim. Vet. Adv.* **2011**, *6*, 572–579. [CrossRef]
101. Upadhyay, N.K.; Kumar, R.; Siddiqui, M.S.; Gupta, A. Mechanism of wound-healing activity of *Hippophae rhamnoides* L. leaf extract in experimental burns. *Evid.-Based Complement. Altern. Med.* **2011**, *2011*, 659705. [CrossRef] [PubMed]
102. Saltman, A.E. D-ribose-l-cysteine supplementation enhances wound healing in a rodent model. *Am. J. Surg.* **2015**, *210*, 153–158. [CrossRef] [PubMed]
103. Dastmalchi, K.; Wang, I.; Stark, R.E. Potato wound-healing tissues: A rich source of natural antioxidant molecules with potential for food preservation. *Food Chem.* **2016**, *210*, 473–480. [CrossRef]
104. El-Ferjani, R.M.; Ahmad, M.; Dhiyaaldeen, S.M.; Harun, F.W.; Ibrahim, M.Y.; Adam, H.; Yamin, B.M.; Al-Obaidi, M.M.J.; Al Batran, R. In vivo assessment of antioxidant and wound healing improvement of a new schiff base derived Co (ii) complex in rats. *Sci. Rep.* **2016**, *6*, 1–12. [CrossRef]
105. Qu, N.; Xu, M.; Mizoguchi, I.; Furusawa, J.; Kaneko, K.; Watanabe, K.; Mizuguchi, J.; Itoh, M.; Kawakami, Y.; Yoshimoto, T. Pivotal roles of T-helper 17-related cytokines, IL-17, IL-22, and IL-23, in inflammatory diseases. *Clin. Dev. Immunol.* **2013**, *2013*, 968549. [CrossRef]
106. Kang, S.S.; Bloom, S.M.; Norian, L.A.; Geske, M.J.; Flavell, R.A.; Stappenbeck, T.S.; Allen, P.M. An antibiotic-responsive mouse model of fulminant ulcerative colitis. *PLoS Med.* **2008**, *5*, e41. [CrossRef]
107. Sellon, R.K.; Tonkonogy, S.; Schultz, M.; Dieleman, L.A.; Grenther, W.; Balish, E.D.; Rennick, D.M.; Sartor, R.B. Resident enteric bacteria are necessary for development of spontaneous colitis and immune system activation in interleukin-10-deficient mice. *Infect. Immun.* **1998**, *66*, 5224–5231. [CrossRef] [PubMed]
108. Couper, K.N.; Blount, D.G.; Riley, E.M. IL-10: The master regulator of immunity to infection. *J. Immunol.* **2008**, *180*, 5771–5777. [CrossRef]
109. Elshibani, F.; Alshalmani, S.; Mohammed, H.A. *Pituranthos tortuosus* Essential Oil from Libya: Season Effect on the Composition and Antioxidant Activity. *J. Essent. Oil Bear. Plants* **2020**, *23*, 1095–1104. [CrossRef]
110. Andjić, M.; Božin, B.; Draginić, N.; Kočović, A.; Jeremić, J.N.; Tomović, M.; Milojević Šamanović, A.; Kladar, N.; Čapo, I.; Jakovljević, V. Formulation and evaluation of helichrysum italicum essential oil-based topical formulations for wound healing in diabetic rats. *Pharmaceuticals* **2021**, *14*, 813. [CrossRef]
111. Genčić, M.S.; Aksić, J.M.; Stošić, M.Z.Ž.; Randjelović, P.J.; Stojanović, N.M.; Stojanović-Radić, Z.Z.; Radulović, N.S. Linking the antimicrobial and anti-inflammatory effects of immortelle essential oil with its chemical composition–The interplay between the major and minor constituents. *Food Chem. Toxicol.* **2021**, *158*, 112666. [CrossRef]
112. Ehrnhöfer-Ressler, M.M.; Fricke, K.; Pignitter, M.; Walker, J.M.; Walker, J.; Rychlik, M.; Somoza, V. Identification of 1, 8-cineole, borneol, camphor, and thujone as anti-inflammatory compounds in a Salvia officinalis L. infusion using human gingival fibroblasts. *J. Agric. Food Chem.* **2013**, *61*, 3451–3459. [CrossRef]
113. Juergens, U.R. Anti-inflammatory properties of the monoterpene 1.8-cineole: Current evidence for co-medication in inflammatory airway diseases. *Drug Res. (Stuttg.)* **2014**, *64*, 638–646. [CrossRef]
114. Juergens, L.J.; Racké, K.; Tuleta, I.; Stoeber, M.; Juergens, U.R. Anti-inflammatory effects of 1, 8-cineole (eucalyptol) improve glucocorticoid effects in vitro: A novel approach of steroid-sparing add-on therapy for COPD and asthma? *Synergy* **2017**, *5*, 1–8. [CrossRef]
115. Murakami, Y.; Kawata, A.; Suzuki, S.; Fujisawa, S. Cytotoxicity and pro-/anti-inflammatory properties of cinnamates, acrylates and methacrylates against RAW264. 7 cells. *In Vivo (Brooklyn)* **2018**, *32*, 1309–1322.

antioxidants

MDPI

Article

Chemical Constituents, In Vitro Antioxidant Activity and In Silico Study on NADPH Oxidase of *Allium sativum* L. (Garlic) Essential Oil

Oscar Herrera-Calderon [1,*], Luz Josefina Chacaltana-Ramos [2], Irma Carmen Huayanca-Gutiérrez [2], Majed A. Algarni [3], Mohammed Alqarni [4] and Gaber El-Saber Batiha [5]

[1] Department of Pharmacology, Bromatology and Toxicology, Faculty of Pharmacy and Biochemistry, Universidad Nacional Mayor de San Marcos, Jr Puno 1002, Lima 15001, Peru
[2] Department of Pharmaceutical Chemical, Faculty of Pharmacy and Biochemistry, Universidad Nacional San Luis Gonzaga, Av. Los Maestros s/n, Ica 11001, Peru; luz.chacaltana@unica.edu.pe (L.J.C.-R.); irma.huayanca@unica.edu.pe (I.C.H.-G.)
[3] Department of Clinical Pharmacy, College of Pharmacy, Taif University, P.O. Box 11099, Taif 21944, Saudi Arabia; m.alqarni@tu.edu.sa
[4] Department of Pharmaceutical Chemistry, College of Pharmacy, Taif University, P.O. Box 11099, Taif 21944, Saudi Arabia; m.aalqarni@tu.edu.sa
[5] Department of Pharmacology and Therapeutics, Faculty of Veterinary Medicine, Damanhour University, Damanhour 22511, Egypt; dr_gaber_batiha@vetmed.dmu.edu.eg
* Correspondence: oherreraca@unmsm.edu.pe; Tel.: +51-956-550-510

Citation: Herrera-Calderon, O.; Chacaltana-Ramos, L.J.; Huayanca-Gutiérrez, I.C.; Algarni, M.A.; Alqarni, M.; Batiha, G.E.-S. Chemical Constituents, In Vitro Antioxidant Activity and In Silico Study on NADPH Oxidase of *Allium sativum* L. (Garlic) Essential Oil. *Antioxidants* **2021**, *10*, 1844. https://doi.org/10.3390/antiox10111844

Academic Editors: Delia Mirela Tit and Simona Bungau

Received: 29 October 2021
Accepted: 18 November 2021
Published: 20 November 2021

Abstract: *Allium sativum* L., also known as garlic, is a perennial plant widely used as a spice and also considered a medicinal herb since antiquity. The aim of this study was to determine by gas chromatography–mass spectrometry (GC–MS) the chemical profile fingerprint of the essential oil (EO) of one accession of Peruvian *A. sativum* (garlic), to evaluate its antioxidant activity and an in-silico study on NADPH oxidase activity of the volatile phytoconstituents. The antioxidant activity was tested using DPPH and β-carotene assays. An in-silico study was carried out on NADPH oxidase (PDB ID: 2CDU), as was ADMET prediction. The results indicated that diallyl trisulfide (44.21%) is the major component of the EO, followed by diallyl disulfide (22.08%), allyl methyl trisulfide (9.72%), 2-vinyl-4*H*-1,3-dithiine (4.78%), and α-bisabolol (3.32%). Furthermore, the EO showed antioxidant activity against DPPH radical (IC_{50} = 124.60 ± 2.3 μg/mL) and β-carotene bleaching (IC_{50} = 328.51 ± 2.0). The best docking score on NADPH oxidase corresponds to α-bisabolol (ΔG = −10.62 kcal/mol), followed by 5-methyl-1,2,3,4-tetrathiane (ΔG = −9.33 kcal/mol). Additionally, the volatile components could be linked to the observed antioxidant activity, leading to potential inhibitors of NADPH oxidase.

Keywords: allyl compounds; molecular dynamic; antioxidant enzyme; oxidative stress; phytochemical study

1. Introduction

Essential oils (EOs) are natural products obtained from aromatic plants and can be extracted from leaves, roots, stems, flowers, and seeds, among others [1]. EOs are widely used in the food, cosmetic, alternative therapy (such as aromatherapy) and pharmaceuticals industries [2]. In terms of volatile chemical composition, EOs are mainly constituted by monoterpenes, sesquiterpenes, phenols, and alcohols. However, allyl structures and phenylpropanoids also constitute the phytochemical profile of some EOs [3]. EOs have been linked to antioxidant activity [4] as free radical scavengers and metal chelators [5], as well as to other biological activities such as anti-inflammatory, analgesic [6], sedative [7], antibacterial [8], antiviral [9], neuroprotective [10], and antifungal properties [11].

The generation of free radicals leads to oxidative stress in cells, which can trigger aging and degenerative diseases. The exposure to environmental stress, UV-radiation, viral and bacterial infection and carcinogenic chemicals, among other things, might also cause severe damage, brought about by the lipid peroxidation of polyunsaturated fatty acids and, consequently, the liberation of toxic metabolites [12]. Studies have revealed that essential oils inhibit lipid peroxidation in foods and in the biochemical process related to oxidative stress [2]. According to Amorati et al., the chemical fingerprint of EOs is mainly comprised of terpenoid and phenylpropanoid compounds [13]. Monoterpenes, free phenols, and allylic alcohols have all demonstrated potential antioxidant capacity in oxidative stress models, while other compounds grouped as sesquiterpenes, and non-isoprenoids have shown low antioxidant activity [14].

Nowadays, computational models such as molecular docking are carried out as a bioinformatic tool to study the inhibition of several enzymes that negatively affect the antioxidant activity, such as xanthine oxidase, nitric oxide synthases (NOS), cytochrome P450 reductase, and nicotinamide-adenine dinucleotide phosphate (NADPH) oxidase, as well as the mitochondrial electron transport chain. Furthermore, it is known that reactive oxygen species (ROS) generation occurs in the mitochondria via oxidative phosphorylation and through the enzyme NADPH oxidase [15]. Additionally, NADPH oxidase-mediated cell proliferation participates in intracellular signaling processes and has been observed in a variety of cancer cells and in tissue repair processes [16]. Thus, the screening of inhibitory molecules on this target could be useful to ameliorate numerous chronical or degenerative diseases and to find selective and non-toxic inhibitors of NADPH oxidases, providing new drugs for the treatment of diseases related to oxidative stress-dependent processes [17].

Allium sativum L. (family Amaryllidaceae), also known as garlic is widespread across the world, being used mainly in gastronomy and for its medicinal properties. Within its medical uses, several pharmacological activities have been evaluated, such as its potential anticancer, hypotensive, hepatoprotective, hypoglycemic, antimicrobial, and immunomodulatory effects [18]. Garlic essential oil contains sulfur compounds as the main volatile phytochemicals, dominated by allyl polysulfides, such as diallyl disulfide, diallyl trisulfide, and diallyl disulfide. Different methods to obtain EOs adopted in some studies which might affect the chemical composition due to enzymatic biotransformation processes occurring during the extraction process, i.e., hydrodistillation compared to the microwave-assisted extraction method [19,20]. The type of apparatus used during the extraction could also influence the results, i.e., use of a Clevenger apparatus vs. industrial extraction. Considering the role of the edaphic factor on the plant growth and chemical composition of different metabolites of interest, the chemical profile of essential oils has been shown to be affected by all these factors [21]. Altitude also plays an important role; to date only the results for garlic collected in low altitudes ranging between 500 and 1000 m.a.s.l. have been reported [22]. Additionally, remarkable qualitative and quantitative differences have been found in EOs extracted after following different drying procedures [23]. In Peru, garlic is used mainly as a food additive, and according to National Institute of Agrarian Innovation (INIA-PERU), Peru has several accessions of garlic such as Purple, Criollo or Napurí, Barranquino, Massone, Pata de Perro, and White Chinese [24], with White Chinese being one of the most consumed and commercialized accessions.

To date, some accounts of the antioxidant capacity of garlic EOs determined by different methods have been reported. However, those garlic samples were cultivated at low altitudes and their chemical compositions could differ to those of garlic cultivated in high altitude zones. In-silico studies allow one to identify molecules with promising inhibitory effects on any antioxidant target such as NADPH oxidase, which is considered an important antioxidant marker in biological systems. The focus on antioxidant phytochemicals contained in EOs is directly linked to their application for the prevention of oxidative damage caused by ROS. Hence, low-molecular antioxidants such as allyl polysulfide structures enhance organism stability under conditions of oxidative stress. Based on all these antecedents, the aims of this research were: (1) To determine the phytochemical constituents of garlic EO by gas chromatography–mass spectrometry (GC–MS) and their antioxidant activity against the 1,1-diphenyl-2-picrylhydrazyl (DPPH) radical and β-carotene bleaching; (2) to evaluate in-silico the inhibitory effect of the volatile phytochemicals of the essential oil from *A. sativum* L. (garlic) on NADPH oxidase (PDB ID: 2CDU).

2. Materials and Methods

2.1. Chemicals

All analytical grade (99.5%) solvents (dichloromethane, chloroform, methanol) and hydrogen peroxide were purchased from Merck (Darmstadt, Germany). 2,2-Diphenyl-1-picrylhydrazyl (DPPH), β-carotene, linoleic acid, Trolox, and rutin were purchased from Sigma Aldrich (St Louis, MO, USA).

2.2. Plant Material

A quantity of 800 g of *A. sativum* (bulbs) cultivated in the Arequipa region of Peru (2.335 m.a.s.l.), was used to carry out the experimental procedures. Bulbs were cleaned and peeled before being placed in a Clevenger equipment to obtain the essential oil by hydrodistillation for 2 h [25]. The essential oil was separated by decantation, then anhydrous Na_2SO_4 was added to eliminate any remaining water drops. Finally, the EO was stored in a sealed amber vial until further use.

2.3. Identification of Volatile Compounds by Gas Chromatography–Mass Spectrometry (GC–MS)

Volatile chemicals were determined with a GC–MS system (7890 Gas Detector and 5975C Mass Spectrometer Detector, Agilent Technologies, Santa Clara, CA, USA). Then, 0.0136 g of the sample was weighed and mixed with 0.5 mL of dichloromethane. Next, 1.0 μL of the working solution was injected into the equipment in splitless mode (split ratio: 20:1). The EO was run on a J&W 122-1545.67659 DB-5ms column (60 m × 250 μm × 0.25 μm, Agilent Technologies). The working conditions were as follows: the temperature program started at 40 °C, with increments of 5 °C/min up to 180 °C, followed by increases of 2.5 °C/min up to 200 °C for 5 min and finally 10 °C/min up to 300 °C, followed by holding for 3 min. The helium flow rate was at 1 mL/min. Volatile chemicals were identified and confirmed by comparing the mass spectrum of the compounds with the NIST20 library data [26].

2.4. Antioxidant Activity

2.4.1. 2,2-Diphenyl-1-picrylhydrazyl (DPPH) Assay

The 2,2-diphenyl-1-picrylhydrazyl (DPPH) method according to Rojas-Armas et al. [27] with slight modifications, was used as an organic radical activity assay. For 0.1 mM of the DPPH solution, methanol was used as a solvent, and 100 μL of this solution was mixed with 900 μL of EO at different concentrations (0–1000 μg/mL). Then, the reaction tubes were incubated at room temperature for 30 min and protected from light. Finally, the absorbance was measured at 517 nm using a Genesys 20 spectrophotometer (Thermo Scientific, Waltham, MA, USA). Trolox was used as an antioxidant control. The inhibitory concentration (IC_{50}) was calculated from the plot of inhibition percentage against the sample concentration.

2.4.2. β-Carotene Bleaching Assay

The lipid peroxidation activity was determined by the β-carotene bleaching method according to Tepe et al. [28]. A stock solution containing 0.5 mg of β-carotene in 1 mL of chloroform was then mixed with 200 mg of Tween 40 and 25 μL of linoleic acid. Then, the chloroform was removed under vacuum at 40 °C for 5 min using a R-100® rotary evaporator (Buchi, Flawil, Switzerland). Subsequently, 100 mL of 3% aqueous hydrogen peroxide solution was added to the residue and mixed for 5 min until an emulsion was obtained. Then, aliquots (2.5 mL) of the emulsion were mixed with garlic EO (350 μL) and rutin. All test tubes were incubated at 25 °C up to 48 h and read at 490 nm using a Genesys 20 spectrophotometer.

2.5. Molecular Docking of the Phytochemical Volatiles of Garlic on the Receptor NADPH Oxidase (PDB ID: 2CDU)

Prior to the docking study, 13 volatile oil compounds of *Allium sativum* (garlic) were drawn in ChemDraw 19.0 (Perkin Elmer, Waltham, MA, USA), and subsequently, the geometry was optimized. The receptor NADPH oxidase was retrieved from the protein data bank (PDB ID: 2CDU). Before performing molecular interaction studies, NADPH oxidase was further curated for missing side-chain residues using the openMM tool. Molecular docking studies were performed with Autodock v 4.2.6 (The Scripps Research Institute, La Jolla, CA, USA). The binding cavity for ligand compound docking in NADPH oxidase was determined from the predefined co-crystallized X-ray structure from RCSB PDB. The residue positions were calculated within 3 Å from the co-crystallized ligand. After the cavity selection in each case, the co-crystallized ligands were removed using the Chimera tool (https://www.cgl.ucsf.edu/chimera/, accessed on 11 September 2021), and subsequently, the energy was minimized using the steepest descent and conjugate gradient algorithm. Then, finally, merging the nonpolar hydrogens, both the receptor and target compound were saved in pdbqt format. A grid box was created with parameters X = 68, Y = 58, and Z = 64 Å with 0.3 Å spacing. Following the Lamarckian genetic algorithm (LGA), docking studies of the protein–ligand complex were performed to achieve the lowest free energy of binding (ΔG). During the molecular docking studies, three replicates were performed, with a total number of 50 solutions computed in each case, with a population size of 500, a number of evaluations of 2,500,000, a maximum number of generations of 27,000, and rest the default parameters were allowed. After docking, the RMSD clustering maps were obtained by the re-clustering command with a clustering tolerance of 0.25, 0.5, and 1 Å, respectively, in order to obtain the best cluster with the lowest energy score and a high number of populations.

2.6. Molecular Dynamic (MD) Studies

The MD simulation studies were carried on the best docked complexes for NADPH oxidase with α-bisabolol and standard diallyl trisulfide using Desmond 2020.1 from Schrödinger, LLC (New York, NY, USA). The OPLS-2005 force field and explicit solvent model with the SPC water molecules were used in this system. Na+ ions were added to neutralize the charge. Moreover, 0.15 M, of the NaCl solution was added to the system to simulate the physiological environment. The NPT ensemble was set up using the Nose–Hoover chain coupling scheme with a temperature of 27 °C, a relaxation time of 1.0 ps, and a pressure of 1 bar, maintained in all of the simulations. A time step of 2 fs was used. The Martyna–Tuckerman–Klein chain coupling scheme barostat method was used for pressure control with a relaxation time of 2 ps. The particle mesh Ewald method was used for calculating long-range electrostatic interactions, and the radius for the coulomb interactions were fixed at 9Å. The RESPA integrator was used for a time step of 2 fs for each trajectory calculating the bonded forces. The root means square deviation (RMSD), radius of gyration (Rg), root mean square fluctuation (RMSF), and number of hydrogen (H-bonds) were calculated to monitor the stability of the MD simulations [29].

2.7. ADMET Prediction

The absorption, distribution, metabolism, excretion, and toxicity (ADMET) properties of the compounds determined in the essential oil of garlic were calculated using the pkCSM online tool (http://biosig.unimelb.edu.au/pkcsm/prediction, accessed on 21 October 2021) [30] and SwissADME (http://www.swissadme.ch/, accessed on 21 October 2021) [31].

2.8. Statistical Analysis

The preparation of essential oil was carried out three times, and the mean and SD of the three independent experiments are presented. The antioxidant assays were repeated three times. The statistical tools employed were Student's *t*-test, two-tailed, and the IC_{50} values were estimated by linear regression statistics. p-Values less than 0.05 were considered statistically significant. GraphPad Prism program version 6.0 (GraphPad Software, La Jolla, CA, USA) was used to develop the statistical analysis.

3. Results and Discussion

3.1. Chemical Profile of the Essential Oil of *A. sativum*

The obtained EO showed a pale yellow color, an extraction yield of 0.78% (v/dry weight), and a density of 0.95 ± 0.01 g/mL at 20 °C. The volatile components of garlic essential oil were analyzed by GC–MS and are presented in Table 1. According to our results, we identified 17 compounds, four of which are of unknown structures, which accounted for 100% of the total composition. The analysis identified diallyl trisulfide (retention time 24.97 min) as the main component (44.21%) of the volatile chemicals, followed by diallyl disulfide (22.08%), allyl methyl trisulfide (9.72%), and 2-vinyl-4*H*-1,3-dithiine (4.78%). According to Figure S1 of the Supplementary Materials, the total time for the evaluation was 50 min.

Table 1. Chemical composition of the volatile oil of *A. sativum* (garlic).

N°	Compound Name (NIST08.L)	Rt (min)	Molecular Formula/ Molecular Mass	% In Sample (Relative Areas)	Chemical Structure
1	Allyl methyl disulfide	12.75	$C_4H_8S_2$ (120.36)	0.47 ± 0.01	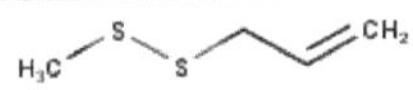
2	3*H*-1,2-Dithiole	14.18	$C_3H_4S_2$ (104.19)	2.41 ± 0.02	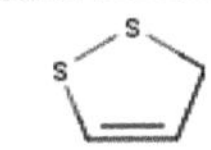
3	Diallyl disulfide	18.11	$C_6H_{10}S_2$ (146.3)	22.08 ± 0.11	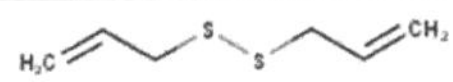
4	1-Propenyl 2-propenyl-(*E*)-disulfide	18.69	$C_6H_{10}S_2$ (146.27)	0.92 ± 0.01	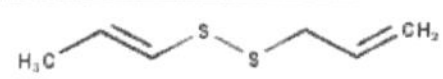
5	Allyl methyl trisulfide	20.06	$C_4H_8S_3$ (152.29)	9.72 ± 0.05	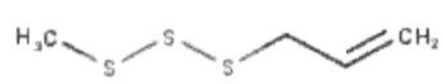
6	Unknown	20.92	$C_6H_{10}S_2$ (146.27)	1.08 ± 0.01	n.d.
7	3-Vinyl-1,2-dithiacyclohex-4-ene	21.71	$C_6H_8S_2$ (144.25)	2.56 ± 0.01	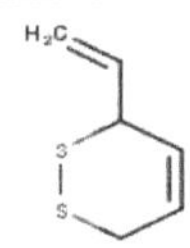
8	4H-1,2,3-Trithiine	22.18	$C_3H_4S_3$ (136.25)	3.07 ± 0.01	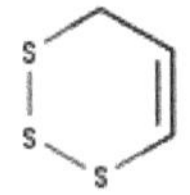
9	2-Vinyl-4*H*-1,3-dithiine	22.46	$C_6H_8S_2$ (144.25)	4.78 ± 0.01	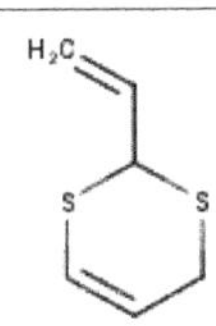
10	Unknown	24.66	$C_9H_{16}O_2S$ (188.29)	0.50 ± 0.01	n.d.
11	Diallyl trisulfide	24.97	$C_6H_{10}S_3$ (178.33)	44.21 ± 0.08	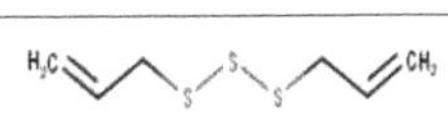

Table 1. *Cont.*

N°	Compound Name (NIST08.L)	Rt (min)	Molecular Formula/ Molecular Mass	% In Sample (Relative Areas)	Chemical Structure
12	1-Allyl-3-propyltrisulfane	25.22	$C_6H_{12}S_3$ (180.34)	1.37 ± 0.01	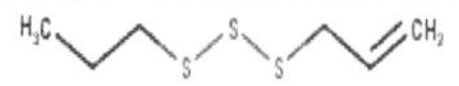
13	5-Methyl-1,2,3,4-tetrathiane	27.20	$C_3H_6S_4$ (170.32)	1.55 ± 0.01	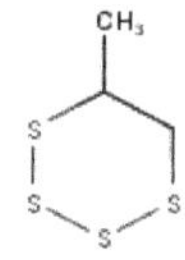
14	Unknown	31.50	$C_9H_{16}S_3$ (220.42)	0.85 ± 0.01	n.d.
15	Diallyl tetrasulfide	31.55	$C_6H_{10}S_4$ (210.39)	0.68 ± 0.01	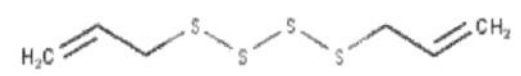
16	α-Bisabolol	35.42	$C_{15}H_{26}O$ (222.37)	3.32 ± 0.03	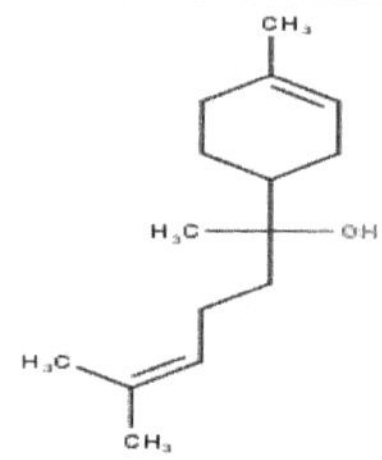
17	Unknown	45.26	$C_{22}H_{42}O_4$ (370.57)	0.43 ± 0.01	n.d.

Rt, retention time; n.d., not determined.

Regarding the phytochemical analysis of garlic essential oil by GC–MS, our results differ to those of other investigations and could be explained by the different extraction methods used to obtain the EO, such as conventional or non-conventional techniques. For example, in sono-hydrodistillation and ultrasound-assisted hydrodistillation, the obtained garlic essential oil has, as a major component, diallyl disulfide, whilst is lower in hydrodistillation [32]. In our study, the most representative molecule was diallyl trisulfide, with 44.21%, followed by diallyl sulfide, with 22.08%. However, in recent studies, as reported by Thuy et al., allyl disulfide was the main component, with 28.44% [9]. In an analysis of garlic EO from Cameroon, diallyl trisulfide (41.62%) and diallyl disulfide (19.74%) were the major components [33]. In a Tunisian study, allyl disulfide (28.0%) and eugenol (15.37%) were the most abundant constituents [34]. In Saudi Arabia, the major component was allyl methyl trisulfide, with 13.10%. [35]. Garlics from Thailand showed values of diallyl disulfide, diallyl trisulfide, and diallyl tetrasulfide, equivalent to 31.67%, 31.56%, and 13.48%, respectively [36]. On the contrary, the drying procedure also affects the composition of sulfur components, i.e., garlic EO from Tunisia obtained by freeze drying contained 45.9% diallyl trisulfide compared to 42.3% for an oven-dried extract and 37.3% for an air-dried extract [37]. The diversity of the chemical composition in garlic EO might be related with external factors such as temperature, soil composition, climate conditions, environmental stress, ecosystem, and altitude. Arequipa is situated at 2335 m.a.s.l., which could be an advantage to produce an EO with a phytochemical profile different to other varieties of garlic found across the world [38].

3.2. Antioxidant Profile of A. sativum Essential Oil

Garlic EO exhibited strong antioxidant activity, as shown in Table 2. Trolox showed better antioxidant activity than the EO. On the contrary, there was a significant difference between EO and Trolox concentrations ($p = 0.0002$). The EO of *A. sativum* showed a good antioxidant response, but other research has reported different values; for example, according to Lawrence et al., the EO of garlic grown in the north Indian plains showed an IC_{50} value of 0.5 mg/mL [39]. Another IC_{50} value that has been reported is 7.67 mg/mL, for an oil obtained by hydrodistillation [40]. In our study the EO showed an IC_{50} of 124.60 ± 2.5 μg/mL, which is different to the findings of Ndoye et al. [33], with an IC_{50} value of 0.19 μg/mL. This is a high value compared to the other results, which could be due to the presence of diallyl trisulfide, diallyl disulfide, and methylallyl di- and trisulfides. Although diallyl sulfide was the main component of the garlic EO produced by hydrodistillation, dialyl disulfide and allyl methyl sulfide did not demonstrate any antioxidant action when tested as inhibitors of the controlled autoxidation of isopropylbenzene or styrene, implying that they are oxidized together with the oxidizable substrate [13]. We detected diallyl trisulfide as the major compound, whilst other sulfur volatiles have been shown to be abundant in other garlic EOs. However, some techniques to extract essential oil might affect its antioxidant activity, according to Boubechiche et al. [32]. The EO obtained by hydrodistillation had a better antioxidant capacity (IC_{50} = 0.96 mg/mL) than that obtained using the ultrasound-assisted (IC_{50} = 1.176 mg/mL) and sono-hydrodistillation processes (IC_{50} = 1.234 mg/mL). After microwave-assisted extraction, the EO showed an inhibition percentage of 72.06% at 500 μg/mL [41]. Factors such as altitude, climate conditions, and chemotype varieties may also be responsible for the differences observed in our study [42]. Regarding the β-carotene bleaching assay, there was a significant difference between rutin and the garlic EO ($p = 0.0012$), being the antioxidant standard more than garlic EO. In a recent study by Ncir et al. [34], an IC_{50} value of β-carotene equivalent to 0.2 ± 0.02 mg/mL was reported, similar to our results but contrary to the DPPH assay, with an IC_{50} value of 0.048 ± 0.007 mg/mL.

Table 2. Antioxidant activities of the *A. sativum* essential oil.

Samples	Antioxidant Activity	
	DPPH IC_{50} (μg/mL)	β-Carotene IC_{50} (μg/mL)
Essential oil of *Allium sativum*	124.60 ± 2.3 **	328.51 ± 2.0 *
Trolox	0.54 ± 0.02	-
Rutin	-	3.5 ± 0.05

Values are reported as the mean ± SD of three experiments. * $p < 0.01$; ** $p < 0.001$. Student's *t*-test, two-tailed.

Another factor considered was the presence of α-bisabolol, which has not been found in other garlic EOs and which could influence the antioxidant activity, as shown in Table 2 in comparison to other garlic EOs from across the world.

3.3. Molecular Docking of the Phytochemical Constituents of the Essential Oil from A. sativum

Molecular docking studies were carried out in order to understand the interaction profile of various volatile oil compounds present in *A. sativum* with NADPH oxidase. Out of the 13 specific compounds found abundant in the chromatography results, α-bisabolol displayed lowest the binding energy, with (ΔG) −10.62 kcal/mol and a predicted inhibitory concentration (Ki) of 0.14 μM (Table 3). The principal residues of NADPH oxidase LYS187 and TYR188 were involved in conventional polar hydrogen bond formation with α-bisabolol (Figure 1). On the contrary, 5-methyl-1,2,3,4-tetrathiane and 4*H*-1,2,3-trithiine also displayed better binding with NADPH oxidase, followed by α-bisabolol, with ΔG −9.33 and −9.05 kcal/mol, respectively (Table 1). However, the predicted Ki was observed to be 1.24 and 1.90 μM, respectively (Table 1). The common residue of NADPH oxidase CYS133 was involved in both the cases, forming polar hydrogen bonds (Figure 1), while other than CYS133, GLY244 also indulged in forming conventional hydrogen bonds with 5-methyl-1,2,3,4-tetrathiane (Figure 1). All of the other ligands' interaction profiles are presented in Table 3 and Supplementary Materials Figures S2–S10.

The inhibition of NADPH oxidase in silico by sulfur components and alpha-bisabolol could be correlated with the in vitro results presented by Schepetkin et al., including the garlic EO and three compounds (diallyl trisulfide, ajoene, and allicin), which inhibited the neutrophil ROS production, with diallyl trisulfide being the major component attributed to ROS inhibition [43]. Additionally, S-allylcysteine showed antioxidant activity, inhibiting gp91^{phox} and gp22^{phox} of NADPH oxidase, where gp91^{phox} was the catalytic subunit and gp22^{phox} is a membrane protein, both of which formed a complex generating superoxide radical, and its over-activation has been linked to several renal diseases [44].

Table 3. Ligand interaction energies and inhibitory concentrations with NADPH oxidase in the molecular docking study.

No.	Ligand Name	ΔG (kcal/mol)	Ki (μM)	Residues Involved in Polar Bonds
1	Allyl methyl disulfide	−6.85	15.0	VAL214
2	3*H*-1,2-Dithiole	−7.45	5.45	CYS133
3	Diallyl disulfide	−7.19	8.48	ILE243, GLY244
4	1-Propenyl 2-propenyl-(*E*)-disulfide	−6.59	23.2	LYS187
5	Allyl methyl trisulfide	−6.65	21.1	MET33
6	3-Vinyl-1,2-dithiacyclohex-4-ene	−8.87	2.78	VAL81
7	4*H*-1,2,3-Trithiine	−9.05	1.90	CYS133
8	2-Vinyl-4*H*-1,3-dithiine	−7.69	3.64	VAL81
9	Diallyl trisulfide	−7.36	6.37	GLY244
10	1-Allyl-3-propyltrisulfan	−7.76	3.32	GLY244
11	5-Methyl-1,2,3,4-tetrathiane	−9.33	1.24	CYS133, GLY244
12	Diallyl tetrasulfide	−6.21	44.4	-
13	α-Bisabolol	−10.62	0.140	LYS187, TYR188

VAL, valine; CYS, cysteine; ILE, isoleucine; GLY, glycine; LYS, lysine; MET, methionine; TYR, tyrosine.

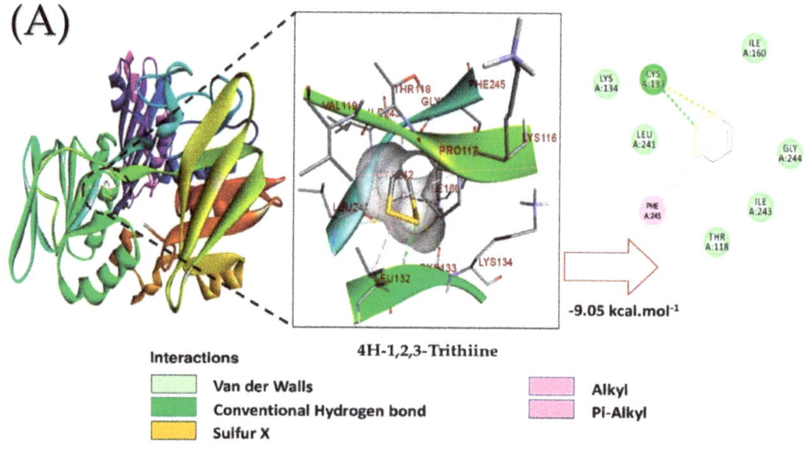

Figure 1. *Cont.*

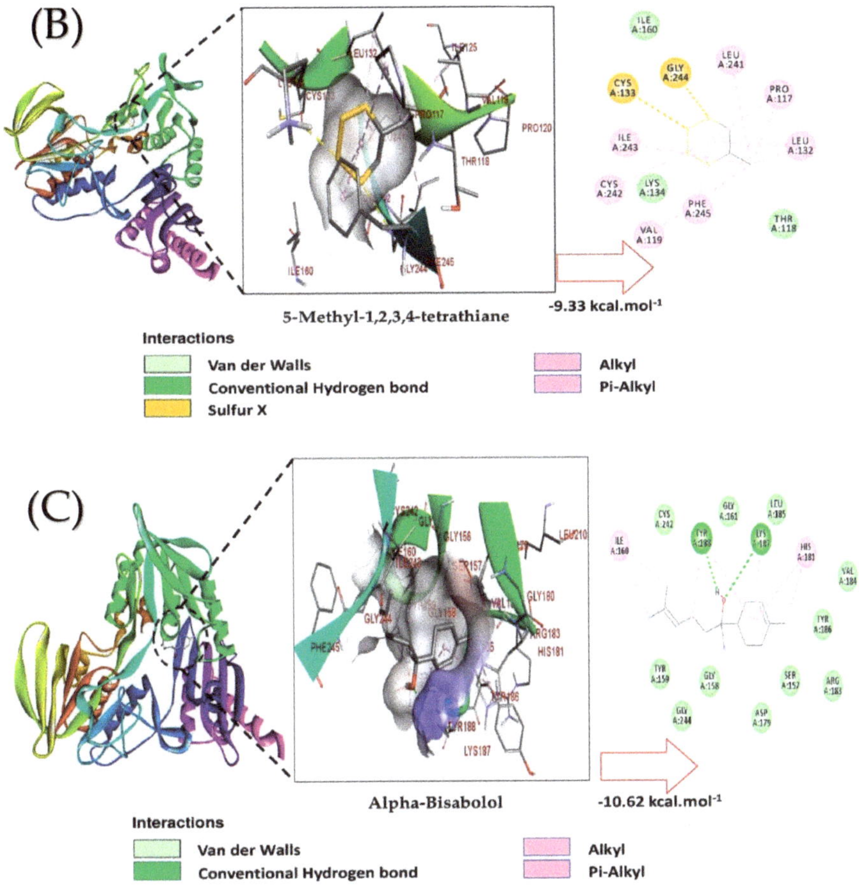

Figure 1. Molecular interaction studies of the most active phytochemical constituents of garlic essential oil with NADPH oxidase (PDB ID: 2CDU). (**A**): 4H-1,2,3-trithiine, (**B**): 5-methyl-1,2,3,4-tetrathiane, and (**C**): α-bisabolol. Surface view (left panel), and 2D (right panel) interactions.

3.4. Molecular Dynamics of the Phytoconstituents of the Essential Oil from A. sativum

In molecular dynamics, the two most common parameters of structural fluctuations are the root mean square deviation (RMSD) and the root mean square fluctuations (RMSF). The RMSD measures the average displacement of the atoms at an instant of the simulation relative to a reference structure, generally the first frame of the crystallographic structure or simulation. The RMSF measures the displacement of a particular atom, or group of atoms, relative to the reference structure, averaged over the number of atoms. The RMSD is useful for analysis of time-dependent motions of a structure. It is frequently used to discern whether a structure is stable during the time-scale of the simulations [45].

Molecular dynamics studies of NADPH oxidase with α-bisabolol and diallyl trisulfide were carried out for a 100 ns simulation time scale. A total of 100 ns of simulation time analysis of the trajectories displayed convergence of the root mean square deviation (RMSD) for α-bisabolol with an average deviation 0.1 Å (Figure 1A, dark cyan). Meanwhile, with diallyl trisulfide, the RMSD showed little more fluctuation compared to alpha-bisabolol, with an average deviation 0.5 Å (Figure 2A, orange). RMSD deviations of the bound complexes were exhibited within an acceptable range, and with alpha-bisabolol, the Cα of NADPH oxidase was more stable. The RMSF of individual amino residues of NADPH oxidase over the function of a 100 ns time scale displayed low fluctuations in α-bisabolol, with an average of 0.1 Å (Figure 2B, drak cyan). Low fluctuations of the amino acid residues indicate higher stability from a converged structure. However, amino acid residues of NADPH oxidase bound to diallyl trisulfide fluctuated more at positions 70 and 370, respectively, as compared to the α-bisabolol-bound complex (Figure 2B, orange).

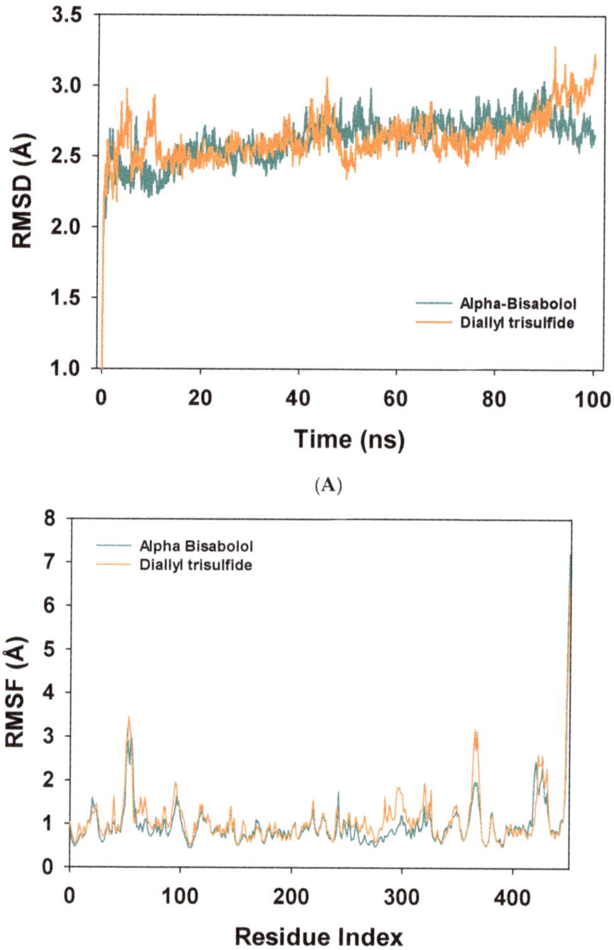

(A)

(B)

Figure 2. (**A**) RMSD plot of Cα atoms of NADPH oxidase displaying deviations from the mean with α-bisabolol (dark cyan) and diallyl trisulfide (orange) over a 100 ns time scale simulation. (**B**) RMSF plot of amino acids of NADPH oxidase bound to α-bisabolol (dark cyan) and diallyl trisulfide (orange) over a 100 ns time scale simulation.

The radius of gyration is the measure of compactness of the protein in the ligand-bound state. The MD simulation of 100 ns of NADPH oxidase Cα atoms complexed with α-bisabolol displayed a slight lowering of the trajectory due to compactness of the complex (Figure 3A, dark cyan); meanwhile, the complex with diallyl trisulfide (orange) displayed more fluctuations and a less compact structure compared to the α-bisabolol-bound complex (Figure 3A, orange).

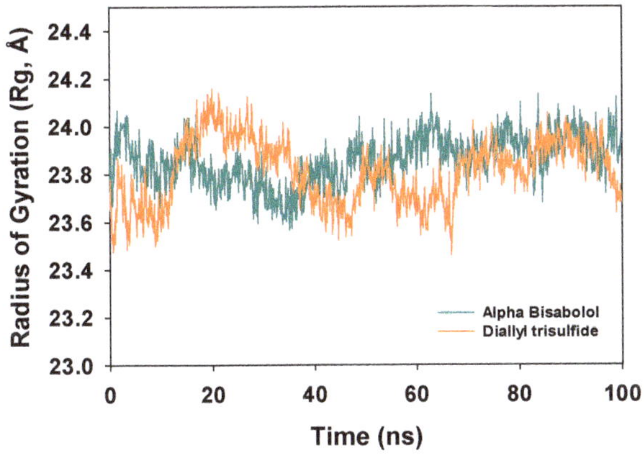

(**A**)

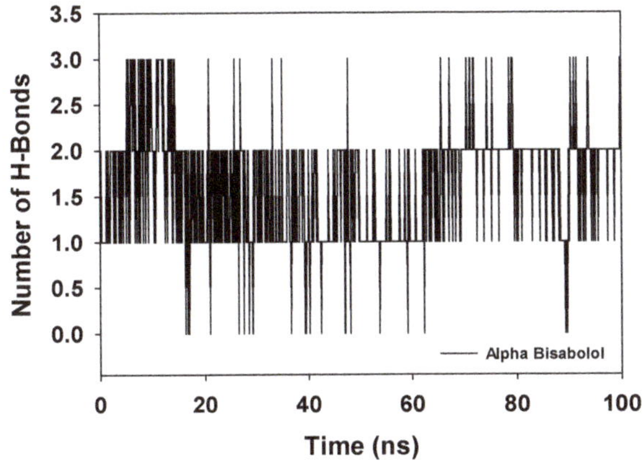

(**B**)

Figure 3. (**A**) Radius of gyration (Rg) plot of Cα atoms of NADPH oxidase complexed with α-bisabolol (dark cyan) and diallyl trisulfide (orange). (**B**) Number of hydrogen bonds formed between NADPH oxidase and α-bisabolol throughout a simulation time of 100 ns.

The number of hydrogen bonds formed between proteins and ligands is an important factor when analyzing a stable complex throughout a simulation. Here, in this case, the number of H-bonds formed between NADPH oxidase and α-bisabolol displayed a consistent interaction, with an average of 1.5 throughout the 100 ns simulation (Figure 2B). In contrast, with diallyl trisulfide, no hydrogen bond formations were recorded. The total energy of the system is another essential function to draw a conclusion regarding the stability of a complex. NADPH oxidase-α-bisabolol displayed an average of −35 kcal/mol, where most of the energies were favorably contributed by the non-bonded Coulomb and van der Waal's forces (vdW) interactions of −5 and −25 kcal/mol, respectively (Figure 4A). On the contrary, the diallyl trisulfide-bound NADPH oxidase complex displayed an average energy of −27 kcal/mol, where the vdW energy contributed more favorably than the Coulomb energy (Figure 4B). The comparative energy plots indicate a more stable and converged structure of α-bisabolol-bound NADPH oxidase than the standard diallyl trisulfide. Therefore, from the simulation studies, it can be suggested that α-bisabolol has the potential to be a better inhibitor against NADPH oxidase and can substitute the application of diallyl trisulfide.

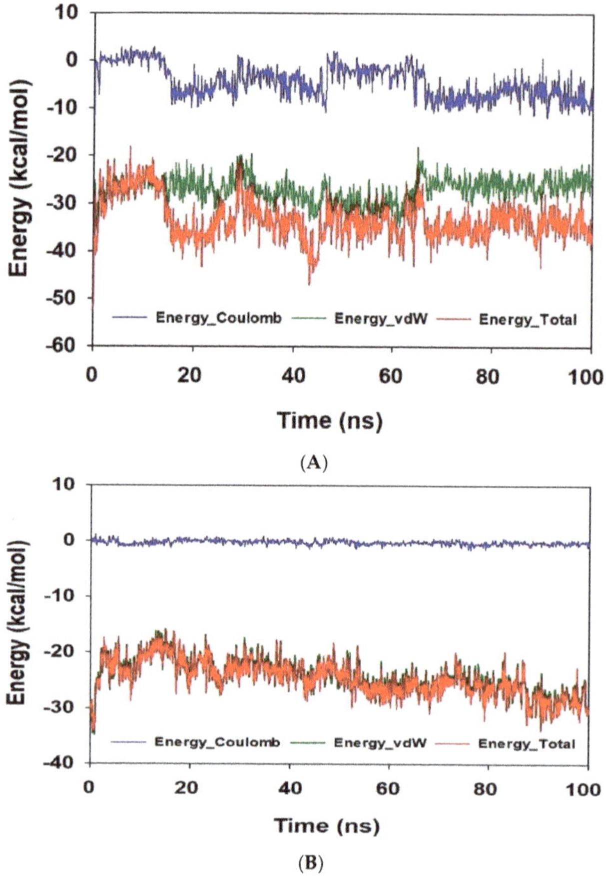

Figure 4. Energy plots of NADPH oxidase complexed with (**A**) α-bisabolol and (**B**) diallyl trisulfide.

3.5. ADMET Profiles

The chemical constituents of the garlic essential oil were analyzed using the online pkCSM tool to predict the absorption, distribution, metabolism, excretion, and toxicity profiles. The results revealed that all compounds had a molecular weight ranging between 100 and 400 g/mol, which is important for penetrability, because the profile to achieve this parameter is for those compounds with values less than 500 g/mol. The Caco-2 permeability had values above 0.90 and a high intestinal absorption (90–95%), which would predict that all compounds of the garlic EO will be absorbed in the small intestine [29]. Due to the lipophilic character of chemical components in garlic EOs, they tend to form micelles and are digested in the small intestine [46]. The skin permeability ranged from −1.29 to −2.232 cm/h (≤2.5), which means that volatile phytochemicals easily penetrate the skin.

The volume distribution (VDss) is acceptable due to its values above −0.15. On the contrary, 11 compounds are probably able to penetrate the blood–brain barrier (BBB) (log BB > 0.3), but 3H-1,2-dithiole and 3-vinyl-1,2-dithiacyclohex-4-ene had a medium value under 0.3, which would result in difficultly accessing the brain. The central nervous system (CNS) permeability achieved between −2.644 and −1.649; therefore, only 5-methyl-1,2,3,4-tetrathiane is able to permeate the central nervous system (logP > −2), as values logP < −3 are unable to penetrate it. Table 4 shows that all of constituents are not able to interfere with CYP2D6 and CYP3A4 and would not be metabolized by either. The excretion parameters showed that the total clearance (0.103 and 1.363 log mL/min/kg) achieved positive values, meaning a quick excretion [15]. Volatile organic compounds can influence the expression and activity of cytochrome P450 enzymes and transferases involved in the metabolism of other drugs in the system. Additionally, drug interactions of essential oil components should be considered if any medication is used together with EOs. Furthermore, all compounds showed no potential contraindication, because they are not substrates of the organic cation transport 2 (OCT2), which is involved in the uptake and secretion of cationic drugs. Regarding the toxicity, the acute oral toxicity in rats (LD_{50}) ranged from 1.739 to 3.035 mol/kg, meaning a low toxicity. Additionally, they were not considered hepatotoxic substances (Table 4).

Table 4. ADMET properties of the chemical constituents of the *A. sativum* essential oil (in the metabolism analysis for CYP 2D6 inhibitors, CYP 3A4 inhibitors, renal OCT2 substrate, and hepatotoxicity, no data were obtained).

	Absorption			Distribution			Excretion	Toxicity	
N°	Caco-2 (Log Papp in 10^{-6} cm/s)	Intestinal Absorption (% Absorbed)	Skin Permeability (log Kp)	VDss (Log L/kg)	BBB Permeability (Log BB)	CNS Permeability (Log PS)	Total Clearance (Log mL/min/kg)	Oral Rat Acute Toxicity (LD_{50} = mol/kg)	Oral Rat Chronic Toxicity (LOAEL = Log mg/kg_bw/day)
1	1.394	94.604	−1.761	0.098	0.332	−2.336	0.444	2.512	1.726
2	1.384	95.015	−2.232	0.159	0.183	−2.599	0.287	2.674	1.733
3	1.406	94.007	−1.374	0.197	0.743	−2.157	0.366	2.433	2.026
4	1.397	92.885	−1.652	0.112	0.437	−2.435	0.347	2.845	1.728
5	1.398	94.877	−1.865	0.241	0.377	−2.439	0.496	2.447	1.777
6	1.394	93.326	−2.088	0.148	0.27	−2.644	0.161	2.956	1.756
7	1.398	94.908	−1.865	0.241	0.378	−2.439	0.467	2.442	1.752
8	1.403	92.573	−1.449	0.216	0.767	−2.309	0.446	2.711	1.857
9	1.403	92.573	−1.449	0.216	0.767	−2.309	0.446	2.711	1.857
10	1.403	92.097	−1.425	0.225	0.757	−2.309	0.389	2.75	1.931
11	1.427	91.045	−1.29	0.165	0.693	−1.649	0.103	2.863	1.812
12	1.406	90.609	−1.552	0.224	0.759	−2.402	0.336	3.035	1.843
13	1.505	93.014	−1.761	0.42	0.605	-2.541	1.363	1.739	1.178

1: Allyl methyl disulfide; 2: 3*H*-1,2-dithiole; 3: diallyl disulfide; 4: 1-propenyl-2-propenyl-(*E*)-disulfide; 5: allyl methyl trisulfide; 6: 3-vinyl-1,2-dithiacyclohex-4-ene; 7: 4*H*-1,2,3-trithiine; 8: 2-vinyl-4*H*-1,3-dithiine; 9: diallyl trisulfide; 10: 1-allyl-3-propyltrisulfane; 11: 5-methyl-1,2,3,4-tetrathiane; 12: diallyl tetrasulfide; 13: α-bisabolol.

The main limitation of this study was in the development of the in vitro and in silico study to demonstrate the antioxidant capacity of the garlic EO, being very important for in vivo models, i.e., mammalian cell lines, as well as for experimental animals. However, these findings pretend to show the probable mechanism of molecular interaction of the volatile components of the EO on NADPH oxidase. In addition, there lacked a comparison of the results determined in the DPPH and β-carotene bleaching assays with other garlic EOs studied elsewhere in the world. Furthermore, there were limitations in showing the chemical profile of Peruvian garlic, which is cultivated in different lands affected by altitude, climate, and temperature, among others. Garlic EO might be a promissory antioxidant agent used in phytotherapy or in the food industry. Furthermore, the chemical structures such as α-bisabolol and sulfur components determined in the EO could be used as templates for the design of new drugs to inhibit NADPH oxidase in the future.

4. Conclusions

We concluded that the essential oil of *A. sativum* cultivated in Peru at 2335 m.a.s.l. has a low percentage yield, and analysis with GC–MS revealed the presence of diallyl trisulfide, with 44.23%, as the main volatile component of the EO, followed by 14 compounds of sulfur structures, with 55.02%, and two oxygenated terpenes, namely, α-bisabolol and one unknown constituent with the formula $C_{22}H_{42}O_4$, which represent 3.75%.

Regarding the antioxidant capacity in vitro, two assay methods were carried out, namely, DPPH and β-carotene bleaching assays, showing a good inhibitory capacity over other types of garlic EOs reported across the world. The presence of two compounds not detected in other garlic EOs across the world, such as α-bisabolol (a sesquiterpene), would be vital in maintaining their high antioxidant power. However, garlic EO were less active than the standard antioxidants like Trolox and rutin. In the in-silico study, the activity of the phytoconstituents of the garlic EO on NADPH oxidase were studied, being α-bisabolol the compound with the best docking score. Furthermore, this compound showed good stability during the molecular dynamics study carried out across 100 ns, whilst diallyl trisulfide was not stable, demonstrating that those ligands with low-energy docking scores have poor stability on the target protein. Additionally, according to the ADMET prediction, all garlic components can be absorbed by oral and transdermal administration, which represent a great advantage against other compounds that experience difficulty in penetrating the gastrointestinal barrier. No toxicity was predicted, but further studies have to be developed to confirm these findings using in vivo models.

Supplementary Materials: The following are available online at https://www.mdpi.com/article/10.3390/antiox10111844/s1, Figure S1: GC chromatogram of the essential oil from garlic, Figures S2–S11: Molecular interaction studies of allyl methyl disulfide, 3*H*-1,2-dithiole, diallyl disulfide, 1-propenyl 2-propenyl-(*E*)-disulfide, allyl methyl trisulfide, 3-vinyl-1,2-dithiacyclohex-4-ene, 2-vinyl-4*H*-1,3-dithiine, diallyl trisulfide, 1-allyl-3-propyltrisulfane, and diallyl tetrasulfide with NADPH oxidase (PDB ID: 2CDU), surface view (left panel), and 2D (right panel) interactions.

Author Contributions: Conceptualization, O.H.-C., I.C.H.-G. and L.J.C.-R.; methodology, O.H.-C.; formal analysis, G.E.-S.B.; investigation, O.H.-C. and M.A.; writing—original draft preparation, O.H.-C.; writing—review and editing, O.H.-C. and G.E.-S.B.; funding acquisition, M.A.A. and M.A. All authors have read and agreed to the published version of the manuscript.

Funding: The authors would like to extend their sincere appreciation to Taif University Researchers Supporting Project (number TURSP-2020/309), Taif University, Taif, Saudi Arabia.

Institutional Review Board Statement: Not applicable.

Informed Consent Statement: Not applicable.

Data Availability Statement: The data is contained within the article and supplementary materials.

Acknowledgments: The authors would like to extend their sincere appreciation to Taif University Researchers Supporting Project (number TURSP-2020/309), Taif University, Taif, Saudi Arabia and the Vicerrectorado de Investigación y Desarrollo (VRID) of the Universidad Nacional San Luis Gonzaga, Ica, Peru.

Conflicts of Interest: The authors declare no conflict of interest.

References

1. Firenzuoli, F.; Jaitak, V.; Horvath, G.; Bassolé, I.H.N.; Setzer, W.N.; Gori, L. Essential Oils: New Perspectives in Human Health and Wellness. *Evid. Based Complementary Altern. Med. Ecam* **2014**, *2014*, 467363. [CrossRef]
2. Tongnuanchan, P.; Benjakul, S. Essential Oils: Extraction, Bioactivities, and Their Uses for Food Preservation. *J. Food Sci.* **2014**, *79*, R1231–R1249. [CrossRef]
3. De Lavor, É.M.; Cavalcante Fernandes, A.W.; de Andrade Teles, R.B.; Pereira Leal, A.E.B.; de Oliveira, R.G.; Gama e Silva, M.; de Oliveira, A.P.; Silva, J.C.; de Moura Fontes Araújo, M.T.; Melo Coutinho, H.D.; et al. Essential Oils and Their Major Compounds in the Treatment of Chronic Inflammation: A Review of Antioxidant Potential in Preclinical Studies and Molecular Mechanisms. *Oxidative Med. Cell. Longev.* **2018**, *2018*, 6468593. [CrossRef]
4. Palmieri, S.; Maggio, F.; Pellegrini, M.; Ricci, A.; Serio, A.; Paparella, A.; Sterzo, C. lo Effect of the Distillation Time on the Chemical Composition, Antioxidant Potential and Antimicrobial Activity of Essential Oils from Different *Cannabis sativa* L. Cultivars. *Molecules* **2021**, *26*, 4770. [CrossRef] [PubMed]
5. Abd-ElGawad, A.M.; Bonanomi, G.; Al-Rashed, S.A.; Elshamy, A.I. Persicaria Lapathifolia Essential Oil: Chemical Constituents, Antioxidant Activity, and Allelopathic Effect on the Weed Echinochloa Colona. *Plants* **2021**, *10*, 1798. [CrossRef]
6. Silva, J.; Abebe, W.; Sousa, S.M.; Duarte, V.G.; Machado, M.I.L.; Matos, F.J.A. Analgesic and Anti-Inflammatory Effects of Essential Oils of Eucalyptus. *J. Ethnopharmacol.* **2003**, *89*, 277–283. [CrossRef]
7. López, V.; Nielsen, B.; Solas, M.; Ramírez, M.J.; Jäger, A.K. Exploring Pharmacological Mechanisms of Lavender (*Lavandula angustifolia*) Essential Oil on Central Nervous System Targets. *Front. Pharmacol.* **2017**, *8*, 280. [CrossRef] [PubMed]
8. Camele, I.; Grul'ová, D.; Elshafie, H.S. Chemical Composition and Antimicrobial Properties of Mentha × Piperita Cv. 'Kristinka' Essential Oil. *Plants* **2021**, *10*, 1567. [CrossRef]

9. Thuy, B.T.P.; My, T.T.A.; Hai, N.T.T.; Hieu, L.T.; Hoa, T.T.; Loan, H.T.P.; Triet, N.T.; van Anh, T.T.; Quy, P.T.; van Tat, P.; et al. Investigation into SARS-CoV-2 Resistance of Compounds in Garlic Essential Oil. *ACS Omega* **2020**, *5*, 8312–8320. [CrossRef] [PubMed]
10. Ayaz, M.; Sadiq, A.; Junaid, M.; Ullah, F.; Subhan, F.; Ahmed, J. Neuroprotective and Anti-Aging Potentials of Essential Oils from Aromatic and Medicinal Plants. *Front. Aging Neurosci.* **2017**, *9*, 168. [CrossRef]
11. Jung, K.-W.; Chung, M.-S.; Bai, H.-W.; Chung, B.-Y.; Lee, S. Investigation of Antifungal Mechanisms of Thymol in the Human Fungal Pathogen, Cryptococcus Neoformans. *Molecules* **2021**, *26*, 3476. [CrossRef] [PubMed]
12. Amiri, H. Antioxidant Activity of the Essential Oil and Methanolic Extract of *Teucrium orientale* (L.) Subsp. *taylori* (Boiss.) Rech. f. *Iran. J. Pharm. Res. IJPR* **2010**, *9*, 417–423.
13. Amorati, R.; Foti, M.C.; Valgimigli, L. Antioxidant Activity of Essential Oils. *J. Agric. Food Chem.* **2013**, *61*, 10835–10847. [CrossRef]
14. Ruberto, G.; Baratta, M.T. Antioxidant Activity of Selected Essential Oil Components in Two Lipid Model Systems. *Food Chem.* **2000**, *69*, 167–174. [CrossRef]
15. Farouk, A.; Mohsen, M.; Ali, H.; Shaaban, H.; Albaridi, N. Antioxidant Activity and Molecular Docking Study of Volatile Constituents from Different Aromatic Lamiaceous Plants Cultivated in Madinah Monawara, Saudi Arabia. *Molecules* **2021**, *26*, 4145. [CrossRef] [PubMed]
16. Takaishi, K.; Kinoshita, H.; Kawashima, S.; Kawahito, S. Human Vascular Smooth Muscle Function and Oxidative Stress Induced by Nadph Oxidase with the Clinical Implications. *Cells* **2021**, *10*, 1947. [CrossRef]
17. Maraldi, T. Natural Compounds as Modulators of NADPH Oxidases. *Oxidative Med. Cell. Longev.* **2013**. [CrossRef]
18. Li, R.; Chen, W.-C.; Wang, W.-P.; Tian, W.-Y.; Zhang, X.-G. Extraction of Essential Oils from Garlic (*Allium sativum*) Using Ligarine as Solvent and Its Immunity Activity in Gastric Cancer Rat. *Med. Chem. Res.* **2010**, *19*, 1092–1105. [CrossRef]
19. Kodera, Y.; Kurita, M.; Nakamoto, M.; Matsutomo, T. Chemistry of Aged Garlic: Diversity of Constituents in Aged Garlic Extract and Their Production Mechanisms via the Combination of Chemical and Enzymatic Reactions (Review). *Exp. Ther. Med.* **2020**, *19*, 1574–1584. [CrossRef]
20. Yamaguchi, Y.; Kumagai, H. Characteristics, Biosynthesis, Decomposition, Metabolism and Functions of the Garlic Odour Precursor, S-Allyl-L-Cysteine Sulfoxide. *Exp. Ther. Med.* **2020**, *19*, 1528. [CrossRef] [PubMed]
21. Fernández-Sestelo, M.; Carrillo, J.M. Environmental Effects on Yield and Composition of Essential Oil in Wild Populations of Spike Lavender (*Lavandula latifolia* Medik.). *Agriculture* **2020**, *10*, 626. [CrossRef]
22. Khalil, N.; El-Jalel, L.; Yousif, M.; Gonaid, M. Altitude Impact on the Chemical Profile and Biological Activities of Satureja Thymbra L. Essential Oil. *BMC Complementary Med. Ther.* **2020**, *20*, 186. [CrossRef]
23. Hazrati, S.; Lotfi, K.; Govahi, M.; Ebadi, M.T. A Comparative Study: Influence of Various Drying Methods on Essential Oil Components and Biological Properties of Stachys Lavandulifolia. *Food Sci. Nutr.* **2021**, *9*, 2612–2619. [CrossRef]
24. Riego, M. *De A. y El Ajo En El Contexto Mundial y Nacional*; Ministerio de Agricultura y Riego: La Molina, Peru, 2020.
25. Liz Leslie Cucho-Medrano, J.; Wesley Mendoza-Beingolea, S.; Máximo Fuertes-Ruitón, C.; Elena Salazar-Salvatierra, M.; Herrera-Calderon, O.; Chemical, O. Chemical Profile of the Volatile Constituents and Antimicrobial Activity of the Essential Oils from Croton Adipatus, Croton Thurifer, and Croton Collinus. *Antibiotics* **2021**, *10*, 1387. [CrossRef]
26. NIST20: Updates to the NIST Tandem and Electron Ionization Spectral Libraries | NIST. Available online: https://www.nist.gov/programs-projects/nist20-updates-nist-tandem-and-electron-ionization-spectral-libraries (accessed on 12 September 2021).
27. Rojas-Armas, J.P.; Arroyo-Acevedo, J.L.; Palomino-Pacheco, M.; Herrera-Calderón, O.; Ortiz-Sánchez, J.M.; Rojas-Armas, A.; Calva, J.; Castro-Luna, A.; Hilario-Vargas, J. The Essential Oil of Cymbopogon Citratus Stapt and Carvacrol: An Approach of the Antitumor Effect on 7,12-Dimethylbenz-[α]-anthracene (DMBA)-Induced Breast Cancer in Female Rats. *Molecules* **2020**, *25*, 3284. [CrossRef]
28. Tepe, B.; Sokmen, M.; Akpulat, H.A.; Sokmen, A. Screening of the Antioxidant Potentials of Six Salvia Species from Turkey. *Food Chem.* **2006**, *95*, 200–204. [CrossRef]
29. Wairata, J.; Sukandar, E.R.; Fadlan, A.; Purnomo, A.S.; Taher, M.; Ersam, T. Evaluation of the Antioxidant, Antidiabetic, and Antiplasmodial Activities of Xanthones Isolated from Garcinia Forbesii and Their In Silico Studies. *Biomedicines* **2021**, *9*, 1380. [CrossRef] [PubMed]
30. Pires, D.E.V.; Blundell, T.L.; Ascher, D.B. PkCSM: Predicting Small-Molecule Pharmacokinetic and Toxicity Properties Using Graph-Based Signatures. *J. Med. Chem.* **2015**, *58*, 4066–4072. [CrossRef]
31. Daina, A.; Michielin, O.; Zoete, V. SwissADME: A Free Web Tool to Evaluate Pharmacokinetics, Drug-Likeness and Medicinal Chemistry Friendliness of Small Molecules. *Sci. Rep.* **2017**, *7*, 1–13. [CrossRef] [PubMed]
32. Boubechiche, Z. Comparison of Volatile Compounds Profile and Antioxydant Activity of Allium Sativum Essential Oils Extracted Using Hydrodistillation, Ultrasound- Assisted and Sono-Hydrodistillation Processes. *Indian J. Pharm. Educ. Res.* **2017**, *51*, S281–S285. [CrossRef]
33. Ndoye Foe, F.M.-C.; Tchinang, T.F.K.; Nyegue, A.M.; Abdou, J.-P.; Yaya, A.J.G.; Tchinda, A.T.; Essame, J.-L.O.; Etoa, F.-X. Chemical Composition, in Vitro Antioxidant and Anti-Inflammatory Properties of Essential Oils of Four Dietary and Medicinal Plants from Cameroon. *BMC Complementary Altern. Med.* **2016**, *16*, 1–12. [CrossRef]
34. Ncir, M.; ben Ali, M.; Sellami, H.; Allagui, M.S.; Lahyani, A.; Makni Ayadi, F.; Boudawara, T.; Allouche, N.; el Feki, A.; Saoudi, M. Protective Effects of Allium Sativum Essential Oil Rich in Disulfides against Deltamethrin Induced Oxidative Stress and Hepatotoxicity in Rats. *J. Food Meas. Charact.* **2020**, *14*, 2667–2675. [CrossRef]

35. Ashraf, S.A.; Khan, M.A.; Awadelkareem, A.M.; Tajuddin, S.; Ahmad, M.F.; Hussain, T. GC-MS Analysis of Commercially Available Allium Sativum and Trigonella Foenum-Graecum Essential Oils and Their Antimicrobial Activities. *J. Pure Appl. Microbiol.* **2019**, *13*, 2545–2552. [CrossRef]
36. Torpol, K.; Wiriyacharee, P.; Sriwattana, S.; Sangsuwan, J.; Prinyawiwatkul, W. Antimicrobia Activity of Garlic (*Allium sativum* L.) and Holy Basil (*Ocimum sanctum* L.) Essential Oils Applied by Liquid vs. Vapour Phases. *Int. J. Food Sci. Technol.* **2018**, *53*, 2119–2128. [CrossRef]
37. Dziri, S.; Casabianca, H.; Hanchi, B.; Hosni, K. Composition of Garlic Essential Oil (Allium Sativum L.) as Influenced by Drying Method. *J. Essent. Oil Res.* **2014**, *26*, 91–96. [CrossRef]
38. Delazar, A.; Bahmani, M.; Shoar, H.H.; Tabatabaei-Raisi, A.; Asnaashari, S.; Nahar, L.; Sarker, S.D. Effect of Altitude, Temperature and Soil on Essential Oil Production in Thymus Fedtschenkoi Flowers in Osko and Surrounding Areas in Iran. *J. Essent. Oil Bear. Plants* **2013**, *14*, 23–29. [CrossRef]
39. Lawrence, R.; Lawrence, K. Antioxidant Activity of Garlic Essential Oil (Allium Sativum) Grown in North Indian Plains. *Asian Pac. J. Trop. Biomed.* **2011**, *1*, S51–S54. [CrossRef]
40. Mnayer, D.; Fabiano-Tixier, A.-S.; Petitcolas, E.; Hamieh, T.; Nehme, N.; Ferrant, C.; Fernandez, X.; Chemat, F. Chemical Composition, Antibacterial and Antioxidant Activities of Six Essentials Oils from the Alliaceae Family. *Molecules* **2014**, *19*, 20034. [CrossRef]
41. Bajpai, V.K.; Sharma, A.; Kim, S.H.; Baek, K.-H. Chemical Composition, Antioxidant, Lipid Peroxidation Inhibition and Free Radical Scavenging Activities of Microwave Extracted Essential Oil from Allium Sativum. *J. Essent. Oil Bear. Plants* **2015**, *18*, 300–313. [CrossRef]
42. Chen, C.; Cai, J.; Liu, S.; Qiu, G.; Wu, X.; Zhang, W.; Chen, C.; Qi, W.; Wu, Y.; Liu, Z. Comparative Study on the Composition of Four Different Varieties of Garlic. *PeerJ* **2019**, *7*, e6442. [CrossRef] [PubMed]
43. Schepetkin, I.A.; Kirpotina, L.N.; Khlebnikov, A.I.; Balasubramanian, N.; Quinn, M.T. Neutrophil Immunomodulatory Activity of Natural Organosulfur Compounds. *Molecules* **2019**, *24*, 1809. [CrossRef] [PubMed]
44. Colín-González, A.L.; Santana, R.A.; Silva-Islas, C.A.; Chánez-Cárdenas, M.E.; Santamaría, A.; Maldonado, P.D. The Antioxidant Mechanisms Underlying the Aged Garlic Extract- and S-Allylcysteine-Induced Protection. *Oxidative Med. Cell. Longev.* **2012**, *2012*, 907162. [CrossRef] [PubMed]
45. Martínez, L. Automatic Identification of Mobile and Rigid Substructures in Molecular Dynamics Simulations and Fractional Structural Fluctuation Analysis. *PLoS ONE* **2015**, *10*, e0119264. [CrossRef] [PubMed]
46. Horky, P.; Skalickova, S.; Smerkova, K.; Skladanka, J. Essential Oils as a Feed Additives: Pharmacokinetics and Potential Toxicity in Monogastric Animals. *Animals* **2019**, *9*, 352. [CrossRef] [PubMed]

MDPI
St. Alban-Anlage 66
4052 Basel
Switzerland
Tel. +41 61 683 77 34
Fax +41 61 302 89 18
www.mdpi.com

Antioxidants Editorial Office
E-mail: antioxidants@mdpi.com
www.mdpi.com/journal/antioxidants